DISCEX'00

Proceedings

DARPA Information Survivability Conference & Exposition

DISCEX'00

25-27 January 2000

Hilton Head, South Carolina

Sponsored by

DARPA

IEEE COMPUTER SOCIETY

Los Alamitos, California

Washington • Brussels • Tokyo

IEEE Computer Society Order Number PR00490
ISBN 0-7695-0490-6
ISBN 0-7695-0492-2 (microfiche)
Library of Congress Number 99-067507

Additional copies may be ordered from:

IEEE Computer Society
Customer Service Center
10662 Los Vaqueros Circle
P.O. Box 3014
Los Alamitos, CA 90720-1314
Tel: + 1-714-821-8380
Fax: + 1-714-821-4641
E-mail: cs.books@computer.org

IEEE Service Center
445 Hoes Lane
P.O. Box 1331
Piscataway, NJ 08855-1331
Tel: + 1-732-981-0060
Fax: + 1-732-981-9667
mis.custserv@computer.org

IEEE Computer Society
Asia/Pacific Office
Watanabe Building,
1-4-2 Minami-Aoyama
Minato-ku, Tokyo 107-0062 JAPAN
Tel: + 81-3-3408-3118
Fax: + 81-3-3408-3553
tokyo.ofc@computer.org

Editorial production by Danielle C. Young

Cover art production by Joseph Daigle/Studio Productions

Printed in the United States of America by Technical Communication Services

IEEE COMPUTER SOCIETY

Table of Contents

DARPA Information Survivability Conference & Exposition (DISCEX'00)

High Confidence Computing

Table of Contents

DARPA Information Survivability Conference & Exposition (DISCEX'00)

Volume II of II

Survivability of Large-Scale Systems

Foreword

During the early 1990's more and more attacks were launched against the networks of the U.S. Department of Defense. In response to these attacks and increasing military reliance on the commercial Internet, DARPA started on a journey into the Information Survivability domain.

A new Information Security program was approved in 1994."Jumpstart" funds, provided in FY 1995, were used to establish and strengthen security components of existing programs. Since DARPA was funding the development of innovative networking and systems technologies, there was an opportunity to insert security into this early research, rather than have a completely separate security program that would have to address these technologies after they had been developed. Early focus areas included network security and electronic commerce.

The first Broad Agency Announcement (BAA) was published in January 1995. This BAA called for research in infrastructure protection, protection of end systems, and assurance. Infrastructure protection called for research to address networking vulnerabilities as well as the creation of new network security services. End-system security called for research on higher-assurance policy-independent firewalls, security management technology, configurable and reusable security policy enforcing modules, and security assessment tools. Assurance called for research in security specification languages, security analysis methods that could apply to complex distributed and networked systems, as well as metrics for security assessment.

During 1995, a DARPA-sponsored Information Science and Technology (ISAT) study on Survivability led to a new program segment aimed at developing means for hardening legacy systems by retrofitting security and survivability technologies. The Assurance segment of the program was changed into the Wrappers and Composition program to satisfy this new requirement. The Infrastructure Protection segment was renamed High Confidence Networking, and End-system Protection was renamed High Confidence Computing. Anticipating an increase in the interest of protecting critical infrastructure, a new program segment Survivability of Large Scale Systems was also added.

Subsequent BAAs (BAA 96-40 and 98-04) called for survivability research inspired by biological models, research in intrusion detection and response in large-scale systems, large-scale intrusion assessment, and survivable highly decentralized systems. Complementing the major thrust in intrusion detection and response were projects that recognized the need for Defense systems to sustain mission critical operations in the face of attack. These projects explored capabilities such as decentralized (e.g., market-based) resource allocation, adaptivity, and artificial diversity that led to the concept of Inherent Survivability and helped form the basis for a follow-on program in Intrusion Tolerant Systems.

The DARPA Information Survivability program consisted of four research areas. A short description of each research area is listed below.

High Confidence Networking

The High Confidence Networking program was aimed at creating network barriers that would prevent network disruption while facilitating secure communication for heterogeneous users and services. This program was designed to develop protocols and services to protect the integrity of internet-based fabrics and guarantee their preferential availability to critical applications as well as the health of the network in time of attack. Higher-level applications need network-wide security services, and this program was designed to supply the security technology for sharing and protecting enterprise facilities and for safe communication among dynamically negotiated sets of peers. This included the development of high assurance methods and algorithms that work with current and emerging network software and hardware.

Activities within this program included the development of secure protocols for key network functions such as routing, name service, mobile internets, data transport, multicast, key management, user authentication and network management. In addition, the development of tools and their application of formal methods for analyzing the security of such protocols were an important part of the program. Research for secure group communication is seen as a basis for virtual private networks, also known as enclaves, and the associated policy negotiation languages are critical for their establishment. Lastly, resource allocation mechanisms, which satisfy quality of service requests of mission-critical priority clients, were developed as part of this program.

High Confidence Computing

The High Confidence Computing program was aimed at developing and demonstrating operating systems (OS) and computing environments with integral security designed to be strong against a powerful adversary and not trail commercial trends in functionality and performance. Its emphasis on policy neutrality and configurability was intended to lead to a technology base of commercially viable secure systems, easily tailored to defensive military needs. As Defense systems must simultaneously satisfy a number of mission-critical properties, High Confidence Computing developed technologies that integrated support for security, dependability, and real-time operation and explored tradeoffs and interactions among these constraints.

Activities within this program included enabling configurable OS security through policy-neutral access control mechanisms, policy languages for ease of specification and management of security rules, and a policy "compiler" that maps these rules into the access control mechanisms. These technologies ensure increased strength against attack by leveraging innovative language and compiler technology to provide low-cost tools for high assurance operating system construction. Development of partitioned OS resource management techniques that provide security through process isolation and resource use limitation were completed as part of this program. Research in security for extensible operating systems led to innovative techniques for assuring the safety and security of mobile code. This program was responsible for incorporating compatible security mechanisms into standard distributed computing environments (e.g., CORBA) to support secure interoperability across administrative domains. This facilitated the survivability of mission-critical applications through technologies which permit dynamic binding of security and fault-tolerance services to applications in response to policy or threat environment changes and are coupled to OS and distributed enforcement mechanisms.

Survivability of Large Scale Systems

The Survivability of Large-Scale Systems program was aimed at developing, evaluating, and demonstrating intrusion detection techniques. This included protocols that allow system elements (such as network managers, firewalls, and filtering routers) to react to detected events, algorithms to redirect resources to the most important tasks and whose behavior is difficult for an adversary to predict, and techniques for reconfiguring to a state not susceptible to the original attack. This program focused on providing detection technologies for local intrusion detection as well as larger-scale tactical indications, assessment, and warning.

Activities within this program included the development of intrusion detection technologies capable of detecting known attacks with high confidence and beginnings at detecting unknown attacks. These techniques offer much improved false alarm rates and higher detection accuracy than the prior existing state-of-the-art. These techniques were also validated in realistic testbeds subject to red-teaming. To address the problem of tactical indications and warning, efficient methods were developed for peer-to-peer cooperative detection and reporting. Probes and sensors were developed for monitoring large-scale systems for attack. This program investigated technologies to estimate the extent of damage, and developed protocols for automated response and reconfiguration. In addition, techniques for resource allocation were developed that assign remaining resources to the most critical tasks. This program explored diversity techniques to ensure that no single attack can completely take out any capability and techniques for recovery to a state not susceptible to the original attack.

Wrappers and Composition

The Wrappers and Composition program was aimed at developing toolkits for integrating security and survivability functionality into legacy systems and developing tools for assessing the security and survivability of components and their composition. Department of Defense systems contain a high percentage of legacy and off-the-shelf components that do not provide the properties needed for survivable systems (e.g., security, robustness). This program was designed to develop the techniques to assess aspects of the trustworthiness of a component and develop tools which transform existing components into ones, which have the properties, needed for survivability. This included the development of formal engineering rules for composing components so as to preserve these properties, and tools for embodying these rules to assist an integrator in assembling components and wrappers in a trustworthy manner.

Activities within this program included the development of tools for automatically generating security and survivability "wrappers." Wrappers encapsulate an existing component by intercepting all input and output to the component, and thereby add security and survivability functionality; e.g., encrypting the traffic or filtering access requests to the component. The demonstration of these wrappers for candidate systems provides security and robustness functionality. Underlying these tools is a rigorous methodology for "survivable composition" to allow an integrator to predict the security and survivability characteristics of compositions of components and wrappers.

These Proceedings

This publication represents the DARPA Information Survivability Conference and Exposition (DISCEX) proceedings and contains research papers describing the projects in the four areas of the Information Survivability program described above. Contributions from 24 academic institutions, 18 industrial research laboratories, and 3 government research laboratories describe progress in the research performed for survivable information systems. These proceedings have been peer reviewed, although not strictly in the traditional sense of a refereed conference or journal, and participating reviewers are noted in the Acknowledgements section. We're confident these proceedings represent high quality papers describing the research carried out at each institution. In addition, we view these proceedings as a technology transfer vehicle for anyone interested in information survivability and the technology available today. We believe these proceedings will be valuable for future information survivability researchers as they see where we've been and they can leverage this previous research as they proceed ahead.

The sections of these proceedings are grouped by program with the papers arranged by topic so similar research projects can be studied together. We are confident these proceedings will provide an excellent source of ideas and insight for the rapidly growing community of survivable information systems. It is our hope that these proceedings will convey the many ideas and results from the information survivability researchers and the results presented will catalyze the discovery of new techniques for securing our nation's information systems.

World Wide Web

Even though the DARPA Information Survivability program is coming to a close, current descriptions of all activities within this program are maintained with links to performers' sites at http://www.darpa.mil/ito/research/is. This will provide the reader with opportunities for contact and collaboration to accelerate the technology transfer process and, hopefully, the progress of research and development of survivable information systems.

Gary Koob
DARPA / ITO
gkoob@darpa.mil

Douglas Maughan
DARPA / ITO
dmaughan@darpa.mil

Sami Saydjari
DARPA / ISO
ssaydjari@darpa.mil

Arlington, Virginia
January, 2000

Acknowledgments

The Defense Advanced Research Projects Agency (DARPA) Information Technology Office (ITO) was pleased to sponsor the DARPA Information Survivability Conference and Exposition (DISCEX) on January 25-27, 2000 in Hilton Head, South Carolina. This conference and exposition was designed to be a showcase for the technologies developed by researchers in the DARPA Information Survivability program over the last five years.

We acknowledge the hard work of several DARPA Program Managers who have guided these programs over the last five years. They are: Michael St. Johns, Teresa Lunt, Gary Koob, Howard Shrobe, Hilarie Orman, Rob Rosenthal, Sami Saydjari, Bob Laddaga, Helen Gill, and Douglas Maughan.

Several volunteers assisted in the review of the papers contained in these proceedings. We acknowledge their assistance in assuring these volumes contain quality papers and thank them for their service to the information survivability research community.

DARPA

Gary Koob, Jay Lala, Douglas Maughan, Michael Skroch, Brian Witten

Air Force Research Laboratory (Rome Labs)

Alan Akins, Richard Butler, Priscilla Cassidy, Vaughn Combs, John Evanowsky, John Faust, Daniel Hague, Brad Harnish, Patrick Hurley, Robert Kaminski, Kevin Kwiat, Thomas Lawrence, Jerry Leppert, Michael Nassif, Robert Paragi, Paul Sierak, Gino Spagnuolo, Walter Tirenin, Jon Valente

SPAWAR Systems Center (San Diego)

Lou Gutman, Ed Howard, Bill Jacobs, Steve Spendlove, Derek Wong

National Security Agency (NSA)

Andrea Colegrove, Eric Harder, Robert Meushaw, Mark Schneider

National Institute of Science and Technology (NIST)

Shu-ien Chang, Kevin Mills, Doug Montgomery

MITRE Corporation

Henry Allen, Donald Faatz, Joshua Guttman, Emily Hawthorn, Charles Howell, Sushil Jajodia, Jay Kahn, Todd O'Boyle

NAI Labs

Lee Badger, David Balenson, Jeffrey Cook, Peter Dinsmore, Kelly Djahandari, Jeremy Epstein, Rich Feiertag, Mark Feldman, Timothy Fraser, Dennis Hollingworth, Peter Kruus, Brian Matt, Russ Mundy, Sandra Murphy, Sue Rho, Daniel Sterne

SPARTA

Jim Demerest, Sharon Downey, Ame Fox, Linda Gilmore, Hugh Harney, Barbara Mayer, Richard McAllister, Carl Muckenhirn, Don Oswald, Thomas Rivers, Howard Weiss, Gene Williams

BBN

Isidro Castineyra, Robert Donaghey, Stephen Kent, David Levin, John Lowry, Charles Lynn, Luis Sanchez, Robert Shirey, John Zao

University of Southern California Information Sciences Institute (USC/ISI)

Dan Boneh, Cliff Neuman, Gene Tsudik, Brian Tung

Carole Kelliher and Sheryl Augustson of Schafer Corporation coordinated the conference. Dave Stenger, also of Schafer Corporation, supported the demonstrations at the exposition. David Ricardo and Dave Stenger assisted with the creation of graphics for the proceedings and web presence. Jeff Nolan, also of Schafer Corporation, did a significant amount of the information gathering during the beginning stages of this conference process.

Lastly, and most importantly, John Drake and Michael Ferguson coordinated the solicitation of papers from the researchers, distribution of submitted papers to the reviewers, collection of comments from reviewers and distribution to the authors, and final solicitation to IEEE. Without their hard work these proceedings never would have been completed.

High Confidence Networking

Retrofitting Security into Internet Infrastructure Protocols

Sandra Murphy
NAI Labs, Network Associates, Inc
Glenwood, MD
Sandra_Murphy@nai.com

Olafur Gudmundsson
NAI Labs, Network Associates, Inc
Glenwood, MD
Olafur_Gudmundsson@nai.com

Russ Mundy
NAI Labs, Network Associates, Inc.
Glenwood, MD
Russ_Mundy@nai.com

Brian Wellington
NAI Labs, Network Associates, Inc.
Glenwood, MD
Brian_Wellington@nai.com

Abstract

There are many different protocols that are critical to the communication of data in the Internet but are not readily visible to the end-users. These infrastructure protocols include the Domain Name System protocol, the various routing protocols, and the Simple Network Management Protocol. The infrastructure protocols were developed early in the history of the Internet, when the risk environment was much more benign, and are not protected against failure or attack. This has left the Internet vulnerable to failure resulting from attack on its infrastructure. Under DARPA sponsorship in the Information Survivability program, Network Associates has designed and implemented enhancements to critical infrastructure protocols, so as to enable the robust operation of the Internet. This paper describes our work on security enhancements to the Domain Name System, the Open Shortest Path First (OSPF) routing protocol, the Border Gateway Protocol (BGP) routing protocol, and the Simple Network Management Protocol (SNMP). [1]

1. Introduction

There are many different protocols that are critical to the communication of data in the Internet but are not readily visible to the end-users. Chief among these protocols, of course, is the well-recognized Internet Protocol (IP) itself. There are other equally critical protocols that are part of the Internet infrastructure and enable scalable communication through a world-wide network like the Internet. These include the Domain Name System protocol, which provides for the use of human-readable names for Internet hosts by translating those names to the machine-readable IP addresses, and the various routing protocols, which distribute the topology information used by IP to forward data packets across networks. Other protocols provide tools necessary for scalable operation of the Internet. In particular, the Simple Network Management Protocol provides mechanisms for remote management of various Internet devices such as routers and hosts.

These infrastructure protocols were developed at a time in the history of the Internet when the threat environment was not yet fully understood and are not designed to protect against failure or attack. In many cases the protocol designers had some idea of the risks to the protocols but had no practical idea of how to secure the protocols. In the last decade, both the threats and network security technologies have become better understood. The lack of protocol security features has left the Internet vulnerable to failure resulting from attack on its infrastructure. Such failures can take the form of spoofing of individual hosts; denial of service to individual hosts, to networks, or to entire regions of the Internet; exposure of sensitive information to unauthorized hosts; network congestion leading to traffic delay or loss; etc. The IP protocol itself has been the subject of intensive study, with the security enhancements called IPSEC developed in the Internet Engineering Task Force. Under DARPA sponsorship in the Information Survivability program, Network Associates has designed and implemented enhancements to some of the critical infrastructure protocols, so as to enable the robust operation of the Internet.

This paper describes our work on security enhancements to the Domain Name System (DNS), the Open Shortest Path First (OSPF) routing protocol, the Border Gateway Protocol (BGP) routing protocol, and the Simple Network Management Protocol (SNMP). Particularly, it describes the use of

[1] This work supported by DARPA as Internet Infrastructure Protection, Contract DABT63-94-C-0001.

cryptography to protect the authenticity and integrity of the information exchanged in these protocols and addresses the provisions for authorization of information in each protocol. We pay particular attention to the features of each protocol that made it difficult to protect after the protocols had been standardized and implemented without security. Section 2 describes the DNS security enhancements. Section 3 describes the security enhancements to routing protocols. Section 4 describes the security enhancements to the SNMP protocol.

2. Domain Name System

The Domain Name System was invented to provide a scalable naming system in the Internet. The IPv4 protocol knows hosts by their 32 bit IPv4 address. Users prefer a human-readable name for hosts. In the early days of the Internet, one file called *hosts.txt* provided a translation from human-readable host names to machine-readable IP addresses. This file was centrally located and maintained at the Network Information Center. However, this centralization of location and administration was not scalable and became unwieldy as the Internet grew. In 1984, DNS was proposed as a scalable substitute.

The DNS is a tree structured name space database, with distribution both of information location and of administrative control. The database is spread over many different *name servers*, each responsible for a portion of the name space tree. DNS *resolvers* query the name servers for information, usually the IP address corresponding to a host name (or vice versa).

The DNS information is organized into a tree, where each subtree is known as a *domain*. Each node of the tree is assigned a label and the *fully qualified domain name* of each node is formed from the labels of that node and each ancestor up to the root of the tree (separated by dots). For example, the root domain contains the **com**, **edu**, and **gov** domains and the **com** domain contains the **acme.com** domain which contains the **coyote.acme.com** domain.

Each domain is administered by some organization that is responsible for ensuring that correct data is entered into that portion of the name space. Any organization may *delegate* administrative control of one of its subdomains to another organization. In this way, authorization to administer the information in the DNS is distributed. A *zone* is the part of a domain that has not been delegated away. If **com** has delegated the **acme.com** domain to ACME, Inc., then **coyote.acme.com** is still in the **com** domain. However, it is not in the **com** zone; it is in the **acme.com** zone. Each zone is associated with a name server that is the *authority* for that zone. The controlling authority for each domain provides the information for that domain through one or more authoritative DNS name servers for the zone.

DNS information is stored in each name server as a set of resource records (RR) which contain data associated with one domain name. Each record specifies type, class (which will normally be the Internet class), and value. Examples of types include A (address), PTR (reverse address, the opposite of A), MX (mailbox where mail addressed to the domain will be forwarded), HINFO (host configuration), and LOC (location). All records of the same type associated with a particular name comprise a set (or RRset).

Two resource records are used to indicate the delegation of part of the domain of a name server – a Start of Authority (SOA) RR and a Name Server (NS) RR for the delegated subdomain. The delegated zone contains both records; the delegating zone contains only the NS record. The NS record gives the fully qualified domain name of the name server for the subdomain. The delegating zone may also contain an A record for the delegated zone's name servers.

Queries from a resolver can result in multiple queries to many different name servers located in different parts of the Internet. If the local name server cannot answer the query, it will ask other name servers, starting with the root name server. Any non-authoritative name server asked will answer with a referral to a name server that has been delegated responsibility for that domain. (To answer a query for **coyote.canyon.acme.com**, the root server will tell where to find a name server for **acme.com**, which will refer to the name server for **canyon.acme.com**, which will authoritatively answer.) The referral includes providing the NS record that indicates the delegation, as well as an A record indicating the corresponding IP address. (This "glue" information, while non-authoritative from the delegating zone, is needed to find the name server which is authoritative.) To improve performance, most name servers will cache responses they receive to avoid redundant queries to the name servers of popular zones.

The distribution of authority for the DNS information coupled with the query response mechanism provides a robust and scalable mechanism for name translation. However, the system is vulnerable to spoofing. There is no protection of the authenticity or integrity of the responses that are received. An external attacker could modify a response or provide false information masquerading as the correct responder. Furthermore, because authority is distributed, it is difficult to know whether to trust the responses received. These is no guarantee that the data received in a response is truly authoritative or that a name server is not maliciously providing false information. A number of attacks using DNS data are described in [2].

Security enhancements for DNS were designed in the DNSSEC working group of the IETF [6] to ensure that responses can be authenticated as being authoritative. Authority in the DNS follows the zone structure, not the domain structure. A zone has complete authority over names in the zone portion of the domain name space and retains no authority over parts of the domain that it delegates away. A zone does have complete authority to determine a delegation, which is indicated by the presence of a NS record in its zone. But authority over the contents of that record is given to the delegated zone.

To prove the authenticity of an authoritative response, the authoritative zone cryptographically protects the data. Using a symmetric cryptographic technique would require creating a separate key for each resolver, server pair. This would clearly not scale to the size of the Internet. The zones have no way of predicting the resolvers that might send a query, so it would not be possible to pre-establish the symmetric keys. Furthermore, symmetric keys would prevent servers from pre-computing and caching protected responses, as the response would be protected differently for nearly every querier. Consequently, the cryptographic protection chosen was a public key digital signature.

NAI Labs provided the leadership of the DNSSEC working group and the first reference implementation, based on Berkeley Internet Name Domain (BIND) software. BIND includes a name server (named), resolvers (dig, nslookup), as well as other DNS related utilities.

DNS Security Extensions DNS Security Extensions [6] adds new record types to incorporate authentication into the DNS. The KEY record stores the public (asymmetric) key for a host or administrative zone, and the SIG record stores a digital signature associated with each set of records. If a zone is signed, it contains a Zone KEY record. The Zone KEY record must be signed by the delegating zone to maintain a chain of trust from the root. In a signed zone, each record set includes a SIG record, which contains the signature of the set as generated by the Zone key. A DNSSEC aware resolver can determine whether or not a zone is signed. In a signed zone, all record sets must be signed so that if a resolver sees an unsigned set when a signed set is expected, it can identify that there is an error.

The following list shows the typical records in an unsigned zone named example:

```
example.      SOA   ...
example.      NS    ...
a.example.    A     ...
d.example.    A     ...
```

The following list shows the KEY and SIG records added to the example zone when it is signed:

```
example.      SOA   ...
example.      SIG   SOA ...
example.      NS    ...
example.      SIG   NS ...
example.      KEY   ...
example.      SIG   KEY ...
a.example.    A     ...
a.example.    SIG   A ...
d.example.    A     ...
d.example.    SIG   A ...
```

When a DNSSEC aware name server responds to a query, it will include SIGs for all signed records in the response. There is no guarantee that a single response will contain enough information to verify all signatures, so the name server or resolver may require additional queries. The DNS message header also indicates if the response (specifically all records in the answer and authority sections) contains authenticated data.

Using the DNS chain of trust Assume a resolver queries a secure DNS server and gets a response. It verifies the signature for each set of records in the response. A verified signature tells the receiver that the message was correctly signed. This establishes data integrity, but the resolver must determine whether it should trust the KEY used to sign the message, and whether the key is permitted to sign that message. The secure resolver must establish a cryptographically verified "path" from a known trusted point in the tree to the point represented by the name in the response. Since a SIG record associated with the KEY set is signed by the parent zone's KEY, the resolver will request security information for the parent zone. This process can continue until the resolver finds a trusted key. In the worst case, this will be the root name server key, which is pre-configured in all DNSSEC aware name servers. This recursive process is referred to as "walking the chain of trust". When the chain of trust includes the full hierarchy from the root name servers, the path validation process is similar to an X.509 certificate environment that is mapped into a tree-structured name space. Additionally, DNSSEC servers can establish trusted agreements with other parties even though the parties may have no hierarchical relationship. If the chain of trust is established from either the DNS root name server key or a trusted party key, the resolver has then established the data origin authentication (often called source authentication) for the information.

To look at authentication another way, consider a record for the host **www.acme.com**. The root key is trusted, since it is pre-configured. So, if the root key signs the **com** KEY and its SIG verifies, we trust the **com** key. If the **com** key signs the **acme.com** KEY and it verifies, it is then trusted. If the **www.acme.com** is signed by the **acme.com** key and the signature verifies, the records for **www.acme.com** can be trusted. Fortunately, an intelligent resolver will cache keys, so once it establishes this chain of trust the first time, the verification process is quicker in future iterations.

Cryptographic authentication also verifies authorization. The key that signs a record must also be authorized to sign the specific set of records. For example, if records in the **bar.com** zone are signed by **foo.com**'s key, a DNS server should consider the information invalid. A key is normally only authorized to sign records for its zone. Thus, **bar.com**'s key may sign any record in the **bar.com** zone but not the **foo.com** zone. Including the DNS name in the security records permits resolvers to determine whether or not the signature is authorized.

Authenticated Denial If a DNS server discovers that the name requested does not have an entry in the DNS, it returns the SOA record for the authoritative zone. However, this record, even when signed, is not adequate to securely deny the existence of a DNS name. The SOA response itself does not indicate the query to which it responds. An attacker could store a prior SOA response and SIG and insert them as a response to any query for any name in that zone.

For authenticated denial of existence of a resource record, another resource record type, NXT ("next") was introduced. One NXT record exists for each DNS name with any associated records. The NXT record's data includes the name of the next DNS entry in the zone, as well as a list of the types of records present for the current name (the determination of "next" is based on a canonical ordering defined in the DNS Security Extensions specification and follows an alphabetical order). The name of the NXT record and the next DNS entry name contained in the NXT record define a range of names that do not exist. The record types listed in the NXT record determine a range of record types that do not exist for the name of the NXT record. When a DNS server responds to a query for which there are no matching records, the DNS server sends the "covering" NXT record. All possible names are thus either present in the DNS or in a range covered by exactly one NXT record.

Transaction security An end system's resolver is often a simple application. Each time this resolver needs DNS information, it builds a query, sends it, and reads the response. This requires little processing and no caching. While this is a reasonable design, it is very inefficient for the main DNS security extensions. Since the resolver does not cache records, the resolver must obtain the entire chain of KEY and SIG records for every query. One solution is to add caching to this type of resolver, but this adds complexity. Additionally, verifying digital signatures is CPU intensive. This may prove impractical since many resolvers may be running on slow machines.

In most cases, this type of resolver communicates with only a small number of DNS servers on the same local network. If the resolver can trust the local server, then it can offload all the security extension processing to the server, leaving the problem of securing communication with the server. Since there are only two participants in the communication between resolver and local server, they can protect the communication using secret key cryptography. Secret key cryptography has the added advantage that operations require less CPU usage than public key cryptography, so the resolver does not need to be especially fast.

A technique has been proposed and implemented to protect the resolver to local server communication. It is presumed that the resolver and server share a symmetric key that is determined and transmitted out of band (the key is not stored in the DNS). DNS messages from the resolver contain a message digest of the request and every message from the server contain a message digest of the transaction. These message digests are generated with the shared key and authenticated by the receiver. The digest is added to the DNS message in a TSIG (Transaction Signature) record, which is included at the end of the additional data section. Note that while SIG records are associated with sets of records, TSIG records are associated with messages; this includes the DNS header in addition to all records.

Other uses of a secured DNS system Those who are interested in the wide availability of a public key infrastructure have shown interest in using a secured DNS. The DNS is used by every host that hopes to communicate over the Internet so software is widely available. The distribution of DNS servers in the Internet means that the public key information could be similarly distributed, with every host within easy reach of a DNS server. The security extensions to DNS mean that information stored in the DNS could be protected for data origin authenticity, integrity and authorization. These are all attributes that are needed in a public key infrastructure.

Besides the KEY records that are part of the DNS security extensions and could hold keys associated with a domain name (referring to a user rather than a host), a new resource record has been defined called a CERT record. This record can store X.509, SPKI or PGP certificates and provide world wide access.

Deployment and Operational Challenges The DNS security extensions have resulted in some significant impacts on both servers and resolvers. The need to effectively operate DNS with a mix of secure and non-secure components has been an important consideration for all DNS security extensions. These considerations are making deployment and operation easier but do not make it simple. In order to facilitate fielding and use of DNS security extensions, NAI Labs is supporting its use on research networks such as the DARPA funded Collaborative Advanced Interagency Research Network (CAIRN) and in workshops for operators that plan on early fielding of DNS security extensions.

Some of the impacts identified include the need for new procedures as well as enhancements to DNS software.

The DNS security authorization structure does not exactly match the zone authorization structure. The zone authorization requirement is that a zone is authoritative for all data content in the zone. Since the security requirement for a DNS chain of trust necessitates a cryptographically verifiable link between zones, the KEY record for the delegated zone must be signed by the parent, delegating zone. Additionally, these requirements also result in the delegating zone having an NS record for the delegated zone which establishes that the zone actually has been delegated. However, this is not the authoritative record for the name server in the delegated zone. The authoritative information for the delegated zone name server must be obtained from the delegated zone. Similarly, the zone authorization requirements dictate that a delegated zone is the authoritative DNS source for its KEY and SIG records. Therefore, the KEY record for a delegated zone must be created by the delegated zone and securely communicated to the parent delegating zone which must create the SIG record for this KEY and return both the KEY and SIG records to the delegated zone.

Creation of secure DNS zone data needs to be accomplished in a secure manner. The zone administrator creates a private/public key pair on a system with sufficient security to protect the private key. In most cases, we expect that this will be done on a system which is not connected to the general Internet nor providing the nameservice function for the zone. The public portion of the key pair will be used in the KEY record that is signed by the parent delegating zone and is part of the authoritative information provided by the zone. The zone signing tool reads in all of the information for the zone, organizes and orders each of the records for the zone, creates all of the NXT records for the zone, creates all of the SIG records using the private key, and creates a file with the full set of information that can be used by the operational name server for the zone. This file then needs to be moved to the name server in a secure manner.

Security aware DNS resolvers need to be able to receive and validly interpret DNS security information. This requires more capability and processing than is needed for merely retrieving DNS data. A secure resolver must judge the security status of retrieved data and may often need to retrieve additional data to verify or deny the DNS chain of trust for an arbitrary name in DNS. This requires the resolver to be able to determine if the requested name is supposed to be in a secure zone, to determine if it has the necessary KEY and SIG records and attempt to retrieve them if it does not, and to verify the additional records. Additionally, resolvers need to be able to securely respond to requests for non-existent names which may involve multiple requests to DNS servers and proper interpretation of NXT records in addition to verification of KEY and SIG records.

The amount of data in the DNS will be increased significantly with the addition of security information. Compared to other DNS records, public keys and signatures are fairly large. This property makes both messages and zone data larger. For example, the size of an A record is 4 bytes plus the record header, but a SIG record generated by a 1024 bit RSA key will be over 100 bytes. Similarly, where an unsigned zone stored by a name server may have only an A record for each name, the signed zone would have an A record, an NXT record, and a SIG record for each RRset. The impact of this size increase on both zone files and communications patterns between servers and resolvers will be an area of continuing study.

In addition to the DNS security extension requirements, the DNS specifications are also being extended to meet other operational requirements such as IPv6 and Dynamic Update. These requirements, coupled with fifteen years of DNS operation and five years of research in DNS security, established the need for a major update to the most widely used Internet name server software, commonly known as BIND. This development, which extends the earlier research funded by the Defense Advanced Research Project Agency (DARPA), is now being jointly funded by the Defense Information Systems Agency (DISA) and various UNIX vendors. NAI Labs and the Internet Software Consortium, which develops and distributes the BIND software, have initial support for DNS security in version 8 of BIND and will have extensive DNS security capabilities in version 9 of BIND.

3. Routing

The Internet carries packets from source to destination by hop-by-hop forwarding through each router along the path. Each router forwards the packet according to its local view of the topology of the network. In the very beginning of the Arpanet, it was possible to construct this topological view by hand for each router. As the number of hosts grew, hand construction of the topology quickly became infeasible. *Routing protocols* were developed to distribute topology information from router to router. In the benign threat environment in which these protocols were developed, the protocols implicitly trust the information received and the peer from which it came.

The lack of protection of data authenticity and integrity in the routing protocols in use when this project began left the Internet vulnerable to failures. Because all the routers cooperate in the collection and distribution of topology information, incorrect routing information propagates over a wide area, resulting in the failure to forward packets properly. Such failures can affect wide areas of the Internet and large numbers of users. Compared to the localized effect of the commonly reported attacks on individual servers, attacks on the routing protocol could be much more damaging in total effect.

The propagation of incorrect routing information can have several different effects on the forwarding of data packets. If the routing information creates loops in the forwarding path, then packets may never be delivered, denying service to every effected source and destination. Loops or unnecessarily long routes through the network also have the effect of increasing network congestion by leaving packets in the network longer than necessary. The resulting application delays can prompt the applications to retransmit data, further increasing the network congestion. Routing information that incorrectly forwards large amounts of traffic over one link may overwhelm the link and delay or prevent the delivery of the traffic. If an overwhelmed link fails completely, legitimate traffic would either not be delivered or would congest other links. Incorrect routing information could result in data traffic transiting networks that would otherwise not see the data. Routing information that incorrectly reports portions of the Internet as unreachable could deny service to all hosts connected to those networks.

This is no idle threat. Many of these failures have been seen in the Internet, resulting from faulty or mis-configured routers. The opportunity exists for these vulnerabilities to be exploited deliberately. Clearly, the routing protocols need to be protected from bogus information.

Several of the existing routing protocols contain provisions for cryptographic protection of the authenticity and integrity of the data as it travels between adjacent routers (e.g., [1], [12]). This typically takes the form of a keyed cryptographic hash of the data. This protection ensures that external attackers cannot cause the damage just described. But this hop-by-hop protection does not ensure the end-to-end protection of the information. As the known incidents have all been caused by routers that are legitimate participants in the routing protocols, clearly end-to-end protection is needed. Besides authentication and integrity, authorization is also important. The router which introduces topology information into the routing protocol should be verified to possess the authorization to do so.

Routing protocols can be characterized as to whether they are used within one autonomous system (AS) (a set of routers under a single technical administration which have a single, coherent routing policy) or between autonomous systems. Those used within one AS are called Interior Gateway Protocols (IGP); those used between AS's are called Exterior Gateway Protocols (EGP). In deciding which routing protocols to address, we focused on those that have are in the standards track of specification within the IETF and that are widely used. We chose to consider protections to OSPF, an IGP, and BGP 4, an EGP.

3.1. OSPF

The OSPF routing protocol [12] is an Interior Gateway Protocol designed for use within a single autonomous system. Consequently, it is assumed that the most important factor in determining the choice of a route is the length of the route and that the shortest will be chosen first (hence the name "shortest path first"). OSPF is a *link state* protocol, which means that each router transmits state information (status and metric) about its links with its neighbors. This link state information is flooded verbatim to the entire network. Each router in the network can therefore collect a detailed view of the network topology and compute the shortest routes to any destination. Because the views are the same in all the routers, the computed shortest routes are loop-free and packets forwarded from router to router are certain to reach the destination.

We have designed and implemented end-to-end protection by adding digital signatures to the OSPF link state routing information. The idea of signing routing information is not new. Foremost, there is the design that Radia Perlman reported in her thesis [14] and in her book [15] for signing link state information and for distribution of the public keys used in the signing. IDPR [18] also recommends the use of public key based signatures of link state information. Kumar and Crowcroft [9] discuss the use of secret and public key authentication of inter-domain routing protocols. Finn [7] discusses the use of secret and public key authentication of several different routing protocols. The design reported here uses the same basic concepts as that reported in [14] and [18]. It should be noted that [14] also presents techniques that provide robust forwarding of data packets, a topic that is not considered here, as we considered it a separate problem from protecting the OSPF routing protocol.

OSPF assumes a network architecture that consists of a two level hierarchy of sets of routers, called *areas*. One area is designated as the backbone area. The other areas connect to the backbone through Area Border Routers (ABR). Within any one area, link state information is fully flooded. However, most link state information does not flood between areas. ABR advertises to the backbone a summary of all the networks it can reach in its attached area. The ABR advertises to its attached area a summary of all the networks it can reach through the backbone. Another group of routers, that can be anywhere in the hierarchy, are known as Autonomous System Border Routers (ASBR) and report routes to addresses external to the Autonomous System.

Each link attached to an OSPF router is the subject of a Link State Advertisement (LSA). Each LSA is transmitted as a single packet. Adjacent OSPF routers learn of each other's existence through a hello protocol, establish adjacency and synchronize their databases of LSAs they have received. Each LSA a router receives is flooded to all adjacent routers.

The LSAs contain the identification of the router that sent them and the link they represent. They also contain a sequence number, a checksum, and an age field. This information is sufficient for the receiving router to compare the new LSA to LSAs in the database; only the most current is stored and propagated. LSAs are periodically retransmitted from their source, with an incremented sequence number. Also, LSAs age as they sit in the database and as they are communicated through the network. Eventually, the age of an LSA will reach a MaxAge value and the LSA will be deleted from the network. This ensures that LSAs originated by a router that goes down will eventually be retired from the network wide collective database. The first router to notice that an LSA has reached MaxAge will flood that LSA through the network. Each receiving router will delete any matching LSA in its database. Although each receiving router would undoubtedly soon have found that the matching local LSA had itself naturally aged to MaxAge and would have then deleted it from its local database, this technique speeds database convergence among the routers.

The originating router also uses the age field, in this case to expunge the collective database of obsolete information. If a router no longer needs to be advertising a link, it issues a new LSA for that link with the age set to MaxAge. Each receiving router will take this prematurely aged LSA as a sign to delete any matching LSA in its database. The prematurely aged LSA will be flooded through the network as any other LSA and will thus expunge the unneeded link information from the collective database.

Cryptographic protection of OSPF To protect OSPF against some of the faults discussed above, it is necessary to ensure the authorization of the router producing the received link state information. This naturally also requires protecting the data origin authenticity and integrity of each message. We rely on cryptographic protection to supply these services.

The OSPF design paradigm is that each router is ignorant of the number or existence of routers that are not directly connected neighbors. This means that cryptographic protection of the OSPF routing information cannot rely exclusively on symmetric keys. When a router is transmitting a new LSA, it cannot know what other routers there are that might receive it. So it cannot know what symmetric key to use. It is necessary to use asymmetric (public key) cryptography to protect the OSPF routing information.

The basis of the cryptographic protection of OSPF is digital signatures added to the OSPF LSAs and certified keys and authorization flooded through the network along with the LSAs. Link state information will be signed by the originator of that information and the signature will stay with the data in its travels via OSPF flooding. This will provide end-to-end integrity and authentication for LSA data.

Authorization is not naturally a part of the OSPF design. Unlike DNS, there is no existing authorization structure that can be used to verify the authority of a router to originate link state information. We must introduce a new component to the OSPF design, an entity that is aware of the link state information that a router is authorized to originate and is known and trusted by all routers to certify that authority. We call this component a Trusted Entity. The Trusted Entity may be on-line or off-line and may be a function performed by any platform in the autonomous system. Because OSPF is an IGP running under a single technical administration, it is possible to identify such an entity and configure its identity into all the routers.

A digital signature attached to an LSA by the source router provides assurance that the data really does come from the originating router. It will provide assurance that the data has not been modified in transit. In the case where incorrect routing data is distributed by a faulty router, the signature provides a way to trace the problem to its source.

Note that the digital signatures protecting the LSAs do not relieve the need for hop-by-hop authentication and integrity protection, as the protocol exchanges like the hello protocol also need to be protected. However, those exchanges are communicated with directly connected neighbors. Symmetric keys shared with the neighbor can be part of the standard configuration information provided a router at startup (its attached networks, its router identification, etc.) and used with a message digest or cryptographic hash algorithm to provide authentication and integrity protection.

Each LSA is signed by the originating router. Each router is given a public, private key pair. The private key is used to sign the LSAs while the public key by the recipients of signed LSAs to verify the signatures.

All fields of the LSA are included in the signature except the age field. (The age field is modified by all the routers the LSA traverses and so it cannot be protected.) This leaves a vulnerability to an attack that sets the age field of an LSA to MaxAge and floods it through the network. As explained above, this will cause that LSA to be deleted from every router's database. However, the originating router has a legitimate need to create an LSA with an age field of MaxAge, when it no longer needs to be advertising a link and wishes to delete the LSA from the collective database. Therefore, we stipulate that when the originating router creates an LSA with an age of MaxAge, the age field will be included in the signature. Similarly, when routers receive an LSA with an age of MaxAge, the age field will be included in the verification.

In order for the LSA signatures to be verified by the recipients, they must have the public key. The public key is distributed to all the routers that need it for verification of signatures by including the key in a new LSA type and distributing it via the normal flooding mechanism. The new LSA type is called a Public Key LSA (PKLSA). The Trusted Entity certifies the router public key by signing the key. The creation time of the key is included to serve as an indicator of the appropriate router public key to use when more than one is present in the network at one time, i.e., during key rollover. The Trusted Entity also certifies the authority of the router to originate certain information by signing the role (ABR, ASBR, etc) of the router and the networks that it is allowed to include in LSAs. These certifications are included in the PKLSA. This PKLSA therefore allows both verification of the data origin authenticity of the signed LSAs and verification of the authority of the router to produce those LSAs. Each router must be configured with the public key of this Trusted Entity in order to verify the certifications in the PKLSA.

It is possible for a signed LSA to be received before the LSA carrying the corresponding public key. We provide that such a not-yet-verified LSA should be retained for an interval to see if the corresponding PKLSA shows up. LSAs that are discarded due to verification failures, that are stored waiting for keys or that are discarded because they are advertising links that are not authorized to the originating router must still be acknowledged to the neighbor. Otherwise the database synchronization process will cause the needless retransmission of the failed LSAs.

A router may flush its keys from routing tables by flooding a PKLSA for that key with the age set to MaxAge. This is called premature aging of the PKLSA and would be used if a key needed to be removed, for example when the key was discovered to be compromised. A key can also be removed from routing tables (superseded) by a PKLSA from the same router, containing a valid certificate for a new key with a more recent creation time. If a key is superseded by a more recent key, it is not necessary to flush the old key with a "MaxAge" PKLSA.

When a router public key changes and it originates a new LSA containing the new key, it must generate a new set of LSAs signed by the new key. To prevent the retention of information signed by the old key, we must insist that this LSA information be deleted from the collective database. However, we would not want key rollover to cause massive routing outages. We therefore specify that LSAs signed with the old key should be artificially aged to within a specified time interval of MaxAge. New LSAs should be originated and signed with the new key within that time interval. This should mean that new LSAs will arrive to take the place of the LSAs signed with the old key before they are deleted from the database. If new LSAs do not arrive then the old LSA data will disappear from the LSDB in a timely fashion.

Operation of the protected OSPF Each router must be configured with the key of the Trusted Entity and with its own private key. It must also be configured with its PKLSA. Generation of the public key and communication to the Trusted Entity for certification (or generation of the key pair by the Trusted Entity and communication of the private key to the router) is a matter of local procedures and out of band communication mechanisms.

The operation of a protected OSPF system faces some difficulties. The OSPF LSA packets are quite small, sometimes 18 bytes long. A digital signature can increase this size by a multiple depending on the size of the key. For an RSA 1024 bit key, the packet size increases more than seven fold. This has significant impact both on the bandwidth needed to transmit the protected LSAs and in the capacity needed to store the database of protected LSAs.

As noted, it is difficult to protect against misuse of the age field in the LSAs. Including the age field in the signature when the value is MaxAge prevents the wholesale deletion of LSA information from the database through a bogus packet. It does, however, prevent the speedup to database convergence mentioned above. If a router notices that an LSA (not one the router has itself originated) has naturally aged in its database to MaxAge and attempts to speed convergence by flooding this MaxAge LSA to the area, then the receiving routers will try to include the age field in the signature verification and will find that the signature on the LSA does not verify. We give up the speedup in database convergence in order to protect against a significant vulnerability.

The presence of areas in an OSPF autonomous system is attractive for scalable configuration. However, the fact that each ABR summarizes data when reporting to the inside of its attached areas means that the signature protection we apply cannot pass area borders. The most we could do to protect against an ABR that is originating bogus data would be to compare its announcements with the announcements of any other ABR attached to the area. The ABR's attached to the same area all have the same data and can therefore double check each other's announcements.

For flexibility, it would be good to be able to handle multiple digital signature algorithms in one OSPF system. However, because LSAs flood throughout their associated areas, each router in the area must understand the algorithm used to sign a protected LSA. Consequently, there is little to gain from allowing multiple algorithms to be used in one area. However, we were able to provide for flexibility between OSPF areas in an autonomous system. Areas may differ in their ability to use digital signature enhancements or in the cryptographic algorithm or key they use.

This flexibility can result in configuration inflexibility in an OSPF system. For example, each ABR sends LSA's to two areas. If the two areas do not use the same signature algorithm, then the ABR must be able to sign and verify LSA's with both algorithms. A more serious difficulty arises for the ASBR's. The routes to external networks announced by the ASBR's do get flooded to all areas of the network. If some areas use a different signature algorithm from the ASBR, the external routes the ASBR announces would not be verifiable in those areas. If a lowest common denominator algorithm is not available, then the ASBR must be placed on the backbone. Each area not using the ASBR's signature algorithm must be given a default route to the backbone to provide connectivity to the external addresses.

3.2. BGP 4

BGP 4 [16] is the most widely used routing protocol between autonomous systems. As an EGP, it presumes that shortest paths are not the most important selection criteria for a route and that routing policy is much more important. Therefore, it uses a *distance vector* technique. Each BGP speaker receives from its peers their view of the topology of the network, expressed as their distance to each reachable network prefix. The BGP speaker summarizes this information, computes its own distance to each network prefix that its routing policy will allow and transmits the result to its neighbors. (Because the distance metric used is the sequence of AS's the routing information has transited, BGP4 is sometimes described as a *path vector* algorithm.)

Because the BGP protocol summarizes the information and transmits its own view, the original source of the information is obscured. It is possible to ensure that the routing information received came from a peer BGP speaker, but it is not possible to ensure that the end source of the routing information is authorized. This has led to serious operational difficulties in the Internet. In particular, it is possible for a faulty or misconfigured BGP speaker to advertise a direct connection to a network that is actually part of another AS. This will disrupt routing to the improperly advertised network, as all traffic will go to the faulty router. In incidents where a number of networks have been so mis-advertised, the effect has spread beyond the mis-advertised networks because the excess mis-directed traffic overwhelms a portion of the Internet and causes problems for other traffic. To make matters worse, it is not always apparent which AS has injected the bogus advertisement into the Internet. Hours are lost as the source of the problem is painstakingly tracked from AS to AS.

No perfect answer exists to this problem. In particular, preventing improper advertisements of networks requires the existence of an authorization structure. One possible authorization structure could be built from the Internet Routing Registry systems. NAI Labs has been working with the Routing Policy System (RPS) working group to help define the security needed in the registries that use the RPS language. Others have suggested using the secured DNS as an authorization structure for network advertisements. The BBN S-BGP work (described briefly below) describes another address allocation authorization hierarchy rooted at the Internet Corporation for Assigned Names and Numbers (ICANN).

Given that an authorization structure exists, it is still necessary to authenticate the end source of the routing information and verify the integrity of the routing messages, in order to verify the authorization. First, as with OSPF, hop-by-hop authenticity and integrity protection are needed to protect the protocol message that do not carry routing information but still effect the behavior of the routing protocol. Second, digital signatures could be used to protect the routing information. Digital signatures are needed because, as with OSPF, there is no way to predict which routers might receive the information. That means that there is no way to preconfigure symmetric shared keys with the necessary routers or choose the appropriate symmetric shared keys to use to protect the message.

There are four different techniques that could provide some protection for the routing information contained in the routing messages. The distances that BGP carries in its routing information are expressed in terms of the sequence of AS's that would carry the traffic (AS_PATH). When BGP chooses the route it would use to forward traffic to a destination network, it prepends its own AS number to the AS_PATH it was provided and transmits the result to its neighbors. This has the advantage to proving authentication and authorization of preserving some idea of the source of the routing information. The four protection techniques provide increasing protection at increasing cost.

- Sign the originating AS
- Sign the originating AS and the predecessor information
- Sign the originating AS and nest signatures of prefixes of the AS_PATH
- Rely on the Internet Routing Registry to confirm the authenticity of the AS_PATH

Sign the originating AS The first technique for protecting the BGP routing information is to authenticate the source of network prefix advertisements. Any router originating an advertisement for a network prefix should digitally sign the advertisement. This information could be carried as a new attribute of the BGP message containing the network advertisement. This gives an authentic identity of the originator of a network advertisement that could be referred to the authorization structure to determine if it is an authorized advertisement. (Even without an authorization infrastructure, this information has the benefit that it would facilitate the discovery of the culprit of an improper advertisement.) The digital signatures would be verified at each receiving router and would be passed on with chosen routes.

The costs related to this technique are the expense of providing a public key infrastructure so that the signatures could be verified, of configuring each router with its public key, of generating and verifying the signatures at the routers, and of transmitting and storing the signatures at all the routers. Since a network infrequently changes its connection point to the routing world, a signed origination advertisement could be cached for a considerable time, reducing the verification processing cost to the receiving routers.

This step provides a solution to a current common failure in the routing world, but it does not completely correct the vulnerabilities in the BGP routing protocol. Knowing that the originator of an advertisement to a network prefix in a route advertisement is authentic and authorized does nothing to ensure that the AS_PATH in the route advertisement is authentic and authorized.

Sign the originating AS and the predecessor information A second technique is suggested in [17]. The router originating an advertisement to a network prefix will digitally sign the advertisement as suggested above. It will include in the signature the identity of the peer to which it will send this information - i.e., its immediate predecessor on the path to the destination advertised. Receiving routers can cache the signed predecessor statements. The AS_PATH in any BGP route can be checked against the database of these signed predecessor statements to ensure that the adjacencies represented by successive AS's in the AS_PATH are legitimate. This works a bit like the signed link state information in the protected OSPF protocol.

The costs related to this technique are all the costs associated with the signing of the originating AS: providing a public key infrastructure; configuring routers; processing, transmission and storage costs of the signatures. Added to this is the cost of signing and verifying the predecessor information and of verifying AS_PATHs against the predecessor database. While the signed origination advertisements can be cached because connection points change infrequently, the adjacencies in the Internet change more often and caching the signed predecessor information may not bring much relief from the signature verification load.

This step provides additional assurance that the BGP routes received are authentic and authorized. It assures that successive AS's in the AS_PATH are indeed peers. However, that does not assure that the path as a whole is valid. BGP peers are free to decide on any basis whether or not they will transit traffic for a network prefix to their peers. Valid adjacencies can be pasted together to make an AS_PATH that would pass the verification check against the predecessor database, but traffic forwarded on that path might not be delivered. Peering arrangements and routing policy for carrying transit traffic are the subject of brisk negotiation in business arrangements between network providers. It is unknown how many of the valid peering adjacencies in the Internet can be certain to carry all of each other's traffic load. So it is unknown how often this technique would identify an AS_PATH as valid when it was not.

Sign the originating AS and nest signatures of prefixes of the AS_PATH A further step in assuring that the BGP routes are authentic and authorized relies on further digital signatures. Each router originating an advertisement to a network prefix will digitally sign the advertisement as suggested above. Each router that adds its AS number to the AS_PATH in a received route and passes the augmented AS_PATH in a new route to its neighbor will sign that new route. It will also include the signature it was passed with the received route. The included signature serves as authentication that the router constructed the AS_PATH from an authentic path. The neighbor will sign the augmentation it adds to the AS_PATH and use this router's signature to authenticate its augmentation. This process generates multiple signatures over nested prefixes of the AS_PATH.

A BGP speaker could snip out a suffix of the data it received as well as the associated signatures and pass those on as the proof that its AS_PATH was based on an authentic path. To prevent this, the signatures generated should cover not only the route information but the intended receiver as well.

For example, suppose AS_1 receives a route to $N_i$from AS_2 containing an AS_PATH of $\{AS_2AS_3...AS_k\}$. Then AS_1 should also receive

- AS_2's signature of "$AS_1, \{AS_2AS_3...AS_k\}, N_i$",
- AS_3's signature of "$AS_2, \{AS_3...AS_k\}, N_i$",
- AS_4's signature of "$AS_3, \{AS_4...AS_k\}, N_i$", etc.

AS_1 would verify the authenticity of AS_2's route by verifying AS_3's signature and the signatures that follow. AS_1 does not need to verify AS_2's signature to verify the authenticity of the route it received (given the hop-by-hop protections discussed above) but it does need to include AS_2's signature in the routes it transmits to demonstrate the authenticity of *its* routes.

AS_1 would compute a path as $\{AS_1AS_2AS_3...AS_k\}$, and pass this path to AS_0 along with its signature of "$AS_0, \{AS_1AS_2AS_3...AS_k\}, N_i$", AS_2's signature of "$AS_1, \{AS_2AS_3...AS_k\}, N_i$", and the other signatures.

The costs related to this technique are all the costs associated with the signing of the originating AS: providing a public key infrastructure; configuring routers; processing, transmission and storage costs. This technique also requires that each router generate a new signature for each route that it produces and passes to a neighbors. Multiple signatures must be verified on each incoming route - as many of the nested signature must be verified as are needed to assure that the route is valid. If it is presumed that there might be no more than K faulty routers in the network, then no more than K signatures need to be verified.

This is quite a costly technique, but it does provide the maximum assurance that the routes are authentic and authorized.

Sign the originating AS and rely on the registry to confirm the authenticity of the AS_PATH Since verifying the authorization of an originated advertisement for a network prefix requires a referral to a routing authorization structure, it would be possible to have that structure verify the entire route as well. Most Internet Routing Registries, for example, collect a sophisticated set of information from their registrants that could be sufficient to verify the AS_PATH as a whole. The communication with the routing authorization infrastructure would itself have to be protected, of course. The assurance derived from this technique would depend on the assurance that the authorization infrastructure maintained the authenticity and authority of the recorded information.

The assurance also depends on the registrants having been completely open about the information they provide. As previously mentioned, information about routing policies is the subject of business agreements that might be considered too sensitive to expose. The engineering problems of ensuring timely transmission to and from the authorization structure would also have to be addressed.

Operational challenges The BGP routing system as described above is a simplified version of actual practice. For scalability, routers frequently combine the routes they receive to several network prefixes and advertise routes to an aggregate network prefix covering them all. The AS's in the AS_PATHs of the combined routes are listed as a set, without maintaining any distinction among them. This aggregation is very important to the efficient operation of the Internet routing system. Unfortunately, it obscures the meaning of the AS_PATH in that traffic to any network address covered by the aggregate prefix is only guaranteed to pass through some, not necessarily all, of the AS's listed in a set. Similarly, the validity checks must be satisfied if any one path through the AS's in the set is authenticated. To assure recipients of the validity of the aggregated route, it would be necessary to pass on the text and the nested signatures of the aggregated component routes. This means an increase in transmission bandwidth at each aggregation point and a similar increase in verification time at each verification point's peers. In the likely case that the signatures dominate the total space required for the aggregated routes, the percentage reduction in the total space would not show as much effect from the reduction in the total network address space as is presently the case.

BBN's S-BGP project [8, 11] [2] is exploring a technique that employs signatures of nested prefixes of the AS_PATH ("route attestations") as suggested here. Their work includes use of IPSEC for hop-by-hop protection of authenticity, integrity and replay of BGP messages. Their work incorporates signed authorization of network prefix originations ("address attestations") into each routing message. Verification of this signature uses an infrastructure of a hierarchy of signed authorizations for address blocks rooted at the ICANN.

In implementing this design, they have reported that the increase in the size of the routing database is daunting. The concomitant cryptographic processing can be manageable if a significant portion of new routing information reinstalls a recently withdrawn route (known as "route flap") and the cryptographic information associated with withdrawn routes is cached instead of being discarded. They are investigating this and other techniques to reduce the processing and storage costs of the technique.

4. Simple Network Management Protocol

Establishing a standards-based network management protocol with effective security has been an elusive requirement for many years. As the principal body that establishes most Internet standards, the IETF has made several attempts at establishing such a standard that until recently have not come to fruition. Previous versions of network management protocols that were IETF standards did not provide useful security. SNMPv1 is the most widely used IETF network management standard but provides only static password protection which, in many cases, are not changed from the well known defaults of "public" and "private" (which are transmitted in the clear). Though the standard is nearly ten years old, only recently has the IETF been able to progress a standard with sufficient security for the threats faced by today's networks. The absence of a standards-based network management protocol with sufficient security has forced both the public and private sector to resort to unique, individual (often functionally similar but not interoperable) approaches for securely managing their networks and applications.

The most recent IETF effort to develop a single network management standard with effective security has made significant progress. The principal objective of this effort

[2] This Internet Draft was released as the paper was being prepared for print, leaving too little time for an in depth review. The statements here are based on previous discussions and may not reflect the current work.

is to provide a network management standard with security for authenticity, integrity and privacy based on cryptographic mechanisms. A SNMP Advisory Team, led by NAI Labs, developed an approach which became the basis for the new network management protocol being defined by the SNMP version 3 (SNMPv3) Working Group. As a result of the work of the SNMPv3 WG, the IETF recently published a set of RFCs as Informational and Draft Standards [3, 4, 5, 10, 19, 20] which provide the principal specifications needed for the new, secure network management standard. NAI Labs led both the SNMP Advisory Team and the SNMPv3 Working Group with the objective of providing the necessary basis for the secure network management needed by current and emerging networks.

The SNMP architecture is described in [19] in terms of modular units that are given distinct functions and can be implemented separately. The major components of the architecture are an SNMP engine containing a Message Processing Subsystem, a Security Subsystem and an Access Control Subsystem, and possibly multiple SNMP applications which provide specific functional processing of management data. A SNMPv3 management system contains:

- several (potentially many) nodes, each with an SNMP entity containing an engine and at least a command responder and a notification generator application which have access to management instrumentation (these are traditionally called agents);
- at least one SNMP entity with an engine and command generator and/or notification receiver application (these are traditionally called a manager); and
- the SNMPV3 management protocol used to convey management information between the SNMP entities.

A less precise but more common description would be that SNMP agents on hosts (or other managed nodes) report information to and respond to directions from their manager using the SNMP protocol across a network which does not provide any guarantee for security of any kind.

Obviously, the correct operation of the SNMPv3 system relies on the protection of the commands and notifications that pass between agents and managers in the system. The principal threats seen to an SNMPv3 system are:

Modification: the risk that an unauthorized entity may modify messages in flight so as to cause unauthorized operations, including causing an object to be set to a false value or causing the false reporting of information.

Masquerade: the risk that an unauthorized principal may assume the identity of an authorized principal and thereby gain the ability to cause some management operation otherwise not available to it.

Message Stream Modification: the risk that the order or timeliness of delivery of a sequence of message may be altered so as to cause an unauthorized or unintended management operation.

Disclosure the risk that sensitive management data might be exposed to an unauthorized entity through eavesdropping on a management exchange.

During the design process, the SNMPv3 Working Group decided specifically not to have the architecture attempt to protect against denial of service and traffic analysis threats.

Responsibility for protection against the defined threats is divided between the Security Subsystem and the Access Control Subsystem. The Security Subsystem is responsible for protecting messages in transmission and receipt between agents and managers. The Access Control Subsystem is responsible for protecting access to sensitive management information. Another way of describing this is the Security Subsystem provides protections "on the wire" between SNMP entities and the Access Control Subsystem provides protection for management information within an particular SNMP entity. Each of these Subsystems has associated management information that can be controlled through use of SNMPv3 mechanisms. Within the architecture, each of the Subsystems needs to have one or more "models" specified. Currently, the SNMPv3 working group has specified only one model for each of the Subsystems. The current Security Subsystem model is the User-based Security Model (USM) [3] while the Access Control Subsystem is the View-based Access Control Model (VACM) [20].

The User-based Security Model (USM) provides cryptographic protection of data integrity, data origin authentication, data confidentiality, and message timeliness and limited replay protection. As the model of communication in an SNMPv3 system is point to point communication and the number of principals who might be permitted to participate in network management is presumably restricted, it is manageable to protect the communication with shared secrets and symmetric cryptography. This presumes that users desiring to participate in the management of a device must first make themselves known so that appropriate keys for the user can be distributed to the agents and managers.

The USM uses a keyed message digest (either HMAC-MD5-96 or HMAC-SHA-96) to protect the data integrity and origin authentication of a message. It uses the DES-CBC symmetric encryption algorithm to protect the confidentiality of a message. Replay and timeliness protection are provided by including timeliness indicators in the message that are compared to the last message received from the same source and by including the generation time in the message that must be a recent time to be accepted.

The origin of a message is the user on whose behalf the message was generated. The user must be known in each

SNMPv3 engine that might generate or receive messages for that user. The attributes of a user (security name and any authentication and privacy keys) are known to each SNMPv3 engine. The user's keys are *localized* to each SNMPv3 engine to which the user is known. The localized key is constructed from a hash of a key computed from the user's password and the identification of the local SNMPv3 engine.

The USM defines three security levels that are associated with each message sent. The levels are: no security (noAuthNoPriv), authentication only(authNoPriv) , and authentication with privacy (authPriv). These three values are ordered such that noAuthNoPriv is less than authNoPriv and authNoPriv is less than authPriv. Every message has an associated security level. All Subsystems (Message Processing, Security, and Access Control) and applications are required to either supply a security level or to abide by the supplied security level while processing the message and its contents. When generating a message, USM will use the key and algorithm appropriate to the user to apply the needed security. It is not possible to request privacy protection alone, so if privacy is needed, authentication will be applied as well. On receipt, if the security level of a message is not supported by the user (e.g., if authentication and privacy are noted but the user supports authentication only) then the message is rejected.

The Access Control Subsystem is responsible for controlling access to the management information stored on an SNMPv3 node. This management information is viewed as a collection of managed objects, residing in a virtual information store, termed the Management Information Base (MIB). This information store is organized in a simple tree structure. Collections of related objects are defined in MIB modules. The View-based Access Control Model defines mechanisms to determine whether access to a managed object should be allowed. It defines a particular access decision function in order to support decisions regarding access rights.

Messages are characterized as belonging to one of five classes:

- Read (retrieve management information, i.e., GET),
- Write (modify management information, i.e., SET),
- Response (sent in response to a request, i.e., a Response or Report),
- Notification (send a notification to a notification receiver application, i.e., Trap or Inform), or
- Internal (exchanged between SNMPv3 engines, i.e., Report)

The VACM defines *groups* that are aggregates of users all of whom have the same access rights to the MIB. Each user is a part of at most one group. The access rights for a group can be separately defined for the three different security levels. For proper protection of the MIB, the access rights of a group is restricted to some subset of the MIB. These subsets are defined as *MIB views* and are defined as a set of managed objects within the MIB. Because the MIB is organized in a tree structure, these MIB views can be conveniently expressed as a set of subtrees of the MIB. A view subtree is defined as all MIB objects that have a common ASN.1 Object Identifier (OID) prefix. A family of view subtrees can be defined by specifying a mask that identifies the significant portions of the OID to be considered in deciding membership in the family.

The VACM divides accesses into read access (which occurs when processing a Read Class message), write access (occurs when processing a Write Class message) and notify access (occurs when processing a Notify Class message). The access rights for each are defined in terms of the view subtrees that are accessible for those accesses. Consequently, these are called the read-view, the write-view and the notify-view. These are stored in the VACM Access Table, indexed by the group name and security level.

The current SNMP architecture provides for several types of protection for management information between SNMP entities and protection within an SNMP entity. The architecture also provides the means for linking these protections so that management information can have protection that is consistent with its sensitivity. Since the management information associated with the Security Subsystem and the Access Control Subsystem is some of the most sensitive on any SNMPv3 entity, that information can require authPriv for all messages carrying such information and restrict access to the information to only the group responsible for the remote management of the entity. For other information that may not be as sensitive, for instance the system name or Internet address of an entity, it may be read by any user with access to the entity and sent in messages without any type of cryptographic protection or authentication.

5. Conclusions

The experience of providing security enhancements to existing protocols has several key lessons.

First, authorization problems subsume authentication problems and are much more difficult to address in a distributed cooperative protocol like DNS or routing. If the protocol has a natural authorization structure like DNS, this can sometimes be put to use to provide the cryptographic authorization. Even then, if the natural authorization structure does not match the security authorization structure, the operational procedures will not adequately handle the difference. Existing operations are difficult to retrofit to the requirements of the security. This is the case for the KEY records in DNS, which must be signed by the delegating zone but are authoritative only in the delegated zone.

When there is no authorization structure to start from, one must be invented. Authentication in a distributed protocol means little if it is impossible to determine if the authentic source of the information has the authority to provide it. Creating a distributed authorization structure for a process already in operation is not a simple task.

Second, end to end authentication is important. In a distributed protocol in a world-wide network, failures of the participants must be expected. Nothing can prevent a failed participant from having an effect on the portion of the protocol operation under its direct control. Any protocol properly designed to be fault tolerant should be ready to deal with this. But if authentication is only hop-by-hop, then any one faulty, mis-configured or subverted participant has the ability to effect the protocol operation over the entire network by inserting bogus information or information masquerading as from an authorized source. As a case in point, the Internet has had hop-by-hop protection between routing peers for a few years now. Even so, outages in the Internet caused by routing problems are a frequent occurrence because routers or autonomous systems themselves fail in interesting ways.

Third, security can be expensive. Cryptographic protections can be expensive in terms of configuration, storage space, bandwidth, and processing time. The design of infrastructure nodes may be impacted as well. For example, processing of cryptographic protections not only takes additional time, it may also force packets to take a different processing path as decisions are delayed waiting for the outcome of a verification or force a packet to be processed twice. Individual users may be willing to make a tradeoff between security and its cost. But the widespread impact of a security failure in infrastructure protocols weighs the tradeoff in the favor of security. Unfortunately, we have seen evidence that individual infrastructure providers may be unwilling to bear this cost alone.

Fourth, protocol design decisions made before security is envisioned can exacerbate the problems associated with retrofitting security later. For example, protocols that decide on small packet sizes over larger aggregated transmissions in order to improve flexibility and responsiveness in operations will be impacted more severely when protections need to be added to each packet. Furthermore, implementation design decisions can interact badly with security provisions. Data may not be stored in a structure amenable to application of cryptographic protections, for example. Our experience here reiterates a common wisdom – security should be considered from the beginning of the design process.

Implementation of the protections discussed here can be found at ftp://ftp.tislabs.com/pub in the DNNSEC and OSPFSIGN directories. An implementation of the SNMPv3 work can be found at http://http://ucd-snmp.ucdavis.edu. For more details of the DNSSEC protections, see [6]. For more details of the SNMPv3 protections, see [19, 3, 20]. For more details of the OSPF protections, see [13].

References

[1] F. Baker and R. Atkinson. Rip version 2. Internet RFC 2082, Cisco Systems, January 1997.

[2] S. M. Bellovin. Using the domain name system for system break-ins. In *Proceedings of the Fifth Usenix UNIX Security Symposium*, June 1995.

[3] U. Blumenthal and B. Wijnen. User-based security model (usm) for version 3 of the simple network management protocol (snmpv3). Internet RFC 2574, IBM T. J. Watson Research, April 1999.

[4] J. Case, D. Harrington, R. Presuhn, and B. Wijnen. Message processing and dispatching for the simple network management protocol (snmp). Internet RFC 2572, SNMP Research Inc., Cabletron Systems, Inc., BMC Software, Inc., IBM T. J. Watson Research, May 1999.

[5] J. Case, R. Mundy, D. Partain, and B. Stewart. Introduction to version 3 of the internet-standard network management framework. Internet RFC 2570, SNMP Research, Inc., TIS Labs at NEtwork Associates, Cisco Systems, April 1999.

[6] D. Eastlake. Domain name system security extensions. Internet RFC 2535, IBM Corporation, March 1999.

[7] G. G. Finn. Reducing the vulnerability of dynamic computer networks. Technical report, University of Southern California Information Sciences Institute, Marina del Rey, California, June 1988.

[8] S. Kent, C. Lynn, L. Sanchez, M. Steenstrup, M. Casagni, and K. Seo. Secure border gateway protocol. http://www.net-tech.bbn.com/sbgp/sbgp-index.html, August 1998.

[9] B. Kumar and J. Crowcroft. Integrating security in inter-domain routing protocols. *Computer Communications Review*, 23(5), October 1993.

[10] D. Levi, P. Meyer, and B. Stewart. Snmp applications. Internet RFC 2573, SNMP Research, Inc., Secure Computing Corporation, Cisco Systems, April 1999.

[11] C. Lynn, J. Mikkelson, and K. Seo. Secure BGP (S-BGP). work in progress. Internet Draft draft-clynn-s-bgp-protocol-00.txt, October 1999.

[12] J. Moy. Ospf version 2. Internet RFC 2328, Ascend Communications, Inc., April 1998.

[13] S. Murphy, M. Badger, and B. Wellington. Ospf with digital signatures. Internet RFC 1771, Trusted Information Systems, June 1997.

[14] R. Perlman. *Network Layer Protocols with Byzantine Robustness*. PhD thesis, Department of Electrical Engineering and Computer Science, Massachusetts Institute of Technology, August 1988.

[15] R. Perlman. *Interconnections: Bridges and Routers*. Addison-Wesley, Reading, Mass., 1992.

[16] Y. Rekhter and T. Li. A border gateway protocol 4 (bgp-4). Internet RFC 1771, IBM Corp. and Cisco Systems, March 1995.

[17] B. Smith and J. Garcia-Luna-Aceves. Securing the border gateway routing protocol. In *Proc. Global Internet'96*, November 1996.

[18] M. Steenstrup. Inter-domain policy routing protocol specification: Version 1. Internet RFC 1479, BBN Systems and Technologies, July 1993.

[19] B. Wijnen, D. Harrington, and R. Presuhn. An architecture for describing snmp management frameworks. Internet RFC 2571, Cabletron Systems, Inc., BMC Software, Inc., IBM T. J. Watson Research, May 1999.

[20] B. Wijnen, R. Presuhn, and K. McCloghrie. View-based access control model (vacm) for the simple network management protocol (snmp). Internet RFC 2575, IBM T. J. Watson Research, BMC Software, Inc., Cisco Systems, April 1999.

Design and Analysis of the Secure Border Gateway Protocol (S-BGP)

Stephen Kent , Charles Lynn, Karen Seo
BBN Technologies

Abstract

The Border Gateway Protocol (BGP), which is used to distribute routing information between autonomous systems (ASes), is a critical component of the Internet's routing infrastructure. It is highly vulnerable to a variety of malicious attacks, due to the lack of a secure means of verifying the authenticity and legitimacy of BGP control traffic. This document describes a secure, scalable, deployable architecture, S-BGP, for a system that addresses most of the security problems associated with BGP. The paper discusses the vulnerabilities and security requirements associated with BGP, describes the S-BGP countermeasures, and explains how they address these vulnerabilities and requirements. The paper also provides a comparison of this architecture with other approaches that have been proposed, analyzes the performance implications of the proposed countermeasures, and reports on prototype implementation experience.

1 Problem Description

Internet routing is based on a distributed system composed of many routers, grouped into management domains called Autonomous Systems (ASes). Routing information is exchanged between ASes in Border Gateway Protocol (BGP) [1] UPDATE messages. BGP has proven to be highly vulnerable to a variety of attacks [2], due to the lack of a scalable means of verifying the authenticity and legitimacy of BGP control traffic. In April 1997, we began work on the security architecture described in this paper. In this section we describe the problem—how the protocol works, the nature of observed BGP traffic in the Internet, the correct operation of BGP, the threat model and BGP vulnerabilities, and the goals, constraints and assumptions that apply to the proposed countermeasures.

1.1 Overview of BGP

The BGP-4 protocol, both message syntax and the route propagation algorithm, is described in [1]. Routers implementing BGP, BGP "speakers," exchange routing information via UPDATE messages. An UPDATE message consists of three parts: a list of address prefixes for destinations that are no longer reachable via the previously specified route; a list of prefixes that are reachable; and the characteristics of the cumulative path and current inter-AS hop, contained in path attributes, that can be used to reach the address prefixes. (A prefix specifies an IP address block, and consists of a count of the initial significant bits in an IP address and the value of those bits.) The attribute used to specify the inter-AS path, the AS_PATH attribute, specifies a sequence of Autonomous Systems (ASes) along the path, each identified by its AS number. When propagating an UPDATE to a neighboring AS, the BGP speaker prepends its AS number to the sequence, and updates certain other path attributes. Since an UPDATE can specify only one path, only prefixes that share that path may be aggregated into the UPDATE.

The backbone routers of the major internet service providers (ISPs) have a route to every reachable IP address. Analysis of BGP UPDATEs recorded during January 1999, showed routing databases that contained about 61,000 IPv4 address prefixes. Each (non-leaf) BGP speaker maintains a full routing table, and sends its best route for each prefix to each neighbor speaker. When a BGP speaker reboots, it receives complete routing tables (via UPDATEs) from each of its neighbors. The worst case arises at Network Access Points (NAPs), where ISPs are connected together via a high speed (100Mb/s) LAN. A BGP speaker at a NAP might have about 30 peers.

On a daily basis, a BGP speaker at a NAP receives about 1425 UPDATEs from each peer, an average UPDATE rate of about 1 per minute per peer. This rate is affected somewhat by Internet growth, but is mostly a function of UPDATEs sent due to link, component, or congestive failures and recoveries. Analysis shows that about 50% of all UPDATEs are sent as a result of route "flaps," i.e., transient communication failures that, when remedied, result in a return to the original route. This sort of routing behavior has long been characteristic of the Internet [3] and the proposed security mechanisms take advantage of this behavior to achieve acceptable

0-7695-0490-6/99 $10.00

performance, as discussed in Section 6. (David Mills, an architect of the NSFNET, confirmed that route flapping has been common in the Internet since the mid-80's.)

1.2 Correct operation of BGP

Security for BGP is defined by the correct operation of BGP speakers. This definition is based on the observation that any successful attack against BGP should result in other than correct operation, presumably yielding degraded operation. Correct operation of BGP depends upon the integrity, authenticity, and timeliness of the routing information it distributes as well as each BGP speaker's processing, storing, and distribution of this information in accordance with both the BGP specification and with the (local) routing policies of the BGP speaker's AS. The following statements characterize the primary correct operation features of BGP.

- Each UPDATE received by a BGP speaker from a peer was sent by the indicated peer, was not modified en route from the peer, and contains routing information no less recent than the routing information previously received for the indicated prefixes from that peer.
- The UPDATE was intended for receipt by the peer that received it.
- The peer that sent the UPDATE was authorized to act on behalf of its AS to advertise the routing information contained within the UPDATE to BGP peers in the recipient AS.
- The owner of an address space corresponding to a reachable prefix advertised in an UPDATE was authorized by its parent organization to own that address space.
- The first AS in the route was authorized, by the owners of the address space corresponding to the set of reachable prefixes, to advertise those prefixes.
- If the UPDATE indicates a withdrawn route, then the peer withdrawing the route was a legitimate advertiser for that route, prior to its withdrawal.
- The peer that sent the UPDATE correctly applied the BGP rules and its AS's routing policies in modifying, storing, and distributing the UPDATE, in selecting the route, and in deriving forwarding information from it.
- The BGP speaker that received the UPDATE correctly applied the BGP rules and its AS's routing policies in determining whether to accept the UPDATE.

The countermeasures developed for S-BGP meet the first six of these criteria, even in the face of subversion of BGP speakers (Byzantine failures). Section 4 provides a detailed analysis of how each countermeasure contributes to correct operation. However, because the local policy features of BGP allow a speaker considerable latitude in determining how to process an UPDATE, these countermeasures cannot meet the last two criteria, i.e., such attacks could be attributed to local policies not visible outside an AS. To address such attacks, the semantics of BGP itself would have to change. Moreover, because UPDATEs do not carry sequence numbers, a BGP speaker can generate an UPDATE based on old information, e.g., withdrawing or reasserting a route based on outdated information. Thus the temporal accuracy of UPDATEs, in the face of Byzantine failures, is enforced only very coarsely by these countermeasures. (Section 5 provides more details on residual vulnerabilities.)

1.3 Threat model and BGP vulnerabilities

BGP has a number of vulnerabilities that can be exploited to cause problems such as misdelivery or non-delivery of user traffic, misuse of network resources, network congestion and packet delays, and violation of local routing policies.

Communication between BGP peers is subject to active and passive wiretapping. BGP uses TCP/IP for transport and this protocol, and its payload, can be attacked. A speaker's BGP-related software, configuration information, or routing databases may be modified or replaced illicitly via unauthorized access to a router, or to a server from which router software is downloaded, or via a spoofed distribution channel. Most of these attacks transform routers into hostile "insiders." Effective security measures must address such Byzantine attacks.

Exploitation of these vulnerabilities allows a variety of attacks. For example, fictitious BGP messages might be injected into a link (spoofing). Authentic BGP messages might be captured and either modified and re-injected into the link, combined incorrectly, or suppressed altogether. A compromised BGP speaker could generate UPDATEs for routes that do not, legitimately, pass through that speaker. All of these attacks are countered by the mechanisms described in Section 3.

UPDATE messages could be generated too frequently by a compromised BGP speaker, or the selection of routes and distribution of UPDATEs could violate the local routing policies. These failures are not addressed by the proposed countermeasures.

Better physical and procedural security for network management facilities, BGP speakers, and communication links; link-level encryption of inter-router (BGP speaker) traffic; and end-to-end encryption of management information would reduce some of these vulnerabilities. However, some aspects of such security approaches are economically unattractive or infeasible. Moreover, accidental (vs. malicious) misconfiguration would not be prevented by such measures and such misconfiguration has proven to be a source of several significant Internet outages in the past. Any security approach that leaves BGP vulnerable to such benign "attacks" violates the "principle of least privilege" and leaves the Internet routing system vulnerable at its weakest link. In contrast, the security approach described here satisfies this principle, so that any attack on any component of the routing system is limited in its impact on the Internet as a whole.

1.4 Goals, constraints, and assumptions

In order to create countermeasures that are both effective and practical, the S-BGP architecture is based on the following goals, constraints, and assumptions.

The S-BGP architecture must handle the projected growth and usage of the Internet in terms of performance (storage, processing, network bandwidth). It should be dynamic (responding automatically to topology changes, including the addition of new networks, routers and ASes) and scalable (able to handle the growth of the Internet in terms of addresses, routes, BGP control traffic, etc.).

The countermeasures must be consistent with the BGP protocol standards and with the likely evolution of these standards. This includes packet size limits, e.g., 4096 byte maximum for UPDATEs, and BGP features, e.g., path aggregation, communities, and multi-protocol support, e.g., multi-protocol label switching (MPLS). For example, to avoid modifying BGP packet formats, we have chosen to employ the BGP optional, transitive path attribute as a mechanism for distributing countermeasures information. Non S-BGP routers should pass this attribute type transparently, without understanding it.

The S-BGP architecture must be deployable. A primary goal of this work is to not only find countermeasures for BGP vulnerabilities but to cause them to be adopted by ISPs and router vendors. To accomplish this, the countermeasures must use available technology that can be incrementally deployed and they should leverage off of the existing infrastructure, e.g., the Internet Corporation for Assigned Names and Numbers (ICANN), and routing registries. In addition, they must avoid dependency loops; the architecture cannot depend on correct operation of inter-AS routing during initialization, e.g., it cannot rely on non-local databases.

2 Prior work

The earliest significant work published on the topic of routing protocol security is Perlman's doctoral dissertation [21]. The S-BGP design shares several features of that work, e.g., we address Byzantine failures and make extensive use of digital signatures. However we differ in many other respects, e.g., our design applies to a standard exterior (vs. interior) routing protocol, and we pay considerable attention to infrastructure and performance implications.

At the time that we began this work, previously published work on improving the security of BGP, and more generally distance-vector protocols, included proposals for adding sequence numbers to BGP messages [4,5,6], authentication of BGP messages [1,5,6], neighbor-to-neighbor encryption of BGP messages [4], and adding information to UPDATE messages to protect against tampering as the UPDATE propagates around the Internet [4,5,7].

None of this work proposed a comprehensive solution to the BGP security problems described above; each focused on one or more aspects of the problem without considering the full range of issues that are critical to a viable solution. For example, none addressed issues associated with the generation and distribution of public key certificates and certificate revocation lists (CRLs) needed to support validation of signed UPDATEs. Some proposals made changes to BGP that are inconsistent with the protocol standards, a reasonable approach only if one were presented with a "clean slate." None of the prior work examined the statistics of BGP operating in the Internet; this sometimes led authors to focus on performance concerns that are not the major impediment to deploying viable solutions. Some of the work developed solutions for distance vector protocols, but erroneously claimed applicability to BGP, which is described as a path vector protocol.

In contrast, the BGP security architecture reported in this paper is comprehensive, including a design for the infrastructure needed to establish and maintain the system. The optional transitive path attribute it employs is consistent with BGP standards and can be safely carried through routers not implementing S-BGP. This architecture incorporates the notion of an *address attestation*, which establishes that a "first hop" BGP speaker is authorized to advertise a route to a destination. No prior work includes an equivalent notion. Finally, the performance of the design presented here has been modeled based on actual BGP statistics, and has been tested via a live BGP feed from an ISP. No other work has been so rigorously analyzed from a performance perspective.

In [7], the scheme proposed is similar to our route attestations (RAs) in that before distributing an UPDATE to an external neighbor, the BGP speaker signs the route. Our approach differs from the scheme proposed in [7] in that we sign a routing data structure that specifies the next hop AS, explicitly indicating that this AS is authorized to advertise the route in question to the identified neighbor. Hence, our approach avoids a vulnerability not addressed in [7], i.e., provides protection against "cut and paste attacks" in which a BGP speaker inserts itself into a route (using a valid UPDATE containing a route which the speaker was not authorized to use).

The approach proposed in [4,5] was developed in response to the perceived communication and computation overhead of schemes such as [7]. In the case of [7] (and our route attestations), each of the signatures must be carried in each UPDATE and verified by each recipient to validate each received UPDATE. The approach proposed in [4,5] includes only a single signature for a route in an UPDATE; this signature covers only the destination and penultimate ASes listed in the path and is generated by a BGP speaker in the destination AS. While the overhead of [4,5] is less than that of [7] or our route attestations, it fails to provide the protection afforded by the iterated signature schemes in an environment with more sophisticated routing policies (e.g., policies other than shortest path), such as those typically supported by BGP. Moreover, our analysis shows that generation, validation, and transmission of digital signatures does not impose an unacceptable computational or communication burden (see Section 6).

3 Proposed countermeasures

The approach adopted to securing BGP route distribution involves two Public Key Infrastructures (PKIs), a new path attribute containing "attestations", and the use of IPsec. These components are used by a BGP speaker to validate the authenticity and data integrity of BGP UPDATEs that it receives and to verify the identity and authorization of the senders. This section discusses in more detail the PKIs and certificates, the attestations, the use of IPsec, and the distribution of this countermeasure information.

3.1 PKIs and certificates

S-BGP uses two PKIs, based on X.509 (version 3) certificates, to enable BGP speakers to validate the identities and authorization of BGP speakers and of owners of ASes and of portions of the IP address space. These PKIs parallel the existing IP address and AS number assignment delegation system and take advantage of this extant infrastructure. Because these PKIs mirror existing infrastructure, their creation avoids many of the "trust" issues that often complicate the creation of a PKI. The two PKIs involve four types of certificates, as illustrated below. In the diagrams:

- The higher node is the issuer for the certificates defined in the tier below it.
- The name of the current tree node (organization, AS, router, etc.) is the subject of the certificate.
- Any additional fields shown in the node, e.g., address block(s), are in an extension in the certificate.
- Other X.509 certificate fields are assumed, but not shown.

Note that the organizations that assign addresses (Registries, ISPs, DSPs, etc.) and the organizations that obtain autonomous system numbers from a Registry may be different. An organization could receive its AS number from a registry and its address block from an ISP. So the Org3_4 shown in the first PKI hierarchy (Figure 1) could correspond to the Org 1 shown in the second PKI hierarchy (Figure 2).

3.1.1 A PKI for address allocation. This architecture calls for a certificate to be issued to each organization that is granted "ownership" of a portion of the IP address space. This certificate is issued through the same chain of entities that, in the existing environment, is responsible for address allocation. The root of this chain is the ICANN, followed by regional address space allocation authorities (e.g., ARIN and RIPE), ISPs, DSPs, and end users. Note that the proposed system does not require that address assignments be certified all the way to the subscriber. If a subscriber's address is allocated from that of a DSP or an ISP with which it is currently affiliated, then the certification process need only be effected as far as the ISP/DSP. The same applies to DSPs that receive their address-space assignments from ISPs. Also note that a subscriber (or a DSP) who does not participate in BGP exchanges (e.g., is singly-homed) need not be issued a certificate if the subscriber's address space is derived from that of an encompassing ISP or DSP[1]. Finally, if an organization owns multiple ranges of addresses, this design calls for assigning a single certificate[2] containing a list of address blocks, so as to minimize the number of certificates needed to validate an UPDATE.

[1] For historical reasons, the chain of issuance described above was sometimes short circuited. A subscriber (or DSP) who has an address-space assignment that has bypassed the normal allocation procedure, who has changed DSPs/ISPs and retained the originally assigned address must also be certified, even if not a BGP user.

[2] If an organization acquires additional address blocks, a new certificate is issued to reflect the increased scope of ownership.

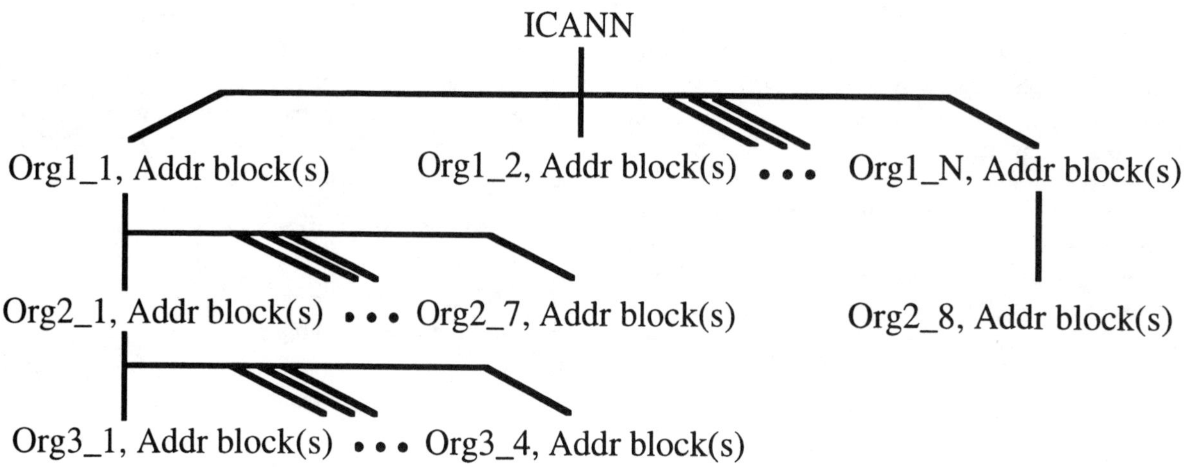

Subject	**Description**
Org1_x	a 1st tier organization (usually a registry)
Org2_x	a 2nd tier organization (usually an ISP or DSP)
Org3_x	a 3rd tier organization (usually a DSP or user organization)

Extension	**Description**
Addr blocks(s)	one or more IP address blocks assigned to the organization

Figure 1: Address Allocation PKI Structure

This PKI reflects the assignment of address blocks to organizations by binding address block(s) to a public key belonging to the organization to which the addresses are being assigned. Unlike a typical X.509 certificate, the identity of the subject is not the primary focus of these certificates; instead, these certificates are used to prove ownership of a block of addresses[3]. Each certificate in this PKI contains a (private) extension that specifies the set of address blocks that have been allocated to the organization. The subject alternate name in each certificate is the DNS name of an organization: an ISP, DSP, or a subscriber. The ICANN, as root, is represented nominally by a self-signed certificate that contains an extension expressing ownership of the entire address space. In Figure 1:

(a) ICANN is the root and issues certificates to the first tier of organizations (Org1_x). Under current practice, Org1_x would be an Internet Registry, although historically it could have been an ISP, an organization, etc. ICANN signs the tier 1 certificates using its private key.

(b) Org1_x then assigns sub-blocks of its address space to ISPs or DSPs. In the diagram, for example, Org1_1 issues a certificate to each of Org2_1 through Org2_7 and Org1_N issues a certificate to Org2_8. Org1_x signs the certificate using the private key corresponding to the public key in the certificate it received in (a).

(c) Org2_x then assigns sub-blocks of its address space to customers, DSPs, etc. In the diagram, Org2_1 issues a certificate to each of Org3_1 through Org3_4. Org2_x signs the certificate using the private key corresponding to the public key in the certificate it received in (b).

(d) And so on....

Table 1 summarizes the Issuer/Subject relationships for the certificates in this PKI.

[3]One could place the address block in an X.509 attribute certificate, linked to an X.509 public key certificate, in a more elegant approach to representing this data. However, because the number of certificates involved is so great, and because attribute certificates are not yet widely supported, we have chosen to add the address block information as a private, v3 extension to a public key certificate.

Certificate Type	Issuer	Subject
Root	ICANN	ICANN
Registry	ICANN	Registry
ISP/DSP	Registry (or ICANN or ISP)	ISP/DSP
Subscriber	ISP/DSP (or Registry or ICANN)	Subscriber

Table 1: Address Allocation PKI Certificate Overview

3.1.2 A PKI for ASes and router IDs. Three types of certificates will be used to support the authentication of ASes and BGP speakers, and the relationship between speakers and ASes. Here too the ICANN is the root and the next tier consists of registries, but the third tier consists of organizations that own ASes, followed by a tier of AS numbers and routers. The result is a broader, shallower certification tree. As before, this tree parallels existing "trust relationships," i.e., the ICANN assigns AS numbers to registries, which in turn assign one or more AS numbers to organizations (e.g., ISPs/DSPs) that run BGP. Each of these organizations is authoritative for identifying routers as representatives (BGP speakers) for the AS(es) that the organization owns. In order to express the ownership of an AS by an organization, each third tier certificate carries an extension that enumerates the ASes assigned to that organization. Validation of fourth tier certificates requires matching asserted AS numbers against these extensions. For each fourth tier AS certificate (b, below) there are typically several router (BGP speaker) certificates (c, below), each specifying the same AS number. (Note that there could be more than one certificate assigned to a BGP speaker if the speaker acts as a proxy for another AS.) As shown in the Figure 2, these 3 types of certificates bind together:

(a) AS numbers and an organization's public key—a registry issues these to organizations and signs them using its private key. The alternate name in the certificate is the DNS name of the organization. An extension contains the (list of ranges of) AS number(s).

(b) an AS number and its public key—An organization issues these and signs them using the private key corresponding to the public key in the certificate described in (a). The issuer alternate name in the certificate is the DNS name of the organization. The subject alternate name is the AS number, represented as a DNS name.

(c) a router (DNS) name, a router id, an AS number, and the router's public key—An organization issues these and signs them using the private key corresponding to the public key in the certificate described in (a). Both the router id (an IP address) and the AS number are extensions in this certificate, and the binding of three items is a critical aspect of this certificate. The subject alternate name is the DNS name of the router corresponding to the router id.

3.1.3 Attestations. An attestation establishes that the subject of the attestation (an AS) is authorized by the issuer to advertise a path to the specified blocks of address space. There are two classes of attestations, address and route, although a single format is employed to represent both. Route attestations are carried in a new type of optional BGP path attribute as part of UPDATE messages:

- Address attestations—here the issuer is the organization that owns the address space and the subject is an AS that may originate it, e.g., the organization's provider. The issuer signs an address attestation using the private key that corresponds to the public key in the certificate (see Figure 1) assigning this address space to the issuer.
- Route attestations—here the subject is a transit AS. A route attestation is signed by the S-BGP speaker (or offline by the management of the AS). The signer uses the private key that corresponds to the public key in the certificate that binds the speaker to the subject AS (see Figure 2).

If an organization has more than one AS, there are separate attestations for each AS rather than just one attestation containing multiple AS numbers. Each AS will have its own set of BGP speakers and its own authentication certificate(s) as well. This applies to both the stub and transit AS cases. Figure 3 summarizes the structure for UPDATEs and route attestations.

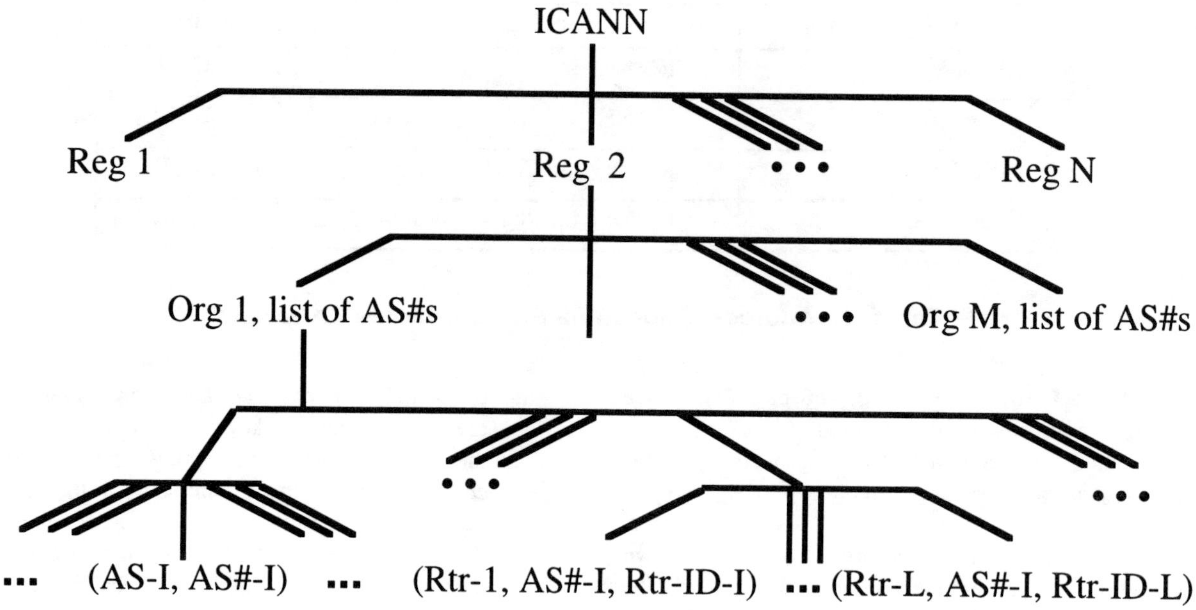

Subject	**Description**
Reg n	DNS name of registry n
Org m	DNS name of ISP/DSP/organization m
AS-i	DNS name of autonomous system i
Rtr-j	DNS name of router j

Extension	**Description**
AS#-i	autonomous system number for AS-i
Rtr-ID-k	router identifier (IP address) for router k

Figure 2: Autonomous System Identification and BGP Speaker PKI

Certificate Type	Issuer	Subject	Extensions
Root	ICANN	ICANN	All AS #s
Registry	ICANN	Registry	AS #s
AS Owner	Registry (or ICANN)	ISP/DSP or Subscriber	AS #s
AS	ISP/DSP or Subscriber	AS number (in DNS format)	AS #
BGP Speaker	ISP/DSP or Subscriber	BGP Speaker DNS name	AS #, Router ID

Table 2: AS and BGP Speaker PKI Certificate Overview

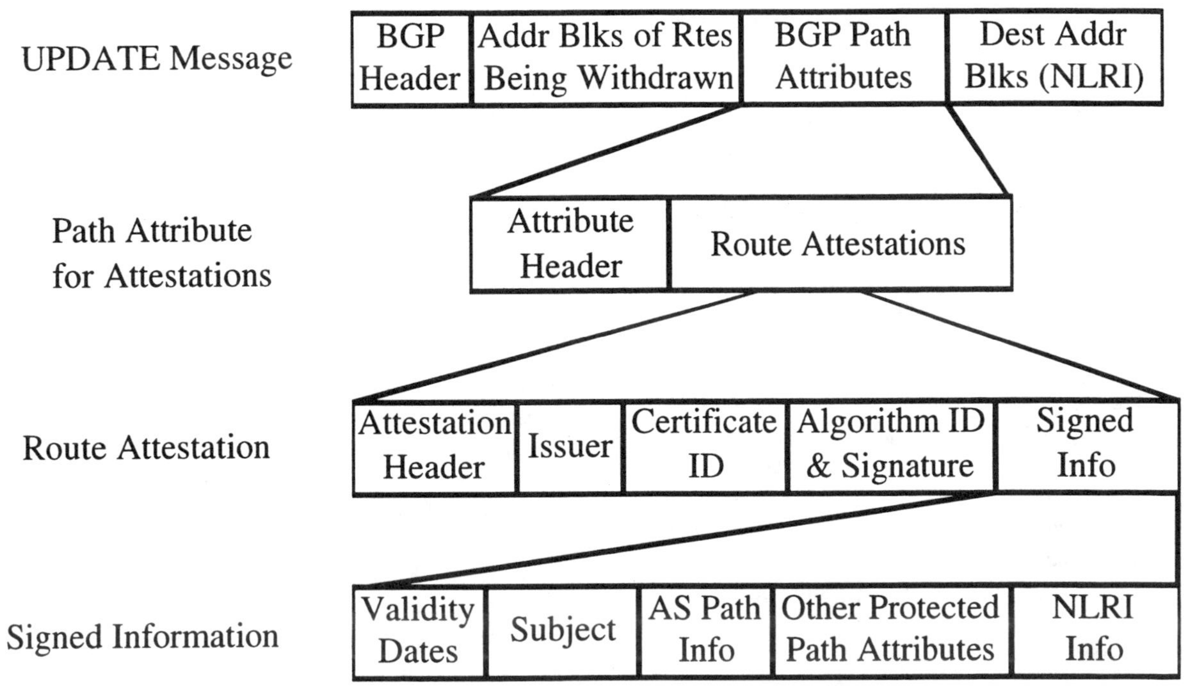

Figure 3: UPDATE Format with Route Attestations

3.2 Route validation

Attestations and certificates are used by BGP speakers to validate routes asserted in UPDATE messages, i.e., to verify that the first AS in the route has been authorized to advertise the address block(s) by the address block owner(s), and that each subsequent AS has been authorized to advertise the route for the address block(s) by the preceding AS in the route. To validate a route received from AS_n, AS_{n+1} needs:

- 1 address attestation from each organization owning an address block or blocks in the NLRI
- 1 address allocation certificate from each organization owning an address block or blocks in the NLRI
- 1 route attestation from every S-BGP speaker (or its AS) along the path (AS_n to AS_1), where the route attestation generated and signed by $router_x$ (or AS_x) specifies the NLRI and the AS_PATH from AS_{x+1} through AS_1
- 1 certificate for each S-BGP speaker along the path (AS_n to AS_1) to check the signatures on the route attestations

and, of course, all the relevant CRLs must have been verified.

This means that for each UPDATE, there must be attestations confirming that all the ASes in the BGP UPDATE are authorized to advertise routes to the destination IP address block(s). This includes ASes that are providing third party advertisements for ASes that are not running BGP.

The attestations are not used for checking withdrawn routes because the authorization of the BGP speaker to advertise those routes was verified at the time they were installed into the local routing information base (Loc-RIB). Moreover, if the BGP speaker has lost the authorization to advertise that route, then the route is by definition no longer valid and should be withdrawn.

Use of IPsec on inter-router communication paths prevents an active wiretapper from spoofing route withdrawals, or replaying valid UPDATEs at times when a BGP speaker would not transmit them, e.g., after a route has been withdrawn and prior to advertisement of the same or a different route.

3.3 Distribution of countermeasure information

This section discusses the mechanisms used to distribute certificates, CRLs, and address and route attestations to the relevant devices performing route validation: S-BGP speakers, route servers, etc.

3.3.1 Distribution of certificates, CRLs and address attestations. Each S-BGP speaker must have access to the public keys required to validate UPDATEs. For non-leaf S-BGP speakers, this amounts to a full set of certificates encompassing all address space owners, AS owners, and some number of S-BGP speakers (plus the ICANN and registry certificates). An X.509 certificate used in this environment is about 450 bytes long, depending on naming conventions and extensions. In the current (June 1999) Internet environment, there are approximately 5,300 autonomous systems, 44,000 organizations that own address prefixes, and 7,500 BGP speakers. The resulting certificate database comprises about 32 Mbytes, and it can be expected to grow each year as more address prefixes, autonomous systems, and BGP speakers are added. The CRL database associated with these certificates adds to this total, but since the certificates are issued to organizations and devices, not people, CRLs probably will not grow significantly.

At first glance it might seem appropriate to transmit certificates as part of each UPDATE. This would ensure that each receiving BGP speaker would receive all the data needed to validate the route attestations in an UPDATE, and it would be easy for each BGP speaker to include its own certificate as part of the forwarding process. However, this would be very wasteful of bandwidth, as each BGP speaker would receive many redundant copies of certificates[4]. More importantly, this approach is infeasible, because BGP UPDATEs are limited in length to 4096 bytes and thus are too small to carry the necessary certificates for most UPDATEs[5]. The introduction of a new type of BGP message for transmission of certificates (and CRLs) could address the packet size problem, but would still tend to be very wasteful of bandwidth and would not be backward compatible.

Instead, this architecture uses out-of-band distribution of certificates and CRLs to all S-BGP speakers. This is an attractive approach to the distribution problem for several reasons. This database is relatively static and thus a good candidate for caching and incremental update. Moreover, the certificates can be validated (and processed against CRLs) and reduced to a more compact format by ISPs/DSPs prior to distribution to S-BGP speakers. (Only the public key, subject and selected extensions need be retained.) This avoids the need for each speaker to perform this processing (entailing tens of thousands of signature validations), and it saves both bandwidth and storage space. Although memory is inexpensive, most currently deployed commercial routers do not possess sufficient memory to store all of these certificates so either additional memory or auxiliary systems will be needed, even with preprocessing.

To address the distribution problem we make use of two tiers of repositories from which one can download the entire certificate and CRL database. The top tier consists of several replicated, easy to access storage sites, e.g., the NAP route servers. The second tier of repositories are operated by the ISPs/DSPs, to provide local access for the S-BGP speakers within each AS. (Per-AS repositories avoid the dependency loop that would occur if one required inter-AS routing in order to access this database.) Bulk transfer of the whole certificate or CRL database, from the first to second tier repositories, can be effected via FTP or TFTP; since certificates and CRLs are signed and carry validity interval information, there is no need for additional integrity mechanisms in the transfer. The second tier repositories also will query top tier repositories to get new CRLs, based on the NEXT UPDATE field in each CRL, and to get new certificates, based on certificate expiration, e.g., using the Lightweight Directory Access Protocol (LDAP). Retrieval of newly issued certificates (in between downloads of the complete database) for newly created ASes, organizations with new address space ownership, etc., could be effected in the same manner, from daily incremental update files.

This allows an ISP/DSP to operate in an anticipatory fashion, retrieving certificates before it needs them. Each second tier repository will validate all of the certificates and CRLs it retrieves, and produce a more compact (locally signed) database ready for consumption by the S-BGP speakers within its administrative purview. If deemed necessary, the top tier repositories also can push CRLs issued prior to scheduled dates.

The same analysis applies to address space attestations. Each address attestation requires about 100 bytes, and this amounts to approximately 4 Mbytes. Carriage of these data items in UPDATEs would usually be redundant and thus is more effectively handled via the same, out-of-band distribution mechanism. Here too, this distribution model allows pre-processing by ISP/DSP NOCs, further eliminating significant signature validation overhead (there would be roughly 44,000 such attestations currently). This model also simplifies revocation of address attestations, i.e., an address space owner can issue a CRL-like, signed data structure and include it in the database for downloading and pre-

[4]The redundancy arises from several sources. A BGP speaker tends to receive routes to the same destination, via each interface, with considerable overlap of ASes in each route. Withdrawal and later re-advertisement of the same routes via the same interfaces results in additional redundancy

[5]Many UPDATEs contain routes for multiple address blocks and some routes contain many AS numbers, and each requires its own certificate.

processing by each ISP/DSP. The total space required for the certificate and address attestation databases can be reduced from about 36 Mbytes to about 11 Mbytes as a result of preprocessing.

3.3.2 Distribution of route attestations. Route attestations are distributed with BGP UPDATEs in a newly defined, optional, transitive path attribute. This approach requires BGP speakers to be upgraded to a BGP release with these countermeasures. When an S-BGP speaker opens a BGP session with a peer, transmitting the advertisable portion of its routing information database via UPDATEs, relevant route attestations are sent with each UPDATE. These attestations employ a compact encoding scheme, to help ensure that they fit within the BGP packet size limits, even when route or address aggregation is employed. (We preserve S-BGP security guarantees in the face of aggregation by explicitly including signed attribute data that otherwise would be lost when aggregation occurs.) The S-BGP speaker receiving an UPDATE caches the associated attestations with the route in its routing information database. (We have not yet determined if a mechanism for revoking route attestations is required, or if a modest attestation lifetime will suffice.) Each BGP speaker generates route attestations based on receipt of UPDATEs from its neighbors, as described in Section 4.2, thus in-band distribution is appropriate. As noted below, the bandwidth required to support in-band distribution of route attestations is negligible (compared to user traffic).

3.4 IPsec and router authentication

BGP is transported over TCP and thus is protected against misordered, lost, or replayed packets, to the extent that the TCP sequence number management facility is secure. BGP-4 provides a means for carrying authentication information in BGP messages, but there is no prescribed key management scheme and there is no facility for sequence numbering of BGP messages, hence this facility is not employed here. Instead, we use the Encapsulating Security Payload (ESP) protocol (with NULL encryption), from the IP security protocol suite (IPsec) [8,9,10,11,12] to provide authentication, data integrity, and anti-replay on a point-to-point basis, i.e., between BGP speakers. (Although the Authentication Header protocol from the IPsec suite could be used here, it is less efficient and thus was not selected.) The Internet Key Exchange (IKE) protocol is used for key management services in support of ESP. The PKI established for router and AS authentication provides the necessary certificates (see Section 3.1.2).

3.5 Other issues

BGP Path Attributes are being standardized to support Communities (to support policy) [13], Confederations (used to transparently split a large AS into several smaller ASes to reduce the n^2 peering requirements) [14], and to support additional protocols (IPV6, MPLS, and multicast) [15]. The Route Attestation mechanism is designed to provide protection for these and other new path attributes, in the same manner in which it protects the current path attributes.

4 How these countermeasures address BGP vulnerabilities

This section describes how the proposed countermeasures reduce the vulnerability of BGP to the attacks described earlier and provide much of the functionality necessary for ensuring the correct operation of BGP.

4.1 Certificates

The certificates described above are used to enable verification of:

- An AS's authorization to "advertise" a block of addresses—The certificates from 3.1.1 are used to verify that an AS is authorized to "advertise" a block of addresses. Specifically, the signature on an address attestation must be verifiable using the public key in a certificate containing the address block(s) that include the address block(s) in the address attestation.
- An organization's ownership of an AS number—The certificates from 3.1.2.a are used to verify that an AS has been assigned to the holder of a particular public key, i.e., an ISP, DSP, or subscriber organization. They are used to validate 3.1.2.b or 3.1.2.a certificates through the AS number linkage.
- An AS's identity—The certificates from 3.1.2.b (or certificate data pre-processed by a NOC and distributed to routers) are used to verify the signature of an AS on a route attestation.
- A BGP speaker's identity and its association with an AS—The certificates from 3.1.2.c are used to verify the signature of a speaker on a route attestation, and in conjunction with 3.1.2.a to make sure that the speaker is authorized to act on behalf of the AS.
- Identity and authorization of a BGP peer—The certificates from 3.1.2.c are used by the BGP speakers when establishing peering sessions, to authenticate each other.

4.2 Address and route attestations

These two countermeasures support validation of the address prefixes and path information in an UPDATE.

Address attestations protect BGP against misbehaving BGP speakers that originate or distribute erroneous UPDATEs and BGP speakers whose advertisable destination addresses have been misconfigured. Whenever an S-BGP speaker advertises itself as the starting point of a route for some address prefix, other S-BGP speakers will verify that the AS represented by that speaker is the subject of an address attestation signed by the owner of the address prefix. Since only the organization that owns the prefix can sign such an attestation, no S-BGP speaker can falsify such an advertisement.

Each S-BGP UPDATE will include a set of route attestations, one per AS listed in the UPDATE, each of which is added to the UPDATE as it propagates among ASes. A route attestation indicates that the signing BGP speaker is authorized to advertise to the neighbor the route constructed thus far, by the organization owning the AS (in which the speaker resides). The route attestation is digitally signed by the S-BGP speaker distributing the UPDATE. It includes the identification of the S-BGP speaker's certificate issued by the owner of the AS, the destination addresses in the route, the list of identifiers of ASes in the route, the identifier of the AS to which the UPDATE is directed, a maximum lifetime, and other transitive data requiring protection. Each recipient of an UPDATE verifies the route attestations contained within it before deciding whether to accept and distribute the UPDATE. Route attestations protect BGP against misbehaving BGP speakers that distribute erroneous UPDATEs, and against misconfigured local routing policies.

4.3 IPsec

IPsec (specifically ESP and IKE) provides the security services needed by the receiving BGP speaker to verify message integrity, the identity of the sender, and the fact that it (the receiver) is the intended recipient of every message. Although the attestations in UPDATE messages protect against a wide range of active wiretap attacks, use of ESP provides protection for all BGP traffic, prevents replay of messages across a link, and protects TCP against various forms of attack, including SYN flooding and spoofed RSTs (resets).

5 Residual vulnerabilities

The S-BGP system (address attestations, route attestations, PKIs, and IPsec for all BGP messages) addresses many of the vulnerabilities of BGP-4. Nevertheless, there exist vulnerabilities that are not eliminated by this system, including the following:

- Suppression of BGP messages by a misbehaving BGP speaker is not addressed. Use of IPsec (and TCP) will detect active wiretap attacks that result in lost or reordered BGP packets. However, a compromised BGP speaker can elect to not transmit BGP messages, even when local policy would call for such transmission, e.g., for route withdrawal. This is undetectable by the proposed countermeasures, although coordinated network monitoring might be able to detect such misbehavior. The substantial flexibility afforded by local policies in BGP appears to preclude countering this vulnerability if such policies are to remain private to ASes (as allowed by the BGP specification and as currently practiced).
- Passive wiretapping to discover network connectivity information is not addressed. These attacks could be countered by enabling the confidentiality feature of ESP, if the risk exceeds the cost to encrypt and decrypt UPDATE messages.
- A BGP speaker may reassert a route that was withdrawn earlier, even if the route has not been re-advertised. This vulnerability exists because BGP UPDATEs do not carry sequence numbers or timestamps that could be used to determine the currency of UPDATEs. However, route attestations do expire, so there is a limit on how long an old attestation can be used for such purposes. The possibility also exists to add a CRL-like function for route attestation revocation, a possibility that will be explored later in our work.
- Verification that the BGP peers that exchanged the UPDATE, correctly applied BGP rules, local policies, etc., is not addressed. As above, BGP affords speakers considerable latitude with regard to local policy and ASes do not usually make public their local routing policies, hence it appears difficult to counter such problems. S-BGP restricts malicious behavior to the set of actions for which the speaker (or AS) is authorized, based on externally verifiable constraints.

6 Performance and operational issues

In developing the S-BGP architecture, we have paid close attention to the performance and operational impact of the proposed countermeasures. Previous work in the area of routing security has often focused almost exclusively on the costs of generating and validating digital signatures. While such costs are an important factor, our analysis suggests that the bandwidth and storage requirements associated with signatures, certificates and CRLs are much bigger problems. The following analysis is based on examination of actual Internet routing data plus simulation of the effects of S-BGP countermeasures.

6.1 Processing

The computation burden for signature generation and validation appears to be tractable in this proposed architecture. We have selected the Digital Signature Algorithm (DSA) to minimize the size of the signatures, specifically for route attestations. DSA yields only a 40-byte signature, vs. the 128-byte signature typical for RSA (using 1024-bit keys). DSA also allows for pre-computation, which permits lower latency in signature generation by S-BGP speakers. Other (S-BGP-specific) techniques (described below) significantly reduce the need for signature validation operations, so the processing asymmetry exhibited by DSA is not a concern here.

The rate at which new UPDATEs are created is not so great that signature generation and validation of route attestations is expected to pose a bottleneck. A BGP speaker at a NAP, peering with about 30 other BGP speakers, receives an average total of about .5 UPDATEs per second. Each route contains an average of 3.6 ASes, and there is one route attestation per AS, yielding a rate of about 1.8 signature validations per second. In contrast, each UPDATE generated by an S-BGP speaker requires just one signature (added by the speaker), thus the signature generation rate is less than one third (1/3.6) of this value. Peak load figures may be about a factor of ten greater, yielding a peak load of about 18 signatures per second. However, analysis of data from NAPs shows that 50% or more of all UPDATEs repeat routes already known to a BGP speaker (e.g., due to link flapping). Thus caching just one route for each address prefix enables a speaker to avoid the need to validate signatures on the vast majority of UPDATEs, reducing the peak load to about 9 signatures per second, a tolerably low computational burden[6].

Upon initialization (reboot), a BGP speaker receives complete routing table updates from each peer. This means that an S-BGP speaker will receive a very large number of route attestations requiring validation, i.e., about 220,000 per peer, in a short time interval. This sort of initialization transient would be unacceptable, even though reboots and installation of new speakers is relatively infrequent[7]. To avoid this problem, we propose the addition of non-volatile storage for validated route attestations, to preserve the cache across reboots. When installing a new BGP speaker in an AS, a NOC could dump the cache from another speaker in the AS and reload it into the new BGP speaker, to seed the cache in the newly installed device.

6.2 Transmission bandwidth

The transmission of countermeasures data in UPDATEs increases the size of these messages to approximately 450 bytes (based on an average of 3.6 route attestations per path), for a typical UPDATE that previously required only 63 bytes. This represents a significant percentage increase in BGP overhead (over 700%), but the transmission of UPDATEs represents a very, very small amount of data relative to subscriber traffic. As noted above, even a speaker at a NAP sees an average rate of less than one UPDATE per second, and such speakers are connected via 100 Mb/s interfaces today, with plans to transition to multi-Gb/s interfaces in the future.

Downloading the certificate, CRL, and address attestation databases contributes an insignificant increment to this overhead. Full database transmission, from a top tier to a second tier repository entails about a 36 Mbyte file transfer for each ISP/DSP. Even if performed on a daily basis, this traffic is swamped by subscriber file transfers. Transfers from a second tier repository to each BGP speaker in its AS are smaller, about 11 Mbytes (due to certificate extraction). Here too, even if performed daily, this would be but a drop in the ocean of subscriber traffic, and use of incremental transfers is a more likely scenario. So the impact on utilization of Internet bandwidth due to transmission of all of the countermeasures data is minimal. Also, the speed of inter-router circuits continues to increase

---

[6] For example, SSLeay software can perform about 40 DSA (1024-bit key) signatures per second on a 450MHz Pentium II processor.

[7] Because of the important service they provide, BGP speakers are usually afforded UPS protection and new software is deployed only after extensive testing, to minimize the likelihood of crashes.

substantially, further minimizing the impact of transmission of additional control traffic.

6.3 Storage/memory

UPDATEs received from neighbors are held by a BGP router in Routing Information Bases (RIBs) and used to generate new UPDATEs for transmission to other BGP routers. The additional memory required for pre-processed certificates and address attestations amounts to about 11 Mbytes. The space required for route attestations is about 26 Mbytes per peer, a modest amount for a typical speaker with 2 or 3 peers, but a significant amount of storage for speakers at NAPs, where each speaker has about 30 peers. This amounts to a large but feasible amount of additional RAM by current standards, where a high end workstation can be configured with hundreds of megabytes of RAM. The routers that act as BGP speakers at NAPs are large, very expensive devices. However the storage capacity of the routers currently used by ISPs/DSPs would not permit storage of S-BGP UPDATEs in their RIBs, if S-BGP were deployed. Thus additional, non-volatile storage is needed in BGP speakers to support these databases. It may be feasible to significantly reduce the storage required here, since the routes (and thus route attestations) received from different peers tend to exhibit significant overlap in their suffixes.

6.4 Deployment and transition issues

Deploying S-BGP raises a number of other issues:

Adoption of S-BGP by several groups— The ISPs, DSPs, and subscriber organizations running BGP will need to cooperate in the generation and distribution of attestations. The first tier ISPs (those connected to the NAPs) must implement the S-BGP security mechanisms in order to offer significant benefit to the Internet community. (Lower level ISPs, DSPs, and subscriber organizations will need to implement the S-BGP security mechanisms only if the expense can be justified.) ICANN and registration authorities will need to expand their operational procedures to support generation of address space and AS number delegation certificates. Finally, router vendors need to provide additional storage in next generation products, or offer ancillary devices for use with existing router products, and revise BGP software to support S-BGP.

S-BGP interaction with other exterior and interior routing protocols—External routes received from external peers need to be redistributed within the AS in order to maintain a consistent and stable view of the exterior routes across the AS. Interior routing protocols will not propagate S-BGP attestations, but if each border S-BGP speaker maintains an iBGP (The acronym "iBGP" denotes intra-AS use of BGP, in contrast to the common inter-AS use of BGP, sometimes referred to as eBGP.) connection with all other transit and border routers within the AS, this problem will be averted.

BGP-4 to S-BGP Transition—The route attestation path attribute is optional for both external and internal BGP exchanges. This allows extensive regression testing before deploying S-BGP on production equipment.

7 Test scenario and results

To verify that the proposed countermeasures work as expected, and to validate our performance projections, we created a prototype implementation of S-BGP and conducted some tests in the DARPA-sponsored CAIRN testbed.

7.1 Testbed characteristics

The testbed routers were configured with FreeBSD 2.2.x as the OS and GateD as the BGP implementation. The hardware platform consisted of Pentium Pro (200-MHz 686-class CPU) with 64 to 256 Mbytes of memory.

Our S-BGP countermeasures were added to the GateD BGP implementation. This software was instrumented to collect performance statistics as described below. We also developed a man-in-the-middle (MITM) tool that was placed between a testbed BGP speaker and a BGP peer in an ISP. This tool was used to intercept BGP control traffic coming from the BGP peer in the Internet, augment it with S-BGP countermeasure data, and then pass it to the S-BGP peer in the CAIRN testbed. In this fashion, we simulated the effect of having deployed S-BGP in the Internet. This tool also can be used to test the effectiveness and robustness of the S-BGP countermeasures, e.g., by injecting bad routes, making the link appear to go up and down, etc. Finally, we developed a tool that accepts a recording of multiple sessions of BGP traffic and replays it to simulate the original order and arrival intervals for the BGP traffic. This latter tool allowed us to conduct tests with deterministic data, to facilitate comparison of results for BGP-4 vs. S-BGP and for optimizations of S-BGP.

For these tests, we used a short duration (10-20 minute) BGP feed, and extrapolated the results to daily operation scenarios. The feed represents a link over which about 2,900 KEEPALIVEs (55 Kbytes) and 1,426

UPDATEs (89 Mbytes) would be transmitted on a daily basis. There were about 58,000 entries in the certificate database (32 Mbytes) and about 44,000 address attestations (4 Mbytes) supporting validation of the route attestations received in these UPDATEs.

7.2 Experiment results

We tested the performance impact of the S-BGP countermeasures, vs. regular BGP, and examined the effect of some optimizations. Where results are a function of the number of peers seen by a BGP speaker, we present them on a per peering session basis. This allows one to estimate the overhead seen by a router with N peers by simply multiplying by N. Also, this provides a good approximation of how a multi-peer speaker behaves, since verification need be performed only for the "best" route. Under typical conditions, a BGP speaker receives one copy of a given UPDATE from every one of its external peers, plus one from every other BGP router in the same AS. In the absence of "best route only" validation, every one of the UPDATEs received via eBGP must be validated, whereas those received via iBGP are assumed to have been validated by the sender.

For example, at a NAP, a router might have approximately 30 eBGP peers. So one would multiply the storage and CPU numbers (which are sensitive to the number of peers) for one peer by 30 plus the number of BGP peers in the router's own AS to obtain the total RA storage and CPU needed. Additional bandwidth would be on a per link basis, so the bandwidth numbers discussed below would not have to be multiplied by the number of peers. Since validation of a route is done only for the route that is selected, the CPU costs approach those of the single peer case. (The router may not receive the "best" route first and hence may have to verify more than one UPDATE, depending on the order of arrival of different paths to the same destination, when the best one arrives, how long one waits before deciding one has the "best" route, etc.)

Bandwidth utilization is not an issue. The bandwidth used to download certificate and Address Attestation (AA) extracts, an operation that might be performed on a daily basis between a NOC and each router in its AS, is tiny compared to user traffic volumes. (The full certificate and AA database used in our experiments amounted to about 343 Mbytes, but was reduced to just 38 Mbytes as a result of the extraction pre-processing.) Even the added overhead for transmission of RAs amounted to only about 1.4Kbs between a pair of routers, again a miniscule amount relative to subscriber traffic and relative to the speed of a typical inter-router link.

Storage utilization was significant, and grew when we employed caching. The certificate extracts occupied about 32 Mbytes of storage, and AAs required about 10 Mybtes. Both figures are higher than the raw database sizes, due to the addition of data structure overhead for efficient access. This AA figure was much higher, due to storage format conventions used in GateD and in S-BGP. The AA storage figure could probably be reduced to less than 5 Mbytes with further attention to data structure design. The most significant overhead is associated with RA storage. Each interface (ADJ-RIB) required almost 17 Mbytes of additional storage, which could become significant for a router at a NAP with over 30 neighbors. When a depth one cache was added (to reduce RA validation and generation costs), the number grew to about 26 Mbytes, a 50% increase. This percentage increase is due to the fact that BGP already requires a depth one cache for inbound UPDATE data, but not for outbound data. Thus a deeper cache would result in larger percentage increases, as both inbound and outbound storage would grow.

The amount of CPU time devoted to RA validation was 123 minutes per day, while generation time was only 16.2 minutes per day. This asymmetry, roughly a factor of 8, arises because of several factors. All inbound UPDATEs were validated in the unoptimtzed case and that requires checking signatures on an average of about 3 RAs per UPDATE. Also, the DSA exhibits an asymmetry of its own, in which signature validation is slower than signature generation. Finally, not every inbound UPDATE resulted in a corresponding outbound UPDATE.

When caching was enabled, the number was largely unchanged, decreasing by only about 2%! We estimate that this is due to two factors. First, the BGP feed we received was from an ISP, and was analogous to what a DSP would receive. This means that most of the route flaps that caching would detect (and use to avoid redundant signature validation) were already filtered out by the ISP. Second, the duration of each test was relatively short, thus minimizing the opportunity for the cache to be filled and then exploited. Thus this test did not verify our thesis regarding the utility of caching. Still, the amount of time devoted to cryptographic processing was sufficiently small as to be tolerable, even using software.

8 Subsequent related work

Since we began this work, there have been several other efforts towards securing BGP. These include an assessment of BGP vulnerabilities and several approaches to addressing some of these vulnerabilities. (S. Murphy's "BGP Security Analysis" surveys this topic. At the time this paper was prepared, the most recent version was <draft-murphy-bgp-secr-03.txt>, June 1999.)

- TCP/MD5 [16]–This defines a TCP option for carrying an MD5 digest. This mechanism offers data origin authentication and data integrity, on a point-to-point basis. It protects the TCP connection used to transport BGP traffic from spoofing attacks and connection hijacking. Lack of an automated key distribution protocol complicates management and encourages overly long term use of symmetric keys. Moreover, because it fails to protect against any attacks that subvert routers or the management of routers, its overall security efficacy is quite limited.
- NLRI Origin Verification [17] – This mechanism proposes adding an address prefix delegation tree to Secure DNS [18]. For each prefix that has been authorized for use, a new resource record specifies the number of the AS that is authorized to originate that prefix. This mechanism does not address route authorization, nor does the proposal describe in detail how this data would be distributed to BGP speakers. It does represent an alternative format and database option for the address attestations developed in our architecture.
- Routing Policy System Security [19] – This approach places authorization information into a small number of databases (Internet Routing Registries) accessible by routers. The information is placed into a database by the organization responsible for the authorization, using some secure access method, e.g., SSL. It proposes that BGP speakers compare the routes that they receive to the routes listed in the database, rejecting any routes not found. No changes to BGP are required. This approach does not appear to be very dynamic, although details of how, and how often, the registries are to be accessed are omitted. Use of such registries would require ISPs/DSPs to publicize what is now local policy information, which most have refused to do. Moreover, the routes stored in the registries are not signed, so attacks against these databases, including malicious or benign errors by an ISP/DSP, could compromise security.
- Hash Chain Signatures [20] – This work describes two protocols, COSP and IOSP, based on hash chain signatures, that offer very rapid signature generation and validation in a routing protocol context. COSP is not applicable because it requires signing messages at fixed time intervals, whereas BGP generates UPDATEs on demand, as a result of topology changes; IOSP is not applicable because its efficiency depends on each router receiving essentially all routing updates, which is not characteristic of the operation of BGP.

9 Future work

Deploying this technology into the Internet will require the creation of the supporting PKIs, rooted at the ICANN, and convincing ICANN, the Internet Registries, the major ISPs, the owners of IP address blocks and the router vendors of the benefits and supportability of these security mechanisms. We hope to be able to pursue a variety of technology transfer activities to facilitate adoption of S-BGP. The prototype implementation and experiments described above are a first step in this direction.

10 Summary

BGP is a critical component of the Internet's routing infrastructure and highly vulnerable to a variety of attacks. The S-BGP countermeasures use IPsec, Public Key Infrastructure (PKI) technology, and a new BGP path attribute ("attestations") to ensure the authenticity and integrity of BGP communication on a point-to-point basis, and to validate BGP routing UPDATEs on a source to (multicast) destination basis. These enhancements will allow Internet Service Providers (ISPs) and their customers to verify that:

- reachability information they receive is from an authentic and authorized BGP peering relationship and has not been modified without authorization;
- the authorization of an organization to claim ownership of a block of IP addresses (a (sub)network) is substantiated by a chain of authorizations rooted at the Internet Corporation for Assigned Names and Numbers;
- an originating AS is authorized to advertise reachability to a block of IP addresses by the organization owning that address block;
- the ASes that processed the routing information en-route from the originating AS are not, either through mis-configuration, internal error, or compromise, advertising reachability information that is inconsistent with nominal topology;

- each AS, and its BGP speakers, that advertise a given route are identifiable and authorized to participate in global Internet routing, by a chain of authorizations rooted at the ICANN.

The significance of this work extends to essentially all Internet users, as almost all depend on the availability of global Internet for access to a wide range of services. A large and growing number of federal government services, including most DoD activities, rely heavily on the Internet for a wide range of (non-tactical) communications, hence this work has significance for national security.

11 Acknowledgements

Many individuals contributed to the design and development of S-BGP. Initial funding was provided by NSA, in April of 1997, yielding a first cut design. DARPA provided later support, under Dr. Hilarie Orman, enabling us to refine, implement and test the design. S-BGP has benefited significantly from the insight and efforts of Martha Steenstrup and Luis Sanchez. As members of the architecture team, their contributions were critical to the design of the attestation and PKI schemes and the associated evaluations of other approaches and performance and operational issues. The authors would also like to thank Michelle Casagni, for her work on the performance analysis and Joanne Mikkelson, Dennis Rockwell, and Nicholas Shectman for their efforts during the ongoing implementation and experimentation phase.

12 References

[1] Y. Rekhter, T. Li, "A Border Gateway Protocol 4 (BGP-4)," RFC 1771, March 1995.

[2] BBN Report 8217, "An Architecture for BGP Countermeasures," November 1997.

[3] C. Villamizar, R. Chandra, R. Govindan., "BGP Route Flap Damping," RFC 2439, November 1998.

[4] Smith, B.R. and Garcia-Luna-Aceves, J.J., "Securing the Border Gateway Routing Protocol," Proceedings of Global Internet '96, November 1996.

[5] Smith, B.R, Murphy, S., and Garcia-Luna-Aceves, J.J., "Securing Distance-Vector Routing Protocols," Symposium on Network and Distributed System Security, February 1997.

[6] Kumar, B., "Integration of Security in Network Routing Protocols," ACM SIGSAC Review, vol.11, no.2, Spring 1993.

[7] Murphy, S., panel presentation (no text) on "Security Architecture for the Internet Infrastructure," Symposium on Network and Distributed System Security, April 1995.

[8] S. Kent, R. Atkinson, "Security Architecture for the Internet Protocol," RFC 2401, November 1998.

[9] R. Glenn & S. Kent, "The NULL Encryption Algorithm and its Use with IPsec," RFC 2410, November 1998.

[10] S. Kent, R. Atkinson, "IP Encapsulating Security Payload (ESP)," RFC 2406, November 1998.

[11] D. Maughan, M. Schertler, M. Schneider, J. Turner, "Internet Security Association and Key Management Protocol (ISAKMP)," RFC 2408, November 1998..

[12] D. Harkins, D. Carrel, "The Internet Key Exchange (IKE)," RFC 2406, November 1998.

[13] R. Chandra, P. Traina, T. Li, "BGP Communities Attribute," RFC 1997, August 1996.

[14] P. Traina, "Autonomous System Confederations for BGP," RFC 1965, June 1996.

[15] T. Bates, R. Chandra, D. Katz, Y. Rekhter, "Multiprotocol Extensions for BGP-4," RFC 2283, February 1998.

[16] A. Heffernan, "Protection of BGP Sessions via the TCP MD5 Signature Option," RFC 2385, August 1998

[17] T. Bates, R. Bush, T. Li, Y. Rekhter, "DNS-based NLRI origin AS verification in BGP," presentation at NANOG 12, February 1998, http://www.nanog.org/mtg-9802.

[18] D. Eastlake, 3rd, C. Kaufman, "Domain Name System Security Extensions," RFC 2065, January 1997.

[19] C. Alaettinoglu, T. Bates, E. Gerich, D. Karrenberg, D. Meyer, M. Terpstra, C. Villamizar, "Routing Policy Specification Language (RPSL)," RFC 2280, January 1998.

[20] K. Zhang, "Efficient Protocols for Signing Routing Messages," Network and Distributed System Security Symposium, March 1998.

[21] R. Perlman, ""Network Layer Protocols With Byzantine Robustness," MIT/LCS/TR-429, October, 1988.

Onion Routing Access Configurations

Paul F. Syverson, Michael G. Reed, and David M. Goldschlag *
Naval Research Laboratory

Abstract

Onion Routing is an infrastructure for private communication over a public network. It provides anonymous connections that are strongly resistant to both eavesdropping and traffic analysis. Thus it hides not only the data being sent, but who is talking to whom. Onion Routing's anonymous connections are bidirectional and near real-time, and can be used anywhere a socket connection can be used. Proxy aware applications, such as web browsing and e-mail, require no modification to use Onion Routing, and do so through a series of proxies. Other applications, such as remote login, can also use the system without modification. Access to an onion routing network can be configured in a variety of ways depending on the needs, policies, and facilities of those connecting. This paper describes some of these access configurations and also provides a basic overview of Onion Routing and comparisons with related work.[1]

Keywords: Security, privacy, anonymity, traffic analysis.

1 Introduction

Preserving privacy means not only hiding the content of messages, but also hiding who is talking to whom (traffic analysis). Much like a physical envelope, the simple application of cryptography within a packet-switched network hides the messages being sent, but can reveal who is talking to whom, and how often. Onion Routing is a general purpose infrastructure for private communication over a public network. It provides anonymous connections that are strongly resistant to both eavesdropping and traffic analysis. The connections are bidirectional, near real-time, and can be used for both connection-based and connectionless traffic. Onion Routing interfaces with off the shelf software and systems through specialized proxies, making it easy to integrate into existing systems. Prototypes have been running since July 1997. As of this article's submission, the prototype network is processing more than 1 million Web connections per month. Connections have come from more than thirty thousand IP addresses in more than sixty countries and in all six main top level domains [12].

Onion Routing operates by dynamically building anonymous connections within a network of real-time Chaum mixes [3].[2] A mix is a store-and-forward device that accepts a number of fixed-length messages from numerous sources, performs cryptographic transformations on the messages, and then forwards the messages to the next destination in a random order. A single mix makes tracking of a particular message either by specific bit-pattern, size, or ordering with respect to other messages difficult. By routing through numerous mixes in the network, determining who is talking to whom is made even more difficult. Onion Routing's network of core onion routers (mixes) is distributed, fault-tolerant, and under the control of multiple administrative domains, so no single onion router can bring down the network or compromise a user's privacy, and cooperation between compromised onion routers is thereby confounded. The prototype network is entirely under our control and connections to it are not protected. Thus, the amount of protection is limited and subject to the trust of our administrative domain. However, arrangements are currently well under way for a network consisting of NRL controlled onion routers as well as onion routers controlled by independent commercial companies. By modifying entrance configurations and exit policies, Onion Routing is completely compatible with a wide variety of policies regarding resistance to traffic analysis and other security needs. This is de-

*Address: Naval Research Laboratory, Center For High Assurance Computer Systems, Washington, D.C. 20375-5337, USA, phone: +1 202.767.2389, fax: +1 202.404.7942, e-mail: syverson@itd.nrl.navy.mil, reed@itd.nrl.navy.mil, david@goldschlag.com.

[1] Versions of portions of this paper have appeared in [7, 13, 15]. This work supported by ONR and DARPA.

[2] This paper provides only a very brief description of Onion Routing system components. For a more detailed description, cf., [13].

scribed below in section 5.

2 Application Support via Proxies

Onion Routing can be used with applications that are proxy-aware, as well as several non-proxy-aware applications, without modification to the applications. Currently supported protocols include HTTP, FTP, SMTP, rlogin, telnet, NNTP, finger, whois, and raw sockets. Proxies are under development for Socks5, DNS, NFS, IRC, HTTPS, SSH, and Virtual Private Networks (VPNs). A proxy has three logical layers: an optional application specific privacy filter that sanitizes the data streams; an application specific layer that translates the data streams into an application independent format accepted by the Onion Routing network; and lastly, an onion building layer that builds and manages the anonymous connections. Because it builds and manages the anonymous connections, the proxy is the most trusted component in the system. Also, to build onions and hence define routes the proxy must know enough about the topology and link state of the network, the public certificates of nodes in the network, and the exit policies of nodes in the network. This information is distributed securely within the network automatically as new nodes come on-line or as the information changes.

3 Anonymous Connections

Onion Routing's anonymous connections are protocol independent and exist in three phases: connection setup, data movement, and connection tear-down. Setup begins when the initiator creates an onion, which defines the path of the connection through the network. An onion is a (recursively) layered data structure that specifies properties of the connection at each point along the route, e.g., cryptographic control information such as the different symmetric cryptographic algorithms and keys used during the data movement phase. The onion is treated as a destination address by onion routers; thus, it is used to establish an anonymous connection. Onions themselves appear differently to each onion router (OR) as well as to network observers. (The same goes for data carried over the connections they establish.) Connection routes can traverse an arbitrary number of ORs. A single onion can set up a route of eleven hops. Longer routes require tunnelling of onions through a connection. Each OR along the route uses its public key to decrypt the entire onion that it receives. This operation exposes the cryptographic control information, the identity of the next OR, and the embedded onion. The OR pads the embedded onion to maintain a fixed size, and sends it to the next OR. After the connection is established, data can be sent in both directions. Data from the initiator is repeatedly pre-encrypted using the algorithms and keys that were specified in the onion. As data moves through the anonymous connection, each OR removes one layer of encryption as defined by the cryptographic control information in the onion defining the route, so the data arrives as plaintext at the recipient. This layering occurs in the reverse order (using different algorithms and keys) for data moving backward. Connection tear-down can be initiated by either end, or in the middle if needed.

All information (onions, data, and network control) are sent through the Onion Routing network in uniform-sized cells. All cells arriving at an OR within a fixed time interval are mixed together to reduce correlation by network insiders. Likewise, the longstanding connections between ORs can be padded and bandwidth-limited to foil external observers. An onion looks different to each OR along a connection because of the layered public-key cryptography. Similarly, the layering of symmetric cryptography over the data phase cells makes them appear different to each OR. This design resists traffic analysis more effectively than any other deployed mechanisms for Internet communication.

4 Connection Overhead

Onion Routing's overhead is relatively small. Connection setup overhead is typically much less than one second and appears to be no more noticeable than other delays associated with normal Web connection setup on the Internet. Computationally expensive public-key cryptography is used only during this connection setup phase. Also, because public-key decryption is much more expensive than encryption, the public-key burden rests mainly upon the onion routers themselves, where the option of dedicated hardware acceleration can be justified. Our modular design is completely compatible with doing the public-key operations in either hardware or software, and we are using both in our test networks. (Aside: Because there is no cryptography in the system code itself, Onion Routing system code has already been approved for unlimited distribution.)

The data movement phase uses only secret-key (symmetric) cryptography, which is much faster. Furthermore, since the symmetric encryption can be pipelined, data throughput can be made as fast as ordinary link or end-to-end encryption. Data latency is affected by the number of ORs along the connection and can vary with route length and the duration of the mix cycles.

5 Access Configurations and Exit Policies

Proxies, onion routers, and other components can be run in a variety of distributed configurations. This allows Onion Routing to mesh well with a wide variety of operational and policy environments. We now consider some of these possibilities for access configurations.

- **Remote-Proxy Access**

  At one extreme, proxies can run remotely. If a user makes an encrypted connection to a trusted remote proxy, Onion Routing's protection can be utilized without installing any software or inducing local computational overhead. If the initiator can trust the remote proxy to build onions, his association with the anonymous connection from the first OR to the responder is hidden from observers and the network. In a similar way, an encrypted connection from an exit funnel (demultiplexor) to a responder hides the association of the responder with the anonymous connection.

  Therefore, if an initiator makes an anonymous connection to some responder, and layers end-to-end encryption over that anonymous connection, the initiator and responder can identify themselves to one another, yet hide their communication from the rest of the world. So, we can build virtual private networks without protected sites.

  Notice, however, that the initiator trusts the remote proxy to conceal that the initiator wants to communicate with the responder, and to build an anonymous connection through the OR network. The next paragraph describes how to shift some of this trust from a remote site to the initiator.

- **Customer-ISP Access**

  Suppose, for example, an Internet Services Provider (ISP) runs a funnel (multiplexor) that accepts connections from onion proxies running on subscribers' machines. In this configuration, users generate onions specifying a path through the ISP to the destination. Although the ISP would know who initiates the connection, the ISP would not know with whom the customer is communicating, nor would it be able to see data content. So the customer need not trust the ISP to maintain her privacy. Furthermore, the ISP becomes a *common carrier*, who carries data for its customers. This may relieve the ISP of responsibility both for whom users are communicating with and the content of those conversations. The ISP may or may not be running an OR as well. If he is running an onion router, then it is more difficult to identify connections that terminate with his customers; however, he is serving as a routing point for other traffic. On the other hand, if he simply runs a funnel to an onion router elsewhere, it will be possible to identify connections terminating with him, but his overall traffic load will be less. Which of these would be the case for a given ISP would probably depend on a variety of service, cost, and pricing considerations. Note that in this configuration the entry funnel must have an established longstanding connection to an OR just like any neighboring OR. In most other cases, where the funnel resides on the same machine as the onion router, establishing an encrypted longstanding connection should not be necessary since the funnel can be directly incorporated into the Onion Router.

- **Island-Unto-Yourself Access**

  If one wants to gain the maximum protection afforded by Onion Routing, it is necessary to have local control of an onion router. Assuming that this OR also serves as an intermediate node for routing of other traffic, not only data and route are hidden but also the time and volume information about connections originating or terminating locally. Of course this additional protection comes at the price of having adequate Internet bandwidth to function in this way.

- **Proxy-and-OR-at-Firewall Access**

  When a proxy and onion router sit on the firewall of a sensitive site, they can serve as an interface between machines behind the firewall and the external network. Connections from machines behind the firewall to the onion router are protected by other means (e.g., physical security). To complicate tracking of traffic originating or terminating within the sensitive enclave, this OR should also route data between other ORs. This configuration might represent the system interface from a typical corporate or government site.

  Connections between machines behind firewall ORs are protected against both eavesdropping and traffic analysis. Since the data stream never appears in the clear on the public network, this data may carry identifying information, but communication is still private. (This feature is used in constructing VPNs via Onion Routing.)

The onion router (more precisely the proxy) at the originating protected site knows both the source and destination of a connection. This protects the anonymity of connections from observers outside the firewall but also simplifies enforcement of and monitoring for compliance with corporate or governmental usage policy.

The use of anonymous connections between two sensitive sites that both control ORs effectively hides their communication from outsiders. Also, by employing a layering of funnels and ORs within firewalls, enclaves can incorporate traffic analysis resistance to their defense-in-depth.

- **Local-Proxy-with-OR-at-Firewall Access**

  It is possible to hide the route and origination of connections originating at an enclave while also protecting the route, application, and data being transmitted from enclave administrators. In this arrangement Onion Routing connected users are visible within the firewall, but not to where they are connected or what applications they are running.

The above discussion describes the various ways that a connection can enter an onion routing network. But, exiting is also important. Unless the responder of a connection is behind a firewall on which the terminal OR resides (e.g., if the responder is some arbitrary Web server) the data stream from the sensitive initiator must also be anonymized to avoid exposing the initiator. For example, an external attacker could simply listen in on the connections to a Web server and identify initiators of any connection to it. This point about exiting an OR network applies equally regardless of the configuration of entrance access.

There are other issues concerning how a connection exits an OR network. Exit points can also set policies for exiting based on where the traffic is going, and what application protocol is being run. Thus, for example, an onion router at a corporate firewall might allow anyone to attempt to remote login, but only to machines behind the firewall. It might allow email traffic to exit to any company site and Web traffic to exit to anywhere. Of course, these exit limitations might be a problem for the proxies attempting to create connections if they did not have the policies available when attempting to build a route.

There is a *database engine* attached to each of the onion routers. These ensure that any changes to this information propagates throughout the entire network to the proxies that construct routes. They also inform neighboring ORs about changes to network topology and link state, as well as ensuring that such information from others propagates throughout the network. In this way, proxies are given the most up-to-date information possible about potential routes. This greatly reduces the chance of bad connections that would then have to be attempted again via another route. The system that generates and distributes this information can be configured to be just as flat as the network itself, and authentication is of the relevant exit point or onion router. Thus, there is much less danger of central failure (or hostile manipulation) of this information. So, it is also more difficult to, e.g., manipulate this information to cause routes to pass through only compromised cooperating onion routers.

6 Background and Comparisons

As mentioned above, Chaum [3] defines a layered object that routes data through intermediate nodes (mixes). These intermediate nodes may reorder, delay, and pad traffic to complicate traffic analysis. In mixes, the assumption is that a single perfect mix adequately complicates traffic analysis, but a sequence of multiple mixes is typically used because real mixes are not ideal. Because of this, mix applications can use mixes in fixed order, and often do. Onion routers differ from mixes in using an indeterminate number of mixes in an indeterminate order, and in at least two other ways: onion routers are more limited in the extent to which they delay traffic at each node because of the real-time expectations that the applications demand of socket connections. Also, in a some Onion Routing access configurations, onion routers are also entry points to the onion routing network, and traffic entering or exiting at those nodes may or may not be visible to outsiders. This can make it hard to track packets, because they may drop out of the network at any node, and new packets may be introduced at each node. While Onion Routing cannot delay traffic to the extent that mixes can, traffic between ORs is multiplexed over a single channel and is link encrypted with a stream cipher. This makes it hard to parse the stream.

Anonymous remailers like Penet [9] strip headers from received mail and forward it to the intended recipient. They may also replace the sender's address with some alias, permitting replies. These sorts of remailers store sensitive state: the mapping between the alias and the true return address. Also, mail forwarded through a chain of remailers may be tracked because it appears the same to each remailer.

Mix based remailers like [4, 8] use mixes to provide anonymous e-mail services. Essentially, the mail message is carried in the innermost layer of an onion-like data structure. Another onion-like structure, used for a return address, can be contained in the message. This makes the return path self contained, and the remailer essentially stateless. Onion Routing shares many structures with Babel [8] but it uses them to build application independent end-to-end connections. This makes anonymous connections accessible to a wide variety of applications.

In [10], mixes are used to provide untraceable communication in an ISDN network. Here is a summary of that paper. In a phone system, each telephone line is assigned to a particular local switch (i.e., local exchange), and switches are interconnected by a (long distance) network. Anonymous calls in ISDN rely upon an anonymous connection between the caller and the long distance network. These connections are made anonymous by routing calls through a predefined series of mixes within each switch. The long distance endpoints of the connection are then mated to complete the call. (Notice that observers can tell which local switches are connected.) This approach relies upon two unique features of ISDN switches. Since each phone line has a subset of the switch's total capacity pre-allocated to it, there is no (real) cost associated with keeping a phone line active all the time, either by making calls to itself, to other phone lines on the same switch, or to the long distance network. Keeping phone lines active complicates traffic analysis because an observer cannot track coincidences.

Also, since each phone line has a control circuit connection to the switch, the switch can broadcast messages to each line using these control circuits. So, within a switch a truly anonymous connection can be established: A phone line makes an anonymous connection to some mix. That mix broadcasts a token identifying itself and the connection. A recipient of that token can make another anonymous connection to the specified mix, which mates the two connections to complete the call.

Our goal of anonymous connections over the Internet differs from anonymous remailers and anonymous ISDN. The data is different, with real-time constraints more severe than mail, but somewhat looser than voice. Both HTTP and ISDN connections are bidirectional, but, unlike ISDN, HTTP connections are likely to be small requests followed by short bursts of returned data. As described in [10], in a local switch, capacity is pre-allocated to each phone line, and broadcasting is efficient. But broadcasting over the Internet is not free, and defining broadcast domains is not trivial. Most importantly, the network topology of the Internet is more akin to the network topology of the long distance network between switches, where capacity is a shared resource. In anonymous ISDN, the mixes hide communication within the local switch, but connections between switches are not hidden. This implies that all calls between two businesses, each large enough to use an entire switch, reveal which businesses are communicating. In Onion Routing, mixing is dispersed throughout the Internet, which improves hiding.

Onion Routing's flexibility with respect to access configurations also make it a natural complement to other services like the Anonymizer [1] and Proxymate [11].

The Anonymizer is a proxy Web site that filters the HTTP data stream to remove a user's identifying information. This makes Web browsing private in the absence of any eavesdropping or traffic analysis. The Anonymizer is vulnerable in three ways: First, it must be trusted. Second, traffic between a browser and the Anonymizer is sent in the clear, so that traffic identifies the true destination of a query, and includes the identifying information that the Anonymizer would filter. Third, even if the traffic between the browser and the Anonymizer were encrypted, traffic analysis could be used to match incoming (encrypted) data with outgoing data. Onion Routing's privacy filters provide a similar function to the Anonymizer. However, the Anonymizer's filters are perhaps the most up-to-date of any readily available filters for the ever changing means by which anonymity can be compromised in the data stream. Also, the high volume that this longstanding service attracts provides some degree of natural cover traffic. An Anonymizer could be used together with Onion Routing as the HTTP proxy front end to provide a nice interface and good filtering for anonymity, with strong resistance to both eavesdropping and traffic analysis. Security is improved because the filtering executes on a machine the user trusts, and communication leaving that machine will resist traffic analysis. Such security in depth removes the central point of failure for network traffic anonymity.

Proxymate (formerly known as LPWA) is a "proxy server that generates consistent untraceable aliases for you that enable you to browse the Web, register at web sites and open accounts, and be 'recognized' upon returning to your accounts, all while still *preserving your privacy.*" Proxymate thus provides various pseudonymy-based services. Like Onion Routing it is designed to handle email in addition to HTTP. And, like Onion Routing, it can be configured so that trusted functions are performed at various locations [2]. However, communication between and from these points is not itself anonymous or resistant to traffic analysis. This makes Proxymate and Onion Routing especially natural complements.

Pipe-net [5] is a proposal somewhat similar to Onion Routing. It has not been implemented, however. Pipe-net's threat model is more paranoid than Onion Routing's: it attempts to resist active attacks by global observers. For example, Pipe-net's connections are permanent and carry constant traffic (to resist timing signature attacks). And, disruptions to any connection are propagated throughout the network. This makes the design impractical for any short lived or large bandwidth connections and implies that the entire network shuts down if even one connection does so. Thus, it is also highly vulnerable to denial-of-service attacks. Pipe-net's design provides the strongest traffic analysis resistance guarantees of any given for connection-based communication infrastructures running over mix-like nodes. But, it accomplishes this at a very high price, so it is not likely to be implemented for large scale Internet use.

Crowds [14] is roughly a distributed and chained Anonymizer, with encrypted links between crowd members. Upon receiving traffic for the first time on a *path* a crowd member flips a weighted coin, and depending on the outcome, continues the path to another randomly chosen crowd member or terminates the path and forwards this (and any future traffic on the path) to its ultimate destination. Crowds is less general than Onion Routing, both in its applications (it is designed only for Web traffic) and its anonymity goals (there is no attempt to hide the ultimate destination of traffic from any node on the path).

Zero-Knowledge Systems [17] has designed a system with many similarities to Onion Routing. Beta versions are available of Freedom, their software for network access akin to the remote-proxy or customer-ISP access described above. Freedom also incorporates local pseudonym management and other features. However, the currently described Zero-Knowledge system appears to limit routes to a fixed length of three hops, which makes connections much more vulnerable to some forms of traffic analysis [16]. Also, the system design does not seen to be as naturally compatible as Onion Routing to enclave level traffic protection, on either the initiator or the responder end.

A natural extension to Onion Routing is the introduction of reply onions. Reply onions allow connections to be made back to an anonymous sender through an onion routing network long after the original connection existed. Reply onions could be used to send anonymous replies in response to a previously received anonymous email. They could also enable novel applications such as anonymous publishing (anonymous URLs) similar to the Rewebber project [6].

7 Conclusion

In summary, Onion Routing is a traffic analysis resistant infrastructure that is easily accessible, has low overhead, can protect a wide variety of applications, and is flexible enough to adapt to various network environments and security needs. The system is highly extensible, allowing for additional symmetric cryptographic algorithms, proxies, or routing algorithms with only minor modifications to the existing code base. Instructions for accessing the prototype network can be found on our Web page along with additional background, pointers to publications, and contact information [12].

References

[1] The Anonymizer. `http://www.anonymizer.com/`

[2] D. Bleichenbacher, E. Gabber, P. Gibbons, Y. Matias, and A. Mayer. "On Secure and Pseudonymous Client-Relationships with Multiple Servers", *Proc. Third USENIX Electronic Commerce Workshop* , Boston, Mass. Sept. 1998, pp. 99–108.

[3] D. Chaum. "Untraceable Electronic Mail, Return Addresses, and Digital Pseudonyms", *Communications of the ACM*, v. 24, n. 2, Feb. 1981, pp. 84-88.

[4] L. Cottrell. "Mixmaster and Remailer Attacks", `http://obscura.obscura.com/~loki/remailer/remailer-essay.html`

[5] W. Dai. Pipe-net, February 1995, post to the `cypherpunks` mailing list.

[6] I. Goldberg and D. Wagner. "TAZ Servers and the Rewebber Network: Enabling Anonymous Publishing on the World Wide Web", *First Monday*, vol. 3 no. 4, April 1998.

[7] D. Goldschlag, M. Reed, P. Syverson. "Onion Routing for Anonymous and Private Internet Connections", *Communications of the ACM*, vol. 42, no. 2, February 1999, pp. 39–41.

[8] C. Gülcü and G. Tsudik.
"Mixing Email with *Babel*", in *Proceedings of the 1996 Symposium on Network and Distributed System Security*, San Diego, February 1996.

[9] J. Helsingius. `http://www.penet.fi/`

[10] A. Pfitzmann, B. Pfitzmann, and M. Waidner. "ISDN-Mixes: Untraceable Communication with Very Small Bandwidth Overhead", in *GI/ITG Conference: Communication in Distributed Systems*, Mannheim Feb, 1991, Informatik-Fachberichte 267, Springer-Verlag, Heildelberg 1991, pp. 451-463.

[11] Proxymate. `http://www.proxymate.com/`

[12] The Onion Routing Home Page. `http://www.onion-router.net/`

[13] M. Reed, P. Syverson, and D. Goldschlag. "Anonymous Connections and Onion Routing", *IEEE Journal on Selected Areas in Communications*, vol. 16 no. 4, May 1998, pp. 482–494.

[14] M. K. Reiter and A. D. Rubin. "Crowds: Anonymity for Web Transactions", *ACM Transactions on Information and System Security*, vol. 1 no. 1, November 1998, pp. 66–92 .

[15] P. Syverson, M. Reed, and D. Goldschlag. "Private Web Browsing", *Journal of Computer Security*, vol. 5 no. 3, 1997, pp. 237–248.

[16] P. Syverson, G. Tsudik, M. Reed, C. Landwehr. "On The Security of Onion Routing", Preprint.

[17] Zero-Knowledge Systems. `http://www.zerokowledge.com/`

Domain Based Internet Security Policy Management

John Zao, Luis Sanchez, Matthew Condell, Charles Lynn, Matthew Fredette, Pamela Helinek, Pajesh Krishnan, Alden Jackson, David Mankins, Marla Shepard, Stephen Kent
< *jzao | lsanchez | mcondell | clynn | mfredette | phelinek | pkrishnan | ajackson | dmankins | mshepard | skent @bbn.com* >
GTE / BBN Technologies Inc.

Abstract. *As security devices and protocols become widely used on the Internet, the task of managing and processing communication security policies grows steeply in its complexity. This paper presents a scaleable, robust, secure distributed system that can manage communication security policies associated with multiple network domains and resolving the policies — esp. those that specify the use of IP-AH/ESP security protocols — into security requirements for inter-domain communication.*

Technology innovation includes a formal model for IPsec policy specification and resolution, a platform independent policy specification language and a distributed policy server system. The formal model consists of a hierarchical domain model for IPsec policy enforcement and a lattice model of IPsec policy semantics. The policy specification language enables users to specify IPsec policies using the formal model regardless of the make of the security devices. The policy servers maintain the security policies in a distributed database, and negotiate the security associations for protecting inter-domain communication. Both the policy database and the policy exchange protocol are protected from passive and active attacks. Several UNIX implementations are available for non-commercial uses.

1. Introduction

Network security technologies are quickly being deployed over the Internet to support electronic commerce and virtual private networking services. In particular, security gateways, commonly known as firewalls, are installed along the perimeters of enterprise internets to enforce data origin authentication and access control, and security protocols, e.g., IP security (IPsec) protocols and Transport Level security (TLS/SSL) protocols, are used to provide end-to-end and hop-to-hop confidentiality, integrity and authentication protection to the Internet traffic. While these security measures have made the Internet a little safer to use, they also create significant management problems. For example, the deployment of security gateways fragments the Internet into security domains, each of which enforces different security policies. The use of security protocols necessitates establishing security associations among communicating end-points and en-route security gateways based on the security requirements of communication. Currently, the Internet Key Exchange protocol [1] can be used to negotiate the necessary security associations; however, a scaleable and general method is still unavailable for deducing the security requirements based on the security policies associated with the end-points and the gateways. In this paper, we present the principles and the design of a distributed policy management system capable of managing the security policies for multiple security domains and resolving them into the security requirements for inter-domain communication.

1.1. Requirement Analysis

We shall begin with a study of the functional requirements of Internet security policy management. Internet security policies are often characterized by the following properties:

- The policies are specified locally but can affect the security requirements of global communications since they are applicable to all communications to and from the network entities managed by the authorities.
- The policies may be specified by several authorities including network administration and users of communication applications. The policies may be classified into *mandatory* and *discretionary* policies respectively.
- The policies may be established for multiple protocol layers including data link, network, transport and application layers.
- The policies may be enforced by different security agents of heterogeneous origins: communication applications, host operating systems, security gateways, secure tunneling and link encryption devices.

Along with these characteristics, there lie the general expectations of safe policy management and assured policy enforcement.

Assured Policy Enforcement. The expectation usually implies the following requirements:

- The specification of security policies should be interpreted consistently by different security devices. This requirement implies the needs for a common policy specification language and a standard policy semantics accepted by both network administrations and application developers.
- The methods of managing the security policies should be applicable to networks of different sizes and topologies. In view of the scale and the growth rate of the global Internet, the management system must be distributed in design with decentralized management and enforcement components.
- The effects of security policy enforcement should remain unaffected by route changes (unless the policies are intended to interact with routing conditions). This requirement is a strong motivation for developing the concepts of domain-based policy management and boundary policy enforcement. By requiring all the security devices along a domain boundary to enforce the same set of security policies, we ensure that the traffic in and out of the domain will receive the same treatment regardless of the route it takes.

Safe Policy Management. This expectation can be translated into the following security requirements:

- *Secure Communication*—Security policy communication should be protected by integrity, data-origin authentication and confidentiality services.
- *Authorized Negotiation*—Policy management agents that participate in policy exchanges must be authorized to represent specific security domains and perform the required operations.
- *Selective Disclosure*—Security policies are sensitive information; even the authorized agents should be allowed to access only the policies necessary to accomplish their objectives.

1.2. Solution Approach

Our solution to the Internet security policy management problem is to develop a distributed system of *security policy servers*, each of which is authorized to maintain the security policies applicable to a cluster of network entities and to negotiate the security services with other servers needed to protect the communication involving those network entities. The security policy servers shall be deployed according to the hierarchical structure of security domains: each cluster of entities subjected to a common set of policies shall be grouped into a security domain; a collection of security domains managed by a common administration shall be associated with a security policy server. The security policies of each domain shall be enforced by the security devices installed along the boundary of the domain. Figure 1 shows the architecture of the domain based policy management system. The architecture consists of four components:

- *Security Policy Domains* — A formal model for security policy specification and enforcement that enables policies to be associated with non-overlapping domains in multiple hierarchies and enforced by the security devices located at domain boundaries; this model significantly reduces the complexity of policy management by transforming it from an inter-device process to an inter-domain process.
- *Security Policy Servers* — A distributed system of servers that maintain, exchange and process security policies associated with security domains in order to determine necessary security associations to protect inter-domain communication.
- *Security Enforcement Agents* — A set of security devices installed along the boundary of a security domain to enforce security policies of the domain.

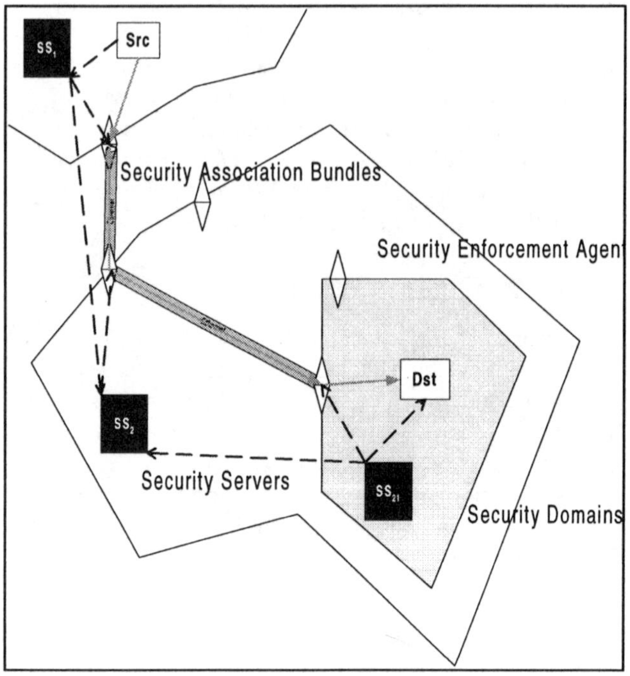

Figure 1. Architecture of Domain Based Security Policy Management System

- *Security Association Bundles* — The collections of security associations derived from the policy exchanges conducted by the security policy servers.

Four new technology components have been created to support this architecture:

- *Security Policy Exchange Protocol* — The protocol used for communication among the policy servers and between them and the enforcement agents
- *Security Policy Specification Language* — The common language for expressing security policies
- *IPsec Policy Algebra* — an algebra for deducing the IPsec actions from the policies
- *Security Policy Management Tools* — Graphic tools for viewing and editing security policies

1.3. Paper Overview

In the remaining sections of the paper, we will discuss the formal concept of hierarchical security domains, the design of the security policy server system and the security policy exchange protocol, the security policy specification language and processing algebra, and finally briefly describe the prototype.

2. Security Domains

2.1. Basic Definitions

An *internet security management domain* (or simply a *security domain*) is a connected cluster of communicating entities on the Internet, referred to as the *members* of the domain, that are protected by a common set of *communication security policies* applied by the *security enforcement agents* (or simply a *security policy*) positioned along the *boundary* of the security domain.

Security domains may form a *hierarchy* based on the following rules:

- A security domain may be a subset of another security domain if all the members of the first domain are also the members of the second domain; in that case, the first domain is considered to be at a higher level of the hierarchy than the second domain.
- A security domain may not overlap partially with another security domain, i.e., a security domain can either be a subset of another domain or share no common member with the other domain. Hence, a communicating entity can only be a *direct member* of a single domain and must be an *indirect member* of all the domains containing that domain.

Conceptually[1], every communicating entity on the Internet must be a member of one or more security domain(s) and subjected to the security policies of the domains to which it belongs directly and indirectly. Any inconsistency among the security policies of these domains that are applicable to a common set of entities is considered a *security policy conflict*.

A *communication security policy* (or simply a *security policy*) is a specification of security services that are required to protect specific network traffic in or out of a security domain under specific conditions. Given these domain-based security policies, one can determine the security services required to protect a communication among two or more network entities by combining the services specified by the security policies of the domains that contain these entities, and the policies of the domains traversed by the communication. We refer to this process as the *resolution* of security policies.

The security policies pertaining to a security domain are enforced solely by the security agents on the boundary of the security domain. We refer to this practice as *boundary policy enforcement*. The practice must be *complete* in the following sense: all communications between the members of the domain and the external Internet must be processed by the security agents according to the security policies of the domain; no communication path can exist between the members of a security domain and the rest of the Internet that can bypass the protection of the security agents.

2.2. Usage Rules

Together, the concepts of security domain hierarchies and boundary security enforcement simplify the management of communication security policies in the following ways:

- It provides a simple but powerful abstraction for reasoning about the deployment and the enforcement of security policies. A security policy will be applied to a communication if and only if the traffic traverses the boundary of the security domain that owns the policy. Hence, in order to protect the communication of some network entities, one must enclose them in a security domain and place security agents at all traffic crossing points on the boundary. Conversely, in order to determine the policies applicable to a communication, one simply has to discover the security domains containing the communication

[1] We shall reconcile the conceptual model with practical network configurations in Section 2.2.

sources and destinations as well as the domains traversed by the traffic.

- It separates the concerns of policy enforcement from those of packet routing: given the practice of boundary policy enforcement is complete then the same set of security policies will be applied to the communication traffic regardless of the routing choice for the traffic.

However, because boundary policy enforcement precludes the application of security policies to the intra-domain communication, the practice shall affect the way we group network entities into security domains. Following are two possible groupings of entities.

2.2.1 Grouping of Local Subnets

The first case corresponds to the typical setting of a firewall protected internet segment: one or more *internet subnets* are surrounded by security gateways that guard all the connections between the subnets and the external Internet. In this setting, the security gateways are the security enforcement agents; the hosts, bridges and routers installed on the subnets are the members of the domain. The security policies specify not the protection to be applied at the communication end-points, but those to be enforced at the security gateways which function as intermediary agents.

In some special settings, the subnets may be surrounded by security gateways that perform different functions and enforce different policies. These settings can be modeled as a collection of overlapping security domains with the same members but different security agents/gateways. Each group of security gateways that enforce the same policies should define the boundary of a distinct domain. As explained in §2.2.3, the overlapping domains shall only be used for formal reasoning; in actual practice, they shall be collapsed into a single physical security domain guarded by multiple groups of security agents.

2.2.2 Grouping of Hosts and Applications

Since security policies can only be enforced at the boundary of a security domain, policies applicable to a communication end-point—be it a layer in the protocol stack, a software application or a hardware module—must be associated with a security domain that encloses the software and hardware components protected by the policies. In this setting, the security enforcement agents are the security modules that perform the services prescribed by the security policies. Network, transport and application layer security can all be modeled in this case. Multiple security domains may also be formed within a single host if different policies may be applied to various user applications and/or at different protocol layers.

2.2.3 Simplification of Domain Hierarchy

The strict definition of security domains [§2.1] seems to imply that a large number of security domains must be established over the entire Internet before domain based policy management can be useful. This is not the case: the system not only can be deployed incrementally but can also be used with a small number of security domains (similar to those established by the network administration). The following techniques can be used to transform the *logical domain model* described above into a *physical domain architecture*.

Collection of end-point domains. A large number of logical security domains may be established when one introduces a policy to a group of communication end-points. Since the policy is expected to protect individual end-points, a domain must be created for each end-point. One way to combine these single-entity domains into one domain is to use a *pointer* for identifying the security enforcement agents. Instead of providing the explicit identity of the security agent, the policy shall use a pointer to refer to the end-point that matches with the policy specification. This will allow all the end-points protected by the same policies to be included in one security domain and use the same pointer as a reference to individual security agents.

Collapse of overlapping domains. Security domains with same members but different security agents [§2.2.1] can be collapsed into one domain by grouping the security agents that enforce the same policies. Again, a pointer must be used to refer to the security agents: the policies that are enforced by the same group of security agents shall use a pointer to refer to any agent inside the group (but not to the ones outside); policies that are enforced by different group of agents shall use different pointers with disjoint coverage. Each of the pointers thus represent a distinct overlapping domain.

3. SECURITY POLICY SYSTEM

3.1. Distributed Server Architecture

The Security Policy Server System (SPS) is a distributed system which provides hosts and security gateways with the policy information required to establish a secure communication end-to-end through possibly multiple security gateways. The Security Policy System provides one or more automated mechanisms for hosts to discover primary and secondary security gateways relevant in an end-to-end communication. Using the Security Policy System, hosts can validate the identity of security gateways and verify that the gateways in question are authorized to represent the source or destination host that they claim to represent.

SPS is comprised of Policy Servers (PS), Policy Clients (PC), Master Files and SPS Databases. Master Files contain local policies and other particular information about a security domain. Local policy information combined with non-local policies (policies outside the boundaries of the security domain) form the SPS Databases. Policy Servers receive request messages from policy clients and other policy servers, process them, and provide the appropriate policy information to the requestor based on the request and access control rules. The servers also maintain the SPS databases by loading local and non-local policy information received through SPS exchanges. Policy Clients generate requests for policy information and transform the replies into the appropriate format required by the application using SPS.

Policy Servers and Clients use the Security Policy Protocol (SPP) to exchange policy information. SPP transports policy information from the SPS Database where this information resides to security gateways and hosts. This protocol provides hosts with the policy information needed to establish security associations with security gateways in the communication path to other hosts, without requiring a-priori knowledge of the identities of the security gateways.

3.2. Policy Databases

In SPS, every security domain must maintain a database containing the policy information for that domain. Security domains could be as small as one host or as large as several networks. This database, called the SPS Database, is comprised of three logical databases: (1) the Local Policy Database, (2) the Cache Database, and (3) the Security Domain Database. The Local Policy Database contains all the policies for the security domain. It is populated with information coming from the Master File of the security domain. The Cache Database contains local and external policies received from other security domains via SPS exchanges.

The Security Domain Database contains a list of all hosts, security gateways, and policy servers that are part of the security domain. Compliant SPS implementations of a policy client and server do not need to implement these three databases separately. However, the information contained in each one of them must exist. The Local Database and the Cache Database must keep a distinct sets of policies since it is not possible to revert cached policy information into Local Database policy information after the cached items expire. While it is not necessary to standardize the format of the database used, the SPS database must contain a minimum set of information.

3.3. Security Policy Protocol (SPP)

In Policy Based Security Management (PBSM), policy clients and servers exchange information using the Security Policy Protocol (SPP) illustrated in Figure 2. SPP defines how the policy information is exchanged, processed, and protected by clients and servers. The protocol also defines what policy information is exchanged and the format used to encode this information.

SPP is flexible and can adapt as policies and topologies change since it does not require a-priori knowledge of the identities of the security gateways along the end-to-end path of a communication. The protocol depends solely upon routing of an appropriate message to discover the identities of the security gateways. In SPP a client may send a query message to a pre-configured local policy server to determine the security policies required for a particular communication. The local server, if it does not have an answer to the query, sends a query to the final destination of the communication.

SPP proxies running on security gateways along the route to the destination will intercept the query and forward it to the policy server for that gateway. That server may then either answer the query only if it has a policy for the communication and if it is authoritative over the final destination. A policy server is authoritative over a host if it asserts policy information for it and can present cryptographic credentials indicating the server's right to make policy statements on behalf of the host. If the policy server is not authoritative over the final destination, it forwards the query to the final destination. This process continues hop-by-hop (server to server) until the query reaches the authoritative policy server for the final destination of the communication.

The policy server ultimately responsible (authoritative) over the final destination sends a reply message destined to the policy server that sent the query. This is repeated on a hop-by-hop basis until the policy server that originated the query receives its reply. When a policy server receives a reply, it must verify a cryptographic signature on the message and the chain-of-trust. The signature proves the integrity and authenticity of the message. The chain-of-trust provides cryptographic proof that every server that has processed this reply is authorized to be involved

```
admin. boundary                           admin. boundary
------------------                   -----------------------------
|                 |                  |             admin. boundary|
|                 |                  |            ----------------|
|       Q1        |       Q2         |      Q3    |              ||
|  H1 ---- SG1 ---- (Internet) --- SG2 ----   | SG3 --- H2   ||
|       R3   |    |       R2         |  |   R1    | |            ||
|            PS1  |                  | PS2        | PS3          ||
|                 |                  |            ----------------|
------------------                   -----------------------------
                           ESP Tunnel
                  |========================|
                           ESP Tunnel
          |=========================   ==============|
                           ESP Transpo:
          |==============================================|

          Legend
          ------
          |==| = security association required by policy
          ---- = connectivity (or if so labeled, administrative boundary)
          Hx   = host x
          SGx  = security gateway x
          PSx  = policy server x
          Qx   = query x
          Rx   = reply x
```

Figure 2. Illustration of SPP Operation

in the communication. The chain-of-trust proves that all the servers involved are either directly or indirectly authoritative over the source or the destination of the communication.

Policy servers communicating through SPP must also resolve their local policies with the policies received in reply messages before accepting and incorporating them into their local policy databases. The resolution process intersects the two sets of policies. SPP implementations perform this process to: (1) find the common policy elements, (2) resolve any ambiguities in how the policies may be enforced; and (3) determine if policies must be sent to a gateway controlled by the server to provide it with the information necessary to allow the communication. The server then replies with the merged policy and sends any necessary signals to its gateways and clients.

All policies exchanged using SPP must be decorrelated. Two policies are decorrelated if there exists at least one selector in both policies for which their values do not overlap or intersect. Due to the local nature of policies, decorrelation is necessary to permit policy servers to properly cache policies.

3.3.1 SPP Message Format

An SPP message consists of a message header and a payload. The SPP header is present in every message. It contains fields identifying the type of message, the status of the message, the number of queries and/or record payloads, and the identity of the host requesting policy information. The header also includes a time-stamp field that provides anti-replay protection. Following the header there might be zero or more SPP payloads. Currently, there are three payload types defined in SPP: Query, Record, and Signature payloads.

SPP has six distinct message types. Query messages contain a specific request for policy information. Reply messages include policy records that answer specific policy queries. Policy messages include policy information and are utilized for up-/downloading security policies to and from a policy server. Policy Acknowledgment messages are utilized to acknowledge corresponding policy messages but do not contain policy information. Policy servers exchange bulk policy information between servers using transfer messages. Finally, policy servers use keep alive messages to inform security gateways and/or other monitoring devices of the status of the server.

SPP messages are authenticated using either IPsec [3] or another security mechanism. SPP provides a basic security mechanism that can be used to provide authentication and integrity to its messages when other security mechanisms are not in use. The SPP authentication is especially useful when traversing heterogeneous domains and the identity of the policy server authoritative for the destination is unknown. These services are provided using digital signatures.

SPP carries signatures in the signature payload. The signature is calculated over the entire SPP message. When this service is used, the entity (host, policy server, or security gateway) verifying the signature must have access to the public key that corresponds to the private key used to sign the SPP message. Certificate fetching is out of the scope of SPP. However, SPP provides a simple certificate fetching mechanism for entities that elect to use it as an alternative to other mechanisms. SPP supports several Public Key certificates formats.

3.3.2 SPP Operation

The following example illustrates how the Security Policy Protocol operates and interacts with IPsec. It describes, step-by-step, the process from the initial application call to the establishment of the necessary security associations. We assume the policy servers have been loaded with policies for their security domains and the policies have been appropriately decorrelated.

1. User application attempts to send a message from H1 to H2 (e.g., finger somebody@H2)
2. IPsec on H1 gets the packet and finds a policy for it in the Security Policy Database (SPD).
3. H1 does not have a security association for the communication and asks the Key Management Protocol to establish one.
4. The policy client in H1 is queried for the policy governing the communication between H1 and H2.
5. H1's policy client creates an SPP query, Q1, and sends it to PS1, its configured policy server.
6. PS1 receives the query. Its security domain database indicates that it is not authoritative over H2 so it checks its cache to see if it has a cached answer. For this example, it does not, so it creates a new SPP query, Q2, with the query and sends it to H2.
7. SG2 intercepts Q2 and forwards it to PS2.
8. PS2 receives the query. Its security domain database indicates that it is not authoritative over H2 and PS2 determines it wants to be involved in the communication. It checks its cache to see if it has a cached answer. For this example, it does not, so it creates a new SPP query, Q3, with the query and sends it to H2.
9. SG3 intercepts Q3 and forwards it to PS3.
10. PS3 receives the query. It checks its security domain database and determines that it is authoritative over H2 so it will send a reply. It checks its cache to see if it has a cached answer. For this example, it does have one cached from previous information sent to it by H2. The cached policy indicates ESP transport must be done with H2 and ESP tunnel must be done with SG3.
11. PS3 creates two messages. One is an SPP reply message, R1, with the policy indicating the required security associations, a security server record indicating PS3 is authoritative over H2, and PS3's certificate in a certificate record. The reply is sent to PS2. The second message is an SPP policy message to signal SG3 that it will need a security association with H1.
12. PS2 receives the R1. It verifies that PS3 is authoritative over H2. It looks up its local policy and resolves it with the policy in the reply. It caches the resolved policy and creates two messages. One is the reply to PS1, R2, that contains the resolved policy, PS3's security server and certificate records and a security server record that states that PS2 is authoritative over PS3 and PS2's certificate. The second message is an SPP policy message to signal SG2 that it will need a security association with H1.
13. PS1 receives the R2. It verifies that PS2 is authoritative over PS3 and PS3 is authoritative over H2, forming a valid chain of trust. It looks up its local policy and resolves it with the policy in the reply. It caches the resolved policy and creates R3, a reply to H1. R3 contains the resolved policy (the security server and certificate records are no longer needed). It also creates a policy message to signal SG1 that it will need a security association with SG2.
14. The policy client receives the R3 and returns it to the application that queried for it. The policy indicates that the three security associations must be established and they must be established in a particular order.
15. The Key Management Protocol is given this information and first creates a security association with SG3 which it can use to set up the security association with H2.
16. Finally, the original message from the application can proceed using the security associations.

4. SECURITY POLICY SPECIFICATION

4.1. Security Policy Specification Language

In order to support vendor and platform independent communication security policy specification, we developed the Security Policy Specification Language (SPSL) [7] by borrowing the syntax from IETF Routing Policy Specification Language (RPSL) [11]. The language was designed originally for specifying IPsec and ISAKMP policies, but its extensible syntax allows new object classes to be added for specifying policies of other security protocols. The language has the following distinct features.

4.1.1 IPsec/ISAKMP Policy Specification

In SPSL, a communication security policy is defined as a binding between a list of communication conditions and a corresponding list of security actions. If a communication matches with one of the conditions then the list of actions must be taken to protect the communication. In an IPsec/ISAKMP policy, a communication condition is specified as a tuple of selector values. SPSL supports an extended selector value set, which includes all the selectors mentioned in IPsec architecture document [5]. The actions, on the other hand, can effect three different kinds of operations:

- They may specify simple packet filtering actions: discard, pass or forward the packet (via tunneling) to a designated network entity.
- They may use IPsec tunnels or transports to pass the packet. The possible security mechanisms for protecting the tunnels and the transports are specified as ISAKMP proposals allowed by ISAKMP-IPsec DOI [10].
- They may also specify security proposals necessary for protecting ISAKMP exchanges.

4.1.2 Node/Domain Based Policy Associations

SPSL supports two different ways to associate security policies with network entities. They are known as node based and domain based policy association.

In *node based policy association*, security policies are bounded to individual network nodes and security devices. The policies associated with a network node specify the protection for the communications to and from the node. These policies are expected to be enforced by the security modules in the node. The policies associated with a security device (formally known as a security enforcement agent) specify the protection for the communications passing through these agents. Either the source or the destination of the communication must be among the nodes that the agent is authorized/expected to protect. Both the network nodes and the security enforcement agents should manage their own policies.

In *domain based policy association*, security policies are bounded to a hierarchical security domain. A security domain is defined as a connected set of network entities that are protected by security enforcement agents (SEA) placed on every communication path going through the domain perimeter. Every security enforcement agent of the domain works to enforce the common set of security policies associated with the domain. The security domains may be completely disjoint or contained in one another, and thus form multiple domain hierarchies. The policies associated with a domain are managed by one or more special agents known as security servers (SS) of the domain. These servers may be distinct network entities or co-located with the nodes or the security enforcement agents of the domain.

SPSL defines special object classes to support the two policy associations: classes node, node-set, gateway and gateway-set are defined to be used in node based policy association while classes gateway, gateway-set and domain are to be used in domain based policy association. A mixed use of two associations is also permitted with the node class objects to be regarded as one-node domain objects.

Multiple Distributed Policy Enforcement Points

SPSL allows explicit selection of security enforcement agent(s) of a security policy. The choices can be specific interfaces of end nodes, en-route security gateways (SG) and/or network nodes identified by IP addresses or DNS names. If the selection is generic such as local/remote security gateways then the exact enforcement agent will be reached by routing. If the selection is specific as by giving IP addresses or DNS names then the enforcement agent may be reached via IP tunneling.

Authentication and Authorization Mechanisms

SPSL defines special object classes for supporting the following security services:

- *Data integrity of policy specification*—Every policy object is protected at least by keyed MD5 data integrity and data origin authentication.
- *Data origin non-repudiation of policy specification*—Policy objects may be protected with public key signatures, which offer non-repudiation proof to the issuer(s) of the policies.
- *Authentication and authorization of policy management entities*—Management objects such as maintainers, person and roles all have public key certificates so that they may issue policies and/or identify themselves to the security management system for access control purposes.

With these services, users of SPSL policy specifications can always verify the integrity and the origin of the policies and allow only authorized personnel to maintain and/or access the policies.

4.1.3 Object Classes

SPSL uses the object programming paradigm. It defines a small set of object classes that maintain policy information in their attributes. There are no executable methods in the object classes. New classes can be created as needed based on a syntactic rule similar to inheritance in object-oriented languages: a new class may have all the attributes of an existing class in addition to its own attributes. Objects in a SPSL file are distinguished and referred to by the unique values of their first attribute, which is known as the key attribute.

The existing classes in SPSL can be divided into the following four categories:

Primitive Data—They contain basic or atomic data elements used in policy specification, e.g., object-name, ipv4-address, integer-range, date, etc.

Management Agents—They contain relevant information of the management entities; the existing classes in this category are maintainer (mntner), person, role, and certificate (cert).

Network Entities—They depict the network elements that are relevant to policy specification; the existing classes are node, node-set, gateway, gateway-set, policy-server and domain.

Policies—They are the classes that contain the policy specification; there are only two classes at the moment: policy class specifies general packet filtering rules and IPsec-policy class specifies IPsec selectors and actions; objects of policy class may appear in two forms for short or long policy specification.

Each class has a set of attributes which store information about the objects of the class. Attributes can be mandatory (man) or optional (opt). A mandatory attribute must be defined for all objects of the class, and an optional attribute may be skipped. Attributes can also be single valued (s-v) or multiple valued (m-v). A single valued attribute may only be used once per object. A multiple valued attribute may be used more than once per object. Each object is uniquely identified by the key attribute of its class.

An SPSL object is textually represented as a list of attribute-value pairs. Each attribute-value pair is written on a separate line. The object's representation ends when a blank line (i.e., a line containing only whitespace characters) is encountered. The default character set is ISO 8859-1 (Latin-1). An object's specification may contain comments. A comment may appear anywhere in an object definition. It starts at the first "#" character on a line and ends at the first end-of-line character. Figure 3 shows a few sample SPSL objects.

4.2. Algebraic Semantics of IPsec Policies

In addition to defining a language for specifying security policies, we also developed an algebra for computing the *composition* and the *difference* of IPsec policies in particular. The algebra can be used in two different ways to process IPsec policies. First, it can be used to perform *policy decorrelation*, a process that can transform a set of policies into a corresponding set of uncorrelated policies. The decorrelated policies have a desirable property that only one of them may match with a specific communication condition; thus, these policies can be processed and used independent from one another. The policy resolution algebra can also be used to build an IPsec policy resolver—a functional module that accepts queries for the security requirements of end-to-end communications and returns the applicable policies to be enforced by specific agents.

4.2.1 Lattice Model of IPsec Policies

The IPsec policy algebra is based upon the set theoretic relations existing among the policy conditions and the partial order relations among the policy actions. As mentioned, every IPsec policy specifies a clause of *communication conditions*, expressed in terms of the values of seven *selectors* (source and destination IP addresses, source and destination port numbers, transport layer protocol identifier, user identifier, and security label) and a clause of *security actions*. Each selector may take a specific value or the "wildcard" value ANY. Source and destination IP addresses may also take ranges of address values. On the other hand, the action clause specifies whether an IP datagram should be *discarded*, *bypassed* without IPsec processing, or *passed* with IPsec processing when it matches with the policy condition. If IPsec processing is necessary then the action clause also specifies a typed list of security mechanisms (such as follows) to be used in datagram processing:

```
(esp (DES HMAC_MD5) or (DES HMAC_SHA))
or ((esp DES), (ah (HMAC_MD5) or
(HMAC_SHA)))
```

An IPsec policy can be modeled as a mapping from its *condition clause*, which is a direct product of seven selector value sets $S_{i=1..7}$, to its *action clause*, which is a typed list of IPsec-ISAKMP descriptors of security mechanisms:

$$P = \underset{i=1}{\overset{7}{\times}} S_i \rightarrow A.$$

```
policy-name:    P11
descr:          incoming TELNET connection to SD1 hosts
domain:         SD1
member-of:      P_SD1
direction:      inbound
dst:            199.100.2.1      # H1 Address
src:            ANY
xport-proto:    TCP
action:         permit
IPsec-proto:    esp transport
IPsec-mech:     esp cipher DES keylen 56 integrity HMACMD5 keylen 128
tech-c:         mcondell
admin-c:        lsanchez
mnt-by:         SS1_MNT
changed:        mcondell@ir.bbn.com 19980529
signature:      <mnter-name><cert-name><signature-alg><signature-value>

sec-serv:       SS1
descr:          primary security server of SD1
name:           sec-server-1.bbn.com
alias:          cotton.bbn.com
ifaddr:         128.89.0.2 masklen 16

sec-agent:      SG1
descr:          security gateway 1 protecting SD1
name:           bbn-firewall.bbn.com
ifaddr:         128.89.0.1 masklen 16
ifaddr:         192.2.1.83 masklen 24

sec-domain:     BBN.COM:SD1
descr:          BBN IR security domain
coverage:       128.89.5/24
sec-serv:       SS1, SS1a
sec-agent:      SG1, SG1a, SG1b, SG1c
sec-policy:     P_SD1
```

Figure 3. Policy Objects in SPSL

Based on the current data model of IPsec policies [6], we deduce that A is a *lattice* with the *top* element T being the NULL action or the least upper bound of conflicting actions and the *bottom* element $\perp$ being the UNDEFINED action lower in order than all actions defined. The lattice model of the policy actions enables us to compose two actions by finding their least upper bound using the *join* operation $\uparrow$ of a partially ordered set.

4.2.2 Composition Operation

The composition of two IPsec policies $(\circ : \wp \times \wp \rightarrow \wp)$ is a *commutative* and *associative* operation that produces a new policy specifying the resultant actions to be taken when the datagram states match with the conditions of *both* input policies. The operation is performed by processing the condition clauses and the action clauses separately. The composition of two policy conditions is performed simply as the *intersection* of the direct products of their selector value sets:

$$\left(\times_k S_{ik}\right) \circ \left(\times_k S_{jk}\right) = \left(\times_k S_{ik}\right) \cap \left(\times_k S_{jk}\right) = \times_k \left(S_{ik} \cap S_{jk}\right)$$

The composition of two action clauses A_i, A_j is performed as the join operation $\uparrow$ defined for the partially ordered set of IPsec policy actions:

$$A_i \circ A_j = A_i \uparrow A_j$$

4.2.3 Difference Operation

The *difference* operation $(-: \wp \times \wp \rightarrow \wp)$ is a *non-commutative* operation that takes *two* IPsec policies and produces a new policy which specifies the actions to be taken when the condition of the *second* policy is excluded from that of the *first* policy. Policy difference can also be performed by processing the condition clauses and the action clauses separately.

Let $P_1 = (C_1 \rightarrow A_1)$ and $P_2 = (C_2 \rightarrow A_2)$ be two policies, and $P_{21} \equiv P_2 - P_1 = (C_{21} \rightarrow A_{21})$ be one of their differences then the formulas to compute the condition C_{21} and the action A_{21} of the difference policy are

$C_{21} = C_2 - C_1$ and $A_{21} = A_2$.

5. PROTOTYPE IMPLEMENTATION

PBSM is comprised of five implementation modules: the parser for the Security Policy Specification Language (SPSL), the Security Policy System (SPS), the Security Management Agent (SMA), the Certificate Infrastructure (CI) and the Policy Management Tool (PMT). The following sections briefly cover key implementation details for each module.

5.1. SPSL Parser

Security Policy Specification Language (SPSL) is a language designed to express security policies, security domains, and the entities that manage the policies and domains. SPSL currently supports policies for packet filtering, IP security (IPsec), and ISAKMP exchanges, however, it may easily be extended to express other types of policies.

SPSL uses the object paradigm although it is neither an object-oriented nor a type based language. The language defines a small set of classes, which can instantiate objects maintaining data relevant to policy specification. The data are contained in the attributes defined in the object classes. There are no executable methods in the classes. The parser is implemented using Bison and Flex.

5.2. Security Policy System (SPS)

The Security Policy System (SPS) is a main component of PBSM. It takes requests from clients (hosts and security gateways) inside/outside a security domain and supplies them with the security policies needed to establish an end-to-end communication across multiple heterogeneous security domains. SPS is comprised of the Security Server Daemon (SSD) and the Security Policy Resolver (SPR).

The Security Server Daemon is the server portion of the Security Policy System and it resides in the security servers. The Security Policy Resolver is the client portion of the Security Policy System and it consists of a set of library calls used to request policy information. Client SPRs at hosts and/or security gateways communicate with Security Server Daemons using the SPP protocol.

The implementation of the SPS consists of developing the library calls that form the Security Policy Resolver, the functions that make the Security Policy Agent, the development of the decorrelation algorithm and the composition algebra processing. The last three components form the Security Server Daemon. The Security Policy Protocol (SPP) is implemented as part of both the Security Policy Resolver and the Security Server Daemon.

5.2.1 Security Server Daemon

The Security Server Daemon (SSD) is the server portion of the Security Policy System and it resides in any host functioning as a security server. The server daemon loads policies for a security domain from its master file into the SPS database. It replies to queries from policy clients and communicates with peer server daemons from other security domains to request policy information related to hosts outside its security domain.

The SSD processing is simple in nature. First, the SSD loads configuration parameters from a configuration file. It then calls the SPSL parser to verify that the policies in the Master files are correct. Upon successful parsing completion the SSD calls the policy decorrelator to eliminate any overlaps among the security policies in the Master File. Once the policies are decorrelated the SSD loads the policies into the SPS database.

After loading the policies, the SSD remains in a loop waiting for messages from client Security Policy Resolvers or peer Security Server Daemons. SSDs establish the validity of incoming messages, make SPS database searches, interface with the Policy Management Agent and generate appropriate replies as needed. The SSD has the ability to identify conflicts and resolve ambiguities between the local and imported policies. It may also take advantage of any ambiguities to help optimize the use of security associations within its security domain.

5.2.2 Security Policy Resolver

The Security Policy Resolver (SPR) is a set of library routines on PBSM compliant hosts and gateways that acts as an interface between the host and its security servers. These routines provide both synchronous and asynchronous interfaces that may be used by applications to request services from the Security Server Daemons. The SPR communicates with the security servers using the Security Policy Protocol.

The Security Policy Resolver is configured with the IP addresses of the Security Servers that it should use to resolve policies. The host on which the SPR resides may be included in this list. At least one Security Server must be configured. Multiple servers may be configured, as in DNS, to allow secondary servers to be accessed in the event that a primary server fails. These configured servers are the default servers that the synchronous interface will use to make policy queries. This information may also be used by applications to determine the appropriate Security Server Daemon to attempt with the asynchronous interface.

5.3. Security Management Agent

The Security Management Agent is an application that intercepts calls from the IPsec engine to a Key Management Protocol (KMP) and vice versa. It serves as the initiator of the policy resolution process in PBSM. When an application requires IPsec the IP datagram is sent to the IPsec engine in the kernel (for native software implementations). The IPsec engines check for an existing Security Association (SA) for that particular communication. If one SA exists in the Security Association Database it uses it otherwise the engine signals the KMP to establish one.

The SMA intercepts the engine signal, asks the SPR for policy information regarding the communication in question and waits for a response from the system. Once the response arrives, the SMA passes such response to the KMP and the process continues as usual. If the SMA doesn't receive a response within a time window it forwards the original request to the KMP and the process continues as usual. As a possible optimization, the SMA can intercept the engine's signal and forward it to the KMP while using it to initiate the policy resolution process with the SPR.

5.4. Certificate Infrastructure

The implementation of the Certificate Infrastructure (CI) consists of the development of the Certificate Infrastructure daemon and the CI database tools. The CI daemon (certd) maintains the database of certificates and CRLs, performs validation and external fetches, and provides certificates or certificate data such as keys, or Certificate Extensions to requesting applications. The CI daemon is a separate process that should be run in the background as root.

The application may request a certificate, key or a set of certificate extensions from the CI. It gives the CI a set of parameters which allows the CI database to search for the requested certificate. If the CI finds one or more certificates that match the input parameters, it returns them all to the application. The more optional input parameters available to the application to configure, the more likely the CI is to be able to identify one correct certificate.

A series of tools were developed for testing and supporting the Certificate Infrastructure daemon or certd. These include tools for inserting certificates and Certificate Revocation Lists (CRLs) into the certificate database, reading certificates from filenames, printing the database contents and validating certificates.

5.5. Policy Management Tools

The Policy Management Tools provide a graphical user interface to create SPSL files. Additionally, they provide an interface to several tools that may be useful for helping system administrators correctly develop their local policies, including tools to interact with the local security server and to provide consistency checks on policies.

6. Future Work and Conclusions

The work presented in this paper represents our first attempt in developing a framework for managing and processing communication security policies. The work may be extended in the following directions.

Policy managing for both network and application layer security. Existing work was focused entirely on the policies governing IPsec AH-ESP protocol uses. It would be meaningful to investigate the use of this system to negotiate the security requirements for both TLS/SSL protection between end hosts and IPsec tunneling among intermediate security gateways.

Algebraic semantics for Virtual Private Networking policies. The algebra discussed in §4.2 specifies the algebraic rules to compose and resolve IPsec policies written according to the simple Pereira-Bhattacharya model. It would be useful to extend the rule set in order to incorporate policy actions relevant to virtual private network (VPN) operation.

Policy management for group communication. The SPP protocol discussed in §3.3 performs only policy negotiation for two-party communication. It would be valuable to extend the protocol for supporting group communication.

Finally, it may be necessary to divide the policy management task into two sub-tasks of *policy discovery and negotiation* and *policy specification and processing.* The first task depends the use of multi-party protocols and distributed databases while the second requires the knowledge of language semantics and program analysis. There are plenty of work remained to be done in both areas.

7. REFERENCES

[1] D. Harkins, D. Carrel. "The Internet Key Exchange (IKE)" *RFC-2409.* Internet Society, Network Working Group, Nov. 1998.

[2] D. Maughan, M. Schertler, M. Schneider, J. Turner. "Internet Security Association & Key Management Protocol (ISAKMP)" *<draft-ietf-IPsec-isakmp-07>,* Internet Society IPsec Working Group, Feb. 97

[3] S. Kent, R. Atkinson. "IP Authentication Header" *RFC-2402.* Internet Society, Network Working Group, Nov. 1998.

[4] S. Kent, R. Atkinson. "IP Encapsulating Security Payload (ESP)" *RFC-2406.* Internet Society, Network Working Group, Nov. 1998.

[5] S. Kent, R. Atkinson. "Security Architecture for the Internet Protocol" *RFC-2401.* Internet Society, Network Working Group, Nov. 1998.

[6] R. Pereira, P. Bhattacharya. "IPsec Policy Data Model" *Internet Draft <draft-ietf-IPsec-policy-model-00>*, Internet Society, IP Security Working Group, Feb. 1998.

[7] M. Condell, C. Lynn, J. Zao. "Security Policy Specification Language" *Internet Draft <draft-ietf-IPsec-spsl-01>.* Internet Society, IP Security Working Group, July 1999.

[8] L. Sanchez, M. Condell. "Security Policy System" *Internet Draft <draft-ietf-IPsec-sps-00>.* Internet Society, IP Security Working Group, Nov. 1998.

[9] J. Zao. "Partially Ordered Composition and Search of IP Security Policies" *To be published in IEEE JSAC special issue on network security*, Feb. 2000.

[10] D. Piper. "The Internet IP Security Domain of Interpretation for ISAKMP" *RFC-2407.* Internet Society, Network Working Group, Nov. 1998.

[11] C. Alaettinouglu, T. Bates, E. Gerich, D. Karrenberg, D. Meyer, M. Terpstra, and C. Villamizer. "Routing Policy Specification Language (RPSL)". *RFC 2280.* Jan. 1998.

Secure Multicast Conferencing

Peter T. Kirstein, Ian Brown and Edmund Whelan

Department of Computer Science, University College London, Gower Street, London WC1E 6BT, UK

{P.Kirstein, I.Brown, E.Whelan}@cs.ucl.ac.uk

Abstract

This paper outlines the architectural principles of secured multicast conferencing. It discusses the general principles of the unsecured activity, giving details of tools, methods of announcing conferences, and relaying between multicast and unicast. It then considers how to secure these activities using techniques being standardised by the Internet Engineering Task Force.

1. Introduction

Multicast conferencing is a rapidly-growing area of Internet use. Audio, video and other media such as shared whiteboard data can be distributed efficiently between groups of conference participants using multicast algorithms that minimise the amount of traffic sent over the network. This is far more effective than systems that maintain a separate link between each participant. However, it also introduces a whole new set of security concerns.

This paper first provides an overview of the mechanisms used to support lightweight multicast conferencing. The protocols used to advertise and invite individuals to conferences, and then communicate between those participants, are described. Active services running in the network to mediate between hosts with different network and processing capabilities are also covered briefly.

We then describe the schemes being developed at University College London under several US Department of Defense and European Commission projects to add security to these services. They are being deployed both to the CAIRN community as part of the privacy aspects of that programme, and to several European communities collaborating in Framework 4 projects. The use of end-to-end encryption at the application and network layers to protect the privacy and integrity of conference data are outlined. Several different schemes for securely announcing and inviting people to conferences, and providing session information to participants, are compared. We also consider the use of smartcards to improve the overall security of these methods. Finally, the security implications of providing in-network code with access to the plaintext of encrypted data are considered.

2. Insecure multicast conferencing

2.1. Basic conferencing architecture

The two defining problems of conferencing are communication between (possibly large) groups of people and real-time delivery of information. In the Internet, these capabilities are supported at a number of levels. The remainder of this section provides an overview of this support.

The humans participating in a conference generally need to have a specific idea of the context in which the conference is happening, which can be formalised as a conference policy. Some conferences are essentially crowds gathered around an attraction, while others have very formal guidelines on who may take part (listen in) and who may speak at which point. In any case, initially the participants must find each other, i.e. establish communication relationships (conference *set-up*, Section 2.3). During the conference, some conference control information is exchanged to implement a conference policy or at least to inform the participants who is present.

A number of different tools can be used to send and receive conference data. *vic*, shown in figure 1, supports video. *rat*, *nte* and *wbd* respectively process audio, shared text and shared whiteboard data. All are available from http://www-mice.cs.ucl.ac.uk/multimedia/software/ Support for multicast conferencing is also now beginning to appear in commercial products such as *RealPlayer* from Real Networks.

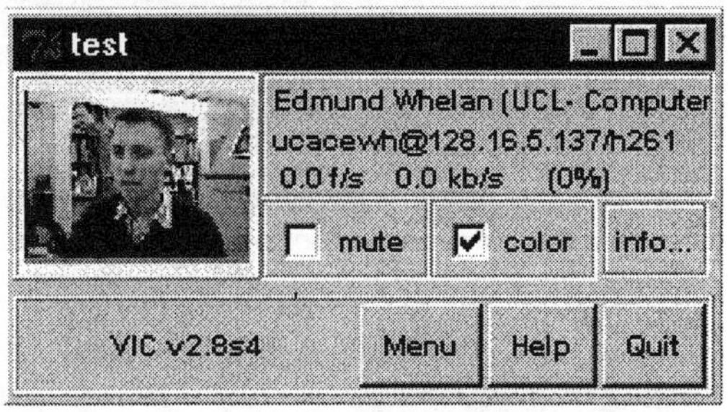

Fig. 1: the VIdeoConferencing tool

0-7695-0490-6/99 $10.00

2.2. Transport mechanism

In a conference, information must be distributed to all the conference participants. Early conferencing systems used a fan-out of data streams, setting up a connection between each pair of participants, which means that the same information must cross some networks more than once. The Internet architecture uses the more efficient approach of multicasting the information to all participants.

Internet Protocol multicast provides efficient many-to-many data distribution in an Internet environment [Dee89]. Multicast is a natural solution for multi-party conferencing because of the efficiency of its data distribution trees, with data being replicated in the network at appropriate points rather than in end-systems. It also avoids the need to configure special-purpose servers to support the session; such servers require support, cause traffic concentration and can be a bottleneck. Receivers do not need to know who or where the senders are to receive traffic from them. Senders never need to know who the receivers are. Neither senders nor receivers need to care about the network topology as the network optimises delivery.

Multimedia conferences require real-time delivery of streamed continuous media (audio and video information) and reliable, near real-time delivery of shared workspace information. In a datagram network, multimedia information must be transmitted in packets, some of which may be delayed more than others. In order that audio and video streams be played out at the recipient with the correct timing, information must be transmitted that allows the recipient to reconstitute the timing.

The Real-time Transport Protocol (RTP) provides all these services [Sch97]. It provides a standard format packet header which gives a media specific timestamp, as well as payload format information and sequence numbering amongst other things. RTP is normally carried using User Datagram Protocol. It does not provide or require any connection set-up, nor does it provide any enhanced reliability over UDP.

Continuous-media tools like *rat* and *vic* do not normally provide error re-transmission facilities; they can reconstruct missing audio and video at the receiver, sometimes helped by Forward Error Correction (FEC) data to correct errors incurred during transmission [Per98]. By contrast shared-workspace tools, like the *network text editor* (NTE), require fully reliable transmission. To achieve this there is a need for a reliable multicast transport. Various techniques to achieve this have been implemented [Flo95]. For RTP to provide a useful media flow, there must be sufficient capacity in the relevant traffic class to accommodate the traffic. How this capacity is ensured is independent of RTP.

Each original RTP source is identified by a source identifier, which is carried in every packet. RTP allows flows from several sources to be mixed in gateways to provide a single resulting flow. When this happens, each mixed packet contains the source IDs of all the contributing sources. RTP media timestamps are in units that are appropriate to the media flow. For example, 8KHz sampled Pulse Code Modulation-encoded audio has a timestamp clock rate of 8KHz. This means that inter-flow synchronisation is not possible from the RTP timestamps alone.

Each RTP flow is supplemented by Real-Time Control Protocol (RTCP) packets. There are a number of different RTCP packet types. RTCP packets provide the relationship between the real-time clock at a sender and the RTP media timestamps so that inter-flow synchronisation can be performed, and they provide textual information to identify a sender in a conference from the source ID.

The penetration of multicast packets can be limited by two mechanisms: *administrative scope* and *time-to-live*. The first ensures that packets for a specific multicast group are constrained to stay inside a set of routers that have been set up to limit the flow. The second ensures that the packets are constrained by the total number of routers the packet can traverse before being discarded

The variety of different network technologies, workstation capabilities and conference system technologies preclude the adoption of a single conferencing system, at a single speed, with homogeneous facilities. Many mechanisms have been suggested, and some even implemented, to address system heterogeneity. At some level these may be independent of any direct intervention inside the network. For example, if streams are layered over different multicast groups, each additional stream providing higher quality, a disadvantaged receiver may just not subscribe to all the relevant multicast groups [Mcc96]. In-network code, such as the Universal Transcoding Gateway described in Section 2.4, can also filter data at points in the network where capacity drops so that the traffic flow matches the network capacity [Amir98].

It is important to include mechanisms for co-ordinating and controlling different streams of data. The Real-Time Streaming Protocol (RTSP) [Sch98] has been defined for controlling activities between systems, while the Message Bus (Mbus) controls activities inside single systems [Per99].

2.3. Session announcements and invitations

Session descriptions provide an advertisement that a session will exist, and also provide sufficient information including multicast addresses, ports, media formats and session times so that a receiver of the session description can join the session. The Session Description Protocol

(SDP) describes the content and format of these descriptions [Han98].

One method of announcing sessions is to regularly send the session description on a well-known multicast address, with a specific scope, using the Session Announcement Protocol (SAP) [Han99a]. The bandwidth consumed by all announcements is limited, so senders must limit the rate that announcements are sent to the total bandwidth divided by the volume taken up by current announcements.

The announcement includes information such as the organiser of the conference, and the session description.

People wishing to participate in a particular conference must then listen for the SAP announcement, and start up their tools with the SDP details provided. Session Directory (SDR) is a tool that can receive these "broadcast" session descriptions, browse through all sessions currently being announced, and then start up the relevant media viewers.

An important feature of SAP is that if the announcement of the message is received, there is a high probability that the session itself can be joined. The distance multicast packets travel can be restricted using the Time-To-Live (TTL) field in packet headers. The TTL is set by the sender and then decreased by each router through which it passes. Once the TTL reaches zero, routers no longer forward the packet. If the TTL of an announcement is set the same as the conference data, recipients who receive the announcement should also be able to receive the conference data. This also applies to regions with limited multicast connectivity to the sender of the data.

Another feature is that each SAP client sees all SAP announcements; for this reason if it is announcing a new session at a particular time, it can check that there has not been a previous announcement for the same multicast group at the same time.

Session announcements can also be used to advertise tightly-coupled sessions, which only requires that additional information about the mechanism used to join the session be given. However, as the number of sessions in the session directory grows, we expect that only larger-scale public sessions will be announced in this manner; smaller, more private, sessions will tend to use direct invitation rather than advertisement. Otherwise either the bandwidth required by SAP, or the interval between announcements, will become too large.

The Session Initiation Protocol (SIP) provides an alternate mechanism whereby a user can be invited to participate in a conference [Han99b]. SIP does not care whether the session is already ongoing, or is just being created. It does not care whether the conference is a small tightly coupled session or a huge broadcast – it merely conveys an invitation to a user in a timely manner, inviting her to participate, and provides enough information for them to be able to know what sort of session to expect. Thus although SIP can be used to make telephone-style calls, it is by no means restricted to that style of conference.

As many users are mobile, it is important that such an invitation mechanism be capable of locating and inviting a user in a location-independent manner. Thus user addresses need to be used as a level of indirection rather than routing a call to a specific terminal. The invitation mechanism also provides for alternative responses, such as leaving a message or being referred to another user.

It is also possible to use off-line mechanisms for providing the information on upcoming sessions. One is to send the information by e-mail; mailer plug-ins can parse messages and start sessions automatically [Hin96].

Alternately the information can be put into a depository known to the potential participants. This mechanism is convenient if potential participants can be expected to access the directory sufficiently often. A combination of the use of ordinary e-mail to announce the existence of a conference, together with a directory mechanism to update session information, will probably be the most popular for a large class of applications. Specialised Web-based tools already allow browsing through lists of conferences, together with client plug-ins that can extract the relevant session description and start up the required tools.

Multicast sessions can be restricted using the TTL field or administrative scoping; hence although an announcement may be retrieved from a directory, it does not follow that the retriever can participate in a particular session. Moreover, because the announcement is only seen by announcers using that directory, there is a chance of two sessions being announced for the same multicast address at the same time. By using random generation of multicast addresses, this danger can be reduced; however if the number of simultaneous sessions increase, the danger of such conflicts also increases. Requiring the use of a depository can also lead to it becoming a single point of failure. It is possible to ameliorate this problem by putting the announcements in several depositories; this is comparable with caching Domain Name System information in nameservers throughout the Internet.

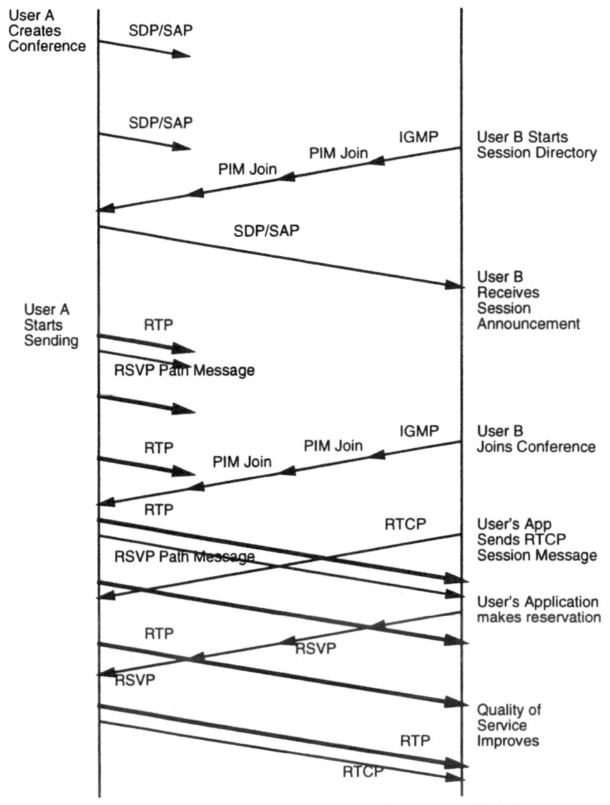

Fig. 2: Joining a lightweight multimedia session

None of these mechanisms will scale well to very large conferences, because of the potential number of messages or depository accesses. However it is not clear that we have the tools to manage really large conferences in any case.

2.4. Active relays

Traditionally, Internet services have been end-to-end, connected over a dumb network. Recently the concept has been advanced of providing active service elements either inside the network or at its edges. One example is an engine that allows workstations with only unicast capability to participate in multicast conferences. Another provides services that permit filtering or transcoding of media payloads – for instance because one region of the Internet has poorer data capacity than another. Some have advocated providing such services at the network level, integrating them with routers; others have proposed the provision of such services at the host level. Three implementations of such services at the host level are m-tunnel [Par96], MeGa [Amir98] and UTG [Kirst98]. The functionality of such devices is largely similar, though there are some differences. In the UCL Transcoding Gateway (UTG), the main functionality is as follows:

- Access to multicast conferences for hosts with only unicast connectivity
- Capability of filtering single streams by the dropping of specific packets. This is an effective way of limiting bandwidth for video by dropping either higher order packets, or even complete frames
- Capability of transcoding media. This is an effective way of reducing bandwidth for some media – particularly audio. Moreover the characteristics of the traffic (e.g. network losses) in one area may be so different from that in another, that transcoding is desirable near the boundary
- Multiplexing or mixing. This is particularly effective where multicast audio comes on different streams but can be multiplexed onto one stream
- Participating in, but terminating requests for retransmission of reliable multicast. This is particularly effective when one region of a network has such poor bandwidth that responding to such requests could strain the infrastructure
- Acting as a cache for Session Announcements

At present the UTG must be placed statically at specific locations in the network, and all its parameters must be set up explicitly. So far we have used the UTG only with unencrypted data; we are now exploring the alternatives with encrypted data. While it normally acts on single streams of data, there is no reason not to treat the total data for a specific conference as a whole; for example, if the total bandwidth available in one area is known (e.g. because of the use of fixed capacity ISDN), then the parameters set in the relay can restrict the traffic on the low capacity side to that known to be available – potentially at the cost of quality or even of suppressing one of the media.

The components in the UTG architecture perform the following tasks:

- The RTSP controller module will provide the control interface to the unicast-only end-system
- The access control module is used to verify that requests for transcoding and gatewaying are from authorised users
- One or more media engines are instantiated to perform transcoding and gatewaying when required
- The Mbus controller provides the necessary glue between all the other modules

3. Security considerations in multicast conferencing

3.1. Introduction

There is a temptation to believe that multicast is inherently less private than unicast communication since the traffic visits so many more places in the network. In fact, this is only the case with broadcast and prune-type multicast routing protocols [Dee91]. However, IP multicast does make it simple for a host to join a multicast group anonymously; it may then receive traffic destined to that group without the other senders' and receivers' knowledge. If the application requirement (conference policy) is to communicate between some defined set of users, then strict privacy can only be enforced through adequate end-to-end encryption.

RTP allows encryption of RTP and RTCP packets using symmetric algorithms. Most of our early work has been done with the Data Encryption Standard (DES) [NIST88] because of the prevalence of implementations and its standardisation in RTP; we are currently moving to more secure ciphers. RTP also specifies a mechanism to manipulate passwords using Message Digest 5 [Riv92] so that the resulting bit string can be used as an encryption key. Passwords can be entered individually into the media tools, but it is more convenient for an application like SDR to provide the session encryption keys needed to encrypt and decrypt the media along with other tool parameters.

The disadvantage of application-level security is that it is necessary to include cryptographic functionality separately in each tool. This was the most convenient method of securing conferences until recently, because standard methods of encrypting the data at a lower layer were not available. The transport layer schemes such as Transport Layer Security [Die99] that have recently been standardised are unsuitable, as they require a reliable unicast connection between parties; this is contrary to our need to use multicast UDP packets. While network-layer IP security extension (IPSEC) proposals have been on the standardisation track for nearly a decade, the relevant standards have only been ratified in the Internet Engineering Task Force over the last year [Tha98]. There are now a number of reference implementations, and many suppliers are about to provide the new functionality in their main production stacks.

These IP extensions provide confidentiality and authentication for IP packets. RTP always intended to use these mechanisms once they became available in preference to its own less flexible schemes. We are investigating where to alter our tools to take advantage of IPSEC mechanisms. This would remove the need to write and maintain encryption code in each application. It would take advantage of fast cryptographic code running in kernel mode, but require all users to have a secure IP driver.

If IPSEC is used in place of application-level security, then the calls to launch the tools are only changed slightly: the media encryption key no longer needs to be supplied. The description will still set up the parameters of the tool; the encryption key will now instead be inserted by SDR in the local Security Association Database, causing IPSEC to decrypt incoming and encrypt outgoing packets. Once standard IPSEC Application Programming Interfaces have been defined we intend each tool to perform this function as it joins the relevant multicast group, removing the need for SDR to run as a trusted application.

The use of IPSEC in multicast environments has not yet been completely defined. In particular, there is no standard way to provide the keys needed to decrypt and authenticate packets to the authorised members of a multicast group. A number of proposals have been made, such as using secure point-to-point connections between a multicast group member and group controller to distribute keys [Can99], but are currently far from implementation. Controllers can also assign unique Security Parameters Index values for a group to avoid collisions on values chosen by receivers as is standard with unicast IPSEC. If individual authentication or replay protection is required, the controller also assigns unique Security Association numbers to each group member.

We intend during the medium term to use the current session announcement and initiation protocols with security extensions to provide this functionality. If the authentication and encryption keys are provided by appropriate mechanisms, then there should be no difficulty in securing the media streams at the network level.

Most current IPSEC implementations concentrate on securing streams at the host level, while conferencing is more concerned with the identity of persons participating. This gap is bridged by the mechanisms used for providing the keys used by IPSEC. For conferencing, these mechanisms should use personal authentication.

Data authentication in IPSEC uses a secret key shared between participants. While this will allow a group member to be sure data came from another member of the group, it does not identify which one. Public-key signatures would allow identification of individual participants, but are too expensive computationally to produce for any moderate stream rate. This problem seems best solved using hybrid signature schemes such as Rohatgi's that also allow low latency signatures, important for interactive multimedia [Roh99].

A potential problem is that IPSEC is designed to operate at the IP level, while media addresses use both IP and port number. It may be desirable to use different cryptographic operations on different conference streams.

With the present specifications, it will be necessary to have any streams with different security policies use different IP addresses. It is not yet clear whether this will cause operational difficulties.

Until multicast key negotiation mechanisms are standardised, setting up secured sessions will remain an application-level function. Security parameters are distributed in a session description using any of the methods described in Section 3.2. A tool such as SDR will insert security parameters from the session description into the local Security Association Database, causing traffic sent to and from that multicast address to be cryptographically processed using the standard IPSEC mechanisms.

3.2. Encrypted/authenticated session descriptions

Announcement and invitation to conference sessions is dependent on passing the session description to the authorised invitees. This information can be passed by many technologies – both in-band and out-of-band.

It is important that announcements and invitations can be authenticated. If making a network reservation results in usage based billing, then some form of authentication is essential. One may wish to be sure that someone who is authorised to do so has called the conference. Finally, there are also mechanisms for modifying session announcements; a simple Denial-of-Service attack is to modify the announced time or location with unauthenticated announcements.

The main reason for constraining access to the Session Description data itself is to ensure that only authorised people may participate in a conference. Since the SD includes the encryption key(s) used to secure the media streams, it is essential that this be passed securely between the persons authorised to participate. This can be achieved by passing the description in an encrypted form, where only authorised parties can obtain the decryption keys. Alternately, it may be accomplished by storing the SD itself in a depository that only authorised persons can access. The session keys may also be kept separate from the rest of the SD. For a conference to be considered private, it must be encrypted; hence as long as the session keys are passed securely, it is less important if the rest of the SD is seen by unauthorised persons.

3.2.1. Authentication of session announcements

Anyone wishing to announce a conference can send out a session announcement. He signs the announcement, allowing it to be verified by recipients. It is impossible to verify his identity without a certification infrastructure, but the main threat is a Denial of Service attack arising from an announcement being contradicted by a subsequent change in another announcement. Even without a security infrastructure, any recipient can check that the changed, received announcement came from the same source as the original announcement. Thus for merely ensuring that only the original announcer can alter a venue, it is not necessary to maintain an infrastructure.

3.2.2. Group management

Whenever conference membership must be restricted, the question of group management arises. It is necessary to either provide a shared secret to a group, or to have some specific state information for each member of a group. One reason for this is that groups are never static; some action needs to be taken to enrol a new member of the group, or to withdraw privileges from an existing member of the group. In all the work that we have done, we assume that each potential participant of a conference has a private/public key pair and a certificate. This has been certified locally, in such a way that everyone is reasonably happy that the certificate is associated with the named individual.

There are many scenarios of how groups should be set up and managed. We now describe one example.

All organisers of conference groups are listed in a directory (which may be just a web page) with their e-mail and public-key certificate. If someone wishes to join a group, they must send a secure e-mail (using S-MIME, for example), to the organiser. The organiser then may either accept or reject her. Assuming she is accepted, her certificate is placed in a list of authorised users for that conference group. The person is also sent to the list of generally authorised conference users, which is maintained by a Security Officer. The list of Conference Organisers is also maintained by the Security Officer. If anybody leaves their organisation, their personnel officers are expected to revoke their certificates. Their appearance on the Certificate Revocation Lists signals that there is a need to revoke their privileges.

The list of authorised participants can be used in many ways. If a shared group key is used in SAP, then this group key can be sent to all the authorised participants by secure e-mail. It can then be used to decrypt the session announcements. If a secure invitation is to be used, the public key of the invitee can be retrieved from the list. If a secured depository is to be used, then the certificate can be made part of the Access Control List for the depository.

We have set up the facilities for group management so that any of these mechanisms could be used.

3.2.3. Distributing session descriptions securely

A number of mechanisms can be used for distributing session descriptions. One is to use secure e-mail; S-MIME [Ram99] or Pretty Good Privacy [Cal98] are the message systems used most commonly. The IETF conferencing community has a strong preference for PGP. A second is secure access to a Web depository, where mechanisms like TLS can be used. Figure 3 shows a prototype Web server we have developed that provides authenticated access to session descriptions. A third is access to a X.500 directory; here access control with strong authentication and encrypted transmission is appropriate. Both the second and the third will be called “access to a secured depository” in the subsequent text. In these last two, an Access Control List, using either a password or a public key certificate, will be used. When a distribution mechanism is suggested, one of the above will be implied. Although the secure DNS (DNSSEC) is appropriate for storing public keys, it is not an appropriate secured depository due to a design decision that there will be no control on access to DNS databases. None of these mechanisms provide for changing keys during a session as might be required in some tightly coupled sessions, but they are sufficient for most usage in the context of lightweight sessions.

Fig. 3: Secure Web access to session descriptions

Security extensions have also been developed for SIP that simply encrypt invitations to users with each user’s public key and sign them with the sender’s private key. For small conferences, each participant could just be invited individually.

It is important that the same SD can be used irrespective of the manner it is transferred. This allows the facility that launches the encrypted media tools to be oblivious of how the SD came to the recipient.

Secure distribution is closely tied to authentication of group members. Conference or session keys can be distributed securely using public-key cryptography on a one-to-one basis. This security is only as good as the certification mechanism used to ensure the conference organiser holds the correct public key for each user. Such mechanisms are not, however, specific to conferencing. Both X.509 [CCITT88] and PGP [Cal98] certificates can be used for this purpose.

There are problems of security policy and a question of how certificates should be used with all these methods.

We have assumed implicitly that the distribution of the shared secret is to known people using their public key certificates. For some purposes, it is not the personal name that is relevant, but the role. Hence both using e-mail and depositories, the access may have to be made to persons using their role rather than their names. Allowing access to depositories may be easier than using e-mail for this case, since it must be assumed that there is local authentication binding the current role to the name.

Another problem will be the question of validation of the certificates used. Certificates may be revoked for many reasons; some will result from changes in roles, but others may be due to private key compromise or other security breaches. Certificate Revocation Lists therefore need to be checked regularly and certificate databases kept up to date.

A further complexity is that people “on the move” may miss session updates sent by e-mail but hidden in a large number of messages. Even with multicast accessibility at their current location, it may be difficult for them to be sure they have up-to-date session information. Clearly it is possible to provide additional plug-ins for mail systems which parse incoming messages for the MIME type which represents a description.

3.2.4. Use of symmetric or asymmetric encryption mechanisms with SAP

A level of indirection can be introduced by multicasting SAP messages encrypted using keys previously supplied to authorised recipients. This may be suitable for stable groups of individuals who hold regular conferences. Because of the availability of PKI toolkits and their use in other projects, we initially implemented this function in SDR using a key pair as the secret required both to encrypt and to decrypt such announcements. Symmetric algorithms with a smaller shared secret could also be used.

There is currently considerable controversy whether session announcements are an appropriate mechanism for announcing private sessions and thus whether there is a place for encrypted announcements at all. The bandwidth taken up by announcements is already quite large. It is typically some 10 minutes before an announcement is refreshed. Another concern is that one of the functions of session announcements is to avoid conflicts in the use of multicast addresses; this avoidance is impractical if the whole announcement is encrypted.

The first concern could be addressed by using session announcement cache proxies [Swa98]; the second by separating out the address allocation functionality from the rest of the announcement mechanism. Both questions are still being discussed in the IETF. The use of proxies will not affect the time taken to refresh the proxies; it will, however, impact dramatically on the time for a workstation to obtain session descriptions from such proxies. But it is still not clear if the bandwidth saving of multicasting an encrypted announcement and separately supplying decryption keys is worth the complexity introduced, particularly regarding group changes that require new keys to be supplied to group members.

3.2.5. Changing encryption keys during sessions

A number of schemes are being investigated to change encryption keys during sessions. In the tightly-coupled sessions of H.323, these are included in the security standard [ITU98]. With multicast conferencing, it is possible to use different techniques – depending on whether the main threat is key compromise due to excessive use, or to a desire to change the set of participants. The first of these can be remedied by sending a new encryption key to the participants encrypted with the old key; at the same time the Session Announcement will need to be updated with the new key to accommodate late joiners. The second can be achieved only by out-of-band techniques – though some can still use largely the existing multicast trees [Wal98]. None of these techniques have been completely standardised, nor have they been implemented in the UCL work; they will not be considered further here.

3.3. Use of smart cards for secure conferencing

All the security mechanisms described so far rely upon each conference participant keeping secret the keys necessary to decrypt session descriptions and encrypted media streams. Most current software stores this information on an individual's workstation, usually in encrypted form. It is possible to streamline many of the operational issues if keys are instead kept on a smartcard. Such a card is easily portable, and can be used on other workstations without the need to transfer keys in software. Secunet[1] has already demonstrated an IPSEC stack that can use a smartcard to decrypt and authenticate data.

It is expected that use of such smartcards will greatly ease the operational deployment of secure conferencing. It should be much easier to automate operations, and to reduce the security constraints on the workstations, with the use of such cards. Users are also relieved of the burden of managing private keys, particularly remembering long passphrases. We certainly plan to demonstrate secure conferencing using smartcards.

4. Secured operation with relays

The relay of Section 2.4 also requires secured operations. This takes several forms; in each it is necessary to decide whether to trust the relay, and how to operate it, for the following purposes:

- To control the relay from individual remote hosts
- To provide encryption keys for secured session descriptions
- To provide encryption keys for secured sessions

Clearly some level of access control is necessary to ensure that only authorised remote hosts can control the relay. Either IPSEC from specific hosts or individuals or TLS from specific people would be good mechanisms to use.

The relay will be used as a SAP proxy, caching SAP announcements from the Mbone. While it could be given decryption keys to allow it to decode all incoming secured announcements, this would be too great a security risk. It would make the relay a very tempting target; if compromised, it could be used to gain access to other communications to that recipient.

Instead, remote hosts will make connections to the SAP proxy. The SAP proxy in the relay passes the SAP announcements to the end-user workstation – whether or not the session is encrypted. Alternately they can be obtained from a trusted depository just as in Section 3.1.2. The user can then invite the relay to join the conference with a secured invitation on behalf of the user of the remote host.

Certain filtering operations can be done without providing the relay with any encryption keys; an example is joining only some multicast groups for layered coding. If application-level encryption has been used, even the multicast-unicast conversion can be done on the encrypted streams.

If more powerful transcoding operations are required, some greater risks must be taken. The recipient should be able to invite the relay to join the conference with the specific keys needed to decrypt data from the conference it is currently filtering. The relay will then do the requisite filtering, and re-encrypt with a new key for the unicast session agreed between the user and the relay. For this last link, an ordinary IPSEC connection would also be appropriate. While this mode of working reduces risk to an acceptable level, the relay must still run in a trusted environment. An attacker should not be able to access the session key or the media data itself. A schematic of the mechanisms just described is shown in figure 4.

[1] http://www.secunet.de/

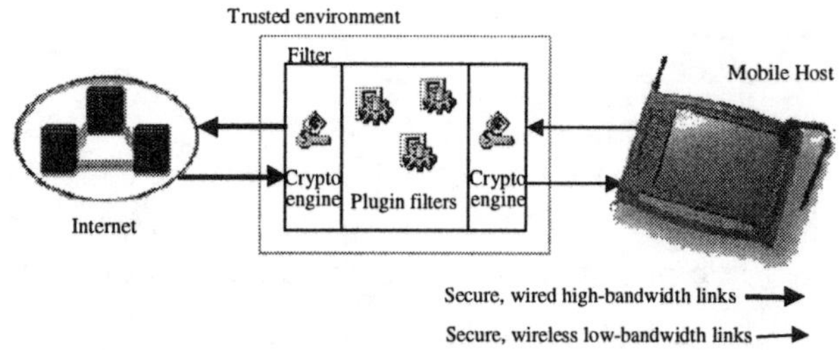

Fig. 4: Operation of secured relay with user-supplied key

This mechanism has the advantage that an attacker and a poorly secured relay may compromise a particular session, but nothing more. Moreover, only those relays that are participating directly in a session need to know any of the session encryption keys.

There is no reason why the relay may not be equipped with a smartcard containing its security environment; as an additional security measure, it could be arranged that sensitive activities could not be done on the relay unless this card was in place.

Relays can also have more than one client. If conference data is encrypted, the cost of decrypting the data for filtering is spread amongst all the clients. While the relay only needs to be given the conference session key once to decrypt the data, it can use knowledge of this key to restrict access to authorised clients who also know the key.

5. Current status of the implementations

We have now secured an adequate number of conferencing tools at the application level[2]. These include *vic* (video), *rat* (audio), *wbd* (shared whiteboard) and *nte* (shared text editor). Sessions can be started with secure SAP as part of a secured *sdr*.

The SAP component of the secured *sdr* uses group keys that must be distributed to all authorised participants. The complete infrastructure for registering public keys of all potential participants, managing groups, and distributing keys securely to groups using both PGP and secure Web pages has been set up. The system is being deployed. Multiple depositories have been set up, to remove single points of failure. The integration of smart cards and IPSEC has not yet been accomplished.

The unsecured UTG has been completed. Implementation of the secure version is being undertaken during the next few months using several of the methods described above.

[2] http://www-mice.cs.ucl.ac.uk/multimedia/software/

6. Conclusions

In this paper we have shown that it is possible to secure the components of multicast multimedia conferencing in several ways – most of which UCL is implementing.

The critical requirement for a secure conference is end-to-end encryption of its media streams. We have described how to use application and network-layer cryptography to achieve this goal, and the relative merits of each. The main tools we are securing are video, audio and shared editor applications; all have been built so that they can use encryption at the application level, or be sent above the IPSEC network level.

We have also discussed the more difficult problem of distributing the session keys used to authorised participants. Until the IETF standardises a method of multicast key distribution, encrypted announcements and invitations and secure e-mail and depository access are perfectly adequate for the needs of lightweight conferences.

Finally, we have stated that there is a need for transcoding relays or gateways in the total architecture. We have summarised the salient aspects of these relays, and discussed how security technology should be used in their protection. Again we have analysed the relative merits of the different mechanisms.

Acknowledgements

Steve Hailes (UCL), Stephen Farrell (SSE) and Colin Perkins (UCL) have made many useful comments on this paper. John Andrews designed and implemented the secured depository based on publicly available code. We gratefully acknowledge the support of DARPA (under order F49620-96-1-0037), the European Office of Aerospace Research and Development (under F61775-98-WE132 and F61775-99-WE006), and the European Commission (under the MERCI, MECCANO, ICETEL and ICECAR projects) for their support of this work.

References

[Amir98] E. Amir et al.: An Active Service Framework and its Application to Real-time Multimedia Transcoding, Proc. ACM SIGCOMM'98, Vancouver, Sep. 1998.

[Cal98] J. Callas et al.: OpenPGP Message Format, RFC 2440, Nov. 1998.

[Can99] R. Canetti et al.: An Architecture for Secure Internet Multicast, IRTF draft, Feb. 1999.

[CCITT88] Consultative Committee on International Telegraphy and Telephony: X.509: The Directory – Authentication Framework, 1988.

[Dee89] S. Deering: Host Extensions for IP

Multicasting, STD 5/RFC 1112, Aug. 1989.

[Dee91] S. Deering: Multicast Routing in a Datagram Internetwork, PhD thesis, Stanford University, Dec. 1991.

[Die99] T. Dierks and C. Allen: The TLS Protocol Version 1.0, RFC 2246, Jan. 1999.

[Flo95] S. Floyd et al.: A Reliable Multicast Framework for Light-weight Sessions and Application Level Framing, Proc. ACM SIGCOMM'95, Cambridge, Massachusetts, Aug. 1995.

[Han98] M. Handley and V. Jacobson: SDP: Session Description Protocol, RFC 2327, Apr. 1998.

[Han99a] M. Handley et al.: SAP: Session Announcement Protocol, IETF draft, Jun. 1999.

[Han99b] M. Handley et al.: SIP: Session Initiation Protocol, RFC 2453, Mar.1999.

[Hin96] E. Hinsch et al.: The Secure Conferencing User Agent: A Tool to Provide Secure Conferencing with Mbone Multimedia Conferencing Applications, Proc. IDMS '96, Berlin, Mar. 96.

[ITU98] International Telecommunication Union: H.235, Security and encryption for H series (H.323 and other H.245 based) multimedia terminals, Feb. 1998.

[Kir98] P. T. Kirstein et al.: Accessing Mbone Sessions over Point-to-Point Connections. Technical report , Department of Computer Science, University College London, Gower Street, London WC1E 6BT, UK, 1998.

[Mcc96] S. McCanne: Scalable Compression and transmission of Internet Multicast Video, PhD thesis, University of California, Berkeley, Dec. 1996.

[NIST88] NIST: Data Encryption Standard, National Institute of Standards and Technology (NIST), FIPS Publication 46-1, Jan. 1988.

[Par96] P. Parnes et al.: mTunnel: a multicast tunneling system with a user based Quality-of-Service model, Proc. European Workshop on Interactive Multimedia Systems and Telecommunication Services, Sep. 1997

[Per98] C. Perkins et a`l.: A Survey of Packet Loss Recovery Techniques for Streaming Media, IEEE Network Magazine, Sep./Oct. 1998.

[Per99] C. Perkins et al.: A Message Bus for Conferencing Systems, IETF draft, Aug. 1998.

[Ram99] B. Ramsdell: S/MIME Version 3 Message Specification, RFC 2633, Jun. 1999.

[Riv92] R. Rivest: The MD5 Message-Digest Algorithm, RFC 1321, Apr. 1992

[Roh99] P. Rohatgi: Hybrid Signature Scheme for Multicast Source Authentication, IRTF draft, Jun. 1999.

[Sch97] H. Schulzrinne et al.: RTP: A Transport Protocol for Real-Time Applications, RFC 1889, Jan. 1996.

[Sch98] H. Schulzrinne et al.: Real Time Streaming Protocol (RTSP), RFC 2326, Apr. 1998.

[Swa98] A. Swan et al.: Layered Transmission and Caching for the Session Directory Service, Proc. ACM Multimedia '98, Bristol, UK, Sep. 1998.

[Tha98] R. Thayer et al.: IP Security Document Roadmap, RFC 2411, Nov. 1998.

[Wal98] D. Wallner et al.: Key Management for Multicast: Issues and Architectures, IETF draft, Sep. 1998.

Policy-Based Security Management for Large Dynamic Groups: An Overview of the DCCM Project[1]

Peter T. Dinsmore, David M. Balenson, Michael Heyman, Peter S. Kruus, Caroline D. Scace, and Alan T. Sherman[2]
Cryptographic Technologies Group
NAI Labs, The Security Research Division
Network Associates, Inc.
{peter_dinsmore, david_balenson, mheyman, pkruus, cscace, asherman}@nai.com

Abstract

The Dynamic Cryptographic Context Management (DCCM) project efficiently provides security for very large, dynamically changing groups of participants. The DCCM system has two novel distinguishing characteristics. First, policy plays a key role in DCCM. Groups at all levels have policies. These policies are represented; they are negotiated; they are managed; and a cryptographic context—an unambiguous set of mechanisms and configuration—is created to make particular interactions possible subject to these policies. Second, DCCM implements a scalable key management system based on One-way Function Trees (OFT) that can handle group sizes up to 100,000 members and can dynamically handle members entering and leaving groups.

1. Introduction

The Dynamic Cryptographic Context Management (DCCM) project [2] efficiently provides security for very large, dynamically changing groups of participants. For example, command and control of tactical military forces requires several types of protection among a very large group of participants, perhaps from different countries or from different Armed Forces units, grouped together under one command for some time period or for a specific exercise. By "large," we mean groups with number of members ranging from 10,000 to 100,000 or more. By "dynamic," we mean new members may be added to the group at any time and existing members may be evicted from the group, thereby requiring immediate changes to some of the security provisions. Members need not be humans; they can be a variety of communicating entities, including sensors, mobile client workstations, server workstations, or network nodes.

The DCCM system has two novel distinguishing characteristics. First, policy plays a key role in DCCM. Groups at all levels have policies. These policies are represented; they are negotiated; they are managed; and a *cryptographic context*—an unambiguous set of mechanisms and configuration—is created to make particular interactions possible subject to these policies. Second, DCCM has a scalable key management system that can handle group sizes up to 100,000 members and can dynamically handle members entering and leaving groups.

This paper presents:

- The DCCM view of a secure group: its organization; lifecycle; and operations;
- Secure group policy and a Cryptographic Context Negotiation Template to represent it;
- Policy negotiation via a Cryptographic Context Negotiation Protocol;
- The One-way Function Tree (OFT) method for large group key management; and
- The architecture of a demonstration system incorporating the DCCM concepts.

2. Secure groups

A secure group is a collection of members who are authorized to access a set of information. Secure group mechanisms enable the members, and only the

---

[1] This work was supported by the Defense Advance Research Projects Agency (DARPA) under Air Force Research Laboratory (AFRL) Contract No. F30602-97-C-0277.

[2] Sherman is also with the Department of Computer Science and Electrical Engineering, University of Maryland, Baltimore County (UMBC).

0-7695-0490-6/99 $10.00 © 1999 IEEE

members, to access the information. DCCM supports group policies that require group confidentiality and/or group integrity of the information.

Group confidentiality protects the group information from disclosure to non-group members. In order to provide confidentiality in a dynamically changing group, security mechanisms must provide forward and backward secrecy. Forward secrecy means that evicted members cannot access future information. Backward secrecy means that new members cannot access past information.

Group integrity protects information from modification by non-group members. Members of a secure group can verify that information has not been modified by anyone outside of the group, and that it originated from an authorized group member. Group integrity does not allow a member to determine which individual group member created a message, *i.e.*, individual source authentication.

DCCM does not provide mechanisms to support individual source authentication of information or to maintain the availability of information.

2.1. Group organization

DCCM recognizes and supports a range of models for the organization of a secure group, from a strict hierarchical model to a broad flat model. A hierarchical group is based on inherent structure of the members forming a group. This structure is typically found in groups created for organizational collaboration, such as an industrial consortium or a military coalition. The security policies and supporting infrastructure for the group also take advantage of existing structure and relationships between members. For example, there may be multiple authentication policies and servers for a group, based on the existing policies within the hierarchy or across multiple hierarchies.

A flat group has no group structure or relationships between group members. It is typically used for individual collaboration or information distribution, such as web broadcasts or pay-per-view. DCCM allows each member to have its own security policy, or to subscribe to a known security policy. In a flat model, certain members may advertise a policy designed to meet certain goals that other members with similar goals can use as well.

The most basic entity in a DCCM secure group is the *participant*. A participant is a single entity that is involved in secure group communication. Typically a participant represents a person, but it can represent a sensor, a Personal Digital Assistant (PDA), or a piece of software. DCCM supports all security services for all participants, and assumes that all participants have the same group communication capabilities. DCCM fully supports the many-to-many model of group communications, meaning all participants can both send and receive information. DCCM can also support the one-to-many model (one sender with many receivers) within its support of the many-to-many model.

A group of participants sharing a secure group communication mechanism for a specified period of time for a common purpose is a *session*. A session implies that all of the participants share the same security mechanisms, share the same security policy for those mechanisms, and share a common security configuration, including cryptographic keys, for the enforcement of the policy. For example, a single lecture broadcast from a semester long class would be a session.

A DCCM *project* is a set of sessions occurring over time. All of the sessions within a project will use the same cryptographic context, or policy, and support the same set of participants. The project is the unit of administration for access control for DCCM. DCCM administers the list of participants in a project and enforces an access control policy between project members and non-members. Within a project, any project participant is free to participate in any session announced for the project. All project members are authorized for all sessions within that project. Project members may choose not to participate, but that is not a security relevant decision. Multiple sessions can occur simultaneously within a project, and a participant can join more than one of these sessions.

The highest level of organization of secure groups in DCCM is a *system*. A system is the supporting infrastructure for a set of related participants that transcends individual projects. The system maintains a single authentication database that is used across multiple projects. The DCCM authentication mechanism utilizes a system base key, or shared secret, that is established when a participant joins the system and authenticates for the first time, typically with a public key mechanism. DCCM amortizes the high cost of the public key authentication over the creation of multiple projects.

2.2. Group lifecycle

All groups go through several phases during their life. The first phase of DCCM, even before any groups are formed, is the induction of participants into a DCCM system. As previously noted, participants authenticate themselves once to a DCCM system and establish a shared key that is used for all future secure operations.

During the next phase, a DCCM project is created from the set of participants authenticated to the system. One of the participants takes on the role of project initiator. The project initiator specifies a list of participants and a proposed security policy for the project. The DCCM system infrastructure uses the proposed policy in negotiation (see Section 4) with each of the proposed participants to derive a cryptographic context for the project.

Once the cryptographic context and membership are established for a project, the key management mechanism establishes the group keys and distributes them to all of the participants (see Section 5, Key management). Participants then use the keys for protecting their group communications. DCCM mechanisms support cryptography at any level in the network stack. Typically, group cryptography is found at the network, transport, or application level.

The group is actively managed during its lifetime (see Section 2.3, Group operations). Participants may be added or removed from the group; keys may expire; and eventually the group is dissolved.

2.3. Group operations

DCCM defines the following group operations for each project:

- **Add**. Participants can be added to a project. There are three security relevant aspects to adding participants to an existing project. First, the project initiator applies an access control policy for the project to determine if the participant can be added. Second, the project policy must specify whether or not new participants are to have access to communications or information generated before they joined the project. If not, a new project key has to be generated when a new participant joins the project. Finally, the project policy must specify how to negotiate the project cryptographic context when new participants are added. Because the negotiation process ensures that the project context is fully compliant with the policies of all participants, a new participant may require a change that is mutually exclusive with an existing member. An even more subtle problem during late additions is that existing members may actually have a different policy depending on the project membership. A new member may require a fresh negotiation for the project cryptographic context.
- **Remove**. Participants can be removed from a project. A remove is used when a participant is no longer participating, but still authorized for a group. There is no change to the group key.
- **Evict**. When participants are evicted from a project, their authorizations are revoked and the group key is changed to prevent them from obtaining further access to any project information.
- **Freeze**. Freezing a participant is a special case of removing a participant. When a participant is frozen, they are not removed from the group keying mechanism, but the system does not send any further key updates to the participant. This operation is useful for a member who knows they will be unable to receive or protect key information for a period of time.
- **Thaw**. Thawing a participant resumes the distribution of key update messages to the participant. A thaw reverses a freeze.
- **Resync**. Resynchronizing a participant resends the latest key update information for that participant only. In addition, it will thaw a frozen participant. A resync operation is used when a participant unexpectedly stops receiving key update information or suspects that its keying material is out of date.

In addition to the project level operations, a participant may also be evicted from the system, which results in an eviction from any projects containing the participant. A participant may choose to join or leave sessions within a project, but those are not security relevant operations. By definition in DCCM, all participants in a project are authorized for all sessions within that project. That authorization is constant, regardless of a participant's choice to exercise it.

3. Policy

A security *policy* is a set of rules specifying how to protect information. Policies can be described by *whom* they cover and by *what* they cover. In DCCM, every organizational entity has a policy. When a participant joins the DCCM system, they bring with them their security policy for the protection of their information. It may be a policy specific to them, such as in a flat group model, or the policy may be an organization policy covering all of the members of some hierarchy.

Policies can cover different security attributes at different levels of abstraction. For example, a high-level policy would state that confidentiality must be protected; a mid-level policy would state that strong encryption must be used; and a low-level policy would specify that triple-DES encryption in CBC mode must be used. Policies can also specify allowable behavior as a range of options. For example, a policy might state that *at least* moderate strength integrity mechanisms must be used.

A policy that contains a range, or set of allowable actions, cannot be enforced by multiple participants with any expectation of interoperability. Interoperability can only be achieved when it can be guaranteed that all of the participants will enforce the policy exactly the same. DCCM accomplishes interoperability by distributing a policy that is completely unambiguous; there are no ranges or options. A policy that specifies a singular instantiation in DCCM is referred to as a *context*, specifically a *cryptographic context.*

There are several problems, however, in creating a cryptographic context that can be enforced by all participants, regardless of their initial policy and their security capabilities. The first problem is specifying a common syntax for representing policy information. There are other policy representations under research, such as SPSL [9] and GSAKMP [12], but neither of them is designed to express group policy including multiple levels of abstraction.

The second problem is specifying common semantics for the interpretation of a policy. Many common security terms and descriptions can be implemented or interpreted in different ways. This situation holds at both high levels and low levels of abstraction. For example, what is *strong* encryption? Does it imply specific algorithms, specific key lengths, specific protocols, or all of them? Is 64 bits of key length strong, or is 128? At low levels of abstraction, does 3DES mean CBC mode and three key EDE?

Without common semantics there is no way to support a common mapping or translation between policy abstractions and implementing mechanisms across the various participants in a project. DCCM addresses this problem with a *Cryptographic Context Negotiation Template* (CCNT) [3].

The CCNT is the common thread for policy representation, distribution, and negotiation in DCCM. It captures policy information at a low level of abstraction to ensure common semantics and interoperability between participants. Mapping from high level constructs to a CCNT can be performed locally; outward representation of policy is made only at the level of the CCNT.

The CCNT represents policy as an *n*-dimensional space. Each axis in the space represents a different aspect of the policy. Each value along an axis is a specific mechanism or configuration for that attribute. For example, *confidentiality* is the vertical axis in Figure 1, and it has the possible values of *3DES, CAST, IDEA,* and *RC4.* DCCM contexts, which are unambiguous sets of mechanisms, are single points in the *n*-dimensional space. Contexts can be proscribed by only a single value on any axis. DCCM policies, which are ranges or sets of allowable mechanisms, are sets of points in this space. Policies are proscribed by ranges of values or separate discrete values on multiple axes. Note that a policy is not required to have an intersection with every axis in the space. Policies can have a "don't care" with respect to that attribute.

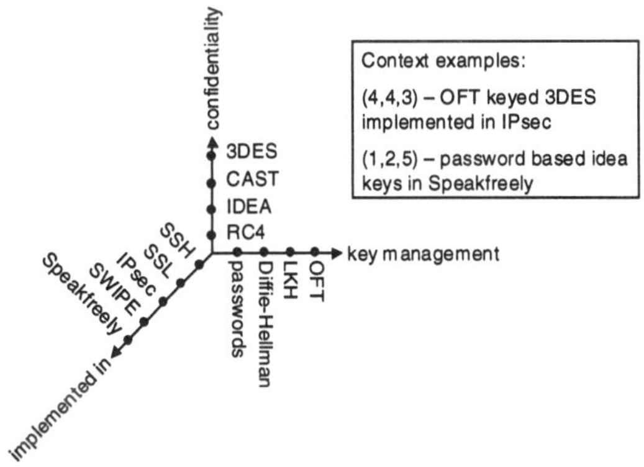

Figure 1: Example cryptographic context negotiation template

The CCNT is designed to represent a broad view of group security policy. Possible axes include:

- Data confidentiality;
- Key management ;
- Implementation mechanism;
- Temporal secrecy (forward secrecy, backward secrecy);
- Key lifetime;
- Key recovery;
- Group authentication;
- Layer for group security;
- Source authentication;
- Integrity;
- Required data throughput;
- Eviction; and
- Error handling.

The axes are not built into DCCM, but are specified in a context file. Axes can be easily added to the policy space by changing the context file; no other changes are required to the system.

4. Policy negotiation

A project context is created through a negotiation protocol during project creation in the DCCM system. The Cryptographic Context Negotiation Protocol (CCNP) [4] creates a fully specified cryptographic context for the project that fulfills the individual policies of the group participants. Viewed graphically in Figure 2, cryptographic context negotiation finds the intersection between the allowable policies of the project participants.

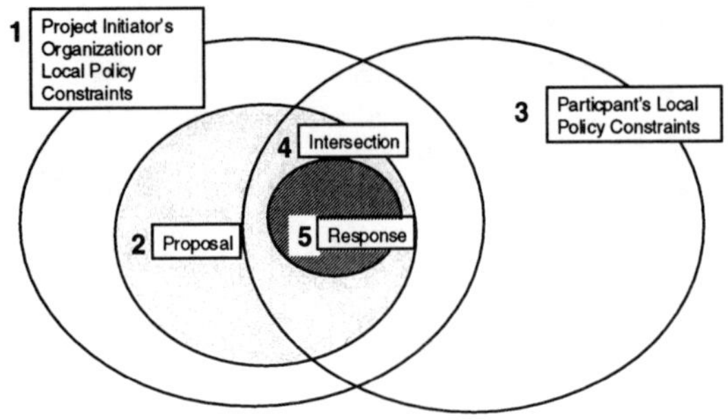

Figure 2: Cryptographic context negotiation

The process starts with the project initiator. The project initiator proposes a policy (area 2) for the project. Whatever is proposed is constrained by the initiator's own local policy (area 1). Each participant finds the intersection (area 4) between the proposed policy and their own local policy (area 3) and chooses a response (area 5) from the intersection. The initiator chooses the final project context from the intersection of the responses.

Careful selection of the proposal by the project initiator can effect different negotiation strategies. If the initiator creates a proposal that is actually a context, *i.e.*, there is only one choice for each axis, then negotiation becomes a directive action. The proposal *is* the policy (context) for the project; accept it as it is or do not join the project. A broad range of policy values (up to the full set of their local policy) proposed by the initiator increases the likelihood of an intersection occurring between a wide range of participants. The smaller the set in the proposal, the higher the chances that some participants will be unable to comply with or enforce the project's policy, and will, therefore, be unable to join the project.

5. Key management

The DCCM project requires a scalable method for establishing session keys for large dynamic groups. The method has to support efficient establishment of a shared secret key, as well as changing this key when group members are added or evicted. The following types of methods were considered:

- *Simple linear methods*, such as a Simple Key Distribution Center (SKDC), scale poorly to large groups, but are easy and straightforward to implement and employ with small-to-moderate size groups.
- *Information-theoretic methods*, such as Blundo's Symmetric Polynomials (BSP) [7] and Chiou-Chen's Secure Lock (CSL) [10], require large amounts of space per user and scale exponentially with the number of members in a group.
- *Multi-party methods based on Diffie-Hellman key agreement*, such as the Group Diffie-Hellman (GDH) method developed under the DARPA-sponsored Cliques project [1], and Burmester-Desmedt methods [8], require slow public-key operations and typically scale linearly with the size of a group. These methods seem better suited for small-to-moderate size groups of 10's or 100's of members. The GDH algorithm is especially attractive when distributed or decentralized control is needed. The Burmester-Desmedt method has the unique quality of constant delay (relative to group size) during initial key establishment.
- *Distributed, fault-tolerant systems* developed by Reiter *et al.* [14], and related *dynamic virtual private network (DVPNs)* developed by Rodeh *et al.* [15] include group key distribution techniques that are better suited for small-to-moderate size groups, but incorporate highly desirable fault-tolerant characteristics.
- *Hierarchical, tree-based methods* include Wallner *et al.*'s Logical Key Hierarchy (LKH) [11][16]. These methods represent group members as the leaves, and the group key as the root of a logical tree, and update keys via encrypting node keys down the tree. Time, space, and broadcast complexity all grow logarithmically relative to the size of a group, and hence these algorithms scale best to very large groups.

For further discussion of related work and additional references, see [2] and [6].

5.1. One-way function trees (OFT)

The hierarchical methods are best suited for the DCCM project. While exploring the use of LKH, the DCCM project developed a new hierarchical method for large dynamic group keying based on the novel application of One-way Function Trees (OFTs) [5][6][13]. The OFT method, like LKH, represents group members as the leaves and the group key as the root of a logical tree. However, rather than "pushing" the group key down the tree, the OFT method "pulls" the group key up the tree, using one-way functions.

5.1.1. OFT structure. As shown in Figure 3, an OFT is a binary tree, each node x of which is associated with two cryptographic keys: a node key k_x and a blinded node key $k'_x = g(k_x)$. The blinded node key is computed from the node key using a one-way function g; it is blinded in the sense that a computationally limited adversary cannot find k_x from k'_x.

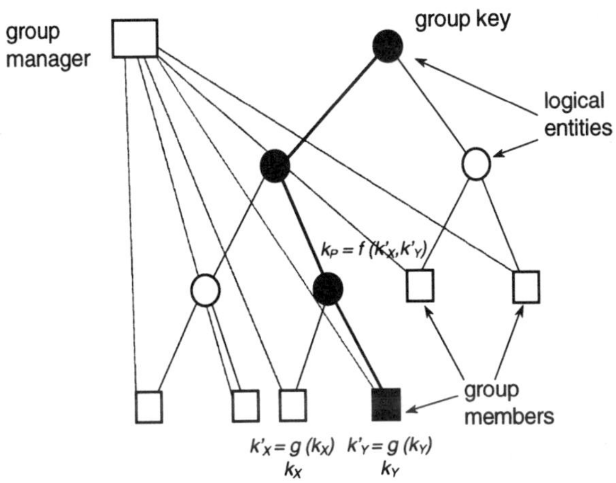

Figure 3: A one-way function tree

A group manager maintains a one-way function tree. Each leaf is associated with a member of the group. The manager uses a symmetric encryption function E to communicate securely with subsets of group members, using unblinded keys as encryption keys as explained below.

A randomly-chosen key is assigned to each member. This key is shared with the manager (via an external secure channel), and the key is assigned as the node key of the member's leaf. A variety of choices are possible governing who chooses the keys. In particular, the key could be chosen by the manager, member, or a combination thereof (see Section 5.1.3).

Each internal node p of the tree has exactly two children. The interior node keys are defined by the rule:

$$k_p = f(\, g(k_x),\, g(k_y)\,), \qquad (1)$$

where x and y denote the left and right child of the node p, respectively. The function g is one-way, and can be based on a cryptographic hash function such as MD5 or SHA-1. The function f does not need to be one-way; it needs to mix its inputs—the bitwise exclusive-or function *XOR* is a fast, simple, and effective choice. The node key associated with the root of the tree is the group key, which the group can use to communicate with privacy among group members and/or authentication of group membership.

The security of the system depends on the fact that each member's knowledge about the current state of the key tree is limited by the following invariant:

> *OFT security invariant* – each member knows the unblinded node keys on the path from its node to the root, and the blinded node keys that are siblings to its path to the root, and no other blinded or unblinded keys.

This invariant is maintained by all operations that add members to the group, and by all operations that delete members from the group.

McGrew and Sherman [13] discuss the properties of f and g in detail and give some preliminary observations regarding the security of OFT. A rigorous proof of the necessary and sufficient properties of f and g needed to satisfy formal security requirements for OFT remains an open problem.

5.1.2. OFT operations. The operations of adding and evicting members rely on the communication of new blinded key values, from the manager to all members, after the node key associated with a leaf has changed. To maintain security, each blinded node key must be communicated only to the appropriate subset of members. If the blinded key k'_x changes, then its new value must be communicated to all of the members who store it. These members are associated with the descendants of the sibling of x, and they know the unblinded node key k_y, where y is the sibling of x. To provide the new value of the blinded key to the appropriate set of members, and keeping it from other members, the manager encrypts k'_x with k_y before broadcasting it to the group. (Here and throughout, we shall use the verb "broadcast" in the sense of "group broadcast"—sending a message from the group manager to all members of the group.)

Adding a member—when a new member joins the group, an existing leaf node x is split, creating new nodes *left(x)* and *right(x)*. The member associated with x becomes associated with *left(x)*, and the new member is associated with *right(x)*. Both members are given new keys. The old member receives a new key because its former sibling knows the old blinded node key and could use this information in collusion with another group member to find an unblinded key that is not on his path to the root. The new values of the blinded node keys that have changed are broadcast securely to the appropriate subgroups, as described in the previous paragraph. The number of blinded keys that must be broadcast to the group is equal to the distance from x to the root plus two. In addition, the new member is given its set of blinded node keys. In order to keep the height h of the tree as low as possible, when a new member is added, the leaf closest to the root is split.

Evicting a member—when the member associated with a leaf x is evicted from the group, the member assigned to the sibling of x is reassigned to the parent p of x and given a new leaf key value. If the sibling y of x is the root of a subtree, then p becomes y, moving the subtree closer to the root. In this case, one of the leaves of this subtree is given a new key (so that the evictee no longer knows the blinded key associated with the root of the subtree). The new values of the blinded node keys that have changed are broadcast securely to the appropriate subgroups, allowing all members to construct the new group key. The number of keys that must be broadcast is equal to the distance from x to the root.

Initialization—group initialization is the process through which the group establishes an initial group communications key. For the OFT method, this process involves two steps. First, the group manager broadcasts some information to the group members needed to apply the OFT key-updating procedures. Second, the members compute a shared group communications key, which is needed to begin secure group communications.

Group initialization is separate from group induction. During group *induction*, each member establishes an individual group base key known only by the member and the group manager. Group *initialization* assumes that each member has already established an individual group base key.

In the first step of OFT group initialization, the manager broadcasts every blinded node key in the OFT to all group members. In this broadcast, each blinded node key is encrypted by the unblinded key of the sibling node, so that only members in the sibling subtree can learn the blinded node key. All members receive the entire broadcast, which consists of a sequence of encrypted blinded node keys.

5.1.3. OFT properties. This section comments briefly on the security, resource usage, and salient characteristic features of the OFT method.

The security properties of OFT stem from the system invariant stated above, from the strength of the component one-way function, and from the random selection of leaf keys. In short, evicted members cannot read future messages, even with collusion by arbitrarily many evicted members, and newly admitted group members cannot read previous messages.

Evicted members have some information about the key tree but not enough to directly compute any unblinded node key. After a member is evicted, the keys along the path from the member's node to the root change. After this change, the evictee knows only the blinded keys of the siblings of the nodes along the path from the evictee to the root. These blinded nodes are insufficient to directly compute any unblinded key.

Interestingly, OFT is a centralized, member-contributory method. OFT is centralized in the sense that the group manager plays a special trusted role. OFT is member contributory in the sense that each leaf can contribute entropy to the group communication key.

With the OFT method, the number of keys stored by group members, the number of keys broadcast to the group when new members are added or evicted, and the computational efforts of group members, are logarithmic in the number of group members. The hierarchical nature of OFT distributes the computational costs of re-keying among the entire group, so that the manager's computational burden is comparable to that of a group member. Table 1 below summarizes the salient resource usage of adding or deleting a member with OFT in terms of time, memory, number of bits broadcast, and number of random bits needed. In the table, n is the group size, K is the size of a key in bits, and h is the height of the OFT ($h = lg\ n$ when the tree is balanced). Either the member or the manager could generate the random bits needed at the leaves. See [6] and [13] for more details, including comparisons with other key establishment methods such as SKDC and LKH.

Table 1: Summary of resource usage of adding or deleting a member with OFT

Resource Measure	Group Member Cost	Group Manager Cost
Time	h	h
Memory	hK	$2nK$
Bits broadcast	0	$hK + h$
Random bits generated	0	K

5.2. Key management summary

OFT was chosen as the group keying method for the DCCM project. While the simple SKDC, Group Diffie-Hellman, and other group keying methods may often be appealing for moderate size groups, many applications will likely demand a method that scales logarithmically in total delay and member memory usage. For such applications, especially if the add-member is more frequent than the evict-member operation, the OFT and other hierarchical methods look attractive for their constant-time add-member. LKH and OFT are similar methods, with LKH offering relatively simpler security semantics, and with OFT requiring fewer bits to transmit for re-keying. If it is critical for the application to minimize the number of bits broadcast, the number of random bits generated, or if a member-contributory method is needed, then OFT may be the method of choice.

6. Implementation architecture

The DCCM concepts described in this paper, in particular OFT, the CCNT, and the CCNP, have been instantiated in a set of toolkits for use in producing demonstrations of providing security for very large groups. Two toolkits were produced, a group key toolkit implementing the OFT algorithm, and a DCCM toolkit implementing the DCCM group policy management.

6.1. Constraints

The design of the DCCM toolkit assumed several constraints. These constraints are a result of the existing structure on the Internet and the state of the current deployment of multicast technologies. The constraints are:

- No individual participant addressing. The design must function without the ability to address a participant individually on the network. Participants may have unique identities, but they may not be network addressable due to private addressing schemes hidden behind a firewall or network address translation box. A participant can address a known server individually and set up a connection, but the participant must initiate the connection.
- Firewalls will block unknown ports. DCCM must not use unknown or nonstandard ports for its communications. If unknown ports are used and participants are protected by a firewall, then the communications will be dropped until specialized configuration of the firewall can be performed.
- Multicast communication is unreliable. Multicast does not provide a connection paradigm. DCCM cannot assume that messages arrive at any point in the DCCM system. DCCM cannot assume that messages arrive in order.
- Multicast communication is subject to fragmentation. Fragments may arrive out of order or not at all. For simplicity, the DCCM design assumes that a message cannot be larger than the standard ethernet MTU of 1500 bytes.

6.2. High-level architecture

The functionality of DCCM is provided through several key software components that can be placed in different parts of an overall system depending on the application and its use of group security mechanisms. As shown in Figure 4, there is a single system component that services multiple instances of the other components.

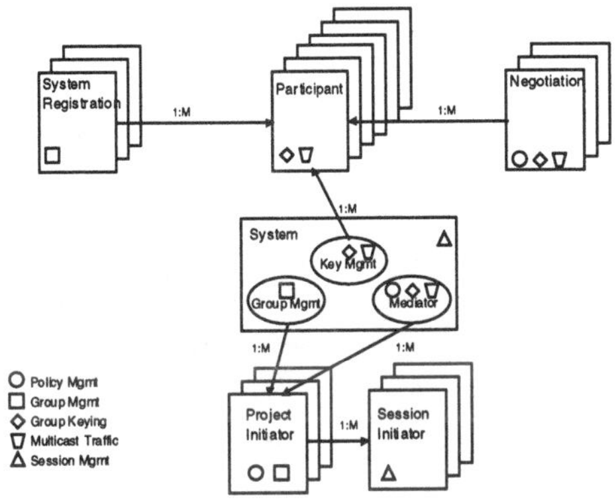

Figure 4: DCCM high-level architecture

The *System* component provides the long-term state of a DCCM system. It maintains the internal DCCM participant list, role information, and authentication shared secret. It enforces project contexts through its implementation of the key management mechanism for a project. The System component also supports the creation and maintenance of individual projects by storing the project membership list and mediating the project context negotiation process.

The *Participant* component provides the long-term state for an individual participant. It manages the participant's registration in a DCCM system, the shared secret, a list of projects for the participant, and the group keys for the projects that include the participant. The participant enforces the cryptographic context, specified in the CCNT, for each project in which it participates.

A participant is represented during policy negotiation by a *Negotiation* component. An instantiation of the negotiation component represents a single policy during project formation. All of the participants with the same policy are represented by the same negotiation component. This procedure limits the number of negotiators involved in a negotiation. Depending on the organization of the members joining a project, the negotiator may be used to represent different policies. In a hierarchical organization, the negotiation component will represent the organization's policy, and a participant will not have a choice of negotiators. In a flat organization with autonomous participants, the negotiators will represent various public policies that participants will select from. If none of the policies offered is acceptable, a participant can be its own negotiator.

A participant joins a DCCM system through a *System Registration* component. There can be multiple registration components, each understanding a specific authentication and enrollment policy. In a DCCM system comprised of previously unrelated organizations, the organization's existing authentication infrastructure can be used by a registration component for each organization. The registration component can also enforce enrollment policy. For example, it can limit the negotiation components used by participants, thus specifying the allowable policies. The registrar can also limit the roles allowed for a participant. Finally, a registrar can remove or evict a participant from the system, and force their removal from all projects as well.

The *Project Initiator* component supports the entity that creates a project, typically one of the participants. It allows the creation of the initial policy proposal, specified in a CCNT; manages the negotiation protocol (CCNP) with the mediator in the system component and the negotiators; and manages the project membership list with the system component. Once the project is created, the project initiator will have the authority to perform group operations.

The *Session Initiator* supports the creation of sessions within a project. The session initiator is the interface between the DCCM keying mechanism and whatever security mechanism is used for the session. The session initiator will communicate the group session key to the mechanism at session establishment, and will then send key updates as they occur. There can be multiple sessions within a project.

6.3. Demonstration system

A demonstration system has been implemented for DCCM utilizing the architectural components. This provides one example of how to utilize the DCCM components in a simple system architecture. As shown in Figure 5, the demonstration system, the components are organized into a single client image and a server. For the simplicity, the demonstration includes a single registrar in the server. All of the roles for a participant are included in the client.

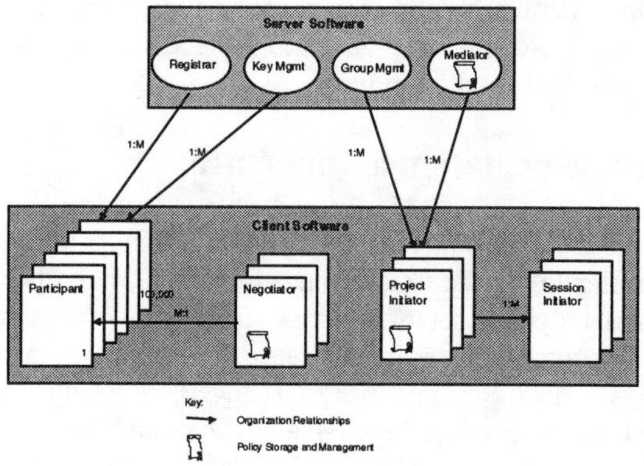

Figure 5: Demonstration architecture

7. Conclusions

Providing policy-based security for large dynamic groups is a challenging problem. It is also a challenge to envision maintaining the security of a group with 100,000 members. There are applications where a secure group of that size is meaningful, and DCCM provides solutions to several important aspects of the problem. DCCM introduces the concept of a group *cryptographic context* that enforces secure interoperability among group members. A Cryptographic Context Negotiation Template is used to represent, distribute, and negotiate group security policies. A Cryptographic Context Negotiation Protocol is used to create a group cryptographic context for a project. During project operation, the One-way Function Tree key mechanism achieves very efficient re-key operations. Its overall cost in terms of time and memory grows logarithmically with the size of the group.

Future work should expand both the policy and key management dimensions of DCCM. Future policy work should focus on representing policy at higher levels of abstraction, formalizing the languages used for policy representation, negotiating group policy at higher levels, and projecting abstract policy onto concrete mechanisms. Policies need to be expanded to represent applications where local autonomy of some mechanisms is desired or necessary; interoperability of security mechanisms may be necessary at the communication level, but not at higher levels.

Future group key management work should provide additional security mechanisms for group security, such as individual source authentication. There is also significant opportunity to enhance group key management techniques to efficiently handle surges in membership and large changes in membership.

8. Acknowledgments

The DCCM project was performed at NAI Labs, The Security Research Division of Network Associates, Inc. (formerly the Advanced Research and Engineering (AR&E) Division of Trusted Information Systems, Inc. (TIS)). Support for the DCCM project was provided by the Defense Advanced Research Projects Agency (DARPA) through Air Force Research Laboratory (AFRL) Contract No. F30602-97-C-0277.

The DCCM system was originally conceived and proposed by Dennis Branstad. During the life of the project the team included David Balenson, Dennis Branstad, Peter Dinsmore, Michael Ferguson, Michael Heyman, Peter Kruus, David McGrew, Matthew Mundy, Caroline Scace, Alan Sherman, and Jay Turner. This paper is based on the work of this talented team, and portions have been excerpted from project reports and briefings. The OFT method for large group key management was developed by David McGrew and Alan Sherman, with contributions from Michael Harding. The demonstration implementation was performed by Michael Ferguson, Michael Heyman, Peter Kruus, Matthew Mundy, and Caroline Scace.

9. References

[1] Ateniese, G., M. Steiner, and G. Tsudik, "Authenticated group key agreement and related protocols," *Proceedings ACM Conference on Computer and Communications Security*, 1998.

[2] Balenson, D., D. K. Branstad, D. A. McGrew, and A. Sherman, *Dynamic Cryptographic Context Management (DCCM) Report 1: Architecture and System Design,* TISR #0709, TIS Labs at Network Associates, Inc. (June 2, 1998).

[3] Balenson, D., D. K. Branstad, D. A. McGrew, J. W. Turner, and M. Heyman, *Dynamic Cryptographic Context Management (DCCM) Report 2, version 2: Cryptographic Context Negotiation Template,* TISR #0745-2, TIS Labs at Network Associates, Inc. (February 24, 1999).

[4] Balenson, D., D. K. Branstad, P. Dinsmore, M, Heyman, and C. Scace, *Dynamic Cryptographic Context Management (DCCM) Report 3: Cryptographic Context Negotiation Protocol,* TISR #0757, TIS Labs at Network Associates, Inc. (February 24, 1999).

[5] Balenson, D., McGrew, D, and A. Sherman, "Key management for large dynamic groups: one-way function trees and amortized initialization," NAI Labs, *Advanced Security Research Journal*, v1n1 (Fall 1998), pp 29-46.

[6] Balenson, D., D. McGrew, and A. Sherman, "Key management for large dynamic groups: one-way function trees and amortized initialization," Internet Draft (work in progress), draft-balenson-smug-groupkeymgt-oft-02.txt, October 1999.

[7] Blundo, C., A. de Santis, A. Herzberg, S. Kutten, U. Vaccaro, and M. Yung," *Advances in Cryptology: Proceedings of Crypto92*, E. F. Brickell, ed., LNCS 740, Springer-Verlag (1992), 471–486.

[8] Burmester, M. and Y. Desmedt, "A secure and efficient conference key distribution system," *Advances in Cryptology: Proceedings of Eurocrypt94*, A. De Santis, ed., LNCS 950, Springer-Verlag (1994), 275–286.

[9] Condell, M., C. Lynn, and J. Zao, "Security Policy Specification Language," Internet Draft (work in progress), draft-ietf-ipsec-spsl-00.txt, October 21, 1998.

[10] Chiou, G. and W. Chen, "Secure broadcasting using the secure lock," *IEEE Transactions on Software Engineering*, 15:8 (August 1989), 929–934.

[11] Harney H. and E. Harder, "Logical key hierarchy protocol," Internet Draft (work in progress), draft-harney-sparta-lkhp-sec-00.txt, March 1999.

[12] Harney, H. and E. Harder, "Group Secure Association Key Management Protocol," Internet Draft (work in progress), draft-harney-sparta-gsakmp-sec-00.txt, April, 1999.

[13] McGrew, D., and A. Sherman, "Key establishment in large dynamic groups using one-way function trees," TIS Report No. 0755, TIS Labs at Network Associates, Inc., Glenwood, MD (May 1998).

[14] Reiter, M., K. Birman, and R. van Renesse, "A security architecture for fault-tolerant systems," TR 93-1354, Cornell University (June 3, 1993).

[15] Rodeh, O., K. Birman, and M. Hayden, "Dynamic virtual private networks," TR 97-1654, Cornell University (1997).

[16] Wallner, D., E. Harder, and R. Agee, "Key management for multicast: Issues and architectures," Internet Draft (work in progress), draft-wallner-key-arch-01.txt, September 15, 1998.

Building Intrusion Tolerant Applications*

Michael Malkin
mikeym@stanford.edu
Stanford University

Thomas Wu
tjw@cs.stanford.edu
Stanford University

Dan Boneh†
dabo@cs.stanford.edu
Stanford University

Abstract

The ITTC project (Intrusion Tolerance via Threshold Cryptography) provides tools and an infrastructure for building intrusion tolerant applications. Rather than prevent intrusions or detect them after the fact, the ITTC system ensures that the compromise of a few system components does not compromise sensitive security information. To do so we protect cryptographic keys by distributing them across a few servers. The keys are never reconstructed at a single location. Our designs are intended to simplify the integration of ITTC into existing applications. We give examples of embedding ITTC into the Apache web server and into a Certification Authority (CA). Performance measurements on both the modified web server and the modified CA show that the architecture works and performs well.

1 Introduction

To combat intrusions into a networked system one often installs intrusion detection software to monitor system behavior. Whenever an "irregular" behavior is observed the software notifies an administrator. In this paper we study a complementary approach we call *intrusion tolerance.* Rather than prevent or detect intrusions after the fact, we provide tools that limit the amount of damage an intruder can cause. Our goal is to ensure that an attacker who penetrates a few system components cannot compromise total system security. To defeat our system an attacker must either penetrate multiple components in a short amount of time or frequently penetrate a certain component. Either way such large scale attacks are much easier to detect than single isolated attacks.

To describe our approach we consider a web server as an example. To enable secure connections to the server one often stores a secret key on the server. Typically the key is used during SSL session key negotiation. An attacker who penetrates the server can expose the private key and can then either masquerade as the server or eavesdrop on connections to the server. Hence, in the case of a bank's web server a single intrusion could result in a compromise of sensitive financial data. In contrast, our approach splits the web server into a small number of components so that a single intrusion does not expose any information about the server's key. Our approach can also be used to safeguard a Certificate Authority's (CA) private key. By splitting the CA into a number of components we can ensure that penetration of a small number of components does not compromise the CA's key.

We protect an application's private key (e.g. a web server or a CA) by sharing the key among a number of *share servers.* An attacker who breaks into a small number of share servers cannot expose the private key. Our main design principle is that long term security information should *never* be located in a single location. Hence, there is no single point of attack at which an attacker can expose critical security information (such as a CA's private key). Shamir's secret sharing scheme [14] is a classic approach for distributing private keys across several sites. Unfortunately, to use the shared key one must reconstruct it at a single location. Thus, secret sharing is inappropriate for our purposes. Instead, we use techniques of Threshold Cryptography [9] to distribute an application's private key among several share servers so that the key can be used without ever reconstructing it.

Our system is designed to be easy to embed into existing applications. For example, by embedding our system into the Apache web server we built a web server whose private key is split across a number of share servers and is never reconstructed in a single location. Whenever the web server receives a request for a secure connection it interacts with the share servers to apply its private key. As we shall see in Section 8 the performance penalty for doing so is small compared to the total time to establish an SSL connection. Consider again the example of a bank's system. By placing the share servers in a secure subnet and ensuring that they only accept connections from the

*Project URL: `http://www.stanford.edu/~dabo/ITTC`

†Supported by DARPA contract #F30602-97-C-0326.

0-7695-0490-6/99 $10.00

bank's web servers one can strengthen the security of the bank's private key. Furthermore, our share servers can service multiple web servers. Hence, a bank maintaining dozens of web servers need not store a private key on each of them. Instead, the private keys can be stored (in shared form) on the share servers. This way a small number of break-ins will not compromise any of the private keys.

As an additional benefit, our system provides high availability of private keys. Even if a few of the share servers crash (by accident or as a result of an attack) and all data stored them is lost, the remaining active share servers automatically compensate for the corrupt ones. Furthermore, the system can recover by redistributing valid shares to the corrupted share servers.

The paper describes the design and implementation of our system including the prototype applications we built. We provide detailed performance measurements on both the modified web server and the modified CA to show that the architecture works and performs well.

2 System components

To protect private keys used by applications such as a web server or a CA, our system distributes shares of private keys among a number of share servers. In this section we describe the system components. We refer to applications using the share servers as *clients*. Web servers and CA's are example clients. Figure 1 illustrates the interaction between the various components.

Share server The share servers are implemented as daemons running on different machines. They hold shares of the private keys belonging to the clients they serve. These shares reveal no information about the client's private key. A share server can manage multiple keys and serve a large number of clients. We envision the total number of share servers being less than ten. Clearly it is desirable that an attacker not be able to compromise the share servers. However, the intrusion tolerant design of our system ensures that even if a few of the share servers are penetrated and the secrets stored on them are exposed or corrupted there is no compromise to overall system security. In other words, an attacker learns nothing from penetrating a few of the share servers. The system identifies corrupt share servers and can be instructed to take appropriate action to refresh the secrets stored on them.

Client. A client is any application that makes use of our client side Threshold LiBrary (the TLB). When a client connects to the share servers it first authenticates itself as an authorized client. It then interacts with the share servers to sign a message or decrypt a given ciphertext using the shared private key stored on the share servers. One concern is that an attacker can penetrate a client and expose the client's authentication key. The attacker can then masquerade as the client and fool the share server into applying a shared key. In Section 6.2 we describe a mechanism for detecting such an attack. At any rate, the attacker cannot expose the shared private key (as is the case when the private key is stored on the client).

Administrator. We provide a central administration utility to administer all the share servers. The administrator utility manages the various keys stored on the share servers. It can shutdown or suspend the servers if necessary, or instruct the servers to take appropriate action if some of the servers have been penetrated. The utility is provided for convenience. If for some reason central administration is not desired the share servers can be controlled locally.

Monitor. We built a monitor utility for testing and demonstrating our system. The monitor is a single daemon that collects information from all key servers and clients. Essentially, the share servers and clients send data to the monitor (via UDP) telling it what they are doing at any given moment. For instance, a client may tell the monitor that it is asking servers number 1, 2 and 4 to apply key number 23 to sign a message. A moment later share servers 1, 2 and 4 will each tell the monitor that they are applying shares of key number 23 to generate a signature, and so on. The monitor collects all this information and sends it to a Java applet that displays it in a human readable form. During actual deployment the monitor can be easily disabled by setting the appropriate switch in the system's configuration file. The administrator can still monitor system behavior by viewing each server's log file.

2.1 Interaction between components

Each share server manages shares of the private key belonging to its clients. These shares are stored on disk encrypted using the administrator's passphrase. When the share server is started (possibly as part of the system's bootup script) it enters *suspended* mode. In suspended mode the share server refuses to accept any connections except a connection from the administrator. When all share servers are started, the admin program can be used to activate the servers by sending the passphrase to all of them at once. Connections between the admin program and share servers are always protected by SSL using mutual authentication. Using a hash of the administrator's passphrase, each server loads all shares available to it and enters *active* mode. The server is then ready to serve its clients.

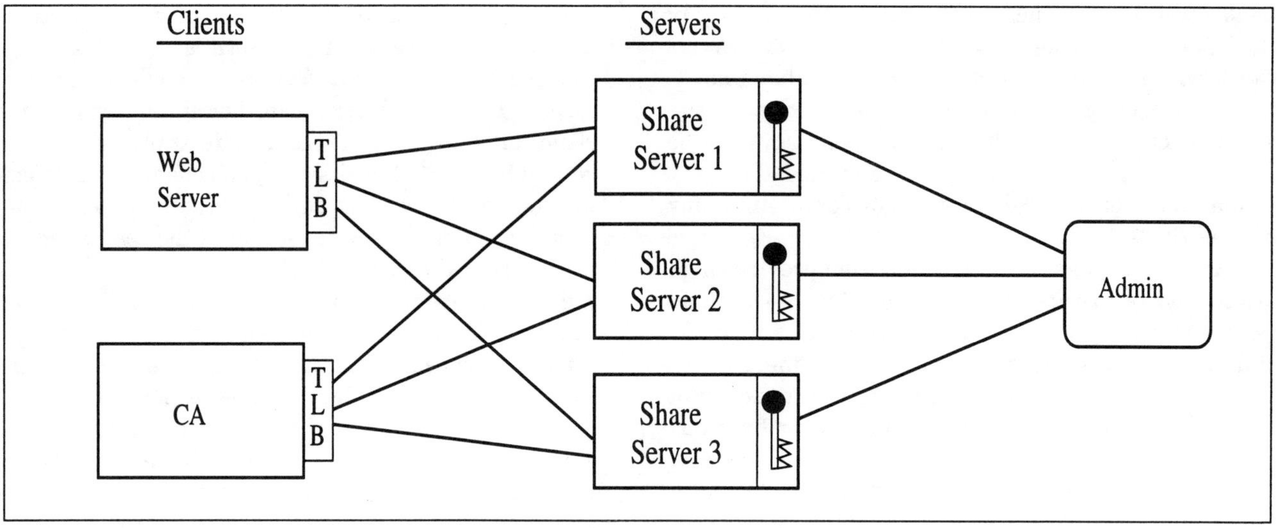

Figure 1: System components

In active mode a share server can accept connections from its clients and the administrator. The first step in any connection to a share server is mutual authentication and key exchange using SSL. The second step is an integrity check of the peer as described in Section 6.2. After a secure link is established, requests to the share server use the following protocol:

Request: COMMAND ⟨*opcode*⟩ ⟨*data*⟩
Response: RESPONSE ⟨*status*⟩ ⟨*data*⟩

The response status is either (1) OK indicating a successful completion, (2) SUSPENDED indicating that the share server is in suspended mode, or (3) a detailed error-code.

To understand the interaction between the system components we describe some of the opcodes supported by the share servers.

Connection from client. Each connection from a client causes the share servers to create a new thread (up to a specified maximum number of threads). These threads last as long as the client keeps open the connection to the share server, reducing the overhead of repeatedly opening and closing connections. A client can send a number of opcodes instructing the servers to apply a shared key. The interaction that takes place as a result of these commands is described in Section 4.

SIGN. Instructs the share server to apply its share of the private key for the purpose of generating a signature. The parameters include a digest of the message to be signed. We note that prior to taking part in the signing protocol described in Section 4 the key-servers verify that the message to be signed follows the PKCS1 format [11].

DECRYPT. Instructs the share server to apply its share of the private key for the purpose of decrypting a ciphertext.

HASH-SIGN. Instructs the share server to apply its share of the private key for the purpose of generating a hash of a signature. The purpose of this opcode is explained in the next section.

Connection from the admin. Some of the commands from the administrator include:

SHUTDOWN, SUSPEND, ACTIVATE. Manage share server modes. Shutdown causes all share server daemons to exit.

GENKEY. The admin can instruct the share servers to generate a new shared key. In our system private keys are *never* constructed in a single location. When an RSA key is generated it is generated in shared form. This uses a protocol due to Boneh and Franklin [3]. We use the implementation of shared RSA key generation described in [10]. The shares of the new private RSA key are stored on the share servers (encrypted with the admin password). The new public key is sent to the admin and is saved on the admin's machine. Authorized clients can then connect to the share servers to generate signatures or decrypt incoming messages using the new key.

A new public key is usually certified by a CA before it can be properly used. To obtain a public key certificate for the new key the admin must generate a *certificate request* to be sent to a CA. The format of a certificate request (specified in PKCS10) includes a *self signature* on the request.

In other words, the newly generated ***private*** key must be used to sign the public key certificate request. The self signature ensures that the entity requesting the certificate has the corresponding private key. Unfortunately, in our case, the admin does not have the private key. In fact, no one has the private key — the private key is *always* stored in shared form on the share servers. Consequently, to generate the certificate request, the admin must connect to the share servers *as a client* and ask them to sign the request. Once the admin obtains the self signed request he can forward it to the CA. The CA will send back the certificate. We provide a utility that enables the admin to generate a self signed certificate request, as described in Section 7.1.

REFRESH. Suppose the administrator suspects the system is under attack and some share servers are compromised. In this case, the admin can instruct all share servers to *refresh* their shares of the private keys. Refreshing the shares does not change the private key. It simply generates a new independent *sharing* of the private key. Suppose an attacker obtains the shares of a private key stored on one of the servers. Once the admin refreshes the shares of the private key the information in the attacker's hands becomes useless.

Refreshing is also used when shares stored on a share server are corrupted or lost. Corruption or loss could happen due to a denial of service attack or a simple server crash. Refreshing the shares causes the uncorrupted servers to generate new valid shares for the corrupted server. Consequently, the system gracefully tolerates corruption or loss of a few shares.

Refresh is described in detail in Section 5.

Both genkey and refresh require all share servers to communicate with all other share servers, so the running time of these operations increase as the square of the number of servers. Since it can take a long time to generate a key among a large number of share servers, it is recommended that the total number of share servers remain less than ten. However, once a key has been generated the total number of share servers does not affect the running time of signatures or decryptions.

3 Key management

We begin by explaining how one shares a private RSA key among a number of share server so that the key can be used without ever having to reconstruct it. We then describe the structure and PEM format used to store these shared keys.

3.1 Sharing RSA keys

Recall that an RSA private key consists of a modulus N and a secret exponent d. The modulus N is a product of two large primes, and d is a positive integer less than N. To decrypt a ciphertext C one computes $C^d \bmod N$. Similarly to sign a message digest M one computes $M^d \bmod N$. Hence, both operations require an exponentiation to the power d modulo N. Without the secret exponent d it is believed to be hard to either decrypt ciphertexts or generate signatures.

We show how a private RSA key can be broken up into a number of pieces (shares). Each share can be stored on a separate server and yet the private key can be used without having to reconstruct the secret. The basic idea, due to Frankel [6], is to pick random numbers d_1, d_2, d_3 in the range $[-N, N]$ so that $d_1 + d_2 + d_3 = d$. We then store share d_i on share server number i, for $i = 1, 2, 3$. Note that an attacker who breaks into any two of the three servers learns nothing about the private key d. All three servers must be compromised to obtain d.

When a client wishes to apply the key to sign a message M it sends M to all three servers. Each server applies its own share d_i to obtain $S_i = M^{d_i} \bmod N$ and sends the result S_i back to the client. The client obtains S_1, S_2, S_3 from the three servers. It multiplies them to obtain the signature $S = S_1 \cdot S_2 \cdot S_3 \bmod N$. Since $d = d_1 + d_2 + d_3$ we have that $S = M^d \bmod N$ as required. Note that the private key d was never reconstructed in order to be used. In addition, there is no communication between the share servers. The only interaction is between the client and each of the servers.

Clearly this approach generalizes to distributing a private RSA key among k servers. Even if $k - 1$ of the shares are exposed, an attacker learns nothing about the private key d. Since all k share servers must be involved in applying to key we call this sharing a k-out-of-k sharing.

3.1.1 t-out-of-k sharing

There are a number of problems with the approach described above. Most importantly, it is not fault-tolerant. If one of the share servers crashes the entire system goes offline. If one of the share servers loses its share, the private key is lost forever. For this reason, we use a t-out-of-k sharing of the secret key; any t of the share servers can be used to apply the key. For instance, if we use a 3-out-of-4 sharing no harm is done

$$d = d_1 + d_2 + d_3$$
$$d = d_4 + d_5 + d_6$$

Server 1	Server 2	Server 3	Server 4
d_1	d_2	d_3	d_3
d_4	d_4	d_5	d_6

3-out-of-4 sharing

$$d = d_1 + d_2$$
$$d = d_3 + d_4$$

Server 1	Server 2	Server 3	Server 4
d_1	d_1	d_2	d_2
d_3	d_4	d_3	d_4

2-out-of-4 sharing

Figure 2: Additive sharings of a private RSA key d.

if one of the share servers is taken offline. In addition, a break-in into any two servers reveals no information about the private key.

Typically a t-out-of-k sharing is achieved using Shamir's classic secret sharing [14]. Unfortunately, when using Shamir's secret sharing the keys must be reconstructed before they can be used. This is inappropriate for our purposes. Instead, we use a combinatorial construction to obtain a t-out-of-k sharing of d from several t-out-of-t sharings of d. We give two examples in Figure 2. In both examples all the d_i's are random integers in the range $[-N, N]$ satisfying the stated equalities. Each server stores multiple d_i's as indicated by the column corresponding to that server. Observe that in the example on the left, any three servers can apply the key while any two servers learn nothing about d. The example on the right is more fault tolerant, but compromising two servers reveals the key. A compromise of a single server reveals nothing about the key. More generally, we implemented an algorithm that constructs tables as above for a t-out-of-k sharing for any t and k. The algorithm is based on ideas from [1, 4].

When a client requests that the servers apply a private key, it specifies which coalition of t servers is used. Based on the coalition, each server locally decides which of its d_i's it will use and sends the resulting S_i back to the client. The client multiplies the t responses and obtains the signature S. The client then uses the public key to check the validity of S as a signature. This is done to ensure that all share servers responded properly. In Section 4 we explain how we deal with corrupt share servers that apply their private shares incorrectly.

3.2 Key structures and key storage

Our system manages three types of keys: (1) a standard RSA public key, (2) a private share stored on each share server, and (3) a public shared key stored on each client. The public key stored on the clients contains the public RSA key plus some additional public information. We describe each of these keys below.

The SSLeay package [17] supports reading and writing both public and private keys in PEM format. Our private shares and public shared keys are represented internally as extensions of the standard RSA key data structure. On disk, we support a PEM-encoded ASN.1 format similar to that used for RSA keys.

Public key: This is a standard RSA public key made up of an RSA modulus N and a public exponent e. It is managed by the standard RSA functions provided in SSLeay. Note that an outsider communicating with the clients is unaware that the corresponding private key is stored in shared form.

Private share: The private share stored on each share server contains the modulus N and a number of private d_i's. For a t-out-of-k shared key the private share file format (as ASN.1) is shows in Figure 3a. Each share server is given w shares $d_1, \ldots, d_w$. Note that the private share file does not contain the optional values $d \bmod p - 1$, $d \bmod q - 1$, or $q^{-1} \bmod p$, normally used to optimize RSA computations, where $N = pq$. After all, none of the parties can construct these values.

Public shared key: The format of the public key file stored on each client is shown in Figure 3b (all arithmetic is done modulo N). Here u is the total number of g^{d_i}'s stored on the client. The values g and $g^{d_i} \bmod N$ are used to detect incorrect (or possibly compromised) private share operations by the share servers, as discussed in the next section. For each share d_i on each share server, there is a g^{d_i} in the public shared key. For example, in the 3-out-of-4 sharing of the previous section the public shared key contains six entries, so $u = 6$.

4 Using a shared key

We are now ready to describe the interaction that takes place when a client requests the share servers to apply a private key to a message M. Throughout the

Figure 3: Shared key file formats.

type:	INT	INT	INT	INT	INT	INT	INT	INT	⋯	INT
data:	version	N	e	k	t	w	d_1	d_2	⋯	d_w

(a) Private key file format.

type:	INT	INT	INT	INT	INT	INT	INT	INT	INT	⋯	INT
data:	version	N	e	k	t	g	u	g^{d_1}	g^{d_2}	⋯	g^{d_u}

(b) Public shared key file format.

interaction the client keeps two bit-vectors `offline` and `corrupt`, indicating which of the share servers are currently offline and which have been found to be corrupt. We assume the client wishes to use a t-out-of-k shared key.

Init The client sets `offline` and `corrupt` to zero.

Step 1: The client picks a *random* coalition of t servers (out of k) that are neither offline nor corrupt.

Step 2: The client sends the following message to each of the servers in the coalition:

COMMAND	SIGN or DECRYPT	M	coalition	key-ID	corrupt	offline

where `key-ID` is the ID of the key the servers should apply. The `key-ID` is simply a 32 bit hash of N and e.

Step 3: Based on the coalition being used, each of the servers extract the appropriate d_i from its private key share of `key-ID`. It then locally computes $S_i = M^{d_i} \bmod N$ and send S_i back to the client. Note that if a signature is being requested, the client first verifies that the message M is in PKCS1 format, and rejects the request if not.

Step 4: The client collects all the S_i's and computes $S = \prod_{i=1}^{t} S_i \bmod N$. If some of the servers were found to be offline, the `offline` bit vector is updated and the process is restarted at Step 1.

Step 5: If $S^e = \pm M \bmod N$ the key was applied correctly and the process terminates.

Step 6: Otherwise, for each server in the coalition the client performs a zero-knowledge test to validate the server's response. The test is described in the next section. All servers that sent incorrect values are marked as corrupt and the process is restarted at Step 1.

If at Step 1 there are less than t servers that are neither corrupt nor offline the process fails. Note that the identity of corrupt share servers is sent to *all* share servers. This enables the servers to notify the administrator, who can take appropriate action to refresh the shares stored on the corrupt servers.

Load balancing: When many different clients use the same share servers, the load on the servers may hurt overall system performance. Fortunately, our choice of a random coalition in Step 1 provides for *automatic load balancing* among the servers. When a 3-out-of-5 key is used each application of the key is likely to make use of a different set of three servers. Furthermore, by using a sufficiently low timeout period clients can avoid waiting for busy servers by switching to a different coalition. Due to the load balancing effect, a 3-out-of-5 sharing will result in a higher throughput than a 3-out-of-4 sharing.

4.1 Identification of corrupt servers

It remains to show how we identify corrupt servers in Step 6. This protocol is only used in the unlikely event one of the servers returns an incorrect response in Step 3. Recall that server i sent $S_i = M^{d_i} \bmod N$ back to the client where d_i is its share of the private key. By examining its own public shared key file the client obtains $V_i = g^{d_i} \bmod N$. To validate sever i's response the client does the following:

Step 6.1: The client picks random a and b in the range $[1, N]$ and computes $Z = M^a g^b \bmod N$.

Step 6.2: It sends the following to server i:

COMMAND	HASH-SIGN	Z	coalition	key-ID	corrupt	offline

Step 6.3: A legitimate server will respond with $A_i = H(Z^{d_i} \bmod N)$ where H is a cryptographic hash function. We use SHA-1.

Step 6.4: The client checks that $A_i = H[S_i^a V_i^b \bmod N]$. If not, the server is declared to be corrupt.

The above protocol is a simplification of a protocol from [7]. The following lemma shows that a share server who sends an invalid S_i will be caught with overwhelming probability. Furthermore, the protocol reveals no information about the server's secret shares. The lemma relies on the Small Order Assumption (SOA): there is no efficient algorithm that, given an RSA modulus $N = pq$, outputs an element x in $\mathbb{Z}_N$, where $x \neq \pm 1$, so that the order of x in $\mathbb{Z}_N$ is less than $N^{1/2}$.

Lemma 4.1 *Assuming SOA, a server who in step 2 sends any value other than* $\pm M^{d_i} \bmod N$ *will fool the above protocol with probability at most* $1/N^{1/2}$. *Furthermore, assuming the hash function used is a random function, the protocol is zero-knowledge (i.e. the client can simulate its interaction with the server).*

For a 1024 bit key, the probability that a server fools the client is less than $1/2^{512}$ — negligibly small. The hash function is used to prevent a malicious client from obtaining signatures on improperly formatted messages (i.e. not in PKCS1 format). For example, in a CA environment the DECRYPT command is disabled. Consequently, the servers can only be used to generate signatures and they only sign properly formatted PKCS1 messages. In this environment, the hash function prevents the client from using the validation test to obtain signatures on arbitrary messages.

We note that when $\gcd(p-1, q-1) = 2$ the SOA assumption is equivalent to assuming the hardness of factoring. However, since our servers cannot test this condition (no single server knows p and q) we use the SOA as the intractability assumption in Lemma 4.1.

5 Refresh

Refresh serves two distinct purposes. First, if some of the shares of the private key are compromised, refresh can be used to secure them. With refresh, an adversary has only a limited ammount of time to penetrate the share servers. If the administrator refreshes before the adversary can break into a threshold of machines, the adversary gains nothing. Second, refresh can be used to generate new shares to replace lost or corrupted shares.

Refreshing a key takes place in three stages:

Step 1: Synchronize on lowest common version

Step 2: Generate new sharing of key

Step 3: Test the new sharing

Step 4: Synchronize on highest common version and send public key to ADM

The first step synchronizes all servers on the lowest version of the key that they all have in common. The second step generates a new sharing of the key, keeping the private key d the same but giving each server a different share of it. In the third step the ADM tests the new sharing to make sure it is valid. The final step tells the servers use the new shares.

5.1 Generating a New Sharing

What follows is a description of how the system generates a new sharing of a key.

Step 1: The ADM encrypts a random message and tries to decrypt it with with all possible coalitions of t servers until one of the coalitions works. Since the coalition can decrypt, the ADM knows that this coalition has a valid sharing of the private key.

Step 2: The ADM sends the servers the command to generate a new sharing. It also sends the key ID, the coalition, the modulus N, the encryption exponent e, the total number of servers k, the threshold t, and the current version of the key.

Step 3: The k servers make a k-out-of-k sharing of the key. If a server participated in the successful decryption of step 1, it contributes the share it used in that decryption. Otherwise, it contributes zero. The new k-out-of-k sharing has the same private key d as the original sharing.

Step 4. The servers take the new k-out-of-k sharing and use it to generate a t-out-of-k sharing. This t-out-of-k sharing has the same private key as the original sharing.

Step 5. All the servers send an acknowledgment to the ADM when they have finished.

5.2 Synchronizing Shares

What follows is an outline of the synchronization part of the refresh algorithm. Synchronization is used to guarantee that all share servers agree to use a particular version of the key.

If synchronization fails, it is automatically attempted again. This is repeated until it has been attempted a maximum number of times, in which case the refresh process is exited with an error.

Step 1: The ADM sends the GET-VERSION opcode to all share servers, along with the ID of the key that is being refreshed.

Step 2: The servers send the ADM the minimum and maximum version that they have for the key. Note that this process assumes that key versions are contiguous. For example, it is assumed that a server can never have versions 2, 3, 5, and 6 of a key.

Step 3: The ADM decides which version of the key to have the servers use.

If this is the first synchronization in the refresh process, then the ADM chooses the lowest version for which a threshold of servers have a share. If there is no such version, an error is given and the refresh process is halted.

A server can have multiple versions of a key only as the result of failed refreshes. The lowest version is chosen because the ADM only gets a shared public key after key generation or a successful refresh, in which case there are no previous versions. Hence, the lowest verison should correspond to the ADM's shared public key.

If this is the second synchronization in the refresh process, the ADM chooses the highest version that all servers agree upon. Since this type of synchronization is done only after the servers have generated a new sharing, they should end up using the new shares. However, if there was an error while generating new shares, the previous shares will be used.

Step 4: The ADM tells the servers which version of the key to use. It sends the servers the ID of the key, the version to use, and and instruction of whether or not to send back the public key. If the ADM is requesting a new shared public key, it backs up the current public key.

Step 5: The servers synchronize on the version by deleting all other versions of the key. They acknowledge this by sending a reply to the ADM. If it was requested, the servers also send the ADM their values for the shared public key.

Step 6: Once the ADM has received acknowledgments from all servers it, knows that the synchronization has finished. If the ADM is receiving a shared public key, it waits until it has the new shared public key and then deletes the backup of the old shared public key.

5.3 Fault-Tolerance of Refresh

If shares were never lost or corrupted, then the two synchronization steps would not be necessary. The first synchronization step ensures that servers will not mix versions when they generate a new shaing of the private key d, because mixing versions when scrambling results in an invalid private key. The second synchronization step causes the servers to revert to a previous version if the scrambling process failed and resulted in an invalid version.

There is fault-tolerance inherent in the structure of the refresh algorithm. The two synchronizations steps ensure that the algorithm can recover from errors or crashes that occur during a refresh. If servers crash during a refresh, they key can be recovered by re-starting the servers and running refresh again. This process can be repeated until there are no crashes and refresh succeeds in generating new shares.

We assume that, in general, the servers are running correctly during a refresh. However, if some of the servers are corrupt, they be able to halt the refresh, but they can't cause the private key to be lost. The shares are tested before the final synchronization, and if they are invalid, the old shares are used.

6 Implementation details

6.1 Certificate formats

The ITTC system uses the Secure Socket Layer (SSL) to authenticate and encrypt all communication. The administrator issues a certificate to each entity (client, share server, and administrative utility) in the system, which must be in a recognized format. The CN (Common Name) field in each certificate contains a string that allows other parties to identify it as described in the following table:

Entity	ID String
Client	`[CLIENT` n`]`
Share Server	`[SERVER` n`]`
Administrator	`[ADMIN` n`]`

When establishing an SSL-secured connection, both peers must send their certificates to establish mutual authentication. Each peer verifies the certificate it receives and then parses it to extract the identifying string to ensure that the peer is authentic. If a certificate is not sent, if verification fails, or if the identity string does not match what was sent in the protocol, the connection fails.

6.2 Sequence numbers

Although client and server certificates thwart direct network snooping and impersonation attacks against the servers, the ITTC threat model requires the system to tolerate and resist successful attacks against individual clients and servers as well. If an attacker, for example, obtains the private key and certificate of a legitimate client, he could use it to obtain access to any shared keys that the client had access to. To limit the amount of damage under these circumstances, ITTC uses sequence numbers in the connection protocols to detect such compromises after the fact. Each client and server keeps a count of how many times it has been accessed by each other entity in the system, and all connections involve an exchange of sequence numbers to verify synchronization. An attacker who uses a stolen key and certificate will cause the sequence number at each server to be incremented without a corresponding change in the sequence number at the "victim" client's system. The next time that client attempts to access a server, the mismatch will be detected.

The actual multi-step sequence number exchange proceeds as follows. Assume that both client and server are currently synchronized at sequence number n initially:

1. The server sends n to the client and sets its own sequence number to $n+1$.
2. The client expects n. It sends $n+1$ back to the server. It sets its own sequence number to $n+1$. (In this step, the client also accepts $n+1$ from the server. If this happens, it sends $n+2$ back to the server and sets its own sequence number to $n+2$. The rest of the protocol proceeds with the client adding one to all its sequence numbers.)
3. The server expects $n+1$. It sends $n+2$ back to the client. It sets its own sequence number to $n+2$.
4. The client expects $n+2$. It sets its own sequence number to $n+2$.

When the protocol completes successfully, both client and server hold sequence number $n+2$. The advantage this protocol has over a simpler single-increment protocol is that it distinguishes between ordinary network failure and a genuine security breach. In a single-increment protocol, it is possible for a network failure to cause one party to increment its sequence number while the other does not. The off-by-one discrepancy in the sequence numbers would be interpreted as a security breach during subsequent interactions between these two parties. This double-increment protocol, on the other hand, increments sequence numbers by two for each successful interaction, while ensuring that the server's sequence number never drifts by more than one relative to the client's number. A network failure would result in at most an off-by-one discrepancy, which would then be resolved in step two of the protocol. A real security breach would cause the numbers to differ by two or more, presumably triggering a legitimate alarm.

6.3 Interface to TLB

To access the underlying ITTC functionality, client programs link against the TLB code and call functions in its external interface. Since it is possible to perform all the standard RSA cryptographic operations (encrypt, decrypt, sign, verify) with an ITTC-style RSA key, existing applications can be rewritten to use these keys by replacing conventional RSA calls with calls to TLB functions.

This approach has some drawbacks, most notably that it requires effort to go through each application to find RSA cryptographic calls and replace them with their ITTC counterparts. Fortunately, the design of SSLeay allows us to implement the TLB interface more elegantly. In SSLeay, RSA keys (type `RSA *`) contain not only the data corresponding to the keys (N, e, d, etc.) but also pointers to functions that perform the four basic RSA operations. This form of polymorphism allows us to reduce the entire TLB interface to a single function call:

```
RSA *ITTC_load_RSA_key(char *ittckeyfile);
```

This function returns an RSA object whose "function table" points to functions that perform ITTC-style decryption and signing (encryption and signature verification are handled just like an ordinary RSA key). Modifying an application to handle ITTC-style keys is now a matter of changing only one piece of code, namely the code that loads RSA keys. Once the key is loaded in this manner, application code cannot distinguish an RSA object obtained through ITTC from a conventional RSA key, and no other code needs to be modified.

Section 7 contains some examples of applications that used this technique to support ITTC-style shared keys. This object-oriented method of encapsulating functionality with data can be carried over to other cryptographic architectures (e.g. Java's JCA or Microsoft CAPI) that support provider interfaces.

7 Example applications

7.1 Intrusion tolerant certification authority

The ITTC-based certificate authority application signs incoming certificate requests with the appropriate shared private key. The application itself is based on Eric Young's `ca` certificate authority program, which is included with SSLeay. As described in Section 6.3, adapting the existing code to handle ITTC keys involved only a minimal change to the code, as seen in Table 1.

```
#ifndef ITTC
  BIO_read_filename(in,keyfile);
  pkey=PEM_read_bio_PrivateKey(in,NULL,NULL);
#else
  pkey=EVP_PKEY_new();
  rsa=ITTC_load_RSA_key(keyfile);
  EVP_PKEY_assign_RSA(pkey,rsa);
#endif
```

Table 1: ITTC Certificate Authority Code Patch

The modified certificate authority (`ittc_ca`), accepts the same inputs and command line arguments as the original program. For example, the command

```
ittc_ca -config ca.cnf -in request.req
    -out newcert.pem -keyfile sharedkey.rsk
```

verifies the self-signed standard-form certificate request in `request.req`, signs it with the shared RSA key `sharedkey.rsk`, and writes the PEM-encoded X509 result in `newcert.pem`. The file `ca.cnf` is an internal SSLeay config file.

In most applications where an ITTC private key is used, it is necessary to obtain a signed certificate for the corresponding public key from a well-known issuer (e.g. when using ITTC to secure a Web server's private key, see Section 7.2). The `ittc_req` program generates correctly-formatted, self-signed certificate requests from ITTC-style shared keys. The command

```
ittc_req -config ca.cnf -new
      -key sharedkey.rsk -out request.req
```

compiles a standard-form certificate request for the shared key `sharedkey.rsk` and applies the shared key once to sign it. The output (`request.req`) is indistinguishable from a certificate request generated from a conventional RSA key and can be sent off to any certificate authority (e.g. Verisign) to be signed.

7.2 Distributed web server

In addition to securing the private key for a certificate authority, ITTC can protect private keys for a Web server. The technique of Section 6.3 was applied to the source code for ApacheSSL 1.2.6 to produce an ITTC-enabled secure server. Since Apache already used SSLeay to handle its public-key cryptography, the entire extent of our modifications to Apache consisted of the three lines of C code shown in Table 2.

```
#ifdef ITTC
   pConfig->prsaKey =
           ITTC_load_RSA_key(szPath);
#else
   pConfig->prsaKey=RSA_new();
   PEM_read_RSAPrivateKey(f,
               &pConfig->prsaKey, NULL);
#endif
```

Table 2: ApacheSSL Code Modification for ITTC

With this change in place, Apache accepts ITTC-style private keys and treats them as if they were conventional RSA keys. An administrator simply generates an ITTC shared key, generates a self-signed certificate request with the `ittc_req` application (see Section 7.1), and sends this request to a recognized certificate authority. Once the signed certificate is received, Apache is configured to use the certificate and corresponding key with the following entries in its configuration file, `httpd.conf`:

```
SSLCertificateFile
     /usr/local/httpd/SSLconf/cert.pem
SSLCertificateKeyFile
     /usr/local/httpd/SSLconf/privatekey.rsk
```

Once the Web server is started up, it will present the certificate `cert.pem` to Web browsers and use the ITTC shared key `privatekey.rsk` to decrypt and/or sign responses.

8 Performance

The following timing measurements were taken on six Intel-based PC's. Two were running Solaris 2.6 at 333Mhz and four were running Windows NT at 450Mhz. The machines were connected via a 10base-T Ethernet. Clearly, a faster local network would improve our performance numbers.

Table 3 shows the latency and throughput for several key-sizes and sharings. Latency is the time that a client has to wait for a request to be serviced. Throughput is the number of requests that can be serviced per second. The table also lists the number of concurrent threads running on the *client*. The more threads run on the client the higher the load on the servers. Load balancing causes each share server to see some fraction of the requests. For example, when 10 threads are used the client makes 10 concurrent requests. If a 3-out-of-5 sharing is in use, each share

	thr.	no sharing		2-out-of-3		2-out-of-4		3-out-of-4		3-out-of-5	
1024	2	0.067s	29.2/s	0.411s	4.86/s	0.403s	4.95/s	0.639s	3.08/s	0.638s	3.13/s
1024	10	0.070s	28.9/s	1.343s	7.14/s	1.180s	7.90/s	1.790s	4.83/s	1.755s	4.90/s
2048	2	0.370s	5.38/s	1.434s	1.38/s	1.268s	1.55/s	1.978s	1.01/s	1.975s	1.00/s

Table 3: Latency and throughput as a function of key size

	2-out-of-3		2-out-of-4		3-out-of-4		3-out-of-5	
0–offline	0.411s	4.86/s	0.403s	4.95/s	0.639s	3.08/s	0.638s	3.13/s
1–offline	0.604s	3.26/s	0.532s	3.72/s	0.979s	2.01/s	0.887s	2.19/s
2–offline			0.763s	2.60/s			0.967s	2.03/s

Table 4: The effect of offline servers (1024-bit key)

server will service an average of $\frac{3}{5}$ of the total requests, or an average of 6 concurrent requests.

If the requests are processed serially, throughput depends only on latency, but when multiple clients connect to the servers at once load-balancing increases throughput. Increasing the number of simultaneous connections slows down each individual request but increases the throughput. Unless stated otherwise, the number of simultaneous connections was set to two.

A larger key corresponds to more communication and longer computations, so as the number of bits in the key increases, the latency increases and the throughput decreases. In every configuration, the 1024-bit key was significantly faster than the 2048-bit key.

Latency is affected by both the number of servers involved in a request and the total number of servers. For a fixed threshold, a larger total number of servers makes load-balancing more effective by distributing concurrent requests over more servers. This keeps the servers from being inundated by requests. When the total number of server is fixed, a higher threshold is likely to increase latency. To see this recall that a request completes only once all the servers in the coalition respond. With a larger coalition it is more likely that a busy server is included in the coalition. Hence, the time until the request completes is likely to increase.

The ratio of servers used to the total number of servers is the dominating factor in throughput. For example, with the 3-out-of-5 sharing, this ratio would be 0.6. As this ratio shrinks, a smaller portion of the servers is required per request, and more simultaneous requests can be processed with the same performance.

Table 3 also compares shared keys versus standard non-shared RSA keys. As expected, latency is larger when using a shared key than when using a non-shared key. The use of the shared key requires communication with share servers which is not required when using a non-shared key. Likewise, throughput is lower when using a shared key, but throughput doesn't degrade as much as latency because of load-balancing and multi-threading of the share servers.

As the number of simultaneous requests (threads) grows, latency also grows, because of the higher load on the servers. However, throughput improves because of load balancing and because the servers can service some threads while other threads are idle.

Table 4 shows what happens when some servers are taken offline. This is implemented by putting the share-servers in "suspended" mode. In this mode, a server can accept connections, but immediately closes them. Performance is identical when a server is simply shut down.

The TLB detects all offline servers in the coalition in one pass. When too many servers are offline, TLB returns an error code. For example, the 2-out-of-3 sharing will return an error code if two servers are offline, since the remaining server cannot service a request by itself.

Taking servers offline increases latency because some requests have to be issued multiple times, with different coalitions. Also, as servers are taken offline, load-balancing forces the remaining servers to service more requests, further increasing latency. Throughput also suffers when servers are taken offline, for the same reasons as latency.

Performance drops more when going from 0–offline to 1–offline than it does when going from 1–offline to 2–offline. 2–offline is worse than 1–offline since in the worst case requests may need to be issued three times until a valid coalition is found. As servers are taken offline, the sharings that use a larger fraction of the total servers degrade the most. For example, taking one server offline affects the 2-out-of-3 and 3-out-of-4 sharings more than the 2-out-of-4 sharing. This is because load balancing is more effective when the number of servers per request is small compared to the total number of servers.

Table 5 shows the effects of corrupt servers. Corrupt servers act just like normal servers, but they return invalid responses. In the tests used to find this data, corrupt servers compute the correct response, then return twice that value. When an invalid response is detected all servers in the coalition are checked for corruption. Hence, all corrupt servers in the coalition are detected in one pass.

Latency grows dramatically when corrupt servers are detected. There are two causes for this growth. First, just as with offline servers, requests may need to be issued multiple times. Second, when a corruption is found, each server in the corrupt coalition must be checked to see if it is corrupt (see Section 4.1). Comparing this chart to the previous chart, it becomes apparent that a corrupt server adds approximately 2.7 times latency penalty as an offline server in the same configuration. As with offline servers, throughput suffers as servers are corrupted, for the same reasons as latency.

Table 6 shows the performance of a web-server and certification authority, both using shared and non-shared private keys of size 1024-bits and 2048-bits. The web-server was tested by repeatedly establishing an SSL connection and issuing a `HEAD` request to a URL on the web-server. The number of threads listed in the table reflects the number of concurrent threads used by the test program when establishing connections to the web server.

In the web-server, both latency and throughput are worse with a shared-key than with a non-shared key, but not by much. With 10 simultaneous connections, latency is only 24% higher for a shared key and throughput is only 17% worse. For most applications, this slowdown is insignificant considering that SSL session establishment is only a small part of the interaction with the web server.

Finally, we mention the performance of refresh. Refresh is a relatively fast operation. For example, refreshing a 1024-bit 2-out-of-3 key takes 10.75 seconds. Most of the time spent by refresh is running the zero-knowledge tests at the end to validate the new shares.

9 Related work

Fray, Deswarte and Powel [8] and Deswarte, Blain and Fabre [5] describe an encrypted file system where file keys are distributed using Shamir secret sharing across several key servers. Keys are reconstructed every time a file is accessed. In contrast, by using threshold RSA, the ITTC system never reconstructs long term private keys in a single location.

The Ω system [12], built at AT&T, also uses threshold cryptography to protect private keys. Ω supported a Certification Authority (CA) used at AT&T. It was the first system to demonstrate the practicality of threshold cryptography. We note that Ω does not support distributed key generation, detection of corrupt servers or the ability to refresh shares in case a share server is compromised.

The proactive security toolkit [2] built at IBM focuses on using proactive security applied to DSS to protect private DSS signing keys. The system shares a DSS key among a number of servers, and proactively refreshes these shares once every predetermined time period (e.g. once a day). Interestingly, DSS and RSA have different sharing properties. RSA keys are easy to share (so that a signature can be generated without reconstructing the key), but are hard to generate distributively (so that none of the participants know the private key). On the other hand, DSS keys are easy to generate distributively, but are harder to use for threshold signatures.

Our particular implementation of threshold RSA is based on one of several possible algorithms. We chose an algorithm for threshold RSA that is best suited to handle a small number of share servers (i.e. less than six). For a larger number of servers one could use a recent algorithm due to Shoup [15].

10 Conclusions

The ITTC system enables applications to store their private keys in an intrusion tolerant fashion. Penetrating a few share servers reveals no information about the private key. Penetrating a client does not expose any shared keys. Using the sequence number mechanism describes in Section 6.2, we force an attacker to repeatedly break into a client within a short period of time.

Our system eliminates single points of failure by never reconstructing a shared private key in a single location. Even key generation is done in shared form. The system detects and corrects offline and corrupt servers. For instance, one of the share servers can be taken offline for maintenance without affecting system behavior. In addition, the system can refresh shares, which recovers lost or corrupted shares and limits the time an adversary has to break into share servers.

ITTC is easy to embed into existing applications. We built a web server and a CA in which private keys are managed using ITTC. Our performance figures show that the cost of using ITTC is reasonable. This is especially true in the case of a web server where session key exchange is only a small fraction of the total

	2-out-of-3		2-out-of-4		3-out-of-4		3-out-of-5	
0–corrupt	0.411s	4.86/s	0.403s	4.95/s	0.639s	3.08/s	0.638s	3.13/s
1–corrupt	0.955s	2.06/s	0.823s	2.40/s	1.478s	1.31/s	1.334s	1.50/s
2–corrupt			1.536s	1.28/s			2.084s	0.95/s

Table 5: The effect of corrupt servers (1024-bit key)

	key-size	thr.	no sharing		2-out-of-3		2-out-of-4		2-out-of-4 1–offline	
web-server	1024	2	0.209s	9.50/s	0.644s	3.10/s	0.515s	3.84/s	0.535s	3.73/s
web-server	1024	10	1.097s	8.25/s	1.362s	6.85/s	1.503s	6.03/s	1.543s	5.66/s
web-server	2048	2	0.389s	5.05/s	1.527s	1.31/s	1.484s	1.34/s	1.509s	1.32/s
CA	1024	2	0.067s	29.2/s	0.411s	4.86/s	0.403s	4.95/s	0.532s	3.72/s
CA	2048	2	0.370s	5.38/s	1.434s	1.38/s	1.268s	1.55/s	1.749s	1.12/s

Table 6: Usage of ITTC in a CA and web server

work performed by the server. Session key exchange is not done too frequently since both the browser and server cache session keys.

Acknowledgments

An earlier version of this paper [16] appeared at the 8th USENIX Security Symposium.

Research supported by DARPA/ITO. We thank Bill Aiello and Sivaramakrishnan Rajagopalan (Raj) for their help throughout the project.

References

[1] N. Alon, Z. Galil, M. Yung, "Dynamic re-sharing verifiable secret sharing against a mobile adversary", in Proceedings of the 1995 European Symposium on Algorithms (ESA), pp. 523–537.

[2] B. Barak, A. Herzberg, D. Naor, E. Shai, "The proactive security toolkit and applications", to appear in the 6th ACM Conference on Computer and Communications Security, 1999.

[3] D. Boneh, M. Franklin, "Efficient generation of shared RSA keys", in Proceedings Crypto' 97, pp. 425–439.

[4] Y. Desmedt, G. Di Crescenzo, M. Burmester, "Multiplicative non-abelian sharing schemes and their application to threshold cryptography", Proceedings ASIACRYPT '94, pp. 21–32.

[5] Y. Deswarte, L. Blain, J Fabre, "Intrusion tolerance in distributed computing systems", Proceedings IEEE Symposium on Security and Privacy, Oakland, 1991, pp. 110–121.

[6] Y. Frankel, "A practical protocol for large group oriented networks", Eurocrypt 89, pp. 56–61.

[7] Y. Frankel, P. Gemmel, P. MacKenzie, M. Yung, "Optimal-resilience proactive public-key cryptosystems", Proceedings FOCS '97, pp. 384–393.

[8] J. Fray, Y. Deswarte, D. Powell, "Intrusion tolerance using fine-grain fragmentation-scattering", Proceedings IEEE Symposium on Security and Privacy, Oakland, 1986, pp. 194–201.

[9] P. Gemmel, "An introduction to threshold cryptography", in CryptoBytes, a technical newsletter of RSA Laboratories, Vol. 2, No. 7, 1997.

[10] M. Malkin, T. Wu, D. Boneh, "Experimenting with shared RSA key generation", Proceedings of the Internet Society's 1999 Symposium on Network and Distributed System Security (SNDSS), pp. 43–56

[11] Public Key Cryptography Standards (PKCS), RSA Labs, available at `http://www.rsa.com/rsalabs/pubs/PKCS/`

[12] M. Reiter, M. Franklin, J. Lacy, R. Wright, "The Ω key management service", Proceedings of the 3rd ACM conference on Computer and Communication Security, 1996.

[13] T. Rabin, "A simplified approach to threshold and proactive RSA", Proceedings of Crypto' 98.

[14] A. Shamir, "How to share a secret", Comm. of the ACM, Vol. 22, 1979, pp. 612–613.

[15] V. Shoup, "Practical threshold signatures", to appear.

[16] T. Wu, M. Malkin, D. Boneh, "Building Intrusion-Tolerant Applications", Proceedings of the 8th USENIX Security Symposium (1999), pp. 79-92

[17] E. Young, SSLeay, `http://www.ssleay.org/`

Automated Trust Negotiation

William H. Winsborough
IBM Transarc Lab
winsboro@us.ibm.com

Kent E. Seamons
IBM Transarc Lab
seamons@us.ibm.com

Vicki E. Jones
North Carolina State University
jones@csc.ncsu.edu

Abstract

Distributed software subjects face the problem of determining one another's trustworthiness. The problem considered is managing the exchange of credentials between strangers for the purpose of property-based authentication and authorization when credentials are sensitive. An architecture for trust negotiation between client and server is presented. The notion of a trust negotiation strategy is introduced and examined with respect to an abstract model of trust negotiation. Two strategies with very different properties are defined and analyzed. A language of credential expressions is presented, with two example negotiations illustrating the two negotiation strategies. Ongoing work on policies governing credential disclosure and trust negotiation is summarized.

Keywords: *Trust Establishment, Trust Management, Agent Negotiation, Access Control, Digital Credentials, Certificates, Privacy.*

1. Introduction

Distributed software subjects face the problem of determining one another's trustworthiness. Current mainstream approaches to establishing trust presume that communicating subjects are already familiar with one another. There are essentially two approaches based on this assumption. The first is identity-based: identifying a subject is often a sufficient basis for doing business. The second is capability-based: subjects obtain capabilities that are specific to the resources they wish to use. Both approaches require that familiarity be established out of band. Identity- and capability-based approaches are both unable to establish trust between complete strangers. Other solutions are needed in open systems, such as the web, where the assumption of familiarity is invalid.

Property-based *digital credentials* [1] (or simply *credentials*) are the on-line analogues of paper credentials that people carry in their wallets. They present a promising approach to trust establishment in open systems. Credentials, which generalize the notion of attribute certificates[19], can authenticate not just the subject's identity, but arbitrary properties of a subject and its relationships with other subjects. To support service authorization, a client can attach appropriate credentials to service requests.

Trust establishment between strangers is particularly important in the context of e-business. Credential exchange between strangers promises to enable software agents to establish trust automatically with potential business partners. For instance, a software agent might be charged with finding new candidate suppliers of commodity goods and services. Even when an automatically generated list of such candidates eventually would be culled by a human, information such as requirements and availability could be sensitive, requiring trust establishment as part of the automated process of identifying candidates.

This paper presents automated trust establishment between strangers through credential exchange when credentials are themselves potentially sensitive. A *sensitive credential* contains private information. For instance, access to a credential containing medical information could be restricted to primary care physicians and HMO staff. Access to a credit card credential could be limited to businesses that are authorized to accept a VISA card and that adhere to guidelines for securing credit card numbers. Prior trust establishment systems based on credential exchange have addressed credential sensitivity only manually, requiring a user at the client to decide which credentials to submit to each new service. In addition to requiring human intervention, this approach provides to the human making the trust decision no assistance in evaluating the trustworthiness of the server.

This paper presents an architecture for client-server applications in which client and server each establishes a *credential access policy* (CAP) for each of its credentials. A credential is disclosed only when its CAP is satisfied by credentials obtained from the opposing software agent. When an agent needs additional credentials, it can request them. Credentials flow between the client and server through a sequence of alternating credential requests and disclosures, which we call a *trust negotiation.*

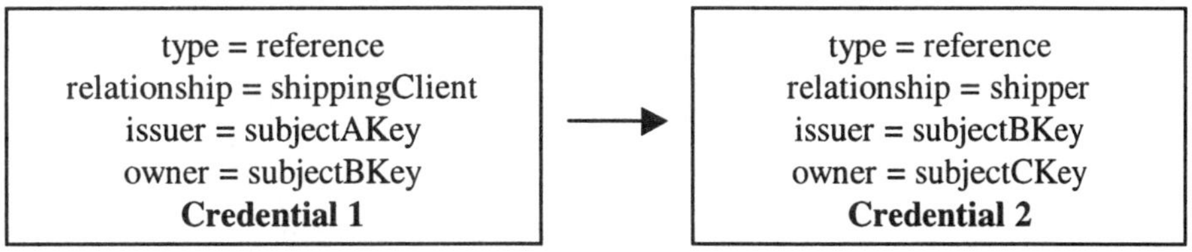

Figure 1. Two credentials forming a chain. Credential 2 was issued by subject B, the owner of Credential 1. In Credential 1, subject A asserts that subject B is a consumer of shipping services. In Credential 2, subject B asserts that subject C is a shipper. If we trust subject A's judgment that subject B is a consumer of shipping, presumably subject B is in a position to know that subject C is a shipper. Additional credentials owned by subject B can be used to engender trust that subject B is a reliable authority on the asserted attributes of subject C.

A negotiation strategy determines characteristics of a negotiation such as which credentials are requested and when the negotiation is halted. We discuss two negotiation strategies with very different properties.

Participants using the first strategy turn over all their credentials as soon as their CAPs are satisfied, without waiting for the credentials to be requested. For this reason we call it an *eager* strategy. It is simple and efficient, and leads to successful negotiation whenever possible. However, it discloses more credentials than necessary to achieve most trust requirements. Participants using the second strategy exchange credential requests that focus the credential exchange, achieving a kind of local minimality of disclosures. For this reason, we call it a *parsimonious* strategy. It is also reasonably efficient, and succeeds whenever possible. However, it has the drawback that the credential requests can disclose sensitive information about the subjects' credentials and properties. This paper discusses these tradeoffs and techniques for managing them.

Section 2 introduces credentials and explains how they can be used to establish trust between strangers. Section 3 introduces credential sensitivity, clarifying the present technical contribution. Section 4 presents a trust negotiation architecture and an abstract model of trust negotiation that is used in Section 5, where eager and parsimonious negotiation strategies are discussed. Section 6 presents two expression languages that can be used within the trust negotiation architecture to express CAPs and credential requests. Section 7 presents two examples of trust negotiation, illustrating the eager and parsimonious negotiation strategies, respectively. Section 8 discusses related work. Section 9 draws conclusions and presents future research directions.

2. Credential-based Trust

A credential is a digitally signed assertion by the *credential issuer* about the *credential owner*. Credentials can be made unforgeable and verifiable by using modern encryption technology: a credential is signed using the issuer's private key and verified using the issuer's public key [12]. A credential aggregates one or more *attributes* of the owner, each consisting of an attribute name/value pair and describing some property of the owner asserted by the issuer. Each credential also contains the public key of the credential owner. The owner can use the corresponding private key to answer challenges or otherwise demonstrate possession of that private key to establish ownership of the credential. The owner can also use the private key to sign another credential, owned by a third subject. In this way, *credential chains* can be created, with the owner of one credential being the issuer of the next credential in the chain.

Credential chains can be submitted to trace a web of trust from a known subject, the issuer of the first credential in the chain (e.g., subject A in Figure 1), to the *submitting subject*, in which trust is needed. The submitting subject is the owner of the last credential in the chain (e.g., subject C) and can demonstrate ownership of that credential, as outlined above. *Supporting* credentials are owned by subjects with whom the submitting subject has a direct or indirect relationship, and, although they are not owned by the submitting subject, the submitting entity does collect, keep, and submit copies of them. Each supporting credential contains the public key whose private-key mate signed the next credential in the chain, enabling reliable verification that the attribute claims made in that next credential were made by the owner of the supporting credential.

The submitted credentials attempt to demonstrate a (possibly indirect) relationship between the submitting subject and the known subject that issued the first credential in the chain. The nature of that relationship can be inferred by inspecting the attributes of the credentials in the chain. Multiple chains can be submitted to establish a higher degree of trust or to demonstrate additional properties of the submitting subject and its relationships with known subjects.

A *credential expression*, ψ, is a logical expression over credentials with constraints on their attributes. A credential expression serves to denote the combinations of credentials, C, that satisfy it. We call those combinations the *solutions* of the expression. For the purpose of trust negotiation, credential expressions can be used to convey requests for credentials between client and server. In this context, credential expressions denote chains of credentials that end with credentials owned by the submitting subject. A credential expression can also be used as a *policy* governing access to a resource. Access to the resource is granted to a subject when a solution is presented that consists of one or more chains ending in credentials owned by the subject. The resource is *unlocked* by the solution.

A policy is *mobile* if it is sent from one subject to another as part of automatic or semiautomatic trust establishment. Mobile policies are used in prior systems to express requirements a client must meet to obtain service. When insufficient credentials accompany a service request, the server returns the *service-governing policy* (SGP). Communicated in this way, the SGP acts as a request for the credentials needed to unlock the resource. Such mobile policies enable clients to select a minimal set of credentials whose submission will authorize the desired service. The client can then issue a second request for service with those credentials attached, and upon verifying the credentials, the server provides the desired service. Policy mobility has two significant advantages. First, it offloads from the server to the client the work of searching the client's credentials. Second, it enables trust to be established in the client without the client revealing irrelevant credentials.

3. When Credentials are Sensitive

A client wishing to do business with a new service may be unwilling to disclose sensitive credentials until some degree of trust has been established in that service. Current credential systems do not address credential sensitivity. The decision to disclose a sensitive credential to a new service is left up to a user at the client. More specifically, client-credential submission policies specify which credentials can be submitted with any request to a specified class of service and which credentials require explicit authorization before they are submitted. This mechanism requires a user be available to make trust decisions when new service classes are contacted. It does not address how the user decides to trust a service.

In [16], Winslett et al. recognize the potential to use server credentials to establish client trust, though they focused mainly on server trust in clients. They present detailed mechanisms for clients to submit credentials to servers. Each service provider establishes a policy governing access to its resources. Upon receiving a request with insufficient credentials attached—often the case on first access—a security agent representing the server sends the client an explanation of the relevant portions of the policy. The client's agent analyzes the server's policy to determine which credentials are needed to support the service request. It makes a decision on which credentials to send to the server (based on a predefined client submission policy and interaction with the user). The client agent then attaches the selected credentials to subsequent requests for the service.

Winslett et al. note the relevance of such machinery for the reverse scenario, in which servers encourage clients by presenting their own credentials, and a client may request credentials from a server. Such a reversal provides a good basis for clients establishing the trust in servers required before disclosing sensitive credentials. However, [16] is unclear about the details of how this kind of trust can be established. The current paper describes recent work on automating the establishment of trust between strangers through *incremental* exchange of sensitive credentials.

One straightforward approach that unfortunately does not work is as follows. The client establishes a policy identifying credentials required from the server prior to the client disclosing any credentials. The client sends this policy to the server as a counter request whenever it receives from the server a credential request, such as a SGP. The problem is that it would then be useless for the server to request client credentials before disclosing its own credentials, as doing so would introduce a cyclic dependence and deadlock. This simplistic approach fails because it governs all client credentials with the same policy, so each request by the server for client credentials leads to an identical counter request from the client.

4. Negotiation Architecture and Model

Section 4.1 presents a high-level architecture that supports fully automated trust negotiation between client and server. Section 4.2 then introduces an abstract model of trust negotiation. The model is used in Section 5 to formalize and analyze two negotiation strategies.

In both architecture and model, each credential is protected by a CAP that controls the credential's disclosure based on credentials presented by the other negotiation participant. Throughout, credentials are disclosed only in observance of these CAPs. Credential requests can also be exchanged to guide the credential exchange.

4.1. Trust Negotiation Architecture

Each of our negotiation participants is represented in trust negotiations by a *security agent* (SA), as in the simple negotiations of Ching et al. [5] and Winslett et al.

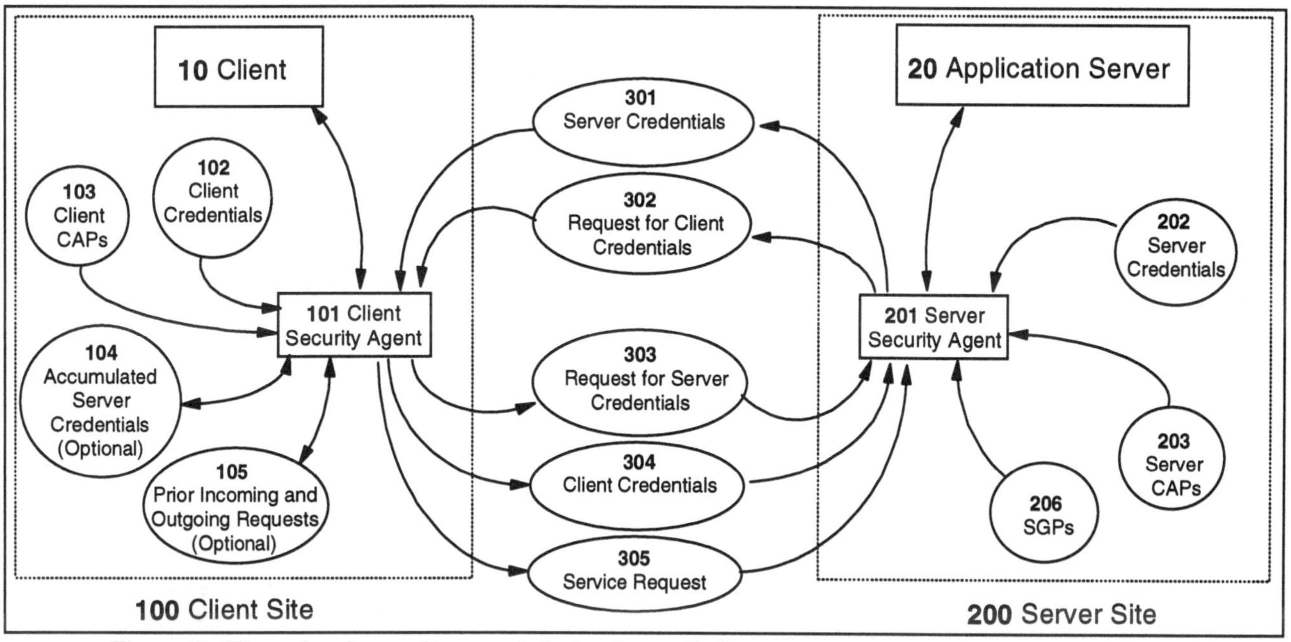

Figure 2. The role of security agents in trust negotiation.

[16]. The role of the SA is illustrated in Figure 2, which depicts the client security agent (101) and the server security agent (201) and several contextual factors that each SA considers during negotiation. The client SA manages the disclosure of client credentials (102) and the server SA manages disclosure of server credentials (202). Like any protected resource, each credential is governed by an access policy (103, 203) that has the same form as a SGP. The CAP identifies credentials from the other negotiation participant that would unlock disclosure of the local credential to that subject.

The client (10) initiates the trust negotiation by making a service request. The client SA intercepts the request and relays it to the server SA. The service (20) is accessible only via the server SA. Upon receiving a request for service (305), the server SA makes an authorization decision based on the appropriate SGP (206). When the client SA is familiar with the SGP, it can attach appropriate credentials (304) to the service request so that the service will be authorized. The server SA determines whether the credentials that arrive with the service request satisfy the SGP. If the policy is satisfied, the trust negotiation has completed successfully; the service is authorized and the request is forwarded to the application server, which provides the service to the client.

Initially, the client is unfamiliar with the SGP, so attaching satisfactory credentials to an initial service request is impractical. A trust negotiation strategy can overcome this problem by using mobile policies. When a server SA receives a request for service without sufficient credentials attached to satisfy the SGP, it sends the SGP to the client SA as a request for client credentials (302). The client SA can then select a combination of credentials that satisfies the SGP, and can attach those credentials (304) to a repetition of the original service request (305).

An important issue in this scenario not addressed in previous trust systems is how to enable the client SA to make independent trust decisions about which credentials to provide to an unfamiliar server. Our SAs use CAPs (103, 203) when selecting credentials to disclose. If the client SA cannot satisfy the SGP by using credentials whose CAPs are unprotected, it can, as the *negotiation instigator*, introduce further stages to the trust negotiation by requesting server credentials (303). These stages seek to build mutual trust through credential exchange, eventually to unlock client credentials that satisfy the SGP. Client credentials are unlocked by incoming server credentials (301); however, as an optimization, the client may also cache and use for this purpose server credentials it received in prior stages (104). (The abstract model of trust negotiation introduced in Section 4.2 does not capture this optimization.)

In each negotiation stage, the active subject responds to an incoming request for credentials either by providing credentials, by formulating a counter request for credentials (302, 303), or both. In some strategies, the client SA can also repeat one of its previous request (105) for credentials that has not yet been satisfied. By exchanging credentials and requests for credentials, the two subjects endeavor to establish trust required to authorize service. Eventually, either the negotiation succeeds or the client SA must abandon the attempt. The negotiation succeeds when the client SA satisfies the original SGP by disclosing sufficient unlocked credentials. At the same time, the client SA repeats the original service request (305), this time with sufficient credentials attached (304) to authorize service.

4.2. Trust Negotiation Model

In this section we introduce the abstract model used in Section 5 to define and analyze two negotiation strategies. That abstract model formalizes a trust negotiation as a sequence of credential disclosures that alternate between the two participants, optionally augmented by a sequence of credential requests that serve to guide the disclosures.

The participants in a trust negotiation are the client and server. Each owns a set of credentials, which we denote by ClientCreds and ServerCreds, respectively. Access to each credential c in ClientCreds or ServerCreds is governed by a policy, denoted $\mathsf{gov}_{\mathsf{client}}(c)$ or $\mathsf{gov}_{\mathsf{server}}(c)$, respectively. If a credential expression, ψ, is satisfied by a set of credentials C, we write $\mathbf{sat}(C, \psi)$. We write $\psi \equiv \psi'$ if for all credential sets C, $\mathbf{sat}(C, \psi)$ *iff* $\mathbf{sat}(C, \psi')$. If $C \subseteq$ ClientCreds and $c \in$ ServerCreds such that $\mathbf{sat}(C, \mathsf{gov}_{\mathsf{server}}(c))$, or if $C \subseteq$ ServerCreds and $c \in$ ClientCreds such that $\mathbf{sat}(C, \mathsf{gov}_{\mathsf{client}}(c))$, we write $\mathbf{unlocked}(c, C)$. If $\mathbf{unlocked}(c, \varnothing)$, c is *unprotected*. Lifting to sets of credentials, C', we write $\mathbf{unlocked}(C', C)$ if $\mathbf{unlocked}(c, C)$ holds for each $c \in C'$.

A *trust negotiation* is formalized by a sequence of credential disclosures, $\{C_i\}_{i \in [0,2n+1]} = C_0, C_1, \ldots, C_{2n+1}$, for some natural number n, that alternates between the two subjects, *i.e.*, $C_{2i} \subseteq$ ClientCreds and $C_{2i+1} \subseteq$ ServerCreds for all i, $0 \leq i \leq n$. Each disclosure corresponds to a message. The number of disclosures is even, modeling a sequence of roundtrips.

The credentials in each disclosure are required to be unlocked by credentials from the other negotiation participant in the previous disclosure, which means that the first disclosure consists entirely of credentials that are unprotected. That is, we have $\mathbf{unlocked}(C_0, \varnothing)$ and $\mathbf{unlocked}(C_{i+1}, C_i)$ for all $0 \leq i \leq 2n$. Any disclosure can be empty, provided the subsequent disclosure consists of unprotected credentials. (This possibility is exercised, for instance, when modeling client/server interactions where the first disclosure is done by the server or the last disclosure is done by the client.)

Both client and server may set *trust requirements* that a trust negotiation may or may not succeed in establishing. Trust requirements are formalized by credential expressions. The trust requirement of primary concern in this paper is the server's policy governing access to a service: the SGP. However, a client might also set a trust requirement that it enforces before doing business with a server.

A trust negotiation is *successful* with respect to a server-set trust requirement, ψ, if the client's final disclosure satisfies ψ, *i.e.*, $\mathbf{sat}(C_{2n}, \psi)$. A trust negotiation is *successful* with respect to a client-set trust requirement, ψ, if the server's final disclosure satisfies ψ, *i.e.*, $\mathbf{sat}(C_{2n+1}, \psi)$. When it is implicit whether ψ is client- or server-set, we just say that $\{C_i\}_{i \in [0,2n+1]}$ satisfies ψ. In negotiation strategies where they provide a goal that focuses the credential exchange, we call trust requirements *trust targets*.

In some trust negotiation strategies introduced in Section 5, disclosures are guided by *credential requests* that are also exchanged by the negotiation participants. Credential requests are formalized by a sequence of credential expressions that accompanies the trust negotiation and that has the same length as the sequence of disclosures. For a given trust negotiation, $\{C_i\}_{i \in [0,2n+1]}$, an accompanying sequence of credential requests has the form $\{\psi_i\}_{i \in [0,2n+1]}$.

5. Negotiation Strategy

In our trust negotiation architecture, the negotiation strategy determines the search for a successful negotiation. The strategy controls which credentials are disclosed, when they are disclosed, and which credentials are requested from the other subject to unlock local credentials. Successful trust negotiation is not always possible. One subject or the other may not possess needed credentials, or subjects may govern their credentials by policies that, together, impose cyclic dependencies. The strategy determines when the negotiation instigator—the client in our architecture—gives up on a negotiation.

Some desirable properties of negotiation strategies are as follows. A strategy should lead to a successful negotiation whenever one exists; that is, it should be *complete*. It should terminate with failure when success is impossible. Ideally, it should avoid disclosing any credentials that are not needed for the negotiation to succeed. Finally, a strategy should be reasonably efficient. We analyze the extent to which these properties are satisfied by using the abstract model defined in Section 4.2.

Within the context of the abstract model, we identify each negotiation strategy with a set of trust negotiations. This high level of abstraction focuses our attention on the essential relationships between CAPs and the disclosures and requests that flow between client and server SAs in each strategy. Some practical matters are not formalized here: what is an effective procedure for constructing counter requests; how to truncate negotiations early when success is impossible; and when do service requests flow? We discuss these matters informally.

This section defines and analyzes two negotiation strategies. The first, our *eager* strategy, is complete and efficient. However, it discloses many credentials needlessly. The second, our parsimonious strategy, begins by exchanging credential requests, but no credentials, exploring (backwards) all possible sequences of disclosures that lead to the satisfaction of a given trust target. The sequence of requests reaches a request that can be satisfied by unprotected credentials at exactly the point where the backwards exploration of fruitful disclosure sequences reaches the beginning of the shortest sequences. Starting at that point, the strategy performs a credential exchange that is minimal in length and that makes a locally minimal disclosure at each step. Unfortunately, because of limits on the degree of cooperation between the two participants in this strategy, a globally minimal disclosure is not guaranteed.

5.1. An Eager Strategy

In this simple strategy, two security agents take turns sending each other every credential they have that is currently unlocked. As credentials are exchanged, more credentials become unlocked. The client terminates a negotiation when it receives a set of credentials from the server that it has already received (no new credentials) or the set it receives unlocks no new client credentials.

Definition. A trust negotiation, $\{C_i\}_{i\in[0,2n+1]}$, is in the *eager strategy* if each C_i is the **maximal** set such that **unlocked**(C_0, $\varnothing$) and **unlocked**(C_{i+1}, C_i) for all $0 \leq i \leq 2n$ and $C_{i+2} \neq C_i$ for all $0 \leq i \leq 2n-2$. We call any negotiation in the eager strategy an *eager negotiation*.

The strategy is reasonably efficient: the number of credential exchanges is bounded by the number of credentials each participant has. A tighter bound is the length of the longest chain of dependencies among credentials possessed by the two participants.

Claim 1 (***Efficiency of eager negotiation***)**:** The length, $2n+2$, of any eager negotiation, $\{C_i\}_{i\in[0,2n+1]}$, is at most 2×**min**(|ClientCreds|, |ServerCreds|).

Claim 2 (***Completeness of eager negotiation***)**:** For any credential expression, ψ, if there exists any trust negotiation satisfying ψ, then there exists an eager negotiation satisfying ψ.

For a given target credential expression, such as the SGP of the service, it is not always necessary to perform an entire eager negotiation. To facilitate this optimization, one may choose to have the server return the SGP as ψ_1. One may modify the eager strategy slightly, making the first two disclosures empty, allowing the service request and SGP to flow first. After those first two messages, if the client has unprotected credentials that satisfy the SGP, it discloses them with a repeat of the service request. If the client does not have such credentials, it terminates the negotiation without sending any credentials. If the client has satisfactory but locked credentials, it continues the negotiation by sending its unlocked credentials and requesting the server's unlocked credentials, checking at each round whether the SGP can now be satisfied with its unlocked credentials. Eventually, either the SGP can be satisfied—in which case the client resends the service request along with the required credentials—or the client determines that negotiation has failed.

An alternative deployment does not use mobile policies: the SGP is never sent to the client. With each of its disclosures, the client repeats the service request, until either service is granted, or no new credentials are disclosed and the negotiation fails.

The strengths of the eager strategy are its simplicity and the fact that no information about policies or credentials is disclosed except where CAPs are satisfied. Its weakness is that it discloses credentials without regard to their relevance to the present negotiation: there is no provision for disclosing on a need-to-know basis.

5.2. A Parsimonious Strategy

Eager negotiations begin exchanging credentials essentially immediately. They make little or no use of credential requests, since all unlocked credentials are disclosed without regard for those requests. This section defines and analyzes the *parsimonious strategy,* which differs from the eager strategy in these respects. An intuitive explanation precedes the formal definition. The numbering in that intuitive explanation corresponds to the numbering in the formal definition below.

1. Requests are exchanged to guide the negotiation toward satisfying a particular trust target, ψ. A trust target is a specific credential expression, such as a SGP or a trust requirement set by the client. To simplify the presentation, we assume the trust target is a SGP. The specification and ensuing results can easily be generalized to cover the case of a trust target set by the client as well. Under this assumption, the first credential request from the server is the trust target.
2. When and if a request is sent that can be satisfied by unprotected credentials, those credentials are disclosed in the next stage.

3. Initial credential disclosures are all empty, until a credential request occurs that can be satisfied by unprotected credentials. If no request is ever sent that can be satisfied by unprotected credentials, the negotiation terminates without disclosing any credentials.
4. Each successive credential request is derived from its predecessor in a manner that makes satisfying that request a necessary and sufficient condition for a disclosure to unlock credentials that satisfy the predecessor.
5. After a request is satisfied by a disclosure of unprotected credentials, the client then resends its prior requests, going through them backwards, at the same time disclosing appropriate credentials to unlock solutions to those requests.
6. As mentioned above in point 2, when and if a request is sent that can be satisfied by a set of unprotected credentials, a **minimal** such set is disclosed in the next stage. Each successive step also discloses a minimal credential set that satisfies a credential request, working backwards through the requests that occurred thus far. The client, which drives the negotiation, will have recorded each of the requests that has flowed. It refers to requests it received from the server when selecting credential disclosures; it resends the requests it sent to the server, as outlined in point 5 above. Each disclosure unlocks the next, until a disclosure satisfying the original trust target is unlocked.

Definition. Let the target credential expression be ψ. To be a *parsimonious negotiation with respect to the trust target*, ψ, a trust negotiation, $\{C_i\}_{i\in[0,2n+1]}$, must be accompanied by a sequence of credential requests, $\{\psi_i\}_{i\in[0,2n+1]}$ and must satisfy the following:

1. $\psi=\psi_1$.
2. If there exists a j, $0 \le j \le 2n+1$, and a C, $C \subseteq$ ClientCreds or $C \subseteq$ ServerCreds, such that **sat**(C, ψ_j) and **unlocked**(C, $\varnothing$), then let k be the least such j and fix $\underline{C}$ to be a **minimal** corresponding C. Otherwise, let $k = 2n+1$.
3. For all i, $0 \le i \le k$, $C_i = \varnothing$.
4. For all i, $0 \le i < k$, ψ_i and ψ_{i+1} have the following relationship:

   For all C, with respectively $C \subseteq$ ClientCreds or $C \subseteq$ ServerCreds, we have **sat**(C, ψ_{i+1}) *iff* there exists a C', with respectively $C' \subseteq$ ServerCreds or $C' \subseteq$ ClientCreds (opposite of C), such that **sat**(C', ψ_i) and **unlocked**(C', C).
5. Prior client requests (which have even indexes) are replayed. If k is even, we require $\psi_{k+j} = \psi_{k-j}$ for even j, $2 \le j < 2n-k$ (if any). If k is odd, we require $\psi_{k+j} = \psi_{k-j}$ for odd j, $1 \le j < 2n-k$ (if any).
6. If $k < 2n+1$, we require that $C_{k+1} = \underline{C}$, and for all j, $1 \le j < 2n-k$, (if any) C_{k+j+1} is a **minimal** set satisfying **sat**(C_{k+j+1}, ψ_{k-j}).

In the following discussion, ClientCreds, ServerCreds, gov_{client}, and gov_{server} are fixed but arbitrary.

Claim 3 (***Determinacy of Requests in Parsimonious Negotiation***): Given two different parsimonious negotiations with the same trust target and with associated credential requests given by $\{\psi_i\}_{i\in[0,2n+1]}$ and $\{\psi'_i\}_{i\in[0,2m+1]}$, we have $\psi_i \equiv \psi'_i$ for all i, $1 \le i \le k$, where k is either the least index such that ψ_k can be satisfied by unprotected credentials, $2n+1$, or $2m+1$, whichever is least.

Claim 4 (***Correctness of Parsimonious Negotiation***): Let $\{C_i\}_{i\in[0,2n+1]}$ be a parsimonious negotiation with respect to trust target ψ with associated credential requests given by $\{\psi_i\}_{i\in[0,2n+1]}$. If there exist k and C with $1 \le k \le 2n+1$ and ($C \subseteq$ ClientCreds or $C \subseteq$ ServerCreds) such that **sat**(C, ψ_k) and **unlocked**(C, $\varnothing$), then there exists a parsimonious negotiation $\{C'_i\}_{i\in[0,2m+1]}$ that satisfies ψ and that is the same as $\{C_i\}_{i\in[0,2n+1]}$, except for possibly being longer.

Claim 5 (***Completeness and Efficiency of Parsimonious Negotiation***): If there exists a trust negotiation that satisfies ψ, then (1) every sufficiently long parsimonious negotiation with trust target ψ also satisfies ψ and (2) no parsimonious negotiation with trust target ψ has length greater than 4×**min**(|ClientCreds|, |ServerCreds|).

Any deployment of the parsimonious strategy should take advantage of the fact that, if a successful negotiation exists, the initial exchange of credential requests will encounter a request that can be satisfied by unprotected credentials within the first $2m$ requests, where m = |ClientCreds|. If such a request has not occurred, the negotiation should be terminated.

An issue faced deploying this strategy is effective derivation of credential requests that satisfy the requirements stated in the construction. As long as the credential language is able to represent logical "and" and "or," these counter requests can be constructed. A "brute force" approach constructs counter requests as follows. (A more efficient approach for the credential expression language presented in Section 6 can be constructed. It is not presented here due to space constraints.) Assume that $\bigwedge$ and $\bigvee$ denote "and" and "or," *i.e.*, that for all C, we have **sat**(C, $\bigwedge\{\psi_0, \psi_1, \ldots, \psi_m\}$) *iff* **sat**($C$, ψ_i) for each i, $1 \le i \le m$ and **sat**(C, $\bigvee\{\psi_0, \psi_1, \ldots, \psi_m\}$) *iff* **sat**($C$, ψ_i) for some i, $1 \le i \le m$. Counter requests can be computed by the server from the incoming request by using $\text{counter}_{\text{server}}(\psi) = \bigvee\{\text{gov}_{\text{server}}(C) \mid \textbf{sat}(C, \psi)\}$, where $\text{gov}_{\text{server}}(C) = \bigwedge\{\text{gov}_{\text{server}}(c) \mid c \in C\}$. Counter requests can be computed analogously by the client.

Example. We illustrate the parsimonious strategy by ignoring several issues. We ignore the fact that credentials are submitted with supporting credentials. We ignore the internal structure of credentials. We use an unrealistic credential expression language that treats credentials as propositional variables.

Let ClientCreds=$\{a, b, c, d\}$, ServerCreds=$\{x, y\}$, $\text{gov}_{\text{client}}=\{a\mapsto\emptyset, b\mapsto\emptyset, c\mapsto x, d\mapsto y\}$, and $\text{gov}_{\text{server}}=\{x\mapsto a\vee b, y\mapsto a\vee b\}$. In the exchanges, we use $\perp$ to represent values that are not determined or used by the parsimonious negotiation. Recall that ψ_1 represents the trust target—an SGP: $\psi_0=\perp$, $\psi_1=(a\wedge d)\vee(c\wedge b)$, $\psi_2=x\vee y$, $\psi_3=a\vee b$, $\psi_4=x\vee y$, $\psi_5=\perp$, $\psi_6=\perp$, $\psi_7=\perp$; $C_0=\emptyset$, $C_1=\emptyset$, $C_2=\emptyset$, $C_3=\emptyset$, $C_4=\{a\}$, $C_5=\{x\}$, $C_6=\{b,c\}$, $C_7=\perp$. The total set of credentials disclosed by the client in this negotiation is not minimal, because the client could have disclosed b in C_4 instead of a. However, if the client had tried this, the server choice in C_5 could instead have been to disclose $\{y\}$, forcing the client to disclose $\{a,d\}$ in C_6, which again is not minimal. This example illustrates the difficulty of devising a negotiation strategy in which the two participants cooperate in exchanging a globally minimal set of credentials.

5.3. Hybrid Strategies

This section sketches informally an approach to combining eager and parsimonious strategies in an effort to use each with the credentials for which it is better suited. CAPs can be made two-part, comprising not only a credential expression, but also a flag to select between parsimonious and eager disclosure. A credential flagged for eager disclosure would be disclosed freely to all sites that present credentials that satisfy the credential-expression component of its CAP. One flagged "parsimonious" would be disclosed only as part of a locally minimal exchange and successful negotiation. A hybrid strategy begins with a phase that uses an eager strategy to attempt to negotiate using only credentials flagged for eager disclosure. If success is possible using just those credentials, the negotiation succeeds during the eager phase. If not, phase two uses a parsimonious strategy to attempt to establish trust by using all credentials. Phase two takes advantage of credentials exchanged during phase one. In a hybrid negotiation, the client determines when phase one has completed unsuccessfully and phase two begins. The client indicates in each request to the server which strategy it is currently employing.

6. Credential Expression Languages

Credential expressions could specify combinations of credentials directly, for instance by representing credential requirements as Boolean combinations of specific credentials. However, by layering levels of abstraction, we aim to provide a basis for overcoming the complexity of authenticating strangers by using credentials obtained for other purposes. Deriving relevant properties from such credentials requires care and expertise. It is important to separate this task from that of defining access requirements for individual services and credentials. For this reason, our credential expression language comprises two parts, separating authentication from authorization. For authentication, credentials are mapped to roles; for authorization, access requirements are expressed as combinations of roles.

The language we present is incomplete and not a formal language-design proposal. However, two language features, introduced and discussed in this section, make or reiterate important points that will guide further work in this area. First, role attributes enhance expressiveness. Second, the monotonic relationship between credentials and access is essential in any context where a subject can withhold disclosure of its own credentials.

This section presents our *Role-based Authorization Language* (*RAL*) used in the example negotiation presented in Section 7. An *authorization policy* defined in RAL consists of a *role-constraint expression*, which expresses requirements for access to the service or credential that it governs. These requirements are expressed in terms of roles of the subject seeking access. These roles are defined by an *authentication policy* written in our *Property-based Authentication Language* (*PAL*), which is also presented in this section. A PAL authentication policy assigns a subject to roles based on subject properties derived from credentials owned by the subject and from the roles of the issuers of those credentials. This assignment is independent of the question of the subject's access to the service. Thus, a PAL role is not a capability, but represents a derived property of the subject.

Section 6.1 introduces PAL, which maps credentials to roles. Section 6.2 then introduces RAL, which uses those roles to define access requirements. Section 6.3 discusses the significance and impact of supporting role attributes.

6.1. Property-based Authentication Language

A PAL authentication policy defines one or more roles. A role is a property of subjects defined in terms of the credentials they possess and the attribute values of those credentials. The notion of assigning subjects to

```
hasCredit(Amount) ← LetterOfCredit : credit,
    creditor(LetterOfCredit.issuer), Amount = LetterOfCredit.amount.
creditor ← Creditor : creditor, self(Creditor.issuer).
knownClient() ← Ref : reference, self(Ref.issuer),
    Ref.relationship = "shipping client".
knownClient() ← Ref1 : reference, Ref2 : reference,
    Ref1.issuer ≠ Ref2.issuer, shipper(Ref1.issuer), shipper(Ref2.issuer),
    Ref1.relationship = "shipping client", Ref2.relationship = "shipping client".
shipper() ← Ref : reference, self(Ref.issuer), Ref.relationship = "shipper".
shipper() ← Ref1 : reference, Ref2 : reference, Ref1.issuer ≠ Ref2.issuer,
    knownClient(Ref1.issuer), knownClient(Ref2.issuer),
    Ref1.relationship = "shipper", Ref2.relationship = "shipper".
```

Figure 3. Six PAL Rules

roles based on credentials goes back to Bina et al. [1]. In Seamons et al. [13] policies are Prolog programs. However, with the exception of a few features, PAL is based on the Trust Policy Language (TPL) of Mass et al. [10]. TPL introduced two intertwined innovations, which we adopt: the role is TPL's sole procedural abstraction; TPL assigns a role not only to the submitting subject, but also to each owner of one of the supporting credentials.

PAL's role attributes enhance TPL, as discussed in Section 6.3. PAL also differs from TPL in that roles are *monotonic* functions of credentials: disclosing more credentials cannot decrease the roles to which a subject authenticates. This is a requirement in trust negotiation because subjects have control over the credentials they disclose, but not the credentials they obtain.

For presentation purposes, we give PAL a concise notation derived from syntax of constraint logic programming languages. A PAL role is a user-defined predicate over the public keys of subjects. Role attributes take the form of additional parameters. A PAL authorization policy consists of a set of *rules*, each of which has the following form:

role(Attributes, ...) ← CredentialVar :
credentialType, ...,
role$_1$(credentialVar.issuer, Attributes$_1$), ...,
credentialConstraints, ...

The *head* of the rule, to the left of the arrow, consists of a single role, applied to an implicit subject key and to zero or more explicit role attributes, illustrated here by **Attributes**. Together, the rules with the same role in their head define that role. The *body* of the rule comprises the portion of the rule to the right of the arrow. One or more credential variables are introduced, each by a variable type declaration, as illustrated by **CredentialVar : credentialType**. The attributes of a credential are denoted with the syntax **CredentialVar.attribute.** Credential variables and role attributes have names that start with capitol letters. Roles, credential types, and credential attributes have names that start with lowercase letters.

Within a given rule, the subject being categorized owns each credential denoted by a credential variable in that rule. The rule states that the subject can authenticate to the role **role** with role attributes **Attributes** if the subject owns a combination of credentials of the designated types such that each of the rule's requirements holds. These requirements are of two kinds. The first requires that the issuer of a credential belongs to a given role. Such an *issuer-role constraint* is an application of a role to a credential-issuer key and role attribute variables, and is illustrated by **role$_1$(CredentialVar.issuer, attributes$_1$)**. Note that the subject key is identified explicitly in role applications that appear in a rule body. Each credential variable must appear in exactly one issuer-role constraint. The second kind of requirement is a *general constraint* on the attributes of the credentials and roles appearing in the rule. These are typically relational expressions, using operators such as =, ≠, and ≤. Although not specified in this paper, user defined functions and operators may also be used. In addition to imposing requirements on the credentials denoted by the credential variables, general constraints serve to define the attributes of the role in the rule head.

Figure 3 shows six PAL rules. The first rule defines one way a subject can be shown to be in the **hasCredit** role. If the full authentication policy contained other rules that defined **hasCredit**, those rules would state other ways to authenticate to the role. The rule here states that a subject is in the **hasCredit** role if it owns a **credit** credential issued by a subject in the **creditor** role. The rule also states that role attribute, **Amount**, is defined by credential attribute, **amount.**

To be satisfied, the rule must be evaluated in the presence of the **credit** credential mentioned above, as well as credentials, owned by the **credit** credential's issuer, that satisfy a rule defining the **creditor** role. The

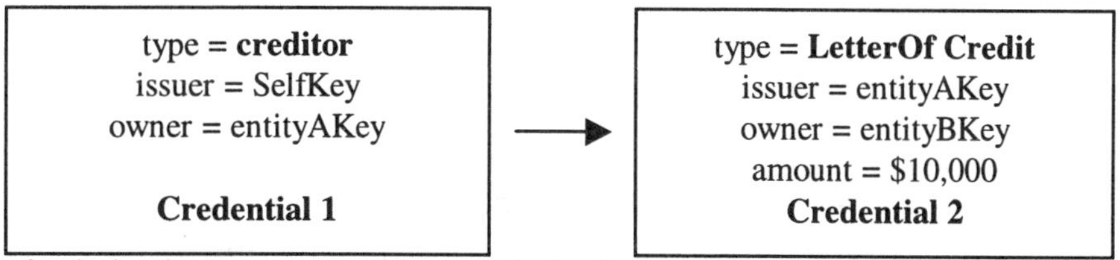

Figure 4. A chain that proves subject A is in the **creditor** role and subject B is in the **hasCredit** role, as those roles are defined in Figure 3

reputation(Rating) ← Membership : businessOrgMember,
    businessOrganization(Membership.issuer), Rating = Membership.rating.
businessOrganization()← ....
clientWithAccount(AccountNumber) ← Account : account,
    self(Account.issuer), AccountNumber = Account.number.
newShippingClient(Pickup, Delivery) ← Destiniation : contract, Warehouse : lease,
    knownBusiness(Destination.issuer), warehouseOwner(Warehouse.issuer),
    Pickup = Warehouse.location, Delivery = Destination.deliveryLocation.
knownBusiness()← ....
warehouseOwner()← ....

Figure 5. Role definitions that, together with those in Figure 3, compose the example's authentication policies. We present the authentication policies defined by the client and the server together here, as some rules are used by both subjects. Note that three rules are incomplete. Their content and the credentials needed to satisfy them are elided from the example presentation to save space.

second rule shown in Figure 3 is such a rule. It states that a subject is in the **creditor** role if it owns a **creditor** credential issued by a member of the **self** role. The **self** role is unlike other roles. It is predefined and its sole member is the owner of the policy, referred to as *self.* A credential issued by the policy owner is a *self-signed credential*, as introduced by Mass et al. Unlike other credentials, self-signed credentials are part of an authentication policy; they define the valid roots of all credential chains. A chain of credentials that supports membership in the **hasCredit** role is shown in Figure 1

By making use of the TPL notion of role as a procedural abstraction mechanism, an authorization policy can trace chains of unbounded length [10]. Rules 3 through 6 in Figure 3 define two roles, **knownClient** and **shipper**. A subject is in the **knownClient** or **shipper** role if it has a corresponding self-signed **reference** credential from the policy owner. A subject is also in the **knownClient** role if it has a reference from two **shipper** members. Similarly, a subject is also in the **shipper** role if it has a reference from two **knownClient** members.

To summarize, an authentication policy has three parts: a set of rules, a set of self-signed credentials, and a set of user defined functions. In this paper we generally exhibit only the rules. However, in any mobile policy, all three parts must flow.

6.2. Role-based Authorization Language

This section introduces the Role-based Authorization Language (RAL). A RAL authorization policy comprises a role-constraint expression, together with a PAL authentication policy defining each role used therein. A role-constraint expression is a logical formula consisting of role applications and attribute constraints combined with the logical connectives AND and OR. (These connectives give exactly the monotonic formulas discussed above.) A role application consists of a role name applied to a subject (in the example, just one of the reserved words **Client** or **Server**) and role attribute variables. Attribute constraints are arbitrary constraints on the role attributes appearing in the rule. An authorization policy that governs a service may also use service parameter names in these constraints. As in PAL, user defined functions may be used.

Example. The following authorization policy could govern a request for a service that schedules shipping. The parameters of the request are **PickupLocation**, **Destination**, and **Tons**.

hasCredit(Client, Amount) AND
    Amount ≥ Tons ×
    costPerTon(PickupLocation, Destination)

This policy requires that the client be in the **hasCredit** role with the **Amount** attribute value at least sufficient to

clientWithAccount(Client, AccountNumber) OR
(reputation(Client,Rating) AND Rating ∈ {good, excellent} AND
(knownClient(Client) OR
(newShippingClient(Client, Pickup, Delivery) AND
PickupLocation = Pickup AND Destination = Delivery)) AND
hasCredit(Client, Amount) AND Amount≥Tons×costPerTon(PickupLocation, Destination))

Figure 6. Policy governing the service, "Schedule Shipping." Parameters of the service request include: PickupLocation, Destination, and Tons.

Client Credentials:

type = contract
deliveryLocation = Vancouver, BC
issuer = Widget Outlet
owner = Acme Widget
Contract

type = lease
location = Hong Kong
issuer = Port of Hong Kong
owner = Acme Widget
Warehouse

type = credit
amount = $15,000
issuer = Mattress Bank
owner = Acme Widget

type = businessOrgMember
rating = excellent
issuer = Better Widget Bureau
owner = A1 Shipping
B-Org-C

Server Credentials:

type = reference
relationship = shipper
issuer = manufacturer 1.
owner = A1 Shipping
Ref-1

type = reference
relationship = shipper
issuer = manufacturer 2.
owner = A1 Shipping
Ref-2

type = businessOrgMember
rating = good
issuer = Intl. Shippers Assoc.
owner = A1 Shipping
B-Org-S

Figure 7. The credentials owned by each negotiation participant. Each credential is labeled by a name used to refer to it in the example.

pay the cost of the desired shipping. The function **costPerTon** is user defined.

When an authorization policy is combined with an authentication policy that defines all the roles it uses, the two together define a set of *solutions*. Each solution is a minimal set of credentials that proves that the subject in question (client or server) belongs to the required roles with the required attributes.

6.3. Role Attributes

The association between attributes and roles enables the policy writer to make credential attributes, exported as role attributes, available in authorization policies. This enables authorization to depend on attributes—such as credit line, rating, or age—without the inconvenience of defining special purpose roles that impose the constraints within the authentication policy.

Role attributes also have at least two other advantages in the context of TPL-style role definitions. First, by exporting credential attributes, they enable attributes of different credentials in a chain to be compared. For instance, they could be used to compare attributes of a credential owner with attributes of the credential's issuer. Second, role attributes can be used to keep track of and to bound the length of a credential chain. Consider the **shipper** role defined in Figure 3. While it is very convenient to let shippers and shipping clients vouch for each other, as the number of subjects in a chain goes up, the policy writer's confidence in the last subject's role may go down. An attribute, **ChainLength**, can be added to each of **shipper** and **knownClient**. By initializing it to zero in the rules that use self-signed credentials, and incrementing it in each of the (mutually) recursive rules, the length can be computed. An authorization policy can then easily impose an upper bound on the length of acceptable chains.

7. Example Negotiations

This section presents two example trust negotiations based on the eager and parsimonious strategies. In the

Client Credential Access Policies:

Contract$_C$ ≡ shipper(Server) AND reputation(Server, Rating) AND Rating ∈ {good, excellent}
Credit$_C$ ≡ reputation(Server, Rating) AND Rating ∈ {good, excellent}
Warehouse$_C$ ≡ shipper(Server)
B-Org-C$_C$ ≡ true

Server Credential Access Policies:

Ref-1$_S$ ≡ reputation(Server, Rating) AND Rating ∈ {good, excellent}
Ref-2$_S$ ≡ reputation(Server, Rating) AND Rating ∈ {good, excellent}
B-Org-S$_S$ ≡ true

Figure 8. The CAPs that govern each credential in the example. Only the authorization portion of each policy is shown. The authentication portion comprises the relevant rules from Figure 3 and Figure 5. The policy by which the client and server, respectively, govern a credential is designated by the credential's name with subscript C and S, respectively.

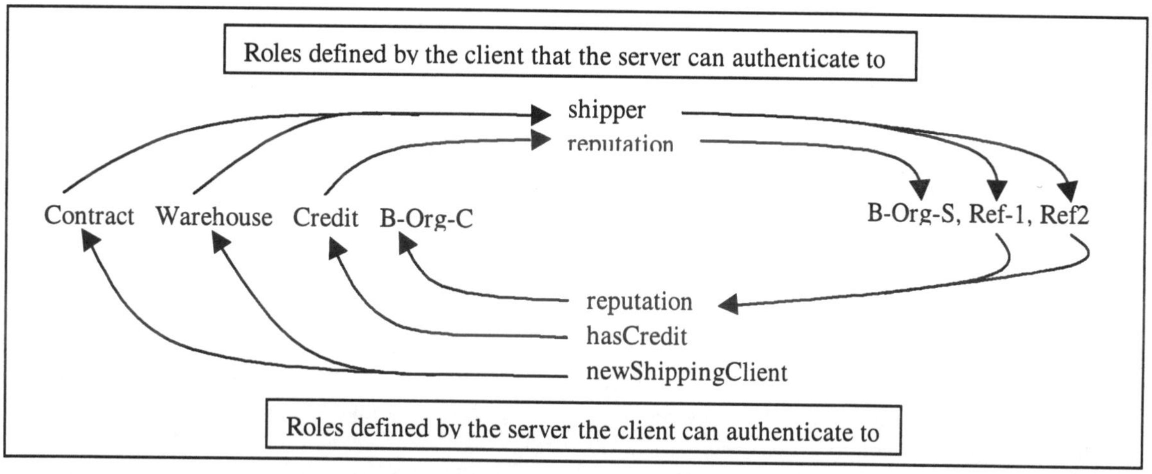

Figure 9 A graph showing the client's credentials to the left and the server's credentials to the right. Incoming arcs to the credentials show the dependency to roles the credential owner can authenticate to using the credentials. Outgoing arcs show the dependency to roles that govern access to the credential. Credentials with no outgoing arc are freely available. Only the roles actually used in the example negotiation are shown.

examples, a negotiation is undertaken to authorize on-line scheduling of shipping services by a small, fictitious widget manufacturer, Acme Widget. The negotiation successfully establishes the trust required to authorize the service request and schedule shipping.

Figure 3 and Figure 5 present the role definitions used by the authentication portions of the SGP and CAPs, as well as the requests for credentials that may flow during the negotiation. To save space, we list each rule only once, although it may have to appear in several places in the various policies and requests. Although not shown, the authentication policies include necessary self-signed credentials. In particular, the server's authentication policies that define **creditor** also include a self-signed **creditor** credential owned by Mattress Bank. Similarly, the client's authentication policies that define **knownClient** also include two self-signed **reference** credentials, one owned by manufacturer 1 and one by manufacturer 2, both with **relationship = "shipping client"**.

Figure 6 presents the authorization policy that governs the service requested. Figure 7 shows the client and server credentials. Figure 8 shows the CAPs of each of those credentials. Figure 9 shows the dependencies between credentials, roles governing access to those credentials, and credentials that authenticate their owner to a given role.

For the negotiation based on the eager strategy, the six stages of the negotiation are described below. Following the initial request for service in stage 1, each stage shows the unlocked credentials that are disclosed.

Stage 1. Client sends (via client SA) request to schedule shipping 5 tons of cargo.

Stage 2. Server SA sends **B-Org-S** to client.

Stage 3. Client SA repeats request to schedule shipping and sends **B-Org-C**, **Credit** to server.

Stage 4. Server SA sends **B-Org-S**, **Ref**$_1$, and **Ref**$_2$.

Stage 5. Client SA repeats service request and sends **B-Org-C**, **Credit, Contract,** and **Warehouse**.

Stage 6. Server SA receives service request and authorizes it, as attached credentials satisfy the SGP.

For the negotiation based on the parsimonious strategy, the eight stages of the negotiation are described below. For each stage where an incoming request for credentials is received, the description shows the solution to the request (a list of local-site credentials) that is selected by the SA. Credentials that are locked appear underlined in the solution. Counter requests combine the CAPs of these locked credentials

Stage 1. Client sends (via client SA) request to schedule shipping 5 tons of cargo.

Stage 2. Server SA returns the SGP.

Stage 3. Client SA's solution to incoming request: **Contract**, **Warehouse**, **Credit**. Outgoing request for credentials: **reputation(Server, Rating) AND Rating** $\in$ **{good, excellent} AND shipper(Server).**

Stage 4. Server SA's solution to incoming request: **B-Org-S**, **Ref**$_1$, **Ref**$_2$. Outgoing request for credentials: **reputation(Client, Rating) AND Rating** $\in$ **{good, excellent}.**

Stage 5. Client SA's solution to incoming request: **B-Org-C.** Outgoing request for credentials (repeating request sent in Stage 3): **reputation(Server, Rating) AND Rating** $\in$ **{good, excellent} AND shipper(Server).** Outgoing credentials (which solve the request sent by server SA in Stage 4): **B-Org-C.**

Stage 6. Server SA's solution to incoming request: **B-Org-S**, **Ref**$_1$, **Ref**$_2$. Outgoing request for credentials: **None.** Outgoing credentials: **B-Org-S, Ref**$_1$, **Ref**$_2$

Stage 7. Client SA repeats service request from Stage 1, attaching credentials that solve the SGP sent by the server SA in Stage 2. Outgoing credentials: **Contract**, **Warehouse**, **Credit.**

Stage 8. Server SA receives service request and authorizes it, as attached credentials satisfy the SGP.

8. Related Work

Credential-based authentication and authorization systems can be divided into three groups: identity-based, property-based, and capability-based systems. The original, public key certificates, such as X.509 [19] and PGP [17], simply bind keys to names (although X.509 version 3 certificates later extended this binding to general properties). Such certificates form the foundation of identity-based systems, which authenticate a subject's identity or name and use it as the basis for authorization. Identity is not a useful basis for our aim of establishing trust among strangers. Property-based credentials were introduced by Bina et al. to incorporate the binding of arbitrary attributes. Trust establishment between strangers can be based on the properties of the subjects without requiring familiarity with the actual subjects. Systems have emerged recently that use property-based credentials to manage trust in decentralized, distributed systems [6][10][13][16]. Capability-based systems [2][3][4][18], discussed further below, manage delegation of authority to operate on a particular application. Capability-based systems are not designed for establishing trust between strangers, since clients are assumed to possess credentials that authorize specific actions with the application server.

Winslett et al. [16] focus on establishing trust between strangers. They present an architecture for using credentials to authorize access to distributed resources. Client and server security assistants manage both the credentials and the policies governing access to sensitive resources. They emphasize the need for credential and policy exchange with little intervention by the client. Seamons et al. [13] continue in this vein, developing policies written in Prolog that use credentials and credential attributes to authenticate clients to roles that have attributes, which can be used in authorization decisions. This work addresses credential sensitivity by using mobile policies to support private client selection of credentials to submit for authorization.

Johnston et al. [6] use both attribute certificates (property-based credentials) and use-condition certificates (policy assertions) to determine access control. Use condition certificates enable multiple, distributed stakeholders to share control over access to resources. In their architecture, the policy evaluation engine retrieves the certificates associated with a user to determine whether all use conditions are met. The certificates are assumed not to be sensitive.

The Trust Establishment Project at the IBM Haifa Research Laboratory [10][11] has developed a system for establishing trust between strangers according to policies that specify constraints on attribute-contents of public-key certificates. Servers use a collector to gather supporting credentials from issuer sites. The system assumes that credentials are not sensitive. The companion Trust Policy Language (TPL) is a special-purpose logic programming language, with XML syntax, that maps certificate holders to roles. TPL policies also map the issuers of each supporting credential to a role. These roles can be used by existing role-based access control mechanisms. The example negotiations presented in Section 7 use a simplified language based on TPL.

The capability-based KeyNote system of Blaze et al. [2][3][4] manages delegation of authority. A KeyNote credential is an assertion that describes the conditions under which one principal authorizes actions requested by other principals. A *policy* is also an assertion that delegates authority on behalf of the associated application to otherwise untrusted principals. Thus an application's policy defines the root of all delegation chains. KeyNote credentials express delegation of authority in terms of actions that are relevant to a given application. KeyNote policies do not interpret the meaning of credentials for the application. This is unlike policies designed for use with property-based credentials, which derive roles from credential attributes, or otherwise bridge the divide between the application and credentials that were issued for unrelated purposes. The IETF Simple Public Key Infrastructure [18] uses a similar approach to that of KeyNote by embedding authorizations directly in certificates.

SSL [6], the predominant credential-exchange mechanism in use on the web today, and its successor TLS [6][7], support credential exchange during client and server authentication. There is no opportunity for the server to authenticate any information about the client before disclosing the server's credential. That is, sensitive server credentials cannot be protected. Furthermore, if the credential disclosed by the server does not satisfy the client, the client has no opportunity to request additional credentials from the server. This presents a serious problem when the client and server are strangers: it is unlikely that any single issuer would be an acceptable authority on all server attributes of interest to all potential clients.

9. Conclusions and Further Work

We have presented a model and an architecture for incrementally establishing mutual trust between clients and servers. The architecture solves the problem of establishing trust sufficient to allow disclosure of sensitive credentials that must be exchanged to authorize application-level operations.

We have discussed two negotiation strategies, one eager and one parsimonious. The eager strategy negotiates efficiently, succeeding whenever possible. It does not exchange credential requests, or otherwise attempt to minimize credential disclosures. There is an advantage in this: credential requests can disclose sensitive information. Credential requests exchanged in the parsimonious strategy can reveal a great deal about which credentials—and properties—a subject has. Although the eager strategy may reveal a credential unnecessarily, it does so only in accordance with the CAP.

The parsimonious strategy conducts a minimal-length exchange in which each disclosure is a locally minimal set. It does this by conducting an exchange of credential requests that in effect considers every possible successful exchange. It remains open how to ensure that the union of each participant's disclosures is minimal. Achieving global minimality remains a goal.

The strategies presented here assume that both participants cooperate in using the strategy. Further research is required to determine whether and how that assumption can be relaxed. A parsimonious negotiation guarantees locally minimal credential disclosure only when both parties "bargain in good faith." This means that each SA assumes the following about the other SA. When the other SA responds to an incoming request by issuing a counter request, if that SA subsequently receives credentials that satisfy the counter request, together with a repetition of the original request, it will return credentials that satisfy that original request. One advantage of a hybrid negotiation strategy, such as the one sketched in Section 5.3, is that the eager phase could be used to establish trust that the other negotiation partner will bargain in good faith before entering a parsimonious negotiation.

Further research is also needed in dimensions of credential-governing policy that determine or influence negotiation strategy. In addition to the credential expression whose satisfaction unlocks credential disclosure, additional policy content may assist in determining credential disclosures and requests. Two orthogonal issues that a credential-governing policy might address are as follows. The first is whether voluntary credential disclosure is permitted whenever the credential is unlocked or whether the credential must first be explicitly requested (presumably implying that its disclosure is necessary for successful negotiation). The second is whether it is permitted to issue counter requests when the credential is requested, potentially disclosing that the credential is held without disclosing the credential itself. The hybrid strategy presented in Section 5.3 assumes that each credential either can be voluntarily disclosed or can have a counter request issued in an effort to unlock it. A natural generalization captures the other two possibilities: that a credential (1) can *both* be voluntarily disclosed and have counter requests issued and (2) can *neither* be voluntarily disclosed nor have counter requests issued. It seems intuitive that some credentials would naturally fall into each of these categories, based, for instance, on whether possession or contents of the credential were more sensitive, and the reasons for the sensitivity.

The potential impact of these policy concepts on hybrid negotiation strategies needs to be examined further. Credentials in the "both" category might enable flexible and focused strategies. Strategies are needed that manage disclosure of credentials whose highly sensitive nature requires the "neither" category. These credentials should only be disclosed to a partner that requests them and that simultaneously provides sufficient credentials to unlock them. For instance, this could be accomplished by a hybrid strategy with an eager negotiation phase followed by a single request for the most sensitive credentials.

Another policy generalization would group credentials into ordered sensitivity classes. This captures a notion of relative sensitivity not captured by other policy constructs above. Strategies could prefer disclosing credentials in lower sensitivity classes.

The architecture presented here addresses some, but not all, of the client's trust needs. It addresses the client's need for trust that enables it to disclose credentials to the server. It does not address the client's need for trust before it requests service. The architecture could assist in negotiating the server's disclosure of credentials that would establish that trust. However, further work is needed to address the question of how the client formulates its request for those credentials. Such a request could be based, for instance, on transaction type or on the contents of service request parameters.

In addition to the above, ongoing work also includes deployment of trust negotiation in a demonstration prototype that uses the eager strategy to negotiate trust between Web clients and servers.

Acknowledgements

The authors wish to thank Marianne Winslett and Sushil Jojodia for their suggestions for improving earlier drafts of this paper. This research project has been supported by DARPA and AFRL under contract F30602-98-C-0222.

10. References

[1] E. Bina, V. Jones, R. McCool, and M. Winslett, "Secure Access to Data Over the Internet," Proceedings of the Third ACM/IEEE International Conference on Parallel and Distributed Information Systems, Austin, Texas, Sept., 1994.

[2] M. Blaze, J. Feigenbaum, J. Ioannidis, and A. Keromytis, "The KeyNote Trust Management System," work in progress, Internet Draft, March 1999.

[3] M. Blaze, J. Feigenbaum, and A. D. Keromytis, "KeyNote: Trust Management for Public-Key Infrastructures," Cambridge 1998 Security Protocols International Workshop, England, 1998.

[4] M. Blaze, J. Feigenbaum, and J. Lacy, "Decentralized Trust Management," *1996 IEEE Conference on Privacy and Security*, Oakland, 1996.

[5] N. Ching, V. Jones, and M. Winslett, "Authorization in the Digital Library: Secure Access to Services across Enterprise Boundaries," *Proceedings of ADL '96 --- Forum on Research and Technology Advances in Digital Libraries*, Washington, DC, May 1996. Available at http://drl.cs.uiuc.edu/security/pubs.html.

[6] T. Dierks, C. Allen, "The TLS Protocol Version 1.0," draft-ietf-tls-protocol-06.txt, Nov. 12, 1998.

[7] S. Farrell, "TLS Extensions for Attribute Certificate Based Authorization," draft-ietf-tls-attr-cert-01.txt, August 20, 1998.

[8] A. Frier, P. Karlton, and P. Kocher, "The SSL 3.0 Protocol," Netscape Communications Corp., Nov., 1996.

[9] W. Johnston, S. Mudumbai, and M. Thompson, "Authorization and Attribute Certificates for Widely Distributed Access Control," *Proceedings of the IEEE 7th International Workshops on Enabling Technologies: Infrastructure for Collaborative Enterprises—WETICE '98.*

[10] Y. Mass, A. Herzberg, D. Naor, Y. Ravid, and J. Mihaeli, "Access Control Meets Public Key Infrastructure, Or: Assigning Roles to Strangers," forthcoming, IBM Haifa Research Laboratory (http://www.hrl.il.ibm.com/), email contact: YOSIMASS@il.ibm.com.

[11] "The Trust Policy Language," IBM Haifa Research Laboratory (http://www.hrl.il.ibm.com/), email contact: YOSIMASS@il.ibm.com.

[12] B. Schneier, Applied Cryptography, John Wiley and Sons, Inc., second edition, 1996.

[13] K. Seamons, W. Winsborough, and M. Winslett, "Internet Credential Acceptance Policies," *Proceedings of the 2nd International Workshop on Logic Programming Tools for Internet Applications.* Leuven, Belgium, July, 1997. Available at http://clement.info.umoncton.ca/~lpnet/proceedings97/

[14] W. Winsborough, K. Seamons, and V. Jones, "Negotiating Disclosure of Sensitive Credentials," Second Conference on Security in Communication Networks '99, Amalfi, Italy, Sept., 1999.

[15] W. Winsborough, K. Seamons, and V. Jones, "Automated Trust Negotiation: Managing Disclosure of Sensitive Credentials," Transarc Research White Paper, May 1999.

[16] M. Winslett, N. Ching, V. Jones, and I. Slepchin, "Using Digital Credentials on the World-Wide Web," *Journal of Computer Security*, **5**, 1997, 255-267. Available at http://drl.cs.uiuc.edu/security/pubs.html.

[17] P. Zimmerman, PGP User's Guide, MIT Press, Cambridge, 1994.

[18] Simple Public Key Infrastructure (SPKI), http://www.ietf.org/html.charters/spki-charter.html.

[19] International Telecommunication Union, Recommendation X.509 - Information Technology - Open Systems Interconnection - The Directory: Authentication Framework, August 1997.

Policy-Based Cryptographic Key Management: Experience with the KRP Project

Dennis K. Branstad and David M. Balenson
NAI Labs, The Security Research Division
Network Associates, Inc.
denny@tislabs.com, david_balenson@nai.com

Abstract

Policy-based cryptographic key management is a powerful, flexible method of creating, distributing, protecting, and destroying cryptographic keys in accordance with an organizational policy governing information security. The Policy-Controlled Cryptographic Key Release project addressed one part of key management. The goals included: (1) developing a formal language for specifying policies indicating to whom and under what conditions a cryptographic key could be accessed; (2) implementing a prototype system for administering (i.e., enforcing) these policies; and (3) experimenting with automated verification tools which analyzed the policies for consistency and completeness. The requirements for the key release policy language and administering systems are identified; the initial language and system design are described; and the lessons learned from the project are summarized. An example key release policy is included.

1. Project overview

Users of modern computer networks want to control and protect their information with high assurance. Military commanders need to share sensitive information with one set of allies in one zone and a different set in another zone. The membership within the sets often changes rapidly. Federal agencies need to disseminate information about their programs to the public, but then control the information obtained in response to this dissemination. Corporations want control over valuable, sensitive information, even when created by employees working on individual projects. Individual computer users want assurance that they have access to their archived personal or financial information, including in times of crises.

As computers have increased in communications and processing capability, applications have also increased in complexity. Cryptography is now being used to protect information in multi-party information-sharing applications as well as two-party communications applications. Corporations are organizing and protecting their information in accordance with specific use policies, often following organizational structures (e.g., accounting, personnel). Military organizations protect their information in accordance with their own national interests and security policies. Security policies historically have been simple (e.g., give to cleared people with a need-to-know), but are now becoming more complex. For example, multi-national task forces are often established to meet specific threats, and the membership of tactical forces often changes dynamically.

The Policy-Controlled Cryptographic Key Release[1] project was initiated to define a new language for describing to whom and under what conditions access to a cryptographic key will be granted in a protective computer system. The set of access rules, including to whom, when, and where access is authorized, is called cryptographic policy for key release. The language for specifying the policy was to be formally defined to eliminate any ambiguities about the conditions under which to grant access to the key. It was also intended to prevent ambiguities on the identity of the key and the person to whom the key is given.

"Give the Toyota key to Jim for tonight" may be perfectly clear within a family context where there is only one Jim and one Toyota. However, there is an implicit policy regarding protection of the key. It is assumed that use of the key will be for a limited time period, that a copy will not be made, and that the key should be

[1] Release is differentiated from recovery in this paper. Key recovery is obtaining a copy of a specific key under emergency conditions (e.g., when the primary key is lost). Key recovery is only emergency access to a *backup* copy of a key by its owner, by the management of the corporate owner, or by law enforcement and national security authorities. While these emergency access privileges may be included in a release policy, key release policies are primarily used to grant authority to use the *original* key.

protected and returned. Access to the Toyota is controlled by access to its key. There is also an implicit, sometimes ambiguous policy, regarding use of the Toyota. However, once Jim has the key to the Toyota, the owner can no longer control its use.

Similar statements can be made about access to sensitive information that has been cryptographically protected (i.e., encrypted). The owner of information often encrypts it with a cryptographic algorithm and a key that is generated specifically to protect that information. The owner must then decide who should get the key and under what conditions. If the encrypted information is to be transmitted to a specific person in a specific location, the access specification is easy, "Give a copy of the decryption key[2] to Jim Thorpe at Internet location XX.YY.ZZ." A key management/distribution system is then called upon to perform this owner-specified action correctly and securely. While not a simple problem, the key distribution system is designed to handle such simple requests correctly and securely.

The policy-based cryptographic key management project was initiated to develop a key management policy language and enforcement system that meets a broad set of requirements, including those of government officials, commercial organizations, and private users. The goal of this project was to go beyond traditional, static rules for key management, and develop a dynamic, policy-based automated system that unified and enforced the rules for using a cryptographic key. These rules are normally specified by the owner (e.g., individual, organization, coalition) of the encrypted information, but may be augmented by other organizations having legal jurisdiction over the information. Figure 1 depicts a very simple view of the system in which a policy for controlling access to a cryptographic key is created using inputs from the creator of the information to be protected. The policy language and administration system also provide a means to get access to the key in emergency situations under the explicit legal authority granted within the law enforcement and national security jurisdiction in which the policy was being administered.

[2] Dr. Stephen Kent, BBN, noted that the word "release" or the phrase "give a copy" might be misnomers. He noted that a modern key management/establishment system for real-time communications applications does not "give a copy" of a key to the people authorized to use the key, but rather that peer entities create a "session" key simultaneously using an appropriate public key algorithm. The KRP Administration System is intended to control the master key used in creating the session key.

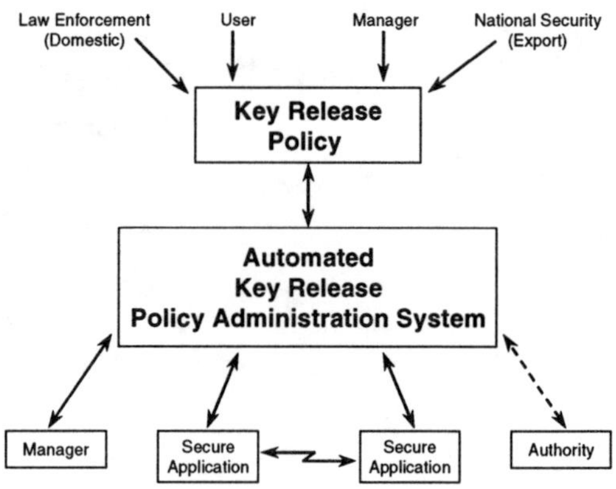

Figure 1. System overview

A fundamental assumption of the project was that cryptographic key management was a broad topic and controlling the release of a key was just one aspect of key management. Release of the key was equated with release of the information. Once a key is released, the person obtaining the key can use (or misuse) the information in any manner. It was hypothesized that an automated system could be created that would enforce a very powerful set of decision rules controlling to whom a key would be given and under what circumstances. These rules would be written in a key release policy language. The project thus became known as the Key Release Policy (KRP) project.

A language is a means to communicate. The written form of a language used for communication among people usually comes long after the initial spoken language has evolved. Attempts to define a formal structure to a natural language usually fail because many exceptions to any general rule about a natural language do exist. However, some languages have been initially defined in a formal structure to avoid ambiguity in their interpretation by a computer. A formal syntax (i.e., language structure) for ALGOL [4] was defined in a syntax specification language called Backus-Naur Form (BNF). While BNF was very good for specifying the acceptable syntax of the new programming language, the semantics, or meaning, of each syntactic element (i.e., word, sentence, phrase) of the new language were not formally defined. This lead to ambiguities in running the same program on different computers.

The key release policy language developed during the project provides a flexible means for users and organizations to describe/select an appropriate policy for releasing the key that is protecting information under their control. It is formally defined in order to describe explicitly to whom and under what conditions a cryptographic key can be released from a protected computer system. This language is intended to prevent ambiguities on the identity of the person to whom the key is to be given and also which key to give. The language design built on relevant technologies available in the research community and extended them in a way that supports multiple security services in multiple security domains.

1.1. Motivation for the KRP project

The dynamic, policy-based key management of the KRP system has many military, commercial, and personal applications. Military applications include:

- Specifying communications security policies that are more flexible than the static ones typically used for cryptographic keying material distributed today;
- Implementing dynamic group-keying applications where participants in large tactical exercises can change very often; and
- Developing IFF (Identify Friend or Foe) codes used to identify rapidly changing members of an allied task force.

Commercial applications range from protecting information within a simple corporation to protecting information within a complex, multi-national consortium of industries joining forces for strategic purposes while still maintaining their competitive structure in order to achieve local, national goals. Personal applications include protecting stored information in accordance with a family oriented access-privilege policy, and medical applications include protecting patient records from unauthorized release.

1.1.1. Military needs. Military organizations are typically hierarchically structured according to a formal organizational structure with personnel organized according to military rank. However, military organizations, like most organizations, also have informal lines of communication through which many of the decisions are made and much of the work of the organization is done. The results of the KRP project are applicable to these organizational needs and can help alleviate problems involving several areas.

Static Keying Problems – One of the problems that the military faces with cryptographic keys is the static keying problem. If a cryptographic key is designated for use by a large number of people for a certain time period, it is difficult to change the key quickly if the key is compromised. The need for dynamic, easy-to-change, cryptographic keying systems have become evident in several recent military operations in which a large number of countries participated as allies.

Group Keying Problems – The group keying problem is yet another dilemma that the military faces. If a key is lost, it may take a month for the key distribution center to notify everyone of this event. Current technology does not easily allow for dynamically updating and changing group keys. This sort of time delay causes this technology to be very ineffective in these types of situations.

Secure Telephone Unit III – During the Desert Storm military action, several examples of military organizations needing improved granularity and flexibility in cryptographic key management were encountered. One of these problems includes the implementation of the Secure Telephone Unit (STU-III), which uses fixed master keys. These units were not deployed to allies, according to an Army General who briefed a STU-III conference, because of concerns of control over keying material (and the technology itself).

Identify Friend or Foe – Also occurring during the Desert Storm military action, Identify Friend or Foe (IFF) codes for identifying members of the allied task force were unavailable or not used properly in several instances, causing losses of both equipment and life in the air and on the ground.

1.1.2. Corporate needs. In addition to military requirements, the KRP administration system is designed to satisfy many commercial requirements. Corporate structures and the information flow among components of the organization are very interesting. Commercial organizations are generally more loosely structured than the military, but they parallel some of the formal and most of the informal military structure very well. There are structure-, role-, and individual-based needs for cryptography in this area.

Corporations have organizational charts that use positions or roles in place of military rank to denote precedence and scope of policy specification. For example, the president has more authority than the vice-president, who in turn has more authority than a secretary. There is a need for cryptographic key management that takes these roles and their respective precedence into account and also allows for the people who fill these roles to change.

1.1.3. Individual needs. Use of personal computers in the home is opening huge markets to many commercial applications. Integration of various home applications such as tax preparation programs, personal financial management programs, and home banking programs are increasing this use. Protection of this information in accordance with a personal access policy will become more important.

Today, personal estate management programs such as financial trust makers and living will creators are available for the personal computer. In a personal will, the will creator itemizes his or her desires for disbursing personal assets to family, friends, charities, and selected organizations. These desires are usually told to a lawyer who codifies them in legal terminology in a will or other legal instrument. The will is filed until the lawyer is notified that the will creator has died and that the estate legal work should be initiated. The provisions and conditions specified in the will are then published, analyzed, interpreted (when necessary), disputed (when potential payoffs are high), and followed in accordance with the law. The lawyer is trusted to have made the provisions legal, consistent, and complete. The lawyer is also trusted to administer the provisions in accordance with professional ethics and probate law.

The KRP system was designed to enhance these types of programs by providing a system that would administer/enforce a personal will. Enforcement of many provisions was assured through the release of cryptographic keys for accessing sensitive information or assets to specified heirs in accordance with a specified time or event schedule. By using an event setting and confirming system, the automated system enforces the provisions of the desires (encoded as a policy) of the will creator.

1.2. Structure of the project

The first phase of the project was to identify the requirements for policy-based key release in unclassified computer applications [1]. Requirements were solicited from individual users, managers, and law enforcement officials. These requirements were refined during system design and language specification. The next phase was to design an initial KRP language and administration system. The final phase was to implement and evaluate a prototype KRP administration system.

1.2.1. Key release policy language. The foundation of a policy-controlled key release system is a language that can specify the exact terms and conditions of the authorized use of a cryptographic key. The language must be easy to understand for the people establishing a policy and must be unambiguously enforceable by a totally automated system. Provisions for emergency access were designed into the language and the enforcement system with very strong auditing features to assure users that unauthorized access to keys was very difficult. Many applications of cryptography (e.g., communications encryption, mass data storage encryption, data element/object encryption, data object signatures, personal identification/authentication) could be supported by the language. The language was based on formal syntax and semantics to the greatest extent possible. The meta-language in which the language itself was specified used an interpreter for implementation.

1.2.2. Key release policy administration system. A design document describing the goals, architecture, and functionality of the policy-controlled key release system was created. A demonstration system was designed for development in software toolkit form. The design included the capabilities of identifying conflicts in multiple policies covering the same information (e.g., combining data objects to form a new object), and attempting to resolve the conflicts (automatically or through manual policy modification). These were used to analyze conflicts within requirements of a single policy. While not completed, the initial implementation, experimentation, and demonstrations helped establish several concepts that are summarized in the latter sections of this paper.

2. Key release policy languages

The KRP language [2] is designed for a creator of a policy to specify the particular policy for releasing a particular key. The language was designed to provide methods for:

- Specifying comprehensive policies for controlling release of cryptographic keys;
- Specifying (and delegating) chains of authority for creating release policies;
- Specifying groups that create, modify, and monitor policies;
- Specifying, setting, and monitoring events/conditions affecting release of a key;
- Specifying groups, including individuals or roles, that can obtain or use keys;
- Specifying and monitoring posting events/ conditions; and
- Specifying a policy for auditing the operation of the KRP administration system.

These goals are achieved by using several languages and subsystems. The primary language is the policy specification language in which a policy component creator specifies the key release policy. Any word processor could be used to create a policy component in this language, similar to using any word processor for creating a computer program in a programming language. However, there is a policy creator support system that uses graphical interfaces that lead a user through logical sequences in creating a policy. Policy Administration Language (PAL) [3] commands are used in defining the policy and administering it.

2.1. Graphical user interface language

The KRP language was implemented in a Graphical User Interface (GUI) that supports the policy creator in performing all the operations that are required in making a policy for which the creator is authorized. This GUI is similar to a word processor and minimizes the time needed by a user to adapt to the GUI and learn its capability.

The KRP GUI has several different components. The main screen is a "launcher" that allows the user to choose between these components. The following options are supported:

- *Set User Name & Password* – this option is usually chosen first, as it allows the user to enter a username, password, and key server.
- *Condition Posting Utility* – this utility shows state variables and current states contained in the current policy. It also allows the user to change state of variables.
- *Encryption Utility* – this service allows the user to encrypt files with selected keys, and also allows the user to create new keys.
- *Command Line Interface* – the command line interface allows a user to enter commands directly into PAL server. The system initializer usually uses this to enter initializing commands to load previously created data into memory.
- *File View Utility* – this simple text-file viewer is used to look at encrypted/decrypted files, and examples of created policies.
- *Group and User Manager* – this GUI utilizes the PAL system commands to display lists of users, create groups and users, and place users into groups. This group and user manager allows the user to easily set up a structure that can be used in creating policies. It also allows the user to enter new keys for new users.

2.2. Specification language

The policy specification language is defined two ways: syntactically and semantically. A formal specification of the syntax defines what "sentences" can exist in the language. A recursive definition language is implemented as part of the syntax specification. Formal syntax specifications are made in the Backus-Naur Form (BNF) "production" language. The formal syntax specification is rigorous and unambiguous, but is difficult to read by human readers. A computer program called a lexical analyzer is used to "recognize" acceptable sentences of the language and acceptable sentences are then translated into semantics of a form (commands to be interpreted) that are administered by the policy administration system.

Policies on most topics are not created in a vacuum. There usually are overriding rules that must be satisfied. For example, parents cannot legally give permission to their son Jim to use the Toyota unless Jim has satisfied his state's policy on driving a motor vehicle and has obtained a driver's license. The parents' policy is, therefore, subordinate to the state's law enforcement policy. Local policies are subordinate to state or federal policies. The KRP language and system are designed to support requirements of higher organizational policies (e.g., law enforcement, corporate) when those policies have precedence over individual policies protecting information.

2.3. Verification language

A key release policy can also be expressed in a verification system language. PVS, the Prototype Verification System from the SRI International Computer Science Laboratory [5], was selected as the KRP verification system language. The specification language and the verification language supported equivalent expressions of a policy and were equivalent to the compiled or interpreted code that enforced the policy.

2.4. Policy administration language

In contrast to the specification language (the language used to express the policy), the Policy Administration Language (PAL) is the language that helps a user create a policy and the system to administer the policy. The policy specification language and the policy administration language are related, but should be considered differently.

The ability to easily change the group authorized to receive a cryptographic key without changing the overall policy is very important. Individuals or people with specific roles in an organization are usually authorized to change a group's membership. For example, the head of the personnel department can change the list of people in the personnel department who are authorized to maintain lists of other roles in an organization. Release of a cryptographic key controlling access to corporate information is often controlled along the lines of the corporation's organizational structure, and is independent of the person filling a specific role. Controlling access to information based on roles rather than individual authorizations is called role-based access control.

2.4.1. Event posting. Release of a key depends on the occurrence of certain events or the existence of certain conditions. Some events are automatic (e.g., time dependent), while others must be monitored and posted by someone. Individuals, people with specific roles, or members of a certain group must be authorized and then required to watch for the occurrence of prescribed events and then to post their occurrence in the KRP system.

2.4.2. Key request. The policy administration language supports users requesting access to keys. To request access to a key, a user must have authorization for the key requester role. Once authorized, a user can request any key. The KRP system evaluates the policy associated with the requested key. The evaluation of a valid key release policy always results in a "permit", "deny", or "conditionally permit." If permission is denied, the root-policy (i.e., the "highest" policy component in a policy domain) can specify if the user is authorized to determine the reason for denial. If evaluation of a policy results in a "permit" condition, a copy of the key is provided to the requester. If the evaluation results in a "conditionally permit", then the conditions required for permission are provided to the requester. The user will be permitted access to the key only after these conditions are satisfied.

A complete policy path must be evaluated when evaluating a request to release a key. A policy path is defined to be the set of nodes from the leaf of a policy tree to its root. The key associated with the leaf will only be released under the permitted conditions specified in the policy. A key requester will be permitted access to the key only if the result of evaluating the policy path explicitly results in the value "permit." Otherwise the result is "deny."

2.4.3. Security event auditing. Policy creators often want to audit the actual release of a particular key in accordance with the policy they created. The posting of certain events may be of utmost importance. While the posting of all events and the release of all keys may be audited by the KRP system administrator, the language supports policy statements that include auditing certain events in the system (unless this release of audit information is denied by a higher authority).

3. Key release system

The previous section described the KRP *languages*. This section outlines the KRP *system* and describes its architecture, methods of system administration, policy specification, and policy enforcement.

3.1. Architectural components

The KRP system provides at least one method for:

- Identifying/authenticating individuals;
- Validating requests for performing a role by an individual;
- Verifying chains of authority in policy components,
- Obtaining policy components created by authorized policy component creators;
- Integrating policy components into a single policy;
- Setting events and conditions prescribed in a policy;
- Auditing security-relevant system events;
- Analyzing completeness and consistency of policies;
- Verifying events and conditions affecting release of a key;
- Protecting keys and policies; and
- Assuring the availability of keys when release is authorized.

Fundamental to any security system is the ability to identify and authenticate individuals. A unique identity (at least within the scope and context of a KRP system) must be provided by, or assigned to, each individual desiring to use the system. The generic term, "user," means any person whose identity and method for being authenticated has been registered in the system. No other authority is granted by simply being a user. Therefore, each user must have specific authorization by a system administrator or a policy creator to perform certain roles or certain acts.

The KRP system also supports storage of policy components that can be integrated to create a policy for release of a key. The present KRP architecture defines an arbitrary number of independent tree structures, each of which has one root, an arbitrary number of nodes at an arbitrary number of layers, and an arbitrary number of leaves. Each of these arbitrary numbers is subject to a maximum established at system initialization. By definition, the number of keys in a KRP system is equal to the total number of leaves of all the trees (i.e., domain structures) specified at any point of time in the lifetime of a system. That is, each leaf of each tree is associated with one, and only one, key. The path from the leaf to its root thus specifies a complete hierarchical policy for the release of that key.

However, if a leaf of a tree is turned into a node (i.e., additional policies can be built subordinated to it), then the policy associated with that leaf must be copied into a new policy component which is subordinated to the appropriate higher node (or root). Therefore, in this situation, a root can never be a leaf.

A state or condition (these are equivalent; a state is explicitly set while a condition is implicitly set simply because a specific event has occurred) must be explicitly created and its name specified along with specifications of who or what can set it. Future additions to these requirements may add levels of protection to the event-posting process.

Additionally, an audit trail (log) processing capability will be present in a KRP system. Authorization to perform audit trail processing is always checked. Report generation capability will then be available.

A method of analyzing the completeness and consistency of policies is provided by the PVS verification system [5]. Consistency verification will entail checking that there is no set of events or conditions under which both a "permit release" and "deny release" of a key to a requester can be the result of evaluating a policy. Completeness is the presence of at least one set of events or conditions under which a key can be released as the result of evaluating a policy.

3.2. System administration

A system initializer role is supported by the KRP system. The first person to start a new KRP system is by definition the system initializer. This person must establish certain parameters for all future operations. The first such parameter is the individual or group of users authorized to be system administrators. Other initial parameters are described in this document.

Other system roles are then established once the system has been started. There is a system administrator, who is authorized to create a new "highest" policy component constituting new domains. Users authorized to audit the KRP system log are also named (may include the system administrators).

Administration information includes the names of the system administrator, system, and policy root managers. It also includes information regarding delegation of authority.

Administration operators are commands that the system administrator can enter into the system. The following commands are supported:

```
PERMIT_READ      UNION
DENY_READ        DIFFERENCE
PERMIT_WRITE     INTERSECTION
DENY             ADD
EQ               REMOVE
LOGIN            DELETE
PASSWORD         LIST_STATE_VARS
KEY              LIST_KEYS
USER             LIST_GROUPS
CREATE_EXPR      TIME_VAR
STATE_VAR        LOAD_SYMTAB
GROUP            DUMP_SYMTAB
```

The policy administration system consists of one or more servers working together as an integrated system and an arbitrary number of client workstations. Each client supports the system in either the policy creator or key requester roles. The server will be able to service an arbitrary number (specified by the system administrator) of clients (policy creators and key requesters) simultaneously.

The system will support the following:

- Generation (or selection) of cryptographic keys;
- Generation (or selection) of a unique ID for a key;
- Creation of a policy for release of a named key within a policy domain tree;
- Explicit establishment of scope of variables to support local policy component definition, policy sub-tree definition, domain-wide definition, and global system definition;
- Explicit release of a key based on set membership, event/state/condition variables, and specified expressions of these;
- Ability to delegate authority to lower nodes in a policy tree and subordinate a policy component to some node in a policy tree;
- Ability to authenticate identities of users, authorizations of users, and occurrence of events; and
- Ability to interrogate/evaluate a policy and determine under what conditions a key can be released and to whom.

3.3. Policy specification

All cryptographic key release policies are established within a security domain within a KRP administration system. Figure 2 depicts two security domains, each having a root policy and several key release policies attached to it. Each security domain has a master policy maker via the mechanism of having a single policy creator for the root policy component. Each intermediate node is a policy component that is subordinate to all policy components connecting it to the root, and is superior to all policy components and policies connecting it to all its leaves.

Each variable has explicit scope. For example, using the former TIS administrative hierarchy, Steve Walker, the president of TIS, is responsible for establishing the base policy. All vice-presidents must adhere to the base policy, but can establish sub-policies for their domain.

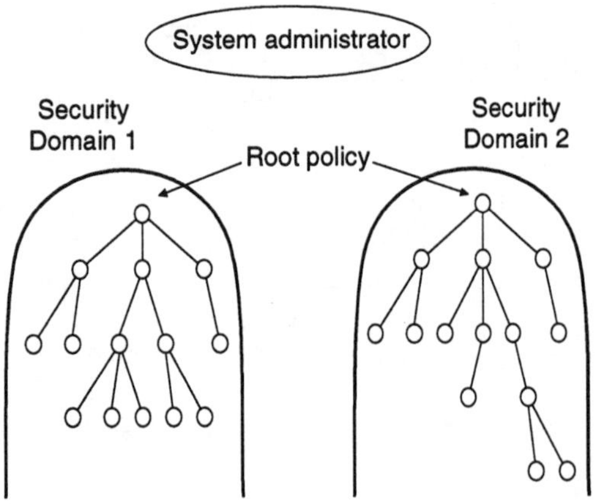

Figure 2. KRP administration system

Project leaders in each domain can establish policies for their project staff, but must conform to higher-level policies. Finally, project staff members can establish a release policy for each of the cryptographic keys that they create but each release policy must conform to all the strictly ordered higher level policies. A simple example of this is included in the appendix.

Policy component precedence – Every policy component has an unambiguous and unique precedence within each domain. The following set of rules is used in the KRP administration system:

- Every KRP administration system has one administrator and many policy domains;
- Every policy domain has one root policy component upon which many policies can be created;
- Every policy (i.e., a leaf in the policy domain tree) has one principal creator and at least one superior policy component;
- Every policy component (i.e., root or node in the policy domain tree) has one creator and at least one subordinate policy;
- Every key in a KRP administration system has exactly one release policy; and
- Each release policy can govern more than one key.

3.4. Policy analysis

The PVS system was developed at SRI to assist in the proof of theorems stated in a formal verification language. A goal of the KRP project was to create a policy specification language that could be automatically translated into an equivalent verification language to be used to verify certain properties about the resulting policy. The goals of policy verification are to assure the policy creator that the created policy is consistent and complete. Formal definitions of these qualities were used in designing the specification language. However, satisfying this goal places restrictions on the language so compromises were made between maximizing the utility of the language while maintaining the ability to verify the consistency and completeness of the resulting policy written in the policy language. For example, a policy in which an action is both permitted and denied is inconsistent.

Completeness is assurance that all possible conditions resulting in a permit can be unambiguously evaluated. Such assurance is very useful in analyzing a policy. Policy enforcement analysis may require answering questions such as "Under what conditions could this key be given to XXX?" or "Under these conditions, who could be given key YYY?" Auditing questions may include "To whom has this key been given in the past?," and "Under what conditions has this key been given and to whom?" There must be explicit answers for all of these questions for a policy to be complete.

4. Lessons learned

The lessons learned from performing a research project should be passed on to others performing similar or related projects. Many lessons were learned during the KRP project regarding policy specification languages, policy enforcement systems, and policy verification.

4.1. System design

Even though the KRP system design and prototype implementation were not due until the second year of the project, the KRP staff decided to start the design and implementation during the first year. We learned that it was difficult to design a policy specification language without understanding how it could and would be enforced. While the major impetus was to have a powerful and flexible language for expressing policies, it was mandatory that policies could be automatically enforced. It was also desirable to have a prototype implementation to demonstrate concepts, approaches, and status.

The prototype system lessons included:

- Many distinct roles and functions needed to be supported, but one person could fill many roles simultaneously and many functions could be performed within a single computing system;
- Interfaces to different functions could be different, but they had to have commonality in order to be easily used by many people;

- Demonstration systems had to be significantly different than operational systems because people had to see much more detail in slow motion during a demonstration than would be needed or tolerated in an operational system;
- Effective demonstration systems must simulate events and produce outputs that would never be needed during operation of a real system; and
- It is difficult to convert a demonstration system to an operational system because the fundamental goals are so different, even though the overall objectives may be the same.

4.2. Language design

The policy specification language design was the major objective of the first year of the project. Three phases of the design were defined. Cook and Rho [2] describe the design after Phase 1, which includes the majority of the language design, but does not include delegation and logging or subordination and precedence.

The KRP language design lessons learned include:

- Training and experience in developing formal computer languages is necessary in order to design a security policy specification language that is useful and enforceable;
- A mathematical foundation is desirable in order to design a language that includes partial ordering, logical operations, set operations, and formal relations;
- Understanding formal and informal organizational structures and communications is desirable in order to design a language that represents reality concerning how information is created, manipulated, withheld, and disseminated in an organization;
- Policies have to be specified in a language understandable to people in order to be useful, and in a language understandable to computers in order to be enforceable;
- Events and conditions have to be identified, quantified, specified, and verified in order to be useful in a KRP system;
- Policies have to be analyzed for consistency in order to be useful and enforceable;
- Managers can describe in general what security policies should be followed, but find it very difficult to state explicitly the conditions and states required for releasing information and to whom to release the information; and
- It is often easier to define a policy than to enforce the policy, or to analyze the policy for completeness.

4.3. Policy verification issues

A major requirement for policy-controlled security is to be able to verify that the policy is unambiguous, consistent, and complete. Specifying a policy in a formal language that is well-structured helps to assure that it is unambiguous. Two aspects must be verified: the syntax (i.e., form) of a policy must be unambiguous and the semantics (i.e., meaning) of a policy must also be unambiguous. The syntax of a language can be verified relatively easily through the use of a syntax analyzer program that tests "sentences" in the language against the formal syntax. Unacceptable formats and sentence fragments can be detected. The semantics of a language is much more difficult to verify. The meaning of sentences can "make sense" to one person, but not to another. The sentences of a computer language can be performed correctly by one computer system, but not by another. A verification program such as PVS is often used to verify the "correctness" of a language by asking the language creator or sentence creator to state assertions about what is intended by a sentence, and then evaluating the assertions in accordance with some predetermined rules.

Consistency of a policy means that different components do not conflict with each other, and that evaluation of a policy does not result in conflicts that cannot be resolved. The KRP policy language, in a consistent policy specified for releasing a key, always results in one of two decisions, either permit or deny release. A consistent policy can never result in both permit and deny results in a policy evaluation.

A complete policy means that all the variables needed to evaluate a policy have a means for determining their value. A complete policy can be evaluated (with difficulty) by setting all variables to all values and verifying that one and only one result can occur for a setting.

One of the lessons we learned from using PVS is that verification systems require significant computing capability to perform even simple verifications useful to a person operating the system in real-time. PVS should be operated by a person with significant experience in automated proof verifiers. We also learned that only simple assertions are easily verified. Currently, the effort of verifying completeness and consistency of key release policies increases exponentially in relation to the complexity of the policies.

5. Acknowledgements

The work described in this paper was supported by the Defense Advanced Research Projects Agency DARPA) under Contract No. DABT63-92-C-0020, DARPA Order 8685. We would like to thank the DARPA program managers Teresa Lunt, Major Michael St. Johns, Dr. Gary Koob, and Robert Rosenthal for their technical interest in, and administrative support of, the KRP project. Teresa Lunt was the first to become interested in creating a powerful, flexible language for creating policies for controlling access to cryptographic keys needed to decrypt sensitive information. Major St. Johns provided funding for the first year of the project. Dr. Koob assisted in developing the statement of work. Robert Rosenthal was the overseer of the technical work and provided project guidance and good suggestions for coordination of results.

We would like to thank Robert Meushaw, Penny Brummit, Elizabeth Ziegler, and Dr. Sylvan Pinsky of NSA for their technical interactions with the KRP staff. We would also like to thank Miles Smid of NIST and Dr. Helen Gill of MITRE for their comments on the KRP project and language.

We would like to acknowledge Dr. Dorothy Denning, Dr. Burt Kaliski, Dr. Warwick Ford, and Mr. Russ Housley for their consultation and recommendations on the project. Dr. Denning provided valuable input on the law enforcement policies. Dr. Kaliski reviewed the cryptographic recommendations and Dr. Ford the policy enforcement mechanisms. Mr. Housley provided inputs on military and commercial requirements for protecting sensitive information. The KRP staff would also like to thank Mr. Mike Gilmore, Federal Bureau of Investigation, for his inputs on federal and state laws governing electronic information access.

We are especially indebted to the staff members of Trusted Information Systems (TIS) who did the majority of the language specification and prototype system development. Sue Rho did the initial policy language design and PVS analysis; Jeff Cook completed the Phase 1 language specification and documentation; Dr. David McGrew did the Policy Administration Language design and prototype implementation.

6. References

[1] Branstad, D., Policy-Based Cryptographic Key Release project (KRP) Requirements, TIS Labs at Network Associates, Inc., May 30, 1998.

[2] Cook, J. and S. Rho, Key Release Policy Language (KRPL): Phase 1, TIS Labs Report #0658D, TIS Labs at Network Associates, Inc., May 30, 1998.

[3] McGrew, D., Policy Administration Language and System Prototype, TIS Labs at Network Associates, Inc., May 11, 1998.

[4] Naur, P., Revised report on the algorithmic language ALGOL 60, CACM 6, 1963, pp. 1-17.

[5] Rushby, J., The PVS Verification System, SRI International Computer Science Laboratory, 1996.

7. Appendix

7.1. Corporate policy example

We now give an example of how a corporation can use the KRP system to define key release policies. It shows how system administrators can use the KRP language and configure the KRP system for their installations. In this example, "Jeff" is the system administrator for his system. He authenticates himself as *system_initializer*, which is the only valid user when the KRP system is initialized. Jeff then creates the set of valid users (*users*) and designates their valid roles (which is represented by *roles*). Next, Jeff removes *system_initializer* from the set of valid users. Finally, he modifies the policy to meet the needs of his installation.

Jeff extends the built-in roles with the following roles:

Roles	Description
President	TIS President
VP	TIS Vice President
KRP_project_manager	The KRP Project Manager
KRP_staff	A KRP Project Staff Member
KRP_user	A User of the KRP System

In addition, he defines the following groups:

Groups	Description
TIS_Management	The TIS corporate management group
TIS	The TIS employee group
KRP	The KRP project group

Although Jeff can submit all of his transaction requests in one step, we split them up as a series of steps for the sake of clarity.

Step 1: Jeff authenticates himself as the *system_initializer* and assumes the (valid) *system_administrator* role. After successful authentication, the KRP system sets the "requestor" variable to *system_initializer* and initializes the (functional) value of the_role.

```
requestor = system_initializer
```

```
the_role(system_initializer) =
  system_administrator
```

Step 2: Jeff adds SteveW, Marty, Denny, David, Jeff, and Sue to the KRP system as valid users.

```
modify((users: var),  users ∪ {
  SteveW, Marty, Denny, David,
  Jeff, Sue
});
```

Step 3: Jeff extends the built-in roles.

```
modify ((roles: var), roles with {
  [SteveW := {President}],
  [Marty :=  {VP}],
  [Denny := {KRP_project_manager,
    KRP_staff, policy_creator}]
  [David := {KRP_staff}],
  [Jeff := {KRP_staff,
    system_administrator,
    policy_creator}],
  [Sue := {KRP_staff}]
});
```

Step 4: Jeff now logs out and back in as himself, assuming the *system_administrator* role.

```
requestor = Jeff
the_role(Jeff) = system_administrator
```

Step 5: Jeff can now void *system_initializer*'s role and then remove the *system_initializer* user.

```
modify((roles: var), roles with
  {[system_initializer := ∅]});
modify((users: var), users \
  {system_initializer});
```

Step 6: Jeff sets up the group structure.

```
modify((groups: var), groups with {
  [SteveW := {TIS_Management,TIS}],
  [Marty :=  {TIS_Management,TIS}],
  [Denny := {KRP, TIS}]
  [David := {KRP,TIS}],
  [Jeff := {KRP,TIS}],
  [Sue := {KRP,TIS}]
});
```

Step 7: Jeff defines an enumeration type, called corporate_states, and a variable, called corporate_status. The policy assigned to corporate_status ensures that only SteveW can modify its value, but all TIS employees can read its value.

```
define((corporate_states: enum),
  (emergency, normal));

create(((corporate_status: var):
  corporate_states),
  normal,
  policy(corporate_state_policy,
    forall (u: user)
      permit((u: user),
      read,(corporate_status: var)),
    permit((SteveW: user),
      modify,(corporate_status: var)),
    forall (u:user)
      deny((u: user),
      modify, corporate_status)
      when (u ≠ SteveW)
  )
);
```

Step 8: Jeff changes his role from *system_administrator* to *policy_creator* (he is still the requestor).

```
requestor = Jeff
the_role(Jeff) = policy_creator
```

Step 9: Jeff can now extend the built-in corporate policy. Marty is permitted to read any key when there is a corporate emergency. Only users whose role is KRP_*staff* are permitted to modify keys. Users who are in the *KRP* group are not permitted to read keys after working hours.

Jeff submits the following request:

```
modify((P0: var),
  policy(... old corporate policy ...,
    forall =(k: key)
      permit((Marty: user),
      read, (k:key))
      when (corporate_status =
        emergency),
    forall (u: user)(k: key)
      deny((u: user),
      modify, (k: key))
      when (the_role(u) ≠
        KRP_staff),
    forall (u: user) (k: key)
      deny((u: user),
      read, (k: key))
      when (the_role(u) =
        KRP_staff ∧
        (current_time ≥ "17:00'' ∨
         current_time ≤ "9:00'' ∨
         current_day = Saturday ∨
         current_day = Sunday))
  )
);
```

7.2. User policy example

We now describe how employees of the corporation can create keys, state variables, and policies. The examples show how employees can specify policies in terms of users, groups, and roles.

7.2.1. Example 1. This example shows how Denny can create variables and keys and define policies for them in terms of users.

Step 1: Denny authenticates himself as the manager of the KRP project.

```
requestor = Denny
the_role(Denny) = KRP_project_manager
```

Step 2: Denny defines an enumeration type, called project_states, and a variable, called project_status.

```
define((project_states: num),
  (emergency, delivery_due,
   on_schedule));

create(((project_status: var):
  project_states),
  on_schedule,
  policy(project_status_policy,
    permit((Denny: user),
      modify, (project_status:var)),
    forall (u: user),
      deny((u: user),
      modify, (project_status: var))
      when (u ≠ Denny),
    forall (u: user),
      permit((u: user),
      read, (project_status : var))
  )
);
```

Step 3: Denny requests to add a key, K1.

```
create(((K1: key) :string),
  "K1's value'',
  policy(K1_policy,
    permit((Marty: user),
      read, (K1: key))
      when ((project_status,Denny) =
        emergency),
    permit((Denny: user),
      {read,delete,modify},(K1:key)),
    permit((Jeff: user),
      {read,delete},(K1: key)),
    permit((David: user),
      read, (K1: key))
  )
);
```

7.2.2. Example 2. This example shows how Denny can define his own roles and define policies in terms of these roles.

Step 1: Denny authenticates himself as the manager of the KRP project.

```
requestor = Denny
the_role(Denny) = KRP_project_manager
```

Step 2: Denny divides the roles assumable by the KRP staff into sub-roles and assigns users to sub-roles.

```
modify(=(user_roles: var),
  user_roles with {
    [Denny := [Denny :=
      {KRP_project_manager}]],
    [Denny := [David :=
      {KRP_system_developer}]],
    [Denny := [Sue :=
      {KRP_language_developer}]],
    [Denny := [Jeff :=
     {KRP_language_developer,
      KRP_system_developer}]]
});
```

Step 3: Denny requests to add a key, K2. This policy for K2 states that users who are in Denny's KRP system developer role may modify, it and users who are in Denny's KRP language developer role may delete it.

```
create(((K2: key): integer),
  34127598464837912487257,
  policy(K2_policy,
    forall (u: user)
      permit((u: user),
      modify, (K2: key))
      when (the_role(u) =
        Denny.KRP_system_developer)
    forall (u: user)
      permit((u: user),
      delete, (K2: key))
      when (the_role(u) =
        Denny.KRP_language_developer)
  )
);
```

The Design of a Group Key Agreement API

Giuseppe Ateniese
Computer Science Department
Johns Hopkins University
ateniese@cs.jhu.edu

Olivier Chevassut
Lawrence Berkeley National Laboratory
University of California
chevassu@george.lbl.gov

Damian Hasse   Yongdae Kim   Gene Tsudik*
Computer Networks Division
USC Information Sciences Institute
{hasse,yongdaek,gts}@isi.edu

Abstract

As collaborative applications grow in popularity the need for appropriate security guarantees, services and mechanisms becomes apparent. This paper describes a protocol suite and an API geared for securing collaborative applications. The API is based on the extensions of Diffie-Hellman key agreement developed in the CLIQUES project. Its core services provide authenticated group key agreement in relatively small (on the order of 100 members) dynamic peer groups.

1. Introduction

Securing group communication is a complex issue. It poses a number of challenges ranging from basic cryptographic algorithms to systems, communication design and secure implementation [4]. The two common group security concerns are: 1) privacy: assuring that intra-group communication remains secret to non-members, and 2) authentication – assuring that legitimate group members can be identified as such.

The standard approach to supporting group security is based on maintaining a secret quantity known only to all current group members The particulars of generating and distributing this secret quantity are known collectively as *group key establishment.* When the latter is achieved with one party generating a group secret, the problem is reduced to *group key distribution.* Whereas, if all group members collectively generate the group secret, the problem is referred to as *group key agreement.* In both cases, only current group members must have access to the group secret. We say *current* since group membership can be highly dynamic. Whenever members join or leave a group, there must be means for securely adjusting the group secret.

This paper discusses the design of an Application Programming Interface (API) for group key agreement. This API, called CLQ_API , is based on the CLIQUES protocol suite and is geared specifically for dynamic peer groups. (Peer groups are relatively small, non-hierarchical groups typically used for replication or collaborative, many-to-many applications.) CLIQUES protocols [14, 15, 3], in turn, are the group extensions of the well-known Diffie-Hellman key exchange [6]. CLIQUES provides authenticated contributory key agreement which guarantees key independence, key confirmation, perfect forward secrecy, and resistance to known key attacks.[1]

CLQ_API separates cryptographic protocols from communication, i.e., the actual group communication is left to the underlying communication subsystem (preferably, a reliable group communication system, e.g., SPREAD[2], TOTEM[11] or TRANSIS[1]). Moreover, network events such as network partitions, failures and other abnormalities are assumed to be taken care of by the same communication system. As described in subsequent sections, this allows the design of a small, concise and communication-independent API.

The rest of this paper is organized as follows. We begin with the notation and the brief description of basic operations in group key agreement. We then describe the CLIQUES protocol suite, which, in turn, consists of the following group operations: join, merge, leave and

*Research supported by the Defense Advanced Research Project Agency, Information Technology Office (DARPA-ITO), under contract DABT63-97-C-0031.

[1] For the definition of the above, the reader is referred to [10].

key refresh. Next, CLQ_API primitives are described in detail along with some examples. Finally, we discuss the efficiency of CLQ_API and conclude with some experimental results obtained with several popular cryptographic packages. (A more detailed description of the API is included in the Appendix.)

2. Group Key Agreement

The following notation is used throughout this paper:

n	number of protocol parties (group members)
i, j	indices of group members
p, q	prime integers
M_i	i-th group member; $i \in [1, n]$
G	unique subgroup of Z_p^* of order q
q	order of the algebraic group
g	exponentiation base; generator in group G
x_i	long-term private key of M_i
N_i	M_i's session random number $\in Z_q$
S_n	group shared key among n members
K_{ij}	long-term shared secret between M_i and M_j
$\|a\|$	bit length of integer a
H	output size of hash function
$inv(a, b)$	multiplicative inverse of a modulo b

2.1. Group key agreement operations

A comprehensive group key agreement API must handle adjustments to the group secret(s) stemming from single- and multiple-member membership population changes. This section describes the purpose of each of these operations.

Single-member changes: include single member additions or deletions. The former occurs when a prospective member wants to join a group and the latter – when a member wants to leave (or is forced to leave) a group. Both operations may be performed by the group controller(s) or by consent of all group members, depending on the local policy.

Multiple-member changes: also include addition and deletion. The former can be further broken into:

- Mass join: multiple disparate new members are brought into an existing group
- Group fusion: two or more groups are merged to form a single group.

The latter includes:

- Mass leave: multiple members must be removed at the same time.
- Group fission: a monolithic group needs to be broken into smaller groups.

Key refresh: is not a membership change operation, however, we discuss it here for the sake of completeness. It has two main purposes:

- Limit the amount of ciphertext generated with the same key.
  Since it is easier to perform cryptanalysis with more ciphertext/plaintext pairs, a routine group key refresh operation is needed. The lifetime of a key is determined by the application-specific policy.
- Recover from the compromise of a current group secret or a member's contribution. (We note that a compromise of a member's contribution can result in disclosure of all group secrets contributed to by this member. Therefore, not only the group shared keys, but also the individual key shares must be periodically refreshed.)

3. CLIQUES Protocols

As mentioned above, CLIQUES is a protocol suite providing authenticated contributory key agreement for dynamic peer groups. The following operations are supported by CLIQUES:

- Join: a new member is added to the group.
- Merge: one or more members are added to the group.
- Leave: one or more members are removed from (or leave) the group.
- Key Refresh: generates a new group shared key.

Each operation is discussed in the remainder of this section. The mapping between the group membership changes and the corresponding key agreement (cryptographic) operations is as follows:

Membership Operation	CLIQUES Operation
Forced leave	Leave
Voluntary leave	Leave
Single join	Join
Mass join	Merge
Group fusion	Merge
Group fission	Leave
Mass leave	Leave
Key refresh	Key Refresh

3.1. Group Controller and Group Key

Group Controller. In the current version of CLIQUES, the last member to join the group becomes a *group controller*. Equivalently, the role of a group controller is always played by the newest member. This behavior can be easily changed if required. In fact, any selection criteria can be used to select a group controller as long as it yields a consistent outcome across all group members. (CLIQUES assumes that the underlying communication system provides a timely and consistent membership view to all group members; this property is referred to as membership *view synchrony.*) One obvious alternative to the present criterion is to let the oldest member be the group controller.

We emphasize that a group controller is not in any way a privileged group member. It exists to prevent contention and should be viewed as a burden (or a chore) rather than a privilege. Specifically, the group controller assists new members in joining the group and initiates operations stemming from departing members or the need for periodic re-keying.

Furthermore, it is important to note that *any* group member can, at any time, initiate a group re-key by unilaterally updating its key share and distributing appropriate partial keys to the rest for the group[2]. (See below.)

Group Key. In the following operations, the group key (also referred to as group secret) has the form $S_n = g^{N_1 \cdots\ N_i\ \cdots\ N_n}$, where n is the group size and N_i $(0 < i \leq\ n)$ is provided by the i-th member, M_i[14]. In all CLIQUES protocols, the last broadcast message is always the set formed by:

$$g^{K_{in} \frac{N_1 \cdots\ N_n}{N_i}} \quad \forall i \in [1, n-1]$$

This set, from the last broadcast, must be retained by each group member. This is necessary since any member can become a controller due to a group partition or a network fault. (In other words, a non-controller can become a group controller if all "younger" members fail or become partitioned out.) Although the entire group receives this message from the current controller, each member uses a different element (key) to compute the new group secret.

Finally, the computation of the pairwise long-term key, K_{ij}, shared between any two members M_i and M_j is assumed to be performed before its use is required in the protocol.

3.2. Join

The join operation adds a new member, M_{n+1}, to the current group of n members. During this process a new group shared key, S_{n+1}, is computed and M_{n+1} becomes the new group controller. Assuming that M_n is the current controller, the protocol runs as follows:

1. M_n generates a new secret and random exponent N'_n and produces the following set:[3]

   $$\mathcal{M} = \{g^{N_1 \ldots N'_n / N_i} \mid i \in [1, n-1]\}$$

   Next, $\mathcal{M}$ is sent to M_{n+1}.

2. Upon receipt of the message, M_{n+1} generates a new secret exponent N_{n+1} and computes:

   $$\mathcal{M}' = \{g^{K_{i,n+1} N_1 \ldots N'_n N_{n+1} / N_i} \mid i \in [1, n]\}$$

   This set is broadcast to the entire group (including itself).

3. Upon receipt of the broadcast, each M_i computes the shared group key as follows:

   $$(g^{\frac{N_1 \ldots N'_n N_{n+1}}{N_i} \cdot K_{i,n+1}})^{K_{i,n+1}^{-1} \cdot N_i} = g^{N_1 \ldots N'_n N_{n+1}} = S_{n+1}$$

Steps 1 and 2 require n modular exponentiations (by M_n and M_{n+1}, respectively) and step 3 requires a single exponentiation by each member. Hence, the protocol requires $(2n + 1)$ serial exponentiations.

3.3. Merge

The merge operation adds $k > 0$ members to the current group of $n \geq 1$ members. Let $m = n + k$. During this process a new group shared key, S_m, is computed and M_m becomes the new group controller. Assuming M_n is the current controller, the protocol runs as follows:

[2]Though we fully support re-keying mechanism, its usage depends on local policy.

[3]This is achieved by exponentiating $N'_n * inv(K_{in}, p) * inv(N_n, q)$ to each element in the last broadcast message.

1. M_n generates a new exponent N'_n and computes: $g^{N_1 \ldots N_{n-1} N'_n}$ by modular exponentiation of the previous group shared key with $(N'_n * inv(N_n, q))$. Then, this value is sent to M_{n+1}, the first new member.
2. Each new (merging) member M_j, $j = n + 1, \ldots, m-1$, generates a new exponent N_j, computes $g^{N_1 \ldots N'_n \ldots N_j}$ and forwards the result to M_{j+1}.
3. Upon receipt of the accumulated value, M_m simply broadcasts it to the entire group.
4. Every group member M_i (old and new) received the broadcast, computes $g^{N_1 \ldots N'_n \ldots N_{m-1}/N_i}$ and sends back to M_m.
5. Having received all responses from the group members, M_m generates a new secret exponent N_m and produces the following set: [4]

   $$\mathcal{M} = \{g^{K_{im} N_1 \ldots N'_n \ldots N_m / N_i} \mid i \in [1, m-1]\}$$

   which it broadcasts to the group.
6. Each member M_i computes the group shared key exactly as in the last step of the Join protocol.

In case of a single-member Merge ($k = 1$), step 2 is not required and the rest of the protocol runs as above.

Steps 1 and 2 require a total of k modular exponentiations while steps 4 and 6 each involve one exponentiation (in parallel by each member). Finally, step 5 needs $(n + k - 1)$ exponentiations. Thus the total number of serial exponentiation for a k-member Merge is $(n + 2k + 1)$.

3.4. Leave

The leave operation removes one or more members from the group of n members. As a result, a new group shared key S_{n-k}, is computed where k is the number of the departing members. Let the set $\mathcal{K}$ denote the leaving members.

There are two cases to consider: 1) the current controller $M_n \in \mathcal{K}$ and 2) $M_n \notin \mathcal{K}$. In the first case, the controller's role is passed on to the most recent remaining member M_d where $d = n - k$. (See Section 4.1 below.) Assuming that members are indexed in the order of joining the group (i.e., M_n is the most recent member to join) the remaining members must be *renumbered* after a Leave. Now, even if the previous controller is not in $\mathcal{K}$, it will acquire a new index and become M_d due to renumbering.

The protocol runs as follows:

1. The controller M_d generates a new exponent N'_d, produces the following set and broadcasts it to the remaining group.

   $$\mathcal{M} = \{g^{K_{i,d} N_1 \ldots N'_d / N_i} \mid M_i \notin \mathcal{K}\}$$

2. Having received $\mathcal{M}$, each M_i computes:

   $$S_d = (g^{\frac{N_1 \ldots N'_d}{N_i} \cdot K_{i,d}})^{K_{i,d}^{-1} \cdot N_i} = g^{N_1 \ldots N_d}$$

Members of the departing set $\mathcal{K}$ cannot compute the new key since the group controller only computes partial keys for the remaining group.

Step 1 requires $n - k$ exponentiations whereas step 2 needs a single exponentiation by each member. In total, the leave operation requires $(n - k + 1)$ serial exponentiations.

3.5. Key refresh

The key refresh operation updates the current group shared key, S_n. It is, in fact, a special case of a Leave with $k = 0$. Its usage depends on the application policy.

4. CLQ_API

CLQ_API is a group key management API based on the CLIQUES protocol suite. It is fairly small, containing only eight function calls.

CLQ_API does not perform any communication. All API calls either accept a token, produce one or both. As mentioned before, CLIQUES (and CLQ_API) requires reliably and sequenced communication and membership view synchrony. Any communication system providing these features can be used in conjunction with CLQ_API , e.g., SPREAD, TOTEM, and TRANSIS might be used.

4.1. CLQ_API Calls

This section describes CLQ_API calls. Each call represents one or more steps in one of the group operations presented in Section 3. Details of the data structures, definitions and secondary function calls can be found in the appendix.

The following terms are used throughout this section:

[4] Note that $\mathcal{M}$ is the same as its counterpart in the Join protocol.

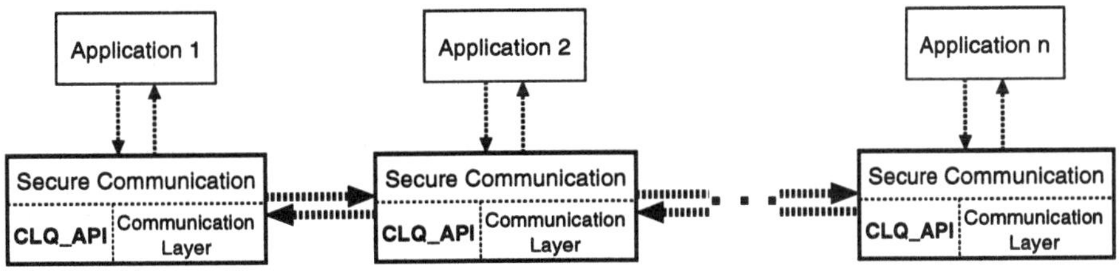

Figure 1. Communication layer and CLQ_API

Context: contains information about each user required by CLQ_API.

Epoch: message sequence number of a token. It is used to keep track of the group secret evolution and to prevent replay attacks.

Timestamp: time when a token is sent. It is used to check the freshness of an incoming token.

Session random: secret key share generated by each member.

Token: basic protocol unit; includes: epoch, timestamp, and a CLIQUES protocol message.

We now describe the CLQ_API calls:

- clq_proc_join: performs *step 1* of the join operation. The controller calls this function to hand over information about the current group to a new member who will eventually become the new controller. The purposes of this function are:
  - generate a new session random for the current controller.
  - remove long term keys and previous session random from the partial keys of all users.
  - add the new session random into the partial keys of all users.
- clq_join: performs *step 2* of the join operation. The new group member calls this function using the token received from the current controller. The purposes of this function are:
  - generate a new session random for the new member.
  - generate long-term keys between the new member and each user (can also be done before calling this function).
  - compute the partial keys for other group members.
- clq_update_ctx: performs the ***last step*** of the join, merge, leave, and key refresh operation. Every member calls this function in order to update the group shared key upon receipt of the token sent by the current controller. The main purpose of this function is to compute group shared key with the caller's session random and the incoming token.
- clq_update_key: performs *step 1 and 2* of the merge operation. This function is called by the current controller and by all (but last) of the new members to add their session randoms to the group shared key.

  The main purposes of this function are:
  - generate new session random.
  - add new session random to the group shared key.
- clq_factor_out: performs *step 4* of the merge operation. Every group member (except the controller) calls this function to factor out its own session random number from the group shared key.
- clq_merge: performs *step 5* of the merge operation. The last new member (new controller) calls this function to add its session random to each of received partial keys from the group members.
- clq_leave: performs *step 1* of the leave operation. Every group member calls this function right immediately after one or more members leave the group. A group member may also become the current controller after calling this function; this can happen if it finds itself to be the newest member within the remaining group. The main purpose of this function is to remove the information of the leaving members. If the controller calls this function, then two more steps need to be performed:
  - generate new session random.
  - compute partial keys for other group members (except, of course, the departed ones).

- clq_refresh_key: performs *step 1* of the key refresh operation. The current controller calls this function, when the group shared key needs to be updated. The main purposes of this function are:
  - To generate new session random number.
  - To compute new partial keys by adding this session random number and removing old one.

4.2. Operations of CLQ_API

Before we explain each of the group operations in the API, we need to define some message types:

- NEW_MEMBER : sent by the controller to a new member, when a new member joins the group.
- KEY_UPDATE_MESSAGE : sent to every member in order to update the group shared key.
- MERGE_KEY_UPDATE : sent to every member to update the group shared key in merge operation.
- MERGE_BROADCAST : broadcast from the last new member when one or more members merge.
- MERGE_FACTOR_OUT : sent by each group member to the new controller in a merge operation.
- MASS_JOIN : sent to the next new member in a merge operation.

We now clarify group operations using the function calls from the previous section.

- Join

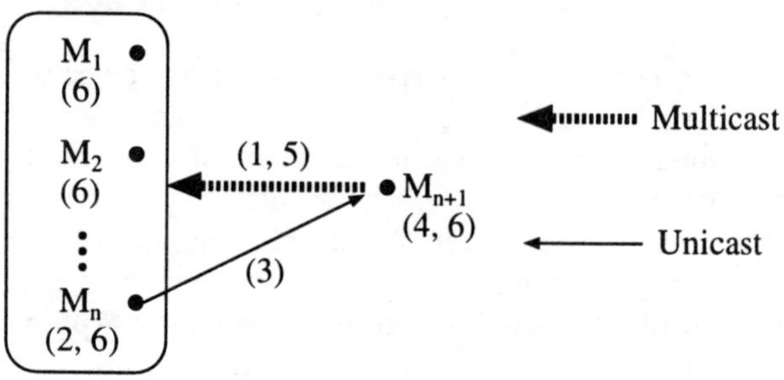

Figure 2. Join operation

(1) New member M_{n+1} calls communication protocol JOIN primitive.

(2) The current controller calls clq_proc_join to generate a token containing NEW_MEMBER message.

(3) The token is sent to the new member.

(4) The new member calls clq_join to generate a token containing KEY_UPDATE_MESSAGE.

(5) The token is broadcast to the entire group.

(6) Every user calls clq_update_ctx to compute the new group shared key.

- Merge (single member)

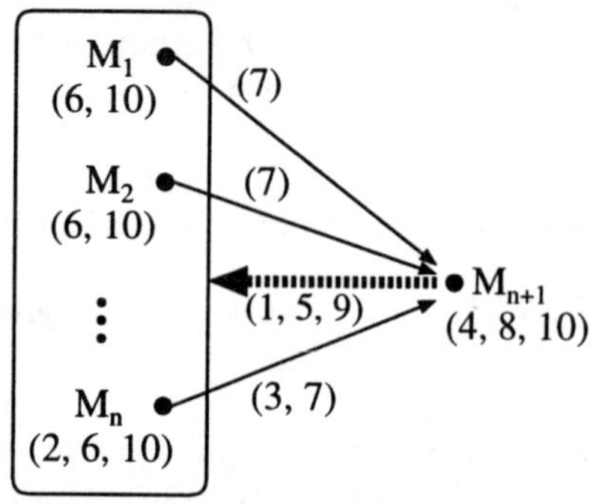

Figure 3. Merge operation

(1) New member M_{n+1} calls communication protocol JOIN primitive.

(2) The current controller calls clq_update_key to generate a token containing MASS_JOIN message.

(3) The token is sent to the new user (who becomes the new controller).

(4) The new controller calls clq_update_key to generate a token containing MERGE_BROADCAST message.

(5) The token is broadcast to the group.

(6) Each member (except the new one) calls clq_factor_out to generate a token containing MERGE_FACTOR_OUT message.

(7) Each member sends the output token back to the new controller.

(8) For each received token, the new controller calls clq_merge to generate a token containing MERGE_KEY_UPDATE message. When the last token is received, MERGE_KEY_UPDATE returns an output token.

(9) The token is broadcasted to the entire group.

(10) Every user calls clq_update_ctx to compute the new group shared key.

- Merge (multiple members)

(1) A MERGE event occurs either at an explicit request or as a result of a multiple members calling communication protocol join to join a group.

(2) The current group controller calls clq_update_key to generate a token containing MASS_JOIN message.

(3) The token is sent to the first merging member.

(4) Upon reception of the token, the next user calls clq_update_key to generate a token containing MASS_JOIN message.

(5) The token is sent to the next new member.

(6) Upon reception of the token, the last user calls clq_update_key to generate a token containing MERGE_BROADCAST message.

(7) The token is broadcasted to the entire group.

(8) Upon reception of the message, each member except the last one calls clq_factor_out to generate a token containing MERGE_FACTOR_OUT message.

(9) The token is sent back to the last new user.

(10) For each output token, the last new member calls clq_merge to generate a token containing MERGE_KEY_UPDATE.

(11) The token is broadcasted to the entire group.

(12) Every user calls clq_update_ctx to compute the new group shared key.

- Leave

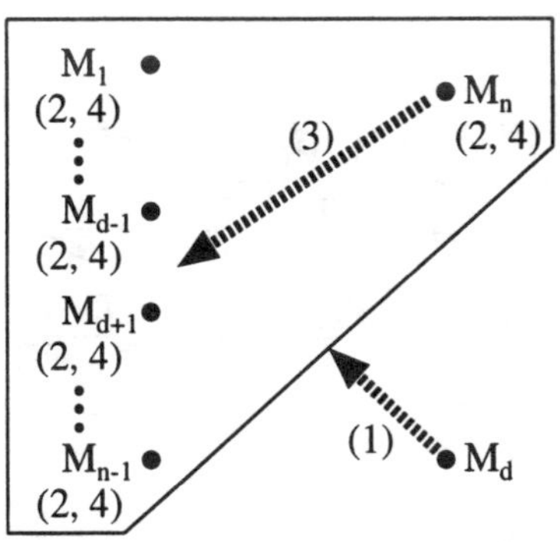

Figure 5. Leave operation

(1) The leaving member M_d calls communication protocol LEAVE primitive. (In case of a member disconnect or network failure the communication system notifies the group of the event.)

(2) All group members call clq_leave; only the group controller obtains the output token containing KEY_UPDATE_MESSAGE.

(3) The token is broadcast to the entire group.

(4) Upon reception of the token, every user calls clq_update_ctx to compute the new group shared key.

- Key refresh

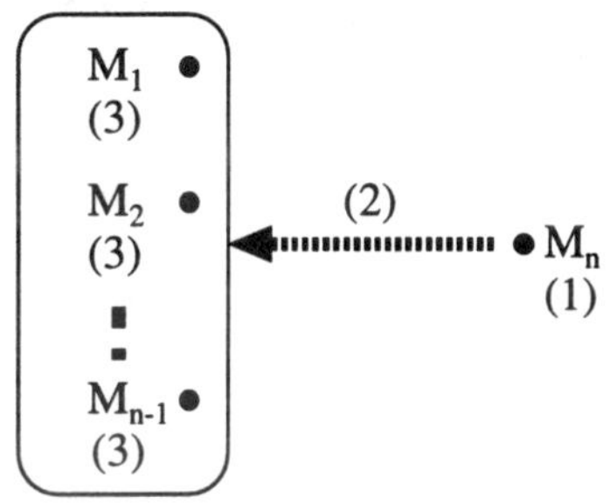

Figure 6. Key refresh operation

(1) The controller calls clq_refresh_key to generate a token containing KEY_UPDATE_MESSAGE.

(2) The token is broadcasted to the entire group.

(3) Upon reception of the token, every user calls clq_update_ctx to compute the new group shared key.

5. Efficiency

The communication overhead is summarized in Table 1.

Table 2 illustrates computation costs. Exponentiation is the most expensive operation as it requires $O(log^3p)$ bit operations in Z_p^*. Given a and p, finding the inverse of $a \in Z_p^*$ requires only $O(log^2p)$ bit operations (using the extended Euclidean algorithm). Similarly, the multiplication of a and b modulo p requires $O(log^2p)$ bit operations. See [9, 10] for more details. Hence, the speed of each operation depends largely on the number of serial exponentiations. (Note that the cost for generating the long term keys is not included in this table.)

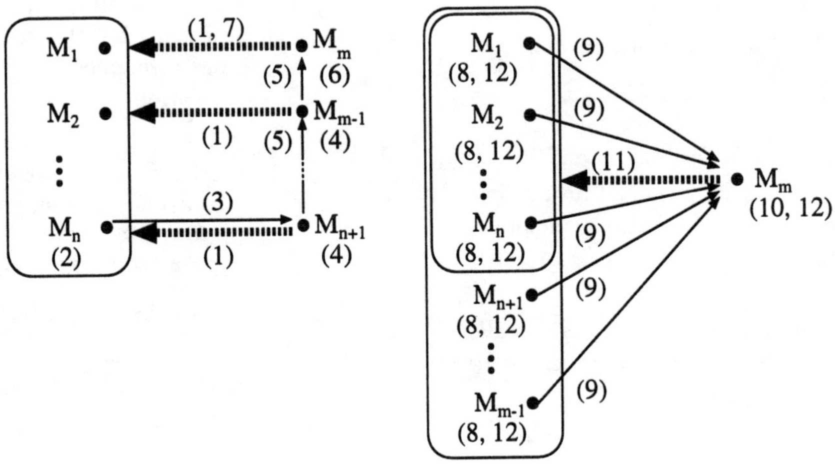

Figure 4. Merge(several members) operation

Table 1. Communication costs

Operations	Join	Merge	Leave	Refresh	k-Merge	k-Leave
Number of users after operation	n+1	n+1	n-1	n	n+k	n-k
Rounds	2	3	1	1	k+2	1
Broadcasts	1	2	1	1	2	1
Reverse broadcast	0	1	0	0	1	0
Total messages	2	n+2	1	1	n+2k	1
Maximum bandwidth	n	n	n-1	n-1	n+k-1	n-k

Table 2. Computation costs

Operations	Join	Merge	Leave	Refresh	k-Merge	k-Leave
Number of users after operation	n+1	n+1	n-1	n	n+k	n-k
Serial exponentiation	2n+1	n+3	n-1	n	n+2k+1	n-k
Total exponentiation	3n+2	3n+2	2n-3	2n-1	3n+4k-2	2n-2k-1

5.1. Exponentiation

As mentioned earlier, modular exponentiation is most expensive operation in CLIQUES protocols. In this section, we compare the performance of modular exponentiation operations using three different cryptographic libraries on three different processors(See Table 3).

The cryptographic libraries are summarized as follows:

- RSAREF[13]
  A cryptographic toolkit for privacy-enhanced mail. We used version 2.0 developed in 1996.

- Crypto++[5]
  A public domain C++ class library of cryptographic schemes published by Wei Dai. Note that, since addition and subtraction are implemented for the Pentium assembler, this package performs better on Pentium than in other microprocessors. We used version 3.1 developed in May 1999.

- OpenSSL[12]
  A successor of SSLeay[8], OpenSSL is a cryptographic toolkit implementing Secure Socket Layer(SSL v.2/3)[7]. (Implements some basic operations in assembler on various platforms.) We used version 0.9.3a developed in May 1999.

We measured the performance of modular exponentiation $y = g^x \pmod{p}$, where p is a random 512-bit prime, g a 512-bit generator for GF(p) of order q(160 bit), and x a random 160 bit integer. Table 4 shows the comparison.

At the time of this writing, OpenSSL appears to be the fastest of all public domain cryptographic libraries. On a Pentium II processor it requires only 2.5 *msecs* for each 512-bit modular exponentiation.

6. Conclusion

This paper describes a protocol suite and an API designed specifically for securing dynamic collaborative applications in unreliable networks. The API is based on the extensions of Diffie-Hellman key agreement developed in the CLIQUES project. It provides core security services (most notably, authenticated key agreement) for relatively small and dynamic peer groups.

References

[1] Y. Amir, D. Dolev, S. Kramer, and D. Malki. A communication sub-system for high availability. In *22nd Annual International Symposium on Fault-Tolerant Computing (FTCS)*, 1992.

[2] Y. Amir and J. Stanton. The spread wide area group communication system. Technical Report CNDS-98-4, The Johns Hopkins University, avalaible at www.spread.org/docs/docspread.html, 1998.

[3] G. Ateniese, M. Steiner, and G. Tsudik. Authenticated group key agreement and friends. In *ACM Symposium on Computer and Communication Security*, November 1998.

[4] R. Canetti and B. Pinkas. *A taxonomy of multicast security issues*, April 1999. draft-irtf-smug-taxonomy-00.txt.

[5] W. Dai. Crypto++, May 1999. available at www.eskimo.com/~weidai/cryptlib.html.

[6] W. Diffie and M. Hellman. New directions in cryptography. *IEEE Transactions on Information Theory*, IT-22(6):644–654, Nov. 1976.

[7] K. E. Hickman and T. Elgamal. The SSL protocol. RFC draft, Netscape Communications Corp., June 1995. Version 3.0, expires 12/95.

[8] T. Hudson and E. Young. SSLeay, Dec. 1998. available at www.ssleay.org.

[9] N. Koblitz. *A Course in Number Theory and Cryptography.* Springer-Verlag, New York, 1987.

[10] A. Menezes, P. V. Oorschot, and S. Vanstone. *Handbook of applied cryptography.* CRC Press series on discrete mathematics and its applications. CRC Press, 1996. ISBN 0-8493-8523-7.

[11] L. Moser, P. Melliar-Smith, D. Agarwal, R. Budhia, and C. Lingley-Papadopoulos. Totem: A fault-tolerant multicast group communication system. *Communications of the ACM*, 39(4), 1996.

[12] OpenSSL Project team. Openssl, May 1999. available at www.openssl.org.

[13] RSA Laboratories. Rsaref2.0(tm): A cryptographic toolkit for privacy-enhanced mail, Jan. 1996. available at www.rsa.com.

[14] M. Steiner, G. Tsudik, and M. Waidner. Diffie-hellman key distribution extended to groups. In *3nd ACM Conference on Computer and Communications Security*, pages 31–37. ACM Press, Mar. 1996.

[15] M. Steiner, G. Tsudik, and M. Waidner. CLIQUES: A new approach to group key agreement. In *IEEE International Conference on Distributed Computing Systems*, May 1998.

Appendix: Data Structures and Definitions of CLQ_API

- CLQ_CONTEXT
  - Type : Structure
  - Description : This structure contains the context of a member, M_i, in a specific group.
  - Contents
    * member_name (String): Name of the user M_i
    * group_name (String): Name of the group

Table 3. Target platform

Machine	CPU speed	Main memory	OS	Compiler
Ultra Sparc I	233 MHz	64 MB	SUN OS 5.5.1	gcc 2.7.2.2
Pentium I	233 MHz	48 MB	Linux 2.0.36	gcc 2.7.2.3
Pentium II	450 MHz	256 MB	Linux 2.2.9	egcs 1.1.2

Table 4. Comparison results

Machine	OpenSSL	Crypto++	RSAREF
UltraSparc I	16.6 msec	60.2 msec	146.3 msec
Pentium I	6.4 msec	N/A	N/A
Pentium II	2.5 msec	7.6 msec	22.2 msec

    * key_share ($|q|$ bit integer): M_i's session random number, N_i
    * group_secret($|p|$ bit integer): Current group shared key
    * group_secret_hash (H bit integer): Hash of group_secret
    * group_members_list (CLQ_GML): List of current group members
    * first (Pointer): Pointer to the first member in the group_members_list
    * last (Pointer): Pointer to the last member in the group_members_list
    * me (Pointer): Pointer to the member M_i in the group_members_list
    * params (CLQ_PARAM): Diffie-Hellman parameters
    * key (CLQ_KEY): Private and public key of M_n
    * epoch (Integer): Last message number used

- CLQ_GML
  - Type : Double linked list of CLQ_GM data structure
  - Description : This structure is a node of the group_member_list.
  - Contents
    * member (CLQ_GM): The current group member
    * prev (Pointer): Pointer to the previous node in the list
    * next (Pointer): Pointer to the next node in the list
- CLQ_GM
  - Type : Structure
  - Description : This structure contains information about a specific member.
  - Contents
    * member_name (String): Name of the member
    * long_term_key($|p|$ bit integer): Long term shared key between myself and member_name, i.e. K_{ij} where i is related to the public key of member_name, and j to my private key
    * last_partial_key($|p|$ bit integer): Last partial key for member_name.
- CLQ_TOKEN
  - Type : Structure
  - Description : Communication token used by CLQ_API, see also CLQ_TOKEN_INFO
  - Contents
    * length (Integer): Size of t_data
    * t_data (Integer array): Contains the following encoded data: group_name, message_type, time_stamp, sender_name, epoch, group_members_list(without long_term_key)
- CLQ_TOKEN_INFO
  - Type : Structure
  - Description : This structure contains information about the token.
  - Contents
    * group_name (String): Name of the group
    * message_type (MSG_TYPE): Type of the message
    * time_stamp (Integer): Time stamp of the message
    * sender_name (String): Name of the sender
- CLQ_PARAM
  - Type : Structure

- Description : Diffie-Hellman, DH, public parameters, i.e. p, q and g
- Contents
  * p ($|p|$ bit integer): DH parameter p
  * q ($|q|$ bit integer): DH parameter q
  * g ($|p|$ bit integer): DH parameter g

- CLQ_KEY
  - Type : Structure
  - Description : Public key and private key of the user
  - Contents
    * priv_key ($|q|$ bit integer): Private key of the user
    * pub_key ($|p|$ bit integer): Public key of the user

Appendix: API Calls of CLQ_API

- clq_join(ctx, member_name, group_name, input_token, output_token)
  - Caller : New member
  - Related to : Join
  - Parameters
    * ctx (CLQ_CONTEXT : Context for the new user, this should be created in this function.
    * member_name (String): Name of the user calling this function
    * group_name (String : Name of the group that has been joined. This name has to match with the one that is included in the input_token.
    * input_token (CLQ_TOKEN): Message received from the controller. It is the output_token generated by clq_proc_join.
    * output_token (CLQ_TOKEN): New key update message to be broadcasted to the group. It will be used as input_token of clq_update_ctx.
- clq_proc_join(ctx, member_name, output_token)
  - Caller : Current controller
  - Related to : Join
  - Parameters
    * ctx (CLQ_CONTEXT): Current group context, ctx will be modified only if the caller is the controller.
    * member_name (String): Name of the new member
    * output_token (CLQ_TOKEN): This token should be used as the input_token for clq_join.
- clq_update_ctx(ctx, input_token)
  - Caller : Every group member.
  - Related to : Join, Merge, Leave, Key Refresh
  - Parameters
    * ctx (CLQ_CONTEXT): Current context of each member
    * input_token (CLQ_TOKEN): Generated by new member or by the current controller, when an update of key is required(i.e. a user join, a user left, or the key has been compromised). It should be the output_token of clq_join, clq_merge, clq_leave, or clq_refresh.
- clq_update_key(ctx, member_list, input_token, output_token)
  - Caller : Current or new controller.
  - Related to : Merge
  - Parameters
    * ctx (CLQ_CONTEXT): Current group context
    * member_list (List of string): List of names of the new members. When a new member calls this function, this list should be null(Since the input_token is valid).
    * input_token (CLQ_TOKEN): Output of clq_update_key by the current controller or previous new member. When the controller calls this function, input_token should be null. (Since the member_list is valid).
    * output_token (CLQ_TOKEN): Contains refreshed group_secret. It will be used as input_token of
      · clq_factor_out by the current group members or
      · clq_update_key by the next new member.
- clq_factor_out(ctx, input_token, output_token)
  - Caller : Every group member (except the last one[5])
  - Related to : Merge
  - Parameters
    * ctx (CLQ_CONTEXT): Current member context
    * input_token (CLQ_TOKEN): output_token of clq_update_key
    * output_token (CLQ_TOKEN): Contains updated last_partial_key by removing each user's key_share. It will be used as input_token of clq_merge.

[5]If the current controller, i.e. the last member, calls clq_factor_out then the output_token will return NULL

- clq_merge(ctx, sender_name, input_token, output_token)
  - Caller : The last new member
  - Related to : Merge, Mass Join
  - Parameters
    * ctx (CLQ_CONTEXT): Current group context. ctx will be modified.
    * sender_name (String): Name of the sender of the input_token
    * input_token (CLQ_TOKEN): output_token of clq_factor_out
    * output_token (CLQ_TOKEN): Contains updated last_partial_key
- clq_leave(ctx, member_list, output_token)
  - Caller : Every group member
  - Related to : Leave
  - Parameters
    * ctx (CLQ_CONTEXT): Current group context
    * member_list (List of string): List of members leaving
    * output_token (CLQ_TOKEN): Updated message to be broadcasted to the group
- clq_refresh_key(ctx, output_token)
  - Caller : Controller
  - Related to : Key Refresh
  - Parameters
    * ctx (CLQ_CONTEXT): Current group context. ctx will be modified.
    * output_token (CLQ_TOKEN): Updated message to be broadcasted to the group
- clq_destroy_ctx(ctx)
  - Description : Frees the space occupied by the current context
  - Parameters
    * ctx (CLQ_CONTEXT): Current group context. ctx will be destroyed.
- clq_destroy_token(token)
  - Description : Frees the space occupied by the token
  - Parameters
    * token (CLQ_TOKEN): Input or output_token
- clq_first_user(ctx, member_name, group_name)
  - Description : clq_first_user is called by the first user who joins a group
  - Main purposes
    * Generates key_share.
    * Generates member context.
  - Caller : The first member in a group
  - Related to : Join
  - Parameters
    * ctx (CLQ_CONTEXT): Current member context. ctx should be created.
    * member_name (String): Name of the first user
    * group_name (String): Name of the group

Trading Off Strength and Performance in Network Authentication: Experience with the ACSA Project[1]

Jamison M. Adcock, David M. Balenson, David W. Carman,
Michael Heyman, and Alan T. Sherman[2]
Cryptographic Technologies Group
NAI Labs, The Security Research Division
Network Associates, Inc.
{jamison_adcock, david_balenson, david_carman, michael_heyman, alan_sherman}@nai.com

Abstract

The Adaptive Cryptographically Synchronized Authentication (ACSA) Project offers a new approach to data authentication in networks by trading off authentication strength and performance. In ACSA, the communicants select among various authentication gears to balance their performance and security needs. These gears include three basic groups: (1) conventional mechanisms that are computationally intensive but considered highly secure; (2) higher-speed, lower-strength mechanisms including Universal Message Authentication Codes (UMACs) and our novel inner-function group (IFG) with bit scattering; and (3) Partial MACs (PMACs) that calculate the authentication tag on only a subset of the message. We are implementing a prototype ACSA System based on the popular IPsec protocols and are demonstrating its effectiveness on high-speed network applications.

1. Introduction

Conventional authentication mechanisms do not operate at speeds fast enough to meet the demands of ultra-fast networks [1][18][23]. This disparity presents a great challenge for high-speed applications that demand network authentication. Much work has focused on devising fast authentication algorithms. The *Adaptive Cryptographically Synchronized Authentication (ACSA)* Project [7] provides a new solution to this challenge by trading off authentication strength and performance to achieve data origin authentication and connectionless integrity.

1.1. Motivation and overview

Some of the applications that motivate our work include real-time high-speed video, high-performance distributed computing, high-speed distributed storage, and 3-D virtual reality. Such applications might run on conventional platforms with high-speed network devices (*e.g.* Gigabit Ethernet or ATM networks on Pentium class machines without cryptographic hardware), on special platforms with conventional cryptographic hardware, or on computationally limited single-processor devices. The ACSA Project offers a practical and flexible solution to high-speed network authentication that can be implemented in software or hardware.

Figure 1 illustrates hypothetical strength-performance tradeoffs that ACSA might achieve. For example, the user can select a lower-speed, high-strength authentication mechanism, a higher-speed, lower-strength mechanism, or a high-speed *Partial Message Authentication Code (PMAC)* that authenticates only a portion of the message. In Section 7.2, we present a concrete instantiation of Figure 1 showing the actual strength-performance levels achieved by our implementation of an ACSA prototype system.

Throughout, we use the term "network authentication" to mean data origin authentication and connectionless integrity, as defined in Internet standards [13]. This type of authentication is without regard to the ordering of the message in a stream of messages.

---

[1] This work was supported by the Defense Advanced Research Projects Agency (DARPA) under Air Force Research Laboratory (AFRL) Contract No. F30602-98-C-0215.

[2] Sherman is also with the Department of Computer Science and Electrical Engineering, University of Maryland, Baltimore County (UMBC).

By "authentication strength" we mean the computational difficulty of producing a forgery. By "performance" we mean computational complexity (CPU time, memory space). The ACSA project aims to lighten the processing loads of the sender and receiver by requiring them to spend less processor resources (especially CPU time) on data authentication.

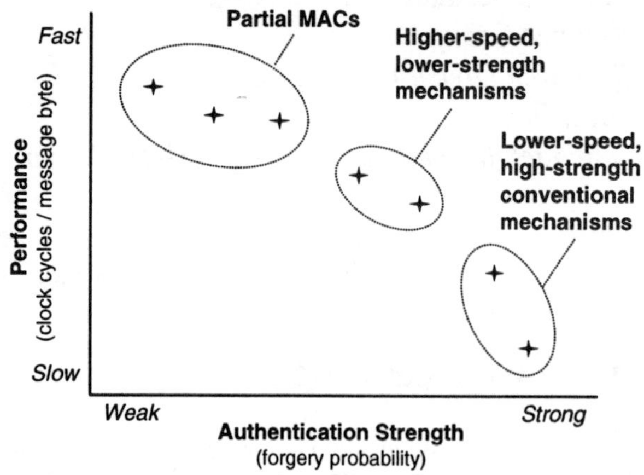

Figure 1. Hypothetical strength-performance tradeoffs for network authentication.

Most previous approaches to network authentication focused on providing high strength, with speed as a secondary concern. By contrast, we adopt the point of view that not all applications require high-strength authentication, and that faster authentication speeds can be achieved by accepting lower-strength (yet adequate) authentication methods. The ACSA System adaptively adjusts its authentication strength and speed to meet current needs based on security policy, observed authentication error rates, alarms from host or network defenses, and processor loading.

The ACSA System embodies a spectrum of authentication mechanisms—like gears of an automobile transmission—that provide various strength-performance levels of network authentication. These authentication gears are organized in the following three classes:

- lower-speed, high-strength conventional mechanisms;
- higher-speed, lower-strength mechanisms; and
- PMACs that authenticate only a portion of the message.

The lower-speed, high-strength mechanisms include algorithms such as HMAC-MD5. The higher-speed, lower-strength ACSA gears include other Message Authentication Codes (MACs) and our novel inner-function groups (IFGs) with bit scattering.

A significant benefit of the ACSA approach is that it supports communicants with the option of using *multiple authentication tags*. For example, with little overhead, the sender can compute two or more authentication tags with different strength-performance levels for the same message packet. Then, the receiver—independently from the sender—could in real time select which tag to verify. This flexibility is useful in unicast and multicast environments when the receiver cannot predict future processor loads, and in multicast environments where there may be many receivers with different processor capabilities.

To demonstrate the strength-performance levels achievable by the ACSA approach, we designed and implemented a prototype ACSA System. We based this prototype on the Internet Protocol Security (IPsec) protocols [13] and Internet Key Exchange (IKE—formerly ISAKMP/Oakley) protocol [11] because they are commonly used and provide the type of connectionless, data-origin authentication that ACSA supports.

Our prototype implementation illustrates the effectiveness of this approach to the IPsec architecture on two 400 MHz Pentium II computers operating Linux connected via a 100 MBit Ethernet network. This implementation is built from our ACSA developer's toolkit, which we are freely distributing.

In comparison with the high-strength HMAC-MD5-96, our prototype achieves speedup factors of greater than a factor of 10 for the higher-speed, lower-strength UMAC-MMX-15, and greater than a factor of 100 for PMAC-64 (using UMAC-MMX-15 as the underlying MAC algorithm). These approximate speedup factors are based on empirical measurements, as described in Section 7.

The ACSA approach is suitable for a wide range of high-speed applications. It is especially suitable for long packets such as video streams in applications that can tolerate lower levels of authentication strength. This approach, however, is not appropriate for all applications—for example, we do not recommend it for financial applications such as banking transactions requiring high levels of authentication security.

1.2. Alternative approaches

Previous work on the challenge of providing high-speed cryptographic network authentication has followed four basic approaches: First, engineers have provided separate special-purpose accelerated cryptographic hardware [1]. Despite the attractiveness of dedicated cryptographic hardware, such hardware is often expensive, supports only a limited set of algorithms and platforms, and is soon rendered obsolete by software operating on rapidly advancing general-purpose CPUs.

Second, software engineers have attempted to speed up algorithms with parallel implementations (for example, see Nahum *et al.* [18]). Whatever gains can be achieved by this strategy can also be applied within and independently of ACSA.

Third, cryptographers have designed fast authentication mechanisms. For example, there have been significant recent advances with the Bucket [21][19], MMH [10], NMH [10], and UMAC [4] algorithms. But, if a communication system depends only on one such fast mechanism, then the system is inflexible and cannot adapt to changing needs. Also, existing communication systems alone cannot easily change the current authentication mechanism in use without a costly renegotiation of the established security association.

Fourth, some researchers have attempted to increase overall security performance by combining confidentiality and authentication functions in a dual-purpose algorithm [9]. Although it can make sense to compute, for example, AES and AES-MAC in a single dovetailed fashion, dual-purpose confidentiality and authentication algorithms risk security weaknesses and forsake the flexible engineering benefits of using separate interchangeable components.

The ACSA project distinguishes itself by taking a broad and flexible engineering approach to providing fast network authentication. Although the ACSA approach can be sped up with hardware, parallelism, and faster algorithms, its gains go beyond these optimizations through offering a fundamentally different strategy of adaptively trading authentication strength for speed.

1.3. Outline

The rest of this paper is organized in seven sections. Section 2 describes the ACSA system architecture, including its components, gears, and adaptive control. Section 3 explains the particular authentication mechanisms used within ACSA—namely conventional MACs, inner-function groups, UMACs, and PMACs. Section 4 analyzes the security of ACSA. Section 5 explains the application of ACSA to IPsec. Section 6 overviews our prototype implementation, including our developer's toolkit. Section 7 reports on our experiences with our prototype implementation, concretely analyzing the system's performance and illustrating achievable risk-performance tradeoffs. Section 8 discusses issues raised by ACSA and summarizes our conclusions. In addition, Appendix A gives pseudocode for our bit-scattering algorithm.

2. Architecture

The ACSA System [2] provides network authentication between a sender and a receiver. For each message (*i.e.* packet), the sender computes an *authentication tag*, which is appended to the message in the data channel. The receiver verifies the received message by recomputing the authentication tag and comparing it with the received tag. If the recomputed and received tags match, the receiver deems the message authentic.

In addition to the data channel, there is also a control channel between the sender and receiver. Using this control channel, the sender and receiver establish a set of *security associations*. These security associations specify a suite of authentication gears and shared authentication keys. Periodically, the sender and receiver exchange control information that can influence dynamic gear changes. All control messages are protected using standard confidentiality and authentication techniques.

2.1. Goals and requirements

The goals of the ACSA System are to make available a spectrum of network authentication mechanisms with various strength-performance tradeoffs, and to support a method for selecting when to apply these mechanisms. In addition, we impose the following two engineering requirements: the ACSA System must be able to operate on a wide range of network devices (*e.g.* firewalls, security gateways, and routers); and the architecture must be consistent with the IPsec and IKE protocols.

2.2. System components

As shown in Figure 2, the ACSA System is organized into five major modules: ACSA controller; local security and resource managers; security association and key management; network application; and security services. Figure 2 shows this view of the ACSA System for the sender; the receiver has a symmetrical organization. In this figure, arrows represent flow of application data and control information.

The main purpose of the controller is to determine which authentication gear to use. In making this decision, the controller relies on inputs from the local security and resource managers and from the network application. Additionally, when the system is performing the sender role for a given connection, the controller may receive inputs from the receiver regarding authentication errors and the receiver's processor load.

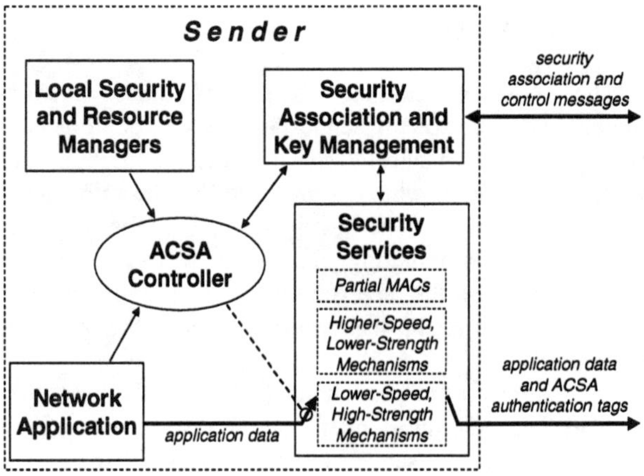

Figure 2. ACSA architecture.

The local security and resource managers represent components that may be provided by a host system. These modules offer information such as security and resource policy, processor load information, and possibly alarms from host or network defenses.

The main purpose of the security association and key management module is to establish and maintain security associations and cryptographic keys between communicants. ACSA further requires that the security association and key management module establish and maintain a suite of security associations that correspond to the ACSA gears supported for a given connection. Control messages are also sent from the receiver to the sender via the security association and key management module. In our prototype implementation, these security associations are negotiated using IKE [6].

In a system that supports ACSA, the network application may communicate to the controller ACSA-specific security and resource policy for its connection.

The main purpose of the security services module in ACSA is to perform the authentication tag computation for the gear specified by the ACSA controller. In the prototype, the IPsec module will perform all the functions of the ACSA security services module. When operating in receiver mode, the security services will also detect and report received authentication errors to the controller.

2.3. Gears, synchronization, and adaptive control

Initially, the sender and receiver agree on a set of acceptable authentication gears to use. Specifically, for consistency with IPsec protocols, for each connection, the sender and receiver initially negotiate a separate security association for each possible gear. To change gears, the controller within the sender system then selects among these agreed upon security associations based on security policy and processor load. The controller within the receiver system can effect gear changes by notifying the sender of observed authentication errors and processor resource limitations in the receiver for the current gear.

For consistency with IPsec, for each packet in our prototype system, the Security Parameter Index (SPI) [12] in the packet header specifies which gear was used to authenticate the packet. By specifying the current gear in the SPI of each packet, no other out-of-band mechanism is needed to achieve gear synchronization between the sender and receiver.

Recovery from dropped packets is performed at the application layer and is not independently guaranteed by the ACSA System. Dropping packets will not cause loss of cryptographic synchronization since the authentication gear and any derived secondary keys (see Section 2.4) depend only on the security association and the packet header.

2.4. Key management

ACSA needs key material for keyed conventional authentication mechanisms, for PMACs, and for our bit-scattering process.

During the negotiation of each connection, the ACSA System relies on the underlying network security protocol (*e.g.* IKE) to establish a common key between the communicants. This key, which we call the *primary authentication key*, is assumed to be valid for the entire lifetime of the connection. Conventional authentication mechanisms use the primary authentication key directly to compute the authentication tag. In our prototype implementation, each primary key is at least 128 bits long.

To meet the needs of additional functionality offered by ACSA, the ACSA security association and key management module derives *secondary authentication keys* from the primary authentication key using a deterministic one-way function. For example, our prototype system derives secondary authentication keys using the SHA-1 algorithm using techniques similar to those described by Matthews [17].

When performing the bit-scattering operation of IFGs or the word-selecting operation of PMACs, the ACSA System derives a secondary authentication key for each packet. By using a shorter-lifetime *secondary authentication key* rather than the longer-lifetime primary authentication key in these instances, we limit loss of any compromised keys due to any use of lower-strength authentication mechanisms.

It is important that these secondary keys be unique for different packets within a security association. Our prototype implementation achieves this objective by seeding the generation of each secondary key for a packet with both the primary key (of the security association) and with the unique packet sequence number available in IPsec to prevent replay attacks.

3. Authentication mechanisms

The authentication gears of the ACSA System are a set of authentication mechanisms that achieve a wide spectrum of strength-performance levels. They are organized in three basic classes: (1) conventional mechanisms, including lower-speed, high-strength algorithms; (2) higher-speed, lower-strength MACs; and (3) Partial MACs (PMACs). By appropriate choice of their operational parameters, PMACs can run at a variety of strength-performance levels, including ultra-fast, lower-strength levels. This section describes these mechanisms, together with ACSA's use of multiple authentication tags.

3.1. Conventional authentication mechanisms

The lower-speed, high-assurance gears of the ACSA System include conventional authentication mechanisms, such as HMACs [3][14]. Our prototype implementation supports the following two IPsec mechanisms: HMAC-MD5-96 [15] and HMAC-SHA-1-96 [16].

These HMACs infuse a key k into a message x by the nested MAC (NMAC) rule:

$$\mathrm{NMAC}_k(x) = f_{k1}(F_{k2}(x)), \qquad (1)$$

where $k1$ and $k2$ are subkeys derived from k; f is a keyed "outer" function; and F is a keyed iterated "inner" function. For the conventional IPsec mechanisms above, both f and F are based on the same underlying hash function (*e.g.* SHA-1).

This NMAC construction enjoys some desirable security properties, as explained by Bellare *et al.* [3]: under suitable security assumptions, the NMAC is at least approximately as strong as the inner function. We also use this NMAC construction in the other gear classes.

3.2. Higher-Speed, lower-strength MACs

Currently, the ACSA System includes two types of higher-speed, lower-strength MACs: inner-function groups (IFGs) and UMACs.

Both IFGs and UMACs gain speed within the NMAC construction by using a higher-speed, lower-strength iterated inner function together with a slower-speed, high-strength outer function. Because the inner function is applied many times over a long message, and the outer function is applied only once, this design offers significant speed-ups over conventional MACs. As pointed out by Bellare, Canetti, and Krawcyzk [3], the security assumptions on the inner function can be significantly relaxed while still maintaining a high level of security for the entire NMAC. The application of this observation in creating higher-speed, lower-strength NMACs is a significant contribution of the ACSA project.

We intentionally use the terms "inner function" and "outer function" to avoid unwanted security connotations associated with the terms "MAC" and "cryptographic hash function." In particular, in an ACSA-style NMAC, the inner function needs to satisfy only a weak collision-resistance property (weaker than that typically required of a MAC), and the outer function might have to satisfy pseudorandomness security properties (stronger than those typically required of a MAC).

Following this NMAC design principle, we developed IFGs with bit scattering. Later, after learning about the recent work of Black *et al.* [4], we also incorporated the UMAC mechanism into ACSA. Although UMACs do not use bit scattering, both UMACs and IFGs exploit the NMAC construction for greater speed. Both UMACs and IFGs are related to the universal hash function paradigm of Wegman and Carter [24]. Wegman and Carter, however, invented their paradigm for strength; we exploit their paradigm for speed and strength.

3.2.1. Inner-function groups with bit scattering. IFGs are based on nested MACs [3] and consist of two or more faster-speed, lower-strength inner functions that feed a high-strength outer function. Additionally, IFGs contain a *bit scattering* function that pseudorandomly selects which bits of the message feed into which inner functions.

The main cryptographic purpose of bit scattering is to prevent an adversary from knowing which parts of the message are processed by which inner functions. Even if an adversary could create collisions in one or more of the lower-strength inner functions when attacked in isolation, the adversary would have difficulty mounting an attack against the IFG because she would not know the inputs to the inner functions. The main performance advantage of IFGs stems from being able to use faster-speed inner functions.

As shown in Figure 3, the sender creates an IFG authentication tag in three steps: bit scattering; compression with inner functions; and outer function computation.

To perform bit scattering, the sender pseudorandomly throws bits of the message in one of two or more temporary buffers called *bins*. Bit scattering is simply a data-moving operation that performs no compression. To enable the sender and receiver to synchronize cryptographically, a function based on a secondary authentication key controls the pseudorandom scattering process.

To scatter bits, ACSA uses the following algorithm. First, the sender pseudorandomly creates a bit mask as long as the longest packet size to be supported. Second, the sender pseudorandomly selects a starting point (which we call a *mask offset*) within the mask. Third, processing pairs of data words at a time, the sender logically ANDs and ORs the data words with the corresponding mask words (and their bitwise complements) to select and interleave bits for the two bins (see Appendix A).

To save computation time, the mask can be reused within a security association. Our prototype generates one mask (as long as the maximum packet size) for each security association. This mask is created by expanding a secondary authentication key in a manner similar to the key expansion algorithm used in IKE [11]. The sender and receiver then use per-packet secondary authentication keys to compute the mask offsets for each packet.

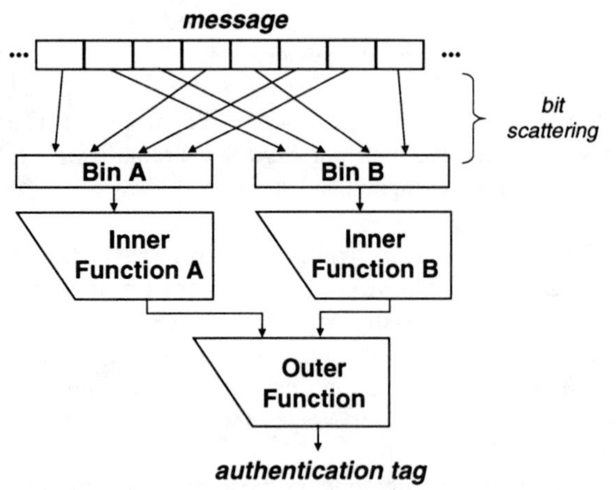

Figure 3. An ACSA NMAC realized by a two-element inner-function group with bit scattering.

Although conceptually simple, scattering bits into bins is nontrivial to perform at high speeds. On a Pentium II, our bit-scattering algorithm runs in less than 1.5 clock cycles per byte. The time to scatter bits is a significant cost of the IFG technique.

After performing bit scattering, a separate inner function is applied to each of the bins. Each inner function is keyed using a secondary authentication key. Candidate inner functions include MD4, AHA [20], MMH [10], NMH [10], Square Hash [8], Bucket [21], Evaluation Hash [22], and Division Hash [22].

The outputs of the inner functions are concatenated and fed into a high-strength outer pseudorandom function (PRF) such as SHA-1 or RC6 [4]. The PRF is keyed using the secondary authentication key as discussed in Section 2.4. The PRF output constitutes the IFG authentication tag.

3.2.2. UMACs. UMAC is a new fast message authentication algorithm by Black *et al.* [4] that runs an order of magnitude faster than high-strength conventional mechanisms such as HMAC-SHA-1. It is based on a new universal family of hash functions called *NH (nonlinear hash),* and the inventors of UMAC have proven security bounds (probability of forgery) of UMAC based on assumed bounds of NH. UMACs fit in very nicely with ACSA as plug-in higher-speed, lower-strength gears.

UMAC follows a slight modification of the aforementioned NMAC formula:

$$\mathrm{UMAC}_k(x) = f_{k1}(\ s,\ \mathrm{NH}_{k2}(x)\ ), \qquad (2)$$

where s is a nonce provided by the sender. Thus, as do IFGs, UMAC achieves fast speeds by using a higher-speed, lower-strength inner hash. Unlike the Bucket, MMH, and NMH hash function descriptions, the UMAC paper [4] provides a complete MAC specification.

The NH function computes a tree of multiplications with increasing word sizes. NH is parallelizable and by design achieves a high performance on MMX™-enabled processors. Black *et al.* propose a family of UMACs of different strengths by truncating the NH tree of multiplications at different heights (word sizes). Although the inventors did not explicitly say so in their paper, their family is an embodiment of an authentication strength-performance tradeoff.

The ACSA System applies a set of UMACs covering a range of strength-performance levels.

3.3. Partial MACs

ACSA uses the novel concept of a *Partial MAC (PMAC)* to achieve even faster authentication speeds than those attained by UMACs and by IFGs. In a PMAC, a UMAC or higher-speed, lower-strength HMAC is applied only to some portion of the message body. Because the message header contains critical information, ACSA always authenticates the entire message header (alternatively, one might always authenticate each header with strong authentication). Care must be exercised when selecting the PMAC gear: PMACs are appropriate only for applications that can tolerate some modified bits in the message body.

We may view each message as a header followed by a sequence of message body parts (*e.g.* 64-bit words). Each PMAC has an associated parameter that specifies with what probability each message body part will be authenticated. We call this parameter the *message-*

selection percentage. Although there are several different ways in which PMACs can be realized, the main effect is the same: applying an authentication algorithm only to selected parts of the message can save significant time.

Abstractly, it is appealing to think of PMACs as follows. For each message body part, the sender tosses a biased coin to decide if that part is authenticated. On Heads, the sender includes that part in the computation of the authentication tag. On Tails, the sender does not include the part in the computation.

The coin flips are generated in a deterministic pseudorandom fashion using a secondary authentication key derived from the primary authentication key. This way, the receiver can synchronize cryptographically with the sender to know what message body parts to include in the authentication tag computation.

PMACs use a very fast inner function (*e.g.* NH) to perform compression on the selected parts of the message. As with UMAC or IFGs, the outer function is still a high-strength conventional PRF. In using the NMAC construction, we do not have to generate any fake authentication tags that might otherwise have been needed.

In our prototype system, we pseudorandomly select words of the message body to authenticate as follows. We start at the beginning. The next word to be selected is determined by a pseudorandomly-chosen offset from the currently chosen word, uniformly chosen from the interval $[1, 2L]$. Here, $L = 1/p$ (rounded to next highest perfect power of 2), where p is the message-selection percentage.

When computing PMACs, our prototype selects 64-bit words and uses UMAC-MMX-15 as the underlying authentication algorithm. We chose to operate on 64-bit words to exploit the Pentium II processor's efficiency at handling that data size.

Finally, note that cyclic redundancy codes are not promising inner-function candidates because they are slower and much weaker than UMAC-MMX-15. For example, we have estimated that CRC-32 executes at approximately one cycle per byte on the Pentium II, at least twice as slow as UMAC-MMX-15.

3.4. Multiple authentication tags

The ACSA System allows and supports the use of *multiple authentication tags.* That is, the sender can compute two or more different authentication tags for the same packet, reflecting different strength-performance levels. This concept can be realized in a variety of ways; its implementation is especially efficient for UMACs and IFGs.

In our prototype implementation, the sender can compute multiple authentication tags for IFGs with relatively little overhead. For example, if an IFG has two inner functions, A and B, then a total of three authentication tags can be separately computed for A, B, and A and B combined, with only three applications of the (non-iterated) outer function. A similar but different strategy could also be applied with UMACs and PMACs.

Multiple authentication tags afford the receiver with increased flexibility in choosing strength-performance levels adaptively. For example, in unicast or multicast communications, the receiver could choose which tag to use in real time based in part on local processor load.

Multiple authentication tags are also useful in multicast environments where different receivers may have different processor capabilities or security policies, and where the sender typically has more powerful computing capabilities than do the receivers. We are not aware of any previous suggestion to use multiple authentication tags in this fashion.

To ameliorate possible security vulnerabilities caused by making multiple authentication tags of the same message available to the adversary, the ACSA System uses a separate secondary authentication key to compute each multiple tag.

4. ACSA vulnerabilities and defenses

The goal of an adversary is to tamper with one or more data words of a packet without detection. Such tampering may include modifying, deleting, inserting, padding, permuting, or replaying bits or data words within a packet or in an authentication tag. An adversary might try attacking the ACSA System by attacking its control mechanisms and by attacking its component functions.

The strength of the system can be analyzed by considering both the computational difficulty of forging an authentication tag, and the probability of modifying a message without detection.

4.1. System attacks and defenses

To attack or influence the ACSA control mechanisms, an adversary might try to spoof, modify, or replay control information. By using strong confidentiality, authentication, and anti-replay services from the underlying network security protocol, the ACSA System protects against such attacks.

An adversary can, nevertheless, block control information. Such interference can result in degradation of service, but not loss of authentication security. The ACSA System does not try to prevent denial-of-service attacks, which are existing vulnerabilities of IPsec and IKE.

A more subtle attack is to modify the communicant's environment, to cause the ACSA controller to use a weaker gear than it might otherwise use. For example, if the adversary could cause the sender's processor load to increase significantly, then the sender may switch to a weaker and faster gear. This vulnerability cannot be solved within ACSA. Each ACSA platform should protect its resources from tampering, not use resource loads to determine ACSA gears, or set an acceptable minimum strength gear.

In our original design of ACSA, for added protection, we had planned to keep the gear choice secret from the adversary. To comply with IPsec protocols, however, the gear selection in our prototype implementation is communicated as part of the SPI, which is authenticated but sent in the clear in each packet header. Fortunately, when IFGs are used, knowing the gear does not enable the adversary to determine which data bits and words contribute to which inner functions. Similarly, knowing the PMAC gear does not enable the adversary to determine which data words contribute to the authentication tag computation.

Because ACSA computes a single tag (or single set of tags in the case of multiple authentication tags) for each packet, it is not possible to use external timing information to determine which data words contribute to which inner functions, nor to determine which words are selected for partial authentication. Timing information can, however, reveal gear choices and the proportional mix of inner functions within an IFG.

4.2. Authentication mechanism security

For authentication mechanisms, the basic measure of security we consider is probability of forgery under the best possible attack. In this section, we estimate and bound the forgery probabilities for authentication mechanisms in each of the three major gear classes by reviewing the best known attacks on these mechanisms.

We also note that the security of bit scattering and PMACs depends crucially on the unpredictability of the secondary authentication keys. These keys are protected through the strong pseudorandom functions that generate them, and through the inclusion of the packet sequence number in their generation.

4.2.1. Conventional mechanisms. The probability of forgery for HMAC-SHA-1 and HMAC-MD5 is directly related to the keyed collision resistance of the SHA-1 and MD5 algorithms, respectively [3]. The best known keyless collision attacks on the underlying hash functions are birthday attacks, so the best known attack on these HMACs has a forgery probability of less than 2^{-64}.

4.2.2. Inner-Function groups with bit scattering. The security of IFGs depends on the security of the inner functions, the security of the outer function, and the effect of bit scattering. Since ACSA uses IFGs with a high-strength outer function and with higher-speed, lower-strength inner functions, for the purpose of the analysis, we may assume that the security of IFGs depends primarily on the inner functions and bit scattering.

Consider an IFG containing two inner functions A and B. Without bit scattering, the adversary would know which message words fed into A and which words fed into B. As a result, the forgery probability of the ensemble would be at least that of the weaker of the two inner hashes. Therefore, bit scattering is a crucial security-enhancing operation for IFGs.

When an IFG employs bit scattering, an adversary has two possible attacks: successfully guess which message bits are processed by which inner functions and attack one of the inner functions; or construct an attack that succeeds regardless of how the message bits are scattered. Since the later possibility seems implausible, it appears that to attack an IFG with bit scattering requires the adversary to guess how the bits to be modified are scattered.

Let n be a typical message length (in bits), and suppose that approximately k of the message bits are scattered into inner hash function A. Suppose further that the adversary wishes to modify t message bits. The chance of correctly naively guessing how these t message bits were scattered is:

$$(k / n)^t. \tag{3}$$

In this sense, the bit-scattering operation appears to enhance authentication strength greatly. We assume that the adversary knows n and k, since these values could be approximated from network and timing analysis. Finally note, that from Equation 2, the scattering process creates significantly greater uncertainty for the adversary when carried out on a bit-by-bit basis rather than on words.

An adversary might try chosen-plaintext attacks in which she observes several message-authentication tag pairs computed from a common mask. However, the per-packet mask offsets, the strong outer function, and the per-packet rekeying of the outer function deter an adversary from garnering information about the mask.

4.2.3. UMACs. Black *et al.* [4] and Rogaway[3] describe forgery probabilities for various UMACs. We summarize these probabilities in Table 1.

Table 1. UMAC probability of forgery.

UMAC version	Probability of Forgery
UMAC-MMX-60	2^{-60}
UMAC-MMX-45	2^{-45}
UMAC-MMX-30	2^{-30}
UMAC-MMX-15	2^{-15}

4.2.4. PMACs. The probability of forgery of a PMAC is based overwhelmingly—not on the strength of the underlying MAC—but on its message-selection percentage. For PMACs with very small message-selection percentages, the probability of forgery is very high (nearly unity).

The utility of PMACs stems not from their ability to detect short modifications of a message; their utility arises from their ability to detect the modification of many bits or words. To this end, consider the probability Ψ_t of successfully modifying at least t message words. This probability is:

$$\Psi_t = (1-p)^t, \quad (4)$$

where p is the message-selection percentage.

Evaluating PMAC security in the context of the number of modified data words is useful for many high-speed applications (*e.g.* high-speed video) that can tolerate some data modifications without a significant loss of fidelity.

5. Application to IPsec

ACSA is applicable to the common network security IPsec protocols. Specifically, ACSA can enhance network authentication service in IPsec by adding higher-speed, lower-strength mechanisms, and by providing an adaptive control system that adaptively switches among IPsec authentication transforms.

Currently, IPsec uses only lower-speed, high-strength authentication algorithms. The two mandatory-to-implement authentication algorithms HMAC-SHA-1 and HMAC-MD5 are so computationally intensive that they must be implemented in special-purpose hardware to support very high-speed network applications. IPsec also specifies HMAC-TIGER as an optional alternative. HMAC-TIGER, however, does not provide significant performance improvement over HMAC-SHA-1 and HMAC-MD5, except on 64-bit processor platforms. Therefore, the higher-speed, lower-strength mechanisms from ACSA offer attractive options for speeding up network authentication in IPsec.

The ACSA control system can be applied in IPsec to select authentication transforms adaptively based on current security policy, detected authentication errors, and processor load. Using the IKE protocol, ACSA initially determines a suite of authentication gears common to the communicants (see Section 2.3). This system of adaptive control gives IPsec a practical strategy for adaptively choosing from among a wider range of strength-performance levels.

6. Prototype system: Design, implementation, and toolkit

The main goals of the ACSA prototype are to demonstrate the feasibility of the ACSA model in enhancing network authentication performance, and to develop a freely available developer's software toolkit that performs and allows others to experiment with and use ACSA functions. At its highest level, the ACSA prototype software comprises an ACSA developer's toolkit, processor-specific optimizations, prototype demonstration software, and third-party network security software.

The ACSA developer's toolkit contains the core functionality that every ACSA implementation must perform. Primarily, the toolkit provides functionality for the controller module. The ACSA developer's toolkit also contains additions designed to integrate easily within most IKE and IPsec implementations.

The ACSA prototype contains processor-specific optimizations for the various supported authentication mechanisms. Optimization of conventional algorithms is important to provide an accurate baseline of the achievable authentication performance of current implementations. Optimization of the proposed authentication mechanisms such as UMACs and PMACs is necessary to demonstrate the performance enhancement that ACSA is capable of providing.

Demonstration software and third-party network security software is also included in the ACSA prototype to provide a fully operational system capable of showing the advantages of ACSA. The third-party software performs the IPsec and IKE protocols. The target platforms for the ACSA prototype are two 400 MHz Pentium II computers operating Linux connected via a 100Mbit Ethernet network connection.

In considering the costs of integrating ACSA, it is important to consider the development cost of adding ACSA to an existing IPsec implementation. Although ACSA controller software is in a self-contained module, significant modifications are required of the existing

[3] Phil Rogaway, personal correspondence (May 4, 1999).

IPsec module to perform ACSA. Specifically, new higher-speed, lower-strength authentication mechanisms must be added, and the ability to handle multiple security associations for a single connection is needed. Modifications to the existing IPsec module can be costly since changes may be required to the operating system's kernel code, or to hardware or firmware in an IPsec co-processor. Modifications to support ACSA must also be made to the IKE module, including additions to interact with the ACSA controller, to negotiate multiple security associations per connection, and to communicate ACSA control messages.

7. System performance

The computational performance of the ACSA system is almost entirely dependent on the performance of its authentication mechanisms. The administrative overhead establishing a gear suite, retrieving local processor load and security condition information, deriving keying material, and determining the appropriate gear is relatively small when amortized over all computations required to create or verify authentication tags for a high-speed network connection.

We give performance measurements for the Pentium II for two reasons. First, in order to give any meaningful detailed high-performance measurements, it is necessary to optimize implementations for some particular processor. Second, we implemented our prototype system on the commonly used Pentium II. Our Pentium II benchmarks calibrate our performance results.

The following two subsections discuss the performance of our prototype implementation on the Pentium II, focusing on the computational costs of performing selected authentication mechanisms and on the achieved strength-performance levels.

7.1. Analysis of system performance

This section describes the performance of each of the three major gear classes in our prototype implementation.

7.1.1. Conventional algorithms. For the large packet sizes that demonstrate the greatest advantages of ACSA, the computational performance of the HMAC-SHA-1 or

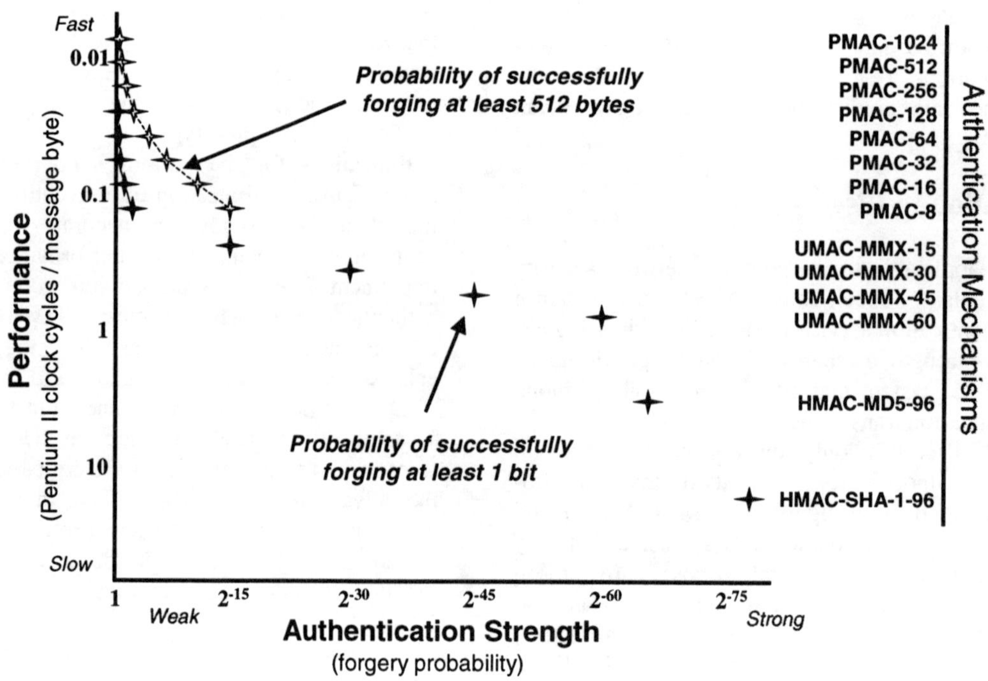

Figure 4. Strength-performance tradeoff measurements for our prototype implementation of the ACSA system.

HMAC-MD5 algorithm is almost entirely dependent on the performance of the underlying SHA-1 or MD5 algorithms, respectively. The fastest assembly language implementation of HMAC-SHA-1 we encountered runs at 12.6 clock cycles per message byte, according to Black *et al.* [4].

As for MD5, Bossalaers [5] reports that his optimized version on the Pentium runs at 5.3 cycles/byte. Since the Pentium II offers increased computational efficiency over the Pentium for most operations, we expect an optimized version of MD5 on the Pentium II to perform even faster.

7.1.2. Higher-Speed, lower-strength algorithms. Since the ACSA prototype will use the Pentium II processor, we may take advantage of the dramatic performance of the UMAC-MMX family of algorithms that use MMX functionality for much of their computations. Black *et al.* [4] report that their implementations of UMAC-MMX-60 and UMAC-MMX-30 run in 0.98 and 0.51 cycles/byte, respectively. Ted Krovetz reports that his implementation of UMAC-MMX-15 runs in 0.32 cycles/byte.[4] These measurements were taken on message data that reside in the Pentium II L1 cache.

Thus, when using an IFG with two separately-keyed inner UMAC-MMX-15 functions, the total speed of IFGs with bit scattering is approximately 1.5 + 2 * (0.32 * ½) = 1.82 cycles/byte on an MMX-enabled Pentium II.

7.1.3. Partial MACs. The performance of PMACs is dependent primarily on selecting message words and compressing the selected words. Although selecting the data words is a linear operation in message length, its dominant per-byte cost is the product of the percentage of data selected and the cost to select each data word. Similarly, the per-byte cost of compressing these words is the product of the percentage of data selected and the per-byte cost of the inner hash function. Thus, the dominant cost is reduced multiplicatively by the selection percentage. For example, in our prototype when selecting one word per 64 message words, the PMAC cost is approximately 0.05 clock cycles per message byte—over 10 times faster than our fastest non-PMAC authentication algorithm, and over 100 times faster than HMAC-MD5-96.

7.2. Strength-Performance tradeoff

Figure 4 summarizes the strength-performance levels achieved in our prototype implementation on a Pentium II. Specifically, Figure 4 shows, for various authentication mechanisms, speed versus the probability of forgery. A notable characteristic of this tradeoff is the large span of security-performance levels over which ACSA provides authentication. Also notable is the dramatic worsening of the forgery probability when PMACs are used. This increase in the forgery probability, however, is less dramatic when evaluated for large amounts of modified data.

8. Discussion and conclusion

ACSA System provides a new approach for meeting performance and authentication demands of high-speed networks. In ACSA, the communicants can choose various strength-performance levels, and they can change their choices within a communication association. This flexibility to choose, and to change within a connection, offers advantages over simply always using one authentication mechanism.

In addition to its overall gear-switching approach, the ACSA Project contributes and applies the novel ideas of Partial MACs (PMACs), inner-function groups (IFGs) with bit scattering, and multiple message authentication tags. Limitations of the ACSA System include lack of provable security (*e.g.* the security of IFGs with bit scattering is now heuristic), possibility of system attacks (*e.g.* an adversary might be able to alter the processor load and thereby cause a gear switch), and additional complexity. For example, in comparison with implementing a single authentication mechanism, it is more costly to implement the ACSA System and it is more difficult to verify the correctness of such an implementation. Nevertheless, such system attacks and implementation problems would likely at worst result in a lower-strength gear being used.

Our immediate future plans for ACSA are to complete and test our prototype system implementation to measure its speeds empirically at various strength levels. We also plan to investigate additional ways of choosing and changing gears adaptively, such as automatically sensing and responding to varying degrees of redundancy in the data stream, and including Quality of Service mechanisms of applications in the security association negotiations. In addition, we plan to write a draft specification (Internet RFC) for implementing ACSA with IKE and IPsec.

Open questions raised by the ACSA Project include the following. (1) What is the security of bit scattering and how much advantage does this operation contribute in relation to its performance cost? (2) Is it possible to extend ACSA ideas to data confidentiality? Although it is possible to combine ACSA with precomputed stream ciphers for fast confidentiality, we do not see how to extend such ideas as IFGs and PMACs to data confidentiality.

[4] Personal email from Ted Krovetz (May 11, 1999).

Using a variety of engineering techniques, the ACSA System offers a practical approach for providing data authentication on high-speed networks by trading off performance and authentication strength.

Acknowledgments

We thank Dennis K. Branstad for originally proposing the ideas of authentication strength-performance tradeoffs, partial message authentication, and gear groups. Phil Rogaway contributed very helpful detailed comments on our preliminary work. Ted Krovetz provided us with assistance in implementing UMAC. Also, we thank Hugo Krawczyk for constructive feedback.

References

[1] S. Abbott, "Cryptographic interfaces: supporting 100mbps and beyond," *Proceedings of the 1999 RSA Data Security Conference.*

[2] D. Balenson, D. Carman, M. Heyman, and A. Sherman, "Adaptive Cryptographically Synchronized Authentication (ACSA): Model and analysis," Revision 1.0 (December 7, 1998). 53 pages.

[3] M. Bellare, R. Canetti, and H. Krawczyk, "Keying hash functions for message authentication" in *Advances in Cryptology: Proceedings of CRYPTO '96*, LNCS 1109, N. Koblitz, ed., Springer-Verlag (1996), 1–15.

[4] J. Black, S. Halevi, H. Krawczyk, T. Krovetz, and P. Rogaway, "UMAC: Fast and secure message authentication" in *Advances in Cryptology: Proceedings of CRYPTO '99*, LNCS 1666, M. Wiener, ed., Springer-Verlag (1999), 216–233.

[5] A. Bosselaers, "Even faster hashing on the Pentium," presented at the rump session of Eurocrypt '97.

[6] D. Carman, J. Adcock, D. Balenson, M. Heyman, and A. Sherman, "Adaptive Cryptographically Synchronized Authentication (ACSA): Prototype system design," Revision 1.0 (May 12, 1999). 56 pages.

[7] DARPA ITO ACSA Web Page, http://www.darpa.mil/ito/psum1998/G374-0.html.

[8] M. Etzel, S. Patel, Z. Ramzan, "Square hash: Fast message authentication via optimized universal hash functions," in *Advances in Cryptology: Proceedings of CRYPTO '99*, LNCS 1666, M. Wiener, ed., Springer-Verlag (1999), 234–251.

[9] V. Gligor, and P. Donescu, "Integrity-aware PCBC encryption schemes," *Security Protocols - 7th International Workshop*, LNCS, B. Chrisianson, B. Crispo, and M. Roe, ed., Cambridge, U.K., Springer-Verlag (April 1999).

[10] S. Halevi and H. Krawczyk, "MMH: Software message authentication in the Gbit/second rates" in *Fast Software Encryption*, LNCS 1233, Eli Biham, ed., Springer-Verlag (1997), 172–189.

[11] D. Harkins and D. Carrel, "The Internet key exchange (IKE)," RFC 2409 (November 1998). ftp://ftp.isi.edu/in-notes/rfc2409.txt.

[12] S. Kent and R. Atkinson, "IP authentication header," RFC 2402 (November 1998). ftp://ftp.isi.edu/in-notes/rfc2402.txt.

[13] S. Kent and R. Atkinson, "Security architecture for the internet protocol," RFC 2401 (November 1998). ftp://ftp.isi.edu/in-notes/rfc2401.txt.

[14] H. Krawcyzk, M. Bellare, and R. Canetti, "HMAC: Keyed-hashing for message authentication," RFC 2104 (February 1997). ftp://ftp.isi.edu/in-notes/rfc2104.txt.

[15] C. Madson and R. Glenn, "The use of HMAC-MD5-96 within ESP and AH," RFC 2403 (November 1998). ftp://ftp.isi.edu/in-notes/rfc2403.txt.

[16] C. Madson and R. Glenn, "The use of HMAC-SHA-1-96 within ESP and AH," RFC 2404 (November 1998). ftp://ftp.isi.edu/in-notes/rfc2404.txt.

[17] T. Matthews, "Suggestions for random number generation," RSA Laboratries' Bulletin, No. 1 (January 22, 1996). 4 pages. http://www.rsa.com/rsalabs/html/bulletins.html.

[18] E. Nahum, S. O'Malley, H. Orman, and R. Schroeppel, "Towards high performance cryptographic software," Technical Report TR95-04, Dept. Computer Science, University of Arizona (1995).

[19] P. Rogaway, "Bucket hashing and its application to fast message authentication" in *Advances in Cryptology: Proceedings of CRYPTO '95,* LNCS 963, D. Coppersmith, ed., Springer-Verlag (1995), 313–328.

[20] P. Rogaway, "Design and analysis of message authentication codes," *Proceedings of the 1996 RSA Data Security Conference* (January 19, 1996).

[21] P. Rogaway, "Bucket hashing and its application to fast message authentication," *Journal of Cryptology,* Vol. 12, No. 2 (Spring 1999).

[22] V. Shoup, "On fast and provably secure message authentication based on universal hashing," *Advances in Cryptology: Proceedings of CRYPTO '96*, LNCS 1109, N. Koblitz, ed., Springer-Verlag (1996), 74–85.

[23] J. Touch, "Report on MD5 performance," RFC 1810 (June 1995). 7 pages. http://www.isi.edu/touch/pubs/rfc1810.html.

[24] M. Wegman and L. Carter, "New hash functions and their use in authentication and set equality," *Journal of Computer and System Sciences*, Vol. 22 (1981), 265–279.

Appendix A. Pseudocode for bit scattering

ScatterBits

Input: message, mask, mask_offset, binA, binB

Result: binA and binB are modified

Description:

Under the control of a pseudorandomly-selected mask, scatter the bits of the message two words at a time into bins A and B. Scatter the next two message words as follows. Twice use the next mask word to mark bits in the next two message words. Interleave the marked bits of the first word with the unmarked bits of the second word, and toss the result into bin A. Interleave the unmarked bits of the first word with the marked bits of the second word, and toss the result into bin B. In the beginning, start with the mask word at the specified mask offset position.

```
Begin

L = length( message )
for X, Y being the next pair of words in message
  with M being the next word in mask
    from mask_offset do

      % 0-fill unmarked X bits; 0-fill marked Y bits
      interleaveA := (X and M) or (Y and (not M))
      push interleaveA into binA

      % 1-fill marked X bits; 1-fill unmarked Y bits
      interleaveB := (X or M) and (Y or (not M))
      push interleaveB into binB
endfor

End
```

Advanced Security Proxies: An Architecture and Implementation for High-Performance Network Firewalls

Roger Knobbe, Andrew Purtell, Stephen Schwab
NAI Labs at Network Associates
{rknobbe, apurtell, sschwab}@nai.com

Abstract

The NAI Labs Advanced Security Proxies (ASP) project is investigating software architectures for high-performance firewalls to enable the secure use of next generation networks. The project objective is to demonstrate an architecture and implementation in which protocol-specific proxies control when data transmission is allowed across the firewall, but which allows the proxy a range of options in determining how that data transits the firewall. By employing proxies that selectively use a range of lower-level protocol stack features, this novel architecture provides higher performance and greater flexibility in determining exactly what information the proxies examine. These decisions are made at the granularity of each proxied connection. We describe the firewall design and implementation and report preliminary experimental results using Fast Ethernet.*

1. Introduction

The Advanced Security Proxies (ASP) project at NAI Labs is experimenting with novel firewall architectures and implementations suitable for use in high-performance network environments. Firewall implementations have traditionally been classified as either packet filtering firewalls [1], in which individual packets are examined and permitted or denied according to policy, or proxy firewalls [2]. In a proxy firewall, the client or process initiating the connection contacts an intermediary, known as a proxy, running on the firewall. After establishing a connection with the proxy, and some additional authentication is performed between the client and the proxy, a second connection is established from the proxy to the server or other remote process listening for a network connection request. Thereafter, the client and server both carry on their normal activities, including sending and receiving application-level protocol messages. However, the messages are relayed through the proxy firewall, which performs all lower-level TCP/IP protocol handling, and delivers the data to the application proxy. The application proxy checks the message, to prevent illegal activity or possible attack, and then forwards the message, once again requiring the firewall to perform all the TCP/IP processing involved in transmitting data.

While proxies allow for the implementation of high assurance firewalls, they have often required that all data flow up and down the protocol stacks, even though much of the data was simply being copied from one side of the firewall to the other side. As network speeds have increased, protocol stacks on general purpose operating systems have not been sufficiently fast to keep pace. As a result, proxy firewalls had become speed bumps on the network. One way of mitigating these performance limitations has been to direct some classes of network traffic through packet filtering software on the firewall hosts. Under proxy control, the packet filters can be programmed to permit packets belonging to approved connections to pass through the firewall. At the completion of a proxied connection, the packet filter rules are dynamically removed, plugging the hole that is temporarily opened to carry the traffic.

In order to support higher speed network technologies with proxy firewalls, we are studying a variety of methods by which a proxy can control the lower-layers of the network hardware or software in order to optimize the flow of data across the firewall. Our research objective is to demonstrate an approach that can keep pace with increasing network speeds without resorting to expensive, overly specialized custom hardware solutions. Our current implementation goal is focused on reaching OC-12 ATM speeds of 622 Mbits/second, which represent a significantly higher performance network than is usually protected using a proxy firewall. Our initial implementation work has utilized 100 Mbit/s FastEthernet, with OC-12 ATM experimentation on the roadmap.

The rest of this paper continues with an overview of the architecture of the ASP firewall. This is followed by sections describing the implementation of the firewall using the Scout operating system, and experimental measurements reporting on the performance of the

* This work was sponsored by the Defense Advanced Research Projects Agency and the Department of Defense and managed by USAF Rome Laboratory under contract number F30602-97-C-0336.

0-7695-0490-6/99 $10.00 © 1999 IEEE

current prototype. Next we describe enhancements and new congestion management functionality added to the Scout TCP implementation which we will utilize in future high-performance experiments, before concluding with a review of related work and an outline of our future directions.

2. Architecture

The ASP firewall architecture is designed to provide multiple fast datapaths across the firewall between the protected and unprotected networks. These network services, accessed through well-defined interfaces, are utilized by ASP firewall proxies to accelerate the traffic transiting the network boundary. Because each proxy retains full control over when traffic is switched to a fast datapath, security functions can be applied to the specific application protocol messages that must be inspected and checked against the security policy enforced by the proxy. For example, an FTP proxy might check that each FTP command including a file pathname uses a bounded length to prevent a buffer overrun attack.

In contrast, network traffic which is not subject to examination by the proxy may be directed over a fast datapath, reducing the delay experienced by each individual packet, and reducing the overall load on the firewall host system. For TCP connections, this will often allow greater throughput, while the firewall as a whole will expend more of its computational resources on performing useful proxy activity rather than just copying data bytes from one network protocol stack endpoint to another endpoint.

The architecture envisions multiple fast datapaths, corresponding to distinct implementation layers and degree of processing of network traffic. In an ATM environment, cells on switched virtual circuits (SVCs) might be the unit of traffic to be forwarded by the network plumbing. In more general environments, the IP datagram, with or without defragmentation support, can be forwarded. A TCP connection forwarder would deliver connection establishment and termination notifications to a proxy while immediately forwarding IP packets on an established connection. Finally, the approach is not limited to TCP connection-oriented traffic. Other classes of traffic including UDP based multimedia protocols or media level signaling protocols (such as ATM UNI or ILMI) can also be handled within this framework.

For a number of reasons, the ASP firewall prototype described below is a software-only implementation of the architecture. One reason is that a suitable hardware platform, upon which we could host firewall software, was not available to us at the time of project inception. More importantly, focusing on software implementation first allowed more flexibility in investigating interesting approaches to the overall problem of developing a high-performance firewall. For example, we anticipated that modifying TCP connections on high-bandwidth networks would aggravate the congestion management mechanisms that are part of the TCP protocol, and reduce throughput of network traffic crossing the firewall. In section 5.1 we describe the TCP congestion management support that we will use to address this issue in the future.

In addition to prototyping in software, we have focused exclusively on TCP-oriented connection traffic and IP-based fast datapaths as the former are the most widely deployed proxy firewalls, and the latter are the most generically supported WAN networks. However, the architecture itself is applicable to a wider range of network technologies and hardware implementations. After describing our implementation and experiments, utilizing the Scout extensible operating system, we also identify requirements for hardware support for the architecture in future work.

3. Implementation

Two separate fast datapath facilities have been implemented in the prototype ASP firewall. Known as IPCUT and NETTAP, both implementations are designed to accelerate the handling of IP packets traversing the firewall. The difference between the two is that IPCUT is designed to carry traffic belonging to a connection that has not been established prior to the proxy issuing commands to IPCUT, while NETTAP is designed to handle traffic belonging to a previously extant connection. In particular NETTAP can switch traffic from the currently proxied connection onto the fast datapath. To evaluate the performance of the firewall acceleration facilities, FTP and HTTP proxies have been implemented that utilize IPCUT and NETTAP respectively. Both proxies and fast datapath mechanisms have been implemented in the Scout extensible operating system.

3.1. Scout Operating System

The Scout OS [1], developed at the University of Arizona, is a modular OS directed toward communication-oriented applications. These include thin-client Java stations, file- and web-servers, QoS routers and firewalls. The key abstraction of Scout is the *path*, which replaces the *process* as the execution-time entity created to service a user request. Unlike a process, a path is explicitly tied to one or more I/O devices. Its purpose is to handle incoming messages across the set of composable modules that comprise the path. The Scout infrastructure provides general facilities for manipulating messages; scheduling threads associated with each path; creating, managing and destroying paths; and demultiplexing messages from an input device to a path capable of processing it.

Paths are composed of sequences of modules (confusingly called routers by the Scout developers) that process messages as they traverse the path. Many of the modules are designed to process one layer of a network protocol stack, and there are modules for TCP, IP, ARP and Ethernet devices (ETH), as well as specific network devices. A Scout kernel is configured by specifying all the modules that will be present in the system, and the router (module) graph indicating which modules have connecting edges. Each module has a number of interfaces, drawn from a small number of types defined by the Scout infrastructure. Interfaces may only be connected to each other if they are of the same type, reducing, though not eliminating, configuration errors. The workhorse type in the system is NET, corresponding to an abstract network interface over which a message between two layers in a protocol stack may be passed. The usual convention for naming protocol stack interfaces is to refer to them by orientation such as UP or DOWN. Thus the TCP module's DOWN:NET interface is connected to the IP module's UP:NET interface in an ordinary protocol stack.

Fig. 1 illustrates a firewall configuration with two protocol stacks. At boot time, each router is initialized, with addresses or routing entries. The Proxy calls pathCreate creating a path from the Proxy to TCP. This passive path waits for a TCP connection request. A software demultiplexing mechanism delivers each incoming packet to the path responsible for processing it. Demultiplexing is accomplished using a distributed approach in which each module attempts to match the fields in its own network layer with one or more possible paths. In the case of TCP, if no established connection exists, the demultiplexing search would first match the incoming packet's IP address with the local host, then the IP packet type (TCP) and TCP port number with the passive path previously created. In response to the connection request, pathCreate is called to construct an active path. Each module along the path initializes per-stage state used by that module in processing packets. TCP control block variables specifying sequence and acknowledgement numbers are stored here, along with congestion control windows and retransmission timers. After the TCP connection closes, a pathDestroy operation frees the stages associated with each module, and also removes the demultiplexing entries for that path.

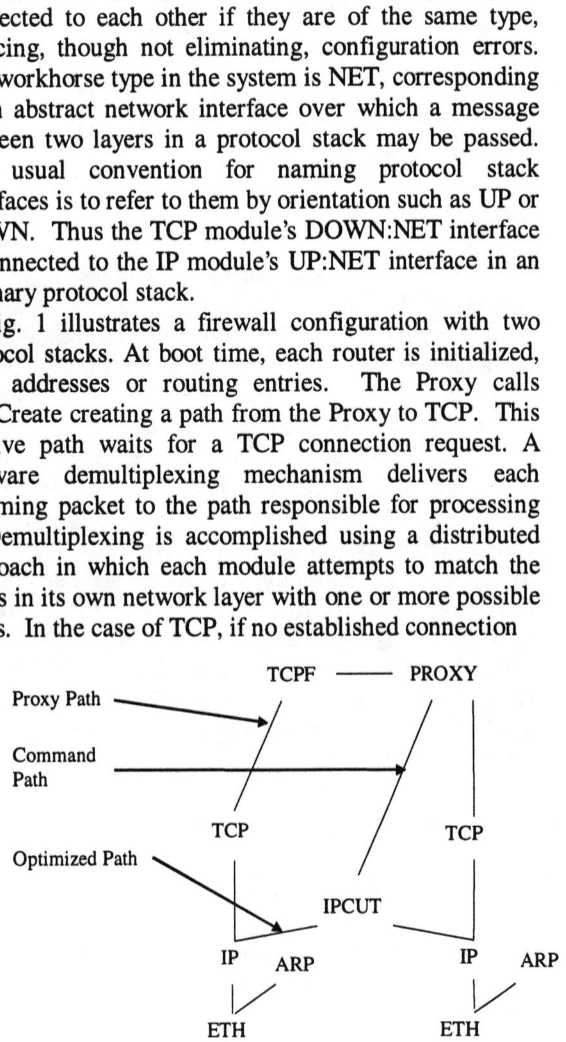

Fig. 1. Configuration of ASP Firewall with IPCUT

3.2. Optimized Firewall Datapaths

Following earlier proxy firewall optimization work, we implemented a pair of identical TCP/IP protocol stacks with one or two modules at the top, inspecting and shuttling application messages from one side of the network to the other. Without any datapath optimization modules, the system performs all the protocol stack processing associated with any other proxy firewall. However, since Scout is a single address space system, data is passed via reference wherever possible. The message handling library takes care to avoid extra byte copying as headers are removed during input processing and later added during output processing. The modules described below provide fast IP forwarding datapaths.

IPCUT, illustrated in Fig. 1, provides an optimized datapath by connecting via its DOWN interface to two different IP modules. In addition, a command interface is provided and used by the FTP proxy to enable the flow of packets over the fast path. IPCUT connects to IP at the same point where a TCP module would connect. An IPCUT command includes parameters describing a TCP connection, consisting of the source and destination IP addresses and TCP port numbers. Optional arguments may include the selection of different initial sequence numbers on each side of the firewall connection. In response to the command, IPCUT creates a path in two steps. First, pathCreate is called to build a path from IPCUT down one protocol stack, through IP and ETH modules. IPCUT uses a hook in the TCP module to add this path to the TCP demultiplexing map. From the Scout point-of-view, IPCUT represents an optimized form of TCP. More importantly, the demultiplexing tree implementation is deterministic, and the TCP module is configured at compile time to demultiplex all IP packets with protocol field equal to 6, the standard value for TCP. Once the first half of the path is established, a pathExtend call is made to repeat the process on the other protocol stack connected to the opposite network interface. PathExtend is functionally equivalent to pathCreate, except the starting point is the existing path, and new stages are added in turn for each module of the new protocol stack.

Once the optimized IPCUT path is established between the two networks, the FTP proxy proceeds to forward the application level message which triggers the file transfer. FTP uses a separate connection for TCP data, which means the original TCP control connection does not need to be optimized. It remains quiescent while the file data moves over the IPCUT path. After the operation completes, the IPCUT module observes the TCP connection teardown by recording the packets with FIN flags, and the respective acknowledgements of these FIN packets by each side. At that point, IPCUT tears down the optimized datapath using pathDestroy, and the proxy is notified of the change.

NETTAP, illustrated in Fig. 2, provides application-level proxies with several additional approaches to establishing and managing fast datapaths. The NETTAP module is inserted directly between the TCP and IP modules, in a position where it can intercept and redirect the IP traffic under proxy control. When using NETTAP, a typical application will first establish two TCP connections and will initially fully proxy data between these connections. Our HTTP proxy currently makes use of the NETTAP module. HTTP traffic is collected and examined until a complete HTTP GET request has been received. The simple proxy implements a policy that filters out URLs, protecting some private URLs on the server, while allowing the rest to be accessed across the firewall. Other policies are easily implemented, although some additional development will be required to tie the proxy-specific policy configuration into the lower-layer policy configuration mechanisms described below in subsection 3.3. Once the proxy determines that the GET request is allowed, it forwards it to the web server, and issues a NETTAP JOIN command, resulting in the HTTP response being carried over an optimized datapath. Unlike the IPCUT module, the NETTAP module switches the same TCP connection used to carry the HTTP traffic onto the fast datapath.

NETTAP affords the proxying application with a range of cost-benefit tradeoffs by operating in one of three selectable modes. The first, JOIN, optimizes host TCP and application processing out of the path completely, in an analogous manner to IPCUT. Once the application has switched a connection to this mode, it cannot receive data on, send data to, or control either connection on the path. The second, SNOOP, establishes a fast datapath but also collects and queues payload data, which is delivered to the application as available CPU cycles permit. In this mode, the application may monitor traffic and, if needed, terminate both connections, but cannot guarantee that some hostile traffic has not already been forwarded before the moment it is detected. The

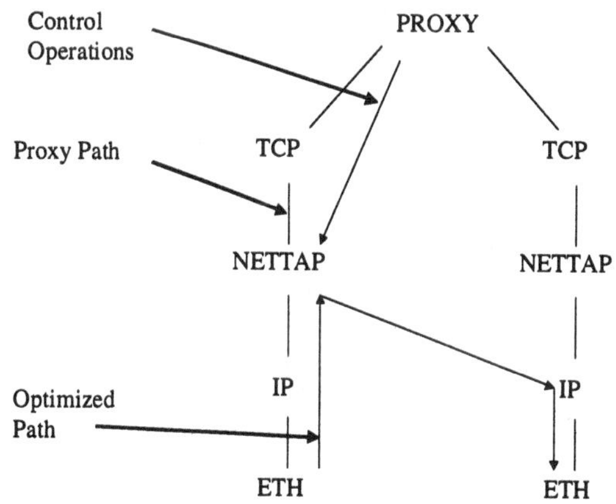

Fig. 2. Configuration of ASP Firewall with NETTAP

third mode, INSPECT, addresses this issue by queuing and delaying the forwarding of traffic until the application has signaled that the collected data may be sent. Of these modes, JOIN has the least impact on system resources, and INSPECT the greatest.

Both IPCUT and NETTAP perform many of the same low-level operations on IP packets. IP addresses and TCP port numbers may be changed to facilitate network address translation and port hiding. In addition, the sequence and acknowledgement fields of the TCP header must be translated between the two halves of the network connection. TCP numbers the bytes transmitted from sender to receiver in order to provide reliable service. Acknowledgements in the TCP header of packets traveling in the opposite direction indicate the greatest sequence number received, and confirm delivery of data to the host. Sequence numbers do not start from any fixed initial value. Instead, each host picks an initial sequence number. In order for a fast datapath to be created, the firewall must modify the sequence and acknowledgement fields by adding or subtracting a constant value corresponding to the difference in the initial sequence numbers. Finally, the IP and TCP checksum fields must be updated to reflect any changes made to the headers. In our implementation, we use an efficient differential checksum algorithm. Instead of examining all the data in the TCP segment to re-compute a fresh checksum, the original checksum is used as a starting point, and is modified based on the old and new values of the header fields that change.

3.3. Policy Infrastructure

We have introduced support for specification and enforcement of dynamic security policies into the ASP firewall prototype through the creation of the INET router, a Scout router module that polices inbound and outbound TCP connections.

Currently we support a subset of the Security Policy Specification Language (SPSL) developed by researchers at BBN [4]. SPSL is a concise policy description language originally designed for use within the Internet routing infrastructure. We call our reduced, upwardly compatible language RSPSL, for Reduced Security Policy Specification Language. An example of an RSPSL policy fragment is presented in fig. 3.

The design of the INET router supports multiple policy languages by separating the policy engine from various parser modules, e.g. the RSPSL parser. A small API is exported which allows parser modules to install policies. A small collection of router control operations allows any component on a path along with the INET router to install or modify policies at any time. The data structures used for communicating policy elements define what concepts may be translated from the policy language into internal representation, and have been designed with extensibility in mind. Currently it is possible to specify complex filtering rules which key on source and/or destination IP addresses or address ranges, source and/or destination port numbers or port number ranges, and IP protocols.

INET enforces policy by intercepting various actions within the Scout kernel. Inbound TCP connection requests travel over special *listening* paths. Outbound TCP connections trigger creation of *active* paths. The compile-time configuration of an INET-enabled Scout kernel places INET stages between TCP and application stages on all paths created in the system which transect TCP.

A policy rule can be defined as *inbound*, *outbound*, or *symmetric* -- applicable to both inbound and outbound connections. If a policy is defined as symmetric, it is applied both on listening paths and during the creation of active paths. If it is defined as *inbound*, the policy is applied only on listening paths. If it is defined as *outbound*, the policy is applied only during creation of active paths. INET rejects inbound connections by dropping the initial TCP SYN packet, which discontinues all further processing associated with the connection. In contrast, INET rejects the creation of active paths for new outbound connections by returning an error during path creation, causing the partially instantiated path to be destroyed and returning a policy-exception error to the application router attempting to establish the new connection. A small number of additional keywords will be defined to specify the policy individual proxies consult in determining when an ASP optimized datapath is to be utilized for a given connection. These policy decisions will be communicated from INET to the proxy router by attaching a newly defined

```
gateway:        sg-foo
 name:          foo-firewall.foo.com
 ifaddr:        172.16.0.1
 ifaddr:        10.0.0.1
 preference:    1

policy-name:    tcp-foo
 association:   sg-foo
 policy:        dst 172.16.0.0-172.16.255.255
                src 10.0.0.0-10.0.0.255
                xport-proto 6 accept
 policy:        dst 172.16.0.0-172.16.255.255
                xport-proto 6 deny
```

Fig. 3. Example RSPSL fragment. Gateway sg-foo is only allowed to forward TCP traffic between the specified subnetworks.

Scout policy attribute to the initial message establishing the connection.

4. Experiments

Initial measurements of firewall performance have been collected using a testbed consisting of 450 Mhz Pentium II PCs with 128 Mbytes of RAM and 100 Mbit/second full-duplex FastEthernet interfaces. A client and server machine are each connected via switches to a firewall machine with two separate network interfaces. Fig. 4 compares the performance of the ASP FTP proxy using IPCUT against a proxy and packet filter firewall. The peak CPU utilization of the ASP firewall was 30%.

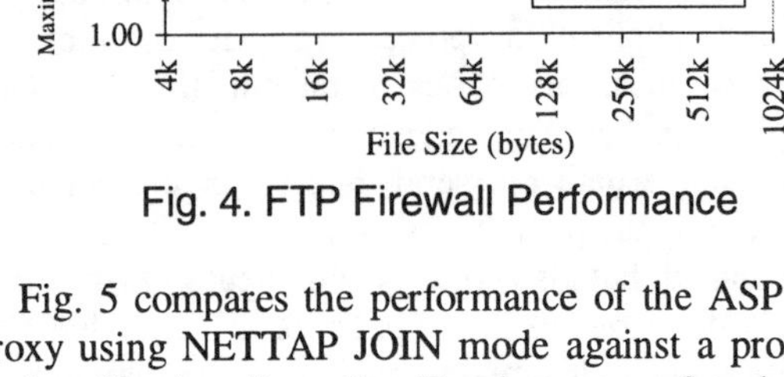

Fig. 4. FTP Firewall Performance

Fig. 5 compares the performance of the ASP HTTP proxy using NETTAP JOIN mode against a proxy and packet filtering firewall. For large transfer sizes, the ASP firewall provides nearly the same throughput as the control case with no firewall between the client and web server. The peak CPU utilization of this configuration

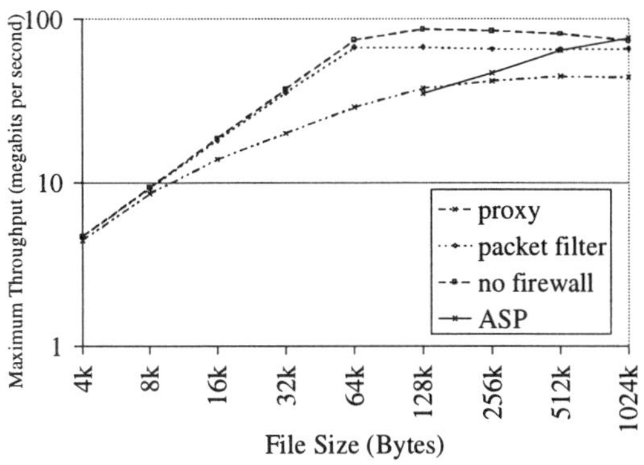

Fig. 5 HTTP Firewall Performance

was 28%, in contrast to 61% with the NETTAP SNOOP mode. Smaller transfer sizes result in performance similar to traditional proxies, as the costs of connection setup and inspecting the HTTP GET messages dominate overall performance in this regime.

5. Incorporation of Recent TCP Results

Recent work in the ASP project has focused on rewriting the TCP implementation in the Scout OS, from an aging code base derived from 4.4BSD, to a more modern implementation incorporating enhancements such as timestamps and scaled windows [5], and TCP NewReno-style fast recovery [6]. In addition, we have constructed a TCP implementation that shares host-pair state among multiple connections [7], described in detail in the following subsection. Although we have not yet conducted a full set of rigorous experiments, preliminary experience indicates that our TCP implementation performs on par with that of modern operating systems, and is roughly twice as fast as the previous Scout TCP implementation. This is due in part to the NewReno extensions, a faster timeout scheduling clock [8][9], separate state management for TIME_WAIT connections [8], and a significant restructuring of Scout TCP path management. The ultimate goal of the TCP enhancements made to the ASP firewall implementation is to enable the system to sustain high traffic throughput. In order to meet this goal, we have also implemented new mechanisms to address congestion management.

5.1. Congestion Management Support

Recent literature, notably [7] and more recently [10], have noted that a distinction can be made between connection-pair state (state which is specific to each endpoint-to-endpoint pair) and host-pair state (state which is specific only to each host-to-host pair). State for each connection is maintained in a data structure commonly termed the TCP Control Block (TCB). The typical TCP implementation usually assigns one TCB to each connection and maintains all state on a per-connection basis.

Per-connection information may be split into macro-state and micro-state, terminology borrowed from RFC2140 [7]. Macro-state describes the finite state machine. Micro-state describes congestion control and sequencing state needed after a connection has been established.

Macro-State

- local and remote host numbers and ports
- connection state: starting, established, closing, etc.
- timers

Micro-State

- local and remote sequence numbers
- send and receive window state
- round-trip time and variance
- congestion window size and threshold
- local and remote MSS

In our experimental TCP implementation, we have moved the elements of micro-state which are host-pair dependent into a data structure shared among connections between such host-pairs. We call this structure a Host TCP Control Block (HTCB). Each HTCB contains the following information:

HTCB-State

- aggregate round-trip time and variance
- aggregate congestion window size and threshold
- local and remote MSS

All connections between a given host-pair reference HTCB-state through a pointer. New connections are initialized from the shared state, gaining advance knowledge of current conditions. Concurrent connections benefit from shared knowledge of network utilization and loss rates. All flows are managed globally. Existing connections may have their congestion windows reduced to free network resources for new connections. Initial slow start can be skipped, maximizing throughput.

Our TCP implementation uses a simple round robin scheme for updating aggregate information. Once established, connections become eligible for nomination as a controlling connection. Each eligible connection is nominated in turn and, once selected, temporarily assumes responsibility for updating transmission timing and congestion control information. A controlling connection relinquishes its control after establishing round-trip time measurement or when it is closed, whichever comes first, thus triggering another nomination cycle. This, in the parlance of RFC2140, is termed ensemble sharing.

The alternative to ensemble sharing is temporal sharing. Each connection maintains its own state and, upon close, averages collected transmission timing and congestion control information into the shared state. As in the ensemble sharing regime, new connections are initialized from shared state. However, we believe temporal sharing is inferior to ensemble sharing because concurrent connections cannot benefit from shared knowledge, an important distinction which eliminates many opportunities for optimizing network utilization among concurrent flows.

Our TCP implementation is currently under active development and we are in the early stages of collecting metrics and verifying conformance to protocol standards. There are many issues that require further analysis. It is unclear whether the simple round robin sharing scheme described above is fair, for example, in the presence of routers that enforce quality of service policies favoring certain flows over others between the same host-pair. Also, the response of our implementation to incidents of congestion or loss when ensemble sharing is enabled is not well known.

6. Related Work

Several other researchers have investigated ways of improving firewall proxy performance by optimizing the underlying network protocol stacks. Spatscheck et. al. [11] describes a TCP connection optimization for Scout paths and a firewall proxy that utilizes it. Maltz and Bhagwat [12] have implemented similar TCP splicing mechanisms in a Unix kernel to support firewall and web caching applications. Fall [13] investigated more general mechanisms for optimizing data copying between both files and network sockets, and introduced a Unix interface to control optimization.

7. Future Direction and Conclusion

There are many additional features we would like to implement in our ASP firewall prototype. While the current Scout system required a simple change to the TCP module to allow IPCUT to insert paths into the demultiplexing map, a more sophisticated Scout infrastructure would allow IPCUT to lookup and access the demultiplexing map dynamically. This would remove a dependency between the IPCUT module on low-level internals of the TCP module. Also, while the path operations are conceptually very simple, the actual interface in Scout requires developers to directly manipulate the pointers linking the various stages of the path together. This code is difficult to debug, extremely dense, and highly prone to coding errors resulting in deformed paths. An enhanced version of the system should provide a higher-level interface to aid developers in the construction and management of paths.

An additional NETTAP command, SPLIT, is currently in development, which will resume the full proxying of a connection by switching the traffic away from the fast datapath and back to the two distinct TCP/IP protocol stacks below the proxy. SPLIT will provide the ability for proxies to periodically re-inspect a connection. A security policy might require re-examination of a connection after a fixed time period, or after a specified number of bytes or packets had traversed the fast path.

Once the ASP architecture is well established, it should be feasible to implement key pieces in hardware, allowing proxy firewalls to scale with increasing network bandwidth. Many hardware routing devices include the capability of updating the IP checksum, but do not have the capability to update the TCP checksum. This is probably due to the fact that all IP routers must decrement the time-to-live field, requiring changes to the IP checksum, but do not change any TCP fields. One clearly identified hardware requirement to support optimized TCP datapaths in firewalls is a method for differential or constant time TCP checksum recalculation. The demultiplexing of packets to different destinations involves the inspection of both IP and TCP fields. The hardware requirements for firewalls vs. routing are not dramatically different, except that TCP flags are also useful as part of the demultiplexing rules. However, the rate at which new TCP connections are established and destroyed may be much greater than the rate at which routing table changes propagate to forwarding engines in routers. Firewalls require an interface to the demultiplexing mechanisms that supports many more changes per second, and with small, bounded latency. Finally, firewalls need to track and record connection termination. A simple state machine for recording the sequence number of packets with the FIN flag, and generating a notification when the corresponding acknowledgement number is seen would also be a requirement for implementing all of the optimized network plumbing in hardware.

As more proxies are implemented that utilize optimized datapaths, the configuration of firewall security policies becomes more complicated. More investigation of the various ways in which security policies exploit dynamic firewall behavior to improve performance without compromising network boundary integrity is needed.

Appendix A. Reduced SPSL

The RSPSL policy language differs from SPSL in that it discards management and certificate information, omits support for domain specifications, and uses only the short form policy definition. The resulting specification language is upwardly compatible with the full SPSL language but is very simple.

Reduced Security Policy Specification Language

Policy:

```
policy-name: <object-name>

association: <node-name> | <node-set-name>
        | <gateway-name> | <gateway-set-name>

policy: dst `*' | any | list of [not] <ip-address>
        | list of [not] <address-range>
        [port `*' | opaque | any | list of [not] <port>
              | list of [not] <port-range>
        [dynamic [<port-range>]]]
        [src `*' | any | list of [not] <ip-address>
              | list of [not] <address-range>
        [port `*' | opaque | any | list of [not] <port>
              | list of [not] <port-range>
         [dynamic [<port-range>]]]]
         [xport-proto `*' | opaque | any
              | list of [not] <proto>
              | list of [not] <proto-range>]
        [direction inbound | outbound [, symmetric]]
        permit [, forward <dest>]
        | deny [, forward <dest>]
        | forward <dest>
```

Node:

```
node:     <node-name>
name:     <dns-name>
alias:    <dns-name>
ifaddr:   <ip-address>
```

Node Set:

```
node-set:   <node-set-name>
members:    list of <node-names>
            and/or <node-set-names>
```

Gateway:

```
gateway:   <gateway-name>
name:      <dns-name>
alias:     <dns-name>
ifaddr:    <ip-address>
[ preference: <integer>     ]
```

Note that the preference attribute is ignored, meaning there are no distinctions made between nodes and gateways in our simplified implementation.

Gateway Set:

```
gateway-set:  <gateway-set-name>
members:      list of <gateway-names>
              and/or <gateway-set-names>
```


Acknowledgements

John Sebes conceived of the original ASP project goals and architecture, and contributed to its success. Hilarie Orman suggested the use of Scout as a firewall platform.

References

[1] B. Cheswick and S. Bellovin. Firewalls and Internet security. Reading, MA, Addison-Wesley, 1994.

[2] M.K. Ranum and F.M. Avolio. A toolkit and methods for Internet firewalls. In Proceedings of the Summer 1994 USENIX Conference, Boston, Massachusetts, pages 37-44.

[3] D. Mosberger and L. Peterson. Making paths explicit in the Scout operating system. In Proceedings of Second USENIX Symposium on Operating Systems Design and Implementation, October 1996, Seattle, Washington, pages 153-167.

[4] M. Condell, C. Lynn, J. Zao, draft-ietf-ipsec-spsl-01.txt: Security policy specification language, IETF IPSec Working Group, July 1999.

[5] Jacobson, et. al., RFC1323: TCP extensions for high performance, IETF Network Working Group, May 1992.

[6] S. Floyd, T. Henderson, RFC2582: The NewReno Modification to TCP's Fast Recovery Algorithm, IETF Network Working Group, April 1999.

[7] J. Touch, RFC2140: TCP control block inter-dependence, IETF Network Working Group, April 1997.

[8] M. Aron, P. Druschel, TCP Implementation enhancements for improving webserver performance, Technical Report 99-335, Department of Computer Science, Rice University, March 1999.

[9] L. Brakmo, L. Peterson, Performance problems in 4.4.BSD TCP, ACM Computer Communication Review, vol. 25, no. 5, October 1995.

[10] H. Balakrishnan, et. al., An integrated congestion management architecture for Internet hosts, In Proceedings of the 1999 SIGCOMM Conference. September 1999, pages 175-187.

[11] D. Maltz and P. Bhagwat. TCP splicing for application layer proxy performance. IBM Research Report RC 21139, March 17, 1998.

[12] K. Fall, A peer-to-peer I/O system in support of I/O intensive workloads, Ph.D. Thesis, 1993, Department of Computer Science, U.C. San Diego.

[13] O. Spatscheck, J. Hansen, J. Hartman and L. Peterson. Optimizing TCP forwarder performance, Technical Report 98-01, Department of Computer Science, The University of Arizona, February 9, 1998.

The Horus and Ensemble Projects

Accomplishments and Limitations

Ken Birman, Bob Constable, Mark Hayden, Jason Hickey, Christoph Kreitz, Robbert van Renesse, Ohad Rodeh and Werner Vogels[1]

Abstract– The Horus and Ensemble efforts culminated a multi-year Cornell research program in process group communication used for fault-tolerance, security and adaptation. Our intent was to understand the degree to which a single system could offer flexibility and yet maintain high performance, to explore the integration of fault-tolerance with security and real-time mechanisms, and to increase trustworthiness of our solutions by applying formal methods. Here, we summarize the accomplishments of the effort and evaluate the successes and failures of the approach.

Index Terms– Reliable multicast, fault tolerance, distributed systems security, distributed computing, automated verification, real-time cluster computing.

I. A BRIEF HISTORY OF FAULT-TOLERANT PROCESS GROUP MECHANISMS FOR RELIABLE DISTRIBUTED COMPUTING

We begin by reviewing the historical trends in distributed computing leading to the present research effort. Brevity prevents us from including a comprehensive bibliography, hence we focus on aspects most directly tied to our work. Readers seeking additional background are referred to [Bir97, Bir99a, Kes97].

Prior to 1985, distributed computing was dominated by the Internet, and the Internet (in turn) was dominated by point-to-point communication mechanisms providing best-effort reliability. Although certain services provided forms of replication (for example, the network news program, the DNS, "yellow pages" (later renamed NIS) and the Xerox Clearinghouse system), application-level support for replicated data was lacking, and even replicated services were typically hardwired to support predetermined sets of clients. Applications built over the Internet employed TCP, UDP and FTP, treating the implementations of these protocols and the mechanisms supporting Internet routing and DNS address resolution as opaque components of the network itself.

During a period from 1983-1985, Cheriton and Zwaenapoel at Stanford extended the basic IP communication suite to support what they called distributed process groups [CZ85]. They used these groups in support of application-initiated broadcast, and proposed a number of group-based services which used broadcast to provide some form of parallelism or accelerated response. Inspired by their work, our fault-tolerant communications effort at Cornell proposed a distributed computing model based on process groups, but extended by the introduction of formal semantics for error handling [BJ87].

The approach that we proposed (soon adopted by several other research groups) became known as reliable process group computing, or virtually synchronous process group computing. In essence, a virtually synchronous process group provides automatically managed membership for application programs, which are permitted to join and leave groups and are informed of membership changes by upcalls. Also provided are multicast interfaces supporting ordered message delivery, and a means of initializing new members when they join the group - we call this the state transfer problem. To ensure that the solution is powerful enough to permit replication of data, membership change is synchronized with respect to multicast sending and delivery, and state transfer is implemented to appear atomic with respect to membership change.

Figure 1 illustrates this model graphically, where time advances from left to right. We see a process group within which multicasts are being used to update the states of members. When new members join, state transfer is shown by a thick arrow. Notice that arrows in the figure are drawn to suggest that events occur synchronously - as if all members that experience the same event, experience it at the same moment in time. Virtual synchrony is "virtual" in the sense that without using a synchronized distributed real-time clock, a running program will be unable to distinguish an actual execution from some closely synchronous one, such as the one in the figure. In reality, however, the events in a virtually synchronous execution may be highly asynchronous. The major benefit of this approach is that by relaxing synchronization in ways that participating processes can't detect, we are able to provide very high performance. Yet the execution model is intuitively simple and makes it easy for application developers to implement very complex, fault-tolerant, distributed services.

The virtual synchrony model was rapidly adopted by research and industry groups world-wide [Bir99a]. Some successes associated with our work on the Isis Toolkit include the overhead display systems that show stock price

[1] Dept. of Computer Science, Cornell University, Ithaca NY 14853. ken@cs.cornell.edu. Hayden is at the Systems Research Center of Compaq Corporation, Palo Alto CA. This work is supported in part by ARPA/ONR grant N00014-92-J-1866, ARPA/RADC grants F30602-96-1-0317 and F30602-98-2-0198, and NSF grant EIA 97-03470

0-7695-0490-6/99 $10.00 © 1999 IEEE

quotes and transactions on the floor of the New York Stock Exchange [Gla98], the entire communications architecture of the Swiss Exchange [PS97], the console clustering architecture used in a new generation of air traffic control technology recently rolled out in France [Bir99a], the control subsystem of several major VLSI fabrication plants (AMD, Seimens, Texas-Instruments), and a number of mobile telephony products.

Military uses of the technology included an intelligence monitoring and reporting technology implemented by NSA, a prototype for the next generation of the AEGIS Naval radar and communications system (called HiperD, this was the basis of the SC-21 standard, which in turn is the basis for DD-21, an important military communications standard expected to have wide-ranging impact during the coming decades) and certain applications associated with Ballistic Missile Command and Control.

These successes can be traced to the ability of the model to support replicated data, to provide high availability (in contrast to database replication methods, which guarantee recoverability but at the cost of sometimes needing to wait for a failed system to recover before the system can resume providing services - a source of potentially long outages), and to support load-balancing within small cluster-styled servers. Yet while these were important accomplishments, virtual synchrony also presented many drawbacks [Bir99a].

Early implementations of the model were monolithic and relatively inflexible: A system like Isis was built from floor to ceiling with one form of communication in mind, and to the degree that one wished to turn features on or off, the technology tended to come with huge numbers of specialized interfaces that the programmer needed to learn and use selectively. For example:

- Isis supported several forms of message ordering. The more costly forms of ordering were also the easiest to use, but the performance hit was considerable. Developers were often forced to tune the choice of message ordering to obtain good performance.
- When an application supports multiple, overlapping process groups, there are many options for the way that events should occur within processes belonging to more than one group. Control over these options was required because no simple default emerged.
- Some applications required that messages be encrypted but this was a significant cost in Isis, so the feature was only enabled as needed.

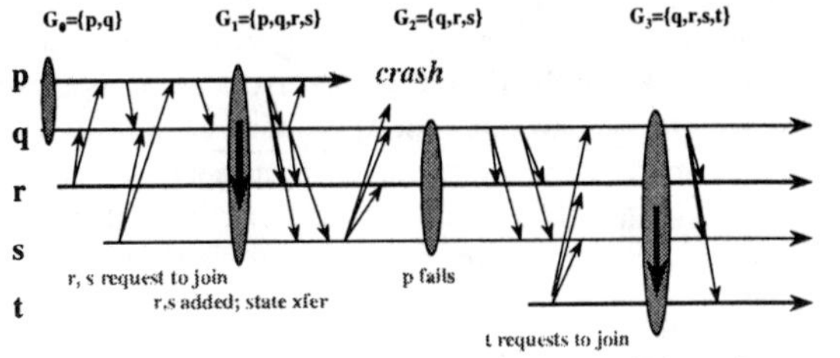

Figure 1: Virtual synchrony model, showing executions of five processes (time advances left to right)

Moreover, at the time Isis was developed, the state of the art for object oriented design was very primitive. CORBA and DCOM/OLE had yet to be introduced, and even RPC had yet to be standardized. As a consequence, early systems like Isis, which used object-oriented designs and concurrent styles of programming were forced to introduce their own solutions. For example, Isis had a widely used threads implementation at the time that CMU first began work on cthreads and pthreads, the threads library that ultimately became standard in Linux. Over time, developments in these areas made Isis less and less compatible with commercial trends.

An additional side-effect of having large numbers of very demanding users was that Isis became more and more complex over time, and it became harder and harder to convince ourselves that the technology itself was free of bugs. Such developments created the concern that process group communication merely invites users to place all their eggs in one basket, and that the basket itself could break, exposing the entire application to mishap. Isis was relatively robust, but it took years to achieve continuous availability with the technology, and when an Isis protocol error surfaced, it could easily bring down an entire distributed system.

II. GOALS OF THE HORUS AND ENSEMBLE PROJECTS

Starting in 1990, we began work on the Horus system [RBM96] to try and overcome these first-generation considerations; Ensemble was subsequently started as a sibling of Horus in 1996. The basic idea underlying both projects is to support group communication using a single generic architectural framework within which the basic group communication interfaces are treated separately from their implementation. One can then plug in an implementation matching the specific needs of the application. To maximize flexibility, each group end-point instantiates a stack of what we call micro-protocols. The developer arranges for the stack used in support of a given group to provide precisely the properties desired from the group. Each micro-protocol layer handles some small aspect of these guarantees. Figure 2 illustrates this idea. Each process in a process group is supported by an underlying protocol stack; the stacks for the various members are identical, but the stacks used in different groups might be very different from one-another.

For example, one layer might overcome message loss by numbering each message and retransmitting lost messages in response to Negative Acknowledgement messages (NAKs). The layer would have an outgoing and an incoming side. Outgoing messages would have a header added to them containing a sequence number, and would be stored pending garbage collection. Incoming messages would be examined to make sure they are in sequence. If not, a NAK message is sent back to the original sender of the message soliciting a retransmission. A separate layer detects when all copies have been delivered (a property called stability) and triggers garbage collection for stable messages.

A trivial example of the opportunity afforded by such an architecture is that the NAK layer might not be needed in some situations - namely, those in which the network doesn't lose messages. At runtime, based on the environment, the Horus system was capable of assisting the user in configuring a stack to provide reliability, inserting a NAK layer if necessary and omitting it if not.

Of course, useful protocol stacks often contain many layers and the kinds of layers one might selectively omit tend to be much more costly and less often needed than NAK. For example, security is often enabled selectively in Horus, the choice of protocol suite implementing virtual synchrony is often of importance (different protocols can give very different performance, and some are much better than others on specific hardware platforms), and different ways of doing multicast ordering are favored under different conditions. Without belaboring the point, the approach provides enough flexibility so that application designers with very different goals can potentially agree on the sharing of a common infrastructure, within which their commonality is captured by layers that they share, and their differences reflected by layers built specifically for their special needs.

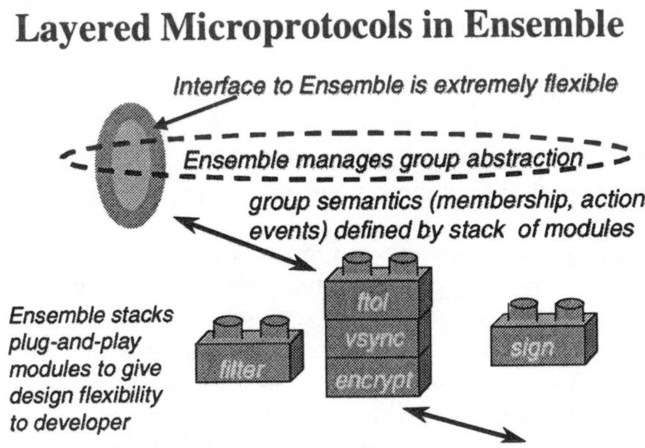

Figure 2: Ensemble and Horus used layered architectures.

Moreover, Horus can potentially support execution models very different from virtual synchrony. Our early hope was that the protocol interfaces could be offered as a standard, and that implementations of such protocols as SRM [Flo95] and RMTP [Pau97], two widely popular scalable protocols with weaker reliability models, might be developed to run on the same platform. Unfortunately, in 1999 as this paper was being written, discussion of possible standards along these lines were still advancing very slowly within the IETF and OMG, two organizations that have shown an interest in developing such standards. (Readers interested in a more detailed discussion of the SRM and RMTP reliability model and a comparison with virtual synchrony are referred to [Bir97, Bir99b].

The Horus effort did more than to simply support a layered stackable architecture. We also wanted to demonstrate that the performance of our architecture could be as good or better than that of a conventional monolithic architecture. We sought to provide real-time features, in response to a requirement coming from the Naval AEGIS application, where there was a need for cluster-style servers able to guarantee real-time event response even under stress (the application involved weapons targeting using radar tracks and had very tight time constants associated with acceptable responses). And whereas Isis was initially focused on securing its own abstractions, Horus was designed to offer security services on behalf of application developers who needed security key infrastructures for purposes of their own.

One can easily imagine that in responding to such varied needs, Horus could become very complicated. Although we managed to control complexity, we did find that the types of transformations we wanted to do on layered protocol stacks exceeded the capabilities of the available C compiler, and hence that quite a bit of hand-coding was needed to obtain high performance and to maximize flexibility.

It was in response to these considerations that Ensemble was developed, starting in the Spring of 1996. As a system, Ensemble is rather similar to Horus, although rewritten using high level programming languages and tools [Hay97]. Our insight was that much of the complexity of Horus came from overcoming inefficiencies associated with stackable protocol layers coded in the C programming language. In contrast, Ensemble's protocol suite is implemented using the O'Caml variant of the ML programming language [Ler97, Mac93]. This language is mathematical in appearance and there are powerful theorem proving tools, notably a system called NuPRL [Con86], available for expressing transformations and other types of operations on programs coded using O'Caml. In our case, we were successful in using O'Caml to code a basic set of protocol layers for Ensemble and then using NuPRL to produce optimized and transformed versions that, in Horus, would have required hand coding and hand optimization. The NuPRL approach is automated and provably correct while the manual Horus approach was a source of bugs.

Moreover, we discovered that NuPRL can potentially do quite a bit more for us.

The remainder of this paper focuses on the successes and limitations of Horus and Ensemble. Both projects are largely at an end now - Ensemble and Horus are both used by modest communities and a number of technology transition efforts should lead to their emergence in products for the mass market within the next few years. Meanwhile, our own effort at Cornell now focuses on what might be seen as third-generation issues that work to move beyond the limitations of the entire process group approach.

For example, we are increasingly convinced that virtual synchrony has some basic scalability limitations that emerge from the model itself, and our new Spinglass project[2] was born out of an insight into a new way to develop a scalable reliability model to overcome these limits. Virtual synchrony, we now believe, is simply better suited to "close grained" cooperation on a scale of tens of members (certainly, less than one hundred group members), while the protocols we are using in Spinglass provide high reliability and steady data delivery to potentially thousands or millions of recipients. We imagine Spinglass as a technology one might use side by side with Ensemble or Horus, because it offers reliability guarantees that are provably weaker than those of the virtual synchrony model, and the virtual synchrony model remains necessary in many situations. An illustration of this arises in our discussion of the Ensemble security work, which combines Ensemble groups with Spinglass protocols.

Similarly, we are continuing to work with NuPRL as a program verification and automated protocol transformation tool of unique power and flexibility. Whereas our initial work focused on using NuPRL to automate some of the protocol stack transformations needed to achieve high performance in stackable architectures, we are now pursuing a more ambitious goal: proving the correctness of the virtual synchrony implementation used in Ensemble, and perhaps of the security key management architecture running over this implementation. But this work, in turn, has revealed yet a third possible goal: the automated generation of provably-correct protocol stacks from relatively high level descriptions of goals. It thus may be possible to see Ensemble much as a compiler used to bootstrap a compilation process - one builds a basic compiler in a first language for a new language, but then implements a second compiler directly in the new language and discards the original one. Our work could yield, within a few years, a completely new and self-supporting infrastructure for building correct group-communication protocols and for optimizing them to achieve extremely high performance.

The remainder of this paper focuses on the accomplishments and limitations of Horus and Ensemble and the major technical challenges we face in transitioning the technology into major commercial product platforms, such as CORBA and COM/OLE.

[2] See http://www.cs.cornell.edu/Info/Projects/spinglass/ for information and references to Spinglass publications.

III. HORUS EFFORT

Our work on the Horus system can be understood in terms of several distinct threads of activity, which were all conducted within the same framework and to a large degree interoperate: Layering and its consequences, real-time issues, security mechanisms, and work on protocol performance and scaling. We consider these in turn.

A. Layered Protocol Architectures in Horus, Performance Issues

The initial focus of our work on Horus concerned its use of layering to simplify the design of the virtual synchrony protocols. When this work was begun, we looked closely at the x-Kernel architecture [PHO89], developed at the University of Arizona by Larry Peterson with similar goals. We found, however, that the x-Kernel was designed with point-to-point TCP-style protocols in mind. For our work on group communications, a more flexible and more standardized interface to each layer was needed.

Accordingly, we developed what we now call the Horus Common Protocol Interface, or HCPI [v*R95*], as a standard interface to and between protocol layers. The interface provides "up", "down" and "control" API's, and operates under a model in which messages and other events travel from the user down the stack to the I/O interface, or from the I/O interface up to the user. For example, an encryption layer might receive outgoing messages from higher layers, use a key to encrypt the body of those messages, and then pass the message to the outgoing message interface of the next layer below. Incoming messages would, similarly, be decoded on arrival and discarded if corruption or tampering was detected.

Over the 7 year period since this interface was first proposed, other groups including the OMG fault-tolerance standards group and the IETF reliable multicast research task-force have proposed creating standard architectural slots similar to the ones occupied by Horus protocol layers. Our effort has offered an updated HCPI interface to these organizations, but until the present, it seems that the aggressive use of layering adopted in Horus remains more advanced that what these organizations might consider.

Layering gives rise to several forms of overhead. A message, traveling down the stack, may be examined by a whole series of layers, most of which do nothing at all to the message. When a layer does add a header to a message, it may need to assume that it is the only layer in the stack, hence to add even a single bit, a layer may need to create a header large enough to hold an integer. By the time a message reaches the wire, it may have many bytes of largely empty headers on it, and may have skipped through as many as 20 or 30 layers that basically took no action.

During 1995, one of us tackled this issue, and developed a methodology for optimizing layers to avoid both forms of overhead [vR96]. Although this work was done separately from Horus, we considered it to be part of the overall technology base. In essence, the approach involves compressing the headers by eliminating wasted space, and also separating headers into different types of data. Header information that remains constant after a stack is established is only transmitted once, and a typical message only carries headers that actually contain changing values - potentially a very small amount of data. Messages are aggregated (packed) to make optimal use of network packets. And, through a decomposition of each layer into data touching and non data-touching parts, it proves possible to short-circuit the path a message takes through the stack, reducing the critical path between the application and the wire to just a few instructions even for a very complex protocol stack.

With these optimizations in place, the Horus Protocol Accelerator set a number of performance records. Running over a zero-copy communications architecture called U-Net [Von95] (similar to the *Virtual Interface Architecture* promoted by the VIA consortium), this version of Horus introduced only a few microseconds of overhead beyond the overhead of the network adaptor and drivers.

B. Real-Time Cluster Computing

Unfortunately, the ability to demonstrate high performance is not enough to achieve real-time responsiveness in some critical applications. Earlier, we noted that our work on Isis was adopted by the Navy for use in its AEGIS architecture. This system includes a number of cluster-style computer systems that are used to compute tracks for airborne objects detected by the AEGIS radar, and serve as the basis of weapons targeting applications. Since threats may be moving at very high speed, real-time response is vital even when failures occur within the cluster. A similar need arises in telecommunications switching architectures, where a co-processor may be asked how a call-establishment request should be routed; the SS7 architecture used in such settings requires 100ms response times even while a failure is handled.

Working with our group, Roy Friedman explored the use of Horus as a technology for cluster control in real-time applications of this sort [FB96]. He considered two styles of solution. In the first, Horus was used to implement small process-groups of two or three processes each, using load-balancing and fault-tolerant RPC mechanisms within these to guarantee that each request would be handled even if one or more failures occurred while the cluster was heavily loaded. With this approach, Friedman was only able to achieve a throughput of a few hundred requests per second, and the Horus failure detection timer (six seconds) emerged as a performance limit: a request might potentially be delayed, if a failure occurred under heavy load, until the detection timer was triggered. For the sorts of applications just mentioned, such delays are totally unacceptable.

Friedman then developed a different solution in which Horus operates as a side-band mechanism for cluster control and data replication, but "offline" from the basic request loop. In this approach, Friedman was able to aggregate batches of requests and used hand-coded, highly optimized protocols for the basic request dispatch and handling communication paths. During the period before Horus discovered a failure, data might pile up, but Friedman used a number of compression schemes to minimize the amount and avoid overloading available buffering.

The approach was a dramatic success: for the SS7 telephone architecture, Friedman now achieved 20,000 requests per second on a 64-node cluster, demonstrated that performance improvements were possible when the cluster size was increased, and was able to sustain 100ms response times even as nodes were taken offline, crashed, or restarted while the switch was under load [FB96].

Friedman's work illustrates, for us, both the power of Horus and a limitation. The benefit of this work was that for the first time, a way to use a cluster of computers in a time-critical fault-tolerance application was demonstrated. Yet the work was technically complex and suggests that unless these methods can be embedded into a very low level of the operating system (for example, into the clustering technology of the NT Clusters system), application developers will have great difficulty exploiting the approach. Horus, viewed from the perspective of this type of real-time application, is a necessary tool, but not sufficient. On the other hand, for high-value applications such as the AEGIS tracking service, it does seem clear that Friedman's work points to a methodology for achieving very high degrees of scalability and real-time responsiveness while tolerating faults.

C. Security in Group Communication Systems

The Horus system was also the setting for our initial foray into security for groups of participants in large networks. Working with Mike Reiter [RBvR94, RB94], we developed a means of securing the virtual synchrony model itself, so that only trusted processes would be allowed to join a process group, and so that group members could obtain a shared group key. This problem involves authentication at the time of the group join, and rekeying when a member joins or leaves (so that prior communication in the group, or subsequent communication, would not be accessible to the new member).

Our work on security can be seen as complementary to work on group security arising directly from the Internet community. Recall that the DNS and routing services of the Internet replicate various forms of data. During the early 1990's it became important to secure the protocols used to update these, and the resulting key distribution and management problem became a classical topic for the security research community. Here, the notion of group membership is much weaker than the one used in the virtual synchrony community, and there is no formal semantics for the execution model. Yet the superficial aspects of the security problem are very similar: we have a group of members, we wish to authenticate joining and leaving, and we plan to use the security keys to encrypt communication within the group.

The Horus security mechanisms have advantages and disadvantages when compared to this more network-oriented form of security. The strong semantics of virtual synchrony groups certainly offers security benefits: within this model, one actually can formalize the question of which processes legitimately belong to a group and which ones do not, and when a process does belong to a group, there are strong guarantees about the state of the data it manages. But there are also disadvantages to the model, notably that it scales poorly beyond about 100 processes. Most experience with Isis was limited to groups of five to ten processes at a time [Bir99a]

and it was only with great care that Isis applications spanning more than about 250 processes were developed successfully. In the Internet, 10,000 members of a DNS service might not be at all unreasonable and one can imagine services containing millions of members. Yet scaling virtual synchrony to this degree seems not to be practical. (Our new project, Spinglass, might well provide this degree of scalability, but it uses a somewhat different execution model).

IV. Ensemble Effort

Earlier, we cited the "eggs in one basket" concern in regard to distributed systems models such as virtual synchrony, or process group security. While such approaches are beneficial to the application designer, whose task is greatly simplified by the strong guarantees of the system, if the model itself is violated as a consequence of a coding error or some unanticipated bug in the protocol itself, the application's correctness or security might be compromised. One can reduce such concerns by exhaustive testing, simulation, or by writing papers in which the protocols employed by the system are presented rigorously and a formal proof of correctness is offered. Yet none of these options yields more than a modest degree of confidence in the ultimate correctness of the running code itself. Even now, more than a decade after the development of Isis, Isis applications that seek to provide continuous availability still exist, and occasionally, one of them reports a bug never before encountered. Such bugs can easily cause the entire distributed application to crash.

As noted earlier, Horus was developed as a partial response to this concern: the technology sought to simplify the monolithic structure of systems like Isis by showing how complex protocols could be broken into simple microprotocols and stacked to match the needs of an environment. Yet Horus offers little to increase the confidence of a skeptic in the ultimate correctness of the protocols and of their implementations.

Ensemble was developed primarily as a response to these concerns. Our fundamental idea was to begin using a new and extremely powerful generation of mathematically rigorous programming and verification tools as a means of moving beyond the hand-coded optimization schemes employed when performing inter-layer optimizations in Horus, and of actually proving the correctness of key components of the system. The technology evolved in several new directions, however, as time passed: we used Ensemble as the basis of initial work on a new protocol suite, and pursued a number of topics involving dynamic adaptation using Ensemble as the base. We also developed a new security architecture within Ensemble, moving well beyond the initial Horus version. This section summarizes each of these threads of research.

A. *Formal Transformation of Protocol Stacks*

Code transformation of the Horus system was impractical in part because of the choice of programming language: by coding Horus in C, we were able to achieve extremely high performance, but this language has limited capabilities for type checking and other types of correctness checking, and such weak mathematical semantics that formally expressed code transformations are largely impossible. Accordingly, a primary reason for building a new system - Ensemble - was to create a version of Horus coded in the O'Caml programming language, a dialect of ML having strong semantics and consequently suitable for analysis and transformation using formal programming tools. This decision was informed by previous success in using NuPRL to reason about a large ML system [AL92] and to reason about hardware [LLHA94]. We were also encouraged by the work at CMU by Harper and Lee on the FOX project to code protocols in SML [HL94].

Our decision to implement Ensemble in O'Caml compelled us to confront an initial challenge of a different nature. The ML family of languages is not traditionally known for high performance, and while O'Caml is compiled, we were concerned that it might not be possible to achieve performance comparable to that of Horus. Yet the verification of a system incapable of the desired level of performance would have been much less satisfying, since we hoped to demonstrate that production-quality distributed software can actually be proved correct. Mark Hayden, who coded the system, undertook a detailed study of this issue, and ultimately developed a methodology for protocol development in O'Caml that overcomes the most common efficiency issues encountered by users of the language. His accomplishment, which involved taking control of garbage collection and using O'Caml's language features very carefully, was reported in [KHH98].

Given an initial version of Ensemble, Hayden, Kreitz and Hickey set out to use a formal mathematical tool called NuPRL ("new pearl") to automate the sorts of optimizations that Van Renesse did by hand in developing Horus [vR96]. They approached this by teaching NuPRL to read Ensemble layers - in effect, NuPRL understands each layer as the "proof" of some property, namely the protocol guarantee implemented by that layer. NuPRL was then able to do several kinds of protocol transformations. For example, because a protocol stack appears as a nested function call to NuPRL, it was possible to request that NuPRL perform an inline function expansion of the code.

The basis for all formal code manipulation is a formal semantics for a large subset of the O'Caml programming language in the logical language of NuPRL. Not long ago, the formalization of such a subset would have been cutting edge research worthy of separate funding and a PhD thesis, but in this case, we were able to build on advances in understanding of formal semantics and on the richness of the NuPRL type theory. Basically the core of O'Caml is a subset of the NuPRL term language, and therefore type theory almost immediately provides a semantics for O'Caml [Kre97]. The method is now called a "shallow embedding". This method has been used to provide a formal semantics for significant extensions of ML in the direction of object orientation, see the work of Crary for example [Cra98].

NuPRL can perform partial evaluation of functions, and this opened the door to a category of optimizations similar to those used by Van Renesse. The approach begins by recognizing that as messages traverse a stack, the code path used may be a very small percentage of the code in the stack as a whole. For example, the virtual synchrony stack treats membership change events very differently from multicasts. If a message is a multicast and no membership change is occurring, the message may be nearly untouched within the stack.

A protocol stack in Ensemble looks like a set of nested function calls. Suppose that x is some form of outgoing event, such as a message to send or a membership change request. Then, Ensemble's job is to evaluate $f_0(f_1(\ldots f_n(x)))$, where each of the f_i is the code implementing some micro-protocol within the stack (f_0 is at the bottom and f_n is at the top). Similarly, for an incoming event, Ensemble can be understood as evaluating the function $f_n(f_{n-1}(\ldots f_0(x)))$. Now, focus on the outgoing case, and suppose we call this entire nested function f. Imagine that we place an if statement in front of it, as follows: "*if(is_a_msg(x)) f(x) else f(x),*" where the predicate *is_a_msg* is true for messages and false for other types of events, such as group membership changes. (Not shown is an additional, implicit argument: the "state" of the protocol stack, which is updated when the stack executes and hence is shared by both function invocations).

Viewing NuPRL as a form of optimizing compiler, the system can be asked to partially evaluate the function under the two cases: "*is_a_msg(x)*" is true, and "*is_a_msg(x)*" is false. Consider the first case: the predicate is true. Under the circumstances just described, very little code needs to be executed for messages, hence the function will collapse to just a few lines of code. "Dead" code branches (those NuPRL can recognize as never being executed) are deleted during the partial evaluation. In effect, we've produced an extremely optimized code path for the common case where we sent a multicast. Yet since the event either is or is not a message, the *behavior* of the original stack is unchanged!

To generate the optimized code while guaranteeing its correctness NuPRL uses two levels of formal optimizations.

- On the first, or *static* level, symbolic evaluation and logical simplification techniques are applied separately to the code of each micro-protocol. They result in formally proven *layer optimization theorems*, which show that the effect of passing an event x through the respective protocol layer, while assuming that a *common case predicate* (CCP) like "is_a_msg(x)" (or some other property of common events) holds, can be expressed by two or three lines of code. These optimizations are executed independently from the application protocol stacks and need to be redone only when the code of a micro-protocol is modified or when new micro-protocols are added to the Ensemble toolkit.

- The second, or *dynamic* level, depends on the particular protocol stacks designed by the application developers. Given the names of the individual micro-protocols occurring in this stack, it composes the corresponding layer optimization theorems into a formal *stack optimization theorem* that describes the effect of passing an event x through the whole stack while assuming that *all* individual CCPs hold. This is not trivial, because Ensemble's composition mechanism allows that events may bounce between layers before leaving the stack instead of passing straight through it. Therefore the technique is based on composition theorems, which abstractly describe the effects of composing common combinations of optimized micro-protocols.

A stack optimization theorem not only describes a fast-path through a protocol stack but also the headers that the stack adds to a message. Since typical messages should only carry headers that actually contain changing values, all constant headers are eliminated before sending the message. For this purpose, the protocol stack is wrapped with code for compressing and expanding headers and then optimized again.

In a final step the stack optimization theorems are converted into O'Caml code, which uses the CCPs as conditionals that select the bypass path in the common case and otherwise the normal stack, as illustrated in Figure 3 (The Transport module below the stack provides marshaling of messages). This program is proven to be equivalent to the original protocol in all cases, but generally more efficient in the common case.

Using this methodology [Kre99], Kreitz, Hayden, Hickey, van Renesse and graduate student Xioaming Liu showed that NuPRL can automatically achieve protocol speedups comparable to the ones that Van Renesse achieved by hand in Horus. Moreover, Liu and Van Renesse developed a version of the method for use in adapting Ensemble to match the protocol stack to the environment. With their work, one can dynamically pick a protocol stack with just the properties needed for a given setting, obtain an optimized version of the stack from NuPRL, compile the resulting O'Caml byte code into machine code, ship this code to a version of Ensemble, and run it as the protocol stack supporting some group. At present, we do the optimization offline, but the methodology could be extended to work on the fly. This work is reported in [Liu99].

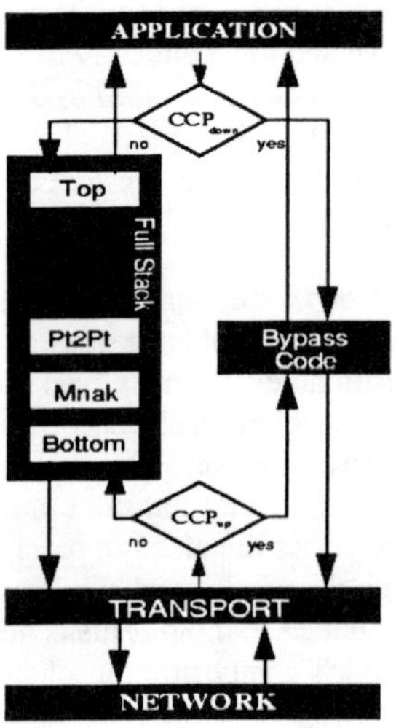

Figure 3: Optimization in the Ensemble Architecture

B. Verification of Ensemble stacks

Formal verification of a system such as Ensemble presents a major challenge because its many micro-protocols (currently over fifty) can be combined in literally thousands of different ways to provide a great variety of services. To verify Ensemble would mean to be able to treat the reasonable combinations of stacks and their services. Moreover, many of the individual protocols are quite sophisticated.

Additionally it is a major challenge to verify the actual system code -- indeed, most verification efforts operate by verifying the correctness of abstracted descriptions of protocol or system components. While verifying these abstractions is an interesting first step, the goal that appealed to us was to work directly with production-quality protocol implementations, and ultimately to prove that the code actually running in the system has the properties required by the user. This goal was a natural extension of our success in transforming the actual O'Caml code of Ensemble stacks.

A key step in verification is the specification of properties. This has been a lively topic in the Horus and Ensemble research. Various temporal logics were tried, including TLA [Kar97], and an axiomatic approach was considered [CHTCB96]. In the end, the ground work for our approach was laid by Hickey and Van Renesse in collaboration with Nancy Lynch, whose research effort at MIT has developed a mathematical programming language called I/O Automata, or IOA [Lyn96]. Jointly with Lynch, they found a way to express virtual synchrony as an IOA, adapted the IOA language itself so that NuPRL could understand such a specification, and extended IOA so that it could be used in compositional settings, such as the Ensemble protocol stacks [HLV99].

We have found a particularly elegant way to formalize IOA using NuPRL class theory [BH98]. We can formalize both services and their implementations in the same style. Moreover the inheritance mechanisms of the formal class theory make it possible to inherit proofs of safety properties of a stack when new layers are added to it. This leads to a methodology that will make it easier to prove properties of stacks from proofs of components. The method seems to scale to stacks composed from many layers from a large base of micro-protocols.

When this work has been completed, we will be able to produce highly optimized executable code from provably correct protocol stacks, dynamically adapting the stack to match the environment where the protocol will be used, and providing guarantees of reliability and security formally verified by a mathematical tool and characterized in a high level notation suitable for use in reasoning about applications built over the resulting process group.

C. Synthesis of layers

Once we know how to prove the correspondence of code fragments to IOA statements, we can also understand this as a means for compiling from IOA into protocols (which we can optimize using our trace driven optimization methodology). So the potential exists to avoid using hand-coded Ensemble protocol stacks in favor of these more automated protocol stacks produced using NuPRL as a compiler.

The basis of this synthesis capability is the fact that global services, abstract protocol specifications and code layers are treated as modules in a common formal language. We can use our formal framework not only for verification purposes but also for the synthesis of protocols from specifications. Because the generated code is correct by construction, our framework supports the development of new communication protocols, which is a notoriously difficult task.

A synthesis of protocol layers that implement a global service will be supported by a small collection of generic methods for transforming specifications of distributed systems. This methodology is not entirely new. Synthesizing algorithms from specifications by applying specification transformations is a well-known principle in program synthesis, and synthesis from proofs has been explored for many years as well [Wal69, BC85,Kre98].The novelty of our approach lies in providing methods for the synthesis of *distributed* systems, which is by far more difficult than synthesizing serial algorithms.

We describe four of the most common generic methods.

1) Replication of Global Values

Applying this method will allow us to represent values locally. It creates a local copy of each value, ensures consistency by making each process broadcast these values, and introduces a unique token. Virtual synchrony (as specified in IOA, *Extended Virtual Synchrony*, or EVS) is a prerequisite for this step; lacking this property, consistency has to be ensured by more complex means.

2) Adding Fault Tolerance

This method, which requires local values as prerequisite, makes the local values fault tolerant. It forbids access to values during a view change and implements merging of multiple copies after a view change.

3) Conversion to Message Passing

This method assumes fault tolerance and generates the typical event-handler interface of Ensemble by converting each operation to a message.

4) Code Generation

This method runs as the last step of a synthesis. It takes a specification that provides the event-handler interface and converts the abstract specification into executable code. Because of the close relation between abstract specifications and the representation of Ensemble's code in NuPRL this step is straightforward.

As an example, we describe the synthesis of a total order layer from the specification of the total order *service*. The total order service has a simple specification. In addition to the properties of EVS, it ensures that all processes in a view receive messages in the same order, and is called ETO.

The total order specification represents this requirement with a global queue that orders all the messages sent in the view. A message can be received only if it is the next message in the global queue.

The ETO specification is not directly implementable because its global queue contains global information, so the first step is to *replicate the global value* so that each process contains a local copy of the queue. The *Replicate* generic method introduces a token to ensure atomic access and consistency of the local copies, but the copies are not robust to failures. The next step is to apply the *Fault Tolerance* generic method, which prohibits access to the queue during a view change, and creates a fresh queue once the view change is completed. At this step, we have partitioned the service into local protocol layers, and in the final step we apply the *Convert to Message* Passing generic method to convert the layer actions (which refer to semantic events like view changes) to explicit message passing style. This final step can be completely automated.

D. Scalable Security Architectures in Ensemble

Ensemble has also been used as a base for an expanded effort to provide security for process group applications and systems [RBD99, RBH98]. As noted earlier, our work on the Horus system resulted in an initial mechanism for securing process group communication and the group abstraction.

In our work on Ensemble, we have focused on expanding the capabilities of this basic idea and using group security keys in innovative ways. With respect to the basic security architecture, Ensemble implements a scheme (developed primarily by Ohad Rodeh, a research in Dolev's Transis group at the Hebrew University in Jerusalem) whereby asymmetric public keys can be "traded" for symmetric point-to-point keys and group keys. These symmetric keys provide a high speed path for signing messages and encrypting their contents, and can also be used by secured applications for application-specific security purposes.

Rodeh's solutions are innovative in two ways. First, he employs a hierarchy of protocols for key management and key refresh, and has a particularly fast solution to the key refresh problem when processes join or leave a group. His algorithm rekeys within milliseconds, permitting a group to (potentially) change keys so rapidly that even if a key were broken, the adversary would gain access to just one or two multicasts before a new key was substituted for the compromised one [RBH98]. Additionally, Rodeh developed a fault-tolerant extension of the well-known Wong-Gouda-Lam tree-based key management architecture [Won98], a protocol that in its original form was centralized and not fault-tolerant. Rodeh's version scales easily within Ensemble's process groups, although these remain limited to perhaps one hundred or two hundred members [RBD99].

The scalability limits of Ensemble prevent Rodeh from using this technique to secure extremely large groups, which might have tens of thousands of members. For this purpose, however, Rodeh is exploring a very simple combination of Ensemble with our new Spinglass system. As mentioned earlier, Spinglass introduces a new and extremely scalable data point in the reliable group communications spectrum, offering a suite of probabilistic protocols that can be used to communicate reliably with huge numbers of processes even in networks subject to the most extreme forms of disruption. Rodeh is linking the secured Ensemble group mechanisms to the Spinglass mechanisms so that Ensemble can control a core group and this, in turn, can manage security keys on behalf of a very large group of leaf nodes. Inspired by the DNS architecture for the Internet, this approach allows typical users to talk to a local representative of the security hierarchy and low cost, while drawing on the strong properties of Rodeh's Ensemble solution to guarantee rapid key refresh and other aspects of a strong security model.

E. Dynamic Adaptation

The third major research topic that has been explored primarily within the context of our work on Ensemble is concerned with dynamic adaptation. As described previously, the adaptation problem arises when an application expresses requirements for a process group that can be satisfied in more than one way, depending upon the environment within which the group runs [Liu99].

Two examples will illustrate this idea. The first is concerned with multicast ordering protocols. Over the past two decades, a great number of ordered, reliable multicast protocols have been developed. Abstractly, it is common to view such protocols as representing optimizations of total ordering. For example, in a process group where we happen to know that only one member will initiate multicasts, a protocol providing fifo ordering would actually be "strong enough" to satisfy a total ordering requirement.

Even in a group where there are multiple senders, one faces such a choice. If the senders are willing to use a token passing or locking scheme, the most appropriate total ordering protocol would be one that exploits the mutual exclusion property - these include causal ordering protocols and token-based total ordering protocols. Lacking a locking scheme, one might still use a token based protocol, but other protocols such as time-stamped ordering protocols now have potential advantages. Broadly, the best choice depends upon the nature of the application, and the nature of the environment within which we run it.

Similarly, the most appropriate security mechanism depends on the setting. Behind a firewall, one may face only a very benign security requirement, and limit "security" to some form of join authentication. Yet when the same application includes a group member running outside the firewall, it may be important to encrypt all group communication, and to authenticate even the messages used within the join protocol itself.

Virtual synchrony offers an appealing way to think about adaptation. Each time the membership of a group changes, the virtual synchrony model announces this through a new process group view, synchronized with respect to ongoing communication among group members. Normally, the view is just a list of group members, ranked in some canonical manner. Our work on adaptation extends the same idea: each time the view changes the new view also includes a new protocol stack for each of the members.

Within Ensemble, we've used this mechanism to change protocol stacks on the fly, with roughly the same (very small) overhead incurred when group membership changes because a process joins, leaves or crashes. What we do is to associate with each stack a small software module responsible for monitoring the environment, watching for conditions under which some other stack would be more desirable than the one currently in use. When conditions change, this module triggers the new-view algorithm, arranging that the new stack be installed at all members in a virtually synchronous manner.

For example, if we are using a total ordering protocol that is appropriate only with a single sender within a group, adaptation might be triggered when a second process first attempts to send: the attempt would cause the group to switch to a new total ordering protocol, at which point the interrupted send request would be allowed to complete. If we are using a security mechanism appropriate only within a firewall, adaptation might occur when a member from outside the firewall joins the group, and so forth.

V. TECHNICAL CHALLENGES ASSOCIATED WITH TRANSITION

One of the most difficult problems we've encountered in our research has been the challenge of technology transition. Broadly, it has been our belief that unless these technologies make the transition from academic demonstrations into broad commercial availability, they will not have the degree of impact that we seek and that DARPA hopes for in projects such as this. Yet in the area of networking and distributed computing, there is tremendous resistance to doing anything other than what the basic Internet supports.

As noted earlier, our first project, the Isis Toolkit, achieved some degree of commercial uptake and had some notable successes. Through this process, Isis became a viable option for military and government projects and had an important influence on technology planning within the services, reflected in the DD-21 architecture.

However, starting a company to commercialize Horus and Ensemble was unappealing to us. Our effort has made both systems available to "all comers" with very few restrictions of any kind, and at no fee. Cornell and the original Ensemble developers - Hayden and Vaysburd - provide some support, also for free. This has resulted in some uptake of the technology, and we have a small user community that includes some large companies (notably, Nortel and BBN) and many small ones. An exciting recent opportunity involves technology transfer into the restructured electric power grid. We are pursuing this topic jointly with a consortium organized by the Electric Power Research Institute, EPRI.

To see a broad-based transition occur, companies of the size of Microsoft and Sun need to become interested in this technology area. There is some good news on this front: the industry as a whole is looking at group communication tools closely, and several standards organizations have been exploring possible standards for reliable multicast and object replication. As these trends advance, one can anticipate a first-class role for reliable multicast in standard operating systems, such as Solaris, Linux and NT, and with that development, technologies such as ours would find a natural home.

Overall, we are encouraged by these developments, but we also see an argument for winding down the effort on Horus and Ensemble in favor of new directions. Accordingly, while continuing work on the NuPRL verification methodology, our overall project has shifted attention to a new technology based on a suite of highly scalable "gossip-based" protocols with probabilistic properties [Bir99b]. The protocols and their behavior take us into a domain rather different from virtual synchrony, while we wait and watch to see what the ultimate impact of our work over the past few years will be.

VI. CONCLUSIONS

Cornell research has placed process group computing, especially with virtual synchrony, or a firm footing. Over the course the three projects we've conducted in this area, we've contributed techniques for achieving high performance, security, real-time guarantees and provable correctness. Although we were not able to scale virtual synchrony to very large settings, the gossip-based protocols developed in our work on Ensemble have taken on a life of their own, and form the basis of a new project (Spinglass) that promises to contribute a new and extremely scalable data point for the spectrum of reliable multicast protocols and applications.

Over the years, the Cornell effort has helped enable some very high profile applications, and we've played an instrumental role in showing that these solutions really work. Down the road we expect that the technologies we pioneered will have a good likelihood of becoming standard in products from major vendors, that IETF standards will emerge for the area, and that CORBA will provide solutions - all traceable to our work and that of related projects.

Our vision of the future is one in which reliable process group computing will reside side-by-side with scalable probabilistic technologies, of the sort now under development in our Spinglass system. Through this approach, we believe that reliable, secure distributed computing can become a commonplace reality on a wide range of systems spanning both large-scale WAN applications based on conventional workstations down to massive networks of very small devices such as might arise in future miniaturized control settings or future sensor networks.

VII. REFERENCES

[AL92] Aagaard, Mark and Miriam Leeser. Verifying a Logic Synthesis Tool in NuPRL. In *Proceedings of Workshop on Computer-Aided Verification.*, 72-83, Springer-Verlag, (1992)

[BC85] Bates, J. L. and Constable, Robert L. Proofs as Programs. In *ACM Transactions on Programming Languages and Systems* 7(1), 53-71 (1985)

[BH98] Bickford, Mark and Hickey, Jason. *An Object-Oriented Approach to Verifying Group Communication Systems* (submitted to CAV), (1998)

[Bir97] Birman, Kenneth P. *Building Secure and Reliable Network Applications.* Manning Publishing Company and Prentice Hall, (1997)

[Bir99a] Birman, Kenneth P. A Review of Experiences with Reliable Multicast. *Software Practice and Experience* 29(9), 741-774 (1999)

[Bir99b] Birman, Kenneth P., Hayden, Mark, Ozkasap, Oznur, Xiao, Zhen, Budiu, Mihai, Minsky, Yaron. Bimodal Multicast. *ACM TOCS* 17:2 (May 1999), 41-88.

[BJ87] Birman, Kenneth P. and Joseph, Thomas A. Exploiting Virtual Synchrony in Distributed Systems. *11th ACM Symp. on Operating SystemsPrinciples,* (1987)

[CHTCB96] Chandra, Tushar D; Hadzilacod, Vassos; Toueg, Sam and Charron-Bost, Bernadette. On the impossibility of group membership. In *15th ACM Symposium on Principles of Distributed computing*, 322-330, (1996)

[Con86] Robert Constable *et. al. Implementing Mathematics in the NuPRL Proof Development System.* Prentice-Hall, 1986.

[Cra98] Crary, Karl. *Type-theoretic Methodology for Practical Programming Languages.* PhD thesis, Cornell University. Department of Computer Science, (1998)

[CZ85] Cheriton, D. and Zwaenepoel, W. Distributed Process Groups in the V Kernel. *ACM TOCS* 3(2), 77-107 (1985)

[Flo95] Sally Floyd, Van Jacobson, Steve McCanne, Ching-Gung Liu and Lixia Zhang. A reliable multicast framework for lightweight sessions and application-level framing. Proc. ACM SIGCOMM, 1995.

[FB96] Roy Friedman and Ken Birman. Using Group Communication Technology to Implement a Reliable and Scalable Distributed IN Coprocessor. Proceedings Telecommunications Information Network Architecture 96, Heidelberg. VDE-Verlag , 25-42, (1996)

[GL99] Goft, G. and Lotem, Y. E. The AS/400 Cluster Engine: A case study. International Workshop on Group Communication (IWGC'99), (1999)

[Gla98] Glade, Bradford. *A Scalable Architecture for Publish-Subscribe Communication in Distributed Systems.* Ph.D. dissertation, Cornell University Dept. of Computer Science, (1998)

[Hay97] Mark G. Hayden. *The Ensemble System.* Ph.D. dissertation, Cornell University Dept. of Computer Science. (1997)

[Hayden-JFP] Hayden, Mark. Distributed Programming in ML. To appear: Journal of Functional Programming, 1999.

[HL94] Harper, Robert and Lee, Peter. Advanced languages for system software: The Fox project in 1994. School of Computer Science Technical Report, CMU-CS-94-104, Carnegie Mellon University, (1994)

[HLv99] Hickey, Jason; Lynch , Nancy and Van Renesse, Robbert. Specifications and Proofs for Ensemble Layers. In *5th International Conference on Tools and Algorithms for the Construction and Analysis of Systems.* LNCS 1579, 119-133, Springer (1999)

[Kar97] Karr, David, A. *Specification, Composition, and Automated Verification of Layered Communications and Protocols.* PhD thesis, Cornell University. Department of computer Science, (1997)

[Kes97] Srinivasan Keshav. *An Engineering Approach to Computer Networking.* Addison Wesley, 1997.

[KHH98] Kreitz, Christoph; Hayden, Mark and Hickey, Jason. A Proof Environment for the Development of Group Communications Systems. In *15th International Conference on Automated Deduction* LNAI 1421, 317-322 Springer, (1998)

[Kre97] Kreitz, Christoph. *Formal Reasoning about Communication Systems I: Embedding ML into Type Theory.* Technical Report TR 97 - 1637 Department of Computer Science, Cornell University, (1997)

[Kre98] Kreitz,Christoph. Program Synthesis. Chapter III.2.5 in *Automated Deduction - A Basis for Applications.* 105-134. Kluwer, (1998)

[Kre99] Kreitz, Christoph. Automated Fast-Track Reconfiguration of Group Communication Systems. In *5th International Conference on Tools and Algorithms for the Construction and Analysis of Systems.* LNCS 1579, 104-118. Springer, (1999)

[Ler97] Xavier Leroy. *The Objective Caml System Release 1.05.* INRIA, France, May 1997.

[Liu99] Liu, Xiaoming; Kreitz, Christoph; Van Renesse, Robbert; Hickey, Jason; Hayden, Mark; Birman, Kenneth and Constable, Robert, L. Building reliable, high-performance communication systems from components. In *17th ACM Symposium on Operating Systems Principles,* (1999)

[LLHA94] O'Leary, John; Leeser, Miriam; Hickey, Jason and Aagaard, Mark. Non-Restoring Integer Square Root: A Case Study in Design by Principled Optimization. In *International Conference on Theorem Proving & Circuit Design.* (1994)

[Lyn96] Nancy Lynch. *Distributed Algorithms.* Morgan Kaufmann Publishers, San Francisco, 1996.

[Mac93] David MacQueen. Reflections on Standared ML. In Peter E. Lauer, Ed. *Functional Programming, Concurrency, Simulation and Automated Reasoning.* Volume 693, pages 32-46, Springer Verlag Lecture Notes in Computer Science, 1993.

[Pau97] Sanjoy Paul, K. K. Sabnani, J. C. Lin, S. Bhattacharyya "Reliable Multicast Transport Protocol (RMTP)", IEEE Journal on Selected Areas in Communications, April 97, Vol 15, No. 3

[PHO89] Peterson, L., Hutchinson, N., O'Malley, S. and Abbott, M. RPC in the *x*-Kernel: Evaluating New Design Techniques. *Proc. 12th ACM Symposium on Operating Systems Principles,* (1989)

[PS97] Piantoni, R. and Stancescu, C. Implementing the Swiss Exchange Trading System. *FTCS 27,* Seattle, Washington, 309-313, (1997)

[RB94] Reiter, M. and Birman, K. How to Securely Replicate Services. *ACM Trans. on Programming Languages and Systems* 16(3), 986-1009, (1994)

[RBD99] Rodeh, O., Birman, K.P. and Dolev, D. Optimized Group Rekey for Group Communication Systems. To appear: 2000 Network and Distributed Systems Security, San Diego (2000)

[RBH98] Rodeh, O., Birman, K.P., Hayden, M., Xiao,. Z., Dolev, D. Ensemble Security. Department of Computer Science Technical Report 98-1703, (1998)

[RBM96] Robbert van Renesse, Kenneth P. Birman, Silvano Maffeis. Horus: A Flexible Group Communication System. *Communications of the ACM* 39(4), 76-83 (1996)

[RBvR94] Reiter, M., Birman, K. and van Renesse, R. A Security Architecture for Fault-Tolerant Systems. *ACM Trans. on Computer Systems* 12(4), 340-371 (1994)

[SWLP94] Stickel, Mark; Waldinger, Richard; Lowry, Michael; Pressburger, Thomas and Underwood, Ian. Deductive Composition of Astronomical Software from Subroutine Libraries. In *12th International Conference on Automated Deduction*, LNAI 814, 341-355, Springer, (1994)

[Vog98] Werner Vogels *et. al.* The Design and Architecture of the Microsoft Cluster Service - A Practical Approach to High Availability and Scalability. Proc 28th FTCS, Heidelberg, (1998)

[Von95] Von Eicken, Thorsten, Basu, Anindya,. Buch V. and Vogels, Werner. U-Net: A User-Level Network Interface for Parallel and Distributed Computing. *Proc. 15th SOSP,* Copper Mountain, CO. Dec. 1995, 40-53.

[vR95] Van Renesse, Robbert, *The Horus Common Protocol Interface (HCPI).* Cornell University Dept. of Computer Science, 1995.

[vR96] Van Renesse, Robbert. Masking the Overhead of Protocol Layering. Proceeding of the 1996 ACM SIGCOMM Conference, Stanford, (1996)

[Wal69] Waldinger, Richard, J. Constructing Programs Automatically Using Theorem Proving. PhD thesis, Computer Science Department. Carnegie-Mellon University, (1969)

[Won98] C.K. Wong, M. Gouda and S. Lam. Secure Group Communication Using Key Graphs. Proc. *ACM SIGCOMM,* Sept. 1998.

Celestial Security Management System *

Chong Xu	Fengmin Gong†	S. Felix Wu
	Ilia Baldine	Zhi Fu
	Chandru Sargor	He Huang
	Frank Jou	
Duke University	MCNC	North Carolina State University
chong@cs.duke.edu	fmg@anr.mcnc.org	wu@eos.ncsu.edu

Abstract

There has been a vast amount of research and development effort aimed at providing solutions and products that address the security needs in the information age. Each solution tends to address only a particular facet of the security problem and only accessible to limited protocols or applications. Moreover, ad hoc deployment of some solutions (e.g., firewalls and IPsec) can hinder our ability to collaborate across networks. A very important question is how any application can discover policy restrictions brought about by these solutions/mechanisms, and make efficient use of them to satisfy the application's security goals. The Celestial project addresses this question by developing a security management architecture that can (1) automatically discover effective security policies and mechanisms along any network path, (2) dynamically configure security mechanisms across protocol layers and across the network, (3) adaptively re-configure these mechanisms to maintain certain levels of security services when the network is under stress. This paper describes the Celestial system design and implementation, and reports the current status of the project.

1 Introduction

In spite of our serious concerns over the information security issues, secure networking is yet to be widely practiced. One reason is the lack of understanding of security issues at a systematic level, *e.g.*, is security protection more important than qaulity of service (QoS) requirement and how to address both security and the traditional QoS in the same service-provisioning framework? Furthermore, there is an overall lack of "secure information-ware", tools for navigating the information space securely and successfully. The Celestial project is addressing this question by developing a security management architecture that can (1) automatically discover effective security policies and mechanisms along any network path, (2) dynamically configure security mechanisms across protocol layers and across the network, (3) adaptively reconfigure these mechanisms to maintain certain levels of security services when the network is under stress. A prototype system (referred to as Celestial) has been implemented to demonstrate the viability and usefulness of this architecture. Main security mechanisms integrated include a cryptographic library in user space, IPsec, and ATM cell encryption. Key components of this system include the Security Management Agent architecture (SMA), a security service API, and an inter-SMA communication protocol. The rest of the paper is organized as follows. The remainder of this section introduces the terminology that will help in the following discussion and describes the problem. Section 2 presents the Celestial architecture and its main components. Section 3 describes the API. Section 4 presents the inter-SMA communication protocol. Section 5 reports the implementation and the status of the system. In Section 6 we review the related work in the area. We discuss the future work in Section 7 and conclude the paper with our thanks to the reviwers.

1.1 Basic terminology

Before discussing the security management problem and our solution to it, a brief introduction of the terms we will be using should be helpful.

Secure communications are made possible through the use of *security services*, such as message confidentiality (encryption), integrity, authenticity, and non-repudiation. Security services are provided by *security mechanisms* that are software or hardware modules by which security functions are implemented. For instance, the Data Encryption Standard (DES) [1] [2] is a security mechanism that provides a data encryp-

*This work is supported in part by DARPA/ITO through Federal Contract # DABT63-97-C-0045

†Please send all correspondence to this author.

0-7695-0490-6/99 $10.00 © 1999 IEEE

tion service. *Security capability* describes a specific implementation of a security mechanism. The *network nodes* refers to end-hosts, routers, and switches. A *secure gateway* is a router or a switch on which security policies are enforced and security mechanisms are provided.

There are many security mechanisms. We will only use IPsec [3, 3, 5, ?] and CellCase [4] throughout this paper to illustrate the heterogeneity of security mechanisms.

1.2 Problem description

Advances in network technology have greatly changed the way we share and exchange information and have boomed many new online activities such as e-commerce. As the Internet gives rise to a new world of global communication, more and more people are becoming concerned with the protection of their online activities. Researchers have been aware of the lack of security in the communication network infrastructure for years. For example, Internet Engineering Task Force (IETF) has several working groups aimed at enhancing the security of network protocols. In the meantime, industry has been quick to catch up by developing many security products.

Different efforts have led to different security solutions, which means different security mechanisms can be used to provide the same security service and they can be employed at different network protocol layers. Meanwhile, different organizations and individuals define security differently by enforcing different security policies. Moreover, different network applications may have varying security service requirements. All of this diversity results in a highly heterogeneous network environment. Figure 1 is an illustration of such an environment. The heterogeneity causes difficulties in setting up secure communication channels because two communication end-points may not share common security mechanisms or the security policies for some nodes on the path may be violated. Let us examine the heterogeneity in closer detail.

Security mechanisms can be applied at different protocol layers. The TCP/IP protocol stack has four layers, i.e., data link layer, network layer, transport layer, and application layer. Network security can be enforced at any of these layers. Enforcing security at different layers has its advantages and disadvantages.

- **Data link layer.** Link layer security mechanisms can operate at link speed, hence they put little performance burden on the network nodes. Encryption is the most common security mechanism applicable at the link layer. One disadvantage of link layer security mechanisms is that authentication, including signatures, can be provided only on a host basis. Furthermore, although message confidentiality can be provided, message integrity can be more difficult to ensure if the link is byte-oriented, rather than frame-oriented.

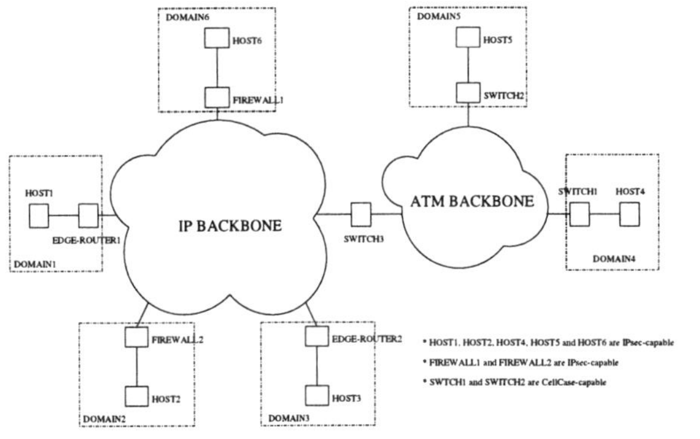

Figure 1: Heterogeneous network environment

- **Network layer.** Many researchers and engineers argue that Internet Protocol (IP) layer is a good place to enforce security because all the packets received by the lower-layer protocols and all the packets sent from the higher-layer protocols and the applications go through it. The biggest advantage of network layer security is its transparency. It can be provided without requiring changes to applications, any other higher layer protocols, or network components that do not need security functions. IPsec [3, 5, 6] is the network layer mechanism being standardized by the IETF.

- **Transport layer.** Providing security at the transport layer permits transport layer peers to communicate securely over insecure networks. The major advantage of providing security mechanisms at the transport layer, as opposed to at the network layer, is that applications can potentially choose different security mechanisms and policies. The reason is that the transport layer typically keeps state that maps to individual applications. The primary drawback of transport layer security is that it is more difficult to enforce a tunneling scheme and the firewall configuration. Examples of transport layer security mechanisms are Secure Socket Layer[7] and Transport Layer Security Protocol[8].

- **Application layer.** Implementing security functions directly at the application layer provides the most flexible way of handling the application-specific requirements. Advantages of providing security functions at the application layer are that they are made available to all the applications, and they can be selectively invoked by the applications when needed, thus limiting the performance cost. However, the ability of the applications to choose whether to utilize the security functions or not makes it harder to enforce the security policies.

Diversity of security policies. The management unit of the Internet is a domain, defined by an autonomous organization. Each organization has its own policies which determine the usage of its resource and data. Security mechanisms employed by different organizations may differ due to their policies. Problems arise when organizations import or export data from one another, while requiring that data continue to be protected in accordance with their policies. Currently, a firewall is the most widely employed mechanism that enforces domain security policies.

Variety of security capabilities. Different network security solutions might not be compatible with each other because they address security problems in different ways. The Internet consists of hosts, routers and switches. Different network nodes may be equipped with different security-related hardware and/or software, which lead to different security capabilities. These security-related hardware/software on one network node can be added or removed without informing others. Moreover, problems arise when two end hosts that do not share the same security capability try to communicate. For example, if one domain signs data to preserve authenticity while the other does not have the algorithm to verify the signature, the authenticity requirement needed in the communication between the two domains is simply not possible.

Different requirements from different applications. Different applications have different security and service requirements. Some applications such as video-conferencing software might need higher speed of encryption while others such as email may have no strict requirement on performance. On the contrary, the email software may need stronger encryption than the video-conferencing software. In this case, static configuration of security mechanisms will not be able to meet every application's requirement.

Survivability over a public network. Talking about secure communications, we have to take into account not only the two end-points of a communication, but also the condition of the path in which the data packets travel. They both influence what policies are to be enforced and what security mechanisms can be used. For example, upon receiving security alarms due to a security breach, the policy of one domain may require stronger encryption, *e.g.*, a longer key, be employed.

From the above discussion, we can see that static configuration of available security mechanisms is impractical as an attempt to cover all the possible situations of today's heterogeneous network environment. As a consequence, **we need a powerful and flexible management infrastructure** that can manage different security mechanisms.

So far, most of the research and engineering efforts have been focused on the individual security mechanisms and protocols. There have been few attempts to address the management of security mechanisms, let alone managing the security mechanisms over different protocol layers. Researchers from BBN proposed Security Policy System [16] which tries to manage IPsec and its policies. Intel is building Common Data Security Architecture [17] that manages heterogeneous security devices. To the best of our knowledge, no other effort has attempted to manage the security mechanisms at different protocol layers and build a fully distributed infrastructure capable of negotiating and dynamically setting up security communication channels.

Our research effort is building a distributed security management infrastructure that manages diverse security mechanisms over different protocol layers for the purpose of secure communications. Our system is capable of discovering the specific mechanisms in common between domains as well as their policy requirements. It negotiates and dynamically sets up secure communication channels utilizing these mechanisms. The decision about the secure channel configuration, which may involve intermediate secure gateways, is made on-the-fly based on the factors such as the domain security policies, the security capabilities of the communication end-points, and the security and service requirements of the application programs.

2 Proposed Solution – Celestial System

We are developing a network security management technology – Celestial system, that addresses the secure communication difficulty introduced by the heterogeneity of the network security environment. The key component of our architecture is the **Security Management Agent** (SMA), a set of software modules located in network nodes that have security requirements and/or provide security services. In order to utilize available security services, end-hosts need to install SMAs. Intermediate, especially backbone, security gateways are not required to have SMAs installed. But we suggest that SMAs be installed on the switches or routers, especially edge routers of a domain, where the security policies need to be enforced. The SMA is implemented as a module residing outside the operating system kernel, thus requiring no kernel change to the systems for it to be installed. Figure 2 illustrates the deployment of SMAs over a network environment.

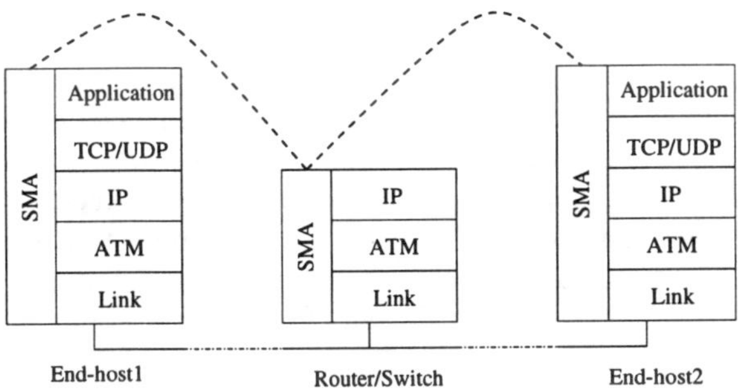

Figure 2: Deployment of Security Management Agent over the Internet

2.1 Deployment of the Security Management Agent

SMA resides logically on the management plane of the network infrastructure. It interacts with the security protocols of each layer of the network protocol stack and manages various management information of the available security protocols. It is authorized to dynamically configure local security mechanisms of all these protocols. To establish a secure end-to-end communication, all the SMAs along the data path have to coordinate in order to set up a security channel.

In addition, an SMA has an interface to the application programs, through which different security services can be provided. Here the application programs include the network application programs and other protocols such as FTP, SMTP that reside on top of TCP/UDP. We refer to them as the applications from now on. This means that secure versions of those higher layer protocols benefit from our system by utilizing existing security mechanisms underneath them.

We will now use the network environment in Figure 1 to explain how the Celestial system works. An application on HOST1 wants to initiate a secure communication with an application on HOST2. It conveys, through an API, its intention to the SMA on HOST1, along with the type of security service needed, say, confidentiality, and the requirements on the service quality. Upon receiving the request from the application, the local SMA generates a request message and sends it to the SMA on HOST2. The message is intercepted by the SMAs on the data path of the communication, i.e., edge Router1 and FIREWALL2. They append their own security capabilities along with special security policy requirements and forward it. The security policies private to the local domains are not appended.

When the SMA on HOST2 gets the request message, it makes a decision on how the secure communication channel should be set up, based on its own policies and the capability and requirement information carried by the request message. The decision includes which intermediate secure gateways will participate in setting up the the secure communication and what security mechanisms will be used between them. In this special case, one plausible decision is as follows:

- FIREWALL2 has to participate (because it is a firewall);
- There will be an IPsec ESP tunnel between HOST1 and FIREWALL2;
- HOST1 and HOST2 will use IPsec ESP to protect data; 3DES will be used as the encryption algorithm.

The reason FIREWALL2 has to participate is determined by the firewall policy requirement. In this case, the application requests a confidentiality service without a specific requirement on the security protocol or encryption algorithm. The SMA decides to use IPsec ESP and its default encryption algorithm, i.e., 3DES.

The decision is sent through the reverse path of the incoming request. Upon receiving the decision, the intermediate SMAs check with their local policies. If there is no policy conflict, the decision is forwarded towards HOST1. Then, the participating SMAs proceed to set up necessary security associations, based on the decision, and exchange keys. After all the security mechanisms are successfully set up, the SMA on HOST1 notifies the application and the application can start initializing handshake with the application on HOST2 and sending data.

Due to the uni-directional nature of IPsec security associations, the SMA on HOST2 will initiate the same procedure by sending a request to the SMA on HOST1 in order to set up the security channel for data to travel from HOST2 to HOST1.

2.2 Structure of the Security Management Agent

With the understanding of how the security management agents are deployed, we now look at what management functions an SMA provides as well as its components. Figure 3 shows the conceptual structure of a security management agent.

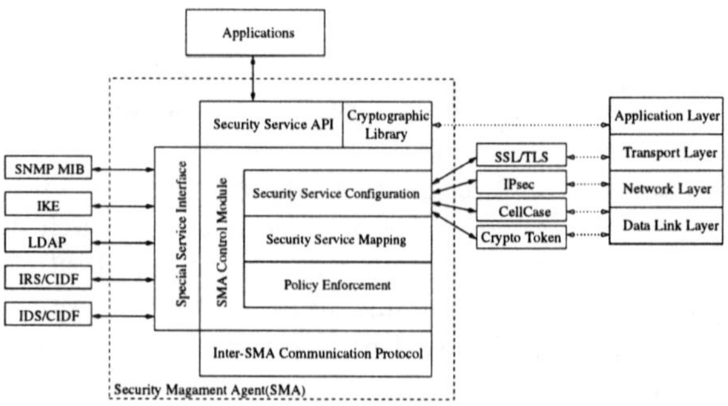

Figure 3: Structure of Security Management Agent

An SMA has eight modules, *Security Service API*, *Inter-SMA Communication Protocol (ISCP)*, *Service Mapping Module*, *Service Configuration Module*, *Policy Enforcement Module*, *Special Service Interface*, *Application-layer Cryptographic Library*, and *Control Module*.

- **Security Service API** is an application programming interface that allows the applications to invoke available security mechanisms in order to receive security services. It is an extension to the BSD-socket API, thus allowing existing network applications to be changed as little as possible in order to use our system. A more detailed description of the API will be given in Section 3.

- **ISCP** is a protocol used to provide communications between SMAs. ISCP itself is a secure and reliable protocol. A summary of ISCP is provided in Section 4.

- **Security Service Mapping Module** is responsible for mapping the security service specifications to locally available security mechanisms.

- **Security Service Configuration Module** is the interface through which SMA manages and invokes different security mechanisms.

- **Policy Enforcement Module** handles the security policies. The policies define the protection requirements for the current domain, for the current local network node, for specific protocols, and for specific applications. Our current desgin assumes that there exist a *policy server* in every security domain where SMA is deployed. When a policy server chooses to delegate policy decisions to individual SMAs in the domain, each SMA will make decisions by its own based on the policies downloaded from the server; otherwise, the policy module will defer the policy decisions to its server by forwarding the contact information for the server to the destination SMA.

- **Special Service Interface** allows us to take advantage of functionality of other infrastructures. For example, we can use IPsec key exchange protocol [9] [10] so that we do not have to develop our own. Another example is that SMA can interface with intrusion detection system. Upon receiving security alarms from an intrusion detection system, SMA will start to enforce stronger security policies and requirements for the later communications.

- **Cryptographic Library** is a library we provide to the application. It contains some basic security mechanisms, regardless of what security mechanisms may be provided in the underlying networks. In the case where no other local security mechanisms, *e.g.*, IPsec, are available, or no common security mechanisms are shared between two communication end-points in lower layers, our system can still support end-to-end secure communications by providing security services at the application layer. Moreover, an application layer cryptographic library gives the applications the flexibility of customizing security functions based on their own needs.

- **Control Module**, aka the SMA *Controller*, is the heart of the SMA that glues other modules together. It maintains a set of data structure, called **security contexts**, each of which describes a secure communication channel setup:

- who are the parties involved in the security service;
- which security mechanisms are used and in what combination;
- the granularity at which a given security protection should be applied;
- the algorithms used for encryption, integrity, and authentication;
- the keys to the cryptographic algorithms.

2.3 Interaction between different SMA modules

The application makes a request for security services through *Security Service API*. Upon receiving this request, the *Controller* first checks, with the *Policy Enforcement Module*, to see whether the request is allowed by its local policies. When getting a positive answer, it gets the local security capabilities from the *Service Mapping Module* and sends them along with the request to the other communication end-point via *ISCP*.

The intermediate SMAs intercept the request. Their *Controllers* check with their own *Policy Modules*, get their local capabilities from their own *Service Mapping Module*, append their capabilities and special policy requirements to the message, and forward the message towards the destination.

When the destination SMA receives the request, its *Controller*, coordinating with its *Policy Module* and *Service Mapping Module*, makes a decision based on the information conveyed in the message and its own security policies and security capabilities. The decision, in the form of security configuration assignments to all the intermediate nodes who need to participate in order to provide an end-to-end secure communication channel, is sent via *ISCP* along the reverse path.

In general, the destination SMA may need to go through several iterations before a conflict-free service configuration can be determined. The destination SMA accomplishes this by examining the policy requirements from those nodes who have already provided this information through the request packet, formulating a configuration, and negotiating with the policy servers for those domains that deferred policy decisions to their policy servers. Clearly, the destination SMA may have to try several alternative configurations before finding one that is conflict free, assuming that one does exist. We should point out that there is a research issue regarding the general strategy for determining service configurations. This strategy should take into account factors such as security, performance for data transfer, and setup time. In-depth study of this issue is outside the scope of the current project.

For all the SMAs, upon receiving the decision, their *Controllers* extract their own assignments. If they need to participate, they check their local *Policy Module* to see whether there is a conflict with the assignment or not. If the assignment is consistent with their local policies, the message is forwarded to the next SMA on the reverse path.

If any intermediate nodes, *e.g.*, firewalls, are required to participate, they will start setting up secure communication channels by configuring security mechanisms through the *Service Configuration Module*, including exchanging session keys through the *Special Service Interface*. When the setup is done, i.e., the *security context* is set up, the two applications can start to communicate.

3 Celestial Application Programming Interface

We have extended BSD-socket API as our *Security Service API*. Currently the *Security Service API* provides TCP support while UDP support is still in development. The following table (Table 1) compares the TCP support between the BSD-socket API and the Celestial *Security Service API*.

BSD-socket API	*Celestial Security Service API*
`socket()`	`sma_socket()`
`bind()`	`sma_bind()`
`connect()`	`sma_connect()`
`listen()`	`sma_listen()`
`accept()`	`sma_accept()`
`read()`	`sma_read()`
`write()`	`sma_write()`
`close()`	`sma_close()`
	`sma_init()`
	`sma_init_sec_context()`
	`sma_accept_sec_context()`
	`sma_release_sec_context()`
	`sma_delete_sec_context()`

Table 1: TCP Support in BSD-socket API versus Celestial Security Service API

As shown in Table 1, the Celestial *Security Service API* is similar to the BSD-socket API. `Sma_socket()`, `sma_read()`, `sma_write()`, etc, are wrappers to the original `socket()`, `read()`, `write()`, etc. They are the same in terms of the syntax, and similar in terms of the semantics. The Celestial *Security Service API* introduces four more calls: `sma_init_sec_context()`, `sma_accept_sec_context()`, `sma_release_sec_context()`, and `sma_delete_sec_context()`, for the security context management. `Sma_init_sec_context()` and `sma_accept_sec_context()` are used to request for security services while `sma_release_sec_context()` and `sma_delete_sec_context()` release and delete security contexts respectively. Setting up a security context can be very time-consuming. Thus, we want a security context to be reused if possible. `Sma_release_sec_context()` releases a security context without deleting it, which gives us an opportunity to reuse the security context in the near future. Security context reuse is very useful for applications like the web client, because the HTTP protocol uses a TCP connection only once to fetch only one object(file) from the web server.

The way the application programs use our *Security Service API* is as follows. The client program first calls `sma_socket()` to create a socket. Then it calls `sma_init_sec_context()` to request the security service and associate the created socket with the secure context being set up. After the successful return of `sma_init_sec_context()`, the client program can use the socket in the same way as the BSD-socket API. That is to say, `sma_write()` and `sma_read()` are used to send and receive data, except that data are sent and received in a secure fashion. After the data exchange, the client program calls `sma_release_sec_context()` or `sma_delete_sec_context()` to release or delete the security context that was used. Then the client program can call `sma_close()` to close the socket. The server program uses our *Security Service API* in a similar way except that `sma_accept_sec_context()` is called instead of `sma_init_sec_context()`. In order to reuse a security context, the web client program calls `sma_init_sec_context()` again. The local *SMA Controller* searches for a reusable security context that has not been torn down. Upon finding one, it simply allows `sma_init_sec_context()` to return without sending out a request to trigger the negotiation process. This is transparent to the web client program.

4 Inter-SMA Communication Protocol

ISCP protocol facilitates the setup and teardown of security contexts between the SMAs. ISCP includes several types of messages that are passed between the SMAs. ISCP also includes several security features that allow for encryption and authentication of ISCP messages. ISCP Protocol number is 137 (122-254 remain unassigned by IANA).

There are two sides to every ISCP communication: the initiator and the responder. The initiator is the SMA-capable node which desires to establish a security association (context) with another SMA-capable node (the responder). SMA-capable nodes that lie on the path between the initiator and the responder are called intermediate nodes.

Each SMA-capable node possesses various security mechanisms, which can be used in the setup of a security context. Examples of such mechanisms could be a) IPSec or other secure tunnelling capability b) SSL (Secure Socket Layer) c) Encryption and Authentication API library (cryptlib) and d) ATM cell encryption

The decision about using particular security mechanisms and involving particular intermediate nodes is left to the responder. It makes these decisions based on the information it receives about the desired properties of the security context supplied by the initiator and the capabilities of the intermediate nodes provided by the SMAs located in those nodes. ISCP facilitates this information exchange and the delivery of decisions.

ISCP was created to be able to work with both IPv4 and IPv6 address spaces. Since ISCP is considered to be a part of security infrastructure, firewalls must be made aware of the ISCP protocol and allow it to pass through in order for the security contexts to be set up. Alhtough ISCP is built to support discovery of and negotiation among SMAs in the Celestial architecture, the discovery and negotiation capabilities can be applied to many other management issues in the emerging networks. For example, firewalls in general can use ISCP for applications to discover their firewall policies so that legitimate applications can traverse the firewall legally and successfully, instead of being blocked blindly. To enable truly dynamic virtual private network (VPN) services using IPSEC, all we need to do is to deploy the Celestial system among security gateways, with IPSEC as the only supported security mechanism; existing applications can be easily adapted to invoke the VPN service through Celestial API. We expect to be able to explore these opportunities and extensions further through evaluation experiments later in the project.

4.1 Message Overview

ISCP supports a number of messages in order to facilitate its functionality:

Pathfinding message: can be sent by the initiator to the responder, when it requires to set up a security context. This message is intended to be intercepted by all SMA-capable nodes that it traverses.

Path Refresh message: Similar in structure to the Pathfinding message, this message is sent by the initiator periodically to the responder in order to maintain the state of the security context in the intermediate nodes and prevent them from timing out.

Security Reservation message: can be sent by the responder to the initiator hop-by-hop through the intermediate nodes (no interception required) with instructions to the intermediate nodes to invoke approriate security services (called assignments from now on) in order to facilitate the setup of a security context between the initiator and the responder.

Reservation refresh message: is sent periodically by the responder in order to maintain the state of the security context.

Confirmation message: can be sent by the responder and the intermedeiate nodes to the initiator in order to confirm the completion of the setup of their local security mechanisms for a particular security context.

Initiator Teardown message: can be sent by the initiator to the responder in order to remove a previously created security context. Also can be sent as a reply to a *Responder Teardown Message*.

Responder Teardown message: can be sent by the responder to the initiator in order to remove a previously created context. Also can be sent in response to the teardown request message from the initiator in order to confirm the teardown. This message is sent hop-by-hop in order to remove the context information from the intermediate nodes.

Acknowledgement message (ACK_CTX): is sent by the initiator to the responder after all the expected confirmation messages have arrived.

Acknowledgement of ACK (ACK_ACK): is sent by the responder in order to confirm the receipt of the ACK message. This is the last message in the initial exchange of messages between the initiator and the responder which precedes every security context setup.

Error message: intended to deliver error information from the intermediate nodes to the initiator.

Bi-directional flow request: is sent by the initiator in order to speed up the connection setup process. ISCP associations are uni-directional in nature because actual IP routes for the two opposite directions may differ. In order for two hosts to have a bi-directional communication, they both must setup contexts for each direction. The bi-directional flow request is sent by the initiator to the responder to inform it that it should initiate a setup of its own.

Reservation acknowledgement (ACK_RES): the responder continues sending reservation messages until it receives the reservation acknowledgement message from the initiator.

Figure 4 describes the typical exchange between SMA nodes during context setup, lifetime and teardown. As can be seen, some of the messages are intercepted using the divert socket functionality, while others are addressed and received directly by indended nodes.

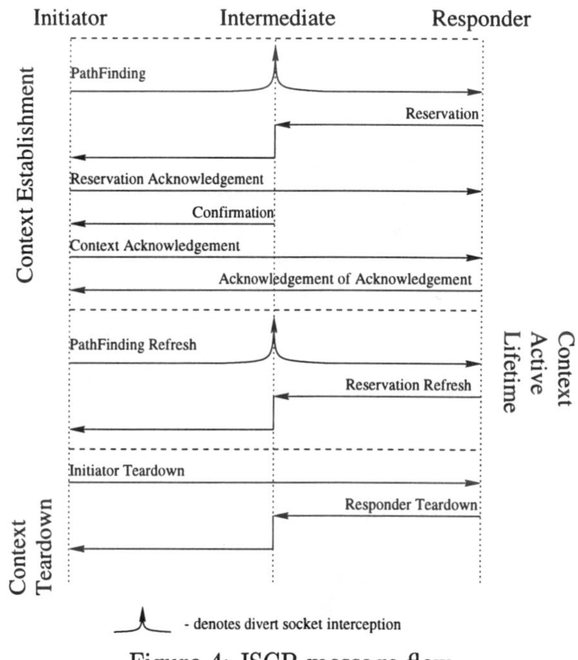

Figure 4: ISCP message flow

4.2 Security Considerations

As any other Internet Protocol, ISCP can be subject to a number of attacks with the purpose of disrupting its service or eavesdropping on the secure traffic.

ISCP takes special care in protecting the integrity and authenticity of its messages. The header of each

ISCP packet is authenticated and the payloads are encrypted to prevent eavesdropping. Separate keys are used for the opposite directions in communications to further improve the security of the protocol. This prevents a "man-in-the-middle" attack from succeeding. Even if a hacker has succeeded in intercepting and reformatting an ISCP message, trying to impersonate an SMA-capable node, he will not be able to decrypt the replies from the responder, as they will be encoded with a separate key.

Two separate mechanisms are used for that purpose. Key distribution is done by using the Diffie-Hellman algorithm run in phone-book mode (with a certificate authority (CA) possessing the public portion of the node's secret, which can be distributed to other nodes on demand). In the absence of a CA, manual configuration can be used for distributing the public component of the key.

In order to encrypt and authenticate its messages, ISCP uses IDEA in CBC mode, with the last 64-bit cypher block used as a signature for authentication. ISCP will be using 128-bit keys. Initially in the message exchange the Diffie-Hellman keys are used. During the initial security context setup new session keys are distributed in order to be used to decrypt and encrypt messages later during the session throughout the lifetime of the context.

5 Implementation and Status

The first prototype system was created based on the version 1.0 of the ISCP design. The system included application level encryption (RC4 and RC5) as the only security mechanisms. It supported both FreeBSD and Solaris operating systems. The system was demonstrated to the DARPA program manager in September 1998 using text-based HTTP client and server programs.

As of this writing, we have completed the design and implementation of ISCP version 2.0. The software structure for the implementation is shown in Figure 5. The current implementation is supported under both FreeBSD and Linux with divert-socket enabled kernels. A Solaris end-host-only version utilizing raw sockets is also planned.

The divert socket mechanism enables IP packet interception and injection on the end-hosts as well as on the routers. Both the interception and the injection happen at the IP layer. The intercepted packets are diverted to sockets in the user space. The intercepted packets can be reinjected into the IP packet flow by writing them to the sockets. The divert socket mechanism supports ISCP operation by intercepting incoming ISCP packets and passing them to ISCP, then receiving them from ISCP and injecting them into the IP packet flow. The Linux divert socket mechanism was implemented in our group and has been made available to the public (`www.anr.mcnc.org/ divert`).

The new prototype system includes the following functional components:

- *Celestial security service API*
- *ISCP version 2.0*
- *SMA Control Module*
- *Service Mapping Modules*
- *Integrated cryptographic library (cryptlib)*
- *Integrated IPsec and IKE capabilities based on FreeS/WAN 1.0*
- *Three adapted applications: text-based web client and server, security-aware GUI web browser based on Arena and libwww, secure Internet audio and intercepter programs.*

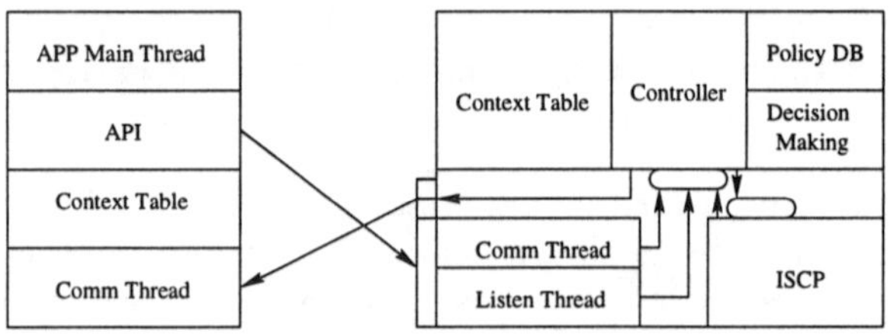

Figure 5: Structure of Multi-Threaded Implementation

Note that the interaction between the application process (including the Celestial API invocation) and the SMA daemon process is achieved through a pair of communication threads. The *Policy* currently allows any type of communications by default. We have examined several candidate technologies for adaptation to provide policy resolution and distribution in the Celestial system.

In the meantime, we will be conducting experiments on our testbed to evaluate the performance and scalability aspects of the system. Celestial architecture allows security management agents to be deployed in an incremental fashion according to the security policies of an application domain. Every SMA-enabled node introduces an access point for security service provisioning and policy enforcement. If more SMAs are deployed, it will allow more flexibility in service provisioning (*e.g.*, multi-hop IPsec tunnels and multi-layer encryptions) but also introduces more latency in end-to-end setup. Such trade-offs can be explored through experiments.

6 Related Work

As we mentioned earlier, most of the research work so far has been focused on individual security mechanisms. There have been few efforts attempting to address the security management over different protocol layers. We discuss the most relevant ones in the next few paragraphs.

6.1 Generic Security Service API(GSS-API)

GSS-API [14] provides a generic interface for security services. It hides all the details of underlying security mechanisms so that the user of GSS-API will not have to worry about the implementation of the security mechanisms. GSS-API, together with a security management system, can provide a similar functionality to that of the Celestial system.

Due to its generic nature, GSS-API is more useful in implementing higher layer security protocols such as Kerberos [15]. It is not tied to an easy-to-use network communication mechanism such as the BSD sockets. Therefore, ordinary network applications will have difficulty adapting to it.

6.2 BBN's Security Policy System

Researchers from BBN have proposed a Security Policy System [16]. In their system, there is a policy server in each domain that controls all the security policies and knows all the capabilities of local network nodes. The policy servers are responsible for making decisions on how to set up security channels.

There are differences between the Policy System and the Celestial system. First, Celestial is a fully distributed system in terms of security management and decision making while Policy System is centralized within a domain. Second, for a policy server to know all the security policies and the security capabilities within its domain is a very strong requirement. For large domains with heterogeneous security capabilities and policies for external connections, it may be necessary to deploy multiple policy servers for both the security isolation and the performance reasons. Third, Policy System only manages IPsec policies, while Celestial focuses on the policy enforcement for all available security mechanisms, not limited to IPsec. Celestial decouples the policy enforcement from the policy distribution/resolution. We are currently considering how BBN's Policy System can satisfy Celestial's policy distribution/resolution needs.

6.3 Intel's Common Data Security Architecture

Common Data Security Architecture(CDSA) [17] focuses on building a security infrastructure that allows different vendors' security products to be plugged-in and utilized.

The key difference between Celestial and CDSA is that Celestial manages security mechanisms and provides security services from the perspective of, but not limited to, the network protocol stack, while CDSA treats security mechanisms more like devices. SMAs of Celestial system can not only negotiate and set up communication channels that may require intermediate secure gateways to participate, but also reconfigure the security communication channels due to factors like security policy changes, intrusion detection alarms, etc. CDSA, on the other hand, aims at end-to-end communications and does not have the notion of a protocol stack and security gateways, thus security mechanisms such as smartcard that can be treated as devices will fit well into CDSA. It is not quite clear how CDSA would manage protocols like IPsec.

Furthermore, CDSA has a GSS-API-like security service API for "applications" to utilize the underlying security services, while Celestial's security service API requires few changes in order for the applications to adapt to the Celestial system because it is an extension to the BSD-socket API.

6.4 Globus Project

Globus project [18] aims at building basic software infrastructure for computations that involve geographically distributed computational and information resources. It implements its own security architecture, using GSS-API.

The major difference between Celestial and Globus is that they have different goals. Globus focuses on resource sharing, thus authentication and access control are the key issues. In this case, GSS-API can be utilized to build its own authentication and access control facility. From this point of view, the security services are used by Globus itself for its own needs.

Celestial focuses on building infrastructure that provides flexible utilization of the underlying secure network transport facilities. It manages all the security protocols and mechanisms that are used for security communications by the applications in a transparent fashion.

7 Future work

We expect to integrate an ATM level encryption capability and an intrusion detection system into the Celestial management system. This will enable us to further develop and demonstrate dynamic reconfiguration of security services. Furthermore, we plan to adapt more applications, such as a video conferencing program to use Celestial Security Service API. It will be interesting to adapt FTP and SMTP protocols from their insecure version by using Celestial Security Service API. These will allow us to show that applications with different security service requirements can utilize the underlying Security Service API as well as the ability of Celestial system to dynamically configure the secure communication channels over an insecure wide-area public network.

8 Acknowledgments

The reviewers provided very careful and constructive comments that helped to improve the quality of this presentation. The authors wish to express their gratitude for the reviewers' effort.

References

[1] W.F. Ehrsam, C.H.W. Meyer, and R.L. Powers, J.K. Smith, and W.L. Tuchman, "Product Block Cipher for Data Security." U.S. Patent #3,962,539, 8 Jun 1976.

[2] Bruce Schneier, "Applied Cryptography." Second Edition, John Wiley & Sons, Inc. 1996

[3] S. Kent, R. Atkinson, "Secure Architecture for Internet Protocol." Internet Engineering Task Force RFC2401, November 1998.

[4] D. Steverson, N. Hillery, G. Byrd, F. Gong and D. Winkelstein, "Design of a Key Agile Cryptographic System for OC-12c Rate ATM." *Internet Society Symposium on Network and Distributed Systems Security*, 1995.

[5] S. Kent, R. Atkinson. "IP Authentication Header(AH)." Internet Engineering Task Force RFC2402, November 1998.

[6] S. Kent, R. Atkinson, "IP Encapsulating Security Payload(ESP)." Internet Engineering Task Force RFC2406, November 1998.

[7] Alan O. Freier, Philip Karlton, Paul C. Kocher, "The SSL Protocol Version 3.0." Available from: http://home.netscape.com/eng/ssl3/draft302.txt.

[8] T. Dierks, C. Allen, "The TLS protocol version 1.0." January 1999. Available from URL ftp://ftp.isi.edu/in-notes/rfc2246.txt.

[9] D. Maughan, M. Schertler, M. Schneider, J. Turner, "Internet Security Association and Key Management Protocol." Internet Engineering Task Force RFC2408, November 1998.

[10] D. Harkins, D. Carrel, "The Internet Key Exchange." Internet Engineering Task Force RFC2409, November 1998.

[11] D. Piper, "The Internet IP Security Domain of Interpretation for ISAKMP." Internet Engineering Task Force RFC2407, November 1998.

[12] R. L. Rivest, "The RC4 Encryption Algorithm." RSA Data Security, Inc., Mar 1992.

[13] R. L. Rivest, "The RC5 Encryption Algorithm." Leuven Workshop on Algorithm'94. Available from URL http://theory.lcs.mit.edu/ rivest/rc5.ps.

[14] J. Linn, "Generic Security Service Application Program Interface Version 2." Internet Engineering Task Force RFC2078, January 1997.

[15] S. P. Miller, C. Nermann, J. I. Schiller, and J. H. Saltze, "Kerberos Authentication and Authorization system." Project Athena Technical Plan Section E.2.1, MIT, 1987. Available from URL http://athena-dist.mit.edu/pub/kerberos/doc/techplan.PS.

[16] L. A. Sanchez, M. N. Condell, "Secure Policy System." Internet Engineering Task Force Internet Draft, November 18, 1998.

[17] Intel, Common Data Security Architecture. Available from URL http://developer.intel.com/ial/security/index.htm.

[18] Globus Project, Available from URL http://www.globus.org/documentation/papers.html.

Multi-Layered Network Survivability – Models, Analysis, Architecture, Framework and Implementation: An Overview

Deep Medhi
Computer Science Telecommunications
University of Missouri-Kansas City
Kansas City, MO 64110-2499
dmedhi@umkc.edu

David Tipper
Department of Information Science
and Telecommunications
University of Pittsburgh
Pittsburgh, PA 15260
dtipper@mail.sis.pitt.edu

Abstract

A major attack can significantly reduce the capability to deliver services in large-scale networked information systems. In this project, we have addressed the survivability of large scale heterogeneous information systems which consist of various services provided over multiple interconnected networks with different technologies. The communications network portions of such systems are referred to as multi-networks. We specifically address the issue of survivability due to physical attacks that destroy links and nodes in multi-networks. The end goal is to support critical services in the face of a major attack by making optimum use of network resources while minimizing network congestion. This is an area which is little studied, especially for large-scale heterogeneous systems. In this paper, we present an overview of our contributions in this area.

1. Introduction

A major attack or a failure can significantly reduce the capability to deliver services in large-scale networked information systems. The drastic effects of communications network failures (even for non-maligned cases) have been demonstrated by several widely publicized network outages in recent years. We broadly classify any type of attack or failure broadly into two groups: physical attack/failure and software attacks. Physical attacks include attacks to destroy switches or transmission systems in some part of a network. Software attacks include attacks such as intruders breaking into systems and destroying, compromising and/or manipulating systems through software change (e.g., injecting false information into network routing tables). In this Darpa project, our work addresses the first group of attacks, i.e., the physical attacks. The obvious question is how to build and evolve network architecture to cope with any major physical attack so that critical information systems can still communicate by utilizing the available network and system resources. It may be noted that a major attack can be caused by a 'single' failure (such as an attack destroying a single high traffic link or node) or 'multiple' failures (such as an attack destroying multiple links and/or nodes). In the rest of the discussion, the terms 'failure' and 'attack' will be used interchangeably.

Most network-based information systems environments consist of a combination of legacy and emerging technologies. Broadly speaking, the environment can be classified into three layers. The top ('application/service') layer is where services such as voice, data, video, multi-cast video, and other distributed services are provided. These services are provided over the middle layer (referred to here as the 'switched network layer') which consists of the networking environment such as circuit-switching, packet-switching (TCP/IP), ATM, and virtual private networks (for special services). Finally, the middle layer is provided over the network transmission environment layer (the bottom 'physical' layer) which typically consists of a mixed technology infrastructure containing fiber and non-fiber wirebased systems as well as wireless components (microwave, cellular, satellite, etc.). Note that an application/service may traverse several interconnected networks with different physical layer and network layer components. For brevity, such heterogeneous multi-layered systems, which takes a resource-directed network architecture view, will be labeled as "*Multi-Networks*".

In this work, we address survivability to provide network design and management procedures towards minimizing the impact of failures on multi-networks. Survivability techniques can be classified into three categories: 1) prevention, 2) network design, and 3) traffic management and restoration. Prevention techniques focus primarily on improving component and system reliability. Some examples

0-7695-0490-6/99 $10.00 © 1999 IEEE

are the use of fault-tolerant hardware architecture in switch design, provision for backup power supplies and transmission equipment, use of frequency hopped spread spectrum techniques to prevent jamming in military radio networks and so on. Network design techniques try to mitigate the effects of system level failures such as link or node failures by placing sufficient diversity and capacity in the network topology. For example, the use of multi-homing nodes so that a single link failure cannot isolate a network node or an access network. Traffic management and restoration procedures seek to direct the network load such that a failure has minimum impact when it occurs and that connections affected by a failure are reconnected around the failure. Survivability goals may be accomplished by designing network infrastructures that are robust to malfunctions of nodes and links, and implementing network control systems that are inherently fault-tolerant and self-healing.

Given the multi-networks environment, survivability and restoration can be addressed at multiple levels. Although it may be possible to address survivability in each layer completely independently, our view is that a multi-layer coordinated and integrated survivability and restoration approach is most desirable to make maximum use of available resources. This should also take into account the underlying technology infrastructure so that evolving survivable network architectures can be generated. With this in mind, our goals in this project have been to address the following inter-related issues:

- Development of network design models/algorithms to provide a quality of service (QoS) specified under any failure condition. This work addresses the problem of intelligently designing and evolving a network topology architecture and configuration starting from an existing architecture and legacy networks to improve survivability.

- Development of network management algorithms (e.g., provisioning of backup routes, virtual circuit rerouting algorithms, etc.) which make optimum use of network resources after a failure (both single and multiple types) in support of critical services. This work concentrates on the design and analysis of multiple priority traffic restoration techniques to provide service continuity while minimizing the network congestion. The restoration algorithms will be suitable for automatic invocation by network components, resulting in a self-configuring system that adapts to the changing fault environment. Since emerging multicasting services such as audio/video conferencing will be critical under an attack, we plan to especially address the issue of the survivability of multi-casting services.

- Emphasis on studying the transient network congestion that occurs after a failure and incorporating its effect into the design of the network and the traffic restoration algorithms. Thus not only will a critical network user be provided service continuity, but the quality of the service can be provided in a graceful manner.

- Functional needs for the network management for multi-networks survivability and demonstration of some of the basic concepts and procedures through a proof-of-concept testbed implementation.

With the above goals in mind, we have divided the work into the following main components:

- Development of design models and tools for survivable networks.

- Development of simulation models and tools that can be used to study of failure.

- Analysis of various network scenarios involving a failure using the tools developed, as well as an evaluation of traffic restoration algorithms.

- A network management framework development for multi-layered network survivability and its proof-of-concept implementation in the MIMIC (Mini Intelligent Multi-networks Information survivability Concepts) testbed at the University of Missouri–Kansas City.

In the rest of the document, we will visit each of these components. It is not possible to cover all of our publications and results in this overview paper. Thus, in most cases, we summarize some of our contributions, while we present a longer description for the network management framework. As applicable, we have indicated the technical reports, theses and publications where more details can be found.

2. Network Design Models

In general survivable network design refers to the incorporation of survivability strategies into the network design phase in order to mitgate the impact of a set of specific failure scenarios. Survivability is typically achieved through, either placing diversity and spare capacity in the network topology (or virtual topology) or adding redundancy to network components (e.g, 1+1 automatic protection switching). In general taking an approach of diversity and spare capacity placement by adopting a mesh type topology with extra capacity is known to be more cost effective and flexible (i.e., can respond to a wider range of failures), than following a redundancy approach. We are interested in network design models that include issues related to network survivability within the framework of the models. In this section, we describe some of the models and tools we have developed.

2.1. Survivable Virtual ATM Network Design

The details of this work are reported in the doctoral dissertation of R. Cotter[6]. In this section, we have highlighted the main contribution of this work.

In a multi-network environment, services required by the upper layer can be provided over a logical network. An example of such a logical network is an ATM-based (Asynchronous Transfer Mode-based) virtual network where virtual paths (VP) are defined for different services based on the demand requirement of the application services level. A critical requirement to address is the survivability of such networks, especially taking into consideration the dynamically rearrangeable capability of ATM VPs. In such an environment, to address for survivability, three different strategies are considered: redesign of the entire network for every possible failure scenario, design of the affected parts (VPs) for any failure scenario, and built-in diversity in the initial design of the network.

While addressing these strategies, along with issues such as consideration of multicast group traffic demand that need to be survivable too, we have been able to develop a generic multi-commodity flow based optimization model that can capture them all very well. This is a major contribution in itself. Further, by looking at the structure of the problem, we have developed an algorithmic framework that can also be used for all of the strategies and scenarios. At the core of this algorithmic framework is a decomposition algorithm which is based on the duality-based subgradient optimization algorithm described in [14].

The design model has been implemented and tested on 96 different test networks ranging from 10-nodes in a network to 100-nodes in a network with multiple service classes and traffic load periods. In addition, a protocol message mechanism that works at the ATM VP level has also been developed. It was geared for ATM VP network survivability.

2.2. Book-ahead Guaranteed Survivable Services

In this work, we have considered a best-effort Internet environment with added capability to provide some guaranteed services that are requested ahead of time. An integral requirement for such guaranteed services may be some level of survivability. A possible approach here is to consider a pair of diverse paths for every survivable demand request. At the same time, to assign bandwidth on the back-up for survivability would mean that other best-effort services can not access this bandwidth if there is *no* failure. Thus, this can lead to consideration of hard requirements and soft requirements.

In addition, such a problem can have conflicting goals; for example, (a) maximization of residual capacity for usage by best effort services, (b) minimization of the cost of the book-ahead survivable provisioning, and (c) possible penalty for not being able to provision for some requests. We have been able to develop an optimization formulation that captures such conflicting goals in a comprehensive manner.

We have done studies on several sample networks and demand requirements to show the effectiveness of the model in meeting different objectives. The details of the model and the results are reported in [23].

2.3. Survivable STM Network Design

STM refers to synchronous transfer mode. Networks based on SONET, and digital cross-connect systems fall under the STM category. In [1, 2], we consider the problem of given a STM mesh type network topology, the normal traffic demand, and the capacity allocation to meet the normal traffic demand, how much spare capacity should be provisioned and where should it be located in order for the network to tolerate a specified set of failure scenarios (e.g., loss of any single link). The term "mesh" does not imply that the network topology is a full mesh, but rather that the network nodes are at least two connected.

Specifically, we present a novel STM survivable network planning technique based on a genetic algorithm formulation of the spare capacity assignment problem for the case of path restoration with link disjoint routes. Genetic algorithms [7] have received considerable attention in recent years for use in solving various combinatorial optimization problems, including the solution of integer programming problems. Genetic algorithms (GA) are stochastic search techniques that mimic the survival of the fittest (or best) paradigm observed in nature.

Our design methodology consists of using the genetic algorithm approach to implement the concept that traffic flows which travel over disjoint routes may be able to share spare capacity on a backup path, since it is unlikely that more than one failure will occur simultaneously. Thus, our approach tries to reduce the cost of spare capacity needed for a particular fault tolerance requirement (e.g., full recovery from any single link failure) by finding a set of backup paths that enables the sharing of spare capacity, which results in reducing the total cost due to the nonlinear economy of scale of spare capacity cost.

The description of the proposed methodology is given in detail in [1, 2] along with a study of numerical results for a variety of network topologies, illustrating the application of the proposed genetic algorithm technique, guidelines for parameter selection and analysis of the computational complexity. Additionally, for the sake of comparison, numerical results for small networks are given for the standard integer programming approaches and a popular heuristic from the literature. It is shown that the GA algorithm is far more computationally efficient while providing near optimal results (2-7%).

3. Simulation Tools

Our interest in studying the effect of a failure on a network and the performance of any newly developed routing and restoration algorithms has led us to the development/enhancement of different simulation tools. Specifically, we have been interested in understanding the implications at the packet-level granularity in the best-effort Internet environment as well as connection/session level granularity in the case of emerging Integrated Services Architecture. Due to our varied interest, it became apparent that just one tool cannot fit every environment. Thus, this led us to the enhancement of the MaRS tool to create the new tool, MoMaRS, which is ideal for understanding packet-level implications due to a failure for multicasting environment. We have also developed extensions to `ns` simulator [33] for the unicast service environment that addresses fault tolerance. However, to address and understand the connection-level effect, we needed a tool that can simulate several thousand connections in a short period of time.

3.1. MoMaRS Tool

MoMaRS is a packet-level simulation tool. The new tool has been created by extending the MaRS tool[3]. MaRS has very good built-in unicast routing components as well as various service level workload components. Its original design allows the study of a link failure. Our interest was to consider multi-cast services and their performance under failure, and we are also interested in considering an environment that addresses Differentiated-Services Internet[4] and Integrated Services Internet[5].

With this in mind, we have made major enhancements to create the MoMaRS tool. In particular, the tool now has two multicast routing components: MKompella[12] and Multicast Shortest Path First (MSPF) routing protocols, as well as multicast workload components for one-to-many and many-to-many communications. We have further added priority based scheduling at the nodes (routers) for emulating a differentiated-services environment, and Resource reSerVation Protocol (RSVP) along with a classifier, a packet scheduler, and an admission controller to emulate the Integrated-Services environment. In addition, a new TCP component is added to allow a user to study interaction between TCP behavior and routing dynamics due to a failure. In particular, this environment allows us to study a single or multiple link failure. We are currently working on a user's manual for this tool and it is expected to be available soon.

3.2. MuSDyR Tool

While MoMaRS is a good tool for understanding packet level impact, it is not geared towards studying session or connection level issues for large-number of sessions/connections. Due to lack of an available tool that suits our needs, we have developed a new simulator called MuSDyR[17] (Multi-service Simulator with Dynamic Routing) for this purpose. MuSDyR is designed to consider connection-level impact along with flow/connection-level quality of service routing components. Due to our special interest in studying a network failure, this tool has been designed from the beginning with the capability to study a failure. In particular, it has a partial restoration capability, allowing the user to do staggered restoration after a failure. In addition, this tool has the capability to re-route a connection in the event of a failure. The tool can easily handle simulation of ten thousand simultaneous connections. To consider the fact that network traffic is non-stationary and changes with time, a dynamic traffic generation module has been included in this tool along with the stationary traffic generation module. In addition, this tool has the built-in capability for reservation-based multicast services.

3.3. Extensions to NS simulator

In order to study the effects of failures and evaluate survivability schemes on unicast services for various next generation Internet architectures, we have developed a simulation tool by extending the Network Simulator-NS (version 2) [33]. NS is widely used in the Internet research community. In order to simulate survivability schemes for unicast services, new modules were added to NS including an admission-control agent, a RSVP agent, a flow-routing agent, a resource agent and a fault-tolerant agent. The admission-control agent determines if a connection requesting QoS will be accepted based on the available resources. The RSVP agent will send reservation messages to setup or tear down the flow along the path given by a flow-routing agent. The flow-routing agent at each node maintains path information (set of candidate paths to other nodes) and routing information once the flow is set up. The candidate path set for each node pair is precomputed and loaded into the simulation before execution. Using the candidate path set the flow-routing agent runs the path selection algorithm to find the path that gives the minimum cost route. A resource agent at each node keeps track of resource levels at all ports. The simulation model is constructed so that all nodes share global information of resource levels. A fault-tolerant agent at each node incorporates the different restoration recovery schemes. Additional modifications to NS include changes to statistics gathering routines to permit the gathering of transient behvior across multiple simulation runs.

4. Some Results

Using simulations, various analyses can be done to understand the impact of a failure, as well as performance, when a new control scheme is introduced to alleviate any problems. In this section, we briefly discuss some preliminary results. We are currently conducting several studies which will be reported elsewhere.

4.1. Transient Behavior with Multicast Services

Using the MoMaRS tool, we have studied the transient network behavior of multicast and unicast connections in both differentiated-service and integrated-service capable Internet architectures under major link failure. Detailed results can be found in [18, 24, 25]. We briefly highlight some of the results here.

Our results show that the performance of multicast routing protocol is essential in reducing the overall network utilization, thereby reducing the overall delay for all packets through efficient network utilization. The priority-based routing algorithm does provide QoS assurance for higher priority traffic. Figure 1 shows a dramatic decrease in the instantaneous delay and jitter for a selected multicast traffic stream when priority-based routing is employed. RSVP along with enhanced queuing and scheduling mechanisms can provide required QoS for real-time traffic in the network with link failure as well as without link failure. Figure 2 shows that in the RSVP-enabled environment, the delay and jitter of the multicast traffic during the failure period remains more or less the same as that during the pre-failure and link recovery periods. This is a very significant improvement in QoS for multicast traffic when compared with base simulations where RSVP is not employed. Simulation results using different routing protocols show that the ability of RSVP to re-establish the affected reserved paths due to network failure depends on the underlying routing protocol.

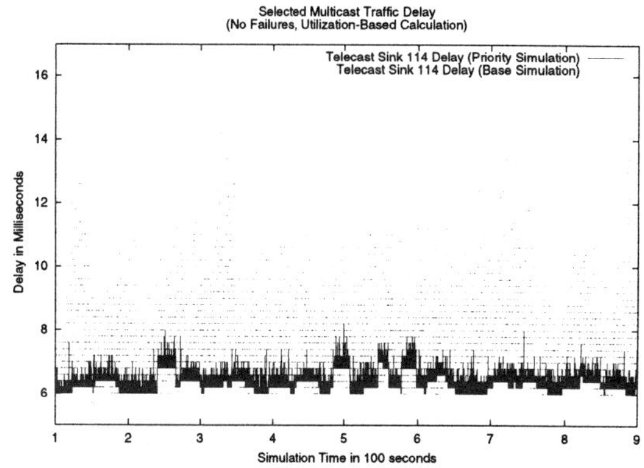

Figure 1. Telecast Sink 114 Instantaneous Delay (No Failure)

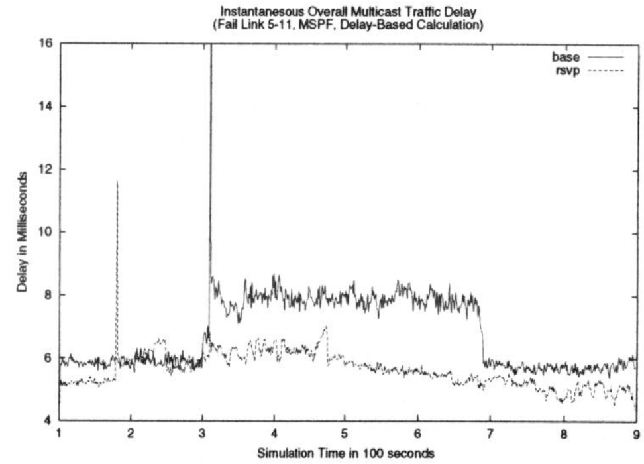

Figure 2. Multicast (Failure, MSPF, Delay-Based)

4.2. TCP Behavior due to Network Dynamics

Using the TCP component of the MoMaRS tool, we have also studied the impact of network dynamics on TCP[21, 22]. Network dynamics refers to the changes in the network state due to route oscillation, link failure and so on. While TCP has been extensively studied over the past decade, surprisingly, very little work has addressed the impact of network dynamics.

Our results show that the impact of network dynamics on TCP depends mainly on its acknowledging (or ACKing) behavior, where the performance of TCP with Non-Delayed acknowledgment (TCP-NDACK) is noticeably affected as compared with that of TCP with Delayed acknowledgment (TCP-DACK). During route oscillations, TCP flow displays asymmetry and false retransmission. Whereas during overlapping link failures, it displays multiple packet loss, retransmission timeouts, false retransmissions, and flow synchronization.

While false retransmissions can be eliminated in most of the cases by sharing congestion information between the header prediction and fast retransmit algorithm, an improved congestion detection mechanism at the nodes/routers and selective acknowledgement in TCP are necessary in order to handle link failures efficiently. We believe that a better acknowledgement technique would make TCP less susceptible and more adaptive to changes in network dynamics.

4.3. Survivable Services in NGI

In this work, we have investigated schemes for the survivability of guaranteed QoS connections that can be applicable in the Next Generation Internet (NGI) network architecture[30] using the extensions done to the `ns` tool. We assume the use of RSVP signaling to reserve resources along a fixed route (explicit path reservation) to provide QoS. A comparative study of the performance of two basic survivability schemes within an NGI architecture is presented. In the first scheme, a QoS connection is provided standby backup resources on a disjoint path by reserving resources on both the working and the backup path. In order to reduce the amount of backup resources required a method for sharing backup resources when the working connections have disjoint routes has been included. In the second scheme a dynamic search for restoration resources is conducted over a preplanned set of alternate paths upon notification of a failure.

A simulation based performance study was conducted to quantify the tradeoff in connection blocking from the guaranteed recovery scheme under normal operations versus the connection blocking after a failure from the dynamic search approach. In addition, the speed of recovery in reconnecting sessions is studied along with the transient network congestion produced by retransmission of lost data. The study shows that the first scheme results in much higher connection blocking under normal operations, faster restoration times, and longer transient congestion times due to non-optimal backup routing.

5. Traffic Restoration Procedures

As noted earlier, survivable traffic management and restoration procedures seek to direct the network load such that a failure has the minimable impact when it occurs while the load affected by a failure is restored. The performance of a traffic restoration procedure will largely depend on the combination of the algorithm used for restoration and the spare capacity allocation in the network. In general, simple restoration schemes will lead to allocating more spare capacity, whereas more sophisticated restoration schemes will require less capacity, but may take longer time to restore connections. In a multi-network environment we contend that distributed dynamic restoration schemes need to play a central role in traffic restoration since such networks will typically be operated by various service providers possibly using different technologies at various network layers and the end-to-end spare capacity planning could be difficult. Our work on traffic restoration has concentrated on the development of algorithms for distributed dynamic traffic restoration to make optimum use of whatever resources are available after a failure, in support of critical services. Here we summarize some of our main results from [9, 10, 34, 35].

5.1. Virtual Circuit Fault Recovery Routing

In [34] and earlier work, we have studied the problem of routing for traffic restoration after a failure in virtual circuit based wide area networks utilizing source node routing. Note that in such networks, a device failure will typically result in several nodes having many virtual circuits to restore and a critical issue in the restoration is the path chosen for rerouting. Another major factor on network performance after a failure in packet switched networks in general is the transient congestion period that results from restored virtual circuits attempting to send out the backlog of packets accumulated during the time delay between the failure occuring and restoration[32]. Standard routing algorithms in packet networks are normally based on minimizing the steady state network delay and such algorithms may be inappropriate for rerouting affected connections after a failure, since at this time, congestion is a paramount issue.

We present an optimization formulation of the rerouting problem by considering residual capacity in the network as well as the decision on whether or not to reconnect a disrupted virtual circuit[34]. Our formulation allows us to consider several routing schemes and fault secnarios in a unified framework. Note that after a failure many virtual circuits will simultaneously need to be restored; thus, we formulate the restoration problem as a bandwidth packing problem. formulation is based on precise information on the network link status and the decision is done in a centralized manner. In an actual implementation in a network, the source-node based routing makes decision in a distributed manner based on delayed information about network link status. We discuss how our proposed routing algorithms can be implemented in a distributed fashion.

The results of a simulation based performance study are reported comparing the performance of both optimal centralized and distributed implementations of five different routing algorithms in terms of network congestion and traditional survivability metrics, such as the call blocking. Through extensive simulation under different network load conditions, our results indicated that while traditional survivability metrics show little differences between the routing algorithms studied, the transient network congestion behavior is noticeably different. Further, it is shown that at low network loads, when there is enough spare resources so that restoration call blocking is low, the network behavior is the best when the load is evenly distributed throughout the network. In contrast, at heavy loads, when there is high amounts of restoration call blocking, one needs to match the routing scheme to the characteristics of the application to ensure the best network performance.

5.2. ATM Fault Recovery Priorities

In [9, 10], we propose a priority scheme for reconnection of virtual circuits (VCs) in ATM networks that have been disrupted by a failure. ATM networks offer several service categories each designed to handle applications with specific traffic characteristics. A failure typically results in *several* nodes being sources for failed virtual circuits with each having *many* virtual circuits in each service category to *simultaneously restore*, possibly on the order of tens of thousands. The way in which the virtual circuits are processed and routed will determine in part, if the connection is restored, the delay in reconnection and the QoS provided after restoration.

In [10], we have developed a restoration priority scheme based in part on the ATM service classes, which aims at minimizing the impact of a failure on the network while providing users the best possible service. The scheme involves both a priority for reconnection among ATM service classes and a rule for ordering and routing VCs within a service class. The proposed scheme is formulated within the context of switched VC routing but is applicable to virtual path restoration as well. The priority traffic restoration technique proposed is based in part on minimizing the number of dropped cells that need retransmission, thus reducing the transient congestion that occurs after restoration.

Numerical results evaluating the performance of the priority scheme show that it significantly reduces the amount of cells needing retransmission after a failure, thereby reducing network congestion at a cost of longer restoration times for lower priority virtual circuits. More recent work [9] formulates the priority restoration problem for ATM within a optimization framework and compares distributed implementations to the benchmark centralized optimization problem solution, suggesting some improvements to our original distributed scheme. This work clearly shows the benefits to both the high priority network user and the network operator in adopting a multi-priority restoration scheme, in terms of speeding up restoration and reducing network congestion.

5.3. Fault Tolerant Connection Oriented Multicasting

In [35] we examine the potential benefits of dedicated backup route(s) to provide survivability for ATM group communications. Specifically we examine the feasibility of providing survivability using working multipoint routes with disjoint dedicated backup multipoint routes where the multipoint routes are setup using either Virtual Rings, shared multicast trees, or VC Mesh groups. We introduce a optimization formulation which identifies a "Disjoint Steiner Ring" within a general graph in order to construct Virtual Rings. Numerical results show that disjoint backup shared multicast trees and disjoint backup VC Mesh groups are not always available while self-healing Virtual Rings exist in all cases tested. In addition, experiments comparing the cost of providing survivability between self-healing Virtual Rings, shared trees, and VC Mesh groups show that self-healing Virtual Rings are lower in cost for the topologies considered.

6. Network Management Framework and Testbed implementation

In the Introduction, we have given the basic notion behind the multi-network framework that is governed by a resource-directed layered network architecture. Towards developing the network management framework for this architectural context, we now elaborate on motivations behind the framework.

In the resource-directed layered network architecture, it may be noted that each layer will have independent policies regarding routing and resource management. However, the granularity and how often the routing changes are different. For example, in today's best-effort Internet, the routing can change at the packet level, while the routing in an ATM Network can be either at the connection or the virtual path level. In digital cross-connect networks, routing is either at the T1 or T3 level and does not tend to change in the scale of minutes or sometimes even days. When we consider the SONET environment, the routing can be at OC-3, OC-12 levels and so on. Further, the time scale of response of routing can be different for different environments. While SONET self-healing rings can reroute quickly (about 50 miliseconds), this is not the case in the SONET mesh network environment. Now, consider such a heterogeneous environment and assume that there is rerouting capabilities in each layer. Some important issues arise such as how a failure at the lower resource layer will affect the overall network, what is the best way for overall networks to respond to a failure, and what functionalities are needed for the failure management.

The administration and management of such a resource-directed multi-layered communication network involve some additional complexities in the routing and resource management strategies during link/node failure situations. Although the traditional management systems, which are designed to manage the network of single administrative domain or of homogeneous technology, do deal with issues related to resource management and survivability, the scope of the management information available to the domain-specific management systems is very localized. For example, the management system of a virtual private network may request an additional capacity on an overflowing point-to-point link. Similarly, a link failure at the physical layer may affect some of the virtual channels of service provider networks at the layers above the physical network. Currently most of these inter-domain issues are handled by human managers. However, with ever-growing multi-layered communications networks, it would be difficult for the human managers to handle the increasing number of such tasks. Another important factor is the desired reaction time for the inter-layer network management issues. For example, when a physical link fails, some of the affected logical channels require immediate attention so that at the user level there will not be any perceivable breakdown of service.

One of the limitations of the domain-specific and technology specific management systems is that the abstraction of the management information is different from one domain to another. Also, each of these management systems will have no knowledge about the abstraction of management information of other domains. The interaction and exchange of management information among the domain-specific management systems are essential in the management systems of multi-layered networks. Hence, the definition of an interface of communication and a common abstraction of management information forms an important factor in the implementation of such management systems for multi-layered networks. The manager at an upper level of hierarchy uses different interfaces of communication to interact with different domain-specific managers at the lower level.

Here, we give an overview of a loosely-coupled hierarchical framework that can facilitate maximal survivability of services in multi-layered networks for various failure situations. More details about this work can be found in [11, 15, 26, 28]. In this discussion, we specifically consider a resource directed two-layered architecture where the networks at the upper layer are called 'user networks', while in the lower layer, we have a 'provider network'. We assume that user networks and the provider network are *not* all completely controlled by the same network administration. In each of the user network or the provider network, we assume that there is at least a domain-specific manager. In this framework, the domain-specific managers are called by the name of the domain for which they are responsible.

All the domain-specific managers will interact with domain independent managers in the upper level of hierarchy. This domain independent manager is called *Across Layer Manager of Managers (ALMoM)*. The communication interface between domain-specific managers and ALMoM specifies the management information that is exchanged and a messaging protocol required for the exchange of information. This architecture is designed to incorporate the management functionalities such as notification of link failures at the provider network and automated expansion of bandwidth of virtual links at the user network. The domain specific manager at the lower layer (provider network) will come to know of the physical link failures almost immediately which may not be the case with domain-specific managers of upper layers (user networks). Depending on the network technology and the exchange due to the management protocol, the managers will realize the failure of logical virtual links (due to failure of physical link) with a certain amount of delay. In such cases, early notification of the link failure from the managers of the physical layer network is useful in initiating recovery mechanisms for affected links.

6.1. Illustrative Examples

We now discuss two simple illustrative examples to show the need for multi-network coordination. More extensive results along this line can be found in [13, 16].

The first example consists of four nodes in the provider network and three nodes in the user network as shown in Figure 3. The links connecting the nodes U0, U1 and U2 in the user network are logical links which are routed through the physical links in the provider network. Each of these logical links is allocated the requested bandwidth in the physical layer network, although the actual utilized bandwidth can be smaller than the allocated bandwidth. In Figure 3, each logical link is marked with both the allocated and the utilized bandwidth. Now let us assume that physical link P2-P3 with a capacity of 60 units fails and hence, logical link U0-U2 with the allocated bandwidth of 60 units also fails. Of the two alternative paths for this logical link in the provider network, the maximum capacity available is only 50 units, which is less than the allocated bandwidth to this link originally. In the absence of utilization information of this channel, the manager of the provider network cannot restore logical link U0-U2 due to insufficient bandwidth. However, as the actual utilization of this link at the time of link failure is only 45 units, the logical channel can be restored using the alternative path P0-P2-P1-P3 with a bandwidth of 45 units allocated, giving the perception of complete recovery. This recovery mechanism between the layers is possible only with the knowledge of both utilization and all the possible alternative paths.

In the second example network shown in Figure 4, when the physical link P1-P2 fails, the logical link U1-U2 with allocated bandwidth of 15 units and 100 % utilization is affected. There is no alternative path for this affected logical link with residual bandwidth of 15 Units. Hence, the manager of the physical (provider) network cannot restore the affected logical link. The user network will then try to route the traffic between nodes U1 and U2 through logical links U1-U0 and U0-U2. Although the unused bandwidth of logical link U0-U2 is sufficient to carry the additional traffic of U1-U2, the bandwidth of U0-U1 is not. For logical link U0-U1 to carry this additional traffic, it requires an additional bandwidth of 5 units to be allocated. When requested by the manager of the user network, the provider network allocates

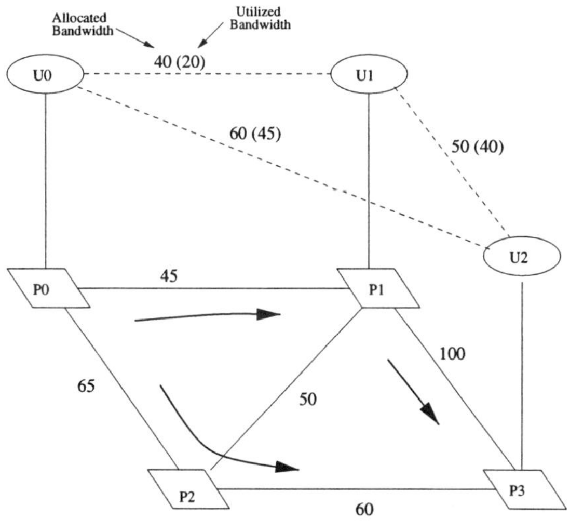

Figure 3. Survivability Mechanism - Example 1

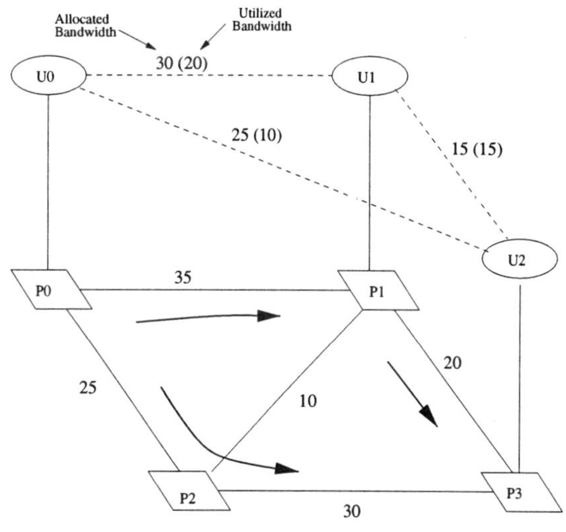

Figure 4. Survivability Mechanism - Example 2

an additional bandwidth of 5 units to logical link U0-U1, so that logical link U1-U2 can be restored. The interaction between the managers of the user network and the provider network is essential in this restoration for two reasons – (1) the manager of the provider network would notify the user network about the failure of the physical link and its inability to restore the affected link, and (2) the manager of the user network negotiates with that of the provider network for the additional bandwidth on logical link U0-U1.

6.2. Architecture of the Multi-layered Network Management System

In the previous section, we have discussed the insufficiency of the domain-specific network managers in a multi-layered network to address a certain level of coordination needed for maximal possible benefits. This type of coordination dictates the need for an integrated management system for multi-layered, multi-domain networks. However, with the increase in size of the network and in number of domains, the amount of management information increases exponentially, hence implementing this integrated management system as a distributed/hierarchical system rather than a single monolithic system is preferable. The choice of a loosely coupled hierarchical distributed architecture is preferred over a flat distributed architecture. In the latter case, each domain specific manager should understand the abstraction of the management information of all other domains, whereas this is not the case with the former.

We propose the loosely coupled hierarchical management system (see Figure 5) where there is an intermediary that interacts with both the upper layer and the lower layer domain-specific managers. We name such an intermediary component as the *Across Layer Manager of Managers (ALMoM)*. The loosely defined framework allows the possibility that ALMoM may not have control over the internal workings (such as routing) within a specific user network. The capabilities between the user network and the provider network can be negotiated at the time of network manager registration through service level agreements. With ALMoM as the coordinating point, each of the domain-specific managers exchange management information, eliminating the need for the domain specific managers to understand the abstraction of management information of all other domains.

We also expect that within an administrative domain, there can be different sub-networks with varying network technologies, with each supporting different management protocols such as SNMP and CMIP. In such a scenario, the preferred architecture for the domain specific managers is again hierarchical with two sub-layers. In the lower sub-layer, the management components are technology-specific and will be responsible for interaction with management agents in a limited number of network nodes. These components will collect the management information from the nodes and translate it into a technology independent abstraction before transferring it to the manager at the upper sub-layer. The manager at the upper sub-layer will perform most of the critical management functions such as processing and archiving of management information and decision making. In accordance with the management functionalities, the management components at lower and upper sub-layers are called *technology-specific* agents and *domain-specific* managers, respectively. In an administrative domain, the number of technology specific agents (i.e. management components

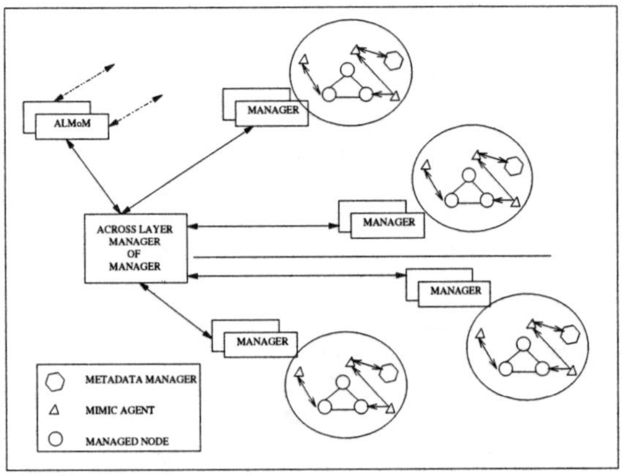

Figure 5. Loosely-Coupled management System for multi-layered Network

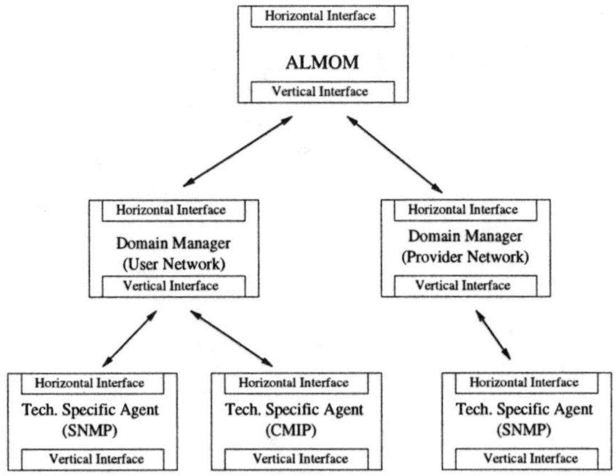

Figure 6. Architecture of Multi-layered Network Management System

at the lower level) depends on the number of subnetworks in the domain and the number of nodes within a sub-network. There can be more than one technology specific agent to monitor a larger sub-network.

The overall architecture of the hierarchical management system for the multi-layered network is shown in Figure 6. This architecture consists of three levels of management components. At the lowest level, technology-specific agents act as proxies between the network nodes and the domain-specific managers at the middle level. The domain-specific managers perform the core of the management functionalities within its administrative domain. The Across Layer Manager of Managers (ALMoM) in the upper-most level is responsible for the inter-domain management functionalities such as survivability and resource management. The exchange of management information between ALMoM and the domain-specific manager is also limited by these functionalities. The role of ALMoM, in this context, can be compared to that of resource trader or broker among the various domains. The interface of communication with management components of the lower level can be of more than one type depending on lower level components. This interface is referred to as the *vertical interface*. Similarly, the interface used for communicating with the management component of the upper level is referred to as the *horizontal interface*. Each management component will interact with more than one management components of the immediately lower level and with only one component in the upper level. The management components in this architecture can also be seen as service providers to the upper level components and as users of service provided by the lower level components. The name *horizontal interface* is used to mean that whichever component in the upper level wants to access the services provided this component, say 'M1', it needs to use the horizontal interface of 'M1'. Similarly, when a management component 'M1' accesses the service offered by a component 'M2' in the lower level, the component 'M1' uses the vertical interface specific to the component 'M2'. In other words, the horizontal interface of component 'M2' will be the vertical interface of another component 'M1' with respect to the component 'M2'.

The definition of an interface (horizontal or vertical) consists of both management information as well as a set of functions, similar to the definition of a class in object-oriented languages such as C++, Java, etc. The definitions of various interfaces in the hierarchical management system mainly are dependent upon the management functionalities implemented in the domain-specific managers as well as ALMoM. Since the context of this paper is survivability strategies, we shall discuss in more detail the definitions of the interfaces specific to these strategies.

Specific to the survivability mechanisms, the definition of the interface also consists of a set of messages exchanged by ALMoM and the domain-specific managers. For example, when a physical link fails in the provider network, the manager of the provider network sends a message to ALMoM. Then ALMoM would determine the set of logical links affected by this failure in various user networks at the upper layers of the network. Depending on the specific implementation of ALMoM, ALMoM may request the current status of the provider network from its manager, while determining the set of affected components. The ALMoM would inform the managers of administrative domains of user networks whose logical links are affected. This will then initiate a string of message exchanges between ALMoM and domain managers for the restoration of affected areas. Similarly, when one of the administrative domains wants the allocated bandwidth of one of its logical links to be increased, the domain manager will send a message to ALMoM to that effect. ALMoM would then determine whether the request can be served. Depending on this determination, ALMoM will initiate a string of message exchanges with domain-specific managers.

Another important aspect of this architecture is that the level of relationship and interaction with ALMoM can be different from one domain-specific manager to another. We envision that each user network may require different levels of survivability requirements. For example, user network-A (UN-A) may require full restoration while user network-B (UN-B) may require partial restoration. What is desirable should be negotiated at the time of registration of the user network domain with ALMoM through a service level agreement (SLA). In such an environment, to address for a failure, ALMoM may be required to correlate notification of failures from different user networks so as to prioritize what restoration to activate in the provider network, so that the user network with the higher priority is restored first, when the bandwidth is limited in the provider network.

Finally, it should be noted that ALMoM, as depicted in Figure 5 does not mean that in practice a single physical entity represents ALMoM. There can be multiple instances of ALMoM for scalability purposes which do distributed networking among themselves, while outside entities such as the user network manager and the provider network manager are transparent to the internal view of ALMoM.

6.3. A Proof-of-Concept Implementation

As a proof of concept, we have implemented our framework for the case of a two layered network with an ATM network as the provider network and an IP network as the user network in the MIMIC testbed at the University of Missouri-Kansas City. Note that this platform is used for a low-cost proof-of-concept demonstration. The basic idea can be implemented in other environment where the user network/provide network paradigm is applicable. In the IP network, we have used FreeBSD routers, while the ATM Network consists of Fore Systems's LE155 switches.

In our current implementation of ALMoM, the survivability mechanism is the only management functionality that is implemented. To make this implementation operating system and location independent, we have used a CORBA-based object management environment and Java programming language. With the CORBA based approach, we have modularized the management system not only at the component level, but also at the management functionality level. For example, the modular implementation of the manager component of the provider network is shown in Figure 7.

An important advantage of modular implementation is that new management functionalities can be added into each

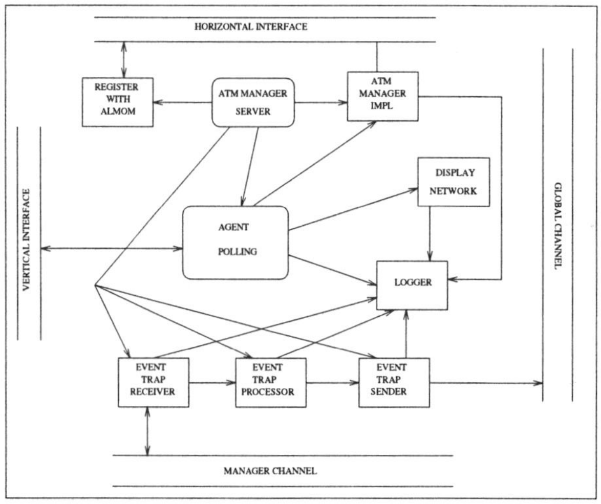

Figure 7. Implementation of the Domain-Specific Manager

of the manager components without much modification to the rest of the implementation. Similarly, *EventChannel* service, which is readily available in the CORBA environment, is used for broadcasting critical events that may have been occuring in the network to various management components. Notably a physical link failure is one such critical event. Depending on the size of the network, there can be more than one *EventChannel* active in the system. In this implementation, we have used two channels, the manager channel and the global channel, providing broadcast service between technology-specific agents and domain-specific managers, and between domain-specific managers and ALMoM, respectively. Each of the events broadcasted in these channels contains enough information, such as event identification, time stamp, origin network and so on,in its header to uniquely identify the event.

7. On-going work

We are currently at the final phase of this project. Several efforts are on-going.

We are currently developing multi-network survivability study tool sets where each of the simulation tools, MoMaRS and MuSDyR, is independently connected with a lower level network optimization-based reconfiguration component. This would then allow us to study a lower layer link failure and its impact on higher-level services (e.g. provider network failure impact on user networks).

In addition, in the MoMaRS tool itself, we are investigating addition of a fault-tolerant multi-cast feature. Similarly, in the MuSDyR tool, we are in the process of adding new routing schemes.

We are also working on the next version of the MIMIC testbed implementation to incorporate some enhancements.

8. Acknowledgments

This work is supported by DARPA and Air Force Research Lab, Air Force Materiel Command, USAF, under agreement No. F30602-97-1-0257. It is also partially supported by NSF Grant NCR-9506652. The U.S. Government is authorized to reproduce and distribute reprints for Governmental purposes notwithstanding any copyright annotation thereon. The views and conclusions contained herein are those of the authors and should not be interpreted as necessarily representing the official policies or endorsements, either expressed or implied, of the Defense Advanced Research Projects Agency (DARPA), Air Force Research Laboratory, or the U.S. Government.

9. Contributors

The work accomplished here would not have been possible without the involvement of various personnel and their dedicated effort. The overall effort at the University of Missouri was led by Deep Medhi while David Tipper took the lead at the University of Pittsburgh.

9.1. University of Missouri-Kansas City

Robert Cotter was instrumental in the development of the survivable virtual ATM network design models, algorithms, and the prototype tool; this is all reported in details in his recently completed Ph.D. dissertation [6].

Several graduate students have been involved in development of MoMaRS simulation tools. Before the project started, Samir Shah started working on the first enhancement to MaRS[3] to consider the multicast shortest path routing component [27]. Then, Gerald Rogers initiated the development of the MKompella multicast routing algorithm and priority-based routing to consider differentiated services environment (reported in his MS thesis [24]; also, see [25]); Suresh Muppala added the RSVP-component to this tool [18] while Ulka Ranadive added the TCP component as well as HTTP workload to this tool [21, 22]. Wen-Jung Hsin as the post-doctoral research associate co-ordinated overall effort related to MoMaRS as well as in the analysis of different network failure scenarios using MoMaRS. Vamsi Valluri and Chengning Lu are currently working on the next version of this tool.

MuSDyR [17] connection-level routing simulation tool to study a failure was designed and developed from scratch by Shishir Ramam, Shankar Subramaniam, and Joshi Sivasankar; their MS theses have reported various aspects of the tool as well as analysis results [20, 29, 31]. Currently, Aekkachai Rattanadilokchai, Loren Rard and Yiming Huang are working on the next version of this tool.

Mark Hieber is finishing up developing a study environment where MoMaRS is coupled with a lower layer network reconfiguration module to allow multi-layer survivability study. Brenda Groskinsky is developing a numerical differential equation model to analyze the impact of reconfigurability in a dynamic traffic environment [8].

The initial set up and testing of the MIMIC testbed, Fred Summa led the effort while Aanand Ramachandran conducted some of the initial testing [19]. In the next phase, when the multi-network management framework was developed and then implemented, Sanjay Jain, Deepa Shenoy and Mukunda Saddi were heavily involved and their contributions have been documented in their respective theses [11, 26, 28]. Currently, Kaushik Deka, Chetan Desai, Jane Zupan and Dijiang Huang are involved in the updated version of the testbed and the framework.

Finally, Post-doctoral fellow T. Srinivasa Rao and Research Associate Wen-Jung Hsin have been key players who plunged into several aspects of the project, specifically new optimization model development, simulation analysis and in the test-bed implementation while Fred Summa has managed to bring sanity to the project by making sure that we have a working environment.

9.2. University of Pittsburgh

Several individuals have contributed to this project at the University of Pittsburgh. In particular Adel Al-Rumaih developed the genetic algorithm based spare capacity planning method and its implementation into a design tool in his Ph.D. dissertation [1]. Graduate student Yu Liu has contributed to the relative comparison study of the genetic algorithm approach and mathematical programming techniques reported in [2]. Ph.D. candidate Anotai Srikitja developed the extended version of the NS simulator and conducted the study on survivable service in NGI reported in [30]. Graduate student Wei-Ping Wang has contributed to the development of fault recovery routing algorithms reported in [34]. Ph.D. candidate William Yurcik developed the survivable ring approach to providing fault tolerant group communications detailed in [35] and his forthcoming Ph.D. dissertation. Visiting scholar Bjorn Jager has contributed to the development of fault recovery routing algorithms and priority restoration schemes as detailed in [34, 10] and his forthcoming doctoral dissertation [9].

References

[1] A. Al-Rumaih. *A Spare Capacity Planning Methodology for Wide Area Survivable Networks*. Ph.D. dissertation, Department of Information Science and Telecommunications, University of Pittsburgh, May 1999.

[2] A. Al-Rumaih, D. Tipper, Y. Liu, and B. Norman. *Spare Capacity Planning for Survivable Mesh Networks*. Technical Report, Department of Information Science and Telecommunications, University of Pittsburgh, September 1999.

[3] C. Alaettinoğlu, A. U. Shankar, K. Dussa-Zieger, and I. Matta. Design and implementation of MaRS: A routing testbed. *Journal of Internetworking — Research and Experience*, 5(1):17–41, March 1994.

[4] Y. Bernet *et. al. A Framework for Differentiated Services*. Internet Draft draft-ietf-diffserv-framework-02.txt, February 1999.

[5] R. Braden, D. Clark, and S. Shenker. *Integrated Services in the Internet Architecture: an Overview*. Internet RFC 1633, June 1994.

[6] R. Cotter. *Network and Protocol Design for Survivable Wide-Area Virtual ATM-based Networks*. Ph.D. dissertation, Computer Networking and Telecommunication Networking, University of Missouri-Kansas City, September 1999.

[7] D. E. Goldberg. *Genetic Algorithms in Search, Optimization, and Machine Learning*. Addison-Wesley, Readings, Mass.,, 1989.

[8] B. Groskinsky, D. Medhi, and D. Tipper. *An Investigation Of Capacity Control Schemes In A Dynamic Traffic Environment*. Technical Report, Computer Science Telecommunications, University of Missouri-Kansas City, July 1999.

[9] B. Jager. *Traffic Restoration in Survivable Wide Area Communication Networks*. forthcoming Ph.D. dissertation, Department of Informatics, University of Bergen, Bergen, Norway, December 1999.

[10] B. Jager and D. Tipper. On fault recovery priority in ATM networks. *Proc. of IEEE ICC 98*, June 1998.

[11] S. Jain. *Design and Implementation of Management Components for Multi-Layered Survivable Networks*. MS Thesis, Computer Science Telecommunications, University of Missouri-Kansas City, May 1999.

[12] V. Kompella, J. Pasquale, , and G. C. Polyzos. Multicast-routing for multimedia communication. *IEEE/ACM Transaction on Networking*, 1:286–291, 1993.

[13] D. Medhi. A unified approach to network survivability for teletraffic networks: Models, algorithms and analysis. *IEEE Trans. on Communications*, 42:534–548, 1994.

[14] D. Medhi. Multi-hour, multi-traffic class network design for virtual path-based dynamically reconfigurable wide-area ATM networks. *IEEE/ACM Trans. on Networking*, 3(6):809–818, December 1995.

[15] D. Medhi, S. Jain, T. Srinivasa Rao, D. Shenoy, M. Saddi, and F. Summa. *A Network Management Framework for Multi-Layered Network Survivability*. Technical Report, Computer Science Telecommunications, University of Missouri–Kansas City, July 1999.

[16] D. Medhi and R. Khurana. Optimization and performance of network restoration schemes for wide-area teletraffic networks. *Journal of Network and Systems Management*, 3(3):265–294, September 1995.

[17] D. Medhi, S. Ramam, R. J. Sivasankar, and S. P. Subramaniam. *Design of MuSDyR: version 1.0*. Technical Report, Computer Science Telecommunications, University of Missouri–Kansas City, July 1999.

[18] S. Muppala. *Network Performance under Failure in the presence of RSVP-based Multicast Services*. MS Thesis, Computer Science Telecommunications, University of Missouri-Kansas City, May 1999.

[19] A. Ramachandran. *On Restoration in ATM Networks and Measurements in a Testbed*. MS Thesis, Computer Science Telecommunications, University of Missouri-Kansas City, June 1998.

[20] S. Ramam. *Multicast Routing with Reservation in Wide-Area Networks*. MS Thesis, Computer Science Telecommunications, University of Missouri-Kansas City, August 1999.

[21] U. Ranadive. *Effect of Network Dynamics on TCP: Observations and Possible Solutions*. MS Thesis, Computer Science Telecommunications, University of Missouri-Kansas City, August 1999.

[22] U. Ranadive and D. Medhi. *Some Observations on the Effect of Network Dynamics on TCP*. Technical Report, Computer Science Telecommunications, University of Missouri–Kansas City, July 1999.

[23] T. S. Rao and D. Medhi. *Book-Ahead Guaranteed Survivable Services in Wide-Area Networks: Problem Formulation and Results*. Technical Report, Computer Science Telecommunications, University of Missouri–Kansas City, July 1999.

[24] G. Rogers. *Analysis of Multicast and Priority based Routing Algorithms with Physical Failures in a Packet Switched Networking Environment*. MS Thesis, Computer Science Telecommunications, University of Missouri-Kansas City, July 1998.

[25] G. Rogers, D. Medhi, W.-J. Hsin, S. Muppala, and D. Tipper. Performance analysis of multicast and priority-based routing under a failure in Differentiated-Services Internet. *Proc. of IEEE MILCOM'99*, October 1999.

[26] M. Saddi. *Network Management System with User Network/Provider Network Paradigm for Multi-Layered Survivable Networks : IP based Implementation*. MS Thesis, Computer Science Telecommunications, University of Missouri-Kansas City, August 1999.

[27] S. Shah and D. Medhi. Performance under a failure of wide-area datagram networks with unicast and multicast traffic routing. *Proc. of IEEE MILCOM'98*, October 1998.

[28] D. Shenoy. *A Network Management Framework for Multiple Layer Survivable Networks: Protocol Development and Implementation.* MS Thesis, Computer Science Telecommunications, University of Missouri-Kansas City, May 1999.

[29] R. Sivasankar. *A Study on Dynamic Traffic in a QoS Routing Environment.* MS Thesis, Computer Science Telecommunications, University of Missouri-Kansas City, August 1999.

[30] A. Srikitja, D. Tipper, and D. Medhi. On providing survivable services in the Next Generation Internet. *Proc. of IEEE MILCOM'99*, October 1999.

[31] S. Subramaniam. *Performance of Dynamic Quality of Service Routing under the Influence of Link Failures.* MS Thesis, Computer Science Telecommunications, University of Missouri-Kansas City, July 1999.

[32] D. Tipper, J. Hammond, S. Sharma, A. Khetan, K. Balakrishnan, and S. Menon. An analysis of the congestion effects of link failures in wide area networks. *IEEE Jrnl. Sel. Areas Comm*, 12:179–192, 1994.

[33] UCB/LBNL/VINT. Network simulator-ns (version 2). *http://www-mash.cs.berkeley.edu/ns/ns.html.*

[34] W. Wang, B. Jager, D. Tipper, and D. Medhi. *On Virtual Circuit Fault Recovery Routing.* Technical Report, Department of Information Science and Telecommunications, University of Pittsburgh, September 1999.

[35] W. Yurcik, D. Tipper, and D. Medhi. Providing network survivability to atm group communications via self healing survivable rings. *Proc. of 7th International Conference on Telcom. Systems*, March 1998.

An Agent Based Architecture for Supporting Application Level Security*

Zhaoyu Liu, Prasad Naldurg, Seung Yi, Tin Qian, Roy H. Campbell, M. Dennis Mickunas
Dept. of Computer Science
University of Illinois at Urbana-Champaign
Urbana, IL 61801
{*zhaoyu*, *naldurg*, *seungyi*, *tinq*, *roy*, *mickunas*}@cs.uiuc.edu

Abstract

The heterogeneous nature of distributed systems raises many security issues and concerns. Traditional systems cannot provide customized security policies and mechanisms for heterogeneous applications. Historically, applications have relied on a static security architecture to provide ad-hoc security guarantees. In this paper we propose a new security architecture based on mobile agents for applications in distributed environments. Our approach allows applications to create and enforce customized policies at run time. These policies and access control requirements can be specified using programs. In addition our framework can handle dynamic requests to change or update these policies and adapt to situational requirements.

1 Introduction

Traditional systems provide security mechanisms to ensure that system resources are used and accessed as intended. They also attempt to detect and prevent accidental or intentional misuse. Typically, these mechanisms tend to be static and it is very difficult to change the security policy or the mechanisms, once the system is installed. Researchers have developed a number of new techniques and mechanisms [7, 5, 10, 9, 15] but very few systems provide support to incorporate these changes.

In a distributed computing environment, applications and users have varying security requirements. In existing systems, these applications or users have very little choice regarding the type of policy or security mechanism and must rely heavily on the underlying infrastructure to provide security guarantees. For example, delegation and security management are severely constrained by static security mechanisms. With systems that support ubiquitous computing devices, it is reasonable to expect that different devices will need different guarantees from the underlying security infrastructure. Traditional static security mechanisms may not be expressive or flexible enough to meet the specific needs of a particular application or device. In order to provide applications (here we mean both users and devices) the ability to customize their security, we need a distributed security architecture that

- is capable of supporting various policies and mechanisms
- can add, replace or revoke policies and mechanisms
- allows applications to specify the kind of security guarantees they want from the system, on the fly
- dynamically enforces these customized policies and mechanisms
- restricts the use of policy to applications and systems that need to know the policy

In this paper we propose an architecture solution that uses mobile agents to provide the required functionality. In our design, these mobile agents are called active capabilities (ACs) [4, 1]. Basically these ACs are signed code fragments that are used to specify policies and mechanisms. Other components of our architecture provide the framework required to evaluate and enforce the policies specified by these ACs and to provide run time revocation, update and enforcement of these policies and mechanisms.

This paper is organized as follows. Section 2 gives a brief overview of the architecture and the trust model and explains each component in detail. Section 3 gives a description of different application scenarios that demonstrate the advantages of using our

*This research is supported by DARPA F30602-98-1-0192 and F30602-97-1-0281

0-7695-0490-6/99 $10.00 © 1999 IEEE

architecture. Section 4 describes two specific implementations of this general architecture. Section 5 presents the preliminary performance results and the discussion on our implementation overhead. The last section presents our conclusions.

2 Architecture Description

This section describes the major components of our architecture and describes the trust model that forms the basis of their interaction. The most important component of our architecture is the active capability(AC). An AC carries a concise representation of security policies and mechanisms, customized or tailored for a particular application or device. The other components in our architecture include AC management, the software framework and the evaluation/enforcement engines. AC Management consists of a distributed network of AC Administrators, software framework component repositories and AC servers. These management entities are trusted. The AC Administrator is responsible for verifying, validating and certifying the code inside the AC, and for signing it.

The AC Administrator can additionally manage the distribution of the ACs using a secure channel. Alternatively, applications may contact the AC server and obtain these ACs and embed them in their code. Trusted applications may be allowed to create their own ACs. Each protection domain typically has one or more AC Administrators that are collectively responsible for the integrity of the ACs.

The AC can use a software framework for context. For example, a software framework may include a hierarchical structure of object-oriented classes of standard security policies. Typically this framework is componentized and arranged so that the components themselves can be downloaded from the software component framework repository using a secure channel. Each node, which is typically a computing device, has an evaluation/enforcement engine in its trusted address space. This engine can also be customized according to the context and its components can be downloaded using the secure channel dynamically. The ACs are evaluated in the sandbox-like environment provided by this engine by instantiating the context of the software framework, which also enforces the result of this evaluation. The subsections that follow give a detailed description of these components.

2.1 Active Capabilities

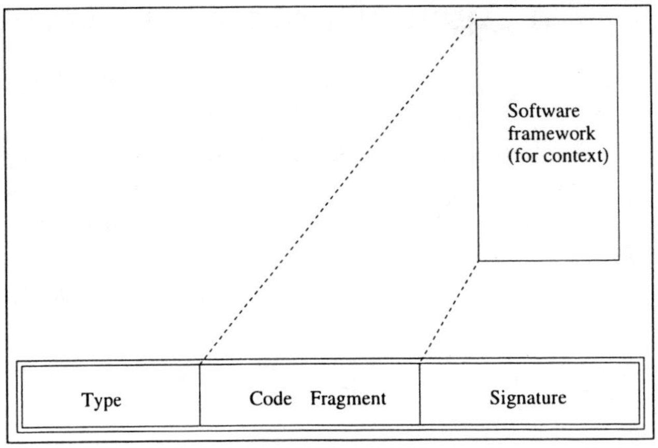

Figure 1: Generic Active Capability

The format of a generic active capability is shown in Figure 1. The first field is similar to a header and contains information about the **type** of the active capability. For instance, the AC type `ACCESS_CONTROL` indicates that the AC encodes an access control policy.

The second field is the most important part of the AC. Ideally it can contain an arbitrary piece of code, that specifies the policies or mechanisms for a particular user. The AC may carry all the code needed to make a policy decision or implement a particular mechanism. However, this approach would be too heavyweight. Instead we advocate the use of a software framework that can provide a context for the code field. The AC code can use the components of this software framework and impose additional constraints on their usage, leveraging the expressive power of the underlying framework.

The next section describes one particular instance of this framework, which implements types of access control policies in a modular and composable fashion [8]. The AC then uses the interface exported by this framework to create a code fragment that concisely represents one particular customized access control policy. In addition, the AC can use the features provided by the underlying language and add conditional processing, based on timestamps, or accumulated credits, to specify timeouts and to impose limits on resource utilization, etc. The flexibility afforded by this approach is limited only by the language syntax.

The last field is the digital signature. Typically the AC is created by an AC administrator or a trusted entity. This entity is responsible for the integrity of the capability, and attests to this by signing the capability. In our distributed architecture, each protection domain has one or a small number of replicated AC administrators. The key distribution and management is simple. If we use a public key infrastructure, we need only one key-pair for AC Management. The administrator(s) can sign the message digest of the AC code using its private key and distribute the corresponding public key within their protection domain. The evaluation/enforcement engines can verify this signature using the public key of the AC administrator. This approach scales well and simplifies trust management.

The AC provides an interface that exports at least the following methods:

- `allowed`
- `revoke`
- `delegate`
- `bind`

The `allowed` method is called by the evaluation/enforcement engine. This method causes the code in the AC to be evaluated. This method returns a boolean value that controls the enforcement of the policy or mechanism requested by the application.

The `revoke` and the `delegate` methods specify interfaces to implement various revocation and delegation strategies. The implementation of these strategies can be customized to suit individual applications. The `bind` method is used to bind capabilities to applications and aids in the retrieval of the context in the evaluation/enforcement engine. This list is not exhaustive and additional methods can be added to extend the functionality and create new AC types.

2.2 Software Framework

This section describes a particular example of a software framework that can be used to provide a context for the active capabilities [8]. Traditional security systems are designed to enforce one particular security policy. In order to provide users more flexibility in terms of access control policy specification, we have implemented a composable and extensible object-oriented policy framework in Java. This framework has a GUI front-end that simplifies the process of specifying the policies. This allows users and commercial organizations to specify access control policies tailored to their specific operational needs. The motivation and the functionality exported by this framework are explained in detail in the subsections that follow.

2.2.1 Access Control Policies

Access control is the mechanism by which a security system exercises control over the access and utilization of shared resources. Historically access control has been defined in terms of $< subject, object >$ tuples and access control matrices. Typically the matrix is indexed by the name of the user (the subject) and by the resource that needs to be protected (the object). The intersection of this pair contains a Boolean value that indicates whether the access is allowed or denied. (The method is usually encoded implicitly in the Boolean value.) For example, Unix file systems use 3 bits to encode various combinations of read, write, and execute permissions for files. However, this matrix method of implementation is not sophisticated or flexible enough and does not scale when the systems serve a large number of users with a large number of resources.

The security policy associated with an access control mechanism refers to the characteristics of its specification, implementation and enforcement. Four different types of access control policies have been defined in literature. They include Mandatory Access Control(MAC), Discretionary Access Control(DAC), Double Discretionary Access Control (DDAC) and Role Based Access Control (RBAC).

The simplest form of access control is DAC. The matrix model is an implementation of this type of policy. Typically a DAC policy implementation maintains an indexed list of allowed $< subject, object, operation >$ triples. DDAC maintains two lists, an "allowed list" similar to DAC and a "denied list". MAC policies use the concept of labeling. MAC is used by trusted operating systems, and every entity in the MAC system is assigned an immutable label. A hierarchy is defined in terms of these labels, and access control is enforced by comparing the labels. Subjects with higher labels have access permissions over objects with equal or lesser labels using a "no read up, no write down" rule [6]. This hierarchy strictly controls the flow of information.

Among these, RBAC is the most flexible type of access control policy [14]. All RBAC subjects are assigned roles. Each role represents a particular set of objects and the allowed operations on each object. The major benefits of this aggregation are the considerable saving in terms of space and simplification in terms of management and enforcement. RBAC allows users to create policies with more sophisticated specifications than simple DAC, DDAC or MAC. A single user may have many different roles, and different permissions depending on the current role. Different constraints related to role and privilege may be enforced in RBAC.

Traditional systems provide a static implementation of any one of these access control mechanisms. Different applications with different access control policies cannot co-exist. Typically, applications cannot be ported across different systems without compromising the security guarantees offered by their access control mechanisms.

2.2.2 Policy framework

The policy framework itself is a hierarchy of classes as shown in Figure 2. It is dynamically configurable and extensible. The classes at the bottom of the framework are mostly abstract, and are mainly used to represent mathematical concepts such as sets and mappings. These classes form the basis for a hierarchy of successively specialized classes representing concepts such as labels and access control lists. Finally, at the top of the framework are classes, which can be used to represent a variety of generic policy forms [13].

Any policy framework that places a heavy burden on its users has never been popular. With this in mind, our policy framework GUI makes the process of creating new policies for ordinary users as painless as possible. Typically, to create new ACs, most users will simply select from a list of predefined policies or will use default settings chosen by a system administrator. However, it is necessary for system administrators and expert users to create and modify policies that respond to specific application needs or security threats. Therefore, our policy framework supports the enforcement of predefined policies efficiently and effortlessly, and also provides a convenient interface for policy authors to create more sophisticated, customized, and situational policies.

The policy framework supports all the following common types of access control: Mandatory(MAC), Discretionary(DAC), Double Discretionary(DDAC), and Role-based(RBAC). It is easy to extend our object-oriented framework to create more fine-grained, application specific policies. ([8] provides several good examples). In our model, we can specify not only the $< subject, object, operation >$ access control triple, but we can also include a resource limit on usage, situational decision rules, constraints and dependencies, e.g., based on current time of day or current role of the principal. The policy framework also lets users specify pre-conditions and post-conditions. Pre-conditions allow necessary security checks to be performed before evaluation takes place, and post-conditions can be used to maintain state and to perform additional checks after evaluation has been completed and when more information becomes available. The central feature of this framework is that the administration of these policies is built into the active capability and the underlying architecture itself. The section on AC management explains this process in detail.

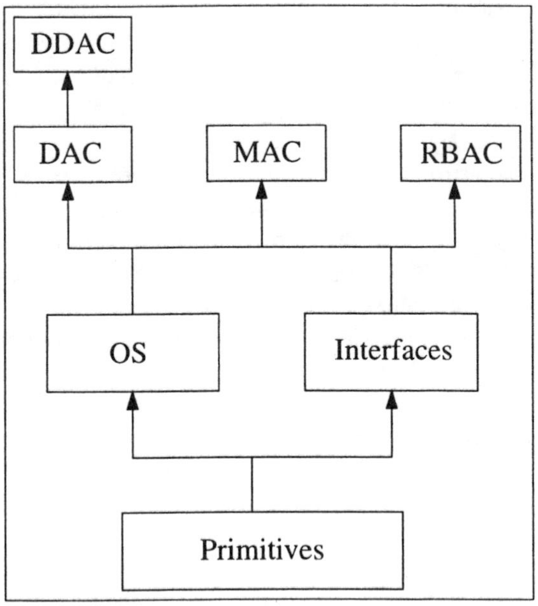

Figure 2: Component-level Map of the Policy Framework

2.3 Evaluation/Enforcement Engine

This section describes the evaluation/enforcement engine component of our architecture. Figure 3 shows a pictorial representation of this component.

The evaluation/enforcement engine consists of an AC cache, run-time resolvable references to customizable AC evaluation sandboxes, and run-time resolvable references to a customizable, componentized software framework. The AC cache is used to cache capabilities that do not change very often and provides a fast processing path for commonly used

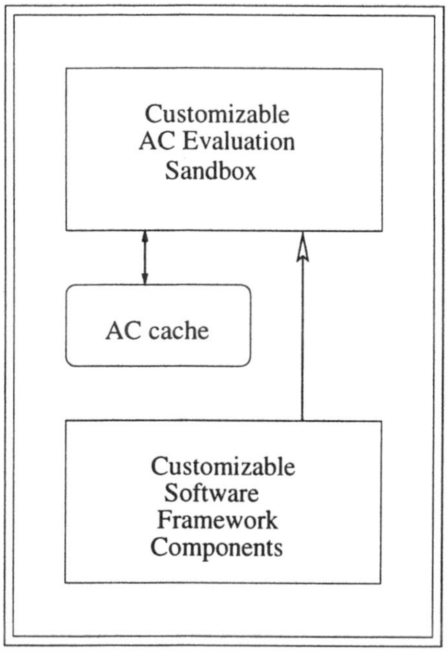

Figure 3: Evaluation and Enforcement engine

or default mechanisms and policies. Different AC types require different contexts. By providing the run-time resolvable references, we can download the software components that form the context from a trusted repository. A sandbox is a restricted execution environment and imposes static and dynamic constraints on the code that runs inside it. A typical example of a sandbox is the Java applet execution environment. This sandbox prevents arbitrary mobile Java bytecode from accessing most of the local files, sensitive data and critical applications.

The sandbox required for the administrator can also customize the evaluation of the ACs or trusted applications. The entire evaluation/enforcement engine needs to be secured in some way. It can run as a process with superuser privilege and create a cryptographically secured channel to communicate with the AC management infrastructure. This channel can be used to obtain the ACs and the downloadable framework and sandbox components. Alternatively it can also be a part of the kernel of a traditional or extensible operating system. The enforcement is done after evaluation. The evaluation/enforcement engine can subsume the concept of a traditional reference monitor. Typically when the application makes a call that accesses a specific resource or requires the use of a specific mechanism, the request is encapsulated and passed to the evaluation engine. The engine builds the context and evaluates the AC associated with the policy or mechanism requested by the application. Depending on the result of this evaluation, the application is either granted or denied the access to the resource or allowed to use the mechanism it requested. To support the enforcement, this engine must export an interface that allows or forces applications to redirect their request to resources and mechanisms through itself. In addition, the engine must either notify applications that the access is denied or forward allowed requests to the appropriate resource, thereby implementing the policy specified in the AC.

2.4 AC Management

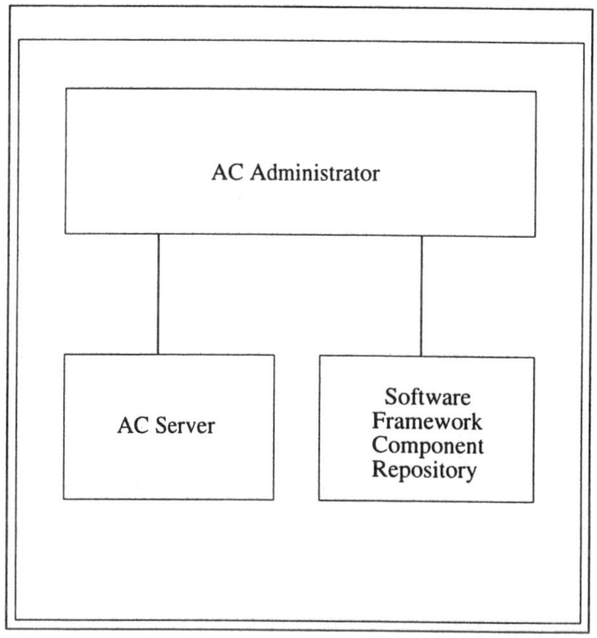

Figure 4: AC management infrastructure

The AC Administrator is equivalent to a trusted third party as in a traditional security model. It is responsible for validating and attesting to the integrity of the active capabilities. For example, using a public key infrastructure(PKI), the AC administrator can sign ACs using its private key, and other entities, like applications and evaluation/enforcement engines, can perform verification using the public key of the AC Administrator. Typically it is also responsible for the creation of the capabilities using the interface provided by the software framework. Although ACs can carry arbitrary code, the creation interface provided by the framework can restrict the capabilities to well-formed expressions and can perform static type checking and verification to make sure that the code in the AC cannot compromise the security of the underlying system. The run-time behavior of ACs is restricted by the sandbox, which limits their access rights and resource utilization. In addition, the communication channel between the administrator and the evaluation/enforcement engine needs to be secure. The AC administrator itself may be implemented as one centralized entity, or its functionality can be distributed throughout the protection domain. There may be multiple instances of the AC administrator to achieve load balancing, scalability and fault tolerance. The AC server acts as a front-end to an AC Administrator's AC repository. This server may be a part of the administrator or may be another entity, closely coupled with the functionality of the AC Administrator.

3 Applications

In this section we describe some application scenarios that highlight the benefits of using our system. One example we have chosen uses ACs to provide dynamic countermeasures against intrusions. In a typical system, when an intrusion or possible intrusion is detected, a security alert is issued. This security alert needs to be issued promptly at runtime without interrupting normal service. This may result in modifying the security policy at the compromised node and its neighbors, activating additional security measures, imposing additional auditing schemes and more restrictive security policies.

Consider the network configuration shown in Figure 5. In this figure, one or more compromised nodes are isolated from the rest of the network and a dynamic firewall is built around them. All nodes that are directly connected to the compromised node are sent ACs to change their existing access control and security mechanisms to minimize the risk of compromising the rest of the nodes in the network.

The traditional way of dealing with intrusion relies on detecting patterns of abnormal or suspicious behavior. Using pattern matching and analysis, or data mining etc., intrusion detection systems define a fixed set of countermeasures that attempt to minimize the damage. One drawback of this approach is that it is very hard to prepare in advance the countermeasures for every kind of attack. Since there is no way to prepare for every kind of intrusion, the fixed set of countermeasures and policies for the firewall nodes needs to be updated very frequently. In

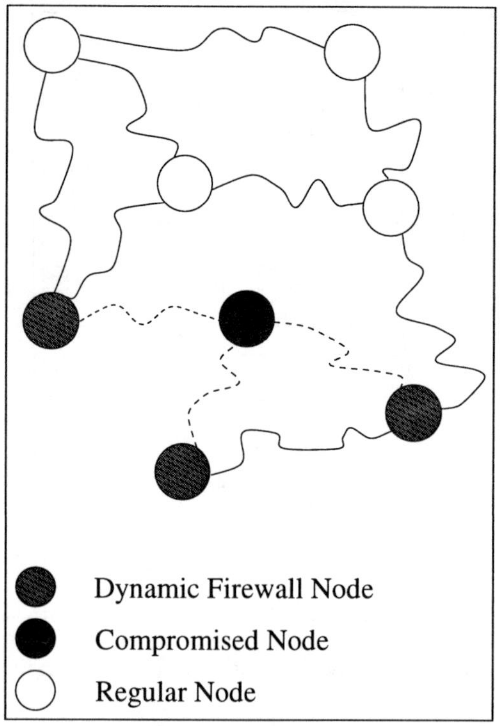

Figure 5: Dynamic Countermeasures for Intrusion

existing systems the overhead associated with this operation is substantial. Using our low-overhead system we can define dynamic policies customized to a particular attack or possible attack, which can be triggered and enforced by suspicious behavior. A static policy that denies all accesses is too restrictive. In many cases it may be worthwhile to adopt customized countermeasures and allow the services that are not compromised to continue. If we detect an intrusion that requires more drastic countermeasures, we can change the security level of the whole system by installing restrictive ACs on each node, for example by changing the current policies to MAC or by disabling a particular service or application.

Active capabilities can also be used to perform distributed computation. The results of this computation can be used to enforce situational policies. Using a weighted trust model, described in [12], applications can assign weights to different entities that reflect upon their "level" of trust in the system. The AC can evaluate the trust level for a specific entity and enforce the policy specific to that trust level. Thus the administrators may dynamically force applications to change their policies and mechanisms based on changing trust levels and force applications to adapt protective measures against nodes on less trustworthy paths.

Another example is a mobile computing environment where users wearing a network of mobile devices enter a smart room that has its own independent security mechanisms and policies. In order to use the services available in that room, these users have to adapt their network's applications and policies and use the underlying security infrastructure. The need to integrate seamlessly the mobile system with the smart room security infrastructure implies that the mobile users must adapt to the security policies of the smart room and vice versa. The administrators use ACs to customize the user, device, network or smart room security profile to enforce these policies at run time.

4 Implementation

Based on the location of the evaluation/enforcement engines we have identified three specific implementation models of our architecture. In the first model, the evaluation/enforcement engines are implemented in the kernel of the system. The engines reside and execute in protected address space. In the second model, the engines reside in the software framework itself. For example, the engine can be built in a JVM(Java Virtual Machine or Java runtime environment) as a sandbox and can manage access to Java objects and Java security policies and mechanisms. This model relies heavily on the security and safety features of the software framework itself. In the third model the engines reside and execute in separate, independent user space, and the system protects them from unauthorized accesses and is responsible for enforcing the behavior of these engines.

In this section we describe two specific implementations of the above three models. The focus of the Cherubim[1] project [2] is to provide dynamic security support in CORBA for mobile computing that allows frequent migration of computers in and out of security enclaves and that facilitates wide area collaboration by creating dynamic sessions that stretch across organizational boundaries. In Cherubim, the ORB is in user space and is assumed to be trusted. This implementation belongs to the third model. The Seraphim[2] project [3] defines a dynamic, interoperable, flexible, composable security architecture for active networks. We implemented a reference monitor in the NodeOS to control NodeOS accesses.

[1]The Cherubim software release is available at http://choices.cs.uiuc.edu/Security/cherubim/software/

[2]Contact authors for the Seraphim software release

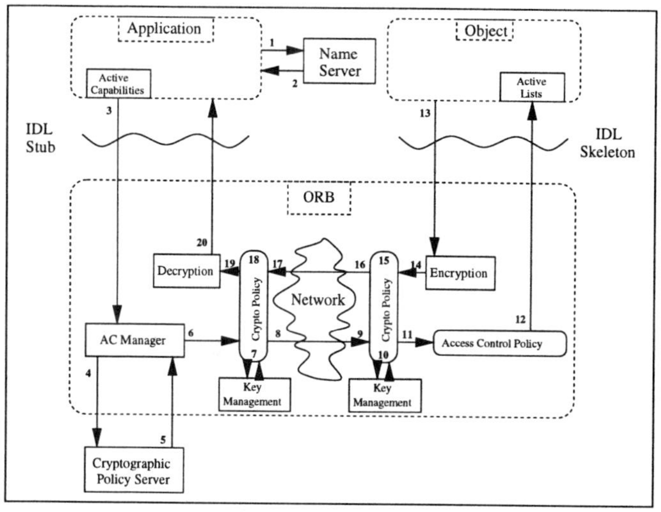

Figure 6: Object Access in *Cherubim*

This implementation belongs to the first model.

4.1 Cherubim

The design of the Cherubim security architecture [11] consists of two major parts: CORBA compliant security services with a security enhanced IDL (Interface Definition Language) providing application level security interfaces, and an agent based dynamic security framework implementing these standard interfaces. Adopting OMG's general security reference model and Security Service Interfaces gave us an open architecture to incorporate a wide variety of security policies and services. In addition, Cherubim aims to provide fast technology evolution and deployment for both user applications and security functions. Using the standard CORBA security services and a security extended OMG IDL provides the necessary basic facilities to achieve this and easily separates security functions from the main application functions. In general, this design provides better extensibility and configurability for both applications and their security features. It also enables wide area software integration with configurable security enforcement. Figure 6 briefly illustrates the process of secure object invocation in Cherubim integrated with CORBA security services. The numbers in the figure show the sequential processing steps for a client request.

As in the OMG model, Cherubim defines a principal as a human user or system entity that is registered in and authenticated to the system. It may have one or several identities, which we call roles in Cherubim. Each role may have different privilege attributes used in making access control decisions. All the information about roles and privilege attributes of a principal is maintained in a secure store object that is called a credential. In Cherubim, credentials may be embedded in the active capabilities to support delegation and to facilitate efficient access control. Objects, policies, and services are organized into security domains, each of which defines a distinct scope with a common set of security policies and mechanisms. In each domain there is a security authority, which we implemented as a policy server with administrative interfaces to security policies in Cherubim. A domain may also be divided into sub-domains to form a security domain hierarchy. To inter-operate between security domains, inter-domain security policies need to be defined and enforced. In addition Cherubim has developed a policy representation framework with role extensions to provide interoperability with RBAC policies while allowing extensibility. One of the major drawbacks of this approach is that OMG CORBA security services are implemented and enforced in user space. It is possible to make stronger guarantees if an ORB can be securely protected.

4.2 Seraphim

Active networks [17] aim to provide a software framework that enables network applications to customize the processing of their data. Applications encapsulate the methods that manipulate the data, with or without the data itself, and inject these capsules into the network. Active routers install and execute these capsules on the data dynamically, thereby facilitating fast protocol and service deployment. Securing this infrastructure against threats and exposures remains a major challenge in this paradigm.

In order to exploit fully the expressive power of the underlying active network, we felt the need for a unified security framework that allows users or applications to create and enforce their own security mechanisms, similar to customizing their own communication protocols. Seraphim [3] is a prototype of a dynamic, fully extensible, inter-operable, security architecture, based on, and built into the underlying active network architecture. This architecture allows active network routers to be configured with only a minimal set of security functions. These functions are recursively used to install and support the secure deployment of new security mechanisms. For instance, sophisticated and application or user specific security functions may be installed at run time using

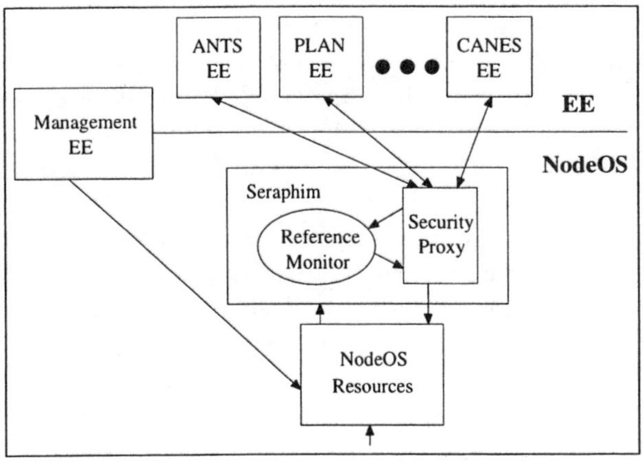

Figure 7: Secure Active Network Node

a secure recursive reconfigurable bootstrap process. Currently, our framework provides mechanisms to dynamically specify, separate and enforce a number of different, often mutually exclusive access control policies. In addition we allow applications to encapsulate credentials and to encode situational policies to alter or revoke dynamically existing access control rules and mechanisms. Applications construct capsules, which use this software environment as context and inject customized security policies into the routers.

The major components of our Seraphim architecture and their interaction, in the context of the active network architecture are shown in Figure 7.

The key component of this architecture is the reference monitor. This is similar to the evaluation/enforcement engine in the general architecture. Currently the reference monitor is implemented as a co-located extension to the NodeOS. Every node has a reference monitor through which all accesses to node resources occur. The policy framework is the software framework component. As mentioned earlier, the policy framework itself is reconfigurable and can be downloaded dynamically when required. Applications or administrators use the interface provided by this policy framework to create ACs that encode the type of access control policy and other constraints used in the access control decision making process.

5 Performance

In this section we describe some experiments we implemented in our Seraphim system. Although our prototype implementation was not built for performance, we did make some preliminary performance

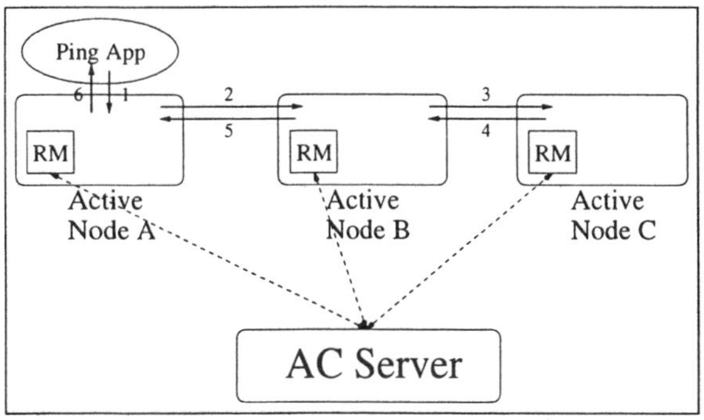

Figure 8: Ping Application Experiment in *Seraphim*

measurements. We also present a discussion on the overheads incurred in our system. In order to write applications for our Seraphim system, we developed SAINTS (Secure Active Inter-operable Network Toolkit System) which is based on the ANTS toolkit [16]. The original ANTS `Node` class was split into distinct NodeOS and EE classes and we added our reference monitor between the EE and the NodeOS. Our SAINTS is backwards compatible with ANTS and can run original ANTS applications. The next few subsections summarize the performance results for Ping, Gnipper and dynamic policy change applications. This is followed by a subsection that discusses strategies for reducing some of the overheads.

5.1 Ping Application

The first experiment measures the performance overhead associated with the modified version of the ANTS Ping application. The experimental setup is shown in Figure 8. The numbers in the figure show the sequence of steps in the transmission of an ANTS Ping capsule. The communication between the AC server and the active node is through TCP. We used three Sun SparcStation 10 machines for the active nodes and a Sun Ultra-60 machine for the AC server. All four machines are on the same 100Mbps Ethernet LAN.

We used four different system configurations to measure the performance. Our measurements exclude the time to load dynamically the active Ping protocol in active nodes. The first configuration, "No RM", is the baseline. In this configuration the Seraphim reference monitor was installed in the NodeOS of all the active nodes, but was bypassed and no access checks were performed. The second configuration, "RM without Cache", was the most straightforward configuration and used the Seraphim reference monitor. In this configuration a reference monitor was installed in all the active nodes, and every time a Ping capsule arrived, one access check was performed. There was no cache in the reference monitor. The reference monitor at each node had to contact the AC server to retrieve the proper AC and to evaluate it for each access check. The third configuration, "RM with Cache", was an improvement over the second configuration. Again, the reference monitor performed one access check for each arriving capsule, but in this case the proper AC was cached inside the reference monitor. Here the reference monitor did not need to contact the AC server at all. The final configuration, "RM with Decisions in Cache", caches the reference monitors evaluation result of the AC. When the AC does not change and the inputs to the AC are the same as to the previous evaluation, a prior cached evaluation value can be used. Each access check in this configuration involved simply a cache lookup, without the dynamic evaluation of the AC.

System Configuration	Ave. RTT (ms)
No RM	10
RM without Cache	1494
RM with Cache	21
RM with Decisions in Cache	10

Table 1: Performance Data for Ping Application

In all configurations, 2000 capsules were sent out from the Ping application and the interval between the Ping capsules was 200ms. We measured the average round trip time (RTT) of the Ping capsules. Table 1 shows the results of our experiment.

We observed that the most inefficient implementation of our architecture is the "RM without Cache" configuration. It has much larger overhead than the base case. When the simple caching scheme was used, the average RTT was approximately twice the base RTT value. When we further assumed that the result of AC evaluation can be cached, average RTT was same as in the base case. This result suggests that by employing suitable optimization techniques, the overhead of our Seraphim architecture can be reduced to an acceptable level. The overhead of checking that the inputs to the AC had not changed and the Ping capsule certificate was the same as that of a previous capsule was negligible. Further research is required on different caching schemes and optimization strategies.

5.2 Gnipper Application

Another experiment we developed in Seraphim is the Gnipper application. This application demonstrates the creation of dynamic protection domains or enclaves. In this application, we create Gnipper vaccine, or an anti-Ping AC. This vaccine is used to disable one user's ability to ping an active node. This vaccine is installed at the reference monitor of the active node that needs the protection. The vaccine dynamically moves one-hop at a time toward the source of the Ping, in response to Ping requests from the original sender. This is best explained with an example presented next.

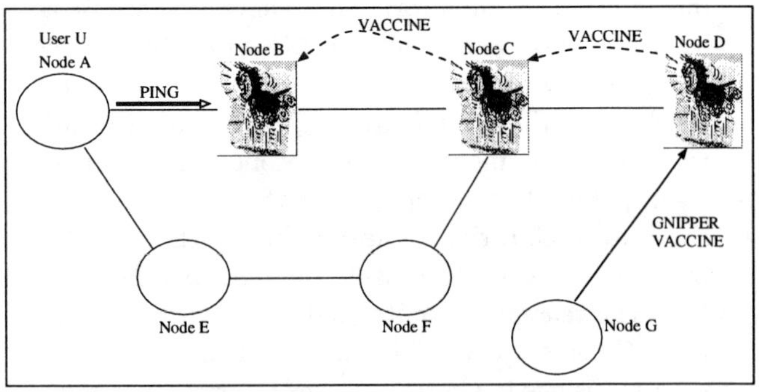

Figure 9: Gnipper Application

The example is shown in Figure 9. We create a vaccine and install it at the reference monitor of Node D. The vaccine is an AC which disables the ability of user U at Node A to ping Node D. User U sends a Ping capsule with destination address of Node D. When the capsule arrives at Node D, Node D drops the capsule and propagates the vaccine to the previous node, Node C. Now Node C is vaccinated. When user U sends another Ping capsule to Node D, the capsule will be dropped at Node C and the vaccine will be installed at Node B.

The exact node which drops the Ping capsule changes dynamically depending on the the number of Ping capsules sent by the source. By installing successively the vaccine at the nodes on the Ping path between Node A and Node D, we have succeeded in building a dynamically growing firewall around D, and also reduced the traffic and moved any denial of service attack away from the intended victim.

This experiment used the same setup as for the Ping application. The average overhead for an application running on a Sun SparcStation 10 machine to create a vaccine and install it at another Sun SparcStation 10 machine on the same local Ethernet LAN was measured as 77ms. Without creation, the average time to send a vaccine and install it under the same setup took 34ms.

This experiment can be extended to build agile and dynamic firewalls that can react to attacks at runtime. When an active node or trusted agent detects attempted attacks, it can send out an active capability carrying a "warning" message with the appropriate vaccine, to build a dynamic line of defense against outside attacks, or to raise the level of security within the domain. When the threat is gone the active node or the trusted agent can send out another active capability to resume normal operation. This can be used as a very powerful security tool in conjunction with intrusion detection and countermeasure systems.

5.3 Dynamic Policy Change

Our policy framework can dynamically change between policy types, for example, from RBAC to MAC. Section 3 mentions an application that can benefit from such a change. We measured the overhead associated with this change using our existing system. For a simple domain with 5 users, 10 objects and 5 operations, changing from RBAC with 5 roles to a MAC with 2 security levels cost about 667ms on an average using a Sun Ultra-60 machine. We have identified several places for optimization and further research is planned to reduce this overhead considerably.

5.4 Secure Multicast Application

A secure multicast application for active networks was implemented in Seraphim. Traditional secure multicast uses session keys to enforce secure data transmission and delivery. When a new member joins a secure multicast group, the sender sends the current session key to the new member so that it can receive data. When a member leaves the group, the session key has to be changed, to prevent eavesdropping. The sender has to generate a new session key and transmit it to all the remaining members.

In our secure multicast example, joining and leaving are both very simple. When a new member joins the group, a new AC is created for this member. The enforcement engine for this new member will get the new AC only when it needs this AC. When a member leaves the group, the AC administrator sends a simple revocation AC to the enforcement engine for this leaving member. The overhead is much smaller than traditional secure multicast systems.

5.5 Discussion

In traditional systems, access control is defined by policy and is enforced by enforcement engines such as reference monitors in operating systems and firewalls in networks. Individual policies can be defined at each enforcement point and managed separately. For example, each firewall in a company can be configured individually to set up a set of rules defined by policy. When there is a policy change, a human administrator can identify the affected firewalls, suspend and reconfigure them to enforce the new policy. If there are many firewalls and frequent policy changes, then this approach is not scalable.

An alternative approach is to have a centralized policy administrator. The policy administrator maintains all policy information for the whole system, and is responsible for distributing policy to individual enforcement engines. When there is a policy update, the policy administrator can distribute the new policy to the enforcement engines. Usually the policy administrator does not know in advance where the new policy is going to be used, so the administrator may have to distribute the new policy throughout the whole system. For example, when a new user is added into the system, the access privileges for this new user needs to be distributed to all enforcement engines. Also when there is a revocation of an existing policy, the revocation information needs to be distributed to all the enforcement engines. In some environments, such as the Cisco Secure Policy Manager for firewalls [5], enforcement engines are suspended in order to update policies. In traditional systems, policy update is often complicated and the overhead is large.

In our architecture, we also have a centralized AC administrator per domain. However this entity may be replicated for fault tolerance and load balancing. The AC administrator keeps track of the location of ACs in the system. When there is a policy change, this causes the ACs to change and the old ACs have to be revoked. The AC administrator sends changed ACs to only those enforcement engines that have the old ACs. When there is a policy change which causes addition of new ACs, the AC administrator does not need to send out anything. When the enforcement engine receives access requests from applications for the first time, it asks for these new ACs from the AC administrator and caches them. Therefore the overhead is for first time use only, and there is no system-wide distribution overhead. An overhead introduced by our approach is the need to maintain information in an AC server about the location of the ACs. Since the centralized computation time usually is smaller than any network delays, the maintenance overhead is much smaller than the extra network overhead of a traditional system.

Sometimes it is difficult for the AC server to keep track of ACs for revocation. In such cases active applications can get ACs from the AC server, and distribute ACs by themselves. For example, in our Seraphim implementation, applications can encode ACs into active capsules, and distribute them through active networks along with active capsules. Applications do not need to know the details of the policy framework and the AC contents. In this way the system can provide very fine-grained customized policies for applications. One simple way to accomplish revocation is to use a time-out. Another solution is to broadcast the revocation information, as in traditional systems. In either case, the overhead is no larger than traditional systems.

6 Conclusions

In this paper we propose an agent based security architecture that allows applications to create customized and situational policies. The motivation for this architecture is that different applications (users or devices) tend to have widely different requirements in terms of security policies and mechanisms. The traditional model of security is very static and cannot support different mechanisms and policies, or change between these policies and mechanisms dynamically.

In our system, applications use the expressive power of our software framework and the flexible nature of our infrastructure to create mobile agents called active capabilities (ACs). These ACs actually carry the customized policy to evaluation/enforcement engines where they are evaluated in a sandbox-like environment and the result of this evaluation is enforced on the applications. The AC management infrastructure defines a trust model for the interaction of various components and manages the creation and distribution of ACs.

The security framework and the evaluation and enforcement engines are composable and extensible and require only a minimal set of functions and mechanisms to be installed. Additional mechanisms and policy implementation components can be downloaded dynamically using a secure communication channel. The overhead of managing policies and mechanisms is very low as this can be handled by the ACs. AC simplifies key management and distribution, making our approach very scalable.

References

[1] Roy H. Campbell, M. Mickunas, Tin Qian, and Zhaoyu Liu. An agent-based architecture for supporting application aware security. In *the Workshop on Research Directions for the Next Generation Internet*, May 1997.

[2] Roy H. Campbell and M. Dennis Mickunas. An agent-based architecture for supporting application aware security. an accepted proposal to DARPA BAA9704, 1997. Also see the web site at http://choices.cs.uiuc.edu/Security/cherubim/.

[3] Roy H. Campbell and M. Dennis Mickunas. Building dynamic interoperable security architecture for active networks. an accepted proposal to DARPA BAA9803, 1998. Also see the web site at http://choices.cs.uiuc.edu/Security/seraphim/.

[4] Roy H. Campbell and Tin Qian. Dynamic agent-based security architecture for mobile computers. In *the Second International Conference on Parallel and Distributed Computing and Networks*, Brisbane, Australia, December 1998.

[5] Cisco Systems, San Jose, CA. *Cisco security manager tutorial, DOC-786905*, 1999. Available at http://www.cisco.com/warp/public/cc/cisco/mkt /security/csm.

[6] D. Denning. *Cryptography and Data Security.* Addison-Wesley Publishing Company, 1982.

[7] David Evans and Andrew Twyman. Flexible policy-directed code safety. In *IEEE Symposium on Security and Privacy*, Oakland, CA, May 9-12 1999.

[8] Tim Fraser. An object-oriented framework for security policy representation. Master's thesis, Department of Computer Science, University of Illinois at Urbana-Champaign, December 1996.

[9] George C. Necula and Peter Lee. The design and implementation of a certifying compiler. In *PLDI '96.*

[10] George C. Necula and Peter Lee. Safe kernel extensions without run-time checking. In *OSDI '96.*

[11] Tin Qian. *Cherubim* agent based dynamic security architecture. Technical report, Department of Computer Science, University of Illinois at Urbana-Champaign, June 1998.

[12] Tin Qian. *Dynamic authorization support in large distributed systems.* PhD thesis, Department of Computer Science, University of Illinois at Urbana-Champaign, December 1999.

[13] Vijay Raghavan. On the design and implentation of a security policy adminstration for a dynamic security system. Master's thesis, Department of Computer Science, University of Illinois at Urbana-Champaign, May 1999.

[14] R. S. Sandhu and E. J. Coyne. Role-based access control models. *IEEE Computer*, 29(2), February 1996.

[15] Dan S. Wallach. *A new Approach to Mobile Code Security.* PhD thesis, Department of Computer Science, Princeton University, January 1999.

[16] D. Wetherall, J. Guttag, and D. Tennenhouse. ANTS: a toolkit for building and dynamically deploying network protocols. In *IEEE OPENARCH'98*, San Francisco, CA, April 1998.

[17] D. Wetherall, U. Legedza, and J. Guttag. Introducing new internet services: Why and how. *IEEE Network Magazine*, July/August 1998.

Recon - A Tool for Incident Detection, Tracking and Response

Juanita Koilpillai
Mountain Wave, Inc.
Juanita@mountainwave.com

John B. Beavers
Mountain Wave, Inc.
Jack@mountainwave.com

Paul Swinton
Mountain Wave, Inc.
Paul@mountainwave.com

Abstract

Recent discussions on the state of Intrusion Detection Systems and Network Security Tools has prompted the notion that what is needed is a solution that

- *can fuse data from heterogeneous distributed network and host sensors*
- *supports sophisticated analysis models and automated responses*
- *provides the user with the appropriate 'situational awareness' so that efforts can be focussed on the right problems*
- is "enterprise-aware".

This paper presents some key concepts for such a solution by describing a tool, code named RECON, which is a result of extensive research and prototyping performed on an adaptive network security management framework funded by DARPA. This research indicates that in order for RECON to satisfy the stated requirements, it would need

- *a standard message format and protocol*
- *an analytical engine such as a rule-base*
- *a flexible and extensible architecture*
- *a graphical user interface that provides a unified view of various levels of information*
- *a language to capture 'enterprise' rules.*

1. Introduction

The intended function for RECON in cyberspace is very similar to that of more traditional RECON units -- those small teams of soldiers who are covertly deployed in forward areas. Human RECON teams move with speed and stealth, and when they detect enemy activity they can react flexibly. In some cases they simply report intrusions to decision-makers so that larger concentrations of power may be applied. Or, if appropriate, a RECON unit might actually begin direct interdiction of a threat, making critical decisions and attacking autonomously. In other instances, RECON teams participate in coordinated assaults against the threats they have discovered. *These are exactly the things that the tool RECON can do for its command in cyberspace.*

RECON can detect intrusions into critical network assets, and interdict those intrusions with strategies and tactics that are *enterprise-aware*. The term "enterprise-aware" implies that adaptations made by RECON are selected with automated consideration of the priorities and needs of the enterprise as related to its electronic assets.

2. Design Premises

RECON is designed around a framework of services that cooperate to detect and react to security incidents in order to minimize or eliminate harm to the enterprise. Selection of such defensive adaptations is based upon automated reasoning about both the state of the network infrastructure and the organization the infrastructure supports.

RECON is also designed to be a general framework for management of enterprise-level functions and assets for application to a large cross-section of potential domains. For example, the basic framework might be applied equally well to the automated management of a constellation of satellites, or to the management of battlefield information streams and assets, or to the management of large-scale suites of enterprise business processes. Concepts of generality necessary for such diverse application pervade the design of all elements of RECON.

The organization of the services has a number of analogies with human organizations. The responsibilities of individual services range from low-level, detail oriented tasks such as information gathering, collation, and storage – to high level decision-making and pattern recognition. Just as in human organizations, the services share common ways (syntax and semantics) of representing and communicating critical information about the current situation and developing events. Services and associated *plug-ins* deal with a common pool of information streams; each service uses this pool to reason independently about what is going on, what might be done about it, and which other services should be notified or consulted with. As an example of the advantage of this approach, if one service or plug-in doesn't know how to deal with a particular message or the

0-7695-0490-6/99 $10.00 © 1999 IEEE

situation, it can simply pass it on to another service or plug-in -- without any translation – for further processing. If a service needs more information about the situation, it can query another service for more information, which is simple to do since both inquiry and representation strategies are common throughout the system.

Plug-ins are the most dynamic, changing components of the framework. The primary purpose of such plug-ins is to make available a wide variety of external information, detection methods, and decision-making techniques. Plug-ins perform the following primary functions:

1. Translate heterogeneous forms of information from the network and other sources and translate it into standard internal formats.

2. Allow for flexible addition to and change of detection, modeling, and decision-making methods used by the Tool to respond to ever-changing environments and challenges.

Some low-level plug-ins simply gather information about the network resources being monitored and protected (e.g. a plug-in to gather SNMP traps). Other high-level plug-ins deal with intrusion detection and decision-making. In some cases, such plug-ins integrate COTS products into the system – products such as existing commercial intrusion-detection software, network management software, rule-based systems, and information management systems. For third-party product integration, the primary function of a plug-in is to provide a simple interface, or wrapper around the COTS product that allows RECON to deal with the foreign software in much the same way as it deals with all of the other plug-ins and components of the system. In other cases, plug-ins can be used to make diverse and highly configurable forms of decision-making available to services that have been programmed by the development team. In all instances, the plug-in aspect of the architecture allows for extremely fast integration of new third-party tools, fast integration of new decision-making methodologies, and fast modification of system behavior that in most instances will not require programming.

Among the most interesting plug-ins to the system are those involved in decision-making. Just as in human organizations, information processing in RECON is partitioned into two basic functional categories:

1. Gathering, organization, and storage of information (Services).
2. Analysis, reasoning, and decision-making associated with the information (Plug-ins).

The first functional category is handled by the RECON's more stable "hard-coded" facilities as represented largely by its major <u>services</u>. Services are analogous to the relatively stable foundations present in most human organizations -- as represented by clerical and office-management staff and by the standard procedures followed by such staff. The second category of information processing is handled largely by decision-making <u>plug-ins</u> that are designed from the ground-up to be changed, extended, or added and removed from the system without the need for change to the foundations – the main services -- of the system.

Most of the high-level detection and decision-making capabilities of RECON are focused in highly-configurable plug-ins. For decision-making, major services send inquiries to "Advisor" plug-ins that review the current situation and direct the services on how best to proceed. These plug-ins share a common interface, and common information exchange protocols, allowing for any of a number of current and future decision-making algorithms to be integrated with the system. Advisor plug-ins apply knowledge to the current situation and generate specific recommendations on responses using standard, well-defined response syntax. Advisors can be implemented using any of a number of specific knowledge processing methodologies ranging from simple table lookups to rule-based systems, Bayesean methods, and neural networks. The only constant is a well-defined interface. As better knowledge processing methods seem more suitable or become available, they may be swapped in without modification to the services that use them. New knowledge processing methods are wrapped to present a consistent interface to services. Examples of the kinds of analysis, reasoning, and decision-making plug-ins that can be readily integrated into the services are:

- Hard-coded (e.g. third-party COTS security products).
- Table-driven.
- Rule-based.
- Fuzzy Logic.
- Bayesean Networks.
- Neural Networks.
- Other methods of pattern recognition.
- Genetic algorithms.
- Etc. – future methods.

Any combination of these approaches might be used simultaneously in the system via its various plug-ins. A service can even employ more than one method (in other words, more than one Advisor) to analyze the same problem simultaneously, particularly if one method of analysis fails to produce a reasonable recommendation.

The fundamental concept derived from the above is this: *analysis, reasoning, and decision-making functions are largely partitioned away from the basic services in order to allow the system to be adapted to a constantly changing security threat environment with speed and sophistication.*

3. Architecture

RECON is based on a flexible and scalable framework in which security software/hardware can be plugged into – entities that represent, monitor, and interdict security threats as depicted in Figure 3.1 (on next page). Such plug-and-play capability is vital for interoperability with any network system, new or old, that needs to be managed securely. This capability is also vital to support rapid expansion and reconfiguration of the system in order to respond to the flood of new threats that are continually being discovered. The underlying technology for building the Tool is the distributed objects paradigm[1].

Organizations like the Object Management Group (OMG) have proposed specifications for communications and collaborations between entities in what is popularly known as the Common Object Request Broker Architecture (CORBA) [2].standard for distributed object computing. OLE/DCOM [3] is a similar standard that has been proposed by Microsoft. In these systems, frameworks define the rules of engagement for 'wiring' components together so that they can play in suites. The Tool uses such industry standard specifications to provide a framework in which security management components can collaborate to deal with threats to an enterprise in real time.

Although the use of standard object computing technologies [4] will support that needs of most security management installations, some applications – in particular Government intelligence and military and financial industry applications, may not always use the platforms and Internet Protocol (IP) networking environment that support the standard distributed object computing tools. For this reason, the design has been layered so that the specifics of a particular distributed object computing environment, such as CORBA, are isolated from the bulk of the code which implements te Tool. This means that the Tool's reach can be readily extended to non-standard system types with unconventional networking environments.

3.1. Services

Text following is a list of services and their intended function in the first releases of the Tool. Since the architecture is based on a distributed objects paradigm, the services can be configured to run in one or more applications depending on the scalability requirements. Multiple instances of the services for true scalability is planned for future releases.

- Audit Service (AS): Assembles, queues and routes events. Among the routing options are
  - Sending messages to the Network Model Service for checking of thresholds.
  - Sending messages to the Detection and Tracking Service for logging, decision on what this event means, and possible declaration of security incident.

- Detection and Tracking Service (DTS): Correlates events into incidents and declares incidents based upon events. Determines whether an event is considered significant using Asset/Threat Service. If so, DTS tries to match the event with an ongoing incident or incident candidate. DTS decides whether to declare a new incident, or whether to report impact on an existing incident to higher-level services. It performs as a database server for incidents and associated events.

- Network Model Service (NMS): Provides a hierarchical representation of network components, computers and of parameters associated with them. Stores initial network configuration and state. Updates configuration, network state changes, and event counts for particular network devices. Reports on parameters exceeding thresholds due to receipt of significant event messages.

- Asset / Threat Service (ATS): Relates classes of assets to classes of threats. This is where various classes of threats must be represented. The hierarchy of data is incident types, threat types and events. A simple decision table represents matching between threats and characteristic attack signatures for various classes of assets.

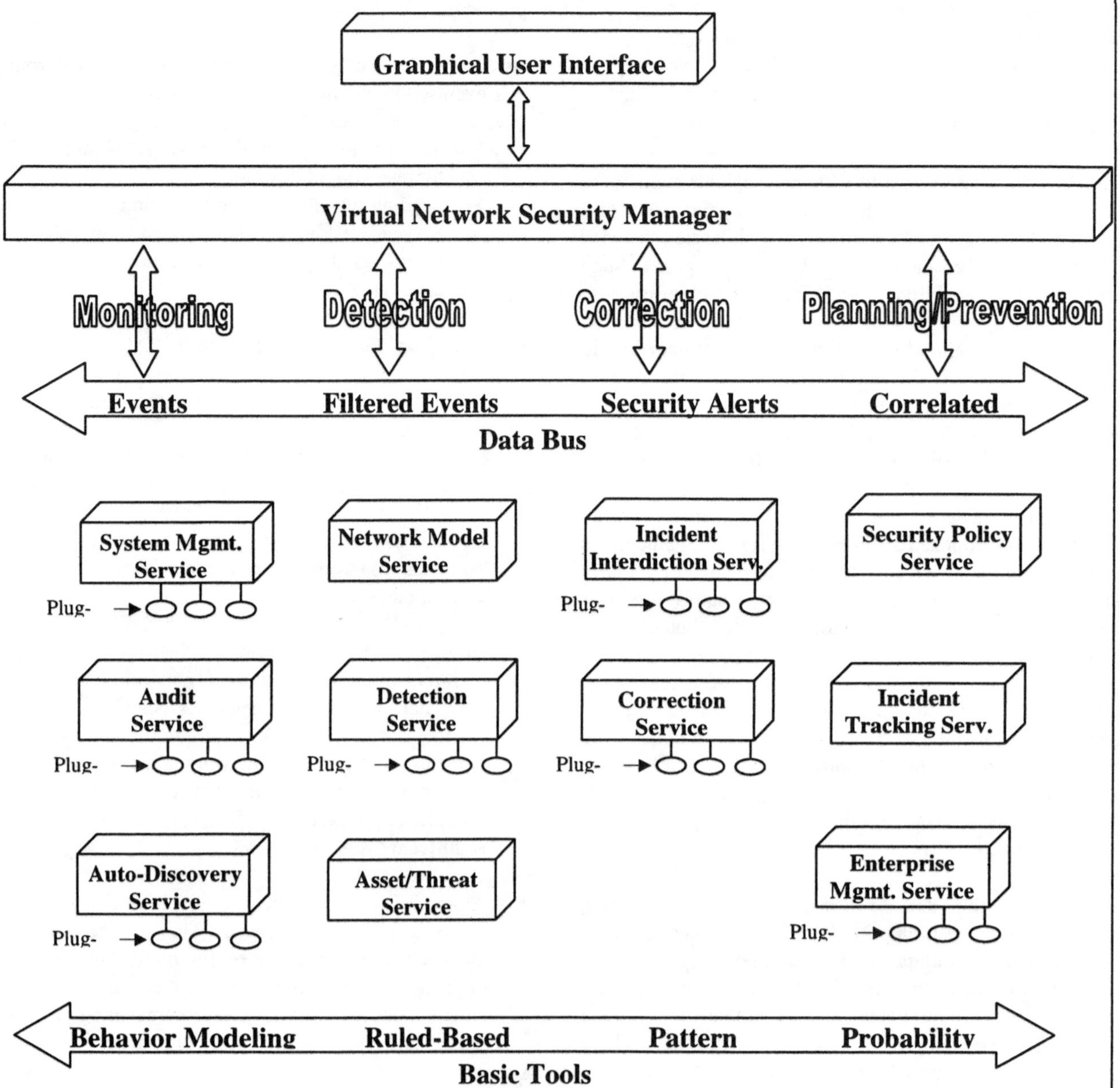

Figure 3.1 Architecture

- Security Policy Service (SPS): Translates high-level policy into general system configuration. Also, represents elements of high-level policy in a form that can be queried by other services such as the Virtual Security Manager. For the initial releases of the tool, this module will only handle simple mappings *from* incident classes and affected asset classes *to* simple adaptive reactions. Implementation of a policy grammar editor and associated parsers is planned for future releases.

- Enterprise Model Service (EMS): Models the enterprise, large-scale enterprise assets and their role in the enterprise, enterprise priorities. This service is to be included in future releases of the Tool.

- Incident Interdiction and Correction Service (IICS): Generates high-level (system independent) response plans and maps them into low-level (hardware/OS-specific) response sequences. Initial releases of the Tool will be capable of carrying out simple adaptive responses.

- System Management Service (SMS): Makes the system easily manageable. For example, it can be used to switch off a particular alert when a valid system administrator is doing maintenance work. It also performs as a system monitor and monitors other services, does health checking, and restarts dead agents. This service is planned for future releases of the tool.

- Virtual Security Manager: This is the highest-level service and it coordinates all adaptive decision-making and attack interdiction.

3.2. Agents and Device Experts

Agents and Device Experts are proxies on local machines throughout the enterprise. The main difference between the two is that Device Experts may have rule-bases or similar analysis capabilities and are thus are not as lightweight as agents, but more sophisticated. They gather information and pass adaptation commands on to the local host.

Agents are capable of, for example, detecting file change events, and sending them on to the relevant services. Agents use translators that provide rudimentary pattern matching to convert relevant information streams into a standardized message set or language. Agents are also able to bring about adaptations on the remote machine by triggering detailed system-specific commands on the local system based on higher-level instructions that have been passed down from higher level services.

Device Experts are Agents that have the ability to read information streams from the particular device that they represent and run filters on these streams to reduce the volume of information received from these devices. The device experts also use translators to convert the information streams into a standardized message set or language. In addition they have some rule processing intelligence to further filter the message streams. They have the ability to store raw information for archive purposes. The Device Experts forward relevant information to higher-level services.

3.3. Messaging

The messaging system used in the tool is layered off so that other emerging messaging standards such as the one defined by the Common Intrusion Detection Framework (CIDF) [5] can be substituted in. The services cooperate both to discern and classify problems and to find feasible and optimal solutions to those problems. To foster this cooperation, all services and agents share a common "language", or interaction protocol. This may be conceptualized as a stream of messages that pass among the services. The information in these messages may be understood by some services and not others, but all services are capable of handling the basic message formatting common to all messages. Services can simply pass on messages that they do not understand. Alternatively, a service may decide to deal with a message, and send messages to other services in order to solicit information or help. As a third message-handling option, a service may elect to add its own information to a message and pass it along to a higher-level or lower-level service for further processing.

In general, any service can query or inform any other service. However, the general strategy for handling information flowing down the "Data Bus" is to pass event data up the chain of services a sequence that follows (very roughly) this sequence:

Agent/Device Expert → Audit → Detection/Tracking → Network Model →Asset/Threat → Virtual Security Manager → Security Policy → Interdiction/Correction → Agent/Device Expert

As messages pass up the "chain of command", the information carried by them become more and more correlated into the current context and behavior of the framework. For example, a message that starts out as a low-level, uncorrelated event may lead to the declaration of a security incident, where not only the impact of the low-level event on the overall enterprise is understood, but a strategy has been developed to deal with the situation.

3.3. Messaging Protocol

The protocol consists of two basic definitions:

1. Structure definition for a simple inter-service message format suitable for communication of virtually any information between modules.

2. A protocol dictionary defining common message types, argument names, allowable argument values, and definitions of standard meanings for all of these.

The second definition defines a common syntax for interaction of services and plug-ins. It also establishes basic semantics associated with the syntax for a common (system-wide) representation of events and current system states. This approach allows heterogeneous elements of the system to reason about what is going on using common representations of the situation (the same language, just as human agents use English to communicate about their world, but tend to reason about it in different ways). This approach, in combination with Advisor-Based / Plug-In Based architecture is what allows the system to be rapidly expanded and adapted to new security challenges and associated response methodologies as they develop

- Message Structure - The generic message element of the Recon framework has the following structure:

  Recipient || Sender || Seq. # || MessageType || Name || Value || Name || Value || etc. ...

A set of well-known message types, argument names, predefined argument values and associated meanings allows for common representation of messages and context between modules. Examples of these are documented below.

- Message format - The general format of a Recon message using this option is:

  Recipient MessageType Name:Value Name:Value etc. ...

where the Sender and Sequence # fields are added by standard inter-process communication methods. Specification of colons is optional, and is allowed in name-value pairs for readability (in long strings it is too easy to lose track of which field is a name and which is a value).

- Some Basic Message Types

  Notify – Notification of an event.
  Adapt – Command to adapt network resources in a particular way, usually in response to an incident discerned by the Recon framework.
  Update – Update a parameter represented by the receiving service. For example, the Model Service accepts update commands to modify its current representation of network configuration and state.
  Request – Request information from a service.

- Some Message Examples

Updating the network model from a plug-in.

English : Update the network model to show a connection reject from the primary router.
Tool : Update Hub1. Raptor. Hub2. PrimaryRouter. TelnetFailedLogins : +1

Notification of event from one service to another

English : A failed connection rejection SNMP trap was generated from Router R2.
Tool : EventNotify EventType:RejectedConnection Source:Router(R2)

Notification of a new security incident from one service to another.

English : Audit Services has decided that there is a 90% chance that router R2 is under a denial of service attack from IP address 223.214.125.11. Therefore Audit Service is telling the Detection/Tracking Service that it is declaring a new incident, incident #213.
Tool : IncidentNotify NewIncidentID:213 IncidentType: DenialOfService Confidence:90 Attacker:223.214.125.11 Asset:Router(R2)

3.4. Rule Engine

A simple but powerful rule engine was developed to understand and prototype the interfaces for advisor plug-ins. Other commercial rule engines were considered but sending and receiving information to these engines proved tedious and inflexible. The other factor was to keep the costs low both is dollars and computing resources and have a functioning tool for agencies and businesses that do not have as much funding for their security efforts. The rule engine consists of a set of rules, a state database, a hierarchical database of objects, a rule processor and an interpreter. The rules are simple if-then statements. The rule processor coordinates the activities of the rule engine. The state database is a collection of facts about the system that eventually reach a goal state and stop the run. The hierarchical database is used to store information about objects which have an arbitrary number of attributes. The interpreter finds rules that must be applied to the current database and conditions within the rules that change the database. The interface developed for the initial releases of the tool is as follows:

Set_Of_Rules CheckRuleBaseForFire (Message)

This call returns a set of fired rules. The service is then free to deal with the fired rules as it wishes. Rules may fire based on checks of the current state database. Rules may also fire based on additional criteria of checking the values of the current input message with the message format described above. This approach allows reasoning about not just the mass of information contained in the rule engine's state database, but also about the particular semantics of an incoming message. It can yield richer opportunities for comprehension of developing situations.

The following is an example of how the rule engine works. Supposing the following message is received from a file monitor plug-in.

FileChange Name://Rebel/dis/dat/index.html user:paul

An example of the kind of rule that will handle this message at a service is:

```
;State Table initial definitions.
//Rebel/dis/dat/index.html->class is CriticalWebServer

;Rules
If Object->class is CriticalWebServerFile and
   Time is Before 6:30 and CurrentTime is After 17:50
   And Object->ChangePermissionToken is Off
then
   ForwardMessage ReconGateway
   Event:CriticalWebServerFileChange
```

The Rule Engine starts with what is known in the database, and applies rules until some goal state is reached by using a forward chaining process which is useful if a lot of information is known in the beginning (for example, patient records where test results are known). Backward chaining is planned for future releases and is useful if few goals and very little information is known up front. In this case, the rule engine will start with a goal hypothesis, and find which rules have this goal as their left-hand side of the condition. It then chooses one rule, and replaces the goal with the left-hand side of the chosen rule. Hopefully the new sub goals will eventually match the database, and be satisfied. When no more sub goals exist, the hypothesis is confirmed.

- Conflict Resolution: Several options for conflict resolution are planned. Among them are :

Specificity - Choose the rule which imposes the most conditions on the database i.e. has the most antecedent / left hand side parts

Recency - Choose the rule which applies to data most recently added to the database (sometimes least recently)

Refractoriness - Don't fire the same rule on the same state of the database twice. Avoids looping.

First - Choose the first of the triggered rules. Likely to lead to looping.

Last - Choose the last of the triggered rules. Likely to lead to looping.

Arbitrary - Choose amongst the triggered rules at random. Likely to lead to strange paths to the goal.

Certain key words are reserved for rules and they are, If, then, else, CurrentTime, Object, and the use of [] brackets which indicates state variables, and the use of { } brackets, which signifies the use of values from input messages.

3.5. Graphical User Interface

At the heart of this interface is a 'control center window' which is the central control station of the Tool. This window is intended to take up as little of the screen as possible, yet present high-level status information and permit control of both the GUI display, such as which windows are being displayed, and overall control of the underlying system throughout its enterprise deployment.

From this window other windows can be launched that display particular information about any of the services running and low-level information that is gathered at the agents or device experts.

These windows can be run in a stand-alone process or can be configured to be displayed in a standard browser.
The graphical user interface is not needed to run the Tool. It can be brought up and allowed to connect to the rest of the system at the user's convenience. The user can choose to view information supplied by any of the services or agents or choose just to have a view into the Incident Situational Display Window.

4. Advantages of Approach

The Tool discussed thus far is best used in environments where best practices and proper configurations of critical enterprise assets need to be monitored and measures have been taken to protect these assets with specialized tools such as Intrusion Detection Systems that also need to be monitored. In such environments the Tool:

- provides the ability to view information from as low-level as log information or as high-level as incident information
- reduces the number of false-positive from Intrusion Detection System
- provides the ability to monitor any deviations from the norm
- provides the ability to correlate information from heterogeneous sources

5. Disadvantages of Approach

In order to use the Tool one needs to set up the appropriate agents and plug-ins, rules to correlate information about the plug-ins, rules to respond to problems detected, and plug-ins to carry out the responses. Although the rules were designed to be as close to English as possible the syntax needs to be learned. However, this syntax can always be changed to a standard, similar to the one proposed by the Common Intrusion Detection Framework (CIDF), that the security industry decides on.

Until agents have been written for every kind of security device that needs to be monitored, new agents need to be written. This problem is alleviated somewhat by the fact that most agents can be written by changing the table-driven rules and filters; compilation is not necessary. However agents that need to monitor more complex tools can take longer to implement.

6. Summary

Next generation Intrusion Detection and Monitoring Systems [6],[7] need to be able to get information from heterogeneous sources, deal not only with events but also alerts and incidents while giving the user a security situational awareness that increases their effectiveness in the fight to secure the enterprise. The concepts proposed in this paper are a huge step in this direction.

With the role of the security administrator still developing with IT departments in the Internet age of increasing cyber-terrorist activities, the best stance might be for them to provide the incident detection, tracking and monitoring capability while assisting the network and system administrators with best practices and configuration guidelines. The Tool will be useful to the security administrator in this role and allow him/her to be part of the team as opposed to being a threat or an after thought.

7. References

[1] Orfali, Robert, Harkey, Dan, and Edwards, Jeri, "The Essential Distributed Objects Survival Guide", John Wiley and Sons, Inc., 1996.

[2] CORBA: Architecture and Specifications (August 1995). The OMG Publications.

[3] Inside OLE2, Second Edition (Microsoft Press, 1995)

[4] Wollrath, Ann, Riggs, Roger, and Waldo, Jim, "A Distributed Object Model for the Java™ System", The Second Conference on Object-Oriented Technologies and Systems Proceedings, Toronto, Canada June 17-21, 1996.

[5] Kahn, Clifford, Porras, Phillip A., Staniford-Chen, Stuart, Tung, Brian, "A Common Intrusion Detection Framework", DRAFT, http://www.isi.edu/~brian/cidf/

[6] Bass,Tim, Gruber, Dave, "A Glimpse into the Future of ID", ;login: Special Issue Intrusion Detection, The USENIX Association Magazine, September 1999.

[7] Forte, Dario, "Intrusion-Detection Systems", ;login:, The USENIX Association Magazine, February 1999, Volume 24, Number 1.

CAPSL Integrated Protocol Environment *

G. Denker and J. Millen
Computer Science Laboratory
SRI International
Menlo Park, CA 94025, USA
{denker || millen}@csl.sri.com

Abstract

CAPSL, a Common Authentication Protocol Specification Language, is a high-level language to support security analysis of cryptographic authentication and key distribution protocols. It is translated to CIL, an intermediate language expressing state transitions with term-rewriting rules. Connectors are being written to adapt CIL to supply input to different security analysis tools, including PVS for inductive verification and Maude for model-checking.

1. Introduction

It is generally understood that high-assurance network security relies upon the proper use of strong cryptography. Cryptographic protocols are the cornerstone of secure electronic banking and commerce, as well as military communications. But a strong cryptosystem is ineffective without safe key management. In modern network environments, keys are distributed using connection establishment protocols, and applied in ways that are supposed to guarantee privacy or authentication of messages.

1.1. Protocol Vulnerabilities

It has been known, at least since the 1981 publication of a replay attack [DS81] on a key distribution protocol in [NS78], that cryptographic protocols are vulnerable to message modification attacks (see Figure 1). These attacks assume that an intruder can eavesdrop on communications and interfere with messages in transit. Message modification or "active" attacks include such tactics as message interception, replay, and address spoofing. Closely related activities on the Internet have been discovered. One notorious example is IP spoofing; the vulnerability was described by Bellovin in [Bel89], and in 1995 a hacker took advantage of it. This confirms the existence of sophisticated active attacks, though without the cryptographic aspect.

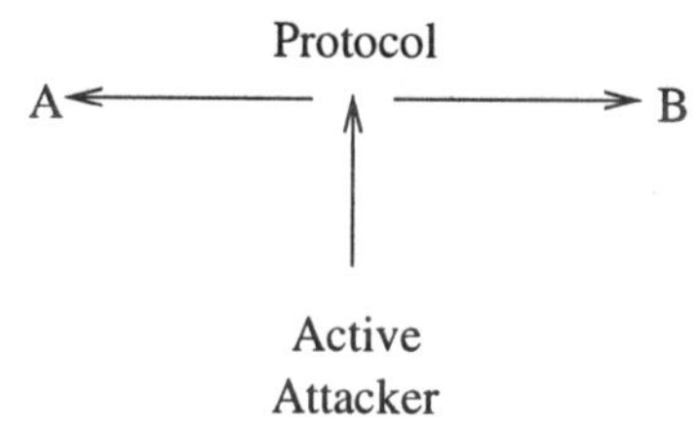

Figure 1. Message Modification Attacks

Some efficient methods have been developed to detect protocol vulnerabilities. Because of the logical, puzzle-like nature of this problem, the analysis is not performed on the executable code, but rather on an abstract model or "formal" specification of the protocol. Specifications of this kind treat cryptographic operations axiomatically, assuming that data cannot be decrypted without the proper key. This assumption distinguishes protocol failure analysis from other concerns, such as possible weaknesses in cryptographic algorithms. Protocol vulnerabilities, once detected, can usually be corrected by relatively small changes in the protocol.

1.2. Analysis Methods

There are three principal protocol analysis approaches: authentication logics, state-space exploration, and inductive proofs. (See Figure 2.) The first and last approaches aim at proving a protocol correct with respect to stated security properties, and the second searches for successful attacks. Notable successes of vulnerability analyses include an attack found with the NRL Protocol Analyzer on a selective broadcast protocol [Mea92] and an attack found with the FDR model-checker on the Needham-Schroeder public-key protocol [Low96]. We will not attempt to survey the history and status of these approaches here, but later on there are

*This work was supported by DARPA through the Air Force Research Laboratory under contract F30602-98-C-0258.

a few references to some of the work that has had a direct impact on the environment we are developing.

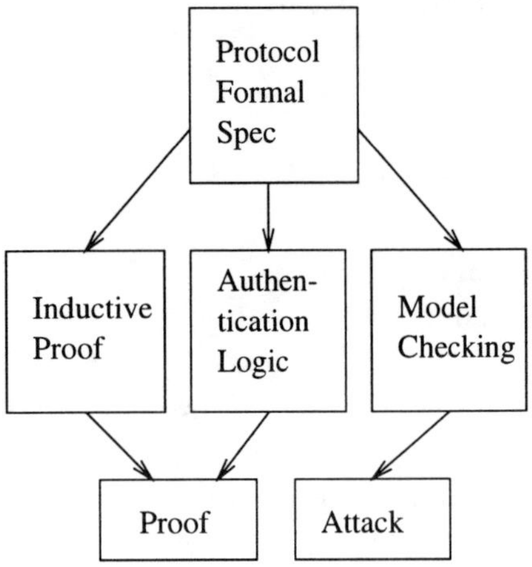

Figure 2. Protocol Analysis Methods

Most formal analysis techniques, even those originally intended to be used by hand, are supported by software tools. The tools require protocols to be expressed in a tool-specific language, with special conventions peculiar to cryptographic protocol analysis. The language may be low-level and based on the general-purpose language that the tool was implemented in or designed to support; or it may be a customized higher-level language. Several researchers realized that it was well worth the trouble to support a higher-level language, if only to save their own time in preparing protocols for analysis.

1.3. The CAPSL Integrated Environment

The acronym "CAPSL" stands for "Common Authentication Protocol Specification Language." CAPSL is a high-level language for cryptographic authentication and key distribution protocols. It is designed to be used as the input format for a variety of formal analysis techniques. The core of a CAPSL specification is a message list like the ones that are often used to present cryptographic protocols in articles and textbooks.

The idea of CAPSL as a single common protocol specification language that could be used as the input format for any formal analysis technique was first presented at the 1996 Isaac Newton Institute Programme on Computer Security, Cryptology, and Coding Theory at Cambridge University. This led to an initial version of CAPSL [Mil97] and later to the current effort.

The general architecture of the CAPSL Integrated Protocol Environment is illustrated in Figure 3. It shows the strategy by which CAPSL can be adapted for use by various protocol analysis tools. The core of this strategy is the use of an intermediate language, CIL (CAPSL Intermediate Language), that is closer to the state-transition representation used by almost all of the inductive proof and model-checking tools. Thus, CIL serves two purposes: to help define the semantics of CAPSL, and to act as an interface through which protocols specified in CAPSL can be analyzed using different tools. We envision customized subtranslators or "connectors" that translate from CIL to whatever form is required for each tool.

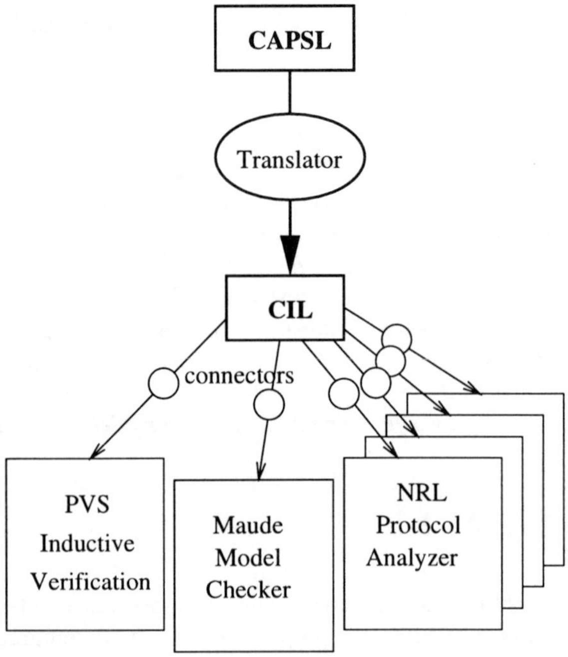

Figure 3. CAPSL Integrated Environment

The translator from CAPSL to CIL deals with the universal aspects of input language processing, such as parsing, type checking, and unraveling a message-list protocol description into the state transitions of the communicating processes.

Translation of CIL to the input language of other tools is viewed as a task for researchers engaged in applying those tools to protocol analysis. At least two connectors will be constructed as part of our own efforts: for PVS and for Maude. PVS is the SRI verification environment, which we are using to support inductive protocol proofs. Maude is an efficiently executable specification language based on rewriting logic which has also been developed at SRI (see Section 5) and which is being used to support model checking. There is a CAPSL interface to the NRL Protocol Analyzer, but it does not use CIL [BMM99].

The integrated protocol environment is still under development. This paper offers an overview of the current state of the CAPSL and CIL design, translator design and implementation, and a snapshot of the current tool support.

2. CAPSL

CAPSL is modular and extensible, and has a number of syntactic features that are unique to protocol analysis. CAPSL syntax is motivated both by user convenience and by the needs of protocol analysis tools.

A CAPSL specification is made up of three kinds of modules: *typespec*, *protocol*, and *environment* specifications, usually in that order. Abstract data type specifications (called *typespecs*) introduce new data types and define cryptographic operators and other functions axiomatically. Standard typespecs are included automatically and others may be supplied by the user.

There may be more than one protocol specification, if there is a top-level protocol that invokes one or more subprotocols. Environment specifications are optional; they are used to set up particular network scenarios for the benefit of search tools.

The following subsections take a closer look at protocol specifications, typespecs, and an environment. Some more advanced CAPSL features are also discussed.

2.1. Protocol Specification

Here is a small example of a protocol specification, illustrating several of the language features. It looks like a strongly typed program, with variable declarations. Unlike an ordinary computer program, the body is a sequence of protocol messages, and it has a list of security goals.

```
PROTOCOL OneMessage;
VARIABLES
    A, B: PKUser;
    K: Skey, FRESH, CRYPTO;
ASSUMPTIONS
    HOLDS A: B;
MESSAGES
    A -> B: {A,K}pk(B);
GOALS
    SECRET K;
    BELIEVES B: HOLDS A: K;
END;
```

This protocol has only one message, in which principal A sends a newly generated key K to another principal B. The key K is concatenated with A and encrypted using B's public key.

Note that this protocol can be instantiated many times with different principals as values for the protocol variables A and B. It helps to keep in mind the distinction between the principal itself, usually a named user (e.g., "Alice" or "Bob"), and the *role* it plays in the protocol. This protocol defines two roles, which could be described as initiator and responder. The name of the initiator role is "A" and the name of the responder role is "B." A principal Alice could play either the "A"-role (initiator) if it is the value of A, or the "B"-role (responder) if it is the value of B.

Protocol variables are typed, and they may also have *properties*. For example, the property FRESH means that the value of the variable is a new value generated for each session, and has not (with cryptographically high probability) been used before by the principal generating it. The property CRYPTO means that the value of the variable is unguessable (because it comes from a large domain and is chosen in an unpredictable way). Properties affect interpretation of the message list and also affect the attacker model.

The HOLDS assumption means that when an A process begins, it is initialized with the identity of a principal B to communicate with.

There are two security goals: that the new key K generated by A is kept secret from any attacker (unless the attacker happens to be either of the named users A or B), and that if the B session completes, B is justified in believing that A holds K.

The belief assertion means that `HOLDS A: K` is interpreted in the context of B's values for A and K. Thus, this is an agreement goal. It fails if there is no "A"-role process for this session with B's value for A and which holds the same value of K. For this protocol, the agreement goal could fail, since the message received by B, encrypted with B's public key, could have been forged by an attacker.

In general, assertions appearing as assumptions or goals are statements about protocol variables and the values held for them by the processes playing the various roles in a protocol session.

2.2. Type Specification

Messages are constructed using cryptographic operators and other functions, such as concatenation and hash functions. Every message field is of type Field, but certain operators require or produce fields of particular subtypes, such as key types.

A portion of the CAPSL type hierarchy is shown in Figure 4, including the type PKUser employed in the "OneMessage" protocol example. The fact that PKUser is a subtype of Principal means that a variable of type PKUser may be used wherever the type Principal is required, such as the source or destination of a message.

All functions used in CAPSL and the data types they operate on must be specified axiomatically in typespecs. The most commonly used types and operators are defined in a standard prelude. Analysts may add new types and operators by providing their own typespecs. Protocol specifications and other typespecs may import user-supplied typespecs.

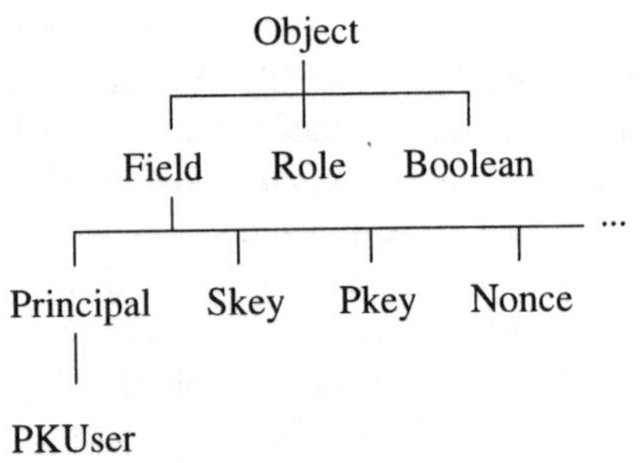

Figure 4. The Object Hierarchy

Here is an example of a typespec that defines the subtype PKUser of Principal. A PKUser has a public/private key pair. (The typespec PPK is actually part of the current prelude.)

```
TYPESPEC PPK;
TYPES PKUser: Principal;
FUNCTIONS
  pk(PKUser): Pkey;
  sk(PKUser): Pkey, PRIVATE, CRYPTO;
VARIABLES
  X: Field;
  U: PKUser;
AXIOMS
  {{X}pk(U)}sk(U)  = X;
  {{X}sk(U)}pk(U)  = X;
  INVERT {X}pk(U): X | sk(U);
  INVERT {X}sk(U): X | pk(U);
END;
```

PRIVATE Functions

The public key of a PKUser is found with the function `pk` and the corresponding private key with `sk`. The secret-key function is declared with the property PRIVATE to indicate that the function call `sk(U)` can be evaluated only by a process running on behalf of U. No other honest process, or the attacker, has direct access to `sk(U)`. This notion of private access is a novel feature of the CAPSL specification language; it is a different concept from the scope restrictions as found in Java and other languages.

Invertibility

The axioms for PKUser include two ordinary reduction axioms and two special invertibility axioms. An invertibility axiom is a special formula used by CAPSL to determine semantic correctness of messages. An axiom `INVERT` *term*: *var* | *termlist* says that a value given by *term* can be inverted to find the value of *var* provided that the value of each term in *termlist* is held. Sometimes invertibility can be inferred from reduction axioms, but sometimes it cannot.

A term like `{X}K` is invertible in the sense that a process holding `{X}K` and the inverse of K is able to extract X. Other terms have other rules for invertibility. The concatenation `[X,Y]` is invertible to extract either X or Y without any additional information. A bitwise exclusive-or `xor(X, Y)` is invertible to extract either X or Y given the other. If *hash* is a one-way function, `hash(P)` is invertible to extract P provided that the domain of the function is a type representing weak passwords (inversion is by guessing and comparing). The invertibility axiom is a uniform way to express all of these properties and any introduced by users.

Bracket Notation

Operators and functions in CAPSL are normally expressed in functional notation, e.g., `sk(Alice)`. However, curly and square brackets ({ } and [ ]) are commonly used in protocol descriptions to signify concatenation with encryption. In CAPSL, an expression like `{A, K}pk(B)` is shorthand for `ped(pk(B),cat(A,K))`. Here, `cat` is a concatenation operator which also can be used by itself (without encryption). The function `ped` is an RSA-like public-key encryption and decryption operator.

Square brackets identify another form of concatenation, with the operator `con`. The difference is that `cat` is associative, `con` is not. A term `{{A,B},C}` is equivalent to `{A,{B,C}}`. Thus, if a pattern $\{X,Y\}$ is matched to a message `{{A,B},C}`, X would be matched with `A`, while if the same pattern is matched to a message `[[A,B],C]`, X would get `[A,B]`.

When a bracket expression is followed by a key, as in `{A,K}Ks`, the encryption operator is assumed to be `ped` if `Ks` is of type Pkey, but it will be interpreted as the built-in symmetric-key encryption operator `se` if the key is of type Skey. If decryption rather than encryption is required, it is written `{A,K}'Ks`.

DENOTES

In protocol specifications, it is common to see names defined for significant expressions, which are then referred to by name in the message list. The DENOTES section in a CAPSL protocol specification can hold definitions like this. A kerberos ticket-granting ticket, for example, could be defined as

```
DENOTES Tgt = {Kct,C,Tgs,T}Ks
```

containing abstracted fields such as the client C, etc. An equation like the one for `Tgt` could have been inserted as an action in the message list. The DENOTES declaration simply features it in an easily found global location, to be referenced implicitly where needed. The symbol `Tgt` and the others are just protocol variables requiring their own type declarations as usual.

2.3. Environment Specifications

When a protocol is being analyzed by a model checker or simulated, the analyst may have to specify which processes are to be run. The analyst may also have to supply other run-specific information such as the initial knowledge of the attacker.

An environment specification defines constants of type Principal and some protocol sessions using them, and creates one or more *runs*, each of which is a scenario involving one or more of the sessions. A run is an interacting collection of sessions, something like a "bundle" in strand-space terminology [THG98]. A *session* is defined by allocating all the roles in a protocol specification to particular constants of type Principal.

For example, suppose we want a run with two sessions, one in which Alice takes the "A" role and Bob takes the "B" role, and another in which a dishonest principal Eve takes the "A" role and Alice takes the "B" role.

We might set that up like this:

```
ENVIRONMENT Test1;
IMPORTS OneMessage;
CONSTANTS
    Alice, Bob: PKUser;
    Eve: PKUser, EXPOSED;
SESSION S1;
    A = Alice;
    B = Bob;
SESSION S2;
    A = Eve;
    B = Alice;
RUN R1;
    EXPOSED sk(Alice);
    ORDER (S1||S2);
END;
```

Note that these session specifications only set up initial states of the processes in each role, and attacker action may prevent some of them from continuing or completing their normal programs.

Sessions are specified by associating roles with constants of type Principal. The role names are the variables of any subtype of Principal that occur in the imported protocol specification. Any other variables that are initially held by any principal must also be assigned in each session to some declared constant.

A principal (Eve, in this case) with the EXPOSED property yields all its PRIVATE values to the attacker. Eve may be thought of as colluding with, or subverted by, the attacker. For each run, there is also optionally a list of individual EXPOSED terms. (The attacker model is dicussed further below.)

The ORDER section can specify a series-parallel sequencing of sessions. Some search tools could benefit from this and save time when such a restriction is assumed. For example, if there were another session `S3`, we might say `ORDER ((S1;S2)||S3)` to mean that session `S2` does not start until `S1` ends, but that sequence runs concurrently with `S3`.

If no run information is given, CAPSL will assume that there is one run "Run1," with no additional terms exposed, in which all declared sessions are run concurrently.

2.4. Other CAPSL Language Features

View Notation

Sometimes a message is viewed differently by the sender and the receiver. The sender and receiver versions of the whole message, or of any individual fields, are separated by the symbol %. In the message `A -> B: {A,Ks}Kas%E` the encrypted term is received as a single value E if B will pass it on rather than decrypt it. This convention was originated by Lowe in Casper [Low98]. If the message were specified simply as `A -> B: {A,Ks}Kas`, it would mean that B was intended to decrypt or check the encrypted content, and CAPSL would report an error if B did not possess `Kas` or its inverse.

Actions

Messages may be preceded and followed by equational actions. An equational action like `X = {Y}K` may be either an assignment to X or a comparison test; it depends on whether or not the acting principal already holds a value for X. The semantics of a comparison, as reflected in the CIL rule output, is that if the test fails, the process cannot take any further state transition.

Interleaved Assertions

Besides their use as assumptions or goals, assertions may appear within a message list, like actions, between messages. An assertion in a message list is prefixed by the keyword ASSUME or PROVE to indicate its status.

There is presently no specific support for message idealizations used in authentication logics. GNY logic, in particular, permits terms in messages to be linked with *message extension* statements [GNY90]. We may add an operator like Brackin's || in ISL [Bra97] for this purpose.

Subprotocols and Selection

In larger protocols it is sometimes convenient to recognize functionally related groups of messages called "exchanges" or subprotocols. In CAPSL, a subprotocol is declared with an ordinary protocol specification, except that it imports a parent protocol to incorporate its variable declarations.

The parent protocol invokes the subprotocol by name where needed, using the form `INCLUDE` *name*.

Conditional selection of protocols is accomplished within a message list by putting `INCLUDE` invocations within an IF-THEN: `IF` *test* `THEN INCLUDE` *subprot1* `ELSE INCLUDE` *subprot2* `ENDIF`.

3. Interface Format For Analysis Tools

3.1. CIL

The challenge for the design of CIL is to make it easy for us and for other researchers to build the connectors needed to translate CIL into the input specifications used by the tools we want to use.

It must be expressive enough to represent a wide range of protocols, and yet at a low enough level to be close to the representation used by most verification or model-checking tools. Many such tools share a specification style that incorporates state-transition rules specified in a pattern-matching style, with symbolic terms to represent encryption and other computations. There is usually a separate and fairly standard attacker model, based on the concept in the early Dolev-Yao paper [DY83].

The meta-notation of Cervesato, et al [CDL$^+$99] was adapted for CIL because it appeared to be the simplest and most general formalism that exhibits these features. Their notation, according to the authors, could be regarded as either an extension of multiset rewriting with a kind of existential quantification, or a Horn fragment of linear logic. The logical style of this formalism made it suitable to serve as the language in which to express the semantics of CAPSL. Furthermore, the term-rewriting aspect corresponded well with the modeling approach taken by Denker, Meseguer, and Talcott with Maude [DMT98]. CIL may be regarded as a notational variant of the linear logic formalism in which certain specific conventions have been used to set up protocol models derived from CAPSL specifications.

3.2. Multiset Term Rewriting

The linear logic formalism uses transition rules of the form

$$F_1, ..., F_k \longrightarrow \exists X_1, ..., X_m : G_1, ..., G_n,$$

where each F_i and G_j is a "fact." Facts are atomic formulas of the form $P(t_1, ..., t_r)$ where P is a predicate symbol and the arguments t_i are terms. A term is constructed from typed constants, variables, and function symbols. Free variables are implicitly universally quantified.

The state of a system can be represented by a multiset of facts. (A multiset is a set in which there may be more than one copy of each element.) A rule is eligible to fire when the facts on the left side of the rule can be matched with facts in the multiset. When a rule fires, the matching facts in the multiset are removed from it and replaced by the facts on the right side of the rule, instantiated according to the substitution required by the pattern match. Removing a fact from the multiset reduces its multiplicity by one, if it was more than one. Facts in the multiset are typically ground terms (no variables) when finite-state search tools are used.

The existential quantifier in linear logic has a special meaning. Quantified variables are instantiated with fresh (unused) constants. This behavior is used to model generation of nonces.

Protocol State Facts

In protocol modeling, facts are used to express the entrance of a process into a state, the transmission of a message, or certain security-relevant events. In CIL, in particular, a state is represented by a fact `state`(*role, num, termlist*) where a role is a constant corresponding to a principal, *num* is a state label, usually an integer, and the terms in the *termlist* are the "memory" of the process in that role and state. A message is a fact `msg`(*sender, receiver, termlist*). A `net`(*term*) fact represents attacker knowledge.

Rules with an empty left side are interpreted as initialization or fact-generating rules. For each role in the protocol, an initial state fact is generated with initially held variables. The rule

$$\rightarrow \mathtt{state}(\mathtt{roleA}, 0, A, B), \quad \mathtt{state}(\mathtt{roleB}, 0, B)$$

creates two facts representing the initial state of two new processes of the protocol, one in role "A" and the other in role "B". The role names are constants taken from the text of the protocol specification, and should be distinguished from the variables A and B, which are free variables whose names could be changed without changing the meaning of the rule. Since A and B are variables of type Principal, this rule can initiate sessions between any pair of principals.

The message `A -> B: {N}sk(A)` would result in two transitions, one for the sender A and one for the receiver B. Assuming that A is in state 0 and N is a nonce originating in this message, the A transition would be:

$$\mathtt{state}(\mathtt{roleA}, 0, A, B) \rightarrow$$
$$\exists N : \mathtt{state}(\mathtt{roleA}, 1, A, B, N), \quad \mathtt{msg}(A, B, \mathtt{ped}(\mathtt{sk}(A), N)).$$

Attacker Facts and Rules

The Dolev-Yao model of the attacker, commonly used in protocol analysis, can be expressed as a set of rules that allow the attacker to eavesdrop on transmitted messages and to forge new messages from available message components.

In principle, the attacker model is very simple. For each message fact $\mathtt{msg}(A, B, X)$, there is a rule

$$\mathtt{msg}(A, B, X) \rightarrow \mathtt{net}(X), \mathtt{msg}(A, B, X)$$

which generates a fact $\mathtt{net}(X)$ asserting that the attacker now holds the content of the message. (Note that the message is repeated on the right so that it will still be available in the multiset for the intended receiver after the attacker reads it.) There are also rules giving the attacker any terms declared as EXPOSED in the environment.

Other attacker rules allow the attacker to analyze complex terms into their components. For example, if `se` is the symmetric-key encryption operator,

$$\mathtt{net}(\mathtt{se}(K, X)), \mathtt{net}(K) \rightarrow$$
$$\mathtt{net}(X), \mathtt{net}(\mathtt{se}(K, X)), \mathtt{net}(K)$$

The attacker can also construct complex terms from the ones it holds, for example,

$$\mathtt{net}(X), \mathtt{net}(Y) \rightarrow$$
$$\mathtt{net}(\mathtt{cat}(X, Y)), \mathtt{net}(X), \mathtt{net}(Y)$$

and to put any term it knows into a message:

$$\mathtt{net}(X) \rightarrow \mathtt{msg}(\mathtt{Spy}, A, X), \mathtt{net}(X)$$

In this message, `Spy` is a constant principal identifying the attacker. Spy is the sender because we have adopted Paulson's convention that the message source and destination arguments are the true source and intended destination of the message, and they are not directly accessible to the receiver of the message. The message is "forged" in the sense that the message content has been constructed by the attacker, and the message can be directed to any principal as its destination. More generally, we could have a predicate `spy( )`, and any principal E such that `spy(E)` is true would have the attacker's properties. If a constant `Eve` is declared as EXPOSED, a fact `spy(Eve)` is generated.

These rules can be extended to deal with messages having several fields.

This attacker model is too simplistic to be useful for model checking. It is noted in [CDL+99], for example, that it is helpful to separate decomposition from composition transitions by using two kinds of facts. However, while every protocol analysis tool needs an attacker model, most of the attacker model is fixed, and protocol-dependent aspects of it are influenced as much by the needs of the tool as by the details of the protocol. For this reason, it does not appear necessary or appropriate for the CIL translator to generate the attacker facts and rules. Our definition of the attacker is needed only to clarify the semantics of the CIL protocol environment, so that analysts can check that their attacker model is consistent with CIL.

3.3. Security Goals

Security goals in CAPSL include SECRET assertions and agreement assertions, expressed as beliefs about HOLDS statements. Other security goals are possible, but we are still in the process of working out their semantics.

The semantics of a security goal consists in an understanding of when it has failed. A secrecy goal has failed when the attacker has learned a value declared to be secret. Since the attacker can always transmit any value it learns in a message, we can recognize a secrecy failure when a secret value appears as the content of a message. That is, if X is secret, the presence of a message fact $\mathtt{msg}(P, Q, X)$ is a secrecy failure, for any principals P and Q.

But how do we know which values are secret? In the multiset model, a value X is declared as secret by generating a fact that says so. If the secrecy goal is `SECRET X: A,B`, the secrecy fact might be of the form $\mathtt{secret}(X, \{A, B\})$. This fact must be generated from some rule whose left side identifies X, A, and B from some state fact. This could be done as an extension to the rule that first transmits X in a message; or it could be done as part of the initialization rule.

When evaluating a protocol for security, we have to keep in mind that if A or B is the Spy, release of the secret X is not considered a failure, since X was intended to be shared with the Spy.

The translator could and will perform the service of adding secret facts, though the present preliminary version does not; it merely passes the SECRET declaration on to the connector.

An agreement goal `BELIEVES B: HOLDS A: X` is violated if B reaches its final state $\mathtt{state}(\mathtt{roleB}, n, B, A, X')$ with a value in the X position that differs from what A actually holds, as indicated by another state $\mathtt{state}(\mathtt{roleA}, m, A, B, X)$ where $X \neq X'$. (It can also fail if there is no corresponding state for `roleA`.)

3.4. Translator Output

A connector from CIL to any analysis tool needs not only the transition rules, but also all the other information supplied in the specification: variable and function declarations, assumptions, goals, and environment. The translator output has all of this in a digested form.

The output of the overall translation is:

1. symbol table and slot table
2. axioms (collected from typespecs, including the prelude, and environments)
3. localized assumptions
4. protocol rewrite rules, i.e., rules for state transitions
5. localized goals
6. environment information

The symbol table lists the types of variables and the type signatures of functions, with entries of the form:

`symbol`(*ident, status, args, val, props*)

where *ident* is the name of the variable, constant, or function, *status* is *var* for dummy variable (from an axiom), *pvar* for a protocol variable, *op* for a constant or function name, or *type* for a type name. The *args* entry is a list of the argument types (empty for a variable or constant), and *val* is the value type. Finally, *props* is a list of the property keywords that were given in the original CAPSL declaration.

The slot table relates state components with protocol variables. The fact that the value held for variable A is the first component in the state of `roleA` (after the required items of the role, state number, and session ID) is indicated by the slot table entry `slot(A,roleA,1)`. State component positions depend only on the role and variable and not on the state number. The reason that we need slot information is to interpret assumptions and goals, which mention protocol variables, in terms of the state facts.

Axioms from typespecs and environment specifications are consolidated into a single list.

Assertions are *localized* by associating them with states, inidicating the role and state number. In particular, assumptions are associated with the initial state, goals are associated with the final state, and interleaved ASSUME and PROVE assertions are associated with the current state where they are placed.

Currently we carry over environment information without change from the CAPSL, except that constant declarations are extracted and incorporated into the single collected set of declarations and that the axioms from the environment specification are consolidated with typespec axioms into a single list.

The form of CIL output is defined with an "abstract syntax" grammar that could be concretized in various ways. Presently, the output is a text file containing the elements described above in a functional notation. There is a technical report with a preliminary description of CIL output [DM99a]; current documentation on the details of CIL output may be found at the TIPE/CAPSL Web site [Mil99a].

Example CIL Output

For the sake of example, and for use in subsequent sections, here is a sample of part of the CIL output derived from the CAPSL modules OneMessage and PPK. The output normally includes additional material from the prelude.

```
symbols(
  symbol(PKUser,type,ids(),Principal,
         props()),
  symbol(pk,op,ids(PKUser),PKey,props()),
  symbol(sk,op,ids(PKUser),PKey,
         props(PRIVATE,CRYPTO)),
  symbol(X,var,ids(),Field,props()),
  symbol(U,var,ids(),PKUser,props()),
  symbol(A,pvar,ids(),PKUser,props()),
  symbol(B,pvar,ids(),PKUser,props()),
  symbol(roleA,op,ids(),Role,props()),
  symbol(roleB,op,ids(),Role,props()),
  ...)
axioms(
  eqn(ped(sk(U),ped(pk(U),X)),X),
  eqn(ped(pk(U),ped(sk(U),X)),X),
  invertible(ped(pk(U),X),X,terms(sk(U))),
  invertible(ped(sk(U),X),X,terms(pk(U))),
  ...)
slots(
  slot(A,roleA,1),
  slot(B,roleA,2),
  slot(B,roleB,1),
  slot(K,roleA,3),
  slot(A,roleB,2)
)
rules(
  rule(
    facts(),
    ids(),
    facts(state(roleA,0,terms(A,B)),
          state(roleB,0,terms(B)))),
  rule(
    facts(state(roleA,0,terms(A,B))),
    ids(K),
    facts(state(roleA,1,terms(A,B,K)),
          msg(A,B,
              terms(ped(pk(B),cat(A,K)))))),
  rule(
    facts(state(roleB,0,terms(B)),
          msg(UNK,B,
              terms(ped(pk(B),cat(A,K))))),
    ids(),
    facts(state(roleB,1,
                terms(B,A,cat(A,K)))))
)
goals(
  loc(
    nodes(node(roleA,1),node(roleB,1)),
    secret(K,ids())),
  loc(
    nodes(node(roleA,1),node(roleB,1)),
    believes(B,holds(A,ids(K))))
)
```

In the last rule, where B receives the message, the sender of the message is designated with a free variable `UNK` that does not have to match A. This is because we have adopted Paulson's convention that the message source and destination arguments are the true source and intended destination of the message, and they are not directly accessible to the receiver of the message. Hence B will accept any message of the given form regardless of the true sender, and will learn values for only those protocol variables that occur in the message content.

3.5. Translator Implementation

The CAPSL-to-CIL translator is presently implemented partly in Java and partly in Maude. The Java portion of the translator parses the CAPSL, performs type checking, and consolidates declarations, axioms, and assumptions. It flags syntax errors or type consistency violations. If there are none, it produces a text file containing the symbol and slot tables, assumptions, goals, message list, and environment.

The Maude portion of the translator converts the message list into transition rules and localizes assertions. It checks that the message list is *implementable* in the sense described in the next section. Essentially, it determines that a process is capable of constructing messages it is supposed to send. This check is analogous to what a compiler does to check that program variables have been initialized before they are used.

The translator can be run with a simple script that calls Java on the main class for type checker and pipes the output to the Maude runtime executable, which interprets a Maude rule-generation program. We plan to re-implement the rule generation in Java, so that the entire translation to CIL can be completed within Java.

4. Translator Algorithm

While the transition-rule part of CIL has its own rewriting-logic semantics, the semantics of CAPSL lies in its translation to CIL. In this section we summarize the main concepts of the message-list translation. More details can be found in [DM99b].

The translator goes through the messages and actions in sequence. Each message or action produces as output a pair of CIL transition rules, and it may also cause slot table entries to be added. In this way the translator keeps track of which variables are held by each role at each point in the protocol.

4.1. Computability and Receivability

While translating messages into rules, the translator checks the implementability of the protocol. Different checks are made depending on what a protocol process has to do, but the key concepts are computability, receivability, and invertibility.

As we noted in the general context of multiset term rewriting, a message causes two state transitions: one for the sender, which generates a rule with a message fact on the right as well as a new state, and one for the receiver, which consumes a message fact on the left.

A equational action can be either an assignment or a comparison. An assignment causes a program variable to become held, so it adds a new role state component, creates a slot table entry, and puts the term on the right into the state. This state change is reflected in a CIL rule. A comparison similarly adds a role state component containing the equation. Then it adds another transition rule to the CIL output which fires when that component contains the value "true."

When the translator creates a rule for the sender, it checks that the content of the message to be sent is *computable*. A term is computable if it can be constructed from the sender's state components, using declared functions if necessary. Functions declared PRIVATE can be employed only when the principal named in the first argument is the sender.

When the translator creates a rule for the receiver, it checks that the content of the message to be received is *receivable*. A term is receivable if it is computable, or if it is a protocol variable that is not yet held, so that receiving the message causes the receiver to add that variable to its state. In that case the translator creates a new slot table entry for that variable and the receiver.

If a term is not computable, it is receivable if it is invertible, and its components are (recursively) receivable. Invertibility is checked using the INVERT declarations. If invertibility requires that some term be held, such as an inverse key, that term must be computable.

Similar considerations apply in the processing of equational actions.

Rule Optimization

The translator produces two rules for each message. Often, successive transitions for the same role can be combined into a single transition.

Consider the two messages

```
A → B: A;
B → A: B;
```

Two transition rules are generated for B:

$$\texttt{state(roleB}, 0, B), \texttt{msg}(A, B, A) \rightarrow \texttt{state(roleB}, 1, B, A)$$

$$\texttt{state(roleB}, 1, B, A) \rightarrow \texttt{state(roleB}, 2, B, A), \texttt{msg}(B, A, B)$$

These can be combined into the rule

$$\texttt{state(roleB}, 0, B), \texttt{msg}(A, B, A) \rightarrow \texttt{state(roleB}, 2, B, A), \texttt{msg}(B, A, B)$$

This does not eliminate any possible attacks, since nothing any other process could do between B's first transition and

B's second transition can affect B; and any transitions taken by other processes before B sent its reply do not depend on B's state, and could still be taken if B's reply is earlier.

Optimizing the rule set is an advantage for verification or model-checking. Fewer rules means fewer transitions, which means a smaller search space. The performance of a finite-state exploration tool can be increased enormously in this way.

5. Analysis of Cryptographic Protocols

Given the CIL output, the next step is to develop a connector to translate the CIL output into a form suitable for analysis with some existing tool or technique. We have been doing protocol analysis using PVS for inductive proofs and Maude for model checking, and connectors for translating a CIL protocol specification into these environments are under development.

This section has some examples constructed by hand of how the translation would work. The connector algorithms are still changing, because some details of the CIL output and also some details of the analysis techniques are undergoing revisions. However, both analysis techniques have been used successfully, and the principal features of the connectors are well established.

5.1. Connector: PVS Inductive Verification

Inductive verification of protocols is the only known way to prove a protocol correct that is deep enough to prove secrecy properties. Our approach uses a trace model based on work by Paulson [Pau98]. To be useful for inductive proof of protocols, a specification and theorem-proving environment should support abstract data types and convenient induction. PVS is such an environment, based on higher-order logic and augmented with a library of decision procedures and the ability to add user-defined proof strategies [ORS92].

A trace model differs from the multiset model because all message facts (called "events" in the trace model) are saved in order, in the trace, instead of being deleted when they are received. A trace model may also save state events. It is easy to reinterpret a CIL state transition rule as a rule for updating the trace.

Before giving an example of how a CIL rule becomes a trace update, it is necessary to provide some background on the structure of a protocol specification in PVS.

PVS Protocol Specification

The PVS specification has some protocol-independent supporting "theories" that correspond roughly to the typespecs in the CAPSL prelude. Theories contain both data type definitions and lemmas about those data types. There is a theory defining traces and messages, with different types of message fields such as keys and principals. Other theories import this one, such as the theory that defines Paulson's set operators *analz* and *synth* and contains lemmas giving their properties. This theory also defines the attacker's ability to construct a forged message from data extracted from a trace. Built on these is a theory defining the strand-space *ideal* from [THG98] and the related notion of a *coideal* defined and used in our own work, such as the proof that a certain protocol has only a parallel attack [Mil99b].

The protocol itself is defined as a set of acceptable traces. The definition is recursive: the empty trace is acceptable, and a new message event or state event can be added to an existing trace if it satisfies some state-transition rule. Thus, transition rules become predicates with two arguments, an event and a trace. The protocol as whole is defined as a global predicate that adds an event to a trace if it satisfies any one of the protocol-rule predicates, or an attacker-rule predicate.

For example, consider the rule:

```
rule(facts(state(roleB,0,terms(B)),
           msg(A,B,terms(A))),
     ids(),
     facts(state(roleB,1,terms(B,A)),
           msg(B,A,terms(B))))
```

This is transformed into a predicate

```
B0(E,T): bool =
  EXISTS (B,A):
   (member(State(roleB,0,B),T)
    AND member(Msg(A,B,A),T)
    AND E = State(roleB,2,Con(B,A))
    ) OR
   (member(State(roleB,1,Con(B,A)),T)
    AND E = Msg(B,A,B))
```

Without going into the details of the PVS syntax, it should be easy to see the close relationship between the CIL and the PVS predicate. This is what makes the connector to PVS easy to write. The complete trace predicate is a disjunction:

```
Protocol(E,T): bool =
  A0(E,T) OR B0(E,T) OR A1(E,T) OR ...
  OR Fake(E,T)
```

where the "Fake" predicate is from a fixed attacker model.

There is an apparently significant difference between the trace model and the multiset model: because old messages and states remain in the trace, a predicate like the one above can be invoked repeatedly to add any number of copies of the same new state and message. From a security verification point of view, this is not a problem, since any attack on the "real" protocol is still possible in this more liberal model. Conceivably one could find an attack in this model that was not real, because it could send information from new message to an old state. Spurious attacks like this are easy to recognize, and they could also be eliminated by modifying the transition rules by putting in an additional test.

Type Declarations

A full protocol theory includes not just the trace predicate but the type declarations of the variables used in it. For this example, that would include

```
E: VAR event
T: VAR trace
A, B: VAR principal
SID: VAR sessid
```

and others. These are easily obtained from the CIL symbol table output. The declarations of the types are in the standard supporting theories, so that the prelude-generated part of the symbol table is ignored by the PVS connector.

At present, we are able to deal with encryption and key-generating functions only if they are already provided by the existing PVS supporting theories. However, there is no reason why we cannot extend support in the future to user-defined functions given in typespecs, since PVS has quite general abstract datatype specification capabilities.

Goals

A security goal, for purposes of inductive proof, is a trace invariant. The general comments about security goals in Section 3.3 apply. A secrecy goal says that if a term X is secret, no message event has X as its content. A term is a secret if an event `secret(X,terms(A,B))` declaring it as such appears in the trace. This is another kind of event, and in the PVS specification, the protocol predicate must be augmented with another rule that generates it. Generally, the state transition that first introduces X in message is the one that is given the new `secret` event.

At present, the PVS connector must look for `secret` goals in the CIL output and use the slot table information to locate the appropriate transition rule.

Agreement goals are also invariants: if the current state event is the last state of B, and B is supposed to agree with A on K, the trace is tested for the presence of an A-state with the same K in the same session. No additional special events are needed.

5.2. Connector: Maude Model Checker

We have had encouraging experiences with Maude [CELM96] for the specification and analysis of cryptographic protocols. The level of specification is detailed enough to capture essential operational aspects. The reflective capabilities of rewriting logic [Mes92, Mes96b] and Maude [CM96] are very helpful, because they support various analysis techniques such as state search, theorem proving, and symbolic simulation. Figure 5 illustrates how this is done in Maude.

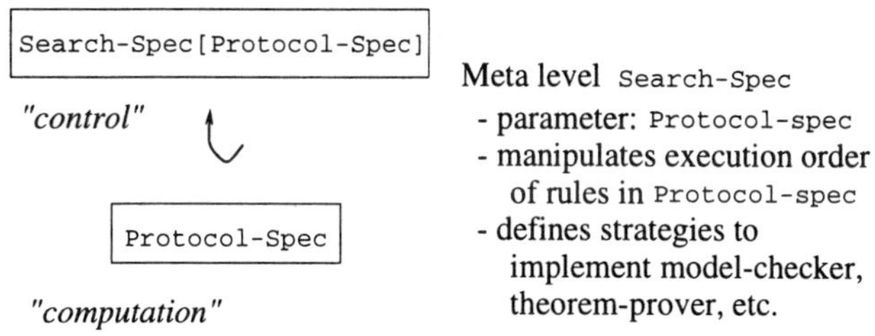

Figure 5. Reflection in Maude

The meta-level specification `Search-Spec` takes `Protocol-Spec` which defines the behavior of the cryptographic protocol as a parameter. At the meta-level the execution of transition rules of the protocol is controlled via user-defined execution strategies. A breadth-first search strategy can exhaustively explore all the executions of a system. Under finite-state or other termination assumptions it is even possible to analyze all executions. This is very helpful in uncovering security flaws under unforeseen attack scenarios.

Maude Basics

There are two types of modules used in Maude to specify cryptographic protocols: *functional* and *system* modules. Functional modules are equational theories used to specify algebraic data types. Cryptographic operators together with their axiomatic behavior can be represented in functional modules. System modules are rewrite theories specifying concurrent systems. Rewrite rules are used in these modules to express the dynamics of a system. All the above modules can be parametrized. This allows vertical and horizontal structuring.

Modeling Protocols in Maude

The CIL specification of a cryptographic protocol is the starting point to derive the corresponding Maude executable specification. The symbol table declarations and the axioms of CIL are directly translated into abstract data type specifications in Maude, protocol rewrite rules correspond to Maude rewrite rules, and environment information is used to declare test scenarios for which the protocol can be executed and analyzed. Figure 6 summarizes the translation.

The CIL declarations of the OneMessage example translate into the following Maude functional module.

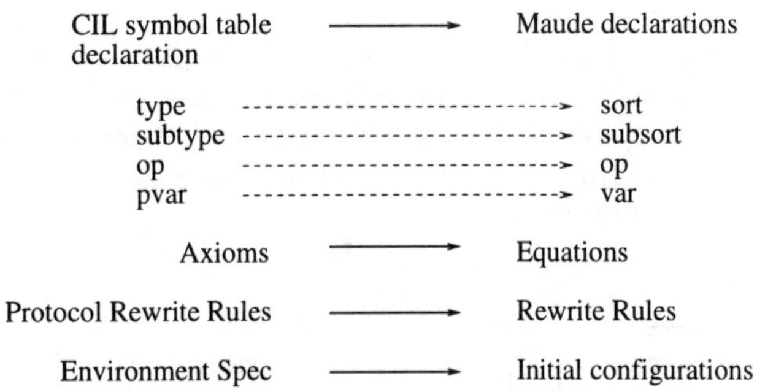

Figure 6. CIL - Maude

```
fmod USER is
  including PRELUDE .
  sort PKUser .
  subsort PKUser < Principal .

  op pk : PKUser -> PKey .
  op sk : PKUser -> PKey .

  var X : Field .
  var U : PKUser.

  eq ped(sk(U), ped(pk(U), X)) = X .
  eq ped(pk(U), ped(sk(U), X)) = X .
  eq invertible(ped(pk(U),X),X,terms(sk(U)))
     = true .
  eq invertible(ped(sk(U),X),X,terms(pk(U)))
     = true .
```

Here, we are assuming that `PRELUDE` is a functional module in which predefined cryptographic types (e.g., `Field`, `Principal`, `PKey`), operators (e.g., `ped`) and predicates (`invertible`) are declared.

In Section 3.4 we presented the CIL rules for the OneMessage protocol. The translation into Maude rewrite rules is very straightforward. For instance, given Maude declarations for state, msg, and fact predicates, the sender CIL rule becomes the Maude rule

```
rl [send] :
facts(state(roleA,0,terms(A,B)), FACTS)
=>
facts(
  state(roleA,1,terms(A,B,mkKey(FACTS))),
  msg(A,B,
      terms(ped(pk(B),cat(A,mkKey(FACTS))))),
  FACTS) .
```

`FACTS` is a variable of type multiset of facts which for a given system configuration is matched with all the remaining facts. `mkKey: FactMultiSet -> PKey` creates a fresh key from a given set of facts. Axiomatically `mkKey` is defined in such a way that the freshness property is guaranteed. In our current implementation we use natural numbers for key creation and we make sure that at any time while executing a specific protocol run (i.e., an ordered set of sessions), no two values of fresh variables are the same.

The Attacker Model in Maude

The following Maude rules describe an attacker that can overhear messages and create fake messages. We assume that principals that are used in scenarios are made available to the attacker during initialization.

```
rl [overhear] :
   facts(msg(A, B,  TERMS), FACTS)
   => facts(msg(A, B,  TERMS),
            net(TERMS), FACTS) .

rl [fake] :
    facts(net(A), net(TERMS),
          FACTS)
    => facts(net(A), net(TERMS),
             msg(Spy, A, TERMS),
             FACTS) .
```

This attacker has the ability to analyze messages via decomposition and decryption. Decomposition is done in one step with a computation like Paulson's *analz* operator [Pau98].

```
rl [decompose] :
   facts(FACTS)
   => facts(closure(unCons(nets(FACTS)),
                    deCrypt(nets(FACTS))),
            noNets(FACTS)) .
```

`nets` and `noNets` are projection functions which split a given multiset of facts into net-facts and other facts. `unCons` and `deCrypt` are one-step decomposition and decryption operators, respectively. That means, given a multiset of facts, all lists are taken apart into their sublists as well as all encrypted fields are decrypted if a corresponding key is part of the attacker knowledge. The `closure` function recursively applies both these functions until no more new fields are gained.

Whereas decomposition of a given set of attacker facts is a terminating task, composition and encryption are not. The attacker may apply arbitrarily many layers of concatenation and encryption in order to construct a message which might result in a successful attack. This makes the search space for a given run of the protocol infinite, even though the number of honest agents and attackers may be finite. Thinking in terms of exhaustive search strategies, the resulting search tree is an infinite data structure. The root of this tree corresponds to the given initial protocol configuration (as given in the environment specification) and each tree node represents one possible extension of the parent node where a transition has taken place. Since we want to make a state-space search feasible, we chose to build a finitely branching search tree. The number of successors of each node is finite, but the tree is arbitrarily deep. For this reason, we chose to implement the attacker in such a way that only one encryption or one composition is allowed per step. Nested encryption and composition is achieved by applying a combination of the composition rules in sequence.

```
rl [composeCons] :
   facts(FACTS, net(FIELD), net(TERMS))
   => facts(FACTS, net(FIELD), net(TERMS),
                   net(FIELD,TERMS)) .

rl [composeEncrypt] :
   facts(FACTS, net(PKEY), net(TERMS))
   => facts(FACTS, net(PKEY), net(TERMS),
                   net(ped(PKEY,TERMS))) .
```

Defining an attack

In order to successfully detect protocol vulnerabilities it is necessary to not only build an abstract model of the protocol and the malicious environment which is amenable to formal analysis techniques, but also to give a formal definition of what constitutes an attack. In other words, when has a protocol security goal failed?

For instance, the secrecy goal in OneMessage can be expressed by introducing a `secret` predicate which in its first parameter states the variable which should be kept secret, and in its second predicate lists the protocol variables which share this secret. An attack situation is characterized by a situation where the secret is known to the attacker.

```
ceq facts(secret(K,terms(A,B)),net(K),FACTS)
    = attack
    if not(spyOf(A) == true)
       and not(spyOf(B) == true) .
```

The appropriate `secret` predicate has to be created along with other facts. As soon as a secret is created or is received by an agent, the secret predicate is created. This can be done by extending the original CIL rules with further predicates. For instance, at the same time that the agent in role "A" in OneMessage creates the secret key in order to send it to the agent in role "B" a corresponding secrecy predicate is created.

```
rule(
  facts(state(roleA,0,terms(A,B))),
  ids(K),
  facts(state(roleA,1,terms(A,B,K)),
        msg(A,B,ped(pk(B),cat(A,K))),
        secret(K,terms(A,B))))
```

Model Checking Via Exhaustive Search Strategies

Our systematic search for an attack using a breadth-first strategy illustrates a form of model checking in which we explore all the possible behaviors beginning with a given state using the reflective capabilities of Maude. Thanks to the reflective nature of rewriting logic [CM96], it is possible to express strategies for executing Maude specifications in Maude itself. This is particularly important because Maude specifications formalize concurrent systems that can be highly concurrent and nondeterministic, and that may not terminate. Therefore, in addition to a default execution strategy that ensures a fair application of the rewrite rules, it is possible to define a very wide variety of rewriting strategies using rewrite rules at the metalevel, and to then execute a specification with a given strategy in Maude. For example, one can program a breadth-first-search strategy to exhaustively determine all possible behaviors from a given initial state using the rewrite rules of a given protocol specification. In this way we can either validate the specification and detect errors, or do formal reasoning about the specification.

To efficiently implement a breadth-first-search strategy, we have chosen to use a bounded depth-first-search strategy which is recursively called with increasing bounds as long as no attack is discovered. The search strategy is defined at the meta-level as a module which includes the protocol as a parameter.

```
fmod SEARCH is
  protecting META-LEVEL[OneMessage] .

  op bdfs : Module Term MachineInt
            LabelList -> Strategy .

  eq bdfs(MODULE,TERM,N,LIST) = ...
```

The bounded depth-first-search operation is equationally specified. The operation is called with the name of the protocol, an initial configuration (several of which are defined in the environment specification), the maximal search depth and a list of rule labels (i.e., all the names of rewrite rules over which the search strategy is defined). The bounded depth-search-first algorithm continues building the path, provided the maximum depth is not reached. For each new step in the path all rewrite rules are tried. Whenever a path is closed, i.e., whenever no other rewrite rules are applicable, or the maximum depth is reached, the algorithm backtracks. Backtracking means that first the same rule as in the previous step is tested with the next possible matching. If this fails, then the next rule in the list of applicable rules is tried. If this also fails, another backtracking step is taken until no more rules can be applied from the initial configuration. If any of the previous two attempts is successful, the bounded depth-first-search algorithm for the new, extended path is called.

After each step in the search it is tested whether an attack situation occurred. In the case where the violation of a goal is defined via an equation, as for instance the secrecy goal of OneMessage, this test is an equality, since the

Maude rewrite engine automatically applies all equations after a rewrite step in order to derive a normal form.

Enhancement of Search Strategies

The search strategies we implemented are designed for reusability. They can be applied to arbitrary protocol specifications. So can the attacker model. The drawback of this generality is poor performance. Model checking through exhaustive state search easily reaches its limits even for small protocols and very limited numbers of participating processes. Therefore, it is highly desirable to enhance the search techniques. We have been experimenting with several possibilities for more efficient, but still complete, search algorithms. Our enhancement efforts can be divided into two main approaches: reordering and pruning of the search tree.

We reordered the search tree to give preference to actions which are more likely to result in an attack. This has no effect on the execution time unless the protocol has flaws, in which case it terminates more quickly. We gave priority to those rules which were generated by the protocol message list. This is due to the fact that their application is very restricted. Attacker rules are more generally applicable, thus they have lower priority. Among those rules, we prefer to apply fake messages over overhearing, and we assigned the lowest priority to composition of messages.

Pruning the search tree cuts down the number of states to be checked for a possible attack. One way to prune the search tree is to test faked messages for their format and receivability. The idea is that only messages which correspond to a valid protocol message are meaningful. Since we are working in a strongly-typed system, we only allow the attacker to fake a message that has the type of one of the messages in the message list. Additionally, a faked message should have an effect on the participating processes. That means that at least one process should be in a state where the faked message could be received, provoking this process to perform the specified transition. We call this kind of pruning "Send-Receive-Dependency".

6. Conclusion

CAPSL, CIL and the translation between them are designed to address important goals in cryptographic protocol specification for analysis purposes. With a common specification language, it becomes possible to harness the combined power of many tools for protocol analysis in a practical way. The components of the CAPSL environment include transportable software for translation of CAPSL to CIL, and connectors to adapt CIL to the input languages of various analysis tools. This software is still under development, but a preliminary CAPSL-to-CIL translator is already available on the Web site [Mil99a]. The site also contains more extensive documentation on CAPSL, including examples of CAPSL specifications, the full grammar, and also abstract syntax specifications for CAPSL and CIL.

With CAPSL, one can express protocols in the simplest accepted message-list form. Type specifications in CAPSL and their use for introducing new operators and subtypes bring an expanding class of protocols within reach. There are plans to broaden the applicability of CAPSL further with extensions for multicast protocols.

CAPSL simplifies what used to be the most awkward aspect of abstract protocol specification, the distinction between short-term session data and the long-term data associated with persistent entities. It was done by applying the general type specification mechanism, together with the novel concepts of private functions and invertibility axioms.

The intermediate language CIL was chosen with an eye toward a clear analysis-level modeling semantics and a universal pattern-matching transition rule style that lends itself both to model checking and inductive proof techniques.

We are developing techniques for inductive protocol proof using PVS and model checking using Maude. In the process, we have confirmed that CIL output is a good match for the specification needs of these tools, and we are in a position to finish up the connectors when those techniques and the CIL translator have stabilized.

References

[Bel89] S. Bellovin. Security problems in the TCP/IP protocol suite. *Computer Communications Review*, 19(2):32–48, April 1989.

[BMM99] S Brackin, C. Meadows, and J. Millen. CAPSL interface for the NRL protocol analyzer. In *IEEE Symposium on Application-Specific Systems and Software Engineering Technology (ASSET '99)*, 1999.

[Bra97] S. Brackin. An interface specification language for automatically analyzing cryptographic protocols. In *Symposium on Network and Distributed System Security*. Internet Society, February 1997.

[CDL+99] I. Cervesato, N. Durgin, P. Lincoln, J. Mitchell, and A. Scedrov. A meta-notation for protocol analysis. In *12th IEEE Computer Security Foundations Workshop*, pages 55–69. IEEE Computer Society, 1999.

[CELM96] M. Clavel, S. Eker, P. Lincoln, and J. Meseguer. Principles of Maude. In Meseguer [Mes96a], pages 65–89.

[CM96] M. Clavel and J. Meseguer. Reflection and Strategies in Rewriting Logic. In Meseguer [Mes96a], pages 125–147.

[DM99a] G. Denker and J. Millen. CAPSL and CIL language design. Technical Report SRI-CSL-99-02, SRI International Computer Science Laboratory, 1999.

[DM99b] G. Denker and J. Millen. CAPSL intermediate language. In *FLoC Workshop on Formal Methods and Security Protocols*, 1999.

[DMT98] G. Denker, J. Meseguer, and C. Talcott. Protocol specification and analysis in Maude. In *Formal Methods and Security Protocols*, 1998. LICS '98 Workshop.

[DS81] D. Denning and G. Sacco. Timestamps in key distribution protocols. *Communications of the ACM*, 24(8), August 1981.

[DY83] D. Dolev and A. Yao. On the security of public key protocols. *IEEE Transactions on Information Theory*, IT-29:198–208, 1983. Also STAN-CS-81-854, May 1981, Stanford U.

[GNY90] L. Gong, R. Needham, and R. Yahalom. Reasoning about belief in cryptographic protocols. In *IEEE Symposium on Research in Security and Privacy*, pages 234–248. IEEE Computer Society, 1990.

[Low96] G. Lowe. Breaking and fixing the Needham-Schroeder public-key protocol using FDR. In *Second Intern. Workshop TACAS'96, Passau, Germany, March 1996*, pages 147–166. Springer, 1996. LNCS 1055.

[Low98] G. Lowe. Casper: A compiler for the analysis of security protocols. *Journal of Computer Security*, 6(1):53–84, 1998.

[Mea92] C. Meadows. Aplying formal methods to the analysis of a key management protocol. *Journal of Computer Security*, 1(1):5–36, 1992.

[Mes92] J. Meseguer. Conditional Rewriting Logic as a Unified Model of Concurrency. *Theoretical Computer Science*, 96(1):73–155, 1992.

[Mes96a] J. Meseguer, editor. *Rewriting Logic and Its Applications, First International Workshop, Asilomar Conference Center, Pacific Grove, CA, September 3-6, 1996*. Elsevier Science B.V., Electronic Notes in Theoretical Computer Science, Volume 4, `http://www.elsevier.nl/locate/entcs/volume4.html`, 1996.

[Mes96b] J. Meseguer. Rewriting Logic as a Semantic Framework for Concurrency: A Progress Report. In U. Montanari and V. Sassone, editors, *Proc. 7th Intern. Conf. on Concurrency Theory: CONCUR'96, Pisa, August 1996*, pages 331–372, 1996. LNCS 1119.

[Mil97] J. Millen. CAPSL: Common Authentication Protocol Specification Language. Technical Report MP 97B48, The MITRE Corporation, 1997.

[Mil99a] J. Millen. CAPSL Web Site. `http://www.csl.sri.com/~millen/capsl`, 1999.

[Mil99b] J. Millen. A necessarily parallel attack. In *FLoC Workshop on Formal Methods and Security Protocols*, 1999.

[NS78] R. Needham and M. Schroeder. Using encryption for authentication in large networks of computers. *Communications of the ACM*, 21(12):993–998, December 1978.

[ORS92] S. Owre, J. M. Rushby, and N. Shankar. PVS: A prototype verification system. In Deepak Kapur, editor, *11th International Conference on Automated Deduction (CADE)*, volume 607 of *Lecture Notes in Artificial Intelligence*, pages 748–752. Springer-Verlag, 1992.

[Pau98] L Paulson. The inductive approach to verifying cryptographic protocols. *Journal of Computer Security*, 6(1):85–128, 1998.

[THG98] J. Thayer, J. Herzog, and J. Guttman. Honest ideals on strand spaces. In *11th IEEE Computer Security Foundations Workshop*, pages 66–78. IEEE Computer Society, 1998.

Automatically Detecting Most Vulnerabilities in Cryptographic Protocols

Stephen H. Brackin *
Arca Systems / Exodus Communications
303 E. Yates St.
Ith :a, NY 14850

Abstract

The Automatic Authentication Protocol Analyzer, 2nd Version (AAPA2) is a fast, completely automatic tool for formally analyzing cryptographic protocols. It correctly identifies vulnerabilities or their absence in 43 of 51 protocols studied in the literature, and it finds errors in previously asserted authentication properties of two large commercial protocols. This paper describes how the AAPA2 finds vulnerabilities in 13 of 16 vulnerable protocols, protocols for which the earlier Automatic Authentication Protocol Analyzer (AAPA) finds vulnerabilities in only 6 of these protocols.

1. Introduction

Cryptographic protocols, called simply *protocols* for the remainder of this paper, are short sequences of message exchanges, usually involving encryption, intended to establish secure communication over insecure networks. Whether they do so, or can be subverted by an active wiretapper who blocks, replays, or modifies messages, is a notoriously difficult problem [1, 18]. The basic issues are *authentication* (i.e., whether participants know with whom they are communicating), and *nondisclosure* (i.e., whether participants reveal information to anyone not meant to receive it).

A *session* is one execution of a protocol. Each message reception begins a new *stage* in a session. A *principal* is a participant in a session. A *field* is either a message field or a non-pair subfield of a message field.

Distortions on a protocol's execution by an active wiretapper (or *attacker*) are called *attacks*. In the worst case, the attacker controls all communication on the network. Five of the many types of attacks that have been studied in the literature follow [13]:

- Freshness — the attacker replaces a field with the same field from an earlier session.
- Type — the attacker puts a field of one type in place of a field of another type.
- Binding — the attacker passes off one principal's public key as another principal's public key.
- Parallel Session — the attacker puts fields from one session into another session, with two or more sessions in progress simultaneously.
- Oracle — the attacker sends a message to a legitimate principal knowing that this principal's response will perform some computation useful to the attacker.

The Automatic Authentication Protocol Analyzer, 2nd Version (AAPA2) is the main product of DARPA's "Complete, Automatic Analysis of Cryptographic Protocols" project, Rome Laboratory contract F30602-97-C-0303 [9, 10]. The AAPA2 finds vulnerabilities to freshness, type, binding, parallel session, and some oracle attacks [8, 7]. The earlier Automatic Authentication Protocol Analyzer (AAPA) [3, 2, 5] finds vulnerabilities only to freshness attacks [6, 4].

For 51 of the 53 protocols in a library by Clark and Jacob of protocols studied in the literature [13], the AAPA2 correctly identifies 13 as failed and 30 as not failed [8]. It misses vulnerabilities in 3 and raises false alarms in 5 others, though 2 of the false alarms obviously arise from its inability to model algebraic properties of the operations used in these protocols [8]. Typos in [13] make it impossible to evaluate the AAPA2's performance on the remaining 2 protocols [8].

The AAPA, by contrast, correctly identifies only 6 of the Clark-Jacob protocols as failed and 33 as not

*This work was supported by the Advanced Research Projects Agency through Rome Laboratory contract F30602-97-C-0303.

0-7695-0490-6/99 $10.00 © 1999 IEEE

failed [6, 8]. It misses vulnerabilities in 10 and raises false alarms in 2 others, though 1 of the false alarms obviously arises from the AAPA's inability to model algebraic properties of the operations used in this protocol [6, 8]. (One of the AAPA "failure identifications" in [6] is actually a false alarm [8].)

The AAPA2 also finds authentication limitations [7] missed by the AAPA [4] in two large commercial protocols from CyberCash, Inc.

The AAPA2 produces its results for the Clark-Jacob library, on a 128-meg Ultra 1, in an average of only 2.6 minutes per protocol [8]. The AAPA is three to six times faster than the AAPA2, but this difference is almost entirely due to how the two are implemented. The AAPA2 uses HOL98 [28], which uses Moscow ML, which is interpreted; the AAPA uses HOL90 [27], which uses Standard ML of New Jersey, which is complied. The AAPA2 uses HOL98 capabilities to find multiple vulnerabilities in single protocols, something that the AAPA cannot do. This new functionality is more important than the loss of speed.

All these AAPA2 and AAPA results are conditional on the assumption that the protocols are implemented so that principals perform the reasonable type, equality, and inequality tests, described in Section 3, that the AAPA2 assumes these principals perform.

Both the AAPA2 and AAPA automatically perform formal analyses based on logics extending the Gong, Needham, Yahalom (GNY) [16] authentication logic, which itself extends the original Burrows, Abadi, Needham (BAN)[12] authentication logic.

The AAPA2's Brackin, Gong, Needham, Yahalom, 2nd Version (BGNY2) logic [9], though, represents a major step toward removing the over-simplifications that have limited the effectiveness of the authentication logics. Unlike the AAPA's Brackin, Gong, Needham, Yahalom (BGNY) logic [3] and all earlier authentication logics, BGNY2 does *not* assume that principals can always correctly identify and interpret the fields they receive. Instead, BGNY2 assumes that principals perform a collection of reasonable type, equality, and inequality checks on the fields they receive, then requires that these checks identify these fields. BGNY2 also imposes *explicitness* requirements to guarantee that the interpretations of these fields are well defined. These additional requirements give the AAPA2 its greater effectiveness.

This paper describes the major differences between the AAPA2 and the AAPA. It also gives detailed descriptions of the 10 cases among the 53 protocols in the Clark-Jacob library [8] for which the results produced by the AAPA2 differ significantly from the results produced by the AAPA.

The remainder of this paper is organized as follows. Section 2 compares and contrasts basic AAPA2 and AAPA operation. Section 3 describes the main sources of the AAPA2's greater effectiveness, its message identification and explicitness requirements. Section 4 gives an example, an ISL2 specification and AAPA2 analysis of the Tatebayashi, Matsuzakai, Newman protocol [29]. Section 5 introduces conventions and notations that Section 6 uses to give very succinct descriptions of AAPA2 protocol analyses. Section 6, the heart of the paper, describes the results produced by the AAPA2 for protocols in the Clark-Jacob library for which these results differ significantly from analogous results produced by the AAPA. Finally, Section 7 gives suggestions for future work.

2. AAPA2 vs. AAPA Operation

This section describes, compares, and contrasts AAPA2 operation and AAPA operation.

Both the AAPA2 and AAPA produce their results by starting with specifications of protocols and their desired properties in simple specification languages. They both translate these specifications into Higher Order Logic (HOL) [17], automatically construct proofs of the protocols' desired properties, and translate their proved and unproved results back into their simple specification languages. They both examine only correct protocol executions, but they both ask whether the *possibility* of attacker interference makes it impossible for legitimate principals to reach their desired authentication conclusions.

Both the AAPA2 and AAPA construct their proofs using *authentication logics* derived from the GNY logic [16]. Authentication logics are collections of rules formalizing authentication inferences that an analyst could validly make on the basis of the information that a principal has received. If the received information does not justify a desired conclusion, then this typically points to a protocol vulnerability that the analyst can exploit to construct an attack on the protocol leading to an authentication violation. The remainder of this paper will follow the usual convention and describe the inferences that an analyst could draw from the data received by a principal as being made by the principal itself rather than by an analyst examining this data.

Authentication logics avoid the combinatorial explosion faced by tools constructing attacks on protocols [19, 23, 21]. They also avoid the extensive expert effort needed to construct inductive proofs that protocols have desired properties [24, 26, 25]. (Cohen is developing resolution-based tools for significantly reducing the amount of effort needed to produce such proofs [14].)

Authentication logics currently miss many failures, though, because they over-simplify the problem. They currently make all their authentication deductions, for instance, by assuming that there have been no nondisclosure failures, even in response to attacks.

While the AAPA2's BGNY2 logic [9] still assumes that there have been no nondisclosure failures, it does *not* assume, as the AAPA's BGNY logic [3] does, that protocol participants can always correctly identify and interpret the fields that they receive. Instead, BGNY2 assumes that legitimate principals perform a computationally feasible set of type, equality, and inequality tests on the fields that they receive, and it requires that these tests uniquely identify every field that is intended to convey a statement. BGNY2 also requires that the statement conveyed by a field is explicit in the field itself, not dependent on this field's context. Section 3 gives a detailed description of BGNY2's message identification and explicitness requirements.

BGNY2's message identification requirements affect the AAPA2's Interface Specification Language, 2nd Version (ISL2) [11]. Because of these requirements, ISL2 differs from the AAPA's Interface Specification Language (ISL) [5] in the following two main ways:

ISL2 allows the user to name different types of plaintext — e.g., name, timestamp, or nonce — and allows the user to assign one of these types to each piece of plaintext. BGNY2 takes this as a user assumption that the protocol will be implemented in such a way that pieces of plaintext of different types will never be mistaken for each other.

ISL2 requires that all keys and passwords be given by functions. (Tables are functions.) This avoids the error, made in the AAPA, of implicitly assuming that a principal that holds a key always knows the role this key plays in a correct session. This correction allows a server, for instance, to hold a key that it shares with the principal who initiates a session without having this server know *which* of the keys that it holds is the one that it shares with the initiator of the session.

Except for these main differences, ISL2 differs from ISL only in providing more conveniences. ISL2 provides a means for letting the AAPA2 label its outputs, and it allows more abbreviations.

An ISL2 specification has six parts:

- The `NAME` section names the protocol. The AAPA2 uses this name to label its terminal output and its output files.
- The `DEFINITIONS` section names and gives the assumed properties of the functions and pieces of plaintext used in the protocol. The `DEFINITIONS` section allows the user to create plaintext type names and assign these types to the different pieces of plaintext.
- The optional `ABBREVIATIONS` section introduces abbreviations for arbitrary ISL2 terms denoting data values.
- The `INITIALCONDITIONS` section is a collection of ISL2 statements defining the protocol's assumed initial conditions.
- The `PROTOCOL` section gives the sequence of message exchanges during a correct session and the ISL2 statements that these messages are intended to convey. The `PROTOCOL` section corresponds to what informal or incomplete descriptions of protocols call "the protocol". (Section 5 introduces notation conventions and assumptions that essentially allow Section 6 to replace ISL2 specifications with these specifications' `PROTOCOL` sections.)
- The `GOALS` section defines the authentication conditions expected to become true at different stages in a correct session.

Informal descriptions of the meanings of the various ISL2 statement constructors follow. These constructors' formal meanings are given via the function `BGNY2` defined in the AAPA2's underlying HOL theory [9]. For brevity, these descriptions say "principal" rather than "principal named by a plaintext field".

- `Believes`: The principal has adequate reason to believe the statement.
- `Conveyed`: The principal was the creator and source, during the current session, of the field.
- `Fresh`: The field was created for the current session.
- `Possesses`: The principal has received the field or is capable of computing this field from fields that it has received.
- `PrivateKey`: The key, for the algorithm, is one of the principal's private keys.
- `PublicKey`: The key, for the algorithm, is one of the principal's public keys.
- `Received`: The principal received the field before the current session, or received this field as, or as part of, some message sent earlier in the current session.

- `SharedSecret`: The field is unguessable, and if the two principals possess it, or come to possess it through secure means, then they are or will be the only ones, other than principals they both trust, who possess it.

- `Trustworthy`: If the principal was the source of a field with an associated statement, then this is adequate reason for believing this statement.

In an ISL2 specification's `PROTOCOL` section, `->` means "sends". Every value sent is implicitly labeled with the names of its purported sender and its intended receiver. In the values sent, `{x}f(y)` denotes the encryption of `x` with the encryption function `f` and the key `y`. The `||` operator binds a statement to a field; BGNY2 assumes that the sender believes this statement and sends the field for the purpose of communicating this statement.

In an ISL2 specification's `GOALS` section, ISL2 statements associated with stage numbers give the authentication conditions, or *user goals*, expected to become true at each stage. Not every stage need have associated user goals.

In proving user goals, both the AAPA2 and AAPA automatically construct and prove *default goals* that capture the basic authentication properties of the protocol that are likely to be of interest, then try to prove the user goals as easy consequences of the proved default goals. The default goals for a stage that begins with the reception of a message include that the message's sender was able to send it, that its receiver can decrypt any encrypted parts of it, that its receiver can determine which principal conveyed each encrypted or hashed field, and that its receiver can believe any statements that the protocol binds to the sending of these parts [10, 2].

Default goals are often false, since neither the AAPA2 nor AAPA can tell what keys a principal is expected to have in its possession at a particular stage. If a default goal fails, both the AAPA2 and AAPA retry it at later stages [10, 2].

Both the AAPA2 and AAPA signal *user-goal failures* if they are unable to prove a user goal. The AAPA2 also signals three additional types of failures if the BGNY2 logic's message identification and explicitness requirements fail. If a principal is unable to uniquely identify a field intended to convey a statement, then the AAPA2 signals an *identification failure.* If the specification binds multiple statements to the same field, then the AAPA2 signals an *ambiguity failure.* If the pieces of plaintext needed to convey the statement bound to a field are not contained in this field, then the AAPA2 signals an *explicitness failure.*

Unlike the AAPA, the AAPA2 gives a *warning* if it finds that a specification has principals do something to check for freshness — e.g., check a timestamp — other than look for fields that they created earlier in the session themselves. Since they can involve messages from different sessions, the AAPA2's identification failures are particularly significant when they occur along with the AAPA2's warnings.

Both the AAPA2 and AAPA produce terminal outputs describing their progress in analyzing a specification, and both produce output files describing the theorems that they proved and the default goals that they were unable to prove. Unlike the AAPA, though, the AAPA2 makes properly labeled *false* assumptions as necessary to allow it to complete its analysis of any specification [10]. This capability allows the AAPA2 to find multiple vulnerabilities in a single protocol — see the example in Section 4 — and is extremely useful for debugging specifications of large protocols [7].

3. Identification and Explicitness

This section describes the BGNY2 logic's identification and explicitness requirements for messages.

The identifications that principals can validly make on the basis of the tests that they perform depend on exactly what these tests are. Neither ISL2 nor BGNY2, though, model protocols in that detail. Instead, BGNY2 assumes that principals perform *all* the type and equality tests that they might reasonably be expected to perform, and some inequality tests.

The identifications that principals can validly make also depend on exactly what pieces of data are available to the attacker, since this determines which possible deceptions principals must avoid. Although it does cover a reasonably large subset of these possibilities, BGNY2 is currently inadequate on this score, which causes the AAPA2 to miss vulnerabilities to oracle attacks for 3 of the 51 Clark-Jacob protocols discussed in Section 1. Section 7 describes a feasible extension to BGNY2 that should correct this problem.

The following three subsections describe the type, equality, and inequality tests that BGNY2 assumes principals perform on all fields they receive that are intended to convey statements. The fourth subsection describes what BGNY2 assumes is the set of all fields that might be used to convey a statement available to the attacker. BGNY2 requires that the tests a principal performs on any field intended to convey a statement uniquely identify this field from among the members of this set. The fifth subsection describes BGNY2's explicitness requirements.

3.1. Type Tests

ISL2 allows the user to name different functions and pieces of plaintext, to define different types of plaintext, and to assign types to all function outputs and to all pieces of plaintext [11]. BGNY2 assumes that the tests principals perform always distinguish objects of different types — BGNY2 takes this as the functional definition of "type". The tests principals perform distinguishing objects of different types need not be explicit. If trying to interpret a network address as a time stamp would cause the software implementing a protocol to produce a core dump, for example, then the user can safely specify network addresses and time stamps as being of different types.

BGNY2 extends these ISL2 type assignments to type assignments for all fields exchanged in a session by making the following assumptions:

- Encrypted and hashed fields are labeled with the names of the functions that produce them, so fields produced by different encryption and/or hash functions are always of different types.
- Tuples of different lengths are of different types.
- All plaintext fields are of different types from all encrypted and hashed fields.
- Algorithms sent as code are of a different type from all other forms of plaintext.
- Passwords, symmetric keys, and public or private keys are of types different from each other and from all user-specified types of plaintext.

These assumptions are reasonable. In implemented protocols, encrypted or hashed information is often labeled with the name and version number of the software used to produce it, so that the recipient can know which software to use for decrypting or recomputing this information. The attacker could falsify the labels on encrypted or hashed values, but not expect legitimate principals to willingly encrypt and send out information having a label other than the expected one. Software implementing a protocol will also malfunction if it does not check that the information it is operating on is of roughly the expected form.

3.2. Equality Tests

BGNY2, like the defaults for the Common Authentication Protocol Specification Language (CAPSL) [22], assumes that principals perform *all* the equality tests that they are capable of performing. If a principal holds a field and receives something that it expects to contain this field, possibly encrypted in a form that this principal can decrypt, then BGNY2 assumes that this principal will test that the field it receives contains the field that it holds.

If a principal holds a field, receives a field that it expects to contain the field that it holds, and finds that the field it receives indeed does contain the field that it holds, though, then this by itself proves nothing. If the field that the principal holds is something that this principal received as plaintext, for example, then the attacker might have replaced what this field should have been with what it is now in order to make the equality test that the principal just performed succeed when it should have failed. (Section 4 gives an example.) Only if the principal can be confident that the field it holds is playing the role that this field is supposed to play does the equality test give evidence that the message field this principal receives is what this message field is supposed to be.

BGNY2 addresses this problem using its `RightRole` construct, which asserts that a field is playing its intended role in the session. A field is playing its intended role if it was placed in that role by a legitimate principal not acting under the influence of an attack.

The BGNY2 rules formally defining `RightRole` [9] involve many HOL concepts extraneous to this paper. These concepts include inductively defined relations and the "deep embedding" of an authentication logic into HOL [3]. These rules' English interpretations, though, which follow, are straightforward:

- A principal can be confident that the fields it initially holds are playing their intended roles.
- If a principal can be confident that a field it receives is from another principal in the current session and is of an expected form, then this principal can be confident that this field is playing its intended role.
- If a principal can be confident that a computed (e.g., encrypted) field is playing its intended role, and is capable of inverting the computation (e.g., decrypting), then this principal can be confident that the result of this inversion (e.g., the plaintext) is playing its intended role.
- If a principal can be confident that a hash code is playing its intended role and is capable of recomputing this hash code, then this principal can be confident that the field hashed is playing its intended role.
- A principal can be confident that a pair of elements is playing its intended role if and only if

this principal can be confident that both elements of this pair are playing their intended roles.

BGNY2 interprets the results of a principal's equality tests by using the `RightRole` statements that this principal believes — i.e., the set of fields held by this principal that this principal can be confident are playing their intended roles. BGNY2 only allows a principal to infer the identify of a field from this principal's equality tests if the subset of these tests consisting of tests against values in the "playing their right roles" set uniquely identify the field.

3.3. Inequality Tests

BGNY2 assumes that principals perform a very limited amount of *inequality* testing. It assumes that if principals receive fields containing what they expect to be other principals' names, and they are capable of extracting these names, then they will object if these names are actually their own names. This is easy to implement, analogous to network software that detects and short-circuits messages sent from a site to itself.

3.4. Fields Available to the Attacker

In considering the possible fields that could be confused with an expected field, BGNY2 only considers fields sent during a correct session. This excludes attacker-constructed fields and fields constructed by legitimate principals in response to oracle attacks. Modeling that level of attacker creativity requires further complicating the logic — see Section 7. (Vulnerability to oracle attacks does often cause explicitness failures, though [8, 7].)

BGNY2 requires that fields be identifiably distinct from fields sent *later* in the session. This models that the attacker can substitute fields from later points in earlier or parallel sessions if the tests that principals perform to determine freshness are inadequate, which they are in 29 of the 53 protocols in the Clark-Jacob library — see Section 6.

For practical purposes, BGNY2's model of the fields available to an attacker is also not so inaccurate as it might seem. The issue of field identification only arises for fields that are intended to convey statements, and only fields whose sources can be reliably determined can reliably convey statements. BGNY2 restricts believing the statement conveyed by a field to fields whose sources can be reliably determined. Fields whose sources can be reliably determined always involve secrets assumed unavailable to the attacker, so the attacker can only obtain these fields from messages sent by legitimate principals. Feasible type tests also prevent what would otherwise be most oracle attacks, so the attacker can often only obtain fields that are sent during the normal execution of the protocol.

3.5. Explicitness Requirements

BGNY2's explicitness requirements protect against attacks that involve tricking the principals who receive fields conveying statements into interpreting these fields in ways that the principals sending these fields did not intend. Such attacks can involve modifying neighboring plaintext fields (e.g., "from" and "to" labels) or redirecting fields to principals not intended to receive them. Binding attacks work in this way.

BGNY2 protects against such attacks by requiring that a principal only believes the statement bound to a field if *every* piece of plaintext in this statement is explicitly contained in the field itself. This prevents misinterpretations arising from attacker-induced distortions in the field's context.

To avoid what would otherwise be a large number of extraneous explicitness failures, though, BGNY2 treats any field encrypted with a principal's private key as containing the name of this principal, and it treats any field encrypted with a symmetric key that is a secret shared by two principals as containing the names of these principals. Since being a private key or a shared secret depends on the current state of the session, this introduces some dependence on context, but not any *new* dependence on context — i.e., any dependence on context not already present in the authentication logics' assumptions about shared secrets.

For the 51 Clark-Jacob protocols discussed in Section 1, BGNY2's explicitness requirements point up potential vulnerabilities in 8 protocols, and for 6 of these protocols there are attacks that successfully exploit these vulnerabilities — see Section 6.

4. Example

This section gives an example of a failed protocol and the AAPA2's analysis of it. This example illustrates ISL2 and AAPA2 operation.

An ISL2 specification and AAPA2 analysis of the Tatebayashi, Matsuzakai, Newman protocol [29] follow. The non-trivial sections of the protocol's ISL2 specification have the following interpretations:

The `DEFINITIONS` section asserts: `A`, `B`, and `S` are pieces of plaintext of a user-defined type `Names` naming the session's principals. `Se` and `Pe` are encryption/decryption functions, `^Kp` is a table of private keys, `Kp` is a table of public keys, and `Ksa` and `Ksb` are tables of symmetric keys. Se is symmetric-key encryption/decryption and every key is its own inverse. Pe is public-key encryption/decryption and encryption with S's public key is inverted by decryption with S's private key.

The INITIALCONDITIONS section asserts:

- A, B, and S can perform all the symmetric- and public-key encryption and decryption operations used, and A and B can generate, or have tables of, symmetric keys.
- A has B's name and S's name — here "name" can be synonymous with "network address" — and knows how it intends to use them in the session. (BGNY2 assumes that every principal knows, and knows the intended uses of, its own name.)
- A and B have S's public key, KpS, and S has its own private key, ^KpS.
- A and B believe that KpS is S's public key, S believes that ^KpS is S's private key, A believes that the symmetric key KsA is unguessable and was not used before the current session, and B believes that the symmetric key KsB is unguessable and was not used before the current session.
- A and B trust each other and S, and S trusts B.

The PROTOCOL section asserts:

- In stage 1, A begins a correct session by sending S the name B of the principal that it wants to communicate with and a symmetric key KsA, encrypted with S's public key KpS. S is to use KsA for returning the symmetric key that S will obtain from B.
- In stage 2, S tells B that A wants to communicate with it.
- In stage 3, B sends S the symmetric key KsB, encrypted with S's public key KpS, for A. In sending this message, B asserts that KsB is suitable for use as a shared secret by A and B.
- In stage 4, S completes the correct session by sending A the key KsB that S obtained from B, encrypted using symmetric-key encryption with the key KsA that S obtained from A. In sending this message, S asserts that KsB is suitable for use as a shared secret by A and B.

The GOALS section asserts: At the end of stage 1, S has received KsA. At the end of stage 3, S has received KsB and believes that it is a suitable shared secret for A and B. At the end of stage 4, A has received KsB and believes that it is a suitable shared secret for A and B.

The protocol's ISL2 specification follows:

```
NAME: "Tatebayashi-Matsuzakai-Newman";

DEFINITIONS:
Names: A,B,S;
ENCRYPT FUNCTIONS: Se,Pe;
PRIVKEY FUNCTIONS: ^Kp;
PUBKEY FUNCTIONS: Kp;
SYMKEY FUNCTIONS: Ksa,Ksb;
Se WITH ANYKEY HASINVERSE Se WITH ANYKEY;
Pe WITH Kp(S) HASINVERSE Pe WITH ^Kp(S);

ABBREVIATIONS:
^KpS = ^Kp(S); KpS = Kp(S);
KsA = Ksa(B); KsB = Ksb(A);

INITIALCONDITIONS:
A Received Se,Pe,Ksa,B,S,KpS;
A Believes
  (Fresh KsA;
   PublicKey S Pe KpS;
   SharedSecret A S KsA;
   Trustworthy B;
   Trustworthy S);
B Received Se,Pe,Ksb,KpS;
B Believes
  (Fresh KsB;
   PublicKey S Pe KpS;
   SharedSecret A B KsB;
   Trustworthy A;
   Trustworthy S);
S Received Se,Pe,^KpS;
S Believes
  (PrivateKey S Pe ^KpS;
   Trustworthy B);

PROTOCOL:
1. A->S: B,{KsA}Pe(KpS);
2. S->B: A;
3. B->S: {KsB}Pe(KpS)
         ||(SharedSecret A B KsB);
4. S->A: {KsB}Se(KsA)
         ||(SharedSecret A B KsB);

GOALS:
1. S Possesses KsA;
3. S Possesses KsB;
   S Believes SharedSecret A B KsB;
4. A Possesses KsB;
   A Believes SharedSecret A B KsB;
```

Despite its simplicity, Lowe and Roscoe have found 10 different attacks on this protocol [20]! In stage 3, for example, S has no evidence that the key it receives is from B, so the attacker can block the message that S sends B in stage 2 and in stage 3 send S a symmetric key of the attacker's choosing, encrypted with the public key KpS, in a message falsely labeled as being from B, and then impersonate B to A. Further, since S has no evidence that the key it receives in stage 1 is from A, the attacker can similarly impersonate A to B, then let A and B send messages to each other with the attacker decrypting and re-encrypting each message.

For the Tatebayashi, Matsuzakai, Newman ISL2 specification above, in a file tmn.isl, the AAPA2 produces the terminal output:

```
Proving default goals, stage 1
Retrying failed default goals, stage 1
Proving user goals, stage 1
Proving default goals, stage 2
Proving user goals, stage 2
Proving default goals, stage 3
Retrying failed default goals, stage 3
Proving user goals, stage 3

User-goal failure, stage: 3!
Goal statement:
 S Believes (SharedSecret A B Ksb(A));
Making the false assumption
  AX1: S Believes (SharedSecret A B Ksb(A));
and continuing analysis of protocol

Proving default goals, stage 4

Explicitness failure, stage: 4!  Term
  {Ksb(A)}Se(Ksa(B))
does not contain the data needed to
communicate statement
  SharedSecret A B Ksb(A)
Making the false assumption
  AX2: {B,A,S} = {A,S};
and continuing analysis of protocol

Proving user goals, stage 4

Failure(s): Tatebayashi-Matsuzakai-Newman
```

The AAPA2 creates file tmn.fail containing the failed default goal:

```
/* ## Failed default goal from stage 3: ## */
S Believes (B Conveyed {Ksb(A)}Pe(Kp(S)));
/* ========== Unproved subgoals: ========= */
S Believes (SharedSecret S B Ksb(A));
S Believes (Fresh {Ksb(A)}Pe(Kp(S)));
/* =========== Proved subgoals: ========== */
S Received {Ksb(A)}Pe(Kp(S));
S Possesses Pe,UNKp(S);
S Believes (PrivateKey S Pe UNKp(S));
```

File tmn.fail also contains a similar failed default goal, from stage 1, with the conclusion

```
S Believes (A Conveyed {Ksa(B)}Pe(Kp(S))).
```

For the Tatebayashi, Matsuzakai, Newman ISL2 specification, the user goal

```
S Believes (SharedSecret A B Ksb(A));
```

fails because S cannot determine the sender of the field meant to communicate this message — i.e., because the default goal

```
S Believes (B Conveyed {Ksb(A)}Pe(Kp(S)));
```

fails. This default goal fails because S has no evidence that this field was generated by B or generated for the current session. No similar user-goal failure occurs for stage 1 only because the protocol's specification does not call for S to acquire any beliefs at this stage.

The failed user goal points to the first of the attacks mentioned earlier, and this goal and the failed default goal from stage 1 together point to the second of these attacks. The failed user goal also points to another attack not given earlier: The attacker can simply replay old messages to get A and B to use an old key that the attacker has been able to determine.

The explicitness failure at stage 4 means that even if S was the source of {Ksb(A)}Se(Ksa(B)) then S might not mean this field to say that Ksb(A) is a suitable shared secret for A and B. This points to still another attack: If A contacts the attacker in its guise as a legitimate principal Z, and if A happens to reuse the key that it uses to protect Z's key to protect B's key — i.e., if Ksa(B) equals Ksa(Z) — then the attacker can replay a copy of {Ksz(A)}Se(Ksa(Z)) at stage 4 to trick A into thinking that a key A shares with the attacker is a key that A shares only with B.

Since this protocol performs no freshness checks at all, the AAPA2 does not output warnings that its freshness checks are inadequate!

5. Abbreviating Assumptions

This section gives assumptions and notation conventions that allow Section 6 to give only partial ISL2 specifications and terse descriptions of AAPA2 outputs without leaving out crucial information. Section 6 also does not list the specified goals for protocols, but notes any goals that are surprising, and gives the goals associated with any AAPA2 user-goal failures. These assumptions and notation conventions represent severe restrictions on ISL2 specifications, but they allow very succinct descriptions.

Section 6 uses the following naming conventions:

- Principal names are single, capital letters.

- Encryption functions are named `E` if the protocol uses only a single encryption function, or `Es` and `Ep` if the protocol uses both symmetric-key and public-key encryption. None of the protocols in Section 6 involve multiple kinds of symmetric- or public-key encryption.

- Names of functions for generating symmetric keys begin with `SK`, names of functions for generating public keys begin with `PK`, and names of functions for generating private keys begin with `^PK`. (Note that the AAPA2 translates `^PK` to `UNPK`.)

- Names of nonces begin with `N`, names of timestamps begin with `T`, and everything whose name begins with `N` or `T` is a nonce or a timestamp.

- If a key-generation function, nonce, or timestamp is initially held by one principal, its name contains the lower-case letter corresponding to this principal's name right after the initial letters that identify it as a key-generation function, nonce, or timestamp. If there is more than one such object, their names end with `1` or `2` to distinguish them.

- If a key-generation function is initially held by all principals or by no principal, its name is simply `SK`, `PK`, or `^PK`.

- `Z` denotes the name of the attacker acting as a principal.

Section 6 makes the following assumptions. If a protocol involves initial-conditions assumptions that do not follow from these general assumptions and the Section 6 naming conventions, Section 6 notes them in its description of the protocol.

- Principal names are computably distinct from all other types of plaintext, but all other plaintext types are indistinguishable.

- Principals initially hold all the data that they use in the session and do not obtain during the session.

- Every public or private key is a function of a single principal name, every symmetric key is a function of two principal names, and these names give the principals who possess these public or private keys or who will use these symmetric keys to communicate. If a principal uses a symmetric key to communicate with itself, this key is a function of the same principal name given twice.

- All encryption functions are there own inverses. Every symmetric-key encryption can be undone using the same key, every public-key encryption can be undone using the corresponding private key, and every private-key encryption can be undone using the corresponding public key.

- Principals initially hold their own private keys and any public keys that they do not obtain during the session. They believe that these keys are what they are — i.e., their own private or others' public keys.

- Principals believe that the symmetric keys they hold are either secrets they share with the principals with whom they use these keys to communicate or that these keys will become such secrets for the principals to whom they send these keys.

- Principals always believe that the nonces and timestamps they initially hold are fresh.

- All legitimate principals trust each other, and no legitimate principal trusts the attacker acting as a principal.

Section 6 describes AAPA2 outputs using the following shorthand:

- "Stage n user-goal failure: *ugoal*; **dgoal*; *sgoal*" means "The AAPA2 finds a user-goal failure at stage n for the goal *ugoal*, which fails because the default goal *dgoal* fails, which fails because the subgoal *sgoal* fails"; *ugoal* and *sgoal* can be multiple bgoals, so *dgoal* is marked with an asterisk.

- "Stage n identification failure: *term*1; *term*2" means "The AAPA2 finds an identification failure at stage n for the field *term*1, which the receiver's default tests do not distinguish from the field *term*2"; here *term*2 can be more than one field.

- "Stage n explicitness failure: *term*1; *stmt*; *term*2" means "The AAPA2 finds an explicitness failure at stage n because the protocol uses *term*1 to communicate *stmt* even though *term*1 does not contain *term*2"; here *term*2 can be more than one missing piece of plaintext.

Using these assumptions, notation conventions, and shorthand expressions, the Tatebayashi, Matsuzakai, Newman protocol in Section 4, and the AAPA2's response to it, can be given simply as

```
1. A->S: B,{SKa(A,B)}Ep(PK(S));
2. S->B: A;
3. B->S: {SKb(A,B)}Ep(PK(S))
         ||(SharedSecret A B SKb(A,B));
4. S->A: {SKb(A,B)}Es(SKa(A,B))
         ||(SharedSecret A B SKb(A,B));
```

Stage-3 user-goal failure:

```
S Believes (SharedSecret A B SKb(A,B));
*S Believes (B Conveyed {Ksb(A)}Pe(Kp(S)));
S Believes (SharedSecret S B Ksb(A));
S Believes (Fresh {Ksb(A)}Pe(Kp(S))).
```

Stage-4 explicitness failure:

```
{SKb(A,B)}Es(SKa(A,B));
SharedSecret A B SKb(A,B); B.
```

6. Protocols with Changed Results

This section gives the AAPA2's results on the 10 of the 53 protocols in [13] for which the AAPA2's results [8] are significantly different from the AAPA's results [6]. The AAPA finds no problems with these protocols.

The AAPA2 signals identification failures for 2 of these 10 protocols and explicitness failures for 8 of them. One of the identification failures is a false alarm arising from the AAPA2's inability to model algebraic properties of the operations used in the protocol. The other identification failure points out a real vulnerability in the protocol. Two of the explicitness failures are apparently false alarms arising from the excess restrictiveness of BGNY2's explicitness requirements. The other six explicitness failures point out real vulnerabilities in the protocols, though one of these is so difficult to exploit that the AAPA2's explicitness failure would probably be taken to be a false alarm if an attack on the protocol were not known.

In addition to the 10 new failures, the AAPA2 signals warnings for the following 29 protocols [8]: ISO Symmetric-Key One-Pass Unilateral Authentication; ISO Symmetric-Key Two-Pass Mutual Authentication; ISO One-Pass Unilateral Authentication with CCFs; ISO Two-Pass Mutual Authentication with CCFs; Denning-Sacco; Wide-Mouthed Frog; ISO Four-Pass Authentication; Woo-Lam Authentication, version Π_f; Woo-Lam Authentication, version Π_1; Woo-Lam Authentication, version Π_2; Woo-Lam Authentication, version Π_3; Woo-Lam Authentication, version Π; Kerberos Version 5; Neuman-Stubblebine, Part 1; Neuman-Stubblebine, Part 2; ISO Public-Key One-Pass Unilateral Authentication; ISO Public-Key Two-Pass Unilateral Authentication; ISO Public-Key Two-Pass Mutual Authentication; ISO Public-Key Three-Pass Mutual Authentication; ISO Public-Key Two-Pass Parallel Mutual Authentication; Needham-Schroeder Public Key; SPLICE/AS Authentication; Hwang-Chen Modified SPLICE/AS; Denning-Sacco Key Distribution with Public Key; CCITT X.509; Davis-Swick Private-Key Certificates, Protocol 1; Davis-Swick Private-Key Certificates, Protocol 2; Davis-Swick Private-Key Certificates, Protocol 3; and Davis-Swick Private-Key Certificates, Protocol 4.

6.1. Otway-Rees

```
1. A -> B: M,{Na,M,A,B}E(SKs(A,S));
2. B -> S: M,A,
           {Na,M,A,B}E(SKs(A,S)),
           {Nb,M,A,B}E(SKs(B,S));
3. S -> B: M,{Na,SKs(A,B)}E(SKs(A,S))
             ||(SharedSecret A B SKs(A,B)),
            {Nb,SKs(A,B)}E(SKs(B,S))
             ||(SharedSecret A B SKs(A,B));
4. B -> A: M,{Na,SKs(A,B)}E(SKs(A,S));
```

Stage-3 explicitness failure:

```
{Nb,SKs(A,B)}E(SKs(B,S));
SharedSecret A B SKs(A,B); A.
```

Stage-4 explicitness failure:

```
{Na,SKs(A,B)}E(SKs(A,S));
SharedSecret A B SKs(A,B); B.
```

These failures are apparently false alarms caused by the AAPA2's inability to model implicitly communicated information. In the session, B sends out the nonce Nb encrypted so that only S can decrypt it, and sends it out as part of an encrypted field that unambiguously associates this nonce with both A and B. S does send Nb out, but only encrypted with the same key that S used to obtain Nb in the first place, so cannot accidentally reveal it to another principal. For S, then, any subsequent message that it sends to B implicitly identifies both A and B. Similar comments apply to A and the nonce Na.

The "attack" on this protocol in [13] requires having the triple M,A,B be mistaken for the key SKs(A,B), which is unrealistic and contrary to the type-checking assumptions made by the AAPA2.

6.2. Amended Needham-Schroeder

```
1. A -> B: A;
2. B -> A: {A,Nb2}E(SKs(B,S));
3. A -> S: A,B,Na,{A,Nb2}E(SKs(B,S));
4. S -> A: {Na,B,SKs(A,B),
             {SKs(A,B),Nb2,A}E(SKs(B,S))
               ||(SharedSecret A B SKs(A,B);
                  Fresh SKs(A,B))
           }E(SKs(A,S))
              ||(SharedSecret A B SKs(A,B);
                 Fresh SKs(A,B));
5. A -> B: {SKs(A,B),Nb2,A}E(SKs(B,S));
6. B -> A: {Nb1}E(SKs(A,B))
             ||(SharedSecret A B SKs(A,B));
7. A -> B: {Minus1(Nb1)}E(SKs(A,B))
             ||(SharedSecret A B SKs(A,B));
```

The protocol assumes S believes SKs(A,B) is fresh.
Stage-6 identification failure:

```
{Nb1}E(SKs(A,B));
{Minus1(Nb1)}E(SKs(A,B)).
```

This failure is an obvious false alarm arising from the AAPA2's limited modeling of the algebraic properties of the operations used. The AAPA2 cannot tell that Minus1 corresponds to subtracting 1, and it cannot automatically apply HOL theorems such as the one that subtracting 1 from a value always changes this value. If the protocol used timestamps rather than nonces, and decrementing a timestamp produced another valid timestamp, then this failure would indicate a real, though minor, vulnerability.

6.3. Wide-Mouthed Frog

```
1. A -> S: {Ta,B,SKa(A,B)}E(SKs(A,S))
             ||(SharedSecret A B SKa(A,B));
2. S -> B: {Ts,A,SKa(A,B)}E(SKs(B,S))
             ||(SharedSecret A B SKa(A,B));
```

This protocol assumes that B and S believe the timestamps Ta and Ts, respectively, are fresh even though they do not create these timestamps; the AAPA2 warns about these beliefs.
Stage-1 identification failure:

```
{Ta,B,SKa(A,B)}E(SKs(A,S));
{Ts,A,SKa(A,B)}E(SKs(B,S)).
```

This failure, particularly in conjunction with the AAPA2's freshness-check warnings, points to a real vulnerability. By alternately pretending to be A and B, and sending S in each stage 1 a copy of the message that S sent out in the previous stage 2, the attacker can keep S updating the timestamp that it puts with the key SKa(A,B) until the attacker has been able to obtain this key or otherwise finds it useful to have A and B use it again.

6.4. Woo-Lam Authentication, version Π

```
1. A -> B: A;
2. B -> A: Nb;
3. A -> B: {Nb}E(SKs(A,S));
4. B -> S: {A,{Nb}E(SKs(A,S))}E(SKs(B,S));
5. S -> B: {Nb}E(SKs(B,S))
             ||(A Possesses Nb);
```

This protocol's ISL2 specification assumes that S believes the nonce Nb is fresh, even though S does not create this nonce; the AAPA2 warns about this belief. The specification makes this assumption to address an AAPA2 limitation: The AAPA2 is incapable of handling conditional statements associated with sending data. In this case, the statement that S actually makes to B in stage 5 is that *if* Nb is fresh *then* A possesses Nb. The assumption allows S to acquire and transmit an unconditional belief.
Stage-5 explicitness failure:

```
{Nb}E(SKs(B,S));
A Possesses Nb; A.
```

This failure points to a real, easily exploited, vulnerability. Since B sends the nonce Nb in the clear, and cannot tell which nonce is in the encrypted field that it receives at stage 3, it cannot guarantee that S does not wind up associating, in a parallel session, the nonce Nb with Z. If this happens, the encrypted nonce Nb that B receives in stage 5 does not represent an statement by S that A possesses Nb, but a statement by S that Z possesses Nb.

Clark and Jacob [13] also give two other "attacks" on this protocol, but one would be prevented by reasonable type checks that the AAPA2 assumes principals perform and the other only works if the encryption operation E is commutative, an algebraic property that the AAPA2 cannot model.

6.5. Bilateral Key Exchange with Public Key

```
1. B -> A: {Nb,B}Ep(PK(A));
2. A -> B: {f(Nb),Na,A,SKa(B)}Ep(PK(B))
             ||(SharedSecret A B SKa(B));
3. B -> A: {f(Na)}Es(SKa(B))
             ||(SharedSecret A B SKa(B));
```

Stage-2 explicitness failure:

```
{f(Nb),Na,A,SKa(B)}Ep(PK(B));
SharedSecret A B SKa(B); B;
```

This failure must be taken seriously, even though the message is to B and the missing information is B's name, because A might have originally encrypted f(Nb),Na,A,Ka1(B) with the public key of some principal acting in league with the attacker, who could then decrypt the message and re-encrypt it using B's public key. In this case, though, the failure is apparently a false alarm caused by the AAPA2's inability to model implicitly communicated information. In the session, B sends out the nonce Nb encrypted so that only A can decrypt it and unambiguously associated with B. A sends Nb out encrypted so that only B can decrypt it, and the protocol does not call for B to do anything further with it, so B cannot accidentally reveal it. Any message from A containing the nonce Nb this implicitly identifies B.

6.6. Needham-Schroeder Public Key

```
1. A -> B: {Na,A}E(PK(B));
2. B -> A: {Na,Nb}E(PK(A))
             ||(SharedSecret A B Nb);
3. A -> B: {Nb}E(PK(B))
             ||(SharedSecret A B Nb);
```

This protocol assumes that the nonces Na and Nb are suitable shared secrets for A and B, though only A uses Na as a shared secret; A uses it to confirm that B is the source of the message sent to A in stage 2.

Stage-2 explicitness failure:

```
{Na,Nb}E(PK(A));
SharedSecret A B Nb; A; B;
```

Stage-3 explicitness failure:

```
{Nb}E(PK(B));
SharedSecret A B Nb; A; B;
```

These failures point to a real, but hard to exploit, vulnerability. As Lowe's attack on the protocol shows [19], if A starts a session with Z, then the attacker can obtain {Na,A}E(PK(Z)) in stage 1 of its session with A, re-encrypt it as {Na,A}E(PK(B)), and send this re-encrypted value to B as a stage-1 message coming from A. A then accepts Nb as a secret that it shares with Z, and sends out {Nb}E(PK(Z)), meaning SharedSecret A Z Nb, to complete its session with Z. Z takes {Nb}E(PK(Z)), re-encrypts it as {Nb}E(PK(B)), and sends it in stage 3 of its session with B. This leaves A thinking that Nb is a secret that it shares with Z and B thinking that Nb is a secret that it shares with A.

In the BGNY2 logic, if A does not trust Z, then when A receives {Na,Nb}E(PK(A)) in stage 2 of its session with Z it will not believe that this message comes from Z or that Nb is a secret that it shares with Z, so will abort the session rather than output {Nb}E(PK(B) in stage 3. If A does this, then Lowe's attack cannot occur.

If one relaxes the BGNY2 assumption that principals do not send messages unless they believe the statements that the protocol binds to them, though, and assumes instead that these principals do send the messages, but send them with only conditional meanings, then the belief that A actually acquires in stage 2 and communicates in stage 3 of its session with Z is that *if* Z is trustworthy *then* Nb is a secret that A shares with Z. Lowe's attack can then occur.

In a correct session, the AAPA2's explicitness failure at stage 2 is a false alarm because nothing in the protocol can cause B to *accidentally* give away Na, but the accidental-disclosure possibility raised by the AAPA2's explicitness failure at stage 3, a possibility confirmed by Lowe's attack, makes it impossible for B to believe that Nb is a secret that it shares with A.

6.7. SPLICE/AS Authentication

```
1. C -> A: S,Nc1;
2. A -> C: {A,C,Nc1,PK(S)}E(^PK(A))
             ||(PublicKey S E PK(S));
3. C -> S: {C,Tc,L,{Nc2}E(PK(S))}E(^PK(C))
             ||(Fresh Nc2;
                SharedSecret C S Nc2);
4. S -> A: C,Ns;
5. A -> S: {A,S,Ns,PK(C)}E(^PK(A))
             ||(PublicKey C E PK(C));
6. S -> C: {S,Plus1(Nc2)}E(PK(C))
             ||(SharedSecret C S Nc2);
```

The protocol's ISL2 specification states as a stage-6 goal that C believes S possesses Plus1(Nc2) rather than that C believes S possesses Nc2. The theorem that these two statements are equivalent is a consequence in the BGNY2 logic of trivial initial-conditions assumptions, but the AAPA2 does not automatically prove such theorems.

This protocol assumes that the nonce Nc2 will be a suitable shared secret for C and S. The protocol also assumes that S believes the timestamp Tc is fresh even though S does not create this timestamp; the AAPA2 warns about this belief.

Stage-2 explicitness failure:

```
{A,C,Nc1,PK(S)}E(UNPK(A));
PublicKey S E PK(S); S.
```

Stage-5 explicitness failure:

```
{A,S,Ns,PK(C)}E(UNPK(A));
PublicKey C E PK(C); C.
```

Stage-5 explicitness failure:

```
{C,Tc,L,{Nc2}E(PK(S))}E(UNPK(C));
(Fresh Nc2; SharedSecret C S Nc2); S.
```

Stage-6 explicitness failure:

```
{S,Plus1(Nc2)}E(PK(C));
SharedSecret C S Nc2; C.
```

These failures point to a real vulnerability. In stage 1, the attacker can substitute Z for S, causing A to return PK(Z) in stage 2. In stage 4, the attacker can substitute Z for C, causing A to return PK(Z) in stage 5. S will then use PK(Z) to recover what it thinks is PK(C), but what is actually PK(Z). S then uses this key and its own private key to recover Nc2, but has no evidence that Nc2 is a secret shared between it and C because

```
{C,Tc,L,{Nc2}E(PK(S))}E(^PK(C))
```

in stage 3 is encrypted with ^PK(C); the attacker can decrypt it using the widely available PK(C), substitute {Nz}E(PK(S)), then re-encrypt the whole thing with ^PK(Z). Finally, the attacker can generate and send a message of the form expected in stage 6.

6.8. Hwang-Chen Modified SPLICE/AS

```
1. C -> A: S,Nc1;
2. A -> C: {A,C,Nc1,S,PK(S)}E(^PK(A))
             ||(PublicKey S E PK(S));
3. C -> S: {C,Tc,L,{Nc2}E(PK(S))}E(^PK(C))
             ||(Fresh Nc2;
                SharedSecret C S Nc2);
4. S -> A: C,Ns;
5. A -> S: {A,S,Ns,C,PK(C)}E(^PK(A))
             ||(PublicKey C E PK(C));
6. S -> C: {S,Plus1(Nc2)}E(PK(C))
             ||(SharedSecret C S Nc2);
```

The protocol's ISL2 specification states as a stage-6 goal that C believes S possesses Plus1(Nc2) rather than that C believes S possesses Nc2. The theorem that these goals are equivalent is a consequence in the BGNY2 logic of trivial initial-conditions assumptions, but the AAPA2 does not automatically prove such theorems.

This protocol assumes that the nonce Nc2 will be a suitable shared secret for C and S.

The protocol also assumes that S believes the timestamp Tc is fresh even though S does not create this timestamp; the AAPA2 warns about this belief.

Stage-5 explicitness failure:

```
{C,Tc,L,{Nc2}E(PK(S))}E(UNPK(C));
SharedSecret C S Nc2; C.
```

Stage-6 explicitness failure:

```
{S,Plus1(Nc2)}E(PK(C));
SharedSecret C S Nc2; C;
```

These failures point to a real, easily exploited, vulnerability. S does not interpret

```
{C,Tc,L,{Nc2}E(PK(S))}E(^PK(C))
```

until stage 5, when it obtains C's public key. At stage 3, the attacker can take this field and replace it with

```
{Z,Tc,L,{Nc2}E(PK(S))}E(^PK(Z))
```

and similarly at stage 6 replace

```
{S,Plus1(Nc2)}E(PK(C))
```

with

```
{Z,Plus1(Nc2)}E(PK(C)).
```

This leaves S believing that it shares Nc2 with Z while C believes that it shares Nc2 with S.

6.9. Denning-Sacco Key Distribution with Public Key

```
1. A -> S: B;
2. S -> A: {A,PKs(A),Ts}E(^PKs(S))
             ||(PublicKey A E PKs(A)),
           {B,PKs(B),Ts}E(^PKs(S))
             ||(PublicKey B E PKs(B));
3. A -> B: {A,PKs(A),Ts}E(^PKs(S)),
           {B,PKs(B),Ts}E(^PKs(S)),
           {{SKa(A,B),Ta}E(^PKs(A))
              ||(SharedSecret A B SKa(A,B))}
            E(PKs(B));
```

This protocol assumes A believes that the timestamp Ts is fresh, and B believes that the timestamps Ta and Ts are fresh, even though they do not create these timestamps; the AAPA2 warns about these beliefs.

Stage-3 explicitness failure:

```
{SKa(A,B),Ta}E(UNPKs(A));
SharedSecret A B SKa(A,B); B.
```

This failure points to a real, easily exploited, vulnerability. The attacker can substitute a copy of a key and timestamp encrypted with A's secret key that A intended to use for communicating with Z for the field {SKa(A,B),Ta}E(^PKs(A)), leaving B believing that a key held by Z is a secret that B shares with A. The attacker needs to do this in two parallel sessions, to keep the timestamp in the field that it substitutes close to the timestamp in the original field.

6.10. CCITT X.509

```
1. A -> B: {Ta,Na,B,Xa,{Ya}E(PK(B))}E(^PK(A))
             ||(Fresh Ya;
                SharedSecret A B Ya);
2. B -> A: {Tb,Nb,A,Na,Xb,{Yb}E(PK(A))}
             E(^PK(B))
              ||(SharedSecret A B Ya;
                 Fresh Yb;
                 SharedSecret A B Yb);
3. A -> B: {Nb}E(^PK(A))
             ||(SharedSecret A B Yb);
```

The protocol assumes that A and B believe that the timestamps Tb and Ta are fresh even though they do not create these timestamps; the AAPA2 warns about these beliefs. This protocol also assumes that A initially believes that Ya is fresh and will be a suitable shared secret for A and B, and that B has similar initial beliefs about Yb. Without these assumptions, without having A and B communicate them in their stage-1 and stage-2 messages, or without assuming that A and B trust each other, the AAPA2 gives user-goal failures for these stages because it cannot prove that B believes A possesses Ya or that A believes B possesses Yb, a problem with the protocol noted in [12].

Stage-2 explicitness failure:

```
{Tb,Nb,A,Na,Xb,{Yb}E(PK(A))}E(UNPK(B));
(SharedSecret A B Ya;
 SharedSecret A B Yb); Ya.
```

Stage-3 explicitness failure:

```
{Nb}E(UNPK(A));
SharedSecret A B Yb; Yb.
```

Since the messages A and B send in stages 1 and 2 are encrypted with private keys, the attacker can decrypt them and re-encrypt them with ^PK(Z). B's only real evidence, then, that A possesses Ya is that B trusts A, A effectively asserts that it possesses Ya in the message that it sends at stage 1, and B accepts the timestamp Ta as dubious proof of freshness; similar comments apply to A's evidence that B possesses Yb.

The first failure is apparently a false alarm, given the strong trust assumptions that the protocol's ISL2 specification makes. A can take Na as implicitly communicating Ya, and even though the attacker can obtain Na there does not seem to be a way for the attacker to induce B into signing a stage-2 message containing both Na and A.

The second failure points to a real, easily exploited, vulnerability. The attacker can block B's stage-2 message, extract Nb from it, use Nb as the attacker's nonce playing the Nb role in a session that the attacker runs with A, then substitute A's final message to the attacker for A's final message to B.

7. Future Work

This section describes possible future work.

The AAPA2's major remaining limitation is that its BGNY2 logic does not accurately model the set of data items available to the attacker. This is responsible for all 3 of the AAPA2's "misses" on the 51 Clark-Jacob protocols discussed in Section 1 [8]. Although it is provably impossible to *exactly* determine the set of data items available to an attacker in less than exponential time [15], it should be possible to incorporate a reasonable *superset* of this set into an extension of BGNY2 without significantly degrading the AAPA2's speed.

An extension of BGNY2 could define a superset of the data items available to an attacker as follows:

- Assume that the attacker has been able to induce an arbitrary number of previous and concurrent sessions of the protocol.
- Assume that every output that could have been produced by a legitimate principal has been produced by this principal — i.e., assume that all possibilities covered by conditional tests have occurred.

Such a superset will be infinite, but the extended logic could still use unification techniques to prove that a fixed set of type, equality, and inequality tests performed by a principal will succeed in identifying a unique member of this set.

The extended logic could also incorporate use of this superset into all of its rules dealing with shared secrets, to require that everything treated as a shared secret is not even potentially made available to the attacker. By doing so, the extended logic would become the first authentication logic to address nondisclosure failure as well as authentication failure.

If this extended logic avoided all of the AAPA2's "misses", then the AAPA2's results indicate that an automatic protocol analysis tool incorporating this logic would find all 16 of the failed protocols among the 51 Clark-Jacob protocols discussed in Section 1, and raise only 5 false alarms in doing so. Further, two of these false alarms would obviously reflect the tool's inability to model algebraic properties of the operations used in the protocols. This 90% to 94% accuracy rate, for a tool capable of producing its results without user guidance and in minutes, would solve the protocol-failure problem for most practical purposes.

References

[1] R. Anderson and R. Needham. Programming satan's computer. In J. van Leeuwen, editor, *Computer Science Today*, number 1000 in Lecture Notes in Computer Science. Springer-Verlag, 1995.

[2] S. Brackin. Deciding cryptographic protocol adequacy with HOL: The implementation. In *Theorem Proving in Higher Order Logics*, number 1125 in Lecture Notes in Computer Science, pages 61–76, Turku, Finland, August 1996. Springer-Verlag.

[3] S. Brackin. A HOL extension of GNY for automatically analyzing cryptographic protocols. In *Proceedings of Computer Security Foundations Workshop IX*, County Kerry, Ireland, June 1996. IEEE.

[4] S. Brackin. Automatic formal analyses of two large commercial protocols. In *Proceedings of the 1997 DIMACS Workshop on Design and Formal Verification of Security Protocols*, Piscataway, NJ, September 1997.

[5] S. Brackin. An interface specification language for automatically analyzing cryptographic protocols. In *Internet Society Symposium on Network and Distributed System Security*, San Diego, CA, February 1997. IEEE.

[6] S. Brackin. Evaluating and improving protocol analysis by automatic proof. In *Proceedings of Computer Security Foundations Workshop XI*, Rockport, MA, June 1998. IEEE.

[7] S. Brackin. Automatically detecting authentication limitations in commercial security protocols. In *Proceedings of the 22nd National Conference on Information Systems Security*, Alexandria, VA, October 1999. IEEE.

[8] S. Brackin. Empirical Tests of the Automatic Authentication Protocol Analyzer, 2nd Version (AAPA2). Technical Report 99006, Arca Systems / Exodus Communications, Ithaca, NY, June 1999.

[9] S. Brackin. A highly effective logic for automatically analyzing cryptographic protocols. Technical Report 99023, Arca Systems / Exodus Communications, Ithaca, NY, May 1999.

[10] S. Brackin. Implementing effective automatic cryptographic protocol analysis. Technical Report 99032, Arca Systems / Exodus Communications, Ithaca, NY, June 1999.

[11] S. Brackin. Using checkable types in automatic protocol analysis. In *15th Annual Computer Security Applications Conference*, Scottsdale, AZ, December 1999.

[12] M. Burrows, M. Abadi, and R. Needham. A logic of authentication. Technical Report 39, Digital Equipment Corporation, Systems Research Center, Palo Alto, CA, February 1990.

[13] J. Clark and J. Jacob. A survey of authentication protocol literature: Version 1.0. A continually updated library, available at `www.cs.york.ac.uk/~jac/`, of protocols analyzed in the literature.

[14] E. Cohen. Towards automatic verification of cryptographic protocols in natural operational models. Unpublished draft, September 1999.

[15] N. Durgin, P. Lincoln, J. Mitchell, and A. Scedrov. Undecidability of bounded security protocols. In *Workshop on Formal Methods and Security Protocols*, Trento, Italy, July 1999.

[16] L. Gong, R. Needham, and R. Yahalom. Reasoning about belief in cryptographic protocols. In *Proceedings of the Symposium on Security and Privacy*, pages 234–248, Oakland, CA, May 1990. IEEE.

[17] M. Gordon and T. Melham. *Introduction to HOL: A Theorem Proving Environment for Higher Order Logic*. Cambridge University Press, Cambridge, 1993.

[18] S. Gritzalis, D. Spinellis, and P. Georgiadis. Security protocols over open networks and distributed systems: Formal methods for their analysis, design, and verification. *Computer Communications Journal*, 22(8):697–709, May 1999.

[19] G. Lowe. Breaking and fixing the Needham-Schroeder public-key protocol using FDR. In *Proceedings of TACAS*, number 1055 in Lecture Notes in Computer Science, pages 147–166, New York, 1996. Springer-Verlag.

[20] G. Lowe and W. Roscoe. Using CSP to detect errors in the TMN protocol. *IEEE Trans. on Software Engineering*, 23(10):659–669, October 1997.

[21] C. Meadows. The NRL Protocol Analyzer: An overview. *Journal of Logic Programming*, 26(2):113–131, 1996.

[22] J. Millen. CAPSL web page. At `www.csl.sri.com /~millen /capsl`, 1998.

[23] J. Millen, S. Clark, and S. Freedman. The Interrogator: Protocol security analysis. *IEEE Trans. on Software Engineering*, SE-13(2):274–288, February 1987.

[24] L. Paulson. Proving properties of security protocols by induction. Technical Report 409, CUCL, Dec. 1996.

[25] L. Paulson. Mechanized proofs for a recursive authentication protocol. Technical Report 418, CUCL, Cambridge, UK, Apr. 1997.

[26] L. Paulson. Mechanized proofs of security protocols: Needham-Schroeder with public keys. Technical Report 413, CUCL, Cambridge, UK, Jan. 1997.

[27] K. Slind. `HOL90.7`. At `research.att.com /dist /ml/hol90/hol90.7.tar.gz`, 1994.

[28] K. Slind. `HOL98`. At `www.cl.cam.ac.uk/Research /HVG/FTP`, 1998.

[29] N. Tatebayashi, N. Matsuzakai, and D. Newman. Key distribution protocol for digital mobile communication systems. In *Advances in Cryptology — CRYPTO '89*, number 435 in Lecture Notes in Computer Science, pages 324–333, New York, 1990. Springer-Verlag.

Open Issues in Formal Methods for Cryptographic Protocol Analysis

Catherine Meadows
Code 5543
Naval Research Laboratory
Washington, DC 20375
meadows@itd.nrl.navy.mil

Abstract

The history of the application of formal methods to cryptographic protocol analysis spans nearly twenty years, and recently has been showing signs of new maturity and consolidation. A number of specialized tools have been developed, and others have effectively demonstrated that existing general-purpose tools can also be applied to these problems with good results. However, with this better understanding of the field comes new problems that strain against the limits of the existing tools. In this paper we will outline some of these new problem areas, and describe what new research needs to be done to to meet the challenges posed.

1 Introduction

The history of the application of formal methods to cryptographic protocol analysis spans nearly twenty years, and recently has been showing signs of new maturity and consolidation. A number of specialized tools have been developed, and others have effectively demonstrated that existing general-purpose tools can also be applied to these problems with good results.

However, with this better understanding of the field comes new problems that strain against the limits of the existing tools. Tools as they exist today are mainly intended to be applied to the problem of showing that a key has been correctly authenticated. This was the usual application for cryptographic protocols in the past, and remains a common one today. But cryptographic protocols are now being applied to new types of problems, such as group communication, financial transactions, and negotiation of algorithms as well as keys. They also face new types of threats, such as denial of service, and older threats, such as traffic analysis, that are becoming more prominent. In many cases the existing tools are not capable of dealing with these problems. However we believe that many of the existing techniques could be adapted so they could be applied successfully. In this paper we will outline some of these new areas, and describe what new research needs to be done to to meet the challenges posed.

Many of the ideas on new directions that are described in this paper were developed during the application of the NRL Protocol Analyzer to the analysis of the Internet Key Exchange protocol and during the development of a set of formal requirements for the Secure Electronic Transactions Protocols, both of which were done as part of the DARPA project Formal Analysis of Internet Security Protocols. One of the purposes of this project was to see how far current tools could be pushed to analyze complex protocols that must satisfy new types of requirements, and also to find out where our tools need to be improved. In this paper we show how this research contributed to our understanding of the areas discussed in this paper.

The rest of this paper is organized as follows. In Section Two we give a brief history and survey of the state of the art in the field. In Section Three we describe the NRL Protocol Analyzer and the IKE and SET protocols that we examined, describing the features that made them a challenge. In Section Four we describe seven different emerging areas: open-ended protocols, denial of service, anonymous communication, high fidelity, composability, negotiation of complex data structures, and getting the products of our research into the real world. Where appropriate, we refer back to our work on IKE and SET for motivation. Section Five concludes the paper.

2 The History and Current State of Formal Cryptographic Protocol Analysis Tools

Cryptographic protocols are protocols that use cryptography to distribute keys and authenticate principals and data over a network. The network is assumed to be hostile, in that it may contain intruders who can read, modify, and delete traffic, and who may have control of one or more network principals. Because of this, such protocols are often subject to nonintuitive attacks which are not easily apparent even to a careful inspector, especially when assumptions about the environment in which the protocol operates change. For example, as is pointed out in [39], the Needham-Schroeder public key protocol [37], which was intended to be used for communication between parties that trust each other, and is secure as long as those assumptions hold, become subject to a man-in-the-middle attack if we assume that one of the communicating parties may be dishonest [24], an assumption which has become more likely in a web-based environment.

For these reasons, it has long been realized that formal methods can be a useful method for analyzing the security of cryptographic protocols. They allow one both to do a thorough analysis of the different paths which an intruder can take, and to specify precisely the environmental assumptions that have been made. Probably the first mention of formal methods as a possible tool for cryptographic protocol analysis came in Needham and Schroeder [37]. However, the first work that was actually done in this area was done by Dolev and Yao [10], and slightly later by Dolev, Even, and Karp [9], who in the late seventies and early eighties developed a set of polynomial-time algorithms for deciding the security of a restricted class of protocols. Unfortunately, it was soon found that relaxing the restrictions on the protocols even slightly made the security problem undecidable [13], and so the work did not go much further than that. Dolev and Yao's work was significant, however, in that it was the first to develop a formal model of an environment in which multiple executions of the protocol can be running concurrently, in which cryptographic algorithms behave like black boxes which obey a limited set of algebraic properties (e.g. the encryption and decryption operations cancel each other out), and which includes an intruder who can read, alter, and destroy traffic, and may also control some legitimate members of the system. Most later work on the formal analysis of cryptographic protocols is based on this model or some variant of it.

Shortly later, work began on developing tools for the analysis of security protocols in general, all of which were based on the Dolev-Yao model or some variant, including the Interrogator [34], the NRL Protocol Analyzer [27], and the the Longley-Rigby tool [23]. Others applied general-purpose formal methods to the problem [20]. Most of this work used some type of state exploration technique, in which a state space is defined and then explored by the tool to determine if there are any paths through the space corresponding to a successful attack by the intruder. Inductive theorem proving techniques were also included in the tool in some cases, as in the NRL Protocol Analyzer, to show that the size of the search space was sufficient to guarantee security. Even during these early stages, much of this work was successful in finding flaws in protocols that had been previously undetected by human analysts, including the use of the NRL Protocol Analyzer to find a flaw in the Simmons Selective Broadcast Protocol [27], and the use of the Longley-Rigby tool to find a flaw in a banking protocol [23].

However, this still remained a fairly esoteric area until the publication of the Burrows, Abadi, and Needham logic [5] brought the problem to the attention of a larger research community. BAN logic uses an approach very different from that of the state exploration tools. It is an example of a logic of knowledge and belief, which consists of a set of possible beliefs that can be held by principals (such as a belief that a message was sent by a certain other principal), and a set of inference rules for deriving new beliefs from old ones. An example would be a rule saying that if A believes that a key K is known only by him and B, and A receives a message encrypted with K, than A believes that that message was sent by B to A, or by A to B. The BAN logic consisted of a very simple, intuitive set of rules, which made it easy to use. Even so, as the BAN paper demonstrated, it was possible to use the logic to pinpoint serious flaws in protocols. As a result, the logic gained wide attention and let to a host of other logics, either extending BAN logic or applying the same concept to different types of problems in cryptographic protocols.

Note that belief logics such as BAN are generally weaker than the state exploration tools since they operate at a much higher level of abstraction. Thus interest in them has waned somewhat as state exploration systems have improved. However, they have an advantage in that they are usually decidable and often even efficiently computable, and thus can be completely automated, as has been shown by Brackin's Automated Authentication Protocol Analyzer [3].

More recently, research has focused on state exploration tools and theorem proving techniques based on the Dolev-Yao model, much of it sparked by Lowe's demonstration that it was possible to use a general-purpose model checker, FDR, to find a man-in-the-middle attack on the Needham-Schroeder public key protocol [24]. Work since then has progressed in applying both model checkers [36, 7] and theorem provers [41, 12] to the problem, as well as in the design of special-purpose model checkers [19, 43, 25] and the use of specialized tools originally intended for somewhat different applications [11].

There has also been a sign of consolidation in the area, an indication that it has been maturing. For example, Millen has been developing CAPSL [35], the Common Authentication Protocol Language, which is intended to provide a common specification language for cryptographic protocol analysis tools. Even more recently, Thayer, Herzog, and Guttman [14] have developed a graph-theoretic interpretation of the Dolev-Yao model, called the strand space model, that brings together many ideas and techniques that have been used in the formal analysis of cryptographic protocols. Because of this, and because of its simplicity and elegance, it has begun to be used both as a basis for new special-purpose tools [43] and as a framework in which to express theoretical results [44]. This trend has promising implications for the integration of future tools and the incorporation of new theoretical results into these tools.

To sum up, at present we are rapidly approaching, if we have not already reached, a state in which we will have a number of different tools available that will be able to verify safety properties such as authentication and secrecy by performing a state space analysis of a protocol specified at the same level of detail that is normally provided in a journal paper using the Dolev-Yao model of a protocol attacker. This is not everything; such tools will not catch errors that arise from implementation details that go beyond the journal-level specification, the Dolev-Yao model leaves out some important attacker capabilities such as cryptanalysis, and, since the Dolev-Yao model assumes an intruder that is capable of blocking any message, it is impossible to prove any kind of liveness property. Nevertheless this is still a great deal; the wide range of attacks found by such tools demonstrate that a number of nontrivial problems occur at this level of specification.

Since the protocol security problem in undecidable, [13, 18, 6], the analysis tools will not be successful all the time, and they may require human intervention at times. But even so, the problem of tool design for this area seems well enough understood by now so that these limitations should not interfere too much with their effective use.

Given that we have reached this plateau, it seems reasonable at this point to ask, what comes next? And indeed, there are a number of related problem areas still to be explored. Some of them have only surfaced in the last few years. Others have been known about for some time, but it was thought more important to concentrate on the basics first. However, now that the basics are well understood, it is time to look more closely at some of these areas. In the remainder of this paper we do this, in each area pointing out what work has already done, and what we believe still remains to be done.

3 IKE, SET, and the NRL Protocol Analyzer

3.1 Overview

In this section we describe the IKE and SET protocols that we analyzed for the DARPA project Formal Analysis of Internet Security Protocols, as well as the tool we applied, the NRL Protocol Analyzer. We concentrate, not on the analyses themselves, which are described elsewhere, but on the challenges that each posed. We identify both those challenges that our tools proved capable of meeting, and those areas where we felt our tools fell short.

3.2 The NRL Protocol Analyzer

The NRL Protocol Analyzer is a formal methods tool for analyzing security properties of cryptographic protocols. It uses automatic invariant generation to limit a potentially infinite search space in combination with exploration of the remaining space to generate attacks on insecure protocols and provide security proofs for secure ones, even in the face of a potentially unlimited number of protocol executions or an unlimited number of intruder actions.

Protocols in the Analyzer model are specified as communicating state machines, one of which is a hostile intruder following the Dolev-Yao model who can read all traffic, modify or delete traffic, perform cryptographic operations, and may be in cooperation with some legitimate users of the system.

The user of the Analyzer attempts to determine whether or not a protocol is secure by specifying an insecure state. This state can be specified not only in terms of values of local state variables and terms known by the intruder, but sequence of events that should or should not have occurred. For example, the user could ask the Analyzer to look for a state in which the same key has been accepted twice by a principal (two events occurring) or a state in which a responder B accepts a key as good for communicating by an initiator A, but in which A never initiated the protocol (one event having occurred and another event not having occurred previously). The Analyzer works backwards from that state until it has explored the search space exhaustively, so that each path produced either begins in an initial state (describing an attack) or an unreachable state. Thus, like the other tools we mentioned earlier, it can be used fo prove safety properties such that no party is authenticated incorrectly, but not liveness properties such that an authentication always completes.

The Analyzer makes no assumptions about limits on the number of protocol executions, the number of principals performing the different executions, the number of interleaved executions, or the number of times cryptographic functions are applied. This results in a search space that is originally infinite. However, the Analyzer provides means for specifying and proving inductive lemmas about the unreachability of infinite classes of states. This allows the user to narrow down the search space so that in many cases an exhaustive search is possible.

We also made use of the NRL Protocol Analyzer Temporal Requirements Language (NPATRL) [48]. NPATRL is a temporal logic language allowing us to specify desirable protocol properties in terms of desirable or undesirable sequence of events. An NPATRL requirement is applied to the NRL Protocol Analyzer by taking the negation of the requirement and using that as an insecure state for the Analyzer to prove unreachable.

The Analyzer has been applied to a number of different cryptographic protocols, and has found flaws in several. In some cases the flaws had not been discovered before. Examples of protocols the Analyzer has been used to examine are the Simmons Selective Broadcast Protocol [27], the Burns-Mitchell Ticket Granting Protocol [32], and an early version of the Encapsulating Security Protocol [46]. The Analyzer has also been used to prove security properties of a number of other protocols, by performing an exhaustive search of the finite space that is left after the necessary lemmas have been proved (see [28, 29]. A more detailed description of the Analyzer is given in [33].

The NRL Protocol Analyzer, as we see, operates within the same paradigm as most specialized tools for cryptographic protocol analysis, and indeed, was one of the first tools to make use of it. That is, it is based on the Dolev-Yao model and it concentrates mainly on proving authentication properties. Thus it is well suited for testing the limits of that paradigm.

3.3 IKE

The Internet Key Exchange protocol (IKE) is a key exchange protocol being developed by the IP Security Protocol (IPSEC) Working Group of the Internet Engineering Task Force (IETF). It is intended to provide the security support for client protocols of the Internet Protocol. As such, it does much more than simply distribute keys; it also is intended to be used to establish *Security Associations* that specify such things as the protocol format used, the cryptographic and hashing algorithms used, and other necessary features for secure communication. Since it is intended to be flexible, it supports a number of different types of key exchange options, including digital signatures, public key encryption, and conventional encryption using shared keys. The Diffie-Hellman algorithm is used to generate shared key material, but is optional in some cases. IKE has evolved from a number of different protocols, including ISAKMP [26], Oakley [38], the Station-to-Station protocol [8], and SKEME [21], the last two of which influenced the development of Oakley.

A typical key establishment protocol proceeds in one phase, in which two parties use master keys to establish shared keying material. IKE, however, proceeds in two such phases. In the first phase, two entities use master keys to agree, not only on keying material, but on the various mechanisms (e.g. cryptographic algorithms, hash functions, etc.), that they will use in the second phase. The keying material and set of mechanisms thus agreed upon is called a *security association.* In the first phase protocol, the initiator proposes a number of possible security associations to the responder, who picks one. In the second phase, the keys and mechanisms produced in the first phase are used to agree upon new keys and mechanisms that will be used to protect and authenticate further communications. The security association established in Phase One is bidirectional, so the initiator in the first phase can be either initiator or responder in the second phase.

At the time we analyzed IKE, it could be used in four different modes: main mode, aggressive mode, quick mode, and new group mode. Main and aggressive modes are used in Phase One negotiations, quick mode is used in Phase Two negotiations, and New Group Mode is used to change the Diffie-Hellman group. Both main and aggressive mode can be implemented using several different types of authentication; there is a different protocol for each one.

The main goals of IKE are the authentication of security associations, and the authentication and generation of keys. However, IKE has some other secondary goals, as well. A good deal of thought has gone into making IKE resistant to denial of service attacks by preventing it from requiring principals to engage in resource-intensive activities until a "reasonable" amount of authentication has occurred. IKE is also designed to be somewhat resistant to traffic analysis, in that in Main Mode identities are encrypted.

In our analysis of IKE, described in detail in [29], we concentrated on verifying that keys were properly authenticated, and that portions of the Security Associations were authenticated. We did not attempt to verify to what degree the protocol resistant to traffic analysis or denial of service; this was beyond the scope of our tool. However, it is clear that such an analysis would have been useful. These issues are discussed in more detail in Section 4.2 on Denial of Service and Section 4.3 on Anonymous Communication.

One of the main challenges in analyzing IKE was the sheer proliferation of related subprotocols. Since we wanted to verify, among other things, that one protocol could not be confused with another, we analyzed as many subprotocols together as we possibly could. The NRL Protocol Analyzer proved adequate to this task, but not before undergoing a major overhaul designed to increase its robustness, and some serious thinking about how to identify possible interactions. These issues are discussed in depth in Section 4.5 on Composability.

3.4 SET

The SET Protocol [31] is a protocol sponsored by major credit card companies and others that is intended to provide a standard for safe, secure credit card transactions over the Internet. ('SET' stands for 'Secure Electronic Transactions'.) As such, it is intended to supply an electronic version of the paper system that exists today. However, there are a number of risks connected with use of the Internet that do not arise in the paper world, or at least are not considered as severe. These arise from the difficulty of identifying participants in transactions and the difficulty of ensuring the private information sent over the Internet remains so. SET is intended to reduce these risks by introducing cryptographic means to protect sensitive information such as credit card numbers and to provide authentication of parties involved in a credit card transaction.

A payment transaction in the SET protocol involves three parties: a customer, a merchant, and an application payment gateway. The customer presents a purchase request to the merchant, which includes credit card information and a proposed purchase amount. The purchase request is identified with a transaction ID. The merchant then passes the request along to the gateway, together with a request that a certain amount (not necessarily equal to the purchase amount) be authorized. The gateway then checks the customer's credit, authorizes a certain amount, and passes this information back to the merchant. The merchant passes this information back to the customer. Either at the same time as the authorization request, or later, the merchant presents a capture request to the gateway for the same transaction, requesting that a certain amount of money be captured. The gateway approves a certain amount which may or may not be equal to the amount requested. The merchant then passes this information back to the customer.

The authentication structure of the SET protocol is complex. Messages between merchant and gateway are always authenticated using digital signatures, as are messages from the merchant to the customer. Digital signatures for authentication for messages from customer to merchant are optional, although they may be made mandatory by a particular application. However, it is in the authentication of forwarded messages that the structure really becomes interesting. The message from customer to merchant includes information that is needed by the gateway but may be hidden from the merchant, such as credit card number (PAN, i.e., Primary Account Number) and expiration date. Also included, when customer digital signatures are used, is a data item called the PANSecret, which is known only by the customer and gateway, but not the merchant. This is not available when customer digital signatures are not used, since it is generated as part of the certificate registration process. This information is protected by the use of a *dual signature.* Two hash functions are computed, one over the the data to be kept hidden from the merchant, and the other over the data to be revealed to it, which includes a hash over the purchase amount and order description that the customer and merchant agreed to offline. The hidden data is encrypted using the gateway's public key. The customer then computes a digital signature (if customer signatures are used) over the two hashes. The signature, the two hashes, and the encrypted and unencrypted information are sent to the merchant. The merchant verifies the signature and forwards the information, including the signature if any, to the gateway. When the gateway receives the message, it verifies the signature, if any, and also verifies the PAN and PANSecret. Whether or not signatures are used, it also verifies the PAN and and the customer's portion of the PANSecret (if any) with the credit card issuer, although this may be done offline.

Authentication of gateway to customer via the merchant is much simpler: there is none. Any information from the gateway that the merchant passes on to the customer is authenticated only by the merchant's signature.

There are also a number of options available. A customer has the option of sending an initialization message prior to its purchase request, which allows it to obtain more up-to-date certificates from the merchant, and allows the merchant to send back a random challenge which it can use to verify the freshness of the customer's subsequent purchase request. When an initialization message is sent, the transaction ID is jointly created by customer and merchant. When no initialization message is sent, the customer may create the transaction ID, or it may be jointly created by the customer and merchant. The gateway also has the option, depending upon the policy followed, of sending the customer's PAN to the merchant (the PANSecret, however, is never sent). There are also protocols for inquiring about the status of an order, cancelling an order, etc.

We have not yet completed our analysis of SET, but we have written a formal specification of the SET requirements using the NPATRL requirements language [31]. This turned out to be a nontrivial task. Although the NPATRL language was expressive enough to state the requirements, their complexity meant that any such specification would be incomprehensible unless we introduced some higher-level framework in which to organize it. We describe the problem and our solution in more detail in Section 4.6 on Negotiation of Complex Data Structures.

4 Emerging Problems

4.1 Open-Ended Protocols

Most of the work on the formal analysis of cryptographic protocols has concentrated on protocols that involve the communication of a fixed number of principals: for example, an initiator and a responder in a key agreement protocol, or a customer, merchant, and bank in an electronic commerce protocol. Most data structures that are used are also closed-ended. That is, in general each message is a fixed structure that is composed of a bounded number fields containing data such as nonces, names, keys, etc. Open-endedness is included in the protocol model, but only with respect to the number of protocol executions that may be going on at the same time, or the number of operations that the intruder may perform to create a message. This means that the protocol models do not need to include such constructs as loops, thus simplifying the model and, one hopes, the analysis.

However, open-ended structures are beginning to show up in a number of different applications. By open-ended, we simply mean the the structure may include an arbitrarily large number of data fields; no precise limit is put on them by the protocol specification. The most obvious is in group communication protocols, in which keys must be shared among members of a group of arbitrary size. Here, it is the group of principals that may be participating in a particular instance of the protocol that is open-ended. However, open-ended structures show up in other types of protocols, as well. For example, anonymous routing protocols make use of an arbitrary number of routers to achieve their goals. Open-ended structures are also used even in protocols in which the number of principals is bounded. For example, the SET protocol allows a merchant to batch transactions for approval by a security gateway. The IKE Protocol offers an even more complex example. One of the purposes of IKE is to agree on a security association (SA), the collection of algorithms and other information used to encrypt and authenticate data. Although there is some information that an SA must include, there is no defined limit on what it can include, so its definition is left open-ended. In addition, an SA is negotiated by having the initiator present a list of SAs to a responder, who then picks one. Thus there are two sources of open-endedness in the use of SAs. Moreover, this open-endedness is security-relevant. For example, recently Zhou [54] and independently Ferguson and Schneier [15] found an attack in which an intruder could trick an initiator into agreeing on the wrong SA by making use of the fact that only part of the SA is actually used in IKE itself.

So far, there has been very little work on applying formal analysis techniques to these kinds of problems in cryptographic protocols. One notable exception has been the work of Paulson [40] on the application of the Isabelle theorem prover to the analysis of a protocol that involves an arbitrary number of principals, and the work of Bryans and Schneider [4] applying the PVS theorem prover to the same protocol. Since all of this work involves general-purpose theorem provers instead of special-purpose tools, we would not be surprised to find that the authors were able to make use of techniques that were not available in tools that were specifically designed for cryptographic protocol analysis. However, it is heartening to note that Bryan and Schneider's work makes use of a construct, the rank function, that Schneider had previously developed for the analysis of cryptographic protocols that involved only a bounded number of participants in a single protocol execution. Thus it may be the case that techniques that were developed to deal with the unboundedness that arises first out of the ability of an intruder to perform an arbitrary number of message operations, and secondly out of the possible execution of an unbounded number of protocol operations in parallel, should be applicable to other types of open-ended protocols as well, although they will probably require some adaptation and expansion.

4.2 Denial of Service

Denial of service was not a threat that was a cause of much concern to the first designers of cryptographic protocols. However, as we have seen from the SYN attacks of TCP/IP, many communication protocols are subject to a particular type of denial of service attack in which the attacker initiates an instance of a protocol and then drops out, leaving the victim hanging. Since the victim must use resources to keep the connection open until the protocol times out, the attacker, by initiating and then dropping enough instances in the protocol quickly enough, can cause the victim to waste enough resources keeping connections open so that it is unable to participate in any more instances of the protocol and is thus effectively cut off from the network.

Strong authentication can both ameliorate and exacerbate this problem. Authentication can be used to identify the source of the attack, allowing the victim to cut off communication with the attacker. But authentication can also be used as a means of launching denial of service attacks, since it is both computation and storage-intensive, and the attacker could launch a denial of service attack on a victim by sending it a series of incorrectly authenticated messages that it would waste its resources verifying.

The approach that has been taken to resolving this problem is to use a tradeoff between resources required of the victim (referred to from now on as the "defender") with resources required of the intruder. Early parts of the protocol require weak authentication that do not require great resources on the part of the intruder to break, but require fewer resources on the part of the defender to verify. More expensive forms of authentication are reserved for later in the protocol when a degree of assurance that the participating parties are legitimate are obtained.

Note that the attacker model used in this strategy is generally weaker than the model used in the verification of traditional authentication goals. Thus the sort of nonintuitive attacks that have been found on these types of goals will not necessarily arise in the case of denial of service, although they are not ruled out entirely. However, the analysis becomes more complicated in a number of other ways. First, the protocol must be analyzed, not only in terms of its final goals, but along each step of the way. Every time a a principal takes part in some action that requires the use of a significant amount of resources, one must check that that an attacker could not fraudulently cause that principal to reach that step without spending a significant amount of its own resources. Secondly, in order to make that verification possible, it is necessary to have a model, not only of principal and intruder actions, but of the cost of those actions. Thus some sort of formal analysis technique would be beneficial, simply in order to to keep track of this complex multi-stage analysis.

Existing protocol analysis tools, although they cannot be applied to the problem directly in their present form, have many features that could be useful if adapted properly. For example, for most it is possible to specify intermediate as well as ultimate goals. Also, although most use a single model of the intruder, most of the weaker intruder models that would be used would be restrictions of this more general intruder model.

We have been working on a framework [30] that could be used to apply existing tools, appropriately modified, to the denial of service problem. We make use of the concept developed by Gong and Syverson [17] of a *fail-stop* cryptographic protocol. Briefly, a protocol is fail-stop if, whenever an attacker interferes with a message, this is detected by the receiving principal and the protocol is halted. We have modified the fail-stop model to include an attacker whose capabilities change as the protocol progresses, and have developed a framework for trading off intruder capabilities against effort expended by the defenders. In this framework the protocol designer specifies a *protocol tolerance relation*, which describes how much effort he believes it should be necessary to expend against an attacker of a given strength. Since we are developing a framework for models instead of a specific model, we do not specify exactly how this effort should be quantified, but examples would include amount of resources expended, amount of time expended, or amount of computational power required. A protocol is then designed so that the effort expended by the defender increases as the protocol executes, and also as each message is verified. The protocol is then analyzed to show that it is fail-stop against an attacker whose capabilities are within the constraints of the desired tolerance relation. That is, at each verification point, the amount of effort required by the attacker to spoof the verification, versus the amount of effort wasted by the defender if the verification is successfully spoofed, should fall within the tolerance relation.

4.3 Anonymous Communication

Anonymous communication is an application that has recently begun to move from the laboratory to the real world, as the ubiquity of the World Wide Web makes even ordinary users more sensitive to the dangers of traffic analysis and of indiscriminately revealing personal information over the Web. Thus systems such as the Onion Router [16], the Anonymizer and Crowds [42], are designed to prevent an onlooker from determining the origination or destination of requests to servers. Basically the way all these systems work is by having requests from users routed through one or more nodes. In the simplest versions, a user proxies a request through a single site, which strips the request of identifying data or otherwise disguises its source, and forwards it to the server. More sophisticated systems, such as Crowds and Onion Routing, have the request routed through a number of nodes. Onion Routing uses cryptographic means to keep each node ignorant of all other nodes in the path except the ones with which it communicates directly.

Other systems provide other types of anonymity. Proxymate distributes pseudonyms for dealing with third parties. The Rewebber supports anonymous publishing. Anonymous remailers support anonymous email. Indeed, we can safely assume that, for any application involving communication over the Internet, there are situations in which one might be concerned about preserving anonymity and preventing traffic analysis.

We note that anonymity has properties that make it challenging to analyze. First of all, it is an emergent property, that is, one that is not apparent in isolation. It would be hard to disguise the source of a request if it is the only request in the network, no matter how many nodes it was routed through. An anonymizing protocol depends upon a mix of traffic to disguise the source and destination of any particular item. Statistical studies, such as the work of Timmerman [51], will be of use here to determine how well this strategy works in different situations. Previous work done on the analysis of covert channels in networks [52] might also be of use here, if applied properly. Next, the more powerful anonymizing protocols require communication between an arbitrary number of nodes, instead of just two or three, whic is the number of principals that are usually assumed to be communicating in most of the protocols that have been analyzed using the existing cryptographic protocol analysis tools. Thus any techniques that are developed for analyzing group communication protocols will probably also be useful in this area. Finally, since an anonymity protocol attempts to preserve secrecy by distributing the data over a wide area, the assumptions made about the capacities of an attacker are very relevant. A ubiquitous attacker will be able to break most anonymity protocols, but ubiquity is not a very realistic assumption. However, attackers who only reside at single nodes and do not communicate are not very realistic either. The work by Syverson and Stubblebine on group principals [47] deals with some aspects of this problem, introducing the notion of a group principal that consists of a certain number of members who are assumed to have certain specified capacities for sharing knowledge.

In summary, we believe that the analysis of anonymity protocols pose a new research challenges which are beginning to be met, at least partially and in different aspects. The main challenge may be tying these different threads together.

4.4 High Fidelity

Most work on the application of formal methods to cryptographic protocol analysis have modeled protocols at a very high level of abstraction. Techniques based on state reachability analysis usually assume that the algorithms used behave like black boxes, with only enough algebraic properties included (e.g. that encryption and decryption cancel each other out) to allow the protocol to function correctly. Techniques based on belief logics are usually even more abstract, forgoing in most cases even a general explicit model of the intruder or of the cryptographic operations; instead goals achieved by the protocol are derived directly from the messages sent.

However, it is well known that many security problems in protocols arise at a much lower level of abstraction. Some come from interactions of the cryptosystem with the protocol, such as a protocol that includes known or chosen plaintext while using a cryptographic algorithm that may be vulnerable to attacks based on the inclusion of this type of plaintext. Others come from problems with other supporting algorithms, such as hash functions or modes of encryption. Some come from other types of low-level implementation details. For example, in our analysis of the Internet Key Exchange protocol, we found an attack that would work if a recipient's decision as to the possible source of a message was implemented in one way, but would fail if it was implemented in another way. Thus, it appears to be well worth our while to take our analysis to a lower level of abstraction.

Some work in this direction already exists. For example, work on the analysis of modes of encryption and chosen and known plaintext has been successful both in finding new problems [45] and reproducing known attacks [46]. From an entirely different angle, work has also been ongoing on introducing some of the polynomial-time reduction techniques used by cryptographers into the framework used by formal methods, making it possible to reason more precisely about the interaction of a protocol with the cryptographic algorithms that is uses [22] Finally, recently new interest has been shown in revisiting the problems of secrecy models in cryptographic protocols, thus going beyond the standard black-box assumptions [53]. Thus we see new interest in the problem of high fidelity from a number of different sides.

4.5 Composability

Most work on the analysis of cryptographic protocols has concentrated on the analysis of protocols that can be described in terms of a single sequence of messages without any choice points or loops. However, most cryptographic protocols are not actually deployed in this fashion. Indeed, many cryptographic protocols as they are actually implemented can be thought of as a suite of "straight-line" sub-protocols (that is protocols that involve no if-then-elses and no loops) along with a number of choice points in which the user may choose which sub-protocol to execute. In this kind of environment, it is necessary, not only that each subprotocol be shown to execute correctly in isolation, but that the subprotocols do not interact with each other in harmful ways. This problem in its general form is known as the composition problem for cryptographic protocols: given that two or more different protocols are executing in the same environment, is it possible that a message or messages from one protocol could be used to subvert the goals of the other?

The composability problem is not only a theoretical concern. Consider, for example, the following attack, described in [1] on a very early version of SSL. The early version included an optional client authentication phase in which the client's challenge response was independent of the type of cipher negotiated for the session, and also of whether or not the authentication was being performed for a reconnection of an old session or for a new one. Moreover, this version of SSL allowed the use of cryptographic algorithms of various strength (weak algorithms for export and stronger ones for home use), and since weakness could be guaranteed by revealing part of the key, it was not always clear by inspection of the key whether weak or strong cryptography was being used. This allowed the following attack (note that in this version of SSL, session keys were supplied by the client):

1. A key K is agreed upon for session A using weak cryptography.

2. Key K is broken by the intruder in real time.

3. The client initiates a reconnection of session A.

4. The intruder initiates a new session B, pretending to be the client, using strong cryptography together with the compromised key K.

5. As part of the connection negotiations for session B, the server presents a challenge to the client. The client should return a digital signature of both K and the challenge. The intruder can't do this itself, but it can pass the server's request on to the client, who will take it to be part of the reconnection negotiations for session A, and produce the appropriate response. The intruder passes the response on to the server as part of the session B negotiations, and the protocol completes.

6. If the client would have been given access to special privileges as a result of using strong cryptography, this could lead to the intruder gaining privileges that it should not be able to have by breaking the key K.

Since this attack involves a confusion of the reconnection protocol with the connection protocol, it is an example of a failure of composition which would not have been found if the two protocols had been analyzed separately.

Early work on composability [18, 17] concentrated on determining under what conditions protocols could be guaranteed to be composable. The early results led to rather stringent requirements: in essence, they required the fail-stop property [17] or something very similar to it [18]. Thus they did not have much practical application.

More recent work has concentrated, not on designing protocols that are guaranteed to be composable, but on reducing the amount of work that is required to show that protocols are composable. In our work in using the NRL Protocol Analyzer to analyze the Internet Key Exchange protocol, we found it useful to take each state transition that required an input message and determine which transitions could produce that output. This information was stored in a database, and only those rules that have a chance of producing that output are consulted when the reachability of an output is being verified. This allowed us to reduce the number of state transitions that had to be tested whenever we had to determine how a message could be produced, thus limiting the state explosion problem. We do not make any claims for the originality of this idea (indeed some sort of rule pre-verification and storage is normally done by rule-based systems that must process a large number of rules), but we were suprised at the dramatic speedup it caused, (at least threefold for the IKE analysis [29]). This is a technique that would be useful for the analysis of any complex protocol, but which was particularly helpful for the analysis of suites of protocols, since only a few state transitions from different protocols had the potential of interacting with each other.

Independently, Thayer, Herzog, and Guttman used a similar insight to develop a technique for analyzing composition properties using their strand space model [49]. Their technique consists of showing that a certain set of terms generated by the first protocol can never be accepted by principals executing the second protocol. This information is then used to prove the full result that the first protocol does not interfere with the second. The techniques used for choosing the set of terms and for using them in the proof are specific to the protocols used in [14], but it is likely that they could be generalized into heuristics that could be applied more widely. We think that it is likely that these techniques will continue to be useful as they are further refined.

We also believe that the last word has not been written on designing protocols that are guaranteed to be composable, or are at least easier to prove composable. Recently some promising work has been done on designing protocols that are verifiable by model-checkers [24, 44]. A model-checker checks that a protocol is secure by simulating its interaction with an intruder. Since simulation is used, only a finite number of executions of the protocol can be examined. Thus it is helpful to know if there is some finite number of executions such that the examination of this finite number is enough to guarantee security. In general, the answer to this is no [13, 18, 6], but it is still possible, as is shown by the work of Lowe [24] and Stoller [44], to put design constraints on protocols that guarantee them to be verifiable through examination of a finite number (and manageable) number of executions. Moreover, the constraints involved are much more realistic than the fail-stop constraint used in the early composability results.

We believe that this work could be used in the analysis of composability in the following way. Consider a protocol that integrates a set of straight-line subprotocols (where a "straight-line" protocol is defined to be one that involves no if-then-elses or loops) by proceeding in two stages. In the first stage, an initiator chooses which of a suite of straight-line protocols it will execute. In the next stage it executes the subprotocol. In order for two subprotocols to interact harmfully, at least two executions of the protocol must have taken place. If we can modify Lowe's and/or Stoller's results so that they would still hold in such an environment, then they could be used to limit the complexity of checking for an interaction by limiting the size of the set of subprotocols that needs to be checked at any one time.

4.6 Negotiation of Complex Data Structures

Key distribution and agreement protocols usually negotiate agreement upon a single field: the key. Most work on formal analysis of cryptographic protocols has concentrated on this problem. But once more structure is introduced into the data to be agreed upon, new problems of maintaining the consistency of the data emerge. For example, Thayer, Herzog, and Guttman [50] recently discovered a new attack on the Otway-Rees protocol, which distributes a key and a key identifier together. In this attack the principals wound up agreeing on the same key identifier but not the same key. Although the Otway-Rees protocol had been carefully studied in the past, this possible behavior had not been uncovered before.

Many of the newer protocols negotiate agreement on much more complex structures than this. As we saw earlier, the Internet Key Exchange Protocol negotiates agreement on a complex data structure that not only includes multiple fields, but is also open ended. SET goes IKE one further by having principals agree on a data structure the contents of whose fields are generated by different members of a three-party protocol as the protocol progresses, so that we are dealing with a data structure that evolves as the protocol completes. Moreover, different parts of the structure may be kept secret from different principals, so it is possible that no one principal knows the contents of all the fields in the data structure, even after the protocol completes.

In spite of this complexity, we found the tool we were using, the NRL Protocol Analyzer, adequate to this task, except for the case of open-ended data structure, described in Section 4.1. The hard part was understanding and specifying the protocol requirements that needed to be satisfied, especially for SET. We were fortunate to have the NPATRL requirements language [48], that had been developed specifically to specify complex protocol requirements for the NRL Protocol Analyzer. However, we still found it necessary to augment it in order to reason about agreement on data structures with possibly unknown or not yet defined fields. We did this in terms of specifying requirements on agreement on projections of the entire data structure. This was not difficult to do, but did require an enhancement to the NPATRL language.

We can see other types of data structures emerging as well. Certificate hierarchies are one example. Since any protocol that manages certificate hierarchies must deal with the revocation of certificates as well, this brings up the problem of introducing negation. There is also a growing body of work on using certificates to implement security policies (e.g. Policymaker [2]). Although there has been a good deal of research in the development of policy definition languages, little of this work has concentrated on examining their interaction with the protocols that implement them. It is still early to tell if there are any problems here the require special attention, but the area bears watching.

4.7 Getting it into the Real World

Throughout most of this paper, we have concentrated on extending the limits of research. But we also need to concentrate on getting the results of our research out to the people who can make best use of it: the designers and evaluators of the cryptographic systems that are being deployed in our networks. For this we want not to concentrate on cutting-edge research problems, but on what we do best now, and what is the best use we can make use of these capabilities.

I think that few would argue that what we do best now is the analysis of straight-line key distribution and authentication protocols, in which the lowest level of abstraction used is a black-box model of a cryptosystem. For these types of protocols there now exist belief logic tools that can do provide a totally automated analysis [3]. On a somewhat deeper level there are a number of state-based analysis tools that can do a more thorough analysis with minimal input from the user. High-level languages like CAPSL [35] also make it easy to specify these protocols in a way usable by the tools.

We have noted, of course, that the reality is often much more complex than the simple protocols that these tools were devised to analyze. But most of the complex protocols had their origins in the simpler journal level protocols; for example the complex Internet Key Exchange protocol is in part derived from the very simple Station to Station protocol. Nor are the "simple" protocols as transparent as they might appear at first glance. Lowe's analysis of the Needham-Schroeder public key protocol and Thayer, Herzog, and Guttman's analysis of the Otway-Rees protocol show that it is possible for possibly harmful properties to go undiscovered for years. Thus, it is important to have the ability to recognize and fix potential problems up front. If we think of the existing tools as providing the potential for animating "back-of-the-envelope" sketches of protocols so that they can be thoroughly examined before the effort is made to developed them into more finished products, I think we will have an idea of how our tools as they exist now can be made immediately useful to protocol designers.

5 Conclusion

In this paper we have given a brief outline of the state of the art of cryptographic protocols and shown many directions in which it could be extended. Most conclusions to papers include suggestions for further research. Since this paper consists of nothing but suggestions for further research, we will forgo that here. However, we do note that we did not intend our list of topics to be exclusive. Indeed, we imagine that there will be many other areas that come to light as research progresses.

6 Acknowledgments

I would like to thank the participants in the FLOC99 Workshop on Formal Methods and Security Protocols for helpful discussions on these topics. This work was sponsored by NSA and DARPA.

References

[1] Josh Benaloh, Butler Lampson, Daniel Simon, Terence Spies, and Bennet Yee. The private communication technology protocol. draft-benaloh-pct-00.txt, October 1995.

[2] M. Blaze, J. Feigenbaum, and J. Lacy. Managing trust in information systems. In *Proceedings of the 1996 IEEE Symposium on Security and Privacy*, pages 164–173. IEEE Computer Socity Press, May 1996.

[3] S. Brackin. Evaluating and improving protocol analysis by automatic proof. In *Proceedings of the 11th IEEE Computer Security Foundations Workshop*. IEEE Computer Society Press, June 1998.

[4] J. Bryans and S. Schneider. CBS, PVS, and a recursive authentication protocol. In *Proceedings of the DIMACS Workshop on Design and Formal Verification of Security Protocols*, September 3-5 1997. available at http://dimacs.rutgers.edu/Workshops/Security/.

[5] Michael Burrows, Martín Abadi, and Roger Needham. A Logic of Authentication. *ACM Transactions in Computer Systems*, 8(1):18–36, February 1990.

[6] I. Cervesato, N. Durgin, P. Lincoln, J. Mitchell, and A. Scedrov. A meta-notation for protocol analysis. In *Proceedings of 12th IEEE Computer Security Foundations Workshop*. IEEE Computer Society Press, June 1999.

[7] Z. Dang and R. A. Kemmerer. Using the ASTRAL model checker for cryptographic protocol analysis. In *Proceedings of the DIMACS Workshop on Design and Formal Verification of Security Protocols*, September 1997. available at http://dimacs.rutgers.edu/Workshops/Security/.

[8] Whitfield Diffie, Paul C. van Oorschot, and Michael J. Wiener. Authentication and Authenticated Key Exchanges. *Designs, Codes, and Cryptography*, 2:107–125, 1992.

[9] D. Dolev, S. Even, and R. Karp. On the Security of Ping-Pong Protocols. *Information and Control*, pages 57–68, 1982.

[10] D. Dolev and A. Yao. On the Security of Public Key Protocols. *IEEE Transactions on Information Theory*, 29(2):198–208, March 1983.

[11] A. Durante, R. Focardi, and R. Gorrieri. CVS: A compiler for the analysis of cryptographic protocols. In *Proceedings of the 12th IEEE Computer Security Foundations Workshop*, pages 203–212. IEEE Computer Society Press, June 1999.

[12] B. Dutertre and S. Schneider. Using a PVS embedding of CSP to verify authentication protocols. In *TPHOLS'97*, 1997.

[13] S. Even and O. Goldreich. On the security of multi-party ping-pong protocols. In *Proceedings of the 24th IEEE Symposium on the Foundations of Computer Science*, pages 34–39. IEEE Computer Society Press, 1983.

[14] F. Javier Thayer Fabrega, Jonathan C. Herzog, and Joshua D. Guttman. Strand spaces: Why is a security protocol correct? In *Proceedings of the 1998 IEEE Symposium on Security and Privacy*, pages 160–171. IEEE Computer Society Press, May 1998.

[15] N. Ferguson and B. Schneier. A security evaluation of IPSec. In M. Blaze, J. Ioannides, A. Keromytis, and J. Smith, editors, *The IPSec Papers*. Addison-Wesley, to appear.

[16] D. Goldschlag, M. Reed, and P. Syverson. Onion routing for anonymous and private internet connections. *Communications of the ACM*, 42(2), February 1999.

[17] Li Gong and Paul Syverson. Fail-stop protocols: An approach to designing secure protocols. In R. K. Iyer, M. Morganti, Fuchs W. K, and V. Gligor, editors, *Dependable Computing for Critical Applications 5*, pages 79–100. IEEE Computer Society, 1998.

[18] N. Heintze and J. D. Tygar. A model for secure protocols and their composition. *IEEE Transactions on Software Engineering*, 22(1):16–30, Jan. 1996.

[19] A. Huima. Efficient infinite-state analysis of security protocols. presented at FLOC'99 Workshop on Formal Methods and Security Protocols, July 1999.

[20] Richard Kemmerer. Using Formal Methods to Analyze Encryption Protocols. *IEEE Journal on Selected Areas in Communication*, 7(4):448–457, 1989.

[21] H. Krawczyk. SKEME: A versatile secure key exchange mechanism for Internet. In *ISOC Secure Networks and Distributed Systems Symposium*, San Diego, 1996.

[22] P. Lincoln, J. MItchell, M. Mitchell, and A. Scedrov. Probabilistic polynomial-time equivalence and security analysis. In J. Wing, J. Woodcock, and J. Davies, editors, *FM'99 - Formal Methods*, pages 776–793. Springer-Verlag LLNCS 1709, September 1999.

[23] D. Longley and S. Rigby. An Automatic Search for Security Flaws in Key Management Schemes. *Computers and Security*, 11(1):75–90, 1992.

[24] G. Lowe. Breaking and fixing the Needham-Schroeder public-key protocol using FDR. *Software - Concepts and Tools*, 17:93–102, 1996.

[25] W. Marrero. A model checker for authentication protocols. In *Proceedings of the DIMACS Workshop on Design and Formal Verification of Security Protocols*, September 1997. available at http://dimacs.rutgers.edu/Workshops/Security/.

[26] D. Maughan, M. Schertler, M. Schneider, and J. Turner. Internet security association and key management protocol (ISAKMP). Request for Comments 2408, Network Working

Group, November 1998. Available at http://www.ietf.org/html.charters/ipsec-charter.html.

[27] C. Meadows. Applying Formal Methods to the Analysis of a Key Management Protocol. *Journal of Computer Security*, 1:5–53, 1992.

[28] C. Meadows. Analyzing the Needham-Schroeder public-key protocol: A comparison of two approaches. In *Computer Security - ESORICS 96*, pages 355–364. Springer-Verlag, September 1996.

[29] C. Meadows. Analysis of the Internet Key Exchange protocol using the NRL Protocol Analyzer. In *Proceedings of the 1999 Symposium on Security and Privacy*. IEEE Computer Society Press, May 1999.

[30] C. Meadows. A formal framework and evaluation method for network denial of service. In *Proceedings of the 12th IEEE Computer Security Foundations Workshop*. IEEE Computer Society Press, June 1999.

[31] C. Meadows and P. Syverson. A formal specification of requirements for payment in the SET protocol. In *Proceedings of Financial Cryptography '98*. Springer-Verlag LLNCS, 1998.

[32] Catherine Meadows. A system for the specification and verification of key management protocols. In *Proceedings of the 1991 IEEE Computer Society Symposium in Research in Security and Privacy*. IEEE Computer Society Press, May 1991.

[33] Catherine Meadows. The NRL Protocol Analyzer: An overview. *Journal of Logic Programming*, 26(2):113–131, 1996.

[34] J. K. Millen, S. C. Clark, and S. B. Freedman. The Interrogator: Protocol Security Analysis. *IEEE Transactions on Software Engineering*, SE-13(2), 1987.

[35] Jonathan K. Millen. CAPSL: Common Authentication Protocol Specification Language. Technical Report MP 97B48, The MITRE Corporation, 1997. See http://www.csl.sri.com/ millen/capsl.

[36] J. Mitchell, M. Mitchell, and U. Stern. Automated analysis of cryptographic protocols using Murϕ. In *Proceedings of the 1997 IEEE Symposium on Security and Privacy*, pages 141–151. IEEE Computer Society Press, May 1997.

[37] R. M. Needham and M. D. Schroeder. Using Encryption for Authentication in Large Networks of Computers. *Communications of the ACM*, 21(12):993–999, December 1978.

[38] H. Orman. The OAKLEY key determination protocol. Request for Comments 2412, Network Working Group, November 1998. Available at http://www.ietf.org/html.charters/ipsec-charter.html.

[39] Susan Pancho. Paradigm shifts in protocol analysis. In *Proceedings of New Security Paradigms Workshop 1999*. ACM, to appear.

[40] L. Paulson. Mechanized proofs for a recursive authentication protocol. In *Proceedings of the 10th Computer Security Foundations Workshop*, pages 84–95. IEEE Computer Society Press, June 1997.

[41] Lawrence C. Paulson. The inductive approach to verifying cryptographic protocols. *Journal of Computer Security*, 6:85–128, 1998.

[42] M. Reiter and A. Rubin. Crowds: Anonymity for web transactions. *ACM TISSEC*, June 1999.

[43] D. Song. Athena: An automatic checker for security protocol analysis. In *Proceedings of 12th IEEE Computer Security Foundations Workshop*. IEEE Computer Society Press, June 1999.

[44] S. Stoller. A bound on attacks on authentication protocols. presented at 1999 Workshop on Formal Methods and Security Protocols, available at http://www.cs.indiana.edu/hyplan/stoller/, July 1999.

[45] S. Stubblebine and V. Gligor. On Message Integrity in Cryptographic Protocols. In *Proceedings of the 1992 Symposium on Security and Privacy*, pages 85–104. IEEE Computer Society Press, May 1992.

[46] S. Stubblebine and C. Meadows. Formal characterization and automated analysis of known-pair and chosen-text attacks. *IEEE Journal on Selected Areas in Communication*, to appear.

[47] P. Syverson and S. Stubblebine. Group principals and the formalization of anonymity. In J. Wing, J. Woodcock, and J. Davies, editors, *FM'99 - Formal Methods*, pages 814–833. Springer-Verlag LLNCS 1708, 1999.

[48] Paul Syverson and Catherine Meadows. A Logical Language for Specifying Cryptographic Protocol Requirements. In *Proceedings of the 1993 IEEE Computer Society Symposium on Research*

in Security and Privacy, pages 165–177. IEEE Computer Society Press, Los Alamitos, California, 1993.

[49] F. Thayer, J. Herzog, and J. Guttman. Mixed strand spaces. In *Proceedings of the IEEE 1999 Computer Security Foundations Workshop*, pages 72–82. IEEE Computer Society Press, June 1999.

[50] J. Thayer, J. Herzong, and J. Guttman. Honest ideals on strand spaces. In *Proceedings of 11th IEEE Computer Security Foundations Workshop*. IEEE Computer Society Press, June 1998.

[51] B. Timmerman. Secure adaptive dynamic traffic masking. In *Proceedings of the 1999 New Security Paradigms Workshop*. ACM, to appear.

[52] B. Venkatraman and R. Newman-Wolfe. Capacity estimation and auditability of network covert channels. In *Proceedings of the 1995 IEEE Symposium on Security and Privacy*. IEEE Symposium on Security and Privacy, May 1995.

[53] D. Volpano. Formalization and proof of secrecy properties. In *Proceedings of the IEEE 1999 Computer Security Foundations Workshop*, pages 92–95. IEEE Computer Society Press, June 1999.

[54] J. Zhou. Fixing a security flaw in IKE protocols. *Electronics Letters*, 35(13):1072–1073, June 1999.

Formal Specification and Analysis of Active Networks and Communication Protocols: The Maude Experience

G. Denker and J. Meseguer
Computer Science Laboratory
SRI International
Menlo Park, CA 94025, USA
{denker | meseguer}@csl.sri.com

C. Talcott
Computer Science Department
Stanford University
Stanford, CA 94305, USA
clt@sail.stanford.edu

Abstract

Rewriting logic and the Maude language make possible a new methodology in which formal modeling and analysis can be used from the earliest phases of system design to uncover many errors and inconsistencies, and to reach high assurance for critical components. Our methodology is arranged as a sequence of increasingly stronger methods, including formal modeling, executable specification, model-checking analysis, narrowing, and formal proof, some of which can be used selectively according to the specific needs of each application. The paper reports on a number of experiments and case studies applying this formal methodology to active networks, communication protocols, and security protocols.

1 Introduction

When a new software technology is first developed, nobody has at first a complete understanding of that new technology. The traditional "build first, learn later" approach is too wasteful and, by providing information too late, does not allow any flexibility to optimize the design and implementation effort. It can also lead to serious integration problems, particularly when different teams are developing different subsystems. All those difficulties apply to Active Networks: it is a novel technology being developed by different teams, and the traditional engineering approaches to deal with its architecture and the integration of subsystems are insufficient. More generally, new methods are needed to properly model and design the increasingly more sophisticated communication systems now in the horizon. We propose that much can be gained from a paradigm shift that uses formal modeling systematically and more aggressively from the earliest phases of system design.

The work presented here focuses on applying a novel, wide-spectrum executable formal specification language, called Maude, to Active Networks and Communication Protocols. Our experience indicates that Maude can substantially help in modeling, symbolically simulating, and analyzing the subsystems of Active Networks and other communication systems, in documenting and ensuring consistency of important parts of the architecture, and in formalizing and analyzing important safety-critical aspects. Part of what makes Maude very well suited for these purposes is its flexible wide-spectrum character: it can deal with very early design phases such as architectures and high-level designs, can be used to quickly develop executable prototypes, and can also be used to generate code. There is also a wide range of options on the kind of analyses that can be performed. One can develop formal models of a system very early, can debug formal specifications—which can be partial and incomplete—by executing them, can do more exhaustive model-checking and symbolic simulation analyses, or, for highly critical subsystems, can in fact do full formal verification using our theorem proving tools. Often much can be gained from the "lightweight" application of our methods through executable specification; the full power of our methods can be reserved for those aspects in which it is more critically needed. The first thing to be gained even from the simplest executable specification uses is a precise documentation of a system's design, that can be used as a clear means of communication between teams, and also to find many bugs and inconsistencies very early on.

The Maude interpreter [7] offers also some useful advantages. The first is efficient executability (1.3 million rewrites per second on a Pentium II for some applications) combined with the ability to trace in detail each execution step. But executability is actually not enough. Since a concurrent system can have many different behaviors, to properly analyze the system it becomes important to explore not just the single execution provided by some default strategy, but many other executions. Under assump-

0-7695-0490-6/99 $10.00 © 1999 IEEE

tions of finite-state or of termination it may even be possible to analyze all executions. The reflective capabilities of rewriting logic and Maude [6] are very helpful in this regard, because they allow user-defined execution strategies that can be formally specified by rewrite rules at the metalevel, including strategies such as breadth-first-search that can exhaustively explore all the executions of a system from a given initial state. This is very helpful in uncovering security flaws under unforeseen attack scenarios or examining the behavior of newly designed algorithms. Maude executables, its manual, and a collection of examples and papers have been available on the web since January 1999 (`http://maude.csl.sri.com`).

In summary, using Maude to formally specify and analyze communication systems offers the following advantages:

- *Early insertion of the formal method.* In this way, maximum benefit can be obtained, since the design can be corrected very early, before heavy implementation efforts have been spent.

- *Simplicity and intuitive appeal of the formalism.* The formalism involved, namely rewriting logic [29], is very simple and it is very well suited for specifying distributed systems, in which local concurrent transitions can be specified as rewrite rules.

- *Modeling Flexibility.* Instead of building in a fixed model of concurrency, rewriting logic allows great flexibility to specify many such models [29, 32], including both synchronous and asynchronous models of communication and a wide range of concurrent object systems.

- *Executability.* Rewriting logic specifications are executable in a rewriting logic language such as Maude [7]. This means that the formal model of the protocol becomes an *executable prototype*, that can be directly used for simulating, testing and debugging the specification.

- *Formal analysis and proof.* Since the behavior of a communications protocol is highly concurrent and nondeterministic, a particular simulation run only exhibits *one* among many possible behaviors. Therefore, although direct execution can already reveal many errors and inconsistencies, a much greater confidence in the correctness of the design can be gained by formal analysis techniques in which *all* possible behaviors—up to termination, or up to a certain depth, or up to satisfaction of a specific state condition—are analyzed in detail. This can be done in Maude by means of exhaustive execution strategies that achieve a form of "model checking" analysis of the state space. In addition, formal proofs of highly critical properties can also be carried out using Maude's theorem proving tools [4].

1.1 Applying Maude to Active Networks and Communication Protocols

In collaboration with other teams working on active networks and on communication and security protocols we have applied Maude to formally specify and analyze active networks protocols and algorithms, security protocols, composable communication services, and distributed software architectures. Our experience so far is quite encouraging:

- In [14] we report joint work with group led by J.J. García-Luna at the Computer Communications Research Group at University of California Santa Cruz in which we used Maude very early in the design of a new reliable broadcast protocol for ANs. In this work, we have developed precise executable specifications of the new protocol and, by analyzing it through execution and model-checking techniques, we have found many deadlocks and inconsistencies, and have clarified incomplete or unspecified assumptions about its behavior.

- We have also applied Maude to the specification and analysis of cryptographic protocols [10] and have shown how our model-checking techniques can be used to discover attacks.

- The positive experience with security protocols has led to the adoption of Maude by J. Millen and G. Denker as the basis for giving a formal semantics to their new secure protocol specification language CAPSL and as the meta-tool used to endow CAPSL with an execution and formal analysis environment [13].

- The paper [44] reports joint work with with Y. Wang and C. Gunter at the University of Pennsylvania in using Maude to formally specify and analyze a PLAN [22] active network algorithm.

- In [11] we present an executable specification of a general middleware architecture for composable distributed communication services such as fault-tolerance, security, and so on, that can be composed and can be dynamically added to selected subsets of a distributed communications system.

- In[7] (Appendix E) we present a substantial case study showing how Maude can be used to execute very high level software designs, namely architectural descriptions. It focuses on a difficult case, namely, *heterogeneous* architectures illustrated by a command and control example featuring dataflow, message passing, and implicit invocation sub-architectures. Using Maude, each of the different subarchitectures can not only be executed, but they can also be interoperated in the execution of the resulting overall system.

- As part of a project to represent the Wright architecture description language [2] in Maude we have developed a representation of CSP ("Communicating Sequential Processes", process algebra notation) [23] in Maude. This is compatible with existing tools for analyzing CSP specifications, complements them by providing a rich execution environment and the ability to analyze non-finite state specifications, and provides a means of combining CSP specifications with other notations for specifying concurrent systems.

1.2 Formal Methodology

The formal methodology underlying our approach can be summarized by stating that *a small amount of formal methods can go a long way*. Approaches requiring full mathematical verification of a system can be too costly. Proof efforts should be used judiciously and selectively, carefully choosing those properties for which a very high level of assurance is needed. But there are many important benefits that can be gained from "lighter" uses of formal methods, without necessarily requiring a full-blown proof effort.

The general idea is to have a series of *increasingly stronger methods*, to which a system specification is subjected. Only after less costly and "lighter" methods have been used, leading to a better understanding and to important improvements and corrections of the original specification, is it meaningful and worthwhile to invest effort on "heavier" and costlier methods. Our approach is based on the following, increasingly stronger methods:

1. *Formal specification.* This process results in a first *formal model* of the system, in which many ambiguities and hidden assumptions present in an informal specification are clarified.

2. *Execution of the specification.* If the formal specification is executable, it can be used directly for simulation and debugging purposes, leading to increasingly better versions of the specification.

3. *Formal model-checking analysis.* Errors in highly distributed and nondeterministic systems not revealed by a particular execution can be found by more a sophisticated model-checking analysis that considers all behaviors of a system from an initial state, up to some level or condition. In this way, the specification can be substantially hardened, and can even be formally verified if the system is finite-state.

4. *Narrowing analysis.* By using symbolic expressions with logical variables, one can carry out a symbolic model-checking analysis in which all behaviors not from a single initial state, but from the possibly infinite set of states described by a symbolic expression are analyzed. Some of these analyses are already a spccial typc of formal proof.

5. *Formal Proof.* For highly critical properties it is also possible to carry out a formal proof of correctness, which can be assisted by formal tools such as those in Maude's formal environment [4].

Up to now, we have used methods 1–3 in the case studies mentioned in Section 1.1, reaping important benefits from this use. In the future we also plan to use methods 4 and 5 for selected purposes.

1.3 Outline of this Paper

Rewriting logic and Maude, including the formal modeling of distributed objects, are summarized in Section 2. To make our methods and ideas concrete, we discuss in detail a case study applying Maude to security protocols in Section 3. Other active network and communication protocol applications are discussed in Section 4. Section 5 presents some concluding remarks and discusses future developments.

2 Rewriting Logic and Maude Basics

Rewriting logic is a very simple logic in which the state space of a distributed system is formally specified as an algebraic data type by means of an equational specification consisting of a signature of types and operations Σ and a collection of conditional equations E. The *dynamics* of such a distributed system is then specified by rewrite rules of the form

$$t \to t'$$

where t, t' are Σ-terms, that describe the *local, concurrent transitions* possible in the system, namely, when a part of the distributed state fits the pattern t, then it can change to a new local state fitting the pattern t'. Therefore, a *rewrite theory* is a triple (Σ, E, R), with (Σ, E) an equational specification axiomatizing a system's distributed state space, and R a collection of rewrite rules axiomatizing the system's local transitions. Rewriting logic has proved to be a general and flexible semantic framework for specifying a wide variety of models of concurrency [29, 33], and in particular for specifying concurrent object systems [30].

Maude [5] is a multi-paradigm executable specification language based on rewriting logic [29]. Maude integrates an equational style of functional specification with an object-oriented specification style for highly concurrent and nondeterministic object systems. Maude modules are rewrite theories whose basic axioms are rewrite rules. The complex concurrent computations of the system so axiomatized then exactly correspond to proofs in the logic. Maude specifications can be efficiently executed using the Maude rewrite engine [5], thus allowing their use for system prototyping and the debugging of specifications.

We briefly summarize the syntax of Maude. There are three types of modules:

- *functional* modules, that are equational theories used to specify algebraic data types; they are declared with the syntax `fmod ... endfm`,
- *system* modules, that are rewrite theories specifying concurrent systems; they are declared with the syntax `mod ... endm`, and
- *object-oriented* modules, that provide special syntax to specify concurrent object-oriented systems but are entirely reducible to system modules; they are declared with the syntax `omod ... endom`.

All the above modules can be *parameterized*, and have an *initial model semantics*. That is, they specify the initial models of their corresponding theories, or, in the case of parameterized modules, the free extension of a model of the parameter theory to a model of the body.

Immediately after the module's keyword, the *name* of the module is given. After this, a list of imported submodules can be added. One can also declare *sorts* and *subsorts* and *operators*. Operators are introduced with the `op` keyword. They can have user-definable syntax, with underbars '_' marking the argument positions, and are declared with the sorts of their arguments and the sort of their result. Some operators can have equational *attributes*, such as `assoc` and `comm`, stating, for example, that the operator is associative and commutative. Such attributes are then used by the Maude engine to match terms *modulo* the declared axioms.

There are three kinds of logical axioms, namely, *equations*—introduced with the keywords `eq`, or, for conditional equations, `ceq`—*memberships*—declaring that a term has a certain sort and introduced with the keywords `mb` and `cmb`—and *rewrite rules*—introduced with the keywords `rl` and `crl`. Functional modules can only have equations and memberships. System and object-oriented modules can have any of the three kinds of axioms. The mathematical variables in such axioms are declared with the keywords `var` and `vars`.

In this paper, given that a concurrent object-oriented style fits communication protocols very well, we will focus on the specification of concurrent object-oriented systems. In object-oriented Maude modules one can declare *classes* and *subclasses*. Each class is declared with the syntax

$$\texttt{class}\ C \mid a_1 : S_1,\ \ldots,\ a_n : S_n$$

where C is the name of the class, and for each $a_i : S_i$, a_i is an attribute identifier, and S_i is the sort over which the values of such an attribute identifier must range. Objects of a class are then record-like structures of the form

$$< O;\ C \mid a_1 : v_1,\ \ldots,\ a_n : v_n >$$

with O the name of the object, $v_1, \ldots v_n$ the current values of its attributes, and with v_i of sort S_i for $1 \leq i \leq n$. Objects can interact with each other in a variety of ways, including the sending of messages. A message declaration

$$\texttt{msg}\ m : p_1 \ldots p_n \rightarrow \texttt{Message}$$

defines the name of the message and the sorts of its parameters. Often, the first parameter refers to the object the message is sent to. The state of a concurrent object system is called a *configuration*. Typically, a configuration is a multiset of objects and messages. The multiset union operator for configurations is associative and commutative so that order and parentheses do not matter, and so that rewriting is multiset rewriting in Maude. It is denoted with empty syntax (juxtaposition). The dynamic behavior of a concurrent object system is then axiomatized by specifying each of its basic concurrent transition patterns by a corresponding labeled rewrite rule. For instance,

```
rl [rule1] :
m1(O1,O3) < O1; C1 | a1: v1, a2: n >
=> < O1; C1 | a1: f(v1), a2: n+1 >
   < O1.n; C2 | b: g(v1) > m2(O3)
```

expresses the result of an asynchronous reception of a message type `m1` by an object of class `C1`, namely, the message is consumed, the object changes its `a1` value to `f(v1)` and has its `a2` value increased by one, a new object of class `C2` with identity `O1.n` is created, and a message of type `m2` is sent to `O3`. In general, a rewrite rule may have several objects in its lefthand side, describing a *synchronization* or *multi-party interaction* between several objects. However, in most of the rules in the communication protocol examples that we shall discuss the communication will be asynchronous and the lefthand sides will consist of either an object and a message, or just an object. By convention, attributes whose value does not change and does not affect the next state of other attributes need not be mentioned in a rule.

3 A Cryptographic Protocol Analysis Application

Cryptographic protocols for key exchange and authentication are the basis for network security. These protocols are often described in a message-list style, i.e., the sequence of messages to be exchanged between the communicating agents is defined as a finite transition system. Though many of the security protocols only involve a very restricted number of messages, it is surprisingly difficult to get the protocols correct. Even under the assumption of "perfect cryptography", i.e., an encrypted message can only be decrypted with the proper key, it has been shown that protocols are vulnerable to message modification attacks [16, 25]. An intruder can overhear or intercept network communications and can manipulate ongoing communications by faked messages. The highly concurrent nature of cryptographic protocols is responsible for the complexity of proving a cryptographic protocol secure with respect to message secrecy or correct with respect to authentication. This suggests that a formal analysis methodology supporting several analysis techniques, such as simulation, model checking and theorem proving, could be of great value for the design of cryptographic protocols.

We have applied Maude to the specification and analysis of several cryptographic protocols. We illustrate our approach with the help of the Needham-Schroeder Public-Key (NSPK) protocol [34]. In Section 3.1 we illustrate the naturalness of an object-oriented NSPK specification in Maude. The level of specification is detailed enough to capture essential operational aspects and to allow rapid prototyping and debugging by directly executing the specifications. In this sense, it can be compared to protocol specifications based on process algebras, CSP, the π-calculus, UNITY, or Petri nets. However, unlike those approaches, rewriting logic does not prescribe a fixed model of concurrency. This translates into additional naturalness and ease of representation.

In order to successfully detect protocol vulnerabilities it is necessary to build not only an abstract model of the protocol but also formal models of the malicious environment, and to define situations in which security goals are violated. It is the combination of the various models of the protocol, the attacker, and the attack that is amenable to formal analysis techniques. In Section 3.2 we describe our formalization of the malicious environment. A breadth-first search strategy and the attack are presented in Section 3.3.

The Maude interpreter allows to efficiently execute the protocol specification. This way we can detect inconsistencies and flaws. But this is not enough to examine all possible behaviors. Our systematic search for an attack using a breadth-first strategy illustrates a form of model checking in which we explore all the possible behaviors beginning with a given state using the reflective capabilities of Maude. We explain our strategies in more detail in Section 3.3.

3.1 Specifying and Executing the NSPK Protocol

The NSPK Protocol [34] is a cryptographic protocol for the authentication of pairs of agents in a distributed computer system. For instance, agents want to be assured of each other's identity before they exchange security-critical data. Thus, an intruder should not be allowed to impersonate another agent. For this purpose, initiator and responder of a communication mutually authenticate each other. NSPK uses public key cryptography, i.e., each agent possesses a public key which can be accessed by all agents and a secret key, which is the inverse of the public key. Moreover, nonces are used in the protocol. Nonces are freshly generated, random numbers to be used in a single run of the protocol. It is assumed that these numbers are generated in such a way that they cannot be guessed by other agents. A textbook-style simplified description of NSPK as for instance given in [25] is:

Message 1 $\quad A \rightarrow B : \{N_a, A\}_{PK(B)}$
Message 2 $\quad B \rightarrow A : \{N_a.N_b\}_{PK(A)}$
Message 3 $\quad A \rightarrow B : \{N_b\}_{PK(B)}$

In this protocol A tries to establish a communication with B. Thus, A plays the role of the initiator and B reacts in the responder role. A encrypts his nonce N_a with the public key of B and sends it along with his identity to B. Now A is in a waiting status. B decrypts the message, creates a new nonce and sends back to A the concatenation of both nonces encrypted with the public key of A. After this, B is waiting for the answer from A. A can decrypt the message sent by B. In case the contents of the message is the concatenation of his own nonce N_a plus another nonce, A can be assured that the message was sent by B, because B is the only agent who can decrypt the message sent by A containing the nonce N_a. Thus, A establishes a communication with B. Moreover, to acknowledge to B the receipt of B's nonce, A separates it from the message and sends it back. If B receives this message from A, he can be assured that he is talking to A and, therefore, of having established a communication with A.

As one can see, there is a lot of implicit knowledge in the textbook-style specification of the NSPK Protocol. Moreover, there are open questions concerning how agents behave if they receive a message which does not have the expected contents, e.g., if A receives a message from B which does not contain his nonce.

We make this implicit knowledge explicit and clarify fuzzy parts of the description by formalizing the protocol in rewriting logic. In addition to the protocol requirements given above, we provide a Maude specification which allows multiple sessions for an agent. Therefore, an agent may simultaneously participate in several runs of the protocol, and may simultaneously play different roles (e.g., initiator or responder) in different runs.

We need to provide some application-specific data types such as fields, nonces, keys, etc. These types are equationally specifiable with the help of Maude's functional modules.

```
fmod DATATYPES is
  sorts Key Field FieldSet Nonce Principal
        Run Role EstabComm .
  subsort Nonce Principal Key < Field .
  subsort Field < FieldSet .
  op keypair : Key Key -> Bool [comm] .
  op mtfield : -> Field .
  op ped : Key Field -> Field .
  op n : Principal Nat -> Nonce .
  op cat : Nonce Nonce -> Field .
  op (_,_,_) : Nonce Principal Nonce -> Run .
  op (_,_,_,_) : Role Nonce Principal Nonce
     -> EstabComm .
  ops i r : -> Role .
  vars sk pk : Key . var f : Field .
  ceq ped(sk,ped(pk,f)) = f
      if keypair(sk,pk) .
  ...
```

In the sort declaration, application-specific data types such as keys, fields, nonces, principals (i.e., identities of agents), etc., are defined. Sorts may be ordered in an inclusion hierarchy by defining subsort relationships. For instance, nonces, principals, and keys are subsorts of `FIELD`. The predicate `keypair` is true for pairs of secret and public keys of one agent. `ped`[1] is the operation to encrypt and decrypt fields. Agents send data of type `Field` in a message. `mtfield` is the empty constant of sort `Field`. `n` is a function to generate fresh, cryptographic nonces. To make theses nonces unique, we implemented this function by using the address of an agent and a number. Concatenation of two nonces is denoted with `cat`. The conditional equation defines the behavior of `ped`. In the complete specification, appropriate equations for all operations are provided. We use the sort `Run` to store information about different sessions of an agent. An agent may concurrently communicate with several other agents by establishing several runs. The first argument of a run is the nonce which is newly created by the agent for the session. The second and third arguments are the identity and nonce of the communication partner. We assume that nonces are always freshly generated and, therefore, can be used to uniquely identify a run. `EstabComm` will be used to store information about successfully established communications, in either of the two roles, initiator or responder. For each established communication an agent stores his own nonce, and for the communication partner it stores the identity and the nonce.

We define an object class `Agent` in Maude. Agents will model the legal participants of a protocol. Later on we will also define a class `Intruder`. Agents have a secret key, two attributes storing information about possible runs (`Run`), one for each possible role, an attribute to store information about established communications, an attribute for desired communications, i.e., a set of object identifiers to identify the agents they want to communicate with, and a counter which will be used to generate nonces.

```
class Agent | e_com: EstabCom, sec_key: Key,
              role_i: Run, role_r: Run,
              d_com: FieldSet, cnt: Nat .
```

Moreover, we need one message to send data between agents. The first two parameters of this message contain the identities of the sender and receiver agents. The third argument is the content to be sent, i.e., an encrypted field.

```
msg from_to_send_ : Principal Principal
    Field -> Message .
```

The Maude specification of NSPK consists of six rules. The first rule `BeginRun` defines the start of a protocol run. The second rule `Message1Rec` defines the behavior of a responder agent to a request of starting a protocol session.

```
vars A B P : Principal . vars RI RR : Run .
vars NI : Nonce . var F : Field .
var S : FieldSet .
vars SKB PKB SKA PKA : Key .
var C : EstabCom .

rl [BeginRun] :
  < A; Agent | role_i: RI, d_com: B U S,
               cnt: J >
  =>
  < A; Agent | role_i: RI U (n(A,J),B,mtfield),
               d_com: S, cnt: J + 1 >
  from(A)to(B)send(ped(pk(B),n(A,J))) .
```

If an agent `A` wants to talk to at least one other agent, say `B` (`U` is the union operator for field sets), then he starts a protocol run by sending to this agent his address `A` and a newly created nonce `n(A,J)`. He deletes the agent `B` from his set of desired communication partners and keeps track of the information of the protocol run in attribute `role_i`. Additionally, the agent encrypts his nonce `n(A,J)` with the public key of his communication partner[2].

```
crl [Message1Rec] :
  < B; Agent | sec_key: SKB, role_i: RI,
               role_r: RR, cnt: J >
  from(A)to(B)send(ped(PKB,F))
  => < B; Agent | role_r: RR U (n(B,J),A,F),
                  cnt: J + 1 >
  from(B)to(A)send(ped(pk(A),cat(F,n(B,J))))
  if keypair(SKB,PKB) and not(F in RR) .
```

[1] "public key encryption and decryption" (borrowed from CAPSL). This operator behaves like RSA encryption.

[2] `pk(B)` is understood as a function which delivers the public key of agent B. In our implementation we used a table of public keys which is accessible by all agents.

If an agent gets a request for a protocol session, he stores the information in attribute `role_r` and sends back the encrypted concatenation of nonces, provided he could decrypt the message. The second part of the condition `not(F in RR)` assures that the nonce is not in use in a responder session of agent `B`. As already mentioned, by default convention one only needs to mention in the righthand side of the rule those attributes which change their value.

The next two rules deal with the receipt of Message 2. In case the initiator gets back his own nonce plus the new nonce of the other agent, he establishes the communication and triggers the last message, i.e., sending Message 3. This behavior is specified in rule `Message2RecCorrect`. Otherwise, the initiator refuses to establish the communication, does not send back any message and just stops the current protocol run (`Message2RecIncorrect`).

```
crl [Message2RecCorrect] :
  < A; Agent | sec_key: SKA, e_com: C
               role_i: RI U (NI,P,mtfield) >
  from(B)to(A)send(ped(PKA,F))
  =>
  < A; Agent | role_i: RI,
               e_com: C U (i,NI,B,rest(F)) >
  from(A)to(B)send(ped(pk(B),rest(F)))
  if keypair(SKA,PKA)
     and (B == P) and NI == first(F)   .

crl [Message2RecIncorrect] :
  < A; Agent | sec_key: SKA,
               role_i: RI U (NI,P,mtfield) >
  from(B)to(A)send(ped(PKA,F))
  =>
  < A; Agent | role_i: RI >
  if keypair(SKA, PKA)  and NI == first(F)
     and B =/= P .
```

`first` and `rest` are operations which are defined on fields. Applied to a concatenation of nonces they deliver the first or last one, respectively. Analogously, there are two rules for the responder, covering the cases for the correct and the incorrect receipt of Message 3 of the NSPK Protocol. Because of space limitations we refrain from presenting these rules. They are very similar to the last two rules above.

A slightly extended version of the specification presented above has been executed on the Maude rewrite engine [5]. The given initial configuration involved two agents `Alice` and `Bob`, where `Alice` wants to talk to `Bob`. Using the default rewriting strategy we already found an unexpected behavior. Analyzing the rewrite path showed that the problem was connected with the ability of having several runs per agent. The condition of the rule `Message1Rec` was not sufficient. If the responder sends Message 2 of the NSPK Protocol, i.e., `from(Bob)to(Alice)send(ped(pk(Alice), cat(NI,n(Bob,J))))`, the rule `Message2RecCorrect` should be applied to indicate the situation where `Alice` already started a run with `Bob` and she waits for his answer. But there is also a matching for another rule, namely `Message1Rec`. Thus, `Alice` sets up a new responder session with `Bob` as initiator without recognizing that there is another run in which the nonce `NI` is already used. This error clarifies a fact which was not clear from the informal specification: in this protocol nonces play the role of session identifiers. In order to reflect this in the specification we have to change the condition for rule `Message1Rec`. It needs to be checked whether the first nonce in the message is already used in another session (`not(first(F) in RI)`). This example illustrates the usefulness of executing specifications in the Maude engine for validation purposes.

3.2 The Intruder

The intruder is part of the system and, therefore, may participate in normal protocol runs. Moreover, an intruder may observe, intercept or fake messages. Introducing a fake message into the system either means replaying a previously observed or intercepted message, or creating a new message using the nonces known by the intruder. In Maude we can give a comprehensive specification of a general intruder who is able to perform all the mentioned actions.

An intruder has usually all attributes an agent has, plus additional ones to store the set of nonces he knows and all the messages he observed or intercepted.

```
class Intruder |
        e_com: EstabCom, sec_key: Key,
        ncs: FieldSet, msgs: FieldSet,
        agents: FieldSet, role_i: Run,
        role_r: Run, d_com: Field, cnt: Nat.
```

Moreover, since the intruder may behave as a normal agent, all agent rules are redefined in such a way that, whenever an intruder creates or receives a nonce, he stores it in the attribute `ncs`. Similarly, he remembers all seen agent identities in `agents` and all seen messages in `msgs`. We do not give all these redefined rules. They are very similar to the normal agent rules, except for being extended, so that the intruder can remember nonces and agent identities. Instead we describe the rules specifying the intrusion behavior.

```
crl [IntruderInterceptMessage] :
  < I; Intruder | sec_key: SKI, ncs: N,
                  msgs: M, agents: S >
  from(B)to(A)send(ped(KEY,F))
  =>
  if keypair(KEY,SKI)
  then < I; Intruder | ncs: N U F,
                       agents: S U A U B >
  else < I; Intruder | msgs: M U ped(KEY,F),
                       agents: S U A U B >
```

```
fi
if A =/= I and B =/= I .
```

If the intruder intercepts a message that is encoded with a key he has, then he can decrypt it and store the nonce in his list of nonces. A similar rule for overhearing messages only differs from the interception rule in the way the messages remain in the configuration. The following two rules express the intruder's ability to fake messages.

```
crl [IntruderReplayMessage] :
  < I; Intruder | msgs: M U MES,
                  agents: S U A U B >
  => < I; Intruder | msgs: M U MES >
     from(B)to(A)send(M)
  if MES =/= mtfield and A =/= I .

crl [IntruderFakeMessage] :
  < I; Intruder | ncs: N U F,
                  agents: S U A U B >
  => < I; Intruder | ncs: N U F >
     from(A)to(B)send(ped(pk(B),F))
  if B =/= I .
```

3.3 Finding an Attack: a Breadth-First-Search Strategy

To prove the correctness of the protocol one has to prove: (1) the authentication of the responder, i.e., that an initiator only establishes a communication with another agent if the agent took part in the protocol run, and (2) the authentication of the initiator. Concerning item (2) there is a well-known attack to the NSPK Protocol proposed by Gavin Lowe [25]. This attack is newer and more subtle than the one proposed by Denning and Sacco in [16] which allows an intruder to replay old, compromised public keys. The attacker takes part in two sessions and exchanges information between them ($I(A)$ stands for the intruder masquerading as agent A).

Message 1	$A \rightarrow I : \{N_a, A\}_{PK(I)}$
Message 1'	$I(A) \rightarrow B : \{N_a, A\}_{PK(B)}$
Message 2'	$B \rightarrow A : \{N_a.N_b\}_{PK(A)}$ /* *intercepted* */
Message 2	$I \rightarrow A : \{N_a.N_b\}_{PK(A)}$
Message 3	$A \rightarrow I : \{N_b\}_{PK(I)}$
Message 3'	$I(A) \rightarrow B : \{N_b\}_{PK(B)}$

At the end of this attack B believes that he talks to A, i.e., the authentication of the initiator fails.

We can take advantage of the Maude engine to analyze the enriched specification "NSPK+Intruder". For this purpose we need to specify an appropriate analyzing strategy. Thanks to the reflective nature of rewriting logic, it is possible to define a very wide variety of rewriting strategies using rewrite rules at the metalevel, and to then execute a specification with a given strategy in Maude. For example, one can program a breadth-first-search strategy to exhaustively determine all possible behaviors from a given initial state using the rewrite rules of a Maude specification. In this way we can either validate the specification and detect errors, or do formal reasoning about the specification. To analyze behaviors starting from a symbolic description of a set of states, the technique of narrowing can instead be used (see Section 5).

For the "NSPK+Intruder" specification we implemented a breadth-first-search strategy (`brFS`) which searches for possible attacks. To efficiently implement this strategy, we have chosen to use a bounded depth-first-search strategy (`bDFS`) which is recursively called with increasing bounds as long as no attack is discovered. A breadth-first-search strategy which starts at a given depth and iteratively calls a depth-first search with increased maximum depth is given as follows:

```
var StartDepth : MachineInt .
op brFS :
   Module Term QidList QidSet MachineInt
     -> Strategy .
op bDFS :
   Module Term QidList QidSet MachineInt
     -> Strategy .
eq brFS(M,T,Labels,StopLabels,StartDepth)
 = if bDFS(M,T,Labels,StopLabels,StartDepth)
        == stop(0, emptyPath)
   then bDFS(
        M,T,Labels,StopLabels,StartDepth+1)
   else bDFS(
        M,T,Labels,StopLabels,StartDepth)
   fi .
```

Both strategies have five parameters: a module with the specification, a term which is the initial configuration, a list of labels which should be applied during the computation (i.e., all rewrite labels in the specification), a set of rewrite labels which define attack situations and for which the search stops, and a depth, which in the case of the breadth-first-search strategy is the depth to start with, and in the case of the bounded depth-first-search strategy is the maximum depth.

For NSPK we defined one stop rule: a successful attack is determined by the fact that nonces have been compromised, i.e., an intruder possesses two nonces with which another agent established a communication. The conditions of the following rule assure that it can only be applied in case an attack happened. The intruder remembers the successful attack by storing the information in the set of established communications.

```
crl [IntruderRecognizeAttack] :
  < I; Intruder | e_com: EC, role_i: RI,
                  role_r: RR,
                  ncs: N U N1 U N2 >
  < A; Agent | e_com: EC' U (ROLE,N1,B,N2) >
  =>
```

```
if ROLE == i
then
  < I; Intruder | e_com: EC U (r,N2,A,N1) >
  < A; Agent | e_com: EC' U (ROLE,N1,B,N2) >
else
  < I; Intruder | e_com: EC U (i,N2,A,N1) >
  < A; Agent | e_com: EC' U (ROLE,N1,B,N2) >
fi
if not((r,N2,A,N1) in EC)
   and not((i,N2,A,N1) in EC)
   and not(N1 in RR)
   and not(N1 in RI)
   and not(N2 in RR)
   and not(N2 in RI) and B =/= I .
```

The intuitive idea is that, starting in the initial configuration, for a rewrite step all rules in the given list of rules are tried until a rewrite is found. In case there is no path containing the rewrite label in the given set, the empty path is returned; otherwise, the first such path found in the search is returned. A rewrite step (`step`) consists of a rewrite label, a number indicating which substitution for the rule with that label was used, and a term that is precisely the righthand side replacement for the given label, rule, and substitution. Thus, a path of the form `path(T1, step(L1,N1,T2), step(L2,N2,T3))` represents a computation sequence which starts in configuration `T1` and in this configuration the rule with label `L1` and matching substitution number `N1` has been applied and resulted in configuration `T2`, from which rule `L2` could be applied with the `N2`th matching substitution resulting in configuration `T3`.

The bounded depth-search-first algorithm continues building the path, provided that none of the rules in the set of stop rules is applied, nor the maximum depth is reached. For each new step in the path all rewrite rules are tried. Whenever a path is closed, i.e., whenever no other rewrite rules are applicable, or the maximum depth is reached, the algorithm backtracks. Backtracking means that first the same rule as in the previous step is tested with the next matching substitution number. If this fails, then the next rule in the list of applicable rules is tried. If this also fails, another backtracking step is taken until no more rules can be applied from the initial configuration. If any of the previous two attempts is successful, the bounded depth-first-search algorithm for the new, extended path is called. Running this strategy delivers the attack given in [25].

Notice that our protocol specification is an infinite state specification. We made no restrictions to fit the specification into a finite state tool. This is one of the strengths of the Maude approach. Moreover, the specification is amenable to further forms of analysis such as narrowing analysis and theorem proving. The NSPK example has been chosen for illustrative purposes to emphasize the wide-spectrum modeling and analysis technique of rewriting logic and Maude.

4 Other Applications

4.1 A Reliable Broadcast Protocol

In [14] we report on an ongoing case study in which a new reliable broadcasting protocol (RBP) [20] currently under development at the University of California at Santa Cruz (UCSC) has been formally specified and analyzed, leading to many corrections and improvements to the original design. Traditionally, the only means to document such protocol designs is pseudo-code. Since this is not suitable for execution or formal analysis, it is only in a post-facto way, after the protocol has been implemented, that flaws and inconsistencies are usually discovered. The process of formally specifying the protocol, and of symbolically executing and formally analyzing the resulting specification, has revealed many bugs and inconsistencies very early in the design process, before the protocol was implemented.

RBP performs reliable broadcasting of information in networks with dynamic topology. Reliable broadcasting is not trivial when the topology of the network can change due to failure and mobility. The aim is to ensure that all nodes that satisfy certain connectedness criteria receive the information within a finite time, and that the source is notified about it. The protocol should furthermore incur as low latency and as few messages as possible.

The full Maude specification tackles the cases of sending and receiving messages and acknowledgments, as well as link deletion and link addition, and can be found in [15]. A major advantage of rewriting logic specifications is that they can be validated immediately by executing test cases to provide quick feedback on the specification. This prototyping possibility comes for free. We used this feature every time the specification was modified, and often encountered errors quite easily on quite simple test examples (such as a network of three nodes). In this validation effort, we executed the test cases using Maude's default interpreter, which simulated an arbitrary run of the protocol for a given initial state of a network. The validation effort helped eliminate errors of syntax and of thought; furthermore, the built-in Maude facilities for tracing an execution were useful for discovering where the error occurred. This validation and correction cycle led to substantial improvements on, and a clear formalization of, the basic ideas of the starting informal protocol.

To substantially increase our confidence in the specification before any costly attempt at a formal proof of correctness, the specification was subjected to close formal analysis using the meta-programming features of Maude to explore all states and behaviors that can nondeterministically be reached from an initial state. Since the specification should be terminating, one could apply a strategy that explores all possible rewrite paths from some given initial state. In particular, we wrote a strategy for finding every non-rewritable state reachable from the initial state. For non-terminating systems, this setting can be modified to give, e.g., every state which is reachable in less than 50 one-step rewrites from some initial state.

We experimented with different, increasingly complex, versions of the protocol. For example, executing the specification using an exhaustive strategy on a clique with three nodes did not produce the hoped-for result, namely a singleton set of irreducible states. Instead, the set of irreducible configurations reachable from the clique of three nodes included a term which indicated a deadlock before the expected end of one round of the protocol. A simple analysis of the output explained the error.

Using Maude's meta-programming features, the user may himself define the rewrite strategies, thereby analyzing the specification in various ways. Although we only needed the quite straightforward exploratory analysis to invalidate our first version, this capability is very useful for analyzing various executions and extracting the relevant information from these automatically. Using the information from the exploratory analysis, the group came up with a new improved version of the protocol.

The effort of formally specifying RBP for dynamic networks using Maude brought to light several weaknesses of the original pseudocode [20]. However, the analysis is not yet finished. For the moment, the following problems were identified and solved (for details see [15]):

1. As described above we could eliminate a deadlock situation in the given protocol.

2. The pseudocode on which we based our first specification of the dynamic RBP was incomplete. In several places it was not clear what are the assumptions about node attributes.

3. A description of initial state was missing, and so was the termination condition for the protocol.

4. Moreover, the original pseudocode was incomplete with respect to how attributes are updated.

5. Some cases were left out in the original specification.

6. Other essential errors were detected using the strategies. For example, we tested a scenario with three nodes a, b, and c where b has the neighbors a and c, a is a source nodes which sends a message and the link between a and b breaks down. Running the protocol with this initial configuration using an exhaustive search strategy delivers three different states of which one is a correct state, the second one reveals an undesired behavior and the third one showed an error in the original pseudocode which has been corrected in the current version.

4.2 Cryptographic Protocol Analysis

As mentioned earlier in this paper, we applied Maude to the specification and analysis of cryptographic protocols and showed how our model-checking techniques can be used to discover attacks.

The TIPE DARPA project focuses on the development of tools for cryptographic protocol analysis. The approach taken is to provide a single common protocol specification language that could be used as the input format for any formal analysis. CAPSL (Common Authentication Protocol Specification Language) [12] has been designed as such a language. It is a modern, easy-to-use specification language for the early specification phase. The main syntactic feature of CAPSL is a message-list style protocol description as can be found in most textbooks published. It is translated to CIL [13], an intermediate language, that expresses the protocol as transition rules in a term rewriting formalism similar to the formalism we used in our experiments with NSPK. The CAPSL Intermediate Language (CIL) serves two purposes: to help define the semantics of CAPSL, and to act as an interface through which protocols specified in CAPSL can be analyzed using a variety of tools. TIPE aims at designing and implementing an integrated protocol environment for the specification and analysis of cryptographic protocols.

Maude has been used in the TIPE project for several purposes. First, the translation from CAPSL to CIL has been carried out in Maude. This way, the formal semantics of CAPSL via its translation to CIL is precisely specified. In the current translator a pre-compiled and type-checked abstract syntax tree is passed to the Maude implementation which determines the corresponding CIL specification. The translator transforms the message list into a set of multiset rewrite rules.

Second, Maude is used as a model-checking tool in the integrated protocol environment and tool-kit. The CIL output of a cryptographic protocol is translated into an executable Maude specification. A metalevel search strategy imports the Maude protocol specification and provides the user with a predefined breadth-first strategy. Search scenarios are automatically derived from the corresponding CAPSL environment specification. The translation from CIL to executable Maude specifications is performed in Maude itself.

4.3 Specifying and Analyzing a PLAN Algorithm

PLAN [22] is a language to program active networks developed at the University of Pennsylvania. In collaboration with Yaw Wang and Carl Gunter at Penn, we have used Maude to formally specify and analyze a PLAN active network algorithm in which active packets scout the nodes of an active network from a source to a destination to find an optimal route relative to a given metric [44]. One interesting aspect of this case study is that it is not clear how to apply traditional network models, such as those used in model checkers, to specify active networks in which active packets carry code that is executed in different nodes and can modify the network. In this regard, the great modeling flexibility of Maude has allowed us to specify the PLAN algorithm in a very direct and simple way that is furthermore executable.

The algorithm has three phases. When a user wants to send a message from a source to a destination, "scout" packets are first sent to try to find an optimal route according to a predefined metric. Then, when a route is found, a flow packet is sent to set up the route. Finally, the desired data packets are sent. Each of these phases is formalized in Maude by appropriate rewrite rules, so that the entire algorithm is specified as a rewrite theory. Since packets and nodes only have local information, it is possible for some data packets to arrive at the destination while some scout packets are still in transit. Also, the route may change while data packets are in transit. Therefore, it is not obvious that the algorithm delivers messages correctly.

The main limitation of testing the algorithm by executing it is that only some behaviors among many are thus explored, while errors may still be lurking in unexplored behaviors. Since the algorithm always terminates, using Maude it is possible to explore all behaviors from a given initial state to check their correctness. This can be done by specifying a suitable strategy at the metalevel. Such a strategy takes the Maude formal specification of the PLAN algorithm and an initial state of the network as inputs, and then explores all the possible behaviors until termination, checking the correct delivery of the messages.

By performing such an analysis we have gained greater confidence of the correct behavior of the algorithm. Of course, even greater confidence could be gained by a narrowing analysis, beginning with a symbolically described family of initial states, or with formal proofs. However, our methodology suggests using those heavier methods only after other errors have already been eliminated by a model checking analysis of the kind described above. Another promising direction would be using Maude to give a formal operational semantics to PLAN. Then, one could use such a formal semantics to analyze and prove properties of PLAN algorithms, that would automatically become formal objects within Maude's logic.

4.4 Composable Distributed Services

In practice NSPK and other protocols provide services designed to ensure desired communication semantics or other properties of the runtime infrastructure. These protocols are not—as it might appear from the usual discussions—the main objectives of the agents involved. It is important to be able to specify and implement services and protocols in a *modular* way, so that they can be installed dynamically and can be composed in many ways to meet complex and changing requirements in open distributed systems. It is also important to harden a system against a variety of threats. If we have protocols protecting against several such threats, how can we modularly combine them to protect the system against a combined attack?

Composition of system properties is in general a nontrivial matter (c.f. [36, 38, 39]) and a modular approach seems particularly promising in this regard. Formally we want operations that compose two or more protocols to form a new protocol, and that install a protocol on a system to obtain a new system. These operations should work on specifications, code, or running systems. Let $\mathcal{S}[\mathcal{C}]$ denote the installation of protocol $\mathcal{S}$ on system $\mathcal{C}$. Two key questions are: if $\mathcal{S}$ ensures property P for $\mathcal{C}$ under what conditions does $\mathcal{S}$ ensure property P for $\mathcal{S}'[\mathcal{C}]$ where $\mathcal{S}'$ is another protocol; if $\mathcal{S}$ ensures property P for $\mathcal{C}$ and $\mathcal{S}'$ ensures property P' for $\mathcal{C}$ under what conditions does $\mathcal{S}$ ensure P' for $\mathcal{S}'[\mathcal{C}]$.

Subclassing is one way to obtain some degree of modularity. An alternative is to use reflection to achieve a highly expressive, modular, dynamic formalism. In particular, reflection can be used as a mechanism for installing protocols by composing a protocol specified as a metalevel activity and a system specified at the base level. Two reflective architectures have been developed for the actor model and used to provide a basis for defining and reasoning about dynamic adaptable distributed systems: the *onion skin* model [1, 3] aims at layered specification of interaction policies for components of distributed systems and allows different protocols for security, reliability, quality of service and other policies to be composed and modified dynamically as the system runs by adding new metalevels; the *two level actor model* [41] focuses on horizontal composition of system-wide services provided by meta-actors that manage resources and control runtime properties of base level application components in an open-distributed system. Services are composed by concurrent operation of service meta-actors and interaction via message passing. This model has been used to specify and reason about services such as migration and distributed garbage collection [43], and QoS based admission control in a multi-media system [42] .

Such reflective models can be naturally represented in Maude. In [11] we present an executable specification of a general middleware architecture for composable distributed communication services based on the onion skin model. We also explain how the compositionality of the resulting middleware can be exploited in proving formal properties about the composed system consisting of the middleware services and given applications. The formal model is based on the rewriting logic axiomatization of concurrent objects specialized to distributed asynchronous objects in the spirit of the actor model. It provides a representation of meta objects and an encapsulation of layers of meta-objects in towers. We show how a system of concurrent objects communicating through asynchronous message passing can be naturally transformed into a behaviorally equivalent system of *meta-object towers of height one*. Composition of meta-objects is very simply expressed by stacking meta-objects on top of each other, with the application meta-object at the bottom of the stack. The behavior of meta-object compositions is then formalized by general rewrite rules governing the communication between the different levels of a stack of meta-objects, as well as by the specific additional rules that govern the behavior of meta-objects providing a particular service. Using the initial semantics of Maude it is now easy to formalize notions such as $\mathcal{S}$ ensures property P for $\mathcal{C}$ (written $\mathcal{S}[\mathcal{C}] \models_{\text{ind}} P$). We show how the formal properties expressing different service guarantees—relative to different hostile environments—can themselves also be stated in a parametric way, so that one can combine in a systematic way the properties of different services when composing them, in order to study whether they hold or not for a given composition of services. We show how to model composable communication services as distributed meta-objects and illustrate the basic ideas with a case study treating fault-tolerance, encryption and authentication services and their composition. Formal specification of service guarantees and combination of guarantees for compositions of these services are also discussed.

4.5 Software Architecture Interoperation

To illustrate the flexibility and ease with which different semantic models can be specified and interoperated in rewriting logic we have carried out a substantial case study as part of the DARPA EDCS initiative. The study, entitled "A Software Architecture Interoperation Example," is documented as Appendix E of [7]. In this study we show how Maude can be used to execute very high level software designs, namely architectural descriptions. It focuses on a difficult case, namely, *heterogeneous* architectures illustrated by a command and control example featuring dataflow, message passing, and implicit invocation sub-architectures. This is accomplished by first defining a general model for *objects* that can interact with each other in a variety of synchronous and asynchronous ways. Then we define a general *dataflow* model as a semantic model for a pipes and filters interaction model as well as a semantic model for *event-based implicit invocation*. We then give an executable high-level specification of a command and control system in which several of these models are combined and interoperated.

To demonstrate the use of Maude to give formal executable and analyzable semantics to software engineering notations, a subset of the Wright architecture description language [2] has been implemented in Maude. This provides an execution environment for Wright specifications as well as the basis for defining additional analysis tools using the reflective capabilities of Maude.

As part of the mapping of Wright to Maude, a mapping of CSP to Maude has been designed and implemented. This mapping is based on concurrent distributed objects (actors), providing as a side benefit a truly concurrent semantics for CSP. The specification of CSP and its machine readable syntax as given in [37] was followed in order to have an independently useful tool for CSP that can interoperate with existing CSP tools and to support interoperation with other notations formalized in Maude such as the CAPSL language for specification of security protocols. This representation of CSP in Maude also provides the starting point for interoperation of CSP specifications and those based on other communication models such as the asynchronous communication of actors, also implemented in Maude.

5 Conclusions and Future Developments

We have presented our experience so far in applying a new formal methodology based on rewriting logic and supported by the Maude language to active networks, communication protocols, and security protocols. The case studies are encouraging and suggest that this methodology, when inserted early in the design process, can uncover many design errors and inconsistencies, can yield unambiguous and executable formal models of the systems being designed, and can be used to reach high levels of assurance about communication systems through a combined series of increasingly stronger techniques.

The case studies discussed in the paper illustrate the use of the techniques 1–3 in the formal methodology of Section 1.2. In the near future we plan to explore more systematically the application of techniques 4–5, namely, the use of narrowing analysis, and of formal proof based on a suitable combination of rewriting and temporal logic. We discuss in more detail each of these techniques in what follows.

5.1 Narrowing

Our systematic search for an attack using a breadth-first strategy and several applications discussed in Section 4 have illustrated a form of model checking in which we explore all the possible behaviors beginning with a given state using the reflective capabilities of Maude. The fact that state-building operators—such as the multiset union of configurations—obey structural axioms like associativity and commutativity allows the rewriting to take place modulo such axioms. This means that we have a *single* representation for the many different expressions describing such a state, and therefore a much smaller search space than if rewriting modulo axioms were not supported.

But what about exploring behaviors starting not just from one state, but from a possibly infinite *set* of states? For object-oriented protocol specifications of the general style illustrated by the NSPK example, this can be accomplished by performing *narrowing* [19] modulo structural axioms such as the associativity and commutativity of the multiset union of configurations. The idea is to describe a set of initial states by a symbolic expression with variables that stands for the set of all its ground substitutions. Then, the exploration of behaviors is entirely similar to the rewriting exploration, except that at each step, instead of *matching* the lefthand side of a rule to a ground term representing a single state, we *unify* (modulo structural axioms such as associativity and commutativity) such a lefthand side with a nonvariable subterm of the symbolic expression representing the current set of states. In principle, such narrowing and unification algorithms can be specified by means of rewrite rules in Maude, but for efficiency reasons it is better to build in the unification algorithm.

We plan to add to Maude built-in unification modulo several equational theories and their combinations, and to exploit such a facility to explore behaviors for sets of states by narrowing. There are some similarities between our proposed use of narrowing and the narrowing analysis performed in the NRL Protocol Analyzer [26, 27, 28]. But in the NRL tool narrowing (with standard unification) is performed only by means of the Church-Rosser *equations* that axiomatize the basic algebraic properties of the underlying cryptography, whereas our proposed use is different, namely, to perform narrowing (modulo structural axioms) using both the (Church-Rosser) equations and the (in general non-Church-Rosser) transition rules specifying the system as a rewrite theory.

5.2 Combining Rewriting and Temporal Logic

Rewriting logic is a logic that we could call *internal* to the concurrent computation of a system that it axiomatizes. Deduction exactly corresponds to computation, and in fact can be used to implement the system axiomatized by the rules. By contrast, temporal logic can be regarded as taking an *external* view of the system. It regards it as a mathematical model—typically some kind of Kripke structure—about which it then makes assertions about its global properties such as safety or liveness. Both levels of description and analysis are useful in their own right; in fact, they complement each other. We therefore plan to use both logics in combination to prove properties about communication protocols and other concurrent systems. The semantic nexus between the two levels of specification is given by the initial model of the rewrite theory specifying the system, which plays at the same time the role of the Kripke structure about which specific temporal logic properties are asserted.

The first issue that must be addressed is the semantic integration between the two logics. Essentially, such an integration is straightforward, because both logics are talking about the same mathematical model. Theories in rewriting logic have *initial models*. The initial model $\mathcal{T}_{\mathcal{R}}$ of a rewrite theory $\mathcal{R}$ is a mathematical model of the concurrent system axiomatized by $\mathcal{R}$. Specifically, it is a category with algebraic structure [29], where the objects correspond to system states, and the arrows correspond to concurrent system transitions. Therefore, $\mathcal{T}_{\mathcal{R}}$ can be regarded as a Kripke structure whose transitions are labeled by the arrows of the category. A variety of different modal or temporal logics can then be chosen to make assertions about such a Kripke structure. We are particularly interested in temporal logics of this kind that provide explicit support for objects. Two candidates that we are considering are the version of the μ-calculus proposed by Ulrike Lechner [24] for reasoning about object-oriented Maude specifications, and the object-oriented Distributed Temporal Logic (DTL) of Ehrich et al. [18, 9], and the temporal logic of Duarte [17] used to axiomatize the actor model. For the second alternative a temporal logic extension of rewriting logic and a technique to transform a rewrite theory into a temporal logic theory have been proposed in [8]. In future work we will investigate how one can reason in the combined calculus and how such reasoning can be supported by tools.

Other approaches to reason about security protocols include Paulson's inductive method [35], Gray and McLean's approach [21], which uses Lamport's TLA notation to specify cryptographic protocols and correctness requirements, and Syverson and Meadows' approach [40], which proposes a logical language for specifying protocol requirements and translates them by hand into the Prolog language of the NRL Analyzer [26, 27, 28]. There is also a large body of work by various authors using *abstraction* techniques to reduce communication protocols—that in principle have an infinite number of states—to finite-state versions for which temporal logic formulas can be decided by model checking. We conjecture that abstraction maps will have a natural high-level specification in rewriting logic as adequate maps between theories.

Our medium-term goal is to reach high assurance for active networks and communication protocols by combining the advantages of Maude executable specification, formal analysis by rewriting-based model checking and narrowing, and temporal logic reasoning by means of both theorem proving and finite-state model checking using abstractions. We expect that rewriting logic and Maude will provide a good logical framework for the integration and implementation of these techniques.

Acknowledgments

This work has been partially supported by DARPA through Rome Laboratories Contract F30602-97-C-0312, by DARPA and NASA through Contract NAS2-98073, by ARPA/AirForce under grant F30602-96-1-0300, by NSF under grant CRR-9633419, and by Office of Naval Research Contracts N00014-96-C-0114 and N00012-99-C-0198.

We thank our fellow members in the Maude team, particularly Peter Ölveczky, Francisco Durán, Narciso Martí-Oliet, Patrick Lincoln, Steven Eker, and Manuel Clavel for their contributions to different aspects of the ideas presented here. We have also benefited from discussions on temporal logic matters with Narciso Martí-Oliet, Hans-Dieter Ehrich, José Fiadeiro, Tom Maibaum, and Carlos Duarte and from many discussions with members of the DARPA Active Networks program, particularly with Kirstie Bellman and Chris Landauer. Finally, we wish to thank Jon Millen at SRI, José García-Luna, Jyoti Rajy, and Brad Smith at UCSC, and Carl Gunter and Yao Wang at UPenn for their important contributions to several of the case studies discussed in this paper.

References

[1] G. Agha. Abstracting interaction patterns: A programming paradigm for open distribute systems. In H. Bowman and J. Derrick, editors, *Formal Methods for Open Object-Based Distributed Systems (FMOODS'97), Volume 2, IFIP TC6 WG6.1 Intern. Workshop, 21-23 July, Canterbury, Kent, UK*, pages 135–153. Chapman & Hall, 1997.

[2] R. Allen and D. Garlan. Formalizing architectural connection. In *Proceedings 16th International Conference on Software Engineering*, 1994.

[3] M. Astley. *Customization and Composition of Distributed Objects: Policy Management in Distributed Software Architectures.* PhD thesis, University of Illinois, Urbana-Champaign, 1999.

[4] M. Clavel, F. Durán, S. Eker, and J. Meseguer. Building equational proving tools by reflection in rewriting logic. In *Proc. of the CafeOBJ Symposium '98, Numazu, Japan.* CafeOBJ Project, April 1998. `http://maude.csl.sri.com`.

[5] M. Clavel, S. Eker, P. Lincoln, and J. Meseguer. Principles of Maude. In Meseguer [31], pages 65–89.

[6] M. Clavel and J. Meseguer. Reflection and Strategies in Rewriting Logic. In Meseguer [31], pages 125–147.

[7] Clavel, M. and Durán, F. and Eker, S. and Lincoln, P. and Martí-Oliet, N. and Meseguer, J. and Quesada, J. *Maude: Specification and Programming in Rewriting Logic.* SRI International, Computer Science Laboratory, Menlo Park, CA, January 1999. `http://maude.csl.sri.com/manual/`.

[8] G. Denker. From Rewrite Theories to Temporal Logic Theories. In H. Kirchner and C. Kirchner, editors, *2nd Int. Workshop on Rewriting Logic and Its Applications (WRLA'98), Pont-A-Mousson, France, September 1-4, 1998.* Elsevier Science B.V., Volume 15 of Electronic Notes in Theoretical Computer Science, `http://www.elsevier.nl/locate/entcs/volume15.html`, 1998.

[9] G. Denker and H.-D. Ehrich. Specifying Distributed Information Systems: Fundamentals of an Object-Oriented Approach Using Distributed Temporal Logic. In H. Bowman and J. Derrick, editors, *Formal Methods for Open Object-Based Distributed Systems (FMOODS'97), Volume 2, IFIP TC6 WG6.1 Intern. Workshop, 21-23 July, Canterbury, Kent, UK*, pages 89–104. Chapman & Hall, 1997.

[10] G. Denker, J. Meseguer, and C. Talcott. Protocol Specification and Analysis in Maude. In N. Heintze and J. Wing, editors, *Proc. of Workshop on Formal Methods and Security Protocols, 25 June 1998, Indianapolis, Indiana*, 1998. `http://www.cs.bell-labs.com/who/nch/fmsp/index.html`.

[11] G. Denker, J. Meseguer, and C. Talcott. Rewriting Semantics of Distributed Meta Objects and Composable Communication Services, 1999. working draft.

[12] G. Denker and J. Millen. CAPSL and CIL Language Design: A Common Authentication Protocol Specification Language and Its Intermediate Language. CSL Report SRI-CSL-99-02, Computer Science Laboratory, SRI International, Menlo Park, CA 94025, 1999. `http://www.csl.sri.com/~denker/pub_99.html`.

[13] G. Denker and J. Millen. CAPSL Intermediate Language. In N. Heintze and E. Clarke, editors, *Workshop on Formal Methods and Security Protocols (FMSP'99), July 5, 1999, Trento, Italy (part of FLOC'99)*, 1999. `http://cm.bell-labs.com/cm/cs/who/nch/fmsp99/`.

[14] Denker, G. and Garcia-Luna-Aceves, J.J. and Meseguer, J. and Ölveczky, P. and Raju, J. and Smith, B. and Talcott, C. Specification and Analysis of a Reliable Broadcasting Protocol in Maude. In B. Hajek and R. Sreenivas, editors, *Proc. 37th Allerton Conference on Communication, Control and Computation*, 1999. url-http://www.comm.csl.uiuc.edu/allerton.

[15] Denker, G. and Garcia-Luna-Aceves, J.J. and Meseguer, J. and Ölveczky, P. and Raju, J. and Smith, B. and Talcott, C. Specifying a Reliable Broadcasting Protocol in Maude. Internal report, Computer Science Laboratory, SRI International, Menlo Park, CA 94025, 1999. `http://www.csl.sri.com/~denker/pub_99.html`.

[16] D. E. Denning and G. M. Sacco. Timestamps in Key Distribution Protocols. *Communications of the ACM*, 24(8):533–536, 1981.

[17] C. H. C. Duarte. *Proof-theoretic Foundations for the Design of Extensible Software Systems.* PhD thesis, Imperial College, University of London, 1999.

[18] H.-D. Ehrich, C. Caleiro, A. Sernadas, and G. Denker. Logics for Specifying Concurrent Information Systems. In

J. Chomicki and G. Saake, editors, *Logics for Databases and Information Systems*, pages 167–198. Kluwer Academic Publishers, 1998.

[19] M. Fay. First-order unification in an equational theory. In *Proceedings of the 4th Workshop on Automated Deduction*, pages 161–167, 1979.

[20] J. J. García-Luna. Reliable Broadcasing in Computer Networks. Manuscript; University of California at Santa Cruz, Computer Science Department, January 1998.

[21] J. W. Gray and J. D. McLean. Using Temporal Logic to Specify and Verify Cryptographic Protocols (Progress Report). In *Proc. of the 8th IEEE Computer Security Foundations Workshop*, pages 108–116. IEEE Computer Society Press, 1995.

[22] M. Hicks, P. Kakkar, J. T. Moore, C. A. Gunter, and S. Nettles. PLAN: A Packet Language for Active Networks. In *Proceedings of the Third ACM SIGPLAN International Conference on Functional Programming Languages*, pages 86–93. ACM, 1998.

[23] C. Hoare. *Communicating Sequential Processes*. Prentice Hall International, 1985.

[24] U. Lechner and C. Lengauer. Modal–μ–Maude — Specification and Properties of Concurrent Objects. In B. Freitag and C.B. Jones and C. Lengauer and H.-J. Schek, editor, *Object Orientation with Parallelism and Persistence*, pages 23–46. Kluwer, 1996.

[25] G. Lowe. Breaking and fixing the Needham-Schroeder public-key protocol using FDR. In *Proceedings of TACAS*, volume 1055 of *Lecture Notes in Computer Science*, pages 147–166. Springer-Verlag, 1996.

[26] C. A. Meadows. Formal Verification of Cryptographic Protocols: A Survey. In J. Pieprzyk and R. Savafi-Naini, editors, *Advances in Cryptology - Asiacrypt '94*, pages 133–150. Springer, 1995. LNCS 917.

[27] C. A. Meadows. Analyzing the Needham-Schroeder Public Key Protocol: A Comparison of Two Approaches. In E. Bertino, H. Kurth, G. Martella, and E. Montolivo, editors, *Proc. of European Symposium on Research in Computer Security (ESORICS'96)*, pages 351–364. Springer, 1996. LNCS 1146.

[28] C. A. Meadows. Language Generation and Verification in the NRL Protocol Analyzer. In *Proc. 9th Computer Security Foundations Workshop*, pages 48–61. IEEE Computer Society Press, June 1996.

[29] J. Meseguer. Conditional Rewriting Logic as a Unified Model of Concurrency. *Theoretical Computer Science*, 96(1):73–155, 1992.

[30] J. Meseguer. A Logical Theory of Concurrent Objects and Its Realization in the Maude Language. In G. Agha, P. Wegner, and A. Yonezawa, editors, *Research Directions in Concurrent Object-Oriented Programming*, pages 314–390. The MIT Press, 1993.

[31] J. Meseguer, editor. *Rewriting Logic and Its Applications, First International Workshop, Asilomar Conference Center, Pacific Grove, CA, September 3-6, 1996*. Elsevier Science B.V., Electronic Notes in Theoretical Computer Science, Volume 4, `http://www.elsevier.nl/locate/entcs/volume4.html`, 1996.

[32] J. Meseguer. Rewriting logic as a semantic framework for concurrency: a progress report. In *Proc. CONCUR'96, Pisa, August 1996*, pages 331–372. Springer LNCS 1119, 1996.

[33] J. Meseguer. Rewriting Logic as a Semantic Framework for Concurrency: A Progress Report. In U. Montanari and V. Sassone, editors, *Proc. 7th Intern. Conf. on Concurrency Theory: CONCUR'96, Pisa, August 1996*, pages 331–372, 1996. LNCS 1119.

[34] R. Needham and M. Schroeder. Using encryption for authentication in large networks of computers. *Communications of the ACM*, 21(12):993–999, 1978.

[35] L. Paulson. The inductive approach to verifying cryptographic protocols. *Journal of Computer Security*, 6(1):85–128, 1998.

[36] J. Peleska. On a unified formal approach for the development of fault-tolerant and secure systems. In H. Rischel, editor, *Nordic Seminar on Dependable Computing Systems, Lyngby, Denmark, August 1994. Technical University of Denmark*, pages 69–80, 1994.

[37] W. Roscoe. *Theory and Practice of Concurrency*. Prentice Hall, 1998.

[38] J. Rushby. Critical system properties: Survey and taxonomy. *Reliability Engineering and System Safety*, 43(2):189–219, 1994.

[39] J. Rushby. Combining system properties: A cautionary example and formal examination. Technical report, Computer Science Laboratory, SRI International, Menlo Park, CA, June 1995. Unpublished project report; available at `http://www.csl.sri.com/~rushby/combined.html`.

[40] Syverson, P. and Meadows, C. A. A Logical Language for Specifying Cryptographic Protocol Requirements. In *Proc. IEEE Computer Society Symp. on Research in Security and Privacy, Oakland, CA*, pages 165–177, May 1993.

[41] N. Venkatasubramanian. *Resource Management in Open Distributed Systems with Applications to Multimedia*. PhD thesis, University of Illinois, Urbana-Champaign, 1998.

[42] N. Venkatasubramanian, G. Agha, and C. L. Talcott. Specifying composable services for qos-based distributed resource management, 1998. submitted.

[43] Venkatasubramanian, N and Talcott, C. L. Reasoning about Meta Level Activities in Open Distributed Systems. In *Principles of Distributed Computation*. ACM, 1995.

[44] B.-Y. Wang, J. Meseguer, and C. A. Gunter. Specification and formal analysis of a PLAN algorithm in Maude. Manuscript, September 1999.

Protecting Privacy in a Decentralized Environment

Andrew C. Myers
Department of Computer Science
Cornell University
andru@cs.cornell.edu

Barbara Liskov
Laboratory for Computer Science
Massachusetts Institute of Technology
liskov@lcs.mit.edu

Abstract

This paper describes the decentralized label model, a new model for controlling information flow in systems with mutual distrust and decentralized authority. The model allows users to share information with distrusted code (e.g., downloaded applets), yet still control how that code disseminates the shared information to others. The model improves on existing multilevel security models by allowing users to declassify information in a decentralized way, and by improving support for fine-grained data sharing. It supports static program analysis of information flow, so that programs can be certified to permit only acceptable information flows and to avoid most run-time information flow checks. In addition to presenting the model, the paper also discusses how the model can be supported in a distributed environment, including an introduction to Jif, an extension to Java that incorporates the model and permits static checking of information flow.

1 Introduction

The common models for computer security are proving inadequate. Security models have two goals: preventing accidental or malicious destruction of information, and controlling the release and propagation of that information. Only the first of these goals is supported well at present, by security models based on access control lists or capabilities (i.e., *discretionary access control*, simply called "access control" from this point on). Access control mechanisms do not support the second goal well: they help to prevent information release but do not control information propagation. For example, if user A is allowed to read B's data, B cannot control how A distributes the information it has read. Control of information propagation *is* supported by existing information flow and compartmental models, but these models unduly restrict the computation that can be performed.

Information flow control is vital whenever there is downloading of distrusted code, as would happen at a router in an active network. Java [GJS96] could be used to download the code and it attempts to prevent the downloaded code from transfering private data to other sites by using a compartmental *sandbox* security model. However, this approach largely prevents applications from sharing data and it does not support fine-grained sharing in which different data manipulated by an application have different security requirements.

This paper describes an approach that provides the needed support. The basis of the work is the *decentralized label model*, a new model of decentralized information flow that relaxes the restrictions of previous models yet is inexpensive in both space and time. The new model supports fine-grained information sharing between distrusted applications, while reducing the potential for information leaks. It allows users to control the flow of their information without imposing the rigid constraints of a traditional multilevel security system. It provides security guarantees to users and to groups rather than to a monolithic organization. It differs from previous work on information flow control by allowing users to explicitly *declassify* (or *downgrade*) data that they own. When data is derived from several sources, all the sources must agree to release the data.

The paper also briefly describes *Jif*, a new programming language that incorporates the model, and the *trusted execution platform*, a runtime environment that ensures that application code is run in a way that avoids information leaks. This platform might run on a router in an active network, or more generally on some node in the Internet. Jif is largely an extension of Java. Jif makes it easy to write programs that control information flow; additionally it allows static checking of code, similar to type checking, to insure that the code does not leak information. Static checking is not just an optimization, however; it is needed to avoid information leaks that can occur when a dynamic check fails.

The paper begins in Section 2 by giving an example of a system that can benefit from our model but that is not well supported by existing models. Section 3 introduces the fundamentals of the decentralized label model and shows

0-7695-0490-6/99 $10.00 © 1999 IEEE

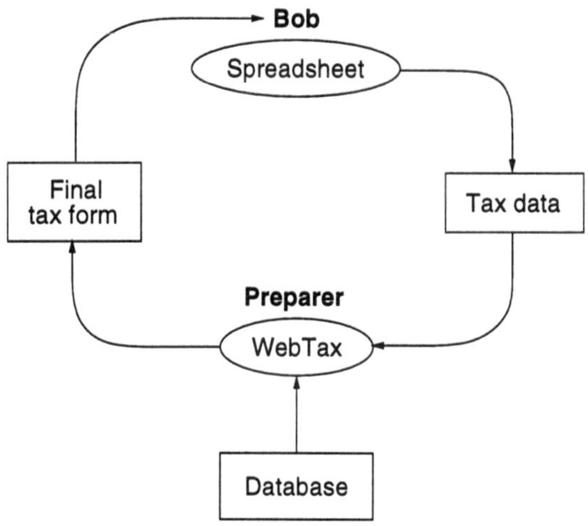

Figure 1. A simple example

how it can be used to support the system described in Section 2. Section 4 briefly discusses the other components of our work. Section 5 describes related work in the areas of information flow models, access control, and static program analysis. We conclude in Section 6.

2 Privacy example

Figure 1 depicts an example with security requirements that cannot be satisfied using existing techniques. This scenario contains mutually distrusting principals that must cooperate to perform useful work. In the example, the user Bob is preparing his tax form using both a spreadsheet program and a piece of software called "WebTax". Bob would like to be able to prepare his final tax form using WebTax, but he does not trust WebTax to protect his privacy. The computation is being performed using two programs: a spreadsheet that Bob trusts and to which he grants his full authority, and the WebTax program, which Bob does not trust. Bob would like to transmit his tax data from the spreadsheet to WebTax and receive a final tax form as a result, while being protected against WebTax leaking his tax information.

In this example, there is another principal named Preparer that also has privacy interests. This principal represents a firm that distributes the WebTax software. The WebTax application computes the final tax form using a proprietary database, shown at the bottom, that is owned by Preparer. This database might, for example, contain secret algorithms for minimizing tax payments. Since this principal is the source of the WebTax software, it trusts the program not to distribute the proprietary database through malicious action, though the program might leak information because it contains bugs.

It may be difficult to prevent some information about the database contents from leaking back to Bob, particularly if Bob is able to make a large number of requests and then carefully analyze the resulting tax forms. This information leak is not a practical problem if Preparer can charge Bob a per-form fee that exceeds the value of the information Bob obtains through each form.

To make this scenario work, Preparer needs two pieces of functionality. First, it needs protection against accidental or malicious release of information from the database by paths other than through the final tax form. Second, it needs the ability to *sign off* on the final tax form, confirming that the information leaked in the final tax form is sufficiently small or scrambled by computation that the tax form may be released to Bob.

It is worth noting that Bob and Preparer do need to trust that the execution platform has not been subverted. For example, if WebTax is running on a computer that Bob controls completely, Bob will be able to steal the proprietary database. Clearly, Preparer cannot have any real expectation of privacy or secrecy if its private data is manipulated in unencrypted form by an execution platform that it does not trust!

However, even if the execution platform is trusted, the scenario cannot be implemented satisfactorily using existing security techniques. With current techniques, Bob must inspect the WebTax code and verify that it does not leak his data; in general, this task is impossible. The techniques described in this paper allow the security goals of both Bob and Preparer to be met without this inspection; Bob and Preparer can then cooperate in performing useful computation. In another sense, this work shows how both Bob and Preparer can inspect the WebTax program efficiently and simply to determine whether it violates their security requirements.

3 Model

This section describes the decentralized label model. The key new feature of this model is that it supports computation in an environment with mutual distrust. The ability to handle mutual distrust is achieved by attaching a notion of ownership to information flow policies. These policies can then be modified by their owners—a form of safe declassification. Declassification is safe because flow policies of other principals remain in force.

The essentials of the decentralized label model are *principals*, which are the entities whose privacy is protected by the model; and *labels*, which are the way that principals express their privacy concerns. In addition, there are rules that must be followed as computation proceeds in order to avoid information leaks, including the safe declassification mechanism. These rules have been shown to be both sound and

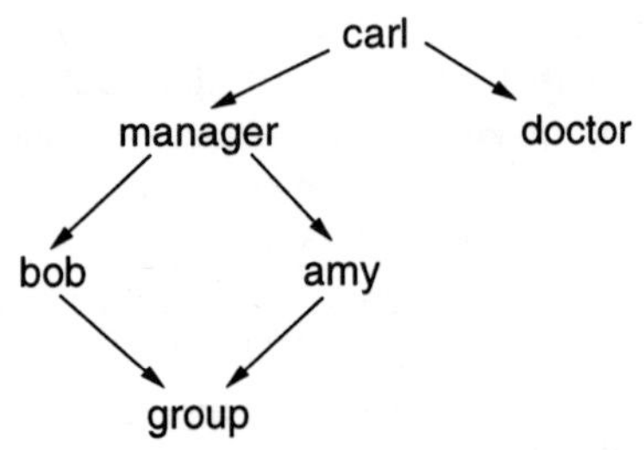

Figure 2. A principal hierarchy

complete with respect to a simple formal semantics for labels: the rule only allows safe relabelings, and it allows *all* safe relabelings [ML98].

In this model, users are assumed to be external to the system on which programs run. Information is leaked only when it leaves the system. The model also contains rules governing the release of information to the external environment, as well as rules for reading information from the environment.

3.1 Principals

In the decentralized label model, information is owned by, updated by, and released to *principals*: users and other authority entities such as groups or roles. For example, both users and groups in Unix would be modeled as principals. A process has the authority to act on behalf of some set of principals.

Some principals are authorized to *act for* other principals. When one principal p can act for another principal p', p possesses all the potential powers and privileges of p'. The *acts for* relation is reflexive and transitive, defining a hierarchy or partial order of principals. For compactness, the statement "p acts for q" is written formally as $p \succeq q$. This relation is similar to the *speaks for* relation [LABW91]; the principal hierarchy also is similar to a *role hierarchy* [San96]. Although the principal hierarchy changes over time, revocation of acts-for relations is assumed to occur infrequently.

The acts-for relation can be used to model groups and roles conveniently, as shown in Figure 2. In this example, a group named `group` is modeled by introducing acts-for relations between each of the group members (`amy` and `bob`) and the group principal. This relation allows Amy and Bob to read data readable by the group and to control data controlled by the group. A principal `manager` representing Bob and Amy's manager is able to act for both `amy` and `bob`. This principal is one of the roles that a third user, Carl (`carl`), is currently enabled to fulfill; Carl also has a separate role as a doctor. Carl can use his roles to prevent accidental leakage of information between data stores associated with his different jobs.

The notion of the acts-for relation permits delegation of all of a principal's powers, or none. In modeling real systems, it is useful to separate out the various powers that may be given by one principal to another. For example, the acts-for relation can be separated into *declassifies-for* and *reads-for* relations between principals. The model presented here can be extended to support finer-grained principal relations in a natural manner by representing them as simple acts-for relations on a more complex principal hierarchy [Mye99b].

3.2 Labels

Principals express their privacy concerns by using labels to annotate programs and data. A label is a set of *policies* that express privacy requirements. A policy has two parts, an owner and a set of readers, and is written in the form *owner*: *readers*. The owner of a policy is a principal whose data was observed in order to construct the value labeled by this policy. The readers of a policy are the principals that this policy permits to read the data. The owner is a source of data, and the readers are possible destinations for the data. Principals not listed as readers are not permitted to read the data.

An example of a label is $L = \{o_1 : r_1, r_2;\ \ o_2 : r_2, r_3\}$. Here, o_1, o_2, r_1, r_2 denote principals. Semicolons separate two policies within the label L. The owners of these policies are o_1 and o_2, and the reader sets of the policies are $\{r_1, r_2\}$ and $\{r_2, r_3\}$, respectively.

The intuitive meaning of a label is that every policy in the label must be obeyed as data flows through the system, so that labeled information is released only by the consensus of all of the policies. A user may read the data only if the user's principal can act for a reader for *each* policy in the label. Thus only users whose principals can act for r_2 can read the data labeled by L.

The same principal may be the owner of several policies; the policies are still enforced independently of one another in this case. If a principal is the owner of *no* policies in the label, it is as if the label contained a policy with that owner, but every possible principal listed as a reader.

This label structure allows each owner to specify an independent flow policy, and thus to retain control over the dissemination of its data. Code running with the authority of an owner can modify the flow policy for a policy with that owner; in particular, the program can *declassify* that data by adding additional readers. Since declassification applies on a per-owner basis, no centralized declassification process is needed, as it is in systems that lack ownership labeling.

Labels are used to control information flow within programs by annotating program variables with labels. The label of a variable controls how the data stored within that variable can be disseminated. The key to secure flow is to ensure that as data flows through the system during computation, its labels only become more restrictive: the labels have more owners, or particular owners allow fewer readers. If the contents of one variable affect the contents of another variable, there is an information flow from the first variable to the second, and therefore, the label of the second variable must be at least as restrictive as the label of the first.

This condition can be enforced at compile time as a kind of type checking, as long as the label of a variable is known at compile time. In other words, the label of a variable cannot change at run time, and the label of the data contained within a variable is always the same as the label of the variable itself. If the label of a variable could change, the label would need to be stored and checked at run time, creating an additional information channel and leading to significant performance overhead. Immutable variable labels might seem like a limitation, but full expressiveness can be obtained by having variables that contain labels.

When a value is read from a variable x, the apparent label of the value is the label of x; whatever label that value had at the time it was written to x is no longer known when it is read. In other words, writing a value to a variable is a relabeling, and is allowed only when the label of the variable is at least as restrictive as the apparent label of the value being written.

Giving private data to an untrusted program does not create an information leak—even if that program runs with the authority of another principal—as long as that program obeys all of the label rules described here. Information can be leaked only when it leaves the system through an *output channel*, so output channels are labeled to prevent leaks. Information can enter the system through an *input channel*, which also is labeled to prevent leaks. It is safe for a process to manipulate data even though no principal in its authority has the right to read it, because all the process can do is write the data to a variable or a channel with a label that is at least as restrictive as the data's label.

3.3 Incremental relabelings

A value may be safely assigned to a variable only if the relabeling that occurs at this point is a *restriction*, a relabeling in which the new label allows the same or fewer information flows than the original. This happens, for example, if the two labels differ only in a single policy, and the readers of that policy in the label of the variable are a subset of the readers of that policy in the label of the value. For example, a relabeling from $\{\texttt{bob} : \texttt{amy}, \texttt{carl}\}$ to $\{\texttt{bob} : \texttt{amy}\}$ is a restriction because the set of readers allowed by `bob` becomes smaller.

There are four ways that a label can be changed without making it less restrictive:

- **Remove a reader.** It is safe to remove a reader from some policy in the label.
- **Add a policy.** It is safe to add a new policy to a label; since all previous policies are still enforced, the policy cannot become less restrictive.
- **Add a reader.** It is safe to add a reader r' to a policy if the policy already allows a reader r that r' acts for. This change is safe because if r' acts for r, it is already effectively considered to be a reader: r' has all of the privileges of r anyway.
- **Replace an owner.** It is safe to replace a policy owner o with some principal o' that acts for o. This change is safe because the new policy allows only processes that act for o' to weaken it through declassification, while the original policy also allows processes with the weaker authority of o to declassify it.

3.4 The relabeling rule

These incremental relabelings capture all of the safe modifications that can be made to a label. A relabeling from a label L_1 to a label L_2 will be safe if there is some sequence of these incremental relabelings that rewrites L_1 into L_2. To allow efficient label checking of programs, a rule is needed to determine whether such a sequence exists, given two labels. This section presents such a rule.

We can define precisely when a relabeling from label L_1 to label L_2 is a restriction with the aid of some additional notation. If the relabeling is a restriction, then either L_1 is as restrictive as L_2, or possibly less restrictive: L_1 is at most as restrictive as L_2, or equivalently, L_2 is at least as restrictive as L_1. The notation $L_1 \sqsubseteq L_2$ is used to denote this relationship.

It is clear that L_2 is at least as restrictive as L_1 if every policy I in L_1 is guaranteed to be enforced by L_2. This guarantee turns out to hold exactly when there is some single policy J in L_2 that is guaranteed to enforce the policy I. For brevity, we will say that J *covers* I. For example, the label $L_1 = \{\texttt{bob} : \texttt{amy}, \texttt{bob}\}$ can be relabeled to the label $L_2 = \{\texttt{bob} : \texttt{bob}; \texttt{amy} : \texttt{carl}\}$ because the single policy $(\texttt{bob} : \texttt{amy}, \texttt{bob})$ in L_1 is guaranteed to be enforced by the first policy of L_2, which has the same owner and fewer readers. The label L_2 contains an additional policy $(\texttt{amy} : \texttt{carl})$, but because information can flow only in accordance with every policy in a label, the addition of this second policy only makes L_2 more restrictive still.

In general, a label may contain two policies one of which covers the other. In this case, the covered policy is redundant and the label is equivalent to one in which the covered policy is removed. This simplification can be important for the performance of label checking, although it never affects correctness.

1. $\{\texttt{amy} : \texttt{bob}, \texttt{carl}\} \sqsubseteq \{\texttt{amy} : \texttt{carl}\}$
2. $\{\texttt{amy} : \texttt{bob}\} \sqsubseteq \{\texttt{amy} :\ \}$
3. $\{\texttt{amy} : \texttt{manager}\} \sqsubseteq \{\texttt{amy} : \texttt{carl}\}$
4. $\{\texttt{manager} : \texttt{bob}\} \sqsubseteq \{\texttt{carl} : \texttt{bob}\}$
5. $\{\texttt{amy} : \texttt{carl}\} \not\sqsubseteq \{\texttt{amy} : \texttt{bob}\}$
6. $\{\texttt{amy} : \texttt{carl}\} \not\sqsubseteq \{\texttt{bob} : \texttt{carl}\}$
7. $\{\texttt{amy} : \texttt{manager}\} \not\sqsubseteq \{\texttt{amy} : \texttt{bob}\}$
8. $\{\texttt{manager} : \texttt{bob}\} \not\sqsubseteq \{\texttt{bob} : \texttt{bob}\}$

Figure 3. Examples of policy relationships

The rule for when one label is more restrictive than another is built upon the notion of one policy covering another. It is important for expressive power that the rule for policies be as permissive as possible while remaining safe. The rule with this property is that a policy I is covered by J (which is written $I \sqsubseteq J$) if two conditions hold. The first condition is that the owner of the policy J must either be the same as the owner of the policy I, or be able to act for the owner of I. The intuition behind this condition is that the owner of the policy is the least powerful principal able to declassify the policy. For the new policy J to be no less restrictive, this principal must either remain the same or increase in power. The second condition is that the readers allowed by J must be a subset of the readers allowed by I, where we consider that the readers allowed by a policy include not only the principals explicitly mentioned by the policy but also any principal able to act for the explicitly mentioned readers. The intuition here is that a principal that can act for an explicitly mentioned reader is able to read the data.

In Figure 3, some examples of this rule are shown, using the principal hierarchy of Figure 2, for labels containing only a single policy each. When two labels each contain only a single policy, the rule for restrictiveness on labels reduces to the rule for restrictiveness on the respective policies, so these examples illustrate the policy restriction rule. In Examples 1 and 2, the reader `bob` has been removed from the reader set, which is safe. In Example 3, the reader `manager` has been replaced by `carl`; in Example 4, the owner `manager` has been replaced by `carl`. These changes are safe because `carl` can act for `manager`.

With the aid of some additional notation, these relationships among labels and policies can be defined formally. The notation $\mathbf{o}(I)$ denotes the owner of a policy I, and the notation $\mathbf{r}(I)$ denotes the set of readers of I. Given a policy I, we can define a function $\mathbf{R}$ that yields the set of principals implicitly allowed as readers by that policy:

$$\mathbf{R}(I) = \{p \mid \exists_{p' \in \mathbf{r}(I)}\ p \succeq p'\}$$

(Recall that $p_1 \succeq p_2$ means that principal p_1 acts for principal p_2 in the principal hierarchy in which the relabeling is being checked.)

This function can be used to define when one label is at most as restrictive as another ($L_1 \sqsubseteq L_2$) and when one policy is at most as restrictive as (that is, covers) another ($I \sqsubseteq J$):

Definition of the complete relabeling rule ($\sqsubseteq$)

$$\begin{aligned} L_1 \sqsubseteq L_2 &\equiv \forall_{I \in L_1} \exists_{J \in L_2}\ I \sqsubseteq J \\ I \sqsubseteq J &\equiv \mathbf{o}(J) \succeq \mathbf{o}(I) \ \wedge\ \mathbf{R}(J) \subseteq \mathbf{R}(I) \\ &\equiv \mathbf{o}(J) \succeq \mathbf{o}(I) \ \wedge\ \forall_{p' \in \mathbf{r}(J)} \exists_{p \in \mathbf{r}(I)}\ p' \succeq p \end{aligned}$$

This rule is called the *complete relabeling rule* because it defines exactly when relabeling from L_1 to L_2 is safe. The difficulty in showing the safety of the relabeling rule is that relabeling is checked statically (perhaps at compile time), and the actual run-time principal hierarchy may differ from that observed at the time of the check (the *static principal hierarchy*).

This proof that the relabeling rule is complete, and a corresponding proof that it is sound, have been presented elsewhere with respect to a simple formal semantics for labels [ML98, Mye99b]. Because the relabeling rule is sound, a relabeling checked at compile time is guaranteed to be a restriction even in the run-time principal hierarchy. Because the relabeling rule is complete, it is the most permissive sound relabeling rule. In other words, the relabeling rule is as expressive as possible while remaining safe; this expressiveness is useful in modeling many information flow scenarios.

3.5 Computing values

During computation, values are derived from other values. Since a derived value may contain information about its sources, its label must enforce the policies of each of its sources. For example, if we multiply two integers, the product's label must be at least as restrictive as the labels of both operands. However, to avoid unnecessary restrictiveness, the label of the product should be the *least* restrictive label having this property.

When a program combines two values labeled with L_1 and L_2, respectively, the result should have the least restrictive label that enforces all the information flow restrictions specified by L_1 and L_2. Recall that a label is simply a set of policies. The least restrictive set of policies that enforces all the policies in L_1 and L_2 is simply the union of the two set of policies. This least restrictive label is the *least upper bound* or *join* of L_1 and L_2, written as $L_1 \sqcup L_2$. (The symbol $\oplus$ also has been used to denote the join of two security classes [DD77, AR80].) Note that $L_1 \sqsubseteq L_1 \sqcup L_2$ for all labels L_1 and L_2: Joining is always a restriction of both input labels. The rule for the join of two labels can be written very compactly by considering L_1 and L_2 as sets of policies:

Labels for Derived Values (Definition of $L_1 \sqcup L_2$)

$$L_1 \sqcup L_2 = L_1 \cup L_2$$

It may be the case that some policies in the union of the two sets are redundant; in this case, they can be removed. For example, the join of the labels $\{\texttt{amy} : \texttt{bob}\}$ and $\{\texttt{amy} : \texttt{bob}, \texttt{carl}\}$ is $\{\texttt{amy} : \texttt{bob}; \texttt{amy} : \texttt{bob}, \texttt{carl}\}$, which is equivalent to the label $\{\texttt{amy} : \texttt{bob}\}$, because the second policy in the union is made redundant by the first, regardless of the principal hierarchy.

3.6 Declassification

Because labels in this model contain information about the owners of labeled data, these owners can retain control over the dissemination of their data, and relax overly restrictive policies when appropriate. This second kind of relabeling is a safe form of *declassification.*

The ability of a process to declassify data depends on the authority possessed by the process. At any moment while executing a program, a process is authorized to act on behalf of some (possibly empty) set of principals. This set of principals is referred to as the *authority* of the process. If a process has the authority to act for a principal, actions performed by the process are assumed to be authorized by that principal. Code running with the authority of a principal can declassify data by creating a copy in whose label a policy owned by that principal is relaxed. In the label of the copy, readers may be added to the reader set, or the policy may be removed entirely, which is effectively the same as adding all principals as readers in the policy.

Because declassification applies on a per-owner basis, no centralized declassification process is needed, as it is in systems that lack ownership labeling. Declassification is limited because it cannot affect the policies of owners the process does not act for; declassification is safe for these other owners because reading occurs only by the consensus of all owners.

The declassification mechanism makes it clear why the labels maintain independent reader sets for each owning principal. If, instead, a label consisted of just an owner set and a reader set, information about the individual flow policies would be lost, reducing the power of declassification.

Because the ability to declassify depends on the run-time authority of the process, it requires a run-time check for the proper authority. Declassification seems to be needed infrequently, so that the overhead of this run-time check is not large. It can be reduced still further using the proper static framework [Mye99a].

Declassification can be described more formally. A process may weaken or remove any policies owned by principals that are part of its authority. Therefore, the label L_1 may be relabeled to L_2 as long as $L_1 \sqsubseteq L_2 \sqcup L_A$, where L_A is a label containing exactly the policies of the form $\{p :\}$ for every principal p in the current authority. The rule for declassification may be expressed as an inference rule:

Relabeling by declassification

$$\frac{L_A = \bigsqcup_{(p \text{ in current authority})} \{p : \} \qquad L_1 \sqsubseteq L_2 \sqcup L_A}{L_1 \text{ may be declassified to } L_2}$$

This inference rule builds on the rule for relabeling by restriction. The subset rule for relabeling L_1 to L_2 states that for all policies J in L_1, there must be a policy K in L_2 that is at least as restrictive. The declassification rule has the intended effect because for policies J in L_1 that are owned by a principal p in the current authority, a more restrictive policy K is found in L_A. For other policies J, the corresponding policy K must be found in L_2, since the current authority does not have the power to weaken them. This rule also shows that a label L_1 always may be declassified to a label that it could be relabeled to by restriction, because the relabeling condition $L_1 \sqsubseteq L_2$ implies the declassification condition $L_1 \sqsubseteq L_2 \sqcup L_A$.

We have now seen three fundamental rules for the manipulation of labels during the execution of a program: rules for relabeling, derived labels, and declassification. Analysis of the safety of a piece of code reduces to analysis of the uses of these rules. As we will see, the rules can be checked at compile time, assuming the programmer is willing to supply some annotations. Code that performs a relabeling that is a restriction can be checked entirely at compile time, but code that performs a declassification requires a run-time test that the process acts for the owner.

3.7 Channels

Input and output channels allow data to enter and leave the domain in which the label rules are enforced. Channels are half-variables; like variables, they have an associated label and can be used as an information conduit. However, they provide only half the functionality that a variable provides: either input or output. As with a variable, when a value is read from an input channel, the value acquires the label of the input channel. Similarly, a value may be written to an output channel only if the label of the output channel is at least as restrictive as the label on the value; otherwise, an information leak is presumed to occur.

Obviously, the assignment of labels to channels is a security-critical operation. It is important that the channel's label reflect reality. For example, if the output of a printer can be read by a number of people, it is important that the label of the output channel to that printer identify all of them,

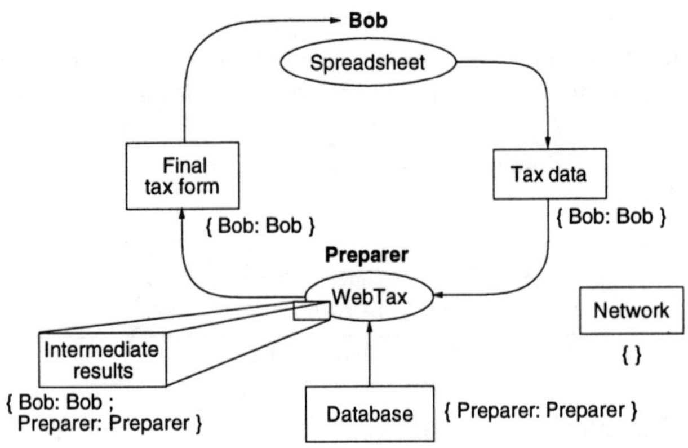

Figure 4. Annotated Tax Preparation Example

because otherwise an information leak is possible. If two computers communicate over channels, it is important that the labels of the matching output and input channels agree; otherwise, labels can be laundered by a round trip.

Labels of input and output channels to devices like displays and printers typically have a single owner (e.g., *root*) that allows as readers principals representing all the actual users of the device. Labels of channels representing communication over a network, however, may have more than one owner.

3.8 Privacy example revisited

The tax preparer example from Section 2 is repeated in Figure 4, except that all data in the example has been annotated with labels to protect the privacy of the principals Bob and Preparer. It can be seen that these labels obey the rules given and meet the security goals set out for this scenario.

In the figure, ovals indicate programs executing in the system. A boldface label beside an oval indicates the authority with which a program acts. In this example, the principals involved are `Bob` and `Preparer`, as we have already seen, and they give their authority to the spreadsheet and WebTax programs, respectively. Arrows in the diagrams represent information flows between principals; square boxes represent information that is flowing, or databases of some sort.

First, Bob applies the label {`Bob: Bob`} to his tax data. This label allows no one to read the data except Bob himself. With this label applied to it, tax data cannot be sent to an untrusted network location, represented as an output channel with label {}, because it is not the case that {`Bob: Bob`} $\sqsubseteq$ {}. Bob can give this data to the WebTax program with confidence that it cannot be leaked, because WebTax will be unable to remove the {`Bob: Bob`} policy from the tax data or any data derived from it.

The WebTax program uses Bob's tax data and its private database to compute the tax form. Any intermediate results computed from these data sources will have the label {`Bob: Bob; Preparer: Preparer`}. Because the reader sets of this label disagree, the label prevents both Bob and Preparer (and everyone else) from reading the intermediate results. This joint label is generated by the rule for join:

$$\{\texttt{Bob} : \texttt{Bob}\} \sqcup \{\texttt{Preparer} : \texttt{Preparer}\}$$

$$= \{\texttt{Bob} : \texttt{Bob} \ ; \ \texttt{Preparer} : \texttt{Preparer}\}$$

Preparer is protected by this label against accidental disclosure of its private database through programming errors in the WebTax application.

Before being released to Bob, the final tax form has the same label as the intermediate results, and is not readable by Bob, appropriately. To make the tax form readable, the WebTax application *declassifies* the label by removing the {`Preparer: Preparer`} policy. The application can do this because the `Preparer` principal has granted the application its authority. This grant of authority is reasonable because `Preparer` supplied the application and presumably trusts that it will not use the power maliciously.

The authority to act as `Preparer` need not be possessed by the entire WebTax application, but only by the part that performs the final release of the tax form. By limiting this authority to a small portion of the application, the risk of accidental release of the database is reduced. Thus the WebTax application might have a small top-level routine that runs with the authority of `Preparer`, while the rest of its code runs with no authority.

4 Supporting the Model

This section briefly describes the other parts of our research: the Jif language, the support for a trusted execution platform, and a way to harden critical information such as the principal hierarchy from malicious attacks.

4.1 The Jif Language

Incorporating the model in a programming language makes it easy for programmers to write code that is label correct and also enables compile time checking of information flow. The model can be incorporated into any programming language; we chose to extend Java because it is becoming widely used. The resulting language is called Jif, which stands for "Java Information Flow."

The idea behind Jif is that annotated programs are checked statically to verify that all information flows within the program follow the rules given in the previous section. Information flow checks can be viewed as an extension to

```
x = 0;
if (b) {
    x = 1;
}
```

Figure 5. Implicit information flow

type checking. For both kinds of static analysis, the compiler determines that certain operations are not permitted to be performed on certain data values. The analogy goes further, because in a language with subtyping one can distinguish between notions of the apparent (static) and actual (run-time) type of a value. Similarly, static label checking can be considered as a conservative approximation to checking the actual run-time labels.

However, there is a difficulty in checking labels at run time that does not arise with type checking. Type checks may be performed at compile time or at run time, though compile-time checks are obviously preferred when applicable because they impose no run-time overhead. By contrast, fine-grained information flow control is practical only with some static analysis, which may seem odd; after all, any check that can be performed by the compiler can be performed at run time as well. The difficulty with run-time checks is the fact that they can *fail*. In failing, they may communicate information about the data that the program is running on. Unless the information flow model is properly constructed, the fact of failure (or its absence) can serve as a covert channel. By contrast, the failure of a compile-time check reveals no information about the actual data passing through a program. A compile-time check only provides information about the program that is being compiled. Similarly, link-time and load-time checks provide information only about the program, and may be considered to be static checks for the purposes of this work.

Implicit information flow [DD77] is difficult to prevent without static analysis. For example, consider the segment of code shown in Figure 5 and assume that the storage locations b and x belong to different security classes {b} and {x}, respectively. (In the Jif language, a label component that is a variable name denotes all of the policies of the label of that variable, so {b} means "the label of b".) In particular, assume b is more sensitive than x (more generally, {b} $\not\sqsubseteq$ {x}), so data should not flow from b to x. However, the code segment stores 1 into x if b is true, and 0 into x if b is false; x effectively contains the value of b. A run-time check can detect easily that the assignment `x = 1` communicates information improperly, and abort the program at this point. Consider, however, the case where b is false: no assignment to x occurs within the context in which b affects the flow of control. The fact of the program's aborting or continuing implicitly communicates information about the value of b, which can be used in at least the case where b is false.

We could imagine inspecting the body of the `if` statement at run time to see whether it contains disallowed operations, but in general this requires evaluating all possible execution paths of the program, which is clearly infeasible. Another possibility is to restrict all writes that follow the `if` statement on the grounds that once the process has observed b, it is irrevocably tainted. However, this approach seems too restrictive to be practical. The advantage of compile-time checking is that in effect, static analysis efficiently constructs proofs that *no* possible execution path contains disallowed operations.

Static analysis would be too limiting for structures like file systems, where information flow cannot be verified statically. Jif defines a new secure run-time escape hatch for these structures, with explicit run-time label checks. Uses of the run-time information flow mechanism are still verified statically, to ensure that they do not leak information.

Jif extends previous work on static flow checking in two ways. First, it contains powerful language features that have not been integrated with static flow checking previously, including mutable objects, subclassing, dynamic type tests, and exceptions. Second, it provides powerful new features that make information flow checking less restrictive and more convenient than in previous programming languages:

- A simple but powerful model of authority allows code privileges to be checked statically, and also allows authority to be granted and checked dynamically.
- *Label polymorphism* permits code that is generic with respect to the security class of the data it manipulates.
- Run-time label checking and first-class label values provide an escape to run-time checks when static checking is too restrictive. Run-time checks are statically checked to ensure that information is not leaked by the success or failure of the run-time check itself.
- Automatic label inference makes unnecessary many of the annotations that otherwise would be required.

4.2 Trusted Execution Platform

To ensure that programs preserve privacy, it is not enough to check their label-correctness at compile time; we also need to ensure that their execution is carried out in a way that is consistent with the labels. Therefore we need to provide a runtime environment that runs programs in a label correct way. We call this environment the *trusted execution platform*.

The platform allows code to be imported and ensures that the imported code follows the rules. For example, it can import the tax preparer's code and also code provided by Bob and ensure that neither program exposes the other's private information.

The platform runs two kinds of code, trusted code and untrusted code. All code that it runs for users, e.g., Bob's code and the tax preparer's code, is untrusted. The trusted computing base, which includes all the trusted code, must include a mechanism that ensures that untrusted code follows the rules of the model. The mechanism might be a compiler that statically checks the information flows in a program and then digitally signs the program. However, having the compiler in the trusted base is not desirable since compilers are very large. Therefore, the platform runs a verifier that checks the output of such a compiler (as annotated byte codes).

The trusted base must also include many of the usual trusted components of an operating system: hardware that has not been subverted, a trustworthy underlying operating system, and a reliable authentication mechanism. We are investigating ways of making this trusted base as small as possible by using the model; that is, all the untrusted code is checked to be sure it is label-correct (and also type-correct). For example, we are planning to decompose the verifier into two parts, a small trusted part that does simple checking, and a larger untrusted part that does inference and whose output is checked by the trusted part. We plan to investigate similar decompositions for other parts of the system, such as the garbage collector and the JVM execution engine.

4.3 Byzantine Fault Tolerance

The correct execution of programs in our model relies on the information in the principal hierarchy: the platform accesses the hierarchy each time it must determine an acts-for relation. If the hierarchy could be corrupted, e.g., to allow the tax preparer to act for Bob, it would not be possible to prevent violations of privacy.

We have been investigating new techniques that can harden critical data bases such as the principal hierarchy from malicious attacks. Our work is leading to a practical way to implement systems that survive byzantine failures, failures in which a failed component appears to be working correctly yet is not. For example, such a component might be corrupted in such a way that it adds erroneous acts-for relations to the principal hierarchy yet it also responds to requests for information about acts-for relations in a way that appears to clients to be correct, i.e., it obeys the protocol.

An introduction to our approach can be found in [CL99]. Our approach is the first to work correctly in asynchronous systems like the Internet; in addition , our mechanism provides excellent performance so that it can be deployed in practical systems.

5 Related Work

There has been much work on information flow control and on the static analysis of security guarantees. The lattice model of information flow comes from the early work of Bell and LaPadula[BL75] and Denning [Den76]. A commonly used label model fitting into the lattice formalism is that of Feiertag et al. [FLR77].

The decentralized label model has several similarities to the ORAC model of McCollum et al. [MMN90]; both models provide some approximation of the "originator-controlled release" labeling used by the U.S. DoD/Intelligence community.Both ORAC and the decentralized label model have the key concept of *ownership* of policies. Both models also support the joining of labels as computation occurs, though the ORAC model lacks some important lattice properties since it attempts to merge policies with common owners. In the ORAC model, as in some mandatory access control models, both process labels and object labels can float upward in the label lattice arbitrarily, a phenomenon called *label creep* that leads to excessively restrictive labels. The absence of lattice properties and the dynamic binding of labels to objects and processes makes any static analysis of the ORAC model rather difficult. Interestingly, ORAC does allow owners to be replaced in label components (based on ACL checks that are analogous to acts-for checks), but it does not support extension of the reader set. The ORAC model also does not support any form of declassification.

Declassification in most information flow control systems lies outside the model. One recent model by Ferrari et al. [FSBJ97] incorporates a form of dynamically-checked declassification through special *waivers* to strict flow checking. Some of the need for declassification in their framework would be avoided with fine-grained static analysis. Because waivers are applied dynamically and mention specific data objects, they seem likely to have administrative and run-time overheads. Also, Abadi [Aba97] has examined the problem of achieving secrecy in security protocols, also using typing rules, and has shown that encryption can be treated as a form of safe declassification through a primitive encryption operator.

One important property of the decentralized label model is that it is amenable to static checking of information flow. However, dynamic enforcement is the approach described in the Orange Book [DOD85] and most systems have followed this approach, leading to difficulties in controlling implicit flows (for example, [Fen73, Fen74, MR92]). Stoughton [Sto81] developed (but did not implement) a purely dynamic model integrating both access control and information flow control, defined formally using denotational semantics.

Static analysis of information flow has a long history,

although it has not been as widely used as dynamic checking. Denning originally proposed a language to permit static checking [DD77], but it was not implemented. Another approach to checking programs for information flows statically has been automatic or semi-automatic theorem proving. Researchers at MITRE [Mil76, Mil81] and SRI [Fei80] developed techniques for information flow checking using formal specifications. Feiertag [Fei80] automatically checks these specifications with a Boyer-Moore theorem prover.

Recently, there has been more interest in provably-secure programming languages, treating information flow checks in the domain of type checking, which does not require a theorem prover. Volpano, Smith and Irvine have developed a simple type system for checking secrecy, encoding Denning's rules in a functional type system and showing them to be sound using standard programming language techniques [VSI96, Vol97]. Heintze and Riecke [HR98] have shown that information-flow-like labels can be applied to a simple language with reference types (the SLam calculus). They show how to statically check an integrated model similar to that of Stoughton and provide soundness results. This approach has been extended as a generalized dependency analysis [ABHR99].

These recent programming-language approaches have the limitation that they are entirely static: unlike Jif, they have no run-time access control, no declassification, and no run-time flow checking. These models also do not provide label polymorphism, support for objects, or tracking of fine-grained information flow through exceptions. The addition of these features in Jifis important for supporting a realistic programming model. However, these features do make the programming language more difficult to treat with the formal tools of programming language theory.

The notion of security enforced by Jif is essentially the same as in the other programming-language approaches; it uses usual definition of correctness for information flow, *non-interference* [GM84]. Intuitively, non-interference says that the low-security outputs of a program may not be affected by its high-security inputs. In Jif, high-security inputs may affect low-security outputs, but only if there is a `declassify` statement along all paths of information flow from the inputs to the outputs. In Java (and Jif), objects may exist both before and after the program runs, so they are effectively persistent, and must be considered to be inputs and outputs themselves.

We have avoided considering covert channels arising from time measurement and thread communication. A scheme for statically analyzing thread communication has been proposed [AR80]; essentially, a second basic block label is added with different propagation rules. This second label is used to restrict communication with other threads. The same technique would remove timing channels, and could be applied to our scheme. It is not clear how well this scheme works in practice; it seems likely to restrict timing and communication quite severely. Smith and Volpano have developed rules recently for checking information flow in a multithreaded functional language [SV98]. The rules they define, while sound, prevent the run time of a program from depending in any way on non-public data, which is arguably impractical.

6 Conclusions

Protecting privacy and secrecy of data has long been known to be a very difficult problem. The increasing use of untrusted programs in decentralized environments with mutual distrust makes a solution to this problem both more important and more difficult to solve. Existing security techniques do not provide satisfactory solutions to this problem.

The goal of this work is to make information flow control a viable technique for providing privacy in a complex, decentralized world with mutually distrusting principals. Information flow control is an attractive approach to protecting the privacy of data because it allows security requirements to be extended transitively towards or away from the principals whose security is being protected. However, it has not been a widely accepted technique because of the excessive restrictiveness it imposes and the computational overhead.

To address these limitations of conventional information flow techniques, this work addresses several problems: the information flow model, its incorporation in a programming langauge, and its runtime support. This paper has concentrated primarily on the first, providing an expressive model of decentralized information flow labels with the ability to express privacy policies for multiple, mutually distrusting principals, and to enforce all of their security requirements simultaneously. The discussions of our solutions to the other problems have been more brief; our work in these areas is still underway. We plan to complete the work on Jif and on the trusted execution platform, including the Byzantine-fault-tolerance techniques for supporting this platform. Then we plan to use this implementation to explore the utility and practicality of the model.

References

[Aba97] Martín Abadi. Secrecy by typing in security protocols. In *Proc. Theoretical Aspects of Computer Software: Third International Conference*, September 1997.

[ABHR99] Martin Abadi, Anindya Banerjee, Nevin Heintze, and Jon Riecke. A core calculus of dependency. In *Proc. 26th ACM*

Symp. on Principles of Programming Languages (POPL), pages 147–160, San Antonio, TX, USA, January 1999.

[AR80] Gregory R. Andrews and Richard P. Reitman. An axiomatic approach to information flow in programs. *ACM Transactions on Programming Languages and Systems*, 2(1):56–76, 1980.

[BL75] D. E. Bell and L. J. LaPadula. Secure computer system: Unified exposition and Multics interpretation. Technical Report ESD-TR-75-306, MITRE Corp. MTR-2997, Bedford, MA, 1975. Available as NTIS AD-A023 588.

[CL99] Miguel Castro and Barbara Liskov. Practical Byzantine Fault Tolerance. In *Proceedings of the Third Symposium on Operating Systems Design and Implementation (to appear)*, New Orleans, LA, February 1999.

[DD77] Dorothy E. Denning and Peter J. Denning. Certification of Programs for Secure Information Flow. *Comm. of the ACM*, 20(7):504–513, July 1977.

[Den76] Dorothy E. Denning. A lattice model of secure information flow. *Comm. of the ACM*, 19(5):236–243, 1976.

[DOD85] Department of Defense. *Department of Defense Trusted Computer System Evaluation Criteria*, DOD 5200.28-STD (The Orange Book) edition, December 1985.

[Fei80] Richard J. Feiertag. A technique for proving specifications are multilevel secure. Technical Report CSL-109, SRI International Computer Science Lab, Menlo Park, California, January 1980.

[Fen73] J. S. Fenton. *Information Protection Systems.* PhD thesis, University of Cambridge, Cambridge, England, 1973.

[Fen74] J. S. Fenton. Memoryless subsystems. *Computing J.*, 17(2):143–147, May 1974.

[FLR77] R. J. Feiertag, K. N. Levitt, and L. Robinson. Proving multilevel security of a system design. *Proc. 6th ACM Symp. on Operating System Principles (SOSP), ACM Operating Systems Review*, 11(5):57–66, November 1977.

[FSBJ97] Elena Ferrari, Pierangela Samarati, Elisa Bertino, and Sushil Jajodia. Providing flexibility in information flow control for object-oriented systems. In *Proc. IEEE Symposium on Security and Privacy*, pages 130–140, Oakland, CA, USA, May 1997.

[GJS96] James Gosling, Bill Joy, and Guy Steele. *The Java Language Specification.* Addison-Wesley, August 1996. ISBN 0-201-63451-1.

[GM84] J. A. Goguen and J. Meseguer. Unwinding and inference control. In *Proc. IEEE Symposium on Security and Privacy*, pages 11–20, April 1984.

[HR98] Nevin Heintze and Jon G. Riecke. The SLam calculus: Programming with secrecy and integrity. In *Proc. 25th ACM Symp. on Principles of Programming Languages (POPL)*, San Diego, California, January 1998.

[LABW91] Butler Lampson, Martín Abadi, Michael Burrows, and Edward Wobber. Authentication in distributed systems: Theory and practice. In *Proc. 13th ACM Symp. on Operating System Principles (SOSP)*, pages 165–182, October 1991. *Operating System Review*, 253(5).

[Mil76] Jonathan K. Millen. Security kernel validation in practice. *Comm. of the ACM*, 19(5):243–250, May 1976.

[Mil81] Jonathan K. Millen. Information flow analysis of formal specifications. In *Proc. IEEE Symposium on Security and Privacy*, pages 3–8, April 1981.

[ML98] Andrew C. Myers and Barbara Liskov. Complete, safe information flow with decentralized labels. In *Proc. IEEE Symposium on Security and Privacy*, Oakland, CA, USA, May 1998.

[MMN90] Catherine J. McCollum, Judith R. Messing, and LouAnna Notargiacomo. Beyond the pale of MAC and DAC—defining new forms of access control. In *Proc. IEEE Symposium on Security and Privacy*, pages 190–200, 1990.

[MR92] M. D. McIlroy and J. A. Reeds. Multilevel security in the UNIX tradition. *Software—Practice and Experience*, 22(8):673–694, August 1992.

[Mye99a] Andrew C. Myers. JFlow: Practical mostly-static information flow control. In *Proc. 26th ACM Symp. on Principles of Programming Languages (POPL)*, San Antonio, TX, USA, January 1999.

[Mye99b] Andrew C. Myers. *Mostly-Static Decentralized Information Flow Control.* PhD thesis, Massachusetts Institute of Technology, Cambridge, MA, January 1999.

[San96] Ravi S. Sandhu. Role hierarchies and constraints for lattice-based access controls. In *Proc. Fourth European Symposium on Research in Computer Security*, Rome, Italy, September 25-27 1996.

[Sto81] Allen Stoughton. Access flow: A protection model which integrates access control and information flow. In *IEEE Symposium on Security and Privacy*, pages 9–18. IEEE Computer Society Press, 1981.

[SV98] Geoffrey Smith and Dennis Volpano. Secure information flow in a multi-threaded imperative language. In *Proc. 25th ACM Symp. on Principles of Programming Languages (POPL)*, San Diego, California, January 1998.

[Vol97] Dennis Volpano. Provably-secure programming languages for remote evaluation. *ACM SIGPLAN Notices*, 32(1):117–119, January 1997.

[VSI96] Dennis Volpano, Geoffrey Smith, and Cynthia Irvine. A sound type system for secure flow analysis. *Journal of Computer Security*, 4(3):167–187, 1996.

High Confidence Computing

Realize: Resource Management for Soft Real-Time Distributed Systems *

P. M. Melliar-Smith, L. E. Moser, V. Kalogeraki, P. Narasimhan
Department of Electrical and Computer Engineering
University of California, Santa Barbara, CA 93106

Abstract

The Realize system simplifies the development of complex applications by separating the application programming from the management of resources for soft real-time CORBA applications, and from the replication of CORBA objects to provide high availability and fault tolerance. Realize uses totally ordered multicast messages to maintain consistency of the states of the object replicas, and adjusts the degree of replication dynamically as resource availability changes. By measuring the usage of the resources and by monitoring the behavior of application objects, Realize allocates objects to processors and migrates objects between processors to balance the load on the processors. Realize employs a least-laxity scheduling strategy that meets soft real-time deadlines for tasks operating across multiple processors. Thus, with Realize, the difficult issues of distribution, real-time scheduling, load balancing, replication, consistency, and fault detection and recovery are hidden from the application programmer.

1 Introduction

Major defense systems are among the most complex systems ever constructed. In some cases the defense application is inherently difficult; in other cases the application is not difficult in itself but is surrounded by so many rules and special cases that the resulting system becomes very complex. Many defense applications must be networked with other defense applications, which further increases the complexity. Each defense application is difficult, and requires considerable time, to program. In a future information war, the speed with which new applications can be developed, and existing applications can be modified, can be critical to success. Moreover, the application must be programmed by experts in the application domain, and such experts are not easy to find.

*This research has been sponsored by the Defense Advanced Research Projects Agency in conjunction with the Office of Naval Research under Contract N00174-95-K-0083.

The situation is even worse. Defense systems must satisfy additional requirements of fault tolerance, real-time performance, and security. Each of these additional requirements is difficult and time consuming to program, and must be programmed by experts in those specific topics; such experts are also scarce. Unfortunately, existing methodologies intermix the mechanisms for fault tolerance and real time with the algorithms for the defense application itself. This causes the applications, which are already quite complex, to become even more complex. The application development timescales increase, because testing for fault tolerance and real time is difficult, which makes it even more difficult to find programmers who are expert both in the application domain and in fault tolerance and real time.

Clearly, what is needed is a methodology that enables application programmers, who are expert in the domain of their defense application, to program that application as cleanly as possible without having to worry about additional considerations. Similarly, the mechanisms that provide fault tolerance and real time capabilities should be programmed and tested separately from the application, once only, by experts in those domains, and then used for a variety of defense applications.

The Realize system [7, 8, 11] achieves that separation. Using Realize, the application programmer programs a fault-tolerant, soft real-time distributed application, without having to address the difficult issues of distribution, real-time scheduling, load balancing, replication, consistency, fault detection and recovery. Realize operates on standard commercial off-the-shelf (COTS) operating systems and with standard unmodified COTS implementations of CORBA [19], preserving the portability, location transparency and interoperability that CORBA provides but augmenting those features with fault tolerance and soft real-time performance.

Thus, the Realize system provides:

- Support for simple programming of soft real-time fault-tolerant distributed applications

0-7695-0490-6/99 $10.00 © 1999 IEEE

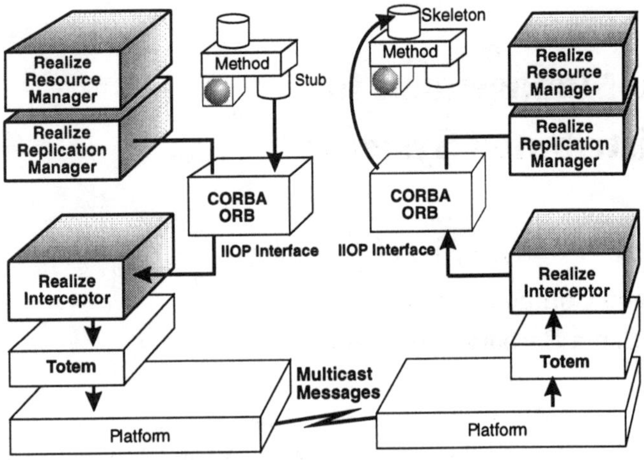

Figure 1: The structure of the Realize system, which includes the Replication Manager, the Resource Manager and the Interceptor.

- Resource management to balance the load on the resources, to schedule the tasks, and to meet soft real-time objectives
- Replication of objects and maintenance of strong replica consistency for high availability and fault tolerance.

The structure of the Realize system is shown in Figure 1. The Object Request Broker (ORB) of CORBA packages the method invocations and responses into messages, formatted according to the Internet Inter-ORB Protocol (IIOP), and subsequently passes those messages to TCP/IP for communication to other CORBA objects over the same ORB or over different ORBs. The Interceptor [14], however, intercepts the IIOP messages before they reach TCP/IP, and diverts them instead to the Replication Manager, which multicasts them to the replicas of the objects using the Totem group communication system [12].

The Resource Manager controls the allocation of resources within the distributed system. It monitors the behavior of the application objects during operation, measuring the utilization of the resources and the durations of the method invocations, and uses this information to balance the loads on the resources by moving objects, and to drive a least-laxity soft real-time scheduler. The Resource Manager is itself implemented as a collection of CORBA objects and, thus, benefits from CORBA's interoperability and Realize's fault tolerance.

The Replication Manager transparently replicates the objects of the application program to provide high availability and fault tolerance with minimal modification of those objects. Both client and server objects can be replicated, and objects can act as both clients and servers. The Replication Manager translates invocations of methods on an object group into invocations of methods on the individual object replicas, and maintains the consistency of those replicas, even in the presence of faults.

2 Resource Management in Realize

The structure of the Realize Resource Management System is shown in Figure 2. A single Resource Manager, replicated for fault tolerance, works in concert with Profilers and Schedulers, located on each processor. The Profilers monitor the behavior of the application objects and measure the current load on the processors' resources. The Resource Manager maintains a repository of data obtained from the Profilers, replicates and distributes the objects across the processors, and moves objects when necessary to maximize the system utility and to maintain a uniform load on the resources. The Schedulers exploit information collected by the Resource Manager to schedule tasks to meet soft real-time deadlines, using a least-laxity scheduling algorithm. The actual replication of objects is provided by the Replication Manager.

The Resource Manager is responsible for assigning objects to processors and for moving objects from one processor to another. As new applications are introduced into the system, the Resource Manager determines, using the Profilers' information, whether it can satisfy their requests and then allocates resources to them accordingly. During operation, the Resource Manager might determine that a resource is overloaded or that a task is not meeting its deadlines, necessitating reallocation of the replicas or reduction of the degree of replication. The Resource Manager determines the number of replicas that are needed to satisfy reliability requirements. If a processor or other resource is lost because of a fault, the Resource Manager might need to reallocate the replicas to maintain a sufficient degree of replication to meet those requirements.

Thus, the soft real-time performance objectives of the Resource Manager are:

- To increase the probability of satisfying soft real-time response time requirements for each application task
- To balance the load of the resources by allocating the application objects to the processors and reallocating them as necessary
- To determine the degree of replication of the objects, as well as the placement of the object replicas, based on the relative importance of the application objects to the tasks.

The Resource Manager operates under the following assumptions:

- The behavior of each method invocation is independent of the context. Thus, the processing time of each method and of methods invoked by a method, and the

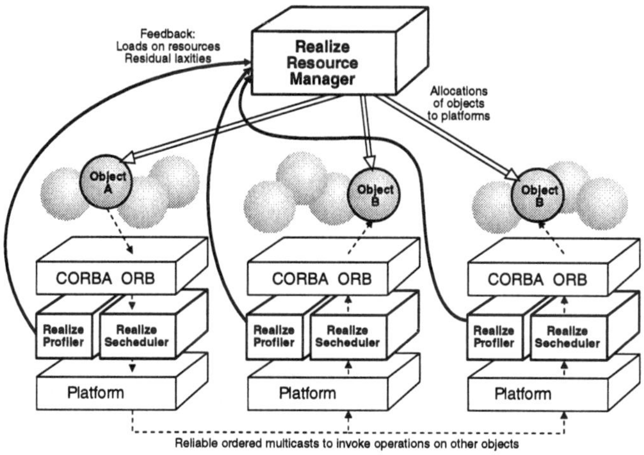

Figure 2: The structure of the Realize Resource Management System, including the Resource Manager, the Profilers and the Schedulers.

communication costs to invoke those methods, do not depend on when or why the method is invoked.

- Each resource is used for multiple activities, performed independently and asynchronously, and each such activity individually contributes only a small proportion of the total load on the resource.
- All resources are lightly loaded, with processor utilizations that are typically less than 0.5 and laxities that are typically at least half the time until the deadline. These margins suffice to accommodate statistical fluctuations in the load with acceptable probability.

The first assumption may make the Resource Manager unsuitable for systems in which nominally equivalent tasks involve substantially different numbers of operations. Well-understood deviations from this assumption can be handled by providing multiple methods to represent different behaviors. In the absence of specific information about the way in which processing times depend on task data, the use of mean processing times and an assumption of independence is the best that can be achieved.

2.1 Feedback Loops

The Resource Manager uses a three-level feedback loop structure:

- The first level (milliseconds), shown in Figure 3, uses measurements of elapsed time. to refine the estimated residual laxity of executing tasks, which are used by the least laxity scheduler.
- The second level (seconds), shown in Figure 3, uses measurements of elapsed time and measurements of resource loads to refine the initial estimates of the laxity for the tasks as they start.
- The third level (several seconds), shown in Figure 4, uses measured resource loads and measured residual laxities to revise the allocation of objects to processors.

2.2 Information Base

Upon initialization, the Resource Manager uses a configuration file to define the physical configuration available to it, including the processors and communication networks along with their specific characteristics (processing speed, memory size, network bandwidth). For each such resource, the Resource Manager periodically receives current load information (processing load, available memory, communication load) from the Profilers on the individual processors. The Resource Manager uses this information to adapt to changing system conditions.

For each application task, the Resource Manager maintains a list of the method invocations required for that task, a deadline for completion of the task, and an importance metric that is used to decide which tasks should be abandoned in the event of system overload.

For each method of each object, the Resource Manager maintains estimates of the computation time and the communication time for invocations of that method. These estimates are used to determine the initial laxity when an application task starts.

2.2.1 Task Metrics

An application task t consists of a sequence of object invocations between an external input (or possibly a timer signal or completion of another operation) and the generation of an external result. Each task t has the following metrics:

- $Deadline_t$: the time interval, starting at task initiation, within which task t should be completed, specified by the application designer
- $Importance_t$: the relative importance of task t, specified by the application designer, that affects the decision of which task should be abandoned in the event of system overload
- $Projected\_latency_t$ or T_t: the estimated amount of time from initiation to completion of task t, measured by the Profiler
- $Laxity_t$: the difference between $Deadline_t$ and $Projected\_latency_t$, a measure of urgency of task t, calculated by the Scheduler and used by the Scheduler to schedule task t
- $Residual\_laxity_t$: the remaining laxity when task t completes, calculated by the Scheduler.

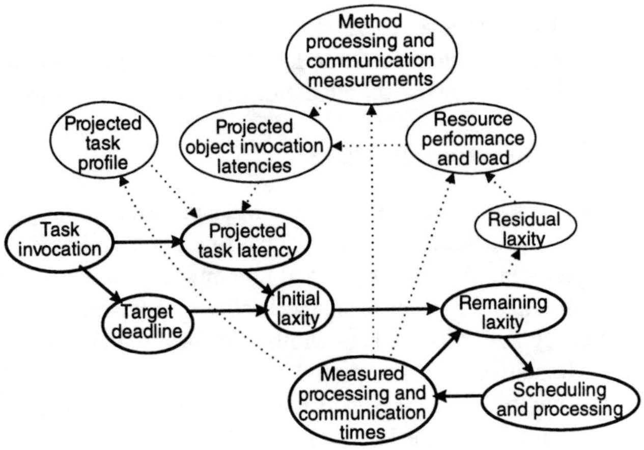

Figure 3: Least laxity scheduling. The first level feedback loop is shown with solid lines, and the second level feedback loop with dotted lines.

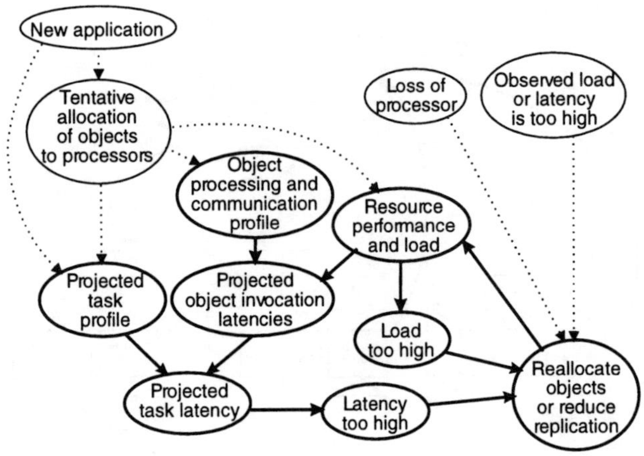

Figure 4: Object allocation and reallocation. The third level feedback loop is shown with solid lines.

2.2.2 Object and Object Replica Metrics

For each object i, the Resource Manager maintains:

- $Methods_i$: the set of methods of object i
- $Replication\_type_i$: the type of replication (active/passive) of object i, determined by the Resource Manager
- $Replication\_degree_i$: the number of replicas of object i, determined by the Resource Manager
- $Host\_names_i$: the set of hosts on which the replicas of object i may be located.

2.2.3 Method Invocation Metrics

For each method m, the Resource Manager maintains:

- $Object_m$: the object of which m is a method
- $Mean\_local\_time_m$: the mean time required, after receipt by a processor of a message invoking method m, for that processor to complete the invocation. This excludes communication time but includes queueing time and the time of embedded invocations of other methods.
- $Mean\_remote\_time_{mn}$: the mean time required for method m to invoke method n remotely, including communication and queueing time and also the time of embedded invocations of other methods
- $Mean\_communication\_time_{mn}$: the mean time to communicate an invocation from method m to method n and to communicate the response back. This is computed as the difference between $Mean\_remote\_time_{mn}$ and $Mean\_local\_time_n$
- $Mean\_execution\_time_{mp}$: the mean time required for method m to execute locally on processor p, including queueing time but excluding the time required for embedded invocations of other methods
- $Mean\_processing\_time_{mp}$ or τ_{mp}: the mean time required for method m to execute on processor p, excluding queueing time and the time required for embedded invocations of other methods
- $Mean\_transmission\_time_{mc}$ or σ_{mc}: the mean time for transmitting an invocation and a response for a method m on link c, excluding queueing time
- $Mean\_invocations_{mn}$: the mean number of invocations that a single invocation of method m makes on method n.

2.3 The Profilers

For complex distributed applications, it is difficult to obtain accurate projections of resource requirements and other behavioral characteristics. Consequently, Realize adopts an approach based on monitoring the utilizations of the resources and the behavior of the application objects using the Profilers to supply feedback information that it uses to adjust the scheduling of tasks and the allocation of objects to processors. The Resource Manager allocates objects to processors, the objects execute and use resources, and the Profilers report resource usage to the Resource Manager, enabling it to adjust the allocations accordingly.

Each processor has its own Profiler, which is implemented between the CORBA ORB and the operating system. Each Profiler reports:

- The current utilization of resources for its processor
- The mean time required for the invocation of each remote method, determined by exponentially weighted averaging

- The mean time and memory required for the invocation of each local method, determined by exponentially weighted averaging
- For each local method, the mean number of invocations of each other method during one invocation of that local method, determined by exponentially weighted averaging
- The mean residual laxity for each task on its completion, determined by exponentially weighted averaging.

Exponentially weighted averaging defines the expectation $E[a]$ for a measurement a, and a constant c, as follows:

$$E_1[a] = a_1$$

$$E_{i+1}[a] = E_i[a] * (1 - c) + a_{i+1} * c$$

where $a_1, a_2, a_2, \ldots$ is the sequence of measurements of a, the most recent measurement is weighted by c, and the prior expectation is weighted by $1 - c$.

2.4 The Resource Manager

The Resource Manager estimates the $Mean\_local\_time_m$ required to invoke method m (processing, embedded invocations, queueing) by calculating the mean of the means $Mean\_local\_time_m$ obtained from the Profilers. Similarly, it estimates the $Mean\_remote\_time_{mn}$ for methods m and n (communication, processing, queueing, embedded invocations and overheads) and also the $Mean\_invocations_{mn}$. In addition, it calculates the $Mean\_communication\_time_{mn}$ by taking the difference of the $Mean\_remote\_time_{mn}$ and $Mean\_local\_time_m$ that it calculated.

Moreover, the Resource Manager estimates the $Mean\_execution\_time_{mp}$ for method m on processor p, which includes queueing time but excludes embedded invocations of other methods. The estimate is derived from the $Mean\_local\_time_m$ for method m, $Mean\_invocations_{mn}$ for each other method n invoked by method m, and $Mean\_remote\_time_{mn}$ for each other method n.

The Resource Manager then estimates the $Mean\_processing\_time$ τ_{mp} for each method m on processor p. Let ρ_p be the load on processor p. In the absence of more specific information about the application's behavior, we assume an M/M/1 queueing model for the execution time. Thus, the $Mean\_processing\_time$ τ_{mp} for method m on processor p is given by:

$$\tau_{mp} \equiv (1 - \rho_p) * Mean\_execution\_time_{mp}$$

Similarly, let ρ_c be the load on communication link c. Then the $Mean\_transmission\_time$ σ_{mc} for method m on communication link c is given by:

$$\sigma_{mc} \equiv (1 - \rho_c) * Mean\_communication\_time_{mc}$$

Using a classical equilibrium flow analysis, the Resource Manager determines the mean number x_{tm} of invocations of method m for one execution of task t.

Given the $Mean\_processing\_time$ τ_{mp}, the $Mean\_transmission\_time$ σ_{mc} for method m on processor p and communication link c, and the mean number x_{tm} of invocations of method m made by task t, the Resource Manager computes the projected latency for the entire task as:

$$Projected\_latency_t \equiv \sum_m \min_{p:m\epsilon p} \left\{ \frac{x_{tm}\tau_{mp}}{1 - \rho_p} \right\} + \frac{x_{tm}\sigma_{mc}}{1 - \rho_c}$$

where $m \epsilon p$ denotes that the object i of which m is a method executes on processor p. The minimum is taken over all processors p that host replicas of which m is a method.

The Resource Manager's estimate of the projected latency for task t, based on the reports of the Profilers, represents the second level of the feedback loop structure, shown by the light arrows in Figure 3. If the Resource Manager's estimate of the $Projected\_latency_t$ for task t is accurate, the $Residual\_laxity_t$ when task t completes will be the same as the initial laxity for task t. The Resource Manager uses the ratio of the $Residual\_laxity_t$ to the initial laxity to adjust its estimates of the $Projected\_latency_t$ for task t.

2.5 The Schedulers

In Realize, the schedulers uses least laxity scheduling, shown as the first level of the feedback loop structure by heavy arrows in Figure 1. Least laxity scheduling is an effective strategy for scheduling tasks across multiple processors provided that the system is not overloaded. In least laxity scheduling, the laxity L_t of a task t represents a measure of urgency of the task, defined as

$$Laxity_t = Deadline_t - Remaining\_projected\_latency_t$$

where the $Remaining\_projected\_latency_t$ is the estimated amount of time from the current point of task t to its completion.

The effectiveness of the least laxity scheduling algorithm compared to the earliest deadline first algorithm in a distributed system can be seen by considering the following example, shown in Figure 5. Assume that tasks A and B arrive independently at processor P. Task A consists of a single method, while task B consists of methods B_1 and B_2.

The earliest deadline first algorithm is driven by the deadlines of the tasks and assigns the highest priority to the task with the earliest deadline. Consequently, task A is scheduled first on processor P and task B starts executing when task A finishes, *i.e.*, at time t = 3 and, thus, task B misses its deadline. In contrast, the least laxity scheduling algorithm takes into consideration the

Task	Deadline	Projected Latency	Laxity t=0	t=1	t=2	t=3	t=4	t=5	t=6	t=7	t=8
A	6	3	3	2	1	1	1				
B_1 & B_2	9	8	1	1	1	1	1	1	1	1	

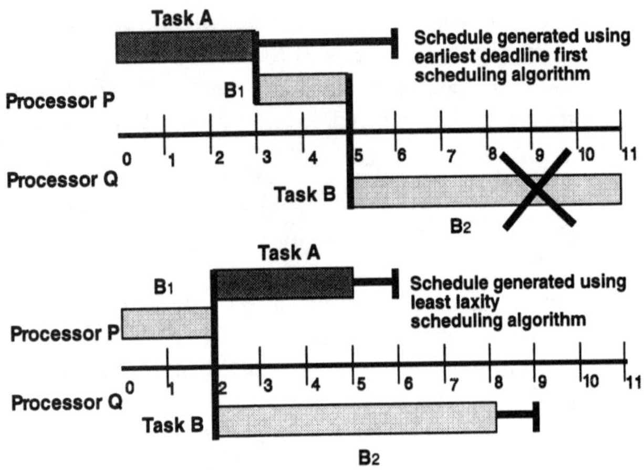

Figure 5: Least laxity scheduling finds an appropriate schedule when earliest deadline scheduling would fail.

execution time (*i.e.*, processing and queueing time) of the tasks on the processor and schedules each method of the task independently according to the task's remaining laxity. Thus, method B_1 of task B is scheduled first on processor P. When method B_1 finishes execution, both tasks have the same laxity values, task A is scheduled on processor P, and method B_2 of task B is scheduled on processor Q. Therefore, both of the tasks meet their deadlines.

As Figure 3 shows, when a task starts, triggered by some external or internal event, the scheduler calculates a target deadline for it, and then subtracts the projected task latency, provided by the Resource Manager, to yield the initial laxity. The strategy that the Resource Manager uses to estimate the projected task latency is described below. As the application executes, the least laxity scheduler schedules the task according to the remaining laxity for that task.

The execution of the task involves the invocation of methods of objects. Each such invocation requires communication to invoke the method, processing for the method, and communication to return the results. The time required for this communication and processing is measured by the Profilers, and is reported to the Resource Manager where it is used to estimate the projected task latency. The actual time to perform the operation may be variable, and increases with the load on the processor and on the communication medium.

When a method is invoked, the time of the invocation is recorded using the operating system's high resolution timer. When the response is received, the time is again noted, and the difference between these two times is the actual time required for the method. This actual time is subtracted from the projected time, and the difference is added to the remaining laxity of the task.

The adjustment of the remaining laxity provides a feedback loop. If a task is delayed, its laxity will diminish and its scheduling priority will increase. All of the computations required to calculate the deadlines and laxities are local to the individual processors. Consequently, overheads are low and priority adjustments can be made on a microsecond-by-microsecond basis. In contrast, the feedback loops used to estimate the projected task latency, from which the initial task laxity is derived, operate more slowly on a timescale of seconds.

The remaining laxity of the task is included in the message that communicates a remote method invocation. This allows the method to be scheduled appropriately on the remote processor hosting the object. There is no need to include a remaining laxity in a response message. As described above, the remaining laxity is adjusted using the measured time at which the response is received. When the task completes, the remaining laxity is recorded as a residual laxity, which is reported back to the Resource Manager as a measure of the current timing margins in the system.

2.6 Allocation of Objects to Processors

The Resource Manager maintains a global view of the application objects in the system and attempts to maintain a uniform load on all of the processors. It tries to meet the requirements of all tasks in the system, but this cannot always be achieved. Moreover, the addition of a new task can result in degradation of the performance of existing tasks. To satisfy the requirements of a new task, the Resource Manager can migrate objects to different processors or reduce the degree of replication of objects [7]. Migration of objects may also be required when a processor fails, when the current load on a processor is too high, or when the latency of a task is too high. The Resource Manager uses a Cooling Algorithm to migrate objects when the load on a processor is too high, and a Hot Spot Algorithm to migrate objects when the latency of a task is too high [8].

2.6.1 Cooling Algorithm

Periodically, the Resource Manager checks whether the load on a processor exceeds a predefined load limit, MAX_LOAD, determined by the application designer, or whether the load on the processor has increased considerably compared to the load on the other processors. If so, the Resource Manager determines whether any of the objects on that processor can be moved to the processor with the lowest load.

The Resource Manager identifies the processor *low* with the lowest load, the processor *high* with the highest load, and the processor $high2$ with the second highest load. It then calculates the contributions to the load on $high$ made by the various objects on $high$. This load is derived from the mean processing time for the various methods of the object and the number of invocations per second of those methods by the various tasks. The objects are considered in the order of their loads. The Resource Manager then determines whether the candidate object i can be moved to the least-loaded processor low without increasing the load on that processor above the allowable load. This continues until the load on processor $high$ is below that of the second most highly loaded processor $high2$ or until all of the objects have been considered.

2.6.2 Hot Spot Algorithm

Periodically, the Resource Manager estimates the projected latency for each task t, to check whether the latency of any task is too high a proportion of its deadline. If so, it searches for the object that is causing the largest queueing delay for task t (the actual processing time cannot, of course, be reduced). This is the object i, executing on some processor p, whose methods m cause, in aggregate, the largest increase in the latency to the completion of task t because of queueing, computed as:

$$\begin{aligned} Queueing\_latency_{tip} &\equiv \sum_{m \epsilon i} \frac{x_{tm} r_{mp}}{(1-\rho_p)} - x_{tm} r_{mp} \\ &= \sum_{m \epsilon i} \rho_p \frac{x_{tm} r_{mp}}{(1-\rho_p)} \end{aligned}$$

where $m \in i$ indicates that m is a method of object i.

The Resource Manager identifies the object i that is causing the largest queueing delay for task t and the processor p on which object i is executing. Once it has identified object i, it selects the processor low with the lowest load and attempts to move object i to processor low. If this object migration would cause the load on processor low to exceed either the predefined load limit MAX_LOAD, or the allowable load on processor p, it attempts to move the next candidate object (causing the next largest queueing delay for task t). This continues until the Resource Manager determines an appropriate object to move. Once it has identified such an object, the Resource Manager moves that object from processor p to processor low.

To estimate the effect of moving an object on the projected task latencies, the Resource Manager considers the effects of increased or decreased processor load on the times required to invoke the methods. It assumes an M/M/1 queueing model for each processor. The estimated time required for invocations of method n on processor p before moving object i containing method m to processor p is:

$$\frac{r_n r_{np}}{1-\rho_p}$$

After moving the object i containing method m to processor p, the estimated time is:

$$\begin{aligned} &\frac{r_n r_{np}}{1-\rho_p - r_m r_{mp}} \\ &\quad = \frac{r_n r_{np}}{1-\rho_p} * \frac{1-\rho_p}{1-\rho_p - r_m r_{mp}} \\ &\quad = \frac{r_n r_{np}}{1-\rho_p} * \left(1 + \frac{r_m r_{mp}}{1-\rho_p} + \left(\frac{r_m r_{mp}}{1-\rho_p}\right)^2 + \ldots\right) \\ &\quad \approx \frac{r_n r_{np}}{1-\rho_p} * \left(1 + \frac{r_m r_{mp}}{1-\rho_p}\right) \end{aligned}$$

where the approximation is justified by the assumption that method m contributes a small proportion of the total load on processor p, and is useful because it is expressed in terms of quantities measured by the Profilers. A similar estimate is made for the reduced time required for method invocations on the processor from which the object is moved.

Figure 4 shows how the projected latencies of Figure 3 are exploited for object allocation and reallocation. Objects are allocated or reallocated when a new application task is introduced into the system, when a processor has been lost due to a fault, when the load on some resource is too high, or when the latency for some task is too high. As the new application task runs, a profile of the method invocations for that task is constructed and is added to the Resource Manager's information base. New objects are tentatively allocated to to the most lightly loaded processors. This increases the load on the processor's resources and requires new projections of the task latencies. If the allocation is acceptable, it is actually performed. However, if a resource load is too high or a latency is too long, then objects are reallocated, or the degree of replication of an object is reduced, or the least important task is suspended. This process is repeated until an acceptable allocation is found, as shown by the solid lines in Figure 4.

3 Replication in Realize

In soft real-time distributed object systems, application objects can be replicated on different processors to provide fault tolerance and high availability. The degree of replication that is provided to an application object depends on the relative importance of that object to the application tasks. Application objects that have critical functionality are expected to be more highly replicated, while less important application objects might not be replicated at all.

In Realize, object replication, multicasting of invocations and responses, maintenance of replica consistency

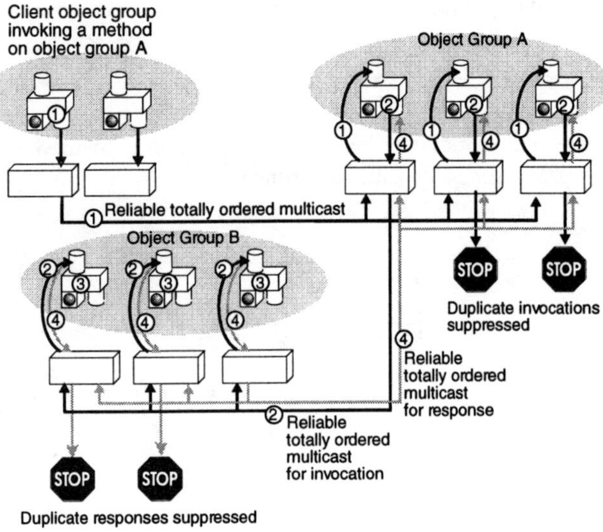

Figure 6: Active replication of objects within the Realize system.

and recovery from faults are transparent to the application objects and to the ORB. The type of replication and the reliability requirement, which implies the degree of replication, are determined for each object individually. Realize operates under the following assumptions:

- Independence of faults in the replicas of an object
- Deterministic behavior of the objects.

3.1 Types of Replication

The Replication Manager supports both active and passive replication, as shown in Figures 6 and 7. The application and the types of faults that must be tolerated dictate the type of replication that is employed.

3.1.1 Active Replication

In active replication, when a client object invokes a method on a server object, Realize multicasts the method invocation to the server object group via Totem and each server replica then executes the method, as shown in Figure 6. The underlying reliable totally ordered multicast protocol ensures that all of the replicas of an object receive the same messages in the same order and, thus, that they execute the methods in the same order. This ordering of method invocations ensures that the states of the replicas are consistent at the end of the operation.

3.1.2 Passive Replication

In passive replication, when a client object invokes a method on a server object, Realize multicasts the method invocation to that group via Totem and only one of the server replicas, the *primary replica*, executes the method, as shown in Figure 7. The Replication Manager at each of the other replicas logs the message containing the invocation so that one of those replicas can execute the method, should the primary replica fail. At the end of the method execution, Realize multicasts the updated state of the primary replica to the nonprimary replicas and multicasts the results to the client object via Totem. During the execution of the method, the states of the nonprimary replicas may differ from that of the primary replica; however, the state transfer achieves replica consistency at the end of the method execution.

3.2 Fault Detection and Recovery

Realize and Totem detect faults by using unreliable fault detectors based on timeouts. The failure of a processor or of the Replication Manager on a processor results in the failure of all object replicas on that processor.

For active replication, if any one of the replicas fails, service is uninterrupted (because the other replicas execute the method) and the results are returned to the invoking object. The replicas must also preserve the state of the object for subsequent transfer to a new or recovered replica.

For passive replication, the effect of a fault depends on whether the failed replica is a primary or nonprimary replica. The failure of a nonprimary replica during the execution of a method is transparent to a client, because Realize removes that replica from the group, while the primary replica continues to execute the method.

If the primary replica fails, Realize detects the fault and determines a new primary replica. The new primary replica must execute the method invocations that it has logged and for which it has not received a response and state transfer from the prior primary replica. Invocations on other objects, and even responses, can be generated by both the original and the new primary replicas. The Replication Manager suppresses such duplicate invocations and duplicate responses.

3.3 Replica Consistency

To ensure that the replicas of an object are updated consistently, Realize exploits the reliable totally ordered message delivery service of an underlying group communication system, in our case Totem [12]. The invocations of methods on the replicas, and the corresponding responses, are contained in multicast messages. Because the messages are delivered in the same total order at each object replica, the methods are executed by each replica in the same order, thereby maintaining consistency of the states of the replicas.

For active replication, this ensures that the invocation of a method, and the corresponding response, is received by all of the replicas of an object or by none of them. Consequently, the method is executed by all of the replicas or by none of them. For passive replication, this ensures that either all of the nonprimary replicas have the updated

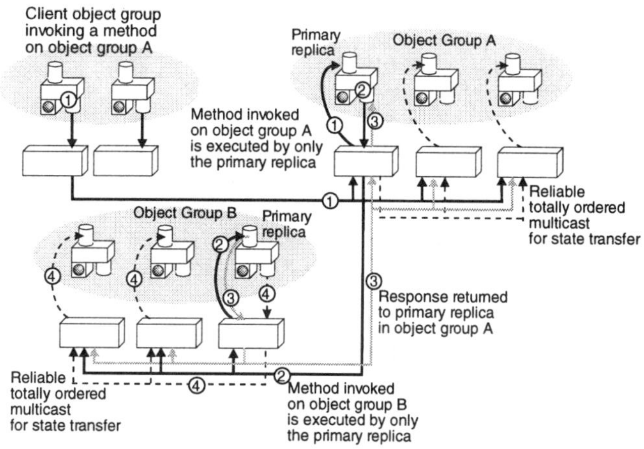

Figure 7: Passive replication of objects within the Realize system.

state of the object or, alternatively, all of the nonprimary replicas know that none of them has the updated state.

A reliable totally ordered message delivery service alone does not suffice to maintain the consistency of the states of the replicas. To maintain replica consistency, Realize also detects and suppresses duplicate operations and transfers state between replicas, as shown in Figure 6. It also ensures that, at any time, only one invocation is being processed within each process, reordering messages as necessary to ensure this. More details of the replica consistency mechanisms used by Realize can be found in [11, 13, 15].

3.4 Degree of Replication

Realize maintains the degree of replication of the application objects to achieve the desired level of reliability, while allocating sufficient resources to the objects to guarantee a certain quality of service. When computing resources are limited, increasing the degree of replication of one application object may adversely affect the quality of service provided to another application object.

Each application object requires a certain minimum amount of resources to perform acceptably. When spare resources are available, those resources can be used for additional object replicas. In such cases, even the small value of one additional replica is greater than the value of an idle resource. On the other hand, when resources are scarce and a large number of application objects are competing for resources, the decision as to which application objects to replicate, if any, is not obvious.

Realize aims to determine the degree of replication of each application object, and to allocate resources to each application task, so as to maximize the system utility while respecting the resource constraints.

As new application tasks are introduced into the system, the Resource Manager determines, using the Profilers' information, whether it can satisfy the requests, determines the replication degree for the application objects, and allocates resources accordingly.

During operation, if insufficient resources remain to run the application tasks or to provide the quality of service they require, the Resource Manager attempts to reallocate the object replicas dynamically to free computing resources by migrating them to other processors in the system; otherwise, it reduces their degree of replication. If a processor or other resource is lost due to a fault, the Resource Manager recomputes the number of replicas for each application object, which may involve reducing the degree of replication of certain objects or reallocating object replicas to adjust to new resource constraints. When a new object is introduced into the system, it may be necessary to remove an existing object replica to allocate resources to the new object.

3.4.1 The Model

The distributed system includes two types of resources: memory and CPUs. Each resource can be used by multiple tasks, and each task individually contributes only a small proportion of the total load on the resources. We let M_p denote the total memory on processor p, L_p the maximum CPU capability of processor p, h_p the proportion of the memory that is allowed to be used, and h'_p the proportion of the CPU that is allowed to be used. For instance, if the utilization of all CPUs should be kept lower than 70%, then h'_p is set to 0.7 for each processor p.

Each application task is a sequence of method invocations of possibly different objects between the acceptance of an external input (or possibly a timer signal or completion of another operation) and the generation of an external result or signal. The tasks arrive independently but their sets of method invocations are not necessarily disjoint. Thus, although each application object is created by one task, its methods may be invoked by several different tasks.

In the ideal case, the system can accommodate at least one replica for the objects of all of the application tasks that require service. The limited amount of system resources requires determination of whether the object of an application task should be replicated, based on the relative importance of the task for the system. The $Importance_t$ of task t is defined as a reward g_t gained from the execution of task t. No penalty is associated with the absence of the execution of a task. An application task completes successfully if each method invoked by the task completes execution of at least one replica of each of the objects successfully. All of the replicas of an object are independent and provide the same functionality.

Assuming that the individual replicas of an object fail independently with equal probabilities, the probability that all replicas of an object fail diminishes as the number of replicas increases. We let q_i denote the probability that a single replica of object i fails and r_i the number of replicas of object i. The $Reward\ G_i$ of object i is the sum of the rewards g_t of all application tasks t that invoke object i:

$$G_i = \sum_{i:i \in t} g_t$$

where $i \in t$ denotes that task t invokes a method of object i.

The $Utility\ U_{i_r}$ of the i_rth replica of object i is defined to be the product of object i's reward and the probability of failure of all but one of the replicas:

$$U_{i_r} = G_i * \binom{i_r}{i_r - 1} * q_i^{i_r - 1} * (1 - q_i)$$

Thus, the utility of the first replica of object i is G_i, the utility of the second replica is the product of G_i and the probability that only one of the two replicas has failed, the utility of the third replica is the product of G_i and the probability that any two of the three replicas have failed, etc. These utilities can be computed quite quickly and stored in a table.

The $System\_utility\ U(\bar{r})$ is the sum of the utilities of the object replicas:

$$U(\bar{r}) = \sum_i U_{i_r} = \sum_i \sum_{i_r} G_i * \binom{i_r}{i_r - 1} * q_i^{i_r - 1} * (1 - q_i)$$

where $\bar{r} = (r_1, r_2, ..., r_i, ...)$ and r_i is the number of replicas of object i.

We let m_{pi} denote the proportion of memory used by object i on processor p and ℓ_{pti} the load that object i invoked by task t imposes on processor p. The load that object i imposes on p is derived from the $Mean\_processing\_time$ τ_{wip} required for a method w of object i executing on processor p, the mean number x_{tiw} of invocations of method w of object i by task t, and the $Projected\_latency$ T_t of task t. Thus, the $Load\ \ell_{pti}$ of an object i on processor p invoked by task t is computed as:

$$\ell_{pti} = \frac{\sum_w \tau_{wip} * x_{tiw}}{T_t}$$

Assuming that it is always possible to recreat at least one replica of each object, the following resource utilization constraints must be satisfied. For each processor p,

$$\sum_{i:i \in p} m_{pi} \leq h_p M_p$$

$$\sum_t \sum_{i:i \in t \wedge i \in p} \ell_{pti} \leq h'_p L_p$$

The quantities in these inequalities are obtained either from measurements and computations (G_i, m_{pi}, ℓ_{pti}, q_i) performed by the Profilers and the Resource Manager or from information (M_p, L_p) specified by the application designer.

We consider the problem of choosing the combination of numbers of replicas that maximizes the $System\_utility$ $U(\bar{r})$, subject to the above resource utilization constraints. For each object, at least one replica is chosen. Then, depending on the availability of resources in the system, additional replicas are chosen with the aim of maximizing the system utility $U(\bar{r})$.

3.4.2 Greedy Algorithm

The above maximization problem is NP hard. An exact solution requires an exhaustive search that considers all possible combinations, which is too expensive for a dynamic allocation algorithm. Furthermore, the values of the importance of the objects and the rewards of the tasks are only estimates. Consequently, the cost of finding an exact solution is unjustified.

Therefore, we employ a greedy algorithm, where the $Load\ \ell_{pi}$ of an object i on processor p for all tasks t executing on processor p is given by:

$$\ell_{pi} = \sum_{t:i \in t, i \in p} \ell_{pti}$$

and the available load $AvailLoad_p$ on processor p is given by:

$$AvailLoad_p = h'_p L_p - \sum_t \sum_{i:i \in t \wedge i \in p} \ell_{pti}$$

and the available memory $AvailMem_p$ on processor p is given by:

$$AvailMem_p = h_p M_p - \sum_{i:i \in p} m_{pi}$$

The greedy algorithm chooses incrementally the unallocated replica with the highest utility and allocates it to the least loaded processor, subject to the resource constraints. In effect, at each step the greedy algorithm determines a local maximum for the system utility function. The final resulting value of the system utility function may not be a global maximum.

The Resource Manager invokes the greedy algorithm when objects are replicated initially and when the system conditions change considerably. Replicas of existing objects may be removed when a new task starts or when resource capacities have been exceeded.

4 Related Work

For a real-time distributed system with two or more processors, it has been shown that no deadline scheduling algorithm can be optimal without complete a priori knowledge of the arrival times, computation times, and deadlines of future tasks [4]. They have shown that least laxity scheduling is an effective strategy for real-time distributed systems, provided that the system is not overloaded. Realize uses least laxity scheduling.

Several efforts are underway to extend CORBA with real-time and fault-tolerance capabilities, including work of the Object Management Group (OMG) Realtime Platform Special Interest Group [17, 18, 20, 21], in which we are participating. As yet, the OMG specifications provide only static fixed priority scheduling.

Gill, Levine, Schmidt and Kuhns [6] have developed a maximum urgency first scheduling algorithm for the TAO real-time ORB [24] that provides scheduling assurances for critical tasks while optimizing the use of scarce resources. Sydir, Chatterjee and Sabata [27] have implemented an end-to-end QoS-driven resource management scheme within the ERDoS ORB, that considers the QoS requirements and resource requirements of each object individually. Wolfe, *et al* [29] have introduced timed distributed method invocations as a real-time CORBA extension to enable the expression and enforcement of timing constraints in client/server interactions by attaching a separate object with timing characteristics to each method invocation.

Nett, Gergeleit and Mock [16] have developed an adaptive system using integrated monitoring, dynamic execution time prediction and scheduling to provide time-awareness for standard CORBA method invocations. Rosu, Schwan and Jha [23] have introduced adaptive resource allocation mechanisms that can adjust resource allocation to changes in the applications' needs and have proposed a set of performance metrics for capturing the impact of those mechanisms on the performance of the applications. Rajkumar, Lee, Lehoczky and Siewiorek [22] have developed the Q-RAM model that focuses on allocating multiple resources to multiple applications such that each application can meet its minimum needs and yet maximize the system utility.

Other researchers have developed middleware that extends CORBA with fault tolerance. The Electra toolkit [9, 10] built by Maffeis on top of Horus provides support for replicated CORBA objects, as does Orbix+Isis [9] on top of Isis. Unlike Realize, Electra and Orbix+Isis are nonhierarchical object systems and support only active replication. Both Electra and Orbix+Isis use an integration approach in that the replication and group communication mechanisms are integrated into the ORB and require modification of the ORB. In contrast, Realize uses the interception approach, which requires no modification to the ORB.

The Distributed Object-Oriented Reliable Service (DOORS) [25] augments CORBA with fault tolerance using passive replication by providing replica management, fault detection, and fault recovery as CORBA service objects above the ORB. The DoorMan management interface monitors DOORS and the underlying system to take corrective action by migrating objects, if their hosts are suspected of being faulty. The OpenDREAMS toolkit [5] also uses the service approach but provides support for both active and passive replication by adding replication and group communication as service objects on top of the ORB. The service approach requires no modification of the ORB, but lacks the transparency and efficiency that Realize provides.

The AQuA framework [3] employs the Quality of Service for CORBA Objects (QuO) runtime, the Proteus dependability property manager, and the Maestro/Ensemble toolkits. QuO [30] extends the Interface Definition Language (IDL) of CORBA with a Quality of Service Description Language (QDL). It provides mechanisms for measuring and enforcing QoS contracts and taking appropriate actions when those contracts are violated. Based on the requirements communicated by the QuO runtime and the faults that occur, the Proteus dependability property manager determines the type of faults to tolerate, the replication policy, the degree of replication, and the location of the replicas, and dynamically modifies the configuration to meet those requirements. The Maestro/Ensemble toolkits [28] uses a gateway to translate CORBA object invocations into messages that are transmitted via the Ensemble group communication system, and detects and filters duplicate invocations (responses). AQuA is substantially less transparent to the application than is Realize.

Shokri, Hecht, Crane, Dussault and Kim [26] have observed that effective fault handling in complex distributed applications requires the ability to adapt resource allocation and fault-tolerance policies dynamically in response to changes in the environment, application requirements and available resources. Bondavalli, Giandomenico and Mura [2] have defined a planner component based on a value-driven admission policy that minimizes the effects that high reliability requirements have on the expected quality of service and achieves the optimal utilization of available system resources.

5 Conclusion

The Realize system simplifies the development of complex distributed defense systems by extending the services of CORBA with resource management, soft real-time scheduling and fault tolerance.

The Resource Manager employs a three-level feedback loop structure, based on the profiling of the application objects, which allocates objects to processors and migrates them between processors to balance the load across the system. It also provides least-laxity scheduling to meet soft real-time deadlines for tasks that span multiple processors.

The Replication Manager replicates objects, using either active or passive replication as appropriate, and maintains the consistency of the states of the replicas using reliable totally ordered multicast messages. Realize determines the appropriate degree of replication of objects, balancing fault-tolerance and real-time performance requirements.

Thus, Realize simplifies the programming of soft real-time fault-tolerant distributed applications by hiding the difficult issues of distribution, real-time scheduling, load balancing, replication, consistency, fault detection and recovery, allowing the application programmer to focus on the application. Realize maintains the portability, location transparency and interoperability that COTS implementations of CORBA provide, and has been demonstrated with seven unmodified commercial CORBA ORBs, incurring an additional overhead of only about 10-15%.

References

[1] R. Bettati and A. Gupta, ''Dynamic resource migration for multiparty real-time communication,'' *Proceedings of the IEEE 16th International Conference on Distributed Computing Systems* (May 1996), Hong Kong, pp. 646-655.

[2] A. Bondavalli, F. D. Giandomenico and I. Mura, ''Value-driven assignment in object-oriented real-time dependable systems,'' *Proceedings of the IEEE 3rd International Workshop on Object-Oriented Real-Time Dependable Systems*, Newport Beach, CA (February 1997), pp. 92-99.

[3] M. Cukier, J. Ren, C. Sabnis, W. H. Sanders, D. E. Bakken, M. E. Berman, D. A. Karr and R. E. Schantz, ''AQuA: An adaptive architecture that provides dependable distributed objects,'' *Proceedings of the IEEE 17th Symposium on Reliable Distributed Systems*, West Lafayette, IN (October 1998), pp. 245-253.

[4] M. L. Dertouzos and A. K. Mok, ''Multiprocessor on-line scheduling of hard-real-time tasks,'' *IEEE Transactions on Software Engineering*, vol. 15, no. 12 (December 1989), pp. 1497-1506.

[5] P. Felber, B. Garbinato and R. Guerraoui, ''Designing a CORBA group communication service,'' *Proceedings of the IEEE 15th Symposium on Reliable Distributed Systems*, Niagara on the Lake, Canada (October 1996), pp. 150-159.

[6] C. Gill, D. L. Levine, D. C. Schmidt and F. Kuhns, ''Evaluating strategies for real-time CORBA dynamic scheduling,'' *International Journal of Time-Critical Computing Systems*, to appear.

[7] V. Kalogeraki, L. E. Moser and P. M. Melliar-Smith, ''Dynamic modeling of replicated objects for dependable soft real-time distributed object systems,'' *Proceedings of the IEEE 4th Workshop on Object-Oriented Real-time Dependable Systems*, Santa Barbara, CA (January 1999), pp. 48-55.

[8] V. Kalogeraki, P. M. Melliar- Smith and L. E. Moser, ''Using multiple feedback loops for object profiling, scheduling and migration in soft real-time distributed object systems,'' *Proceedings of the IEEE 2nd International Symposium on Object-Oriented Real-Time Distributed Computing*, Saint Malo, France (May 1999), pp. 291-300.

[9] S. Landis and S. Maffeis, ''Building reliable distributed systems with CORBA,'' *Theory and Practice of Object Systems*, vol. 3, no. 1, pp. 31-43.

[10] S. Maffeis, ''Electra --- Making distributed programs object-oriented,'' *Proceedings of the USENIX Symposium on Experiences with Distributed and Multiprocessor Systems*, vol. IV, San Diego, CA, pp. 143-156.

[11] P. M. Melliar-Smith, L. E. Moser, V. Kalogeraki and P. Narasimhan ''The Realize middleware for replication and resource management,'' *Proceedings of the IFIP International Conference on Distributed Systems Platforms and Open Distributed Processing, Middleware '98*, The Lake District, England (September 1998), pp. 123-138.

[12] L. E. Moser, P. M. Melliar-Smith, D. A. Agarwal, R. K. Budhia and C. A. Lingley-Papadopoulos, ''Totem: A fault-tolerant multicast group communication system,'' *Communications of the ACM*, vol. 39, no. 4 (April 1996), pp. 54-63.

[13] L. E. Moser, P. M. Melliar-Smith and P. Narasimhan, ''Consistent object replication in the Eternal system,'' *Theory and Practice of Object Systems*, vol. 4, no. 2, pp. 81-92.

[14] P. Narasimhan, L. E. Moser and P. M. Melliar-Smith, ''Exploiting the Internet Inter-ORB protocol interface to provide CORBA with fault tolerance,'' *Proceedings of the USENIX 3rd Conference on Object-Oriented Technologies and Systems*, Portland, OR (June 1997), pp. 81-90.

[15] P. Narasimhan, L. E. Moser P. M. and Melliar-Smith, ''Replica consistency of CORBA objects in partitionable distributed systems,'' *Distributed Systems Engineering*, vol. 4, no. 3 (September 1997), pp. 139-150.

[16] E. Nett, M. Gergeleit and M. Mock, ''An adaptive approach to object-oriented real-time computing,'' *Proceedings of the IEEE 1st International Symposium on Object-Oriented Real-Time Distributed Computing*, Kyoto, Japan (April 1998), pp. 342-349.

[17] Object Management Group, ''Fault Tolerance for CORBA,'' OMG Technical Committee Document orbos/98-10-08 (October 1998).

[18] Object Management Group, ''Realtime CORBA,'' OMG Technical Committee Document orbos/99-02-12 (February 1999).

[19] Object Management Group, ''The Common Object Request Broker: Architecture and Specification,'' 2.3 edition, OMG Technical Committee Document formal/98-12-01 (June 1999).

[20] Object Management Group, ''Fault Tolerant CORBA,'' OMG Technical Committee Document orbos/99-10-05 (October 1999).

[21] Object Management Group, ''Dynamic Scheduling,'' OMG Technical Committee Document orbos/99-10-06 (October 1999).

[22] R. Rajkumar, C. Lee, J. Lehoczky and D. Siewiorek, ''A resource allocation model for QoS management,'' *Proceedings of the IEEE 18th Real-Time Systems Symposium*, San Francisco, CA (December 1997), pp. 298-307.

[23] D. I. Rosu, K. Schwan and R. Jha, ''On adaptive resource allocation for complex real-time applications,'' *Proceedings of the IEEE 18th Real-Time Systems Symposium*, San Francisco, CA (December 1997), pp. 320-329.

[24] D. C. Schmidt, D. L. Levine and S. Mungee, ''The design of the TAO real-time object request broker,'' *Computer Communications*, vol. 21, no. 4 (April 1998) pp. 294-324.

[25] J. Schonwalder, S. Garg, Y. Huang, A. P. A. van Moorsel and S. Yajnik, ''A management interface for distributed fault tolerance CORBA services,'' *Proceedings of the IEEE 3rd International Workshop on Systems Management*, Newport, RI (April 1998), pp. 98-107.

[26] E. Shokri, H. Hecht, P. Crane, J. Dussault and K. H. Kim, ''An approach for adaptive fault-tolerance in object-oriented open distributed systems,'' *Proceedings of the IEEE 3rd International Workshop on Object-Oriented Real-Time Dependable Systems*, Newport Beach, CA (February 1997), pp. 298-305.

[27] J. J. Sydir, S. Chatterjee and B. Sabata, ''Providing end-to-end QoS assurances in a CORBA-based system,'' *Proceedings of the IEEE 1st International Symposium on Object-Oriented Real-Time Distributed Computing*, Annapolis, MA (May 1998), pp. 10-17.

[28] A. Vaysburd and K. Birman, ''The Maestro approach to building reliable interoperable distributed applications with multiple execution styles,'' *Theory and Practice of Object Systems*, vol. 4, no. 2 (1998), pp. 73-80.

[29] V. F. Wolfe, L. C. DiPippo, R. Ginis, M. Squadrito, S. Wohlever, I. Zykh, and R. Johnston, ''Expressing and enforcing timing constraints in a dynamic real-time CORBA system,'' Technical Report, Department of Computer Science and Statistics, University of Rhode Island (1997).

[30] J. Zinky, D. Bakken and R. Schantz, ''Architectural support for quality of service for CORBA objects'' *Theory and Practice of Object Systems*, vol. 3, no. 1 (1997), pp. 55-73.

Survivability through Customization and Adaptability: The Cactus Approach

Matti A. Hiltunen, Richard D. Schlichting, Carlos A. Ugarte, and Gary T. Wong
Department of Computer Science
The University of Arizona
Tucson, AZ 85721, USA
E-mail: hiltunen/rick/cau/gary@cs.arizona.edu.

Abstract

Survivability, the ability of a system to tolerate intentional attacks or accidental failures or errors, is becoming increasingly important with the extended use of computer systems in society. While techniques such as cryptographic methods, intrusion detection, and traditional fault tolerance are currently being used to improve the survivability of such systems, new approaches are needed to help reach the levels that will be required in the near future. This paper proposes the use of fine-grain customization and dynamic adaptation as key enabling technologies in a new approach designed to achieve this goal. Customization not only supports software diversity, but also allows customized tradeoffs to be made between different QoS attributes including performance, security, reliability, and survivability. Dynamic adaptation allows survivable services to change their behavior at runtime as a reaction to anticipated or detected intrusions or failures. The Cactus system provides support for both fine-grain customization and dynamic adaptation, thereby offering a potential solution for building survivable software in networked systems.

1. Introduction

A survivable system is one that is able to complete its mission in a timely manner, even if significant portions are incapacitated by attacks or accidents [3]. Systems connected to large networks such as the Internet are faced with additional challenges—they must prevent unauthorized use, maintain confidentiality, and provide adequate service to proper users [19]. While survivability builds on research in security, reliability, fault-tolerance, safety, and availability, it is more than just a combination of these—it is also about the interaction of these different properties [61]. A survivable system must be able to continue providing service despite attacks or accidents. To do so, it must have facilities to protect against such threats, detect threats when they occur, and react to counteract threats and recover from damage. Thus, building a survivable system architecture requires not just exploiting techniques from security, fault-tolerance, safety, and real time, but it also requires developing new techniques.

We contend that *fine-grain customization* and *dynamic adaptation* are both key enabling technologies for survivability in networked systems. Fine-grain customization allows a service behavior to be tailored in fine detail to the requirements of the service user and the execution environment. Our approach to customization is based on *configurability*, which is the ability to easily construct customized service variants out of a set of software modules. Dynamic adaptation is the ability to change the behavior of a service by modifying the service configuration quickly at runtime. Such adaptation may involve, among other things, loading new code modules dynamically. A system based around these technologies can support customized survivability solutions, as well as evolve these solutions in both the short and long term to react to threats or to take advantage of new techniques.

Specific advantages of using fine-grain customization and dynamic adaptation in this context include the following:

- Services can be configured from existing collections of modules realizing different techniques for protection, detection, and reaction, rather than built from scratch.
- Configuration choices can be made depending on the perceived threats at a given time, with explicit control over tradeoffs such as the cost of an intrusion detection technique versus its coverage.
- Different survivability strategies for peer services can be adopted depending on the system scale, e.g., a closely-coupled strategy for collections of servers on

0-7695-0490-6/99 $10.00 © 1999 IEEE

a LAN versus a highly decentralized strategy for a corporate intranet.

- Multiple protection and detection techniques can be used in combination within a single service to further enhance survivability, e.g., successive authentication layers to protect sensitive data.
- Multiple customized services can be used in combination to protect sensitive aspects of system operations.
- Artificial diversity can be achieved easily by making different choices for a service configuration on different machines.
- Systems and individual services can adapt dynamically to respond to changing circumstances, such as a detected threat or a heightened state of alert (e.g., a wartime situation).
- New survivability techniques can be introduced dynamically into services as developed.

Done properly, a survivable system based on customization and dynamic adaptation could be a unifying framework that leverages both existing and future survivability techniques.

The Cactus Project at the University of Arizona is developing an infrastructure to support the fine-grain customization and dynamic adaptation capabilities needed to construct this type of survivable system. This paper introduces the Cactus model for configurable and dynamically adaptive distributed services, describes Cactus services relevant to survivability, and outlines potential future applications of Cactus in the survivability context.

2. The Cactus approach

The Cactus system is a framework for constructing highly customizable and dynamically adaptable middleware services for networked systems. Cactus provides an integrated approach that addresses a wide range of functional properties, as well as quality of service (QoS) attributes such as reliability, timeliness, performance, and security. The fine-grain customization provided by the systems allows application-specific control over these attributes and their potential tradeoffs.

2.1. The Cactus model

The Cactus model provides tools for building highly configurable distributed protocols and services. In Cactus, different properties and functions of a service are implemented as software modules that interact using an event-driven execution paradigm. A service in Cactus is implemented as a *composite protocol*, with each semantic variant of a QoS attribute or other functional component within the composite protocol implemented as a *micro-protocol*. A micro-protocol is, in turn, structured as a collection of *event handlers*, which are procedure-like segments of code that are executed when a specified *event* occurs. Events are used to signify state changes of interest, such as "message arrival from the network". When such an event occurs, all event handlers bound to that event are executed. Events can be raised explicitly by micro-protocols or implicitly by the Cactus runtime system. Event handling operations are implemented by the Cactus runtime system that is linked with the micro-protocols to form a composite protocol. The use of events provides indirection between micro-protocols that facilitates configurability and adaptability.

The model provides a number of operations for controlling events and handler execution. The two most important are:

- *bid* = **bind**(*event, handler, order, static_args*): Specifies that *handler* is to be executed when *event* occurs. *order* is a numeric value specifying the relative order in which *handler* should be executed relative to other handlers bound to the same event. When the handler is executed, the arguments *static_args* are passed as part of the handler arguments.
- **raise**(*event, dynamic_args, mode, delay*): Causes *event* to be raised after *delay* time units. If *delay* is 0, the event is raised immediately. The occurrence of an event causes handlers bound to the event to be executed with *dynamic_args* (and *static_args* passed in the **bind** operation) as arguments. Execution can either block the invoker until the handlers have completed execution (*mode* = SYNC) or allow the caller to continue (*mode* = ASYNC).

Other operations are available for unbinding handlers from events, creating and deleting events, halting event execution, and canceling a delayed event. Execution of handlers is atomic with respect to concurrency, i.e., each handler is executed to completion before the execution of the next handler is started. The binding of handlers to events can be changed at runtime.

In addition to a flexible event mechanism, Cactus supports two other features useful for configurable and adaptive service: service variables and the Cactus message abstraction. Service variables are variables and data structures shared by all micro-protocols within a given composite protocol. Such variables can be used to store those portions of the service state that need to be accessed by multiple micro-protocols. For example, a communication service would typically have a data structure storing messages that have been transmitted, but cannot be discarded yet. The atomic execution of event handlers makes it easy to use the service variables since it eliminates most concurrency problems.

The Cactus message abstraction generalizes the typical message format consisting of a message body and a header. In particular, a Cactus message consists of a body and a dynamic set of named attributes that have scopes corresponding to the composite protocol (*local*), the protocols on a single machine (*stack*), and the peer protocols at the sender and receiver (*peer*). The local attributes correspond to local variables associated with the message and the peer attributes correspond to normal header fields in a message. The stack attributes allow controlled information sharing across protocol layers when necessary.

Micro-protocols can add, read, and delete message attributes using operations `setAttr()`, `getAttr()`, and `deleteAttr()`. Two micro-protocols can share an attribute if they agree on the name of the attribute. A `pack()` operation adds any peer attributes to the message body, while an analogous `unpack()` operation strips those attributes from the body. The default pack operation includes in the message body all the necessary information for unpacking, including attribute names and lengths. However, the pack and unpack operations are completely customizable so if the message format is known *a priori* by the sender and receiver, this additional information can be omitted. The custom packing and unpacking also makes it possible to create standard message formats in case a composite protocol is interacting with a peer not implemented using Cactus. Finally, this facility makes it possible to use other header processing techniques such as header compression transparently to the micro-protocols that use the attributes.

The Cactus message abstraction makes it easy to manage message headers for highly configurable services. Typically, each combination of micro-protocols would require its own header format that includes those fields needed by the chosen micro-protocols. The message abstraction eliminates the need to explicitly construct the appropriate header format since micro-protocols can simply add the needed attributes to a message. This feature also makes it easy for the message format to change under different operating modes and to piggyback information on messages. In particular, a micro-protocol that wants to piggyback information can do so by adding a peer attribute with an appropriate attribute name at the sender. At the receiver, the corresponding micro-protocol can check if the information is piggybacked on a message simply by invoking `getAttr()` with the same attribute name as an argument. If the attribute is not piggybacked on this particular message, `getAttr()` returns an error indication. This flexibility also allows a message to carry only those attributes—i.e., header fields—that are required for that message.

The Cactus model promotes a *two-level composition* approach to constructing subsystems: a service is constructed out of micro-protocols and a subsystem is constructed out of services. This approach is illustrated in Figure 1. This figure depicts a communication subsystem consisting of composite and traditional protocols. In the middle is a composite protocol, which contains shared service variables, including a bag of messages, micro-protocols, and some events. The boxes in the middle represent micro-protocols, while to the right are events with binding between handlers and events represented by arrows.

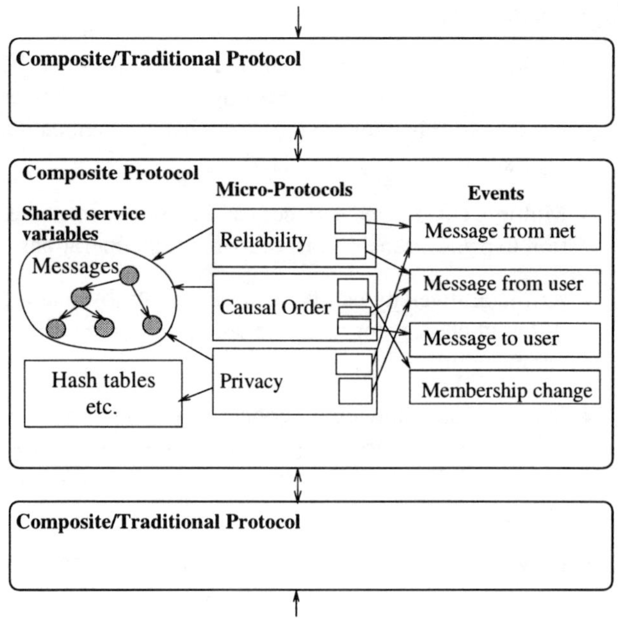

Figure 1. Composite protocol.

The Cactus model does not dictate the external interface provided by a service implemented as a composite protocol. For example, the service may export a CORBA object interface with methods. In this example, the service is a communication protocol that resides in the protocol stack on top of a lower level communication service such as UDP. In such a case, interaction between protocols is either via a standard Uniform Protocol Interface (UPI) such as that supported by the *x*-kernel [35] or via a separate systemic interface that supports vertical integration of support for Quality of Service (QoS) requests and modifications.

Four major prototype implementations of Cactus are currently available or under development. One implementation runs on OpenGroup/RI Mach MK 7.3 using C and CORDS, a variant of the *x*-kernel [59]. A separate version of this implementation called CactusRT supports construction of configurable service with real-time constraints [34]. Another version called Cactus++ is a C++ based implementation that runs on Linux and Solaris. Finally, a Java implementation of the model called Cactus/J runs on numerous operating systems, including Windows NT, Linux, and Solaris. All released Cactus prototypes and services are available through the Cactus home page http://www.cs.arizona.edu/cactus/.

2.2. Configurability in Cactus

All the mechanisms provided by the Cactus model are designed to support configurability by promoting independence between micro-protocols. For example, a micro-protocol may bind an event handler to an event without knowing which micro-protocol will raise the event, or raise an event without knowing which micro-protocol, if any, has handlers bound to this event. Similarly, the message abstraction allows micro-protocols to add new message attributes to a message without coordination with other micro-protocols. As a result, micro-protocols typically do not need to know about one another, which facilitates construction of customized variants of a service by configuring together the desired set of micro-protocols.

Our design approach to constructing highly-customizable services is illustrated in Figure 2. First, the useful abstract properties or execution guarantees of a service are identified by exploring existing implementations of the service. Second, the useful abstract properties are implemented as configurable modules. Note that there may be multiple alternative implementations of a property, where each implementation is optimized for certain execution environment. The properties of a service, as well as the modules that implement the properties, have fundamental relations that dictate which combinations are feasible. The goal of the implementation is not to introduce any additional restrictions on configurability beyond those found in these relationships [29].

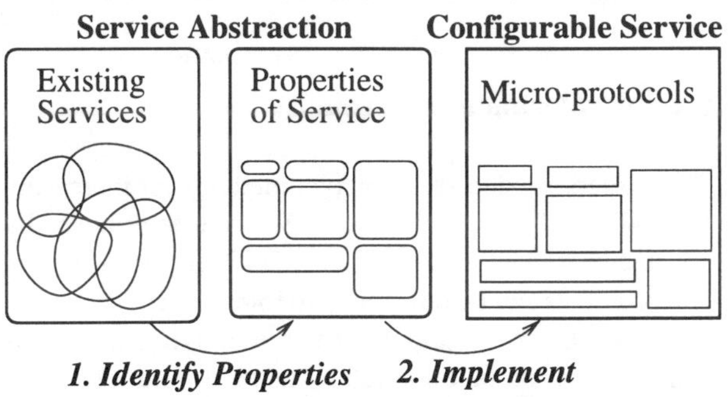

Figure 2. Design of a configurable service

Cactus services with survivability features are discussed below in section 4, but a number of other configurable services have also been implemented. A group membership service provides a range a customization options related to event ordering, synchrony, partition handling, and failure detection [33]. A group remote procedure call service allows customization of message ordering, call synchrony, atomicity, and reply collation [8]. Finally, a real-time communication service allows customization of message ordering, reliability, message deadlines, and the number and roles of communication participants [34]. Other services under development include a distributed shared memory service and a distributed file service.

2.3. Adaptability in Cactus

In addition to fine-grain configurability, the Cactus model offers features for supporting adaptive solutions. The indirection provided by the Cactus mechanisms also makes it easier to construct adaptive services that change their behavior as a result of changes in the execution environment or in the service requirements. In particular, the event mechanism allows new micro-protocols to be activated by simply binding their event handlers to the appropriate events or old micro-protocols to be deactivated by unbinding their handlers. Moreover, service variables make it easy to transfer state from old to new micro-protocols, while the message abstraction makes it trivial to modify the message header format due to an adaptive change from one algorithm to another.

The Cactus model supports different methods of adaptation, ranging from changing the execution parameters of a micro-protocol to actually switching the set of micro-protocols used to implement a service. Two approaches are available for changing the set of micro-protocols: event-handler rebinding and dynamic code loading. In the case of rebinding, the composite protocol is compiled with the alternate behaviors implemented as different micro-protocols, and a mode switch only involves unbinding the old event handlers of the deactivated micro-protocols and binding the handlers of the activated micro-protocols. In the case of dynamic code loading, new micro-protocols are loaded into a running composite protocol. After the new code is loaded, it can be activated in the same manner as above. The Cactus model makes such runtime code modification easier and allows more coordination than simply loading procedures into an ordinary running program because of the decoupling between micro-protocols provided by the event-driven execution model. We have experimented with dynamic code loading using dynamic libraries on Solaris, as well as using Java facilities for class loading in Cactus/J.

The ability to change the behavior of a running program is only the first step in adaptation. In particular, in the case of networked systems, the more difficult problem is often the coordination of the adaptation at the different sites involved in the computation. To coordinate this process, we have developed a three phase model for adaptations consisting of change detection, agreement, and action [32]. The agreement phase, where the different sites decide if the adaptation is necessary and what the adaptation should be, can be relatively complicated. To simplify the implementation of this phase, we have developed a library of consensus micro-protocols. The current set of consensus protocols are fault-tolerant, but do not currently address other aspects of survivability. However, new consensus micro-protocols could easily be added with other features such as authentication or tolerance to Byzantine failures.

We are in the process of defining a Cactus-based software architecture for constructing highly-adaptive configurable services [11]. In this architecture, a configurable service consists of adaptive and static (non-adaptive) components, where each component implements a QoS attribute, property, or function of the service. The service components interact using events and service variables. Each adaptive component is implemented as a collection of micro-protocols, including a number of adaptation-aware micro-protocols (AAMPs) and a component adaptor micro-protocol that coordinates adaptations between the AAMPs (Figure 3). Each AAMP provides a different implementation of the component functionality, with the component adaptor switching between the alternative AAMPs depending on the state of the execution environment and the service requirements. For example, a component that provides communication security may switch between different cryptographic micro-protocols depending on the level of security provided by the underlying network. Since the service components do not interact directly—e.g., by invoking methods on one another—the AAMPs can be switched without any effect on other system components.

The AAMPs are somewhat different than normal micro-protocols. In particular, they export an adaptation interface that consists of operations for evaluating the fitness of the micro-protocol for the current execution environment, and operations for activating and deactivating the micro-protocol. Adding this interface does not require any changes in the Cactus model, however, since an adaptation aware micro-protocol can simply be defined as a subclass of an ordinary micro-protocol with the new adaptation operations.

The software architecture for adaptive systems has two major design goals: to minimize the overhead of adaptability during normal operation and to minimize the impact of the adaptive process itself on the application using the adaptive service. The event mechanism provides a natural solution to the first goal. Since the new AAMP can simply bind its event handlers to the appropriate events, no additional redirection is required. Thus, the only overhead during normal operation is the system monitoring needed to detect changes in the execution environment. The second goal is achieved by allowing the adaptation to occur in phases, where both the old and new AAMP are partially active during certain phases. The Cactus model facilitates this approach, since any number of event handlers may be bound to an event at the same time. Furthermore, the consensus protocols execute in the background without halting normal processing. As a result, an application using an adaptive service should experience at most a slight temporary slowdown in service during adaptation.

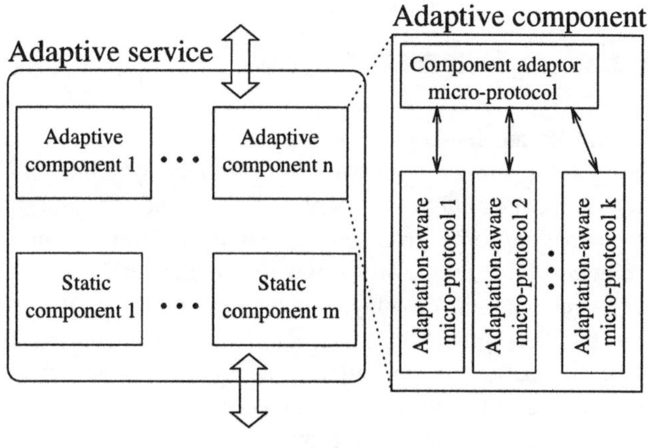

Figure 3. The structure of an adaptive service.

3. Survivability through customization and adaptability

The fine-grain QoS customization and dynamic adaptability supported by Cactus provide powerful mechanisms for survivability. This section outlines how mechanisms supported by Cactus can be utilized in the survivability context.

3.1. Fault tolerance and security

Fault tolerance and security, two methods for increasing survivability [39], are two of the QoS attributes provided by many customizable Cactus services. Different fault-tolerant micro-protocols can be developed to tolerate different types of failures. Omission failures of the underlying communication network can be tolerated by using different types of retransmission micro-protocols. Examples of such micro-protocols include traditional positive and negative acknowledgment based protocols, multiple transmission protocols, and atomicity protocols for group communication where the group members collectively ensure that if at least one member receives a message, all members will eventually receive it. Message corruption failures can be tolerated using either traditional message checksums or cryptographic checksums. The latter have the ability to detect intentional message modifications. Processor failures can be tolerated by using micro-protocols that implement object or process replication (active or passive) or message logging. Alternative micro-protocols can also be developed to tolerate different failure models ranging from crash to arbitrary failures [30].

Similar to fault tolerance, different types of security micro-protocols can be provided to deal with various situations. For communication security, security micro-protocols can provide privacy, authenticity, and message integrity using cryptographic techniques. Micro-protocols can also protect from replay attacks by adding sequence numbers, timestamps, or other unique nonces to messages. Even non-repudiation can be provided by having a micro-protocol log messages. Similar techniques can also be applied to realize other types of data security. For example, files in a file system can be encrypted for privacy or protected from tampering by using cryptographic checksums.

Customization allows each user or system administrator to determine how much security and fault tolerance is required at any given time. An increase in fault tolerance or security requires more resources, and increases the response time and reduces the throughput of a service. Thus, it is important that the levels of fault tolerance and security are set based on the perceived threats at any given time, as well as the requirements of the current applications and users. For example, the security requirements for an email service used by the military can differ based on whether the country is currently involved in a conflict, what information is being transmitted, and who is sending and receiving the information.

Two different approaches can be used for providing fault tolerance and security guarantees using Cactus: integrating the code implementing the enhanced QoS with an existing service or providing it as a separate fault tolerance or security service. In the integrated approach, an existing Cactus service such as RPC or group communication is enhanced to provide additional fault tolerance or security by developing new micro-protocols. For example, fault tolerance through replication can be added to the RPC service as a micro-protocol that sends each request to a collection of servers instead of one. Similarly, security can be added to any communication service by adding encryption, authentication, or message integrity micro-protocols.

The second alternative implements fault tolerance and security guarantees as a separate service. An example of such a fault-tolerance service is the group membership described in [33], while an example of a security service is the secure communication service described in [31]. Such specialized services provide customizable fault-tolerance or security abstractions for higher level services and applications.

The choice between these alternatives is case specific. The integrated approach allows fault tolerance and security micro-protocols to be totally customized for the particular service. For example, the security requirements may be different for different message types used by the service. However, since the integrated approach increases the number of micro-protocols, it makes the service more complicated. The flexibility of the two-level composition model in Cactus allows the use of the design alternative that best fits the requirements.

3.2. Diversity

Diversity methods attempt to make it difficult for an intruder to exploit known security vulnerabilities by making different instances of a program or an operating system different enough that one method of breaking the service might not work on others [13, 14]. Diversity methods include wrapper programs that filter inputs to a potentially buggy program [60], inserting additional checks into programs, and permutations of program code and data layout.

The configurability supported by Cactus provides a natural mechanism for constructing diverse programs and services. Different configurations of Cactus services have some degree of natural diversity given that users can customize the service to their exact requirements and execution environment. Additional diversity can be provided by implementing alternative micro-protocols for different service properties and QoS attributes. For example, communication privacy can be provided using numerous different encryption methods (e.g., DES, RSA, IDEA, Blowfish) or combinations of encryption methods. If the intruder does not know which method or combination of methods is used for the encryption, it becomes more difficult to break the encryption.

Similar diversity techniques can be applied to other security services. For example, a configurable intrusion detection service [16] could be constructed using Cactus, where new intrusion detection methods are added as new micro-protocols and various combinations could be used together. Similarly, the authentication service of an operating system could be customized for each OS installation.

3.3. Adaptability

Adaptability has numerous current and future applications in survivability, including automated reaction to intrusions. For example, methods for stopping an intrusion after it has been detected by terminating a suspected connection or by preventing the intrusion from spreading to other computers [9] can be viewed as adaptive behavior. Automatic actions to restore the system state after an intrusion and to prevent future exploitation of the same security holes can be considered another type of adaptation. Traditional recovery methods used for fault tolerance can often be utilized for recovery of this type as well, although the additional problem of identifying data and programs that may have been modified must also be solved. Moreover, most adaptations related to intrusion tolerance are currently not even automated. For example, responses to CERT advisories must be implemented manually, leading to delays and non-uniform deployment.

Adaptability can also be used as a preventive mechanism by allowing a service to change its behavior dynamically based on user requirements and perceived or detected threats. For example, if an intrusion detection service suspects an intrusion, other services can adaptively increase their security levels to prevent the intruder from gaining more information or doing further damage. Such preventive mechanisms must naturally be combined with corrective actions if damage was done before detection.

We are currently exploring adaptability related to a number of QoS attributes in the context of Cactus. Some of this work focuses on performance adaptations, that is, changing algorithms within a service to increase performance given the current execution environment [11]. We have also explored using adaptability to implement fault tolerance, ranging from considering failure repair as an adaptive action [32] to considering algorithms that adapt to a change in the failure models exhibited by the underlying system [10]. Future work will include adaptive security to accommodate changed user requirements and detected or perceived threats, and adaptive real time to accommodate changes in available resources and system workload.

Both adaptive security and adaptive real time are relevant for survivability. The use of adaptation to increase security when an intrusion is detected is obvious, but other types of security adaptations can also be used to enhance survivability. For example, if an intrusion reduces available resources, a service may adaptively weaken its security guarantees to reduce CPU resource requirements. The weakened security can then be compensated for by frequently switching different low cost security mechanisms. Note that this type of adaptation is an example of a trade off between two different QoS attributes, security and performance.

Adaptive security introduces a number of challenges that still need to be solved. For example, the consensus algorithms used to determine if an adaptation is required and which adaptation should be made must be made intrusion tolerant using message authentication or Byzantine methods. It may also be important to disguise the adaptations from the intruder to hide the original detection. Current research is addressing these issues.

Adaptive real time is also important for survivability, since a survivable system is often required to provide a timely response in spite of failures and intrusions [3]. If failures or intrusions reduce the available set of resources, adaptive services must either reallocate resources or reduce the resource usage of some of its constituent parts. The adaptive functionality provided by Cactus supports this process in several ways. One is by facilitating construction of services that can provide a reduced level of service when resources are reduced. Another is by making it easy to effect tradeoffs between different resources, e.g., by changing to a compression algorithm that requires more CPU time in order to reduce the network bandwidth required. Of course, a primary challenge in this area is to design and implement adaptive actions that can execute in bounded time in order to meet the application's real-time constraints.

3.4. Enhancing survivability transparently

It is often necessary to increase the survivability of existing legacy or off-the-shelf applications. This can be accomplished either by replacing underlying communication and operating system services with survivable versions, or by transparently inserting new middleware services between the application and the underlying services.

We have experimented with the base functionality need to realize the first approach for communication services on both MK 7.3 and Linux. On MK 7.3 with CORDS, all or part of a protocol stack to be inserted into the kernel, which can be used to replace existing communication services provided by the operating system. On Linux, kernel loadable modules can be used to achieve similar functionality. Currently we are experimenting with the use of loadable modules on Linux to simulate the characteristics of a wireless network on top of a wired Ethernet, but similar techniques could be used to insert survivable behavior into the kernel.

There are a number of options for replacing other operating system services with Cactus equivalents depending on the OS platform. One option is to modify the operating system kernel directly if the source code is available (e.g., Linux). Other options include instrumented connectors in NT [2] and, again, loadable kernel modules. While kernel modifications are difficult for the intruder to circumvent, the other options may not provide 100 % protection from intruder circumvention.

We are also exploring the transparent insertion of new middleware services. In the context of CORBA, our goal is to enhance the QoS for existing CORBA clients and servers without modifying either their code or the underlying ORB. We are working on accomplishing this goal on the Orbix ORB by inserting composite protocols into smart proxies at clients and filters at servers, and on the Visibroker ORB using smart stubs at clients and wrappers at servers. We are also working on inserting new middleware services by intercepting signals on Linux and Solaris. This approach is being used in a distributed shared memory service in which the shared memory abstraction is implemented by intercepting segmentation fault signals related to access attempts.

4. Survivable Cactus Services

A number of configurable services built using Cactus have important survivability features. This section discusses some existing services, and their current and planned survivability features.

4.1. Secure communication service

SecComm, a secure communication service implemented using Cactus, allows fine-grain customization of a range of security attributes including privacy, authenticity, message integrity, replay prevention, and non-repudiation [31]. A secure connection created through SecComm is customizable in the sense that only the required security attributes are guaranteed. Furthermore, the strength of guarantee associated with each attribute can be customized at a fine grain level. For example, the service offers the choice of the cryptographic techniques and key lengths used to implement each attribute, and each attribute can be guaranteed using arbitrary combinations of security algorithms. Using combinations of algorithms can make it harder for intruders to break system security because they do not know the method that needs to be broken, although methods based on diversity and secret methods such as this are not uniformly accepted by the security community. Nonetheless, SecComm allows each user to choose between using proven methods such as DES or RSA, using secret combinations of secret methods, or using combinations of proven and secret methods to further increase security.

SecComm exemplifies and extends a current trend in network security design that allows customization of security attributes for explicitly managing the cost/benefit tradeoff. For example, IPSec [37], a collection of protocols being developed by the IETF to support secure packet exchange at the IP layer, provides two security options. The *authentication header* (AH) option does not encrypt the data contents of the packet, but provides optional authenticity, integrity, and replay prevention by adding an AH that contains a cryptographic message digest. The *encapsulating security payload* (ESP) option provides privacy by encrypting the data contents of the packet and optional authenticity, integrity, and replay prevention using a message digest. Naturally, due to the enhanced support for configurability provided by Cactus, SecComm provides more flexibility and extensibility than existing secure communication services.

Figure 4 illustrates the main micro-protocol classes and their interactions through events in SecComm. Note that each component in this figure represents a class of micro-protocols, e.g., the privacy class includes DESPrivacy, RSAPrivacy, and OneTimePadPrivacy micro-protocols.

In the current version, a custom secure communication session is created by selecting the desired set of micro-protocols. Future versions could include higher-level configuration interfaces in which the user specifies only the required level of assurance for each property. The micro-protocol combination that satisfies these requirements would then be created automatically using information on, for instance, relative strengths of different encryption methods.

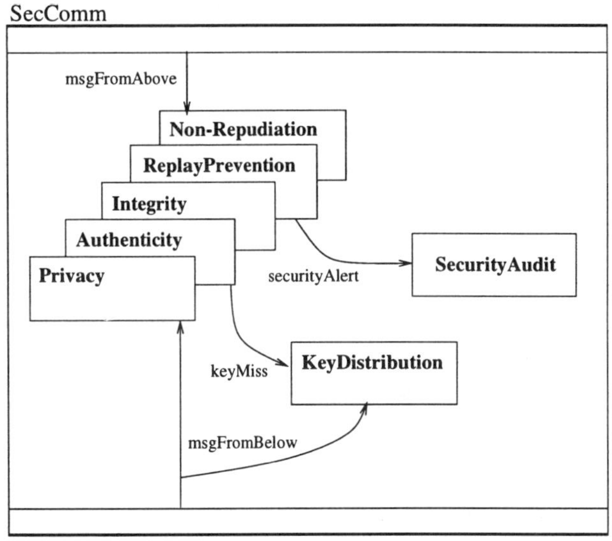

Figure 4. Micro-protocol classes and their interactions.

Future work related to SecComm will focus on adding adaptability features to the protocol. This type of adaptive security service must be able to increase the level of security when requested by the user or when an intrusion attempt is detected, and it must be able to reduce the resource usage of the service when performance is not satisfactory. An interesting challenge will be determining if adaptation can be inexpensive enough that reasonably high level of security can be provided by switching frequently between different inexpensive cryptographic methods. A special instance of this type of protocol switch would be altering the cryptographic key at runtime.

4.2. GroupRPC service

The GroupRPC service allows a client to execute a procedure call on one or more servers. Such facility can be useful for a number of reasons, but the main survivability aspect of such facility is that it can provide fault tolerance through replication. Unfortunately, providing fault tolerance of this type can be expensive. One important factor determining this cost is the type and number of server failures to be tolerated. Typically, the type of failures to be tolerated is expressed in terms of *failure models* that range from relatively benign crash or omission failures to arbitrary (or Byzantine [41]) failures. In terms of the amount of redundancy and execution time required, tolerating a processor crash is much cheaper than tolerating Byzantine failures. Furthermore, the cost is determined not only by the failure model, but also by the number or frequency of the failures expected to occur. Any realistic system can in principle exhibit failures in any of the failure models, but typically, failures in the more benign classes are more frequent than severe failures. Thus, the choice of failure model for a particular task should be based on the frequency of different types of failures in the given execution environment, as well as the criticality of the task. The tradeoff, of course, is that making stronger assumptions about failures improves the performance of the system, but lessens the degree of fault coverage provided by the system and thus the reliability of the task [48].

We have implemented a configurable GroupRPC service that allows the failure model and number of failures to be tolerated to be customized depending on the needs of the application [30]. The service includes micro-protocols for each failure model that implement the appropriate detection and membership algorithms. For example, if the service is configured to tolerate Byzantine failures, the servers run Byzantine agreement protocols to agree on the set of requests to be processed, as well to agree as any membership changes of the server group. Thus, any faulty or compromised server or client cannot cause the server group to become inconsistent. In addition to the algorithms used, the choice of failure model dictates how many failures can be tolerated, that is, how many server replicas are required. Thus, in order to tolerate M server failures, we need M+1 replicas for crash, send omission, and late timing failures, 2M+1 replicas for receive omission and early timing failures, and 2M+1 replicas for Byzantine failures assuming message authentication [30].

Failure Model	Clients	Servers	Time
None	1	1	3.6
	2	1	5.9
Crash	1	2	6.2
	2	2	13.4
Send omission and late timing	1	2	6.6
	2	2	13.8
Receive omission and early timing	1	3	10.5
Byzantine	1	3	18924

Table 1. Average response time (in ms)

Table 1 illustrates how the change in the failure model affects the response time [30]. Each of the configurations listed tolerates one failure of the given type for configurations that guarantee atomicity and total ordering for the calls. The large increase of response time in the Byzantine case is due to slow RSA-based message authentication and an agreement algorithm that operates in synchronous rounds. This concrete example illustrates the need for customization between reliability and performance, and thus, survivability and performance.

4.3. System monitoring services

Two distributed system monitoring services have been constructed using Cactus. One, implemented using Cactus++, runs on Solaris, while the other, implemented using Cactus/J, runs on numerous operating systems including Linux and Windows NT. Both services are structured as one centralized global monitor that provides the user a GUI for monitoring and controlling the system and a number of local monitors executing on each of the machines being monitored. In the current versions, the local monitors are user processes that use standard operating system interfaces to collect information about the system load, CPU and memory usage of processes, current users, etc. Each aspect of the system behavior is monitored and reported by a separate micro-protocol, which makes it easy to configure a service that monitors only the necessary information. This structure also makes the service easily extensible since new micro-protocols can be added to monitor other aspects of the system behavior. The Cactus++ version is extensible at configuration time, i.e., the monitoring service must be recompiled and relinked to include any new micro-protocols. The Cactus/J version, on the other hand, is adaptive, i.e., new micro-protocols can be added dynamically at runtime. This functionality is implemented using a specialized micro-protocol class loader.

These monitoring services can easily be extended to provide monitoring oriented towards security and survivability. For example, the tools can be used to detect denial of service attacks by constructing micro-protocols that collect resource utilization information and produce warnings when thresholds are exceeded. A resource being fully utilized—such as CPU, memory or network buffers—might indicate that a denial of service attack is underway.

Monitoring micro-protocols can also be used to oversee other security-critical aspects of system operation. For example, one can monitor the password file and report any changes. Such a distributed monitoring architecture can also make it more difficult for intruders to cover their tracks by modifying log files, since the monitor can record any changes in the log files and store them on the global monitor. Although a considerable amount of useful information can be collected by just monitoring the file system and using operating system calls, more powerful survivability monitoring would require instrumenting existing code in the operating system as well as services such as ftp, telnet, and http.

Naturally, changes are also needed in the monitoring service itself to make it survivable. For example, it must be impossible for an intruder to modify or replay messages sent by a local monitor or to replace a genuine local monitor by one that does not report any suspicious activity. Message modification and replay can easily be prevented using authentication and message integrity methods similar to those used in SecComm, or the monitoring service could just be built on top of SecComm. Local monitor replacement can be prevented, or at least made very difficult, by having the global monitor give the local monitor a private authentication key at startup. Similarly, the intruder must not be able to impersonate the global monitor to the local monitors. This would be particularly dangerous in the Cactus/J version since it allows new micro-protocols to be loaded at runtime. The natural solution is again authentication of all messages sent by the global monitor. Note that the configurability provided by Cactus makes it easy for each installation to customize the type and strength of authentication used.

5. Future Survivability Services

Cactus could be used as a framework to implement a number of other survivability services with customizability and adaptability features. In the following, we discuss some potential survivability services and their implementation using Cactus.

5.1. Intrusion detection services

Research on intrusion detection has traditionally focused on auditing system logs and more recently, on automated intrusion detection [16]. Intrusion detection systems analyze audits provided by the computing environment to detect attacks or misbehavior occurring in the environment. Two major approaches can be identified. The *misuse-detection approach* detects known signatures or symptoms of attacks, while the *anomaly-detection approach* detects deviations from an assumed normal behavior of the system. Misuse-detection is generally more accurate, but can only account for known attacks. Tools such as statistics [43], expert systems [21], signature analysis [40], neural networks [15], genetic algorithms, and model based reasoning [26] have been used to implement detection of either approach. Detection may focus on analyzing information from one computer (host-based), e.g., [28, 62], inspecting packets passed through a network (network-based), e.g., [49, 53], or some combination, e.g., [46]. Recently, research on intrusion detection has focused on integrating existing intrusion detection systems by providing a Common Intrusion Detection Framework (CIDF) that provides protocols for interaction and collaboration [57].

The customization abilities provided by the Cactus framework could be used as the basis for implementing a highly-configurable intrusion detection service (IDS). The micro-protocol approach would make the service easily extensible in the sense that new micro-protocols could be added to collect information from new sources (e.g., log files, operating system, other system components), to analyze information, and to react to any suspected intrusions. The new analysis micro-protocols could include ones developed for newly recognized misuse patterns. The analysis could also use a combination of different methods (e.g., signature analysis, neural networks, model based reasoning) and use voting or agreement to determine if an intrusion should be suspected. The reaction micro-protocols could range from ones that notify an operator to others that automatically attempt to stop the intrusion.

A configurable IDS designed in this way has a number of other advantages, such as allowing tradeoffs to be customized. For example, it becomes possible to control the tradeoff between the coverage of the detection and its performance and resource utilization. It also allows customization of detection coverage versus the number of false detections, and thus, the potential inconvenience caused to legitimate users. A configurable IDS could also be customized depending on the perceived threat and the current mode of operation. For example, the system could be designed to react automatically to suspected intrusions that occur during nighttime hours when system operators are not at work and few real users are using the system. During daytime operation, the service could be more conservative in its actions and pass responsibility for most reactions to the system operators.

The detected or suspected intrusions detected by an IDS are fundamental for initiating adaptation in other survivability services, but an IDS could also adapt itself. Quick adaptations could occur at the time a threat is suspected, for example, to add additional detection mechanisms focused on the suspected activities. Furthermore, an IDS could adapt in the long term in the sense of evolving to incorporate countermeasures to newly detected attack methods. In particular, newly discovered intrusion signatures might be added to detection micro-protocols, and new detection micro-protocols could be introduced into the IDS at runtime. The changes between different modes of operation could also occur adaptively, rather than through reconfiguration. For example, an IDS could adapt by increasing its sensitivity when system administrators suspect that an intrusion might occur.

The primary goal of developing a Cactus-based IDS would not be to develop new intrusion detection methods, but rather to leverage existing and future methods into a configurable adaptive framework. That is, individual intrusion detection methods could be implemented as micro-protocols, which would allow different variants of the service to be created by configuring chosen subsets of the methods. Any coordination required between the different detection methods, such as weighted voting, could easily be implemented as a separate coordinator micro-protocol that is designed to handle any detection configuration.

5.2. Survivable data storage services

Survivability can be very important for data storage services such as file systems and database systems. A survivable data storage service (SDSS) must not only ensure confidentiality and integrity of data, but also provide data availability in spite of, and after, failures and intrusions. Techniques such as data fragmentation and replication can be used to achieve both confidentiality and availability even if some of the computers are successfully compromised [17, 22, 38]. If the data cannot be physically protected—that is, only authorized access is allowed in all cases—confidentiality and integrity can be implemented using cryptographic techniques similar to those used for secure communication. Furthermore, other techniques such as checkpointing and logging of changes can be applied to allow restoration of data after an intrusion. Finally, a part of the information may be stored outside of the SDSS to enhance survivability. For example, decryption keys or file checksums may be stored on a smartcard that the user provides to the system on demand. Note that a SDSS can use an existing underlying data storage service such as a file system for actually storing the data. Thus, in the case of a distributed file system such as NFS, the actual file server does not need to be modified.

Configurability can be applied to an SDSS as easily as it can be applied to the secure communication service—the different options for service guarantees can be implemented as separate micro-protocols. Note that one SDSS may store data with different survivability requirements. This means that the service must be configured with a set of micro-protocols that is sufficient to satisfy all the guarantees for all the data currently in the system. For each data item accessed, only the required micro-protocols will be activated. The information of which micro-protocols are used for each data item could be stored transparently in the SDSS itself, or this information can be provided by the user when they try to access a data item.

Adaptability can be used for SDSSs in the form of corrective actions if an intrusion and data destruction or modification is detected. Changing the survivability methods on the fly is more difficult for storage services since it could be expensive, for example, to decrypt all files and then encrypt them again using a new method.

5.3. Access control and authentication services

An authentication service (AS) is used to ensure the identity of a principal (i.e., a user or a program operating on a user's behalf), and an access control service (ACS) is used to determine for each file and other system resource if a principal is allowed to use it in some specified manner. Each operating system implements its own authentication (e.g., password, one time password, smartcard, and physical characteristics) and access control methods (e.g., access control lists, capabilities, and Unix-style permission bits).

The authentication and access control methods provided by operating systems are relatively rigid and do not provide much support for survivability. In particular, intruders that manages to pass the authentication test (e.g., guess root password) are free to do any amount of damage. Thus, the higher levels of customization provided by Cactus would be a useful feature for survivable systems.

Allowing such services to be customizable would open a range of possible solutions with better survivability characteristics than traditional systems. For example, authentication could be customized to be one time only at login time, periodic at fixed time intervals, or usage based depending on what files and other resources the user attempts to access and use. Such combinations of access control and authentication could have many interesting applications. For example, accessing some important system files might require that the user is executing as root and that they know an additional password required to modify the file. Such additional security boundaries could also be specified for directories, resulting in arbitrary sequences of successive security boundaries for accessing important system resources. Another approach would be to specify a level of authenticity required for accessing each system resource and executing additional authentication steps when a user attempt to access something higher than the current level. The authenticity level of a user could also reduce over time to at least partially handle the problem of someone else using a session left open by a legitimate user. Since authentication and access control are typically services provided by the operating system, it is not easy to replace them using Cactus equivalents. Some options described in section 3.4 could be applied here depending on the platform.

Survivability adaptations useful for an access control service include revoking all access rights from a compromised user identity and increasing authentication requirements on system resources if the system is attacked. Adaptations in the case of an authentication service include reauthenticating users that have used authentication challenges that have been compromised, as well as replacing authentication methods that have been compromised. Furthermore, the overall required authentication level of the whole system may be increased, in which case all current users are reauthenticated. The signal to perform for such adaptations would typically come from other system components such as the intrusion detection service.

6. Related work

6.1. Configurability

Configurability or extensibility has gained increased attention during the last decade. In the area of communication protocols, notable examples are *x*-kernel [35], Adaptive [56], Horus [51], and Ensemble [27]. The *x*-kernel presents a model for hierarchical composition of modules with identical uniform protocol interfaces (UPIs). The Horus and Ensemble projects adopt the same hierarchical model, but extend the UPI to include a larger set of operations. Adaptive uses a service specific back-end with slots for different functions of the service, where each function can be implemented using a range of alternative modules. Extensibility has also been used in operating systems, e.g. SPIN [6], Scout [45], and Exokernel [20], and database systems, e.g. RAID [7] and Genesis [4].

Customizability is beginning to gain greater attention in the area of communication security. A number of Internet standards under development include some customization features, including IPSec [37], Secure Socket Layer (SSL) [25], Transport Level Security (TLS) [18], Secure HyperText Transfer Protocol (S-HTTP) [52], and Privacy-Enhanced Mail (PEM) [42]. These protocols typically allow a choice of privacy and message integrity methods, as well as optional authentication methods. A number of research projects have also addressed customizable security. For example, the *x*-kernel model has been used to construct configurable protocol stacks for secure communication [47]. Recently, TIS Labs has worked on composable, replaceable security services [58], and the Ensemble project has added a choice of customizable cryptographic techniques for data communication [54].

Customizability is naturally very useful also in system monitoring services, and most services offer some way of modifying the service behaviors. For example, Pulsar [23] supports extensibility by allowing new Tcl scripts to be written for collecting additional information, satool [44] uses Unix scripts, and CARD [1] uses Perl scripts. Some monitoring tools, such as Rscan [55], have been designed to detect security loopholes. Many intrusion detection tools have similar type of extensibility features, including Emerald [46]. In general, however, the use of configurability in the survivability context has been limited and most examples have focused on secure communication and system monitoring, including intrusion detection.

In contrast with service-specific frameworks that allow some extensibility, Cactus is a general framework specifically designed for fine-grain configurability and extensibility. Thus, the possibilities for extensibility are much greater. For example, it would be easy to add security and reliability features into our system monitoring tool, whereas it would probably be difficult to add such features to many other monitoring tools where extensibility is limited.

6.2. Adaptability

Adaptability has been proven to be a useful feature in particular for networking protocols. For example, the Transmission Control Protocol (TCP) of the Internet protocol suite uses adaptive mechanisms for flow control, retransmission, and congestion control [36]. Other examples include adaptive media access control, e.g. [12], adaptive encoding and compression, e.g. [24], and adaptive routing algorithms, e.g. [5]. However, adaptability in such protocols is typically limited to changing some parameters that control the execution the protocol, e.g., timeout values or transmission windows, rather than switching the protocol. To our knowledge, the only other work addressing runtime algorithm switching is work done in Ensemble [27, 50]. In contrast with our work that allows each protocol or service component to adapt with minimal impact on other parts of the system, Ensemble requires that a new protocol stack be created with the new protocol. Message transmission must then be interrupted while the adaptation is in progress.

7. Conclusions

This paper has outlined potential uses of the Cactus approach for implementing various aspects of survivability. In particular, we claim that the fine-grain customizability and dynamic adaptability supported by Cactus can have a significant impact on the survivability of networked systems. Fine-grain customization allows each user or system administrator to make individual survivability versus performance tradeoffs, as well as provides a foundation for using artificial diversity for survivability. Adaptability allows services to react to attacks and failures when they do occur, and facilitates their evolution when new attacks or countermeasures become available.

References

[1] E. Anderson and D. Patterson. Extensible, scalable monitoring for clusters of computers. In *Proceedings of Eleventh Systems Administration Conference (LISA '97)*, pages 9–16, San Diego, CA, Oct 1997.

[2] R. Balzer. Instrumented connectors for mediating architecture interactions. DARPA Windows NT Workshop, Seattle, WA. http://www.dyncorp-is.com/darpa/meetings/-win98aug/Balzerinstruments.html, Aug 1998.

[3] M. Barbacci. Survivability in the age of vulnerable systems. *IEEE Computer*, 29(11):8, Nov 1996.

[4] D. Batory, J. Barnett, J. Garza, K. Smith, K. Tsukuda, B. Twichell, and T. Wise. GENESIS: An extensible database management system. *IEEE Transactions on Software Engineering*, SE-14(11):1711–1729, Nov 1988.

[5] P. Bell and K. Jabbour. Review of point-to-point network routing algorithms. *IEEE Communications Magazine*, 24(1):34–38, 1986.

[6] B. Bershad, S. Savage, P. Pardyak, E. Sirer, M. Fiuczynski, D. Becker, C. Chambers, and S. Eggers. Extensibility, safety, and performance in the SPIN operating system. In *Proceedings of the 15th ACM Symposium on Operating Systems Principles*, pages 267–284, Copper Mountain Resort, CO, Dec 1995.

[7] B. Bhargava, K. Friesen, A. Helal, and J. Riedl. Adaptability experiments in the RAID distributed database system. In *Proceedings of the 9th Symposium on Reliable Distributed Systems*, pages 76–85, 1990.

[8] N. Bhatti, M. Hiltunen, R. Schlichting, and W. Chiu. Coyote: A system for constructing fine-grain configurable communication services. *ACM Transactions on Computer Systems*, 16(4):321–366, Nov 1998.

[9] P. Brutch, T. Brutch, and U. Pooch. Electronic quarantine: An automated intruder response tool. In *Proceedings of the Information Survivability Workshop 1998*, pages 23–27, Orlando, FL, Oct 1998.

[10] I. Chang, M. Hiltunen, and R. Schlichting. Affordable fault tolerance through adaptation. In J. Rolin, editor, *Parallel and Distributed Processing, Lecture Notes in Computer Science 1388*, pages 585–603. Springer, Apr 1998.

[11] W.-K. Chen, M. Hiltunen, and R. Schlichting. Software architecture for building adaptive software. Technical report, Department of Computer Science, University of Arizona, Tucson, AZ, 1999. In preparation.

[12] M. Choi and C. Krishna. An adaptive algorithm to ensure differential service in a token-ring network. *IEEE Transactions on Computers*, C-39(1):19–33, 1990.

[13] C. Cowan and C. Pu. Immunix: Survivability through specialization. In *Proceedings of the 1997 Information Survivability Workshop*, Feb 1997.

[14] C. Cowan and C. Pu. Survivability from a Sow's ear: The retrofit security requirement. In *Proceedings of the Information Survivability Workshop 1998*, pages 43–47, Orlando, FL, Oct 1998.

[15] H. Debar, M. Becker, and D. Siboni. A neural network component for an intrusion detection system. In *Proceedings of the IEEE Symposium on Research in Security and Privacy*, pages 1–11, Oakland, CA, 1992.

[16] D. Denning. An intrusion-detection model. *IEEE Transactions on Software Engineering*, SE-13(2):222–232, Feb 1987.

[17] Y. Deswarte, J.-C. Fabre, J.-M. Fray, D. Powell, and P.-G. Ranea. Saturne: A distributed computing system which tolerates faults and intrusions. In *Proceedings of the Workshop on Future Trends of Distributed Computing Systems*, pages 329–338, Hong Kong, Sep 1990.

[18] T. Dierks and C. Allen. The TLS protocol, version 1.0. Request for Comments (Standards Track) RFC 2246, Certicom, Jan 1999.

[19] R. Ellison, D. Fisher, R. Linger, H. Lipson, T. Longstaff, and N. Mead. Survivable network systems: An emerging discipline. Technical Report CMU/SEI-97-TR-013, Nov 1997.

[20] D. Engler, M. Kaashoek, and J. O'Toole. Exokernel: An operating system architecture for application-level resource management. In *Proceedings of the 15th ACM Symposium on Operating Systems Principles*, pages 251–266, Copper Mountain Resort, CO, Dec 1995.

[21] M. Esmaili, R. Safavi-Naini, and J. Pieprzyk. Computer intrusion detection: A comparative survey. Technical Report Ref. 95-07, Center for Computer Security Research, University of Wollongong, Australia, May 1995.

[22] J.-C. Fabre, Y. Deswarte, and B. Randell. Designing secure and reliable applications using fragmentation-redundancy-scattering: an object-oriented approach. In *Proceedings of the 1st European Dependable Computing Conference*, pages 21–38, Berlin, Germany, Oct 1994.

[23] R. A. Finkel. Pulsar: An extensible tool for monitoring large Unix sites. *Software Practice and Experience*, 27(10):1163–1176, 1997.

[24] A. Fox, S. Gribble, E. Brewer, and E. Amir. Adapting to network and client variation via on-demand, dynamic distillation. In *Proceedings of the 7th ASPLOS Conference*, Oct 1996.

[25] A. Freier, P. Karlton, and P. Kocher. The SSL protocol, version 3.0. Internet-draft, Netscape Communications, Nov 1996.

[26] T. Garvey and T. Lunt. Model-based intrusion detection. In *Proceedings of the 14th National Computer Security Conference*, Washington, DC, Oct 1991.

[27] M. Hayden. The Ensemble system. Technical Report TR98-1662, Department of Computer Science, Cornell University, Jan 1998.

[28] L. Heberlein, K. Levitt, and B. Mukherjee. A method to detect intrusion activity in a networked environment. In *Proceedings of the 14th National Computer Security Conference*, pages 362–371, 1991.

[29] M. Hiltunen. Configuration management for highly-customizable software. *IEE Proceedings: Software*, 145(5):180–188, Oct 1998.

[30] M. Hiltunen, V. Immanuel, and R. Schlichting. Supporting customized failure models for distributed software. *Distributed Systems Engineering*, 1999. To appear.

[31] M. Hiltunen, S. Jaiprakash, and R. Schlichting. Exploiting fine-grain configurability for secure communication. Technical Report 99-08, Department of Computer Science, University of Arizona, Tucson, AZ, Apr 1999.

[32] M. Hiltunen and R. Schlichting. Adaptive distributed and fault-tolerant systems. *Computer Systems Science and Engineering*, 11(5):125–133, Sep 1996.

[33] M. Hiltunen and R. Schlichting. A configurable membership service. *IEEE Transactions on Computers*, 47(5):573–586, May 1998.

[34] M. Hiltunen, R. Schlichting, X. Han, M. Cardozo, and R. Das. Real-time dependable channels: Customizing QoS attributes for distributed systems. *IEEE Transactions on Parallel and Distributed Systems*, 10(6):600–612, Jun 1999.

[35] N. Hutchinson and L. Peterson. The *x*-kernel: An architecture for implementing network protocols. *IEEE Transactions on Software Engineering*, 17(1):64–76, Jan 1991.

[36] V. Jacobson. Congestion avoidance and control. In *Proceedings of the SIGCOMM '88 Symposium*, pages 314–332, Aug 1988.

[37] S. Kent and R. Atkinson. Security architecture for the internet protocol. Request for Comments (Standards Track) RFC 2401, BBN Corp, Home Network, Nov 1998.

[38] H. Kiliccote and P. Khosla. Borg: A scalable and secure distributed information system. In *Proceedings of the Information Survivability Workshop 1998*, pages 101–105, Orlando, FL, Oct 1998.

[39] P. Krupp, J. Maurer, and B. Thuraisingham. Survivability issues for evolvable real-time command and control systems. In *Proceedings of the 1997 Information Survivability Workshop*, Feb 1997.

[40] S. Kumar and E. Spafford. A pattern matching model for misuse intrusion detection. In *Proceedings of the 17th National Computer Security Conference*, pages 11–21, Oct 1994.

[41] L. Lamport, R. Shostak, and P. M. The Byzantine generals problem. *ACM Transactions on Programming Languages and Systems*, 4(3):382–401, Jul 1982.

[42] J. Linn. Privacy enhancement for internet electronic mail: Part I: Message encryption and authentication procedures. Request for Comments RFC 1421, Feb 1993.

[43] T. Lunt, R. Jagannathan, D. Anderson, C. Dodd, F. Gilham, C. Jalali, H. Javitz, P. Neuman, A. Tamaru, and A. Valdez. System design document: Next-generation intrusion detection expert system (NIDES). Technical report, SRI International, Mar 1993.

[44] T. Miller, C. Stirlen, and E. Nemeth. satool — A system administrator's cockpit, an implementation. In *Proceedings of Seventh Systems Administration Conference (LISA '93)*, pages 119–129, Monterey, CA, Nov 1993.

[45] A. Montz, D. Mosberger, S. O'Malley, L. Peterson, T. Proebsting, and J. Hartman. Scout: a communications-oriented operating system. In *Proceedings of the 1st Symposium on Operating Design and Implementation (OSDI)*, page 200, Nov 1994.

[46] P. Neumann and P. Porras. Experience with EMERALD to date. In *Proceedings of the 1st USENIX Workshop on Intrusion Detection and Network Monitoring*, Santa Clara, CA, Apr 1999.

[47] H. Orman, S. O'Malley, R. Schroeppel, and D. Schwartz. Paving the road to network security or the value of small cobblestones. Technical Report 94-16, Department of Computer Science, University of Arizona, Tucson, AZ, May 1994.

[48] D. Powell. Failure mode assumptions and assumption coverage. In *Proceedings of the 22nd IEEE Symposium on Fault-Tolerant Computing*, pages 386–395, 1992.

[49] M. Puldy and M. Christensen. Lessons learned in the implementation of a multi-location network based real-time intrusion detection system. In *Proceedings of the 1st International Workshop on the Recent Advances on Intrusion Detection*, Louvain-la-Neuve, Belgium, Sep 1998.

[50] R. v. Renesse, K. Birman, M. Hayden, A. Vaysburd, and D. Karr. Building adaptive systems using Ensemble. *Software Practice and Experience*, 28(9):963–979, Jul 1998.

[51] R. v. Renesse, K. Birman, and S. Maffeis. Horus, a flexible group communication system. *Communications of the ACM*, 39(4):76–83, Apr 1996.

[52] E. Rescorla and A. Schiffman. The secure hypertext transfer protocol. Internet-draft, Terisa Systems, Inc., Jun 1998.

[53] L. Ricciulli, S. De Capatini di Vimercati, P. Lincoln, and P. Samarati. PNNI global routing infrastructure protection. In *Proceedings of the Information Survivability Workshop 1998*, pages 123–126, Orlando, FL, Oct 1998.

[54] O. Rodeh, K. Birman, M. Hayden, Z. Xiao, and D. Dolev. The architecture and performance of security protocols in the Ensemble group communication system. Technical Report TR98-1703, Department of Computer Science, Cornell University, Dec 1998.

[55] N. Sammons. Multi-platform interrogation and reporting with rscan. In *Proceedings of Ninth Systems Administration Conference (LISA '95)*, Monterey, CA, Sep 1995.

[56] D. Schmidt, D. Box, and T. Suda. ADAPTIVE: A dynamically assembled protocol transformation, integration, and evaluation environment. *Concurrency: Practice and Experience*, 5(4):269–286, Jun 1993.

[57] S. Staniford-Chen, B. Tung, and D. Schnackenberg. The common intrusion detection framework (CIDF). In *Proceedings of the Information Survivability Workshop 1998*, pages 139–143, Orlando, FL, Oct 1998.

[58] R. Thomas and R. Feiertag. Addressing survivability in the composable replaceable security services infrastructure. In *Proceedings of the Information Survivability Workshop 1998*, pages 159–162, Orlando, FL, Oct 1998.

[59] F. Travostino, E. Menze, and F. Reynolds. Paths: Programming with system resources in support of real-time distributed applications. In *Proceedings of the IEEE Workshop on Object-Oriented Real-Time Dependable Systems*, Feb 1996.

[60] W. Venema. TCP WRAPPER: Network monitoring, access control, and booby traps. In *Proceedings of the 3rd Usenix UNIX Security Symposium*, pages 85–92, Baltimore,MD, Sep 1992.

[61] J. Voas, G. McGraw, and A. Ghosh. Reducing uncertainty about survivability. In *Proceedings of the 1997 Information Survivability Workshop*, Feb 1997.

[62] A. Wespi, M. Dacier, H. Debar, and M. Nassehi. Audit trail pattern analysis for detecting suspicious process behavior. In *Proceedings of the 1st International Workshop on the Recent Advances on Intrusion Detection*, Louvain-la-Neuve, Belgium, Sep 1998.

Sigma: Security and Interoperability for Heterogeneous Distributed Systems

Terry Vickers Benzel
NAI Labs, Network Associates, Inc.
(tbenzel@nai.com)
Pierre X. Pasturel
NAI Labs, Network Associates, Inc.
(pierre_pasturel@nai.com)
Deborah Shands
NAI Labs, Network Associates, Inc.
(dshands@nai.com)

Daniel F. Sterne
NAI Labs, Network Associates, Inc.
(dan_sterne@nai.com)
Gregg W. Tally
NAI Labs, Network Associates, Inc.
(gregg_tally@nai.com))
E. John Sebes
Kroll-O'Gara Information Security Group
(ejs@securify.com)

This research was funded by DARPA under contract F30602-97-C-0268.

Abstract

The DARPA ITO sponsored Sigma project was a three-year research effort (September 1995 - September 1998) which investigated the integration and interoperation of security technologies into CORBA-based distributed computing environments. The focus of the effort was on architectures and technologies that allowed controlled, selective exchange of object-oriented services among enclaves, which differ in security policy, mechanism, and assurance. The project produced several results that through technology transfer efforts are being used in multiple DARPA research efforts and a commercial product.

1. Project Overview

1.1. Background

The Common Object Request Broker Architecture (CORBA) is an emerging international standard for interoperable object-oriented distributed computing. CORBA provides an architecture for Object Request Broker (ORBs) which enable diverse applications to interoperate in a uniform, platform independent manner. ORBs are primarily designed to provide object management infrastructure for one enclave, one independent distributed system operating under one administrative domain. Security services within an ORB were only beginning to be minimally addressed at the time of the project initiation and security services for interoperability between enclaves was an open research issue. The *Sigma* project produced results which significantly advanced the state of the art for both within enclave security and between enclave security. The project encompassed architecture studies and prototyping efforts including the following: 1) provision of CORBA Services in a high-assurance trusted multi-level enclave, 2) development of object-oriented access control mechanisms for CORBA based on our Domain and Type Enforcement (DTE) research that are suitable for incorporation into operating systems, Object Request Broker (ORB), and object servers; and 3) development of ORB gateways that selectively filter object requests entering and exiting an enclave and interpret security attributes from other enclaves.

1.2. Objectives

The *SIGMA* project had three principal objectives:

Develop security mechanisms for protecting an enclave by applying access control to interoperation with other enclaves. Every enclave needs security mechanisms which restrict access to enclave-internal CORBA resource by parties outside the enclave. Firewall-like gateway security mechanisms are essential to turning away unwanted CORBA traffic before it reaches internal hosts. Basic elements of firewall functionality for CORBA must be developed, but these are not enough. Support for outsiders' authentication technology, and for object-oriented access control, is needed as well.

Improve the state of the art of security mechanisms for object-oriented distributed systems. Security for CORBA systems even within one enclave is an area of security in a nascent state. Development of end-system (e.g., server-

0-7695-0490-6/99 $10.00 © 1999 IEEE

side) CORBA security mechanisms must proceed along with inter-enclave CORBA security mechanisms, so that the same security model, access control techniques, and policy management approach apply to both intra-enclave and inter-enclave CORBA security.

Extend interoperability access controls to apply to heterogeneous security mechanisms and disparate policies of different enclaves. When enclaves interoperate to share resources via CORBA, different enclaves may be using different security mechanisms (particularly for distributed authentication) and may be enforcing different security policies. Therefore, inter-enclave security must allow for heterogeneity of mechanism, while avoiding confusion from disparate policies. That is, each enclave must be able to enforce its own inter-enclave security policy given whatever security technology is used by remote clients.

1.3. Results Overview

The *SIGMA* project produced research results and proof of concept prototypes corresponding to each of the above goals.

Development of ORB Gateways that selectively filter object requests entering an enclave. The *Sigma ORB* Gateway is deployed on or behind firewalls where it intercepts CORBA's Internet Inter-ORB Protocol (IIOP) traffic addressed to servers in the enclave. The Gateway performs access control in accordance with a DTEL++ policy before forwarding requests to the servers. If a Gateway denies a request, the request is not forwarded. The Gateway does not require modifications to CORBA object references or IIOP messages.

Object-oriented access control mechanisms for CORBA, based on our Domain and Type Enforcement (DTE) research. OO-DTE extends DTE operating systems concepts to distributed object systems, providing both scalability and fine-grained access control for CORBA applications. Two prototype implementations were developed: a DTE kernel-based implementation using the ILU ORB (Xerox PARC), and an above-kernel implementation using Orbix (Iona) and Solaris (Sun). Both implementations support role-based policies, security unaware applications, and optional per-object access control in security aware applications. The above-kernel implementation was transitioned to two other DARPA programs.

Use of distributed authentication technologies to support object-oriented access controls. The ORB Gateway supports multiple different mechanisms for distributed authentication, through the use of multiple security protocols for SSL, DCE, and IPsec. Because of the multiplicity of distributed security technologies, we separated message handling, authentication, and domain derivation into separate components. The authentication component included a framework for insertion of software to support various kinds of credentials, to validate them, extract information from them, and pass the information to the domain deriver. We were able to successfully demonstrate a variety of authentication mechanisms, and trust and domain policies that represent the use of each individually and in combination.

Our results in each of these areas were based on a common approach to specifying security policies for distributed object applications. We designed an OO policy model (OO DTE) and specified a policy language (DTEL++, based on earlier DTE research) for expressing security policy for CORBA objects. The language has several benefits in terms of scalability, flexibility, and capability for fine-grained specification exactly when needed. We developed software tools for DTEL++ (a compiler and a runtime system) and used them for prototypes in the above areas of work. In particular, OO DTE enforcement mechanisms were embedded in demonstration prototypes of an OS, an ORB plug-in, and the ORB Gateway.

2. Project Results

2.1. Architecture

The initial phase of our research focused on a series of analyses of architectural elements. The major areas of architectural study were: Access Controls for Interoperable Object Management, Domain and Type Enforcement for Distributed Object Control, and Security of Client/Server Communication. Our analysis and study led to a number of proposed architectures for providing security for CORBA-based interoperability. The report documenting the architecture analysis recommended phase 2 research and prototyping activities for the effort in three parallel and complimentary areas: Inter-Enclave Interoperability via ORB Gateways, DTE in a Secure Object System, and Secure Distributed Communication and Authentication. Details of the preliminary architecture study can be found in [ARCH96].

2.2. ORB Gateway

The *Sigma* project developed the architectural concept of an ORB Gateway, a new security component that provides security for CORBA-based computing between enclaves communicating over open networks such as the Internet. In developing this concept we studied the functional requirements, identified several design issues for providing inter-enclave security and investigated alternative approaches to resolving these issues. These preliminary results and the rationale for selecting alternatives to prototype are described in [INTER]

There are a number of issues in developing Inter-Enclave Network Boundary mechanisms. In general, distinct enclaves can interoperate via CORBA applications communicating over open networks such as the Internet. In such cases, each enclave has a network boundary, and some security mechanisms deployed at the network boundary, e.g., a firewall. Boundary security is needed to avoid the requirement that every internal host be carefully configured to be resistant to attack from outside networks. Particularly important when the outside network is the Internet, firewalls enforce a policy that describes what kinds of network communication are allowed.

When enclaves need to interoperate via CORBA, each enclave's firewall certainly has a role in regulating CORBA network communication, i.e. network traffic in the IIOP protocol. For example, a firewall might regulate which internal server hosts can be accessed by external clients. In fact, the *Sigma* project developed an early prototype of an IIOP proxy for a traditional firewall, and the IIOP proxy did exactly this.[1]

However, the capabilities of a conventional firewall proxy proved to be too limited to handle the challenges of fine grained access control for inter-enclave CORBA. The all-or-nothing per-host traffic regulation is not flexible enough to enforce security on object-oriented applications except at the coarsest level of granularity. Also, inter-enclave CORBA traffic is very frequently transported by some security protocol (such as SSL) which proxies typically do not handle. In fact, CORBA is specified to allow secure transport and authentication via multiple possible security technologies. If single-security-protocol capability is not part of conventional firewall technology, support for multiple security protocols for one application protocol (IIOP) is even further out of reach for firewalls. Furthermore, inter-enclave secure transport and authentication brings up issues of inter-organizational trust that are rarely handled even by systems that do use security technology such as SSL.

[1] After the *Sigma* project completed, project staff developed an IIOP proxy for Network Associates Gauntlet product. This is discussed further in our description of technology transfer.

Prototyping and investigation into CORBA firewall issues was valuable, however, because of exposure of fundamental issues of CORBA and firewalls. At the outset of the *Sigma* project, there was no standard interoperable method of handling IIOP by firewalls, even in the simplest conventional manner. By the midpoint of the project, *Sigma* project outreach into the OMG resulted in the development of a CORBA/Firewall specification process which addressed a critical issue: how clients use object references to communicate with objects on the other side of a firewall. The overall approach was to extend the interoperable object reference (IOR) format to include information for clients on how to communicate with the referenced object via one or more proxies.

This general, multi-proxy approach was a key enabler for two important approaches for inter-enclave CORBA. The first approach is a near-term use of conventional firewall techniques to perform host-level restrictions and IIOP protocol verification on CORBA communication between the protected network and the Internet. IOR data directs clients' ORBs to the firewall, so that the client connections, rather than being blocked by the firewall as attempts to bypass the firewall IIOP proxy, are instead processed by the IIOP proxy and permitted if in accordance with the firewall's security rules.

The second approach enables more complex boundary security mechanisms to be deployed on a separate security gateway, behind the firewall. This separate gateway can be the network component for the complex CORBA security functionality that is beyond the scope of conventional firewall proxies. The *Sigma* ORB Gateway is a prototype of exactly such a system. In fact, early prototypes of the ORB Gateway were the model for the more advanced functionality described in the CORBA Firewall specification. The general approach is for the firewall to direct all outside IIOP traffic to the ORB Gateway, which provides fine-grained OO security and support for multiple authentication protocols and for inter-enclave trust management. The firewall is the boundary target for outside clients (and may also provide a protocol verification function to protect the gateway for protocol-level attacks), but otherwise delegates control of inter-enclave CORBA traffic by requiring all such traffic to pass through the gateway behind the firewall.

In summary, the network boundary issues were how is inter-enclave CORBA traffic to be handled securely,

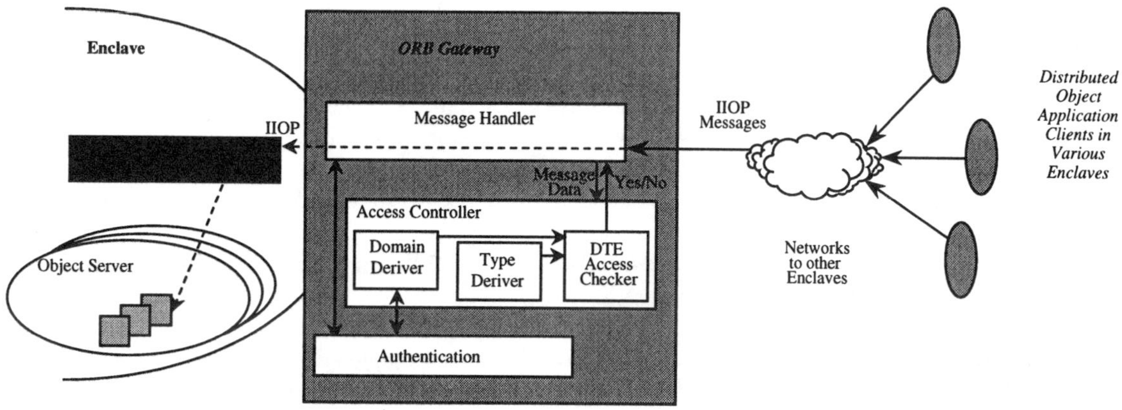

Figure 1: ORB Gateway Architecture

especially since much of the required functionality is beyond the scope of conventional firewall technology?

We rejected the alternative of pushing the envelope of conventional firewall technology, because that would constrain the technology transfer path to the cases where a firewall vendor chose to incorporate into a firewall product a great deal of new technology unfamiliar to all except a handful of CORBA security experts. We also rejected the alternative of downscaling inter-enclave security requirements to compatibility with current firewall products, because that would not meet the objectives of the projects. However, we did demonstrate the kind of IIOP proxy functionality that is consistent with current firewalls, and how that functionality is complementary with the broader set of functions of a behind-firewall CORBA security gateway. We chose the alternative of focusing efforts on such an ORB Gateway, first pioneering the concept and then using it as a model for CORBA/Firewall work in the OMG. We used that OMG experience to validate our initial assumption that inter-enclave access control is the difficult problem area to focus on. The ORB Gateway architecture and concept of operations is shown in Figure 1.

2.3. Distributed Authentication

In order for any useful access control to be possible, the ORB Gateway must be able to determine, for each message, who is making the request. In other words, the ORB Gateway must support distributed authentication mechanisms, and use them to provide data to the OO DTE access control machinery embedded in the ORB Gateway. This requirement generated the three sets of issues described below. These issues and the design and proof of concept prototypes to address these are described in more detail in [INTER].

Authentication Protocols

Because of the need to support multiple different mechanisms for distributed authentication, the ORB Gateway's protocol handling components support multiple security protocols. Each protocol support module implements a security protocol, collects the credential data transported by the protocol, and passes the credential data to another component of the ORB Gateway, the authentication framework. The authentication framework handles various kinds of credentials and performs computation on them in support of access control decisions.

Authentication Support for Access Control

The ultimate purpose of all authentication-related functionality in the ORB Gateway is to support access control decisions on requests for inside objects from outside clients. The relationship between authentication and access control depends significantly on the model for specifying access controls. There are several common models (e.g., access control lists, capabilities, labels, etc.) and we chose a policy model with proven properties of scalability and flexibility— Domain and Type Enforcement (DTE). Sigma's object-oriented extension— OO DTE— is summarized below, but there is one critical difference for OO DTE deployment in the ORB Gateway as compared with other kinds of DTE. When DTE or similar kinds of access control schemes are used in settings within a single security domain (e.g., a network under a single security administration), it can be assumed that the domain of each user is available to access control mechanisms. A domain is a set of users (or other principals that can access resources) that all have the same rights to access various kinds of resources. When an access control mechanism decides whether a particular request should be granted, the decision is based in part on what domain the requester is in.

The ORB Gateway operates in networked environment with requesters that operate in other enclaves. A key part of an enclave's inter-enclave security policy is the definition of which outside users are in which domains. These rules, together with other rules that specify what resources are accessible to members of various domains, comprise the enclave's policy on access to internal CORBA-based resources by outside parties. None of those outside parties directly authenticates as the member of any domain. Authentication credentials (e.g., X.509 certificates, DCE tickets, IPsec-authenticated IP source addresses) are from other enclaves that are unaware of the access control policy enforced by the ORB Gateway. Therefore, a critical part of the ORB Gateway's security functionality is *domain derivation*, that is, computing an OO DTE domain from one or more authentication credentials.

<u>Domain Derivation</u>

Domain derivation is performed as a computation from various attribute/value pairs to a domain. The attributes are extracted from a credential by software specific to a kind of credential. The domain deriver is an ORB Gateway component that compares the attribute values from a particular credential with a set of rules about membership in domains.

2.4. Multi-Level Security

One of the early goals of the *Sigma* project was the investigation of security for distributed object interoperability between trusted and untrusted systems. In this area we studied cases that included applications executing on systems that must provide data separation based on sensitivity or other content attributes. This is an area which is not addressed by current ORB technology. In this case enclave data must be protected by strong label-based separation mechanisms, and application service must be provided for several operational environments, one for each label. Requirements of this sort are exhibited in both military and commercial computing systems, as distributed computing interconnects previously isolated systems. However, previous technology to meet these requirements – multi-level secure operating systems – have not been widely accepted and are not currently part of the COTS technology base of CORBA systems. Under the *Sigma* project we developed an approach to supporting enclaves with multi-level security, but using only standard OS and object service interfaces. In cases where multi-level requirements can be met by polyinstantiated applications, these applications run on COTS ORBs on polyinstantiated Unix execution environments. In cases where MLS applications requires an MLS system base, the object services and object brokerage are provided by an MLS TCB. In both cases, clients access objects in the same way, without regard for whether objects are local or remote, or whether services are polyinstantiated or not. Details of the architectures and prototypes for multi-level security CORBA environments can be found in [ACSAC96] and [MILCOM97].

2.5. DTE In a Secure Object System

A common approach to security mechanisms for object-oriented distributed systems was needed to meet all three goals of the *Sigma* project. We chose to develop object-oriented access control mechanisms for CORBA based on our previous research in Domain and Type Enforcement research [BADG]. DTE is an authorization scheme that is based on the scalable definition of equivalence classes of resources to be accessed and entities that access them. A *domain* is an attribute of a process. Processes running in the same domain share a common set of access rights or privileges. A *type* is an attribute of a system resource. Resources with the same type share a common set of rules about who is authorized for access. Domains and types are defined scalably and flexibly by making use of the natural groupings of accessors and resources, and implicit inclusion in domains or type by virtue of such grouping.

A modest number of role, domain, and type assignments can define access controls for large numbers of resources and principals. The levels of indirection in the relationships create equivalence classes of resources and principals. The equivalence classes reduce the complexity and size of the DTE policy without prohibiting fine-grained control when needed.

OO DTE extends the DTE concepts to distributed object systems by expanding the definition of "resources" to include operations on remote objects. *Sigma* research activities have focused on CORBA, which is the Object Management Group's architecture for interoperable access to distributed objects. The DTE policy language, DTEL, provides the ability to bind types to resources and to define domains in terms of their permitted access to types. DTEL++ is the policy language for OO DTE. It is a superset of DTEL that provides additional constructs to specify policy for distributed object systems.

The extensions to DTEL apply to two main areas: domain definitions and type assignments. Domain definitions have been extended to allow *invoke* and *implement* privileges on types. Invoke and implement privileges are only effective in the context of types assigned to CORBA operations. Type assignments for CORBA operations are specified in several new constructs that have a lexical structure similar to CORBA IDL. Both DTEL++ and IDL have nested modules which provide name scoping for interfaces and other modules. IDL interfaces provide the bottom of the lexical scope within which are defined operations, types, exceptions, and the like.

In DTEL++, however, the content of interfaces is not a set of operation definitions, but rather a set of DTE statements that assign DTE types to the operations of the interface. Specific operations may be assigned, but default assignments are used to implicitly assign a type to entire interfaces and modules. DTEL++ provides an additional powerful means for implicit typing by using inheritance relationships. When an assignment is made to an interface (or individual operations in an interface), that assignment may apply recursively for all the interfaces lower down in the inheritance hierarchy defined in IDL. More specific assignments lower down in the hierarchy may be used to override higher assignments. Various flags are defined to control overridability and limit or expand the scope of assignment statement with respect to inheritance.

Finally, DTEL++ has facilities for customizing operation/type assignments on a per-object basis. DTEL++ assignments can be grouped together into "templates" to form a set of operation/type assignments that represent an authorization policy. These templates can then be applied to specific named objects so that different objects can be subject to different policies. As with original DTEL, a hierarchical name space of CORBA objects is used so that a modest number of assignment statements can be used to govern a large number of objects.

The access control decision in OO DTE is based on a process's domain and the operation's type. A client is permitted to call an operation if the client's domain has the *invoke* privilege for the type assigned to the operation. Similarly, a server is permitted to perform an operation if the server's domain has the *implement* privilege for the operation's type.

A full description of the DTEL++ policy language can be found in [TALL].

2.6. OO DTE Intra-Enclave

Kernel-based OO DTE

Two prototype implementations of OO DTE were developed during the project. The first implementation was based on a DTE OS kernel using the ILU ORB from Xerox PARC. The initial prototype was developed earlier in the project when the then-current state of CORBA security was unsatisfactory in a number of ways. The OMG specification on CORBA security was not final, there were no CORBA security implementations, and there was little understanding of how to use CORBA "interceptor" functionality for plug-in authorization. This prototype served three very useful purposes. First, it allowed us to experiment with the integration of OO DTE into an ORB, at a time when COTS ORBs were not capable of such integration. (ILU was capable of such integration, by means of modification of its freely available source code.) Second, it allowed us to validate our first attempts at OO DTE capability, which was a valuable antecedent to the complete definition of DTEL++, development of language tools, and development of full-scale OO DTE authorization capability. Third, it allowed us to experiment with the use of OS kernel level DTE to support and strengthen the OO DTE security mechanisms of the modified ORB. This initial basis of OO DTE on the previously developed DTE OS was a valuable first step for OO extension of previously developed DTE technology.

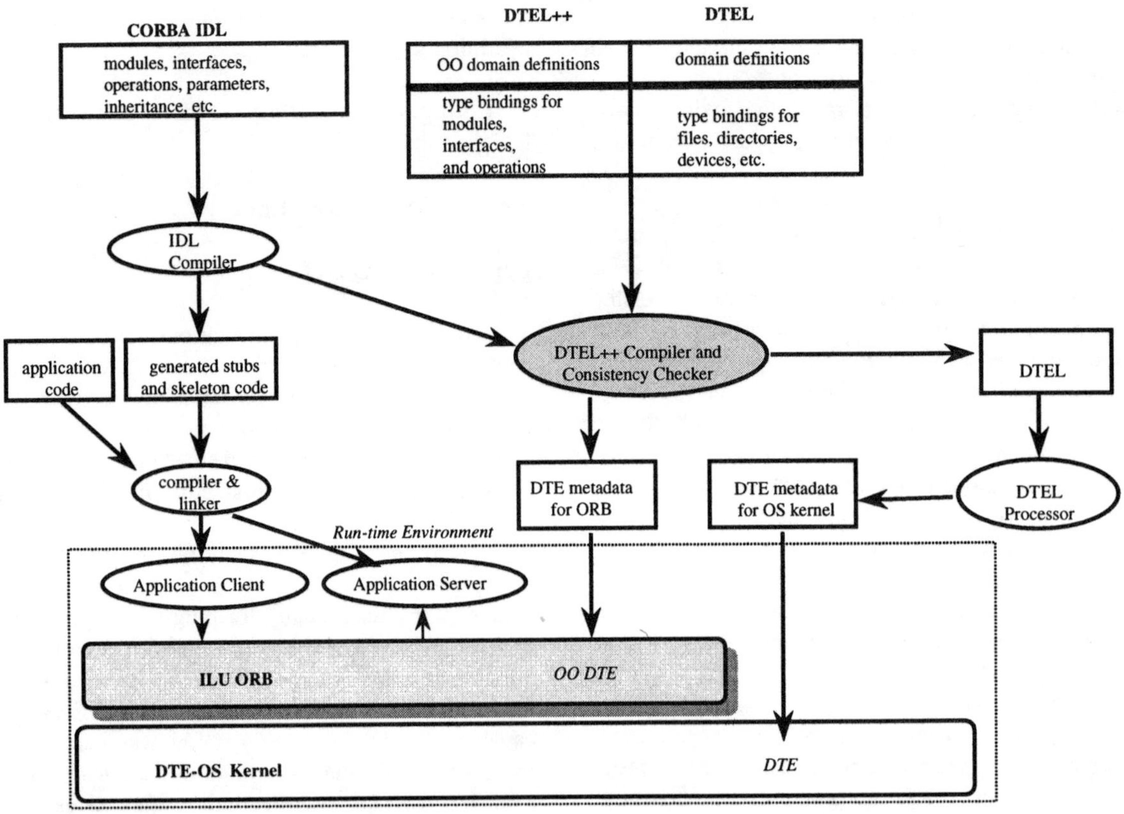

Figure 2: ILU/DTE-OS Prototype

Figure 2 shows how IDL and DTEL++ processing fit with the OO DTE access checker that is part of the ILU/DTE-OS prototype. In this prototype, both the ORB and the OS kernel have a role in access checks. OO DTE metadata is produced by DTEL++ compilation as in Figure 2, but in addition there is also kernel-level DTE metadata. The kernel-level data is produced by compilation of DTEL, the DTE OS policy specification language. Both DTEL++ and DTEL are combined in a single policy specification. The DTEL++ compiler processes the DTEL++ statements, and leaves the DTEL statements for later processing by the DTEL compiler. Both kinds of metadata are used at runtime to perform OO DTE access checks on CORBA object requests.

These access checks are also shown in Figure 2, in the client/server application software composed of IDL-generated code and application-specific logic. Clients invoke object requests that flow through the ORB and the DTE OS kernel. The ORB uses the Type Deriver to derive the DTE type of the operation from ORB-level data and from the OO DTE metadata. The ORB passes the type to the kernel. The kernel extracts the client's domain from OS-level data, and performs the comparison of the domain and type that determines whether the client/server communication is permitted. If so, the OS proceeds with the system-level services that the ORB uses to transport the object request from client to server. If not, the OS does not perform these services; client/server communication is blocked, and the object request is denied.

Above-Kernel OO DTE

The second prototype attempted to provide a more viable platform for using OO DTE. It was implemented in a COTS ORB (Orbix) on a standard operating system (Solaris) and is referred to as the 'above-kernel OO DTE' implementation. COTS ORBs provide a "plug in" interface for interception and examination of each object request. The second prototype focused on using ORB plug-in interfaces to insert an OO DTE authorization module into a COTS ORB. Existing OO DTE authorization software is encapsulated in the manner required for Orbix, with the hope of porting to other ORBs in the future. The resulting module is integrated with the ORB to support experimentation with ORB plug-in capabilities and interaction with CORBA security capabilities.

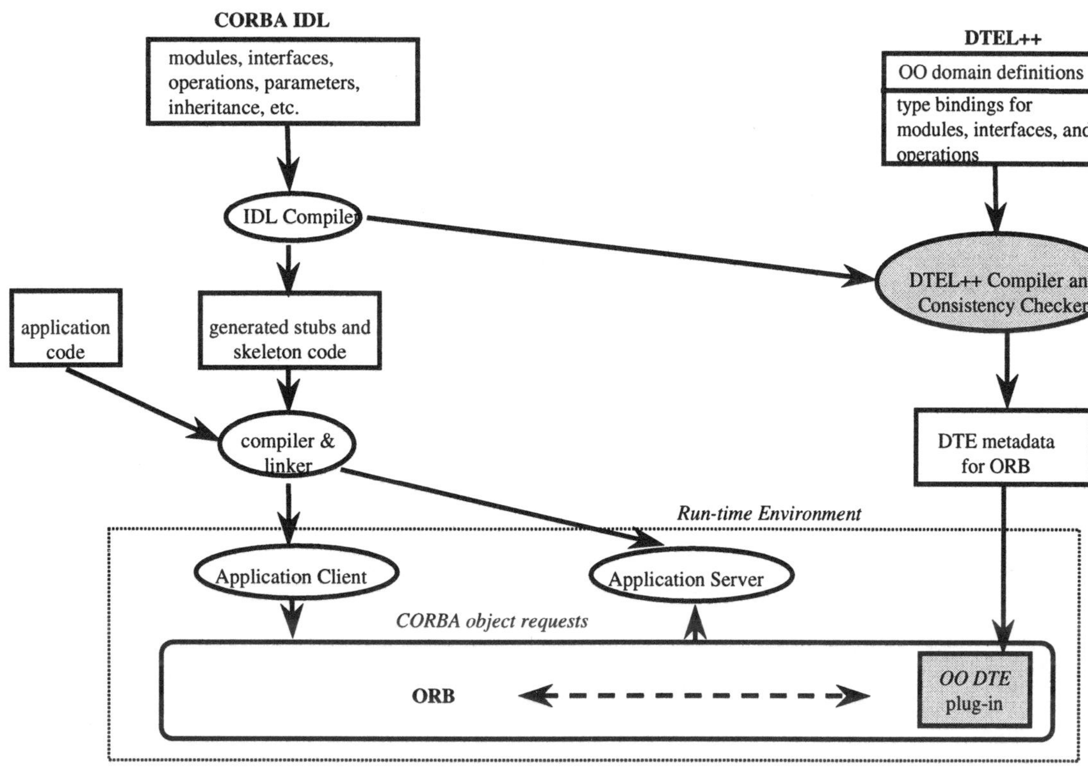

Figure 3: ORB Plug-in Prototype

Figure 3 provides a different, detailed view of Figure 2, in order to show how IDL and DTEL++ processing fit with an OO DTE access checker that is an ORB plug-in. Figure 3 shows a COTS ORB in which there is a plug-in implementing OO DTE. The main difference from Figure 2 is that there is no additional DTEL processing for the DTE-enhanced kernel. The policy specification is purely OO and the enforcement is entirely in the ORB. As with the ILU/DTE OS prototype, a significant component of the plug-in is the Type Deriver, which consumes the output of DTEL++ compilation described above. Information about each object invocation is passed to the Type Deriver, which returns a type. The domain is extracted from CORBAsec authentication data that the ORB provides along with the object invocation data. The domain and type are compared to produce the boolean result of the authorization check. As in the previous figures, Figure 3 shows client-side and server-side application software that is the result of IDL compilation. As part of the ORB's transport of object requests between client and server, the ORB invokes CORBA security mechanisms on each object request-- including the invocation of the OO DTE authorization checker.

Policy Distribution and Synchronization

Policy management and administration is a key element of the above-kernel OO DTE implementation. OO DTE provides the ability to centrally administer and distribute the DTEL++ policy to all hosts within an enclave. The policy distribution and synchronization mechanism supports not only OO DTE applications, but also the ORB Gateway.

There are four major components for policy distribution within the *Sigma* project: the policy directories, the policy server processes, the policy distribution library routines that are linked into each CORBA application, and the administration program, these are shown in figure 4.

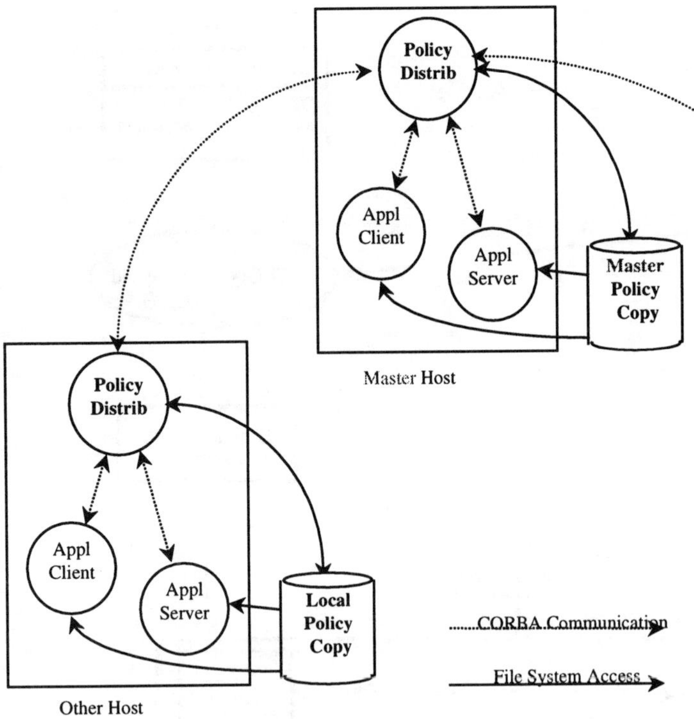

Figure 4: Policy Distribution Overview

The first major component is the policy directory. A policy directory resides on each host in the enclave. One host within the enclave is the master host, as specified in a configuration file, and contains the repository for the policy that governs the enclave. Each of the other hosts within the enclave contains a local copy of the policy retrieved from the master.

The second component is the policy server processes. A policy server runs on each host in the enclave. The policy server running on the master host is the master policy server, and is responsible for notifying the other policy servers of any changes to the security policy. Each policy server is responsible for keeping its policy directory up to date, which means copying the latest policy files and configuration files from the master policy server. Each policy server is also responsible for informing the CORBA applications on its host when the policy changes.

The third major component is the policy distribution library routines that are linked into each CORBA application. These routines read the policy and configuration files at application startup, register with the policy server running on their host, and accept notifications from the policy server when the policy has changed, so that they can re-read the policy and configuration files.

The final major component is the administration program, which runs as a CORBA application on the master host. The administration program is used to communicate policy changes to the master policy server, and to shutdown policy servers and applications.

The policy distribution mechanism is separate from the policy enforcement mechanism. The distribution mechanism knows only what files and directories comprise the policy being distributed. Those names are specified to the policy server through entries in a configuration file. The policy files could be OO-DTE files, metadata, both, or neither. The policy distribution mechanism only makes certain that those files and directories, what ever their contents may be, reside in the policy directory on each host, and that applications are notified when their contents are modified.

OO-DTE technology and the various prototypes along with design issues addressed during development are described in more detail in [USENIX].

The policy distribution mechanism uses CORBA capabilities to notify applications when a policy change occurs and to distribute the policy when it changes. This allows the full use of the OO-DTE mediation and SSL authentication to control policy updates and policy distribution.

2.7. OMG Activities

An important ongoing technology transfer aspect of this project involved working closely with the Object Management Group, the organization responsible for defining the CORBA standards. Project personnel regularly attend OMG Technical Committee meetings, reviewed RFPs and proposals, and provided feedback on security issues. Project members helped develop the OMG specification for CORBA Firewall services and were consulting members on all of the major vendor submission responses to this specification. In addition project members were active in the OMG ORBOS (the primary technical task force), the Security SIG, and CORBAmed (the medical community task force).

During the *Sigma* project, we initiated the standardization process for allowing IIOP traffic to pass through firewalls safely and wrote the RFP. We also partnered with several ORB vendors to help write the final joint submission, which is in the final stages of approval by the OMG. Experience with the ORB Gateway was instrumental in developing the specification. As the project concluded we began exploring the possibility of working with ORB vendors to test the interoperability of their Firewall/IIOP compliant implementations with the ORB Gateway.

3. Demonstrations

During the life of the *Sigma* project we conducted a series of demonstrations showing progressive capabilities in the three key areas of research:

- Multiple applications demonstrated with ORB GW
  - Flight Scheduler with GUI Interface
  - JTF Application IA IFD 1.1
  - IA Challenge Application IA IFD 1.2
- OO-DTE access control for multiple COTS ORBs
  - Orbix
  - Visibroker
- Support for multiple distributed authentication services
  - SSL
  - IPSec
  - DCE

These experimental prototypes demonstrated the feasibility of enforcing access controls on inter-enclave CORBA traffic using Sigma's enclave security gateway for CORBA, the ORB Gateway. The use of OO DTE policy specifications for concisely and flexibly stating an access control policy for objects of CORBA applications of moderate complexity using two commercial ORBs was demonstrated and the enforcement of policy by the ORB Gateway prototype was shown. The project culminated with a demonstration of all of the mechanisms in an enhanced interoperable demonstration in September 1999.

The *Sigma* project engaged in early technology transfer by transferring some of the early research results and prototypes to the DARPA ISO IA program. Under the auspices of the IA program, OO DTE technologies and the ORB Gateway were demonstrated during the Information Assurance program IFD 1.2 in August 1998. All of the demonstrations used the 'Challenge Problem' application, which simulated an image analysis system running in multiple enclaves at multiple security levels. OO DTE was demonstrated using Orbix with a security plug-in (filter). The OO DTE demonstration showed:

- An above-kernel implementation of OO DTE on a COTS ORB: Access control was implemented in an Orbix filter without modifications to the ORB itself. Both the clients and servers were implemented in Orbix with OO DTE filters.
- SSL for authentication and secure invocations – X.509 certificates contained a domain identifier. Clients and servers exchanged certificates as part of the SSL handshake. This established their respective domains for the duration of the connection. All operation requests and replies were encrypted.
- Per-object access control – Image objects were created and named such that 'infrared analysts' could update an 'infrared image analysis', but not 'radar image analysis'. All of the images shared the same CORBA interface.
- Automatic policy distribution and synchronization – The OO DTE policy was changed 'on the fly' from a master policy server. The policy change propagated to each client and server and took effect upon receipt, without the need to restart the applications or take any other manual steps.

The ORB Gateway demonstration showed:

- Automatic policy distribution and synchronization – The same distribution mechanism used by OO DTE also provided policy distribution for the ORB Gateway.
- Interoperation with a COTS ORB (Visibroker) using SSL – The clients and servers were non OO DTE applications written in Java using Visibroker and SSL. The ORB Gateway extracted the domain information from the client certificate and made an access control decision based on the OO DTE access control policy.

4. Technology Transition

Technology transfer was an important aspect of the *Sigma* project. This was in part motivated by the increased focus on CORBA security in both the government and industry and in project directions which strongly favored rapid technology transfer.

The DARPA IA program initial user community was oriented towards the use of CORBA which played a central role in the DARPA ISO sponsored JTF ATD program as well as several other ISO programs. The IA program expressed interest in *Sigma* research results for use in JTF ATD and IA programs in January of 1997, approximately half way into the project period of performance. Under on-going contracts from the IA program the *Sigma* ORB GW and OO-DTE technology were integrated and enhanced for use in the IA program.

Similarly, ORB GW and OO-DTE technology are being transferred for integration and extension in the DARPA ITO Quorum program. OO-DTE, DTEL++ and aspects of the ORB Gateway have been provided to BBN for use in the Quorum Object Integration (QuOIN) project. The initial integration of OO-DTE with BBN's Quality Definition Language (QDL) went very smoothly and further extensions of *Sigma* security technology with BBN Quality of Service technologies are being planned.

Technology transfer to the OMG and COTS product communities was also an important product of the project. This was due to the tremendous need for research results to advance the state of the art in a rapidly emerging technology space. As described above in 2.7 OMG Activities, the *Sigma* project played a leadership role in the development of the OMG specification for CORBA and firewalls.

In the year immediately after the project completed, components of the ORB Gateway, OO-DTE technology and initial research prototypes were enhanced and further developed under internal corporate funding to provide a CORBA IIOP proxy for the COTS Gauntlet Firewall. This proxy includes the following key features:

- Compatible with version 1.1 of CORBA's Internet Inter-ORB Protocol (IIOP).
- Passes only well-formed IIOP messages in accordance with the policy specified in the Gauntlet administration tools.
- Operates transparently for both inbound and outbound connections.
- Supports "callback" requests from servers to clients using bi-directional IIOP.
- Interoperates with Orbix v2.3, Orbix Web, VisiBroker v3.3 (C++ and Java).

In addition to the very successful technology transfer results the project produced three journal quality published papers.

5. Summary

The *Sigma* project objective was to perform research and advanced proof of concept development in three key areas in order to advance the state of the art of security technologies for interoperable heterogeneous distributed systems. We choose the emerging standard, CORBA, which embodies architecture for interoperable object-oriented distributed computing as the basis of our research. The project produced significant results in each of the three key areas: an ORB Gateway that selectively filters object requests entering an enclave, object-oriented access control mechanisms for CORBA, and mechanisms for the application of multiple distributed authentication technologies to support object-oriented access controls. The accomplishments in these areas were based on a common approach to specifying security policies for distributed object applications and common enforcement engines along with policy distribution approaches.

The *Sigma* project was highly successful in its technology transfer goals producing results for use in two major DARPA programs, industry standards, and a COTS product. All of these results continue to be used and extended one year after the project's completion.

6. References

[ACSAC96] T. Benzel and E.J. Sebes, *SIGMA*: "Security for Distributed Object Interoperability between Trusted and Untrusted Systems", "Proceedings of the 12th Annual Computer Security Applications Conference, pp. 158-168, December 1996

[ARCH96] T. Benzel, et.al., "*SIGMA* Security and Interoperability for Heterogeneous Distributed Systems, Architecture Study Report", August 1996

[BADG] L. Badger, D. F. Sterne, D. L. Sherman, K. M. Walker. "*A Domain and Type Enforcement UNIX Prototype,*" USENIX Computing Systems Volume 9, Cambridge, MA, 1996.

[INTER] T. Benzel, et.al., "Interoperability Results and Recommendations Report, Trusted Information Systems, September 1997.

[MILCOM] T. Benzel and E.J. Sebes, "Multi-Level Security for CORBA Applications in DoD Computing Environments" Milcom, October 1997

[TALL] G. Tally, D. F. Sterne, C. D. McDonell, "Sigma Project:DTEL++ Language Specification", Trusted Information Systems, September 1998.

[USENIX99] D. Sterne, G. Tally, C. McDonell, D. Sherman, D. Sames, P. Pasturel and E. J. Sebes, "Group Scalable Access Control for Distributed Object Systems", USENIX paper, Summer 1999.

Security Policy Realization in an Extensible Operating System

Dennis Hollingworth, Timothy Redmond, and Robert Rice
NAI Labs at Network Associates
{dholling,tredmond,rrice}@nai.com

Abstract

We describe the practical application of security to extensible operating systems, focusing on the SPIN extensible OS. Because SPIN and other extensible operating systems support the loading of user-developed extensions directly into the kernel, they require the application of traditional user-level security in addition to kernel-level security, to prevent kernel extensions from subverting kernel functionality. In order to show that such an approach to extensible OS security is possible and practical, we augmented the existing security architecture of SPIN, maintaining the separation between policy mediation and enforcement. Several well-known user-level policies, described here, are supported in the architecture. In addition, we have introduced kernel-level security through the separation of kernel code into segments and the enforcement of a domain/type policy on threads as they progress through system code. Finally, we have devised a useful domain/type policy extensibility strategy that is conservative, but preserves the existing policy.

1. Introduction

Over the last decade there has been a shift in operating systems research focus away from large, monolithic operating systems with static code bases to coarsely extensible, microkernel-based, server-oriented operating systems such as Mach [13], Chorus [4], and Spring [10]. This shift from monolithic operating systems to server-oriented operating systems was motivated primarily by the desire for increased system flexibility and robustness. Unfortunately, server-oriented operating systems tend to suffer from performance problems due to fundamental architectural limitations involving context-switching overhead. For example, this was a particular problem with the Mach operating system, despite a substantial effort on improving system performance toward the end of the Mach OS development effort.

More recent research in operating systems has tended to focus on extensible operating systems that support dynamic kernel modification--systems such as SPIN [2], VINO [6], and the Exokernel [7] architecture. This research on dynamic kernel extensibility is largely motivated by a desire to significantly improve overall application performance by better satisfying application-specific computing and communication requirements. A secondary motivation is the desire to develop a more flexible platform for exploring new OS principles and innovative concepts in operating system services.

This paper describes a practical approach to extensible operating system security (the EXOS project) including the rationale upon which that approach is based. It focuses specifically on the SPIN extensible operating system. SPIN introduces a level of operating system extensibility not previously encountered in operating system development efforts or other research efforts. SPIN core functionality is characterized as little more than a mechanism for loading and executing extensions.

Our approach addresses the security needs surrounding the loading of user-developed extensions directly into the kernel. It also addresses the problem of least privilege among service components of the SPIN system that extend the functionality of the base operating system. Our approach integrates traditional security requirements with SPIN operating system development objectives, specifically high performance and fine-grained extensibility.

This paper has four major areas of discussion. The first of these, Part 2 of the paper, briefly addresses what we mean when we use the term *extensible operating system* and how this concept interacts with the traditional notion of the kernel as part of the Trusted Computing Base. The second, Part 3 of the paper, explores the components and emphasis of a candidate security policy model for extensible operating systems, including basic concepts and practical policy model support for user-level policies. It also establishes a notion of a *kernel-level policy* to augment University of Washington security support for kernel extensions and implements it by extending domain and type enforcement to software segments. The third area of discussion, Part 4 of the paper, discusses the SPIN security architecture developed at the University of Washington as well as architectural enhancements introduced under the EXOS project associated with the production of this document. The

0-7695-0490-6/99 $10.00 © 1999 IEEE

final major area of discussion, Part 5 of the paper, examines how the identified system-level policy based on domain and type enforcement might be correctly extended in a *conservative* fashion to support system, middleware, and application-level services that must extend the operational security policy to support their security objectives. We end the paper with conclusions.

2. Security in extensible systems

2.1. Extensible operating systems and protection mechanisms

Before discussing the security requirements of extensible operating systems, it is useful to clarify what we mean by the term extensible operating system as it is used in this report. As discussed in [17], most OSs are extensible to some degree. They allow for static or even dynamic linking of specific types of system components such as protocol stacks or device drivers to pre-defined interfaces. An *extensible operating system*, as we employ the term, allows much more extensive alteration of core system functionality than previously supported. An extensible operating system allows the run-time loading and execution of operator- or application-supplied extensions to the kernel that extend or replace basic kernel functionality, such as virtual memory support.

SPIN employs kernel extensibility as a mechanism for improving OS performance in support of application software. A significant detractor from the performance of traditional operating systems is the overhead involved with supporting hardware-enforced execution contexts and switching between them. SPIN depends primarily on language features rather than hardware mechanisms for its safety features. In SPIN, procedure calls replace hardware context switches, reducing the overhead of separate contexts.

An additional problem with hardware mechanisms is their coarse granularity. For example, hardware protection may define access enforcement at page boundaries, when what is required is access enforcement on non-contiguous memory regions. Software protection mechanisms, as employed in SPIN, can easily provide protection at such a granularity.

In spite of the advantages of software protection mechanisms, they are likely to raise concerns among security professionals, who have been relying on hardware protection mechanisms for three decades. A convincing argument is required that the granularity, performance, and utility of software protection outweigh the capabilities of the better understood and more accepted hardware protection mechanisms.

2.2. The trusted computing base in extensible operating systems

The shift from monolithic to server-based systems stretches but still preserves the concept of a trusted computing base (TCB) that controls the security of the system. The movement to dynamically extensible kernels challenges the concept at its foundations, primarily because these systems are oriented toward user-supplied, fine-grained, run-time extensions to executing kernel code.

Such systems constitute an apparent deviation from the traditional interpretation of operating system security principles and established high-security architectural concepts. They replace the concept of a relatively static kernel with that of a dynamic kernel specialized at run-time to match system and user requirements. They introduce the problem of code that lies outside a trust boundary but inside the kernel.

From an OS security perspective, a major architectural issue raised by this new OS research trend is that of address space isolation and security policy enforcement. Common security practice for high-security systems includes the operating system kernel in the definition of the trusted computing base, isolating it from user-level code through virtual address space management and hardware-enforced separation. However, these new research operating systems support dynamic, run-time alteration or augmentation of kernel functionality through untrusted or trusted extensions to a minimal "core" kernel code base that does not include much of the functionality we have come to expect in the traditional kernel code base.

The fine-grained extensibility supported by these research operating systems is fundamentally different than the traditional model in scale and approach. Extensibility extends into traditionally core OS services, including fundamental access control mechanisms such as the virtual memory support subsystem. Moreover, extensibility may be supported at as fine a granularity as the customization of an internal kernel procedure call. Individual procedures are potentially replaceable or extensible via user-supplied code in addition to system library code. This raises the issue of how one defines and preserves an encompassing and non-bypassible security policy for the system. Also at issue is what security policies can be enforced by and on such systems.

3. Security policy in the SPIN extensible operating system

Highly extensible systems manifest classical security requirements as well as introduce new ones. With respect to classical security requirements, user-level security needs still exist and must be addressed. Access control between user-level subjects and objects must be supported by and captured within a policy model that reflects the security objectives of the system.

Extensible systems introduce a new wrinkle on security policy support. Extensible operating systems force us to reflect upon the diffusion of user-level security requirements across the user/kernel boundary as well as requiring what we refer to as kernel-level security. They require that we distinguish and clarify the boundary between the two concepts, focus of each, and degree of entanglement that exists.

Extensions loaded into the kernel may retain their association with user identity and user-level security attributes, such as security labels or type assignments. At the same time they become part of the OS kernel code footprint. As they are moved into the kernel's address space, barring the use of mechanisms like those described in this paper, they acquire privileges associated with system-level functionality kernel code, despite the fact that they are of user-space origin and may be untrustworthy. This requires additional protection strategies to ensure that they do not violate both the explicit and implicit policies that apply to the user security context from which the code was introduced. These extensions, running in kernel space, must be prevented from bypassing or otherwise subverting core kernel mechanisms that support policy. To support this requirement, the EXOS project has identified the need for kernel-level policy support in addition to user-level support. The EXOS project is investigating the means by which the analogue of user virtual address space protection might be efficiently applied to the SPIN kernel. There are two aspects to this kernel-level support: confinement and least privilege.

Confinement separates application/system components into protected compartments not based on user identity. The purpose of confinement is to separate functionality into independent system components so that failures in one component are confined to a specific region of code and system functionality. It allows the developers to ensure that the operation of critical code segments is not subject to inadvertent or deliberate tampering by misbehaving code segments. In the case of extensible operating systems, we believe that it is especially important to implement some sort of extension confinement mechanism to ensure the integrity of system components.

The ease of extending the SPIN kernel underscores the need to restrict privilege of extensions to only that necessary to support their intended functionality. System structuring to support *least privilege* is, therefore, a significant security requirement for extensible system security architecture. The objective is to provide a partitioning on the privileges available to software components so that each system component operates with only the privileges required to support the intended service.

A fundamental concept explored in the EXOS project involves the use of security domains to provide a supplemental access control mechanism that is orthogonal to the user or authority that spawns a task. In traditional systems, a common model is that of a thread executing on behalf of a user request. Access control decisions are made based on the user's identity. However, in the SPIN extensible OS security architecture, there are the additional security objectives of least privilege and confinement. As the thread executes, it may progress through several different extensions, each of which has different security requirements. For example, a thread may execute device driver code that has special privilege and needs to be protected from tampering, or it may execute server code that has no special privilege but still needs to be protected from outside tampering. Hence, there is a requirement to provide security separation between these extensions as well as the core system functionality itself. Security domains can be utilized to constrain the allowed and denied access modes associated with the thread as it follows its execution flow.

3.1. Policy models

While there has recently been increasing interest in non-specific security policy expression languages such as Adage [20], there is significant value in the existing policy-specific models when it comes to stating the security policies that hold for a given system. The existence of a body of knowledge associated with specific policy models aids understanding their implications in a given system.

Existing policy models reflect established conceptual protection needs and provide useful abstractions for defining and understanding system security requirements. These policy models are widely recognized as addressing important government as well as commercial security requirements. Consequently, we believe that major user-level policy models such as MLS, RBAC, and Chinese Wall should be explicitly supported by the SPIN security architecture. Part of the EXOS project work involves selecting and supporting specific well-known policy models to establish how important security policies can be supported in the kernel through policy specification and enforcement mechanisms.

3.2. Basic concepts and primitives

Prior to discussing specific policy models, it is useful to define selected fundamental security policy primitives required for the expression of a security policy in the models that will be discussed. The primitives we describe are security objects, subjects, and domains.

3.2.1. Security objects. Security objects are identifiable, distinguished entities, including code segments and data containers, to which an active security policy applies and that can be recorded with a policy manager. They must have an associated set of security attributes by which enforcement mechanisms mediate access by security subjects. Not all system objects are necessarily security objects.

A security type is assigned to every security object. The security type of an object represents certain security attributes that are associated with an object. This type assignment groups objects that will be treated identically for access control purposes.

3.2.2. Security subjects. A security subject is a security policy entity for which access to security objects is mediated by active security policy elements. The traditional definition of a security subject as defined in the "TCSEC Orange Book" is that a subject is a process-domain pair [5]. The nearest equivalent of a process in the SPIN architecture is a SPIN thread, so we define subjects to be thread-domain pairs. As in the traditional case, a thread acts on behalf of some user. Threads move between security domains, the access to which may be constrained by the active security policy. In the SPIN security architecture, a thread initiates domain transitions by either executing a procedure call in kernel space or by executing or returning from a system trap to move from user space to kernel space.

3.2.3. Security domains. A security domain may be both a constituent part of the security subject, as discussed above, and a security mediation target, depending on its use during any given mediation event. In the former case, it is characterized as a set of access permissions to objects. In the latter case, it is the mediation target of an access check event involving a subject's right to extend the domain or transfer control into that domain.

We use security domains to enforce the isolation of system-level components from one another, confining the effects of potentially misbehaving or subverted system components. In the SPIN security architecture, security domains are used to help preserve the integrity of the operating system by structuring the OS so as to minimize privilege and isolate components from one another.

3.3. Policy model support

For extensible operating system security we are interested in two types of policy models--those that are based on the identity of the user associated with a thread of execution and those that are based on the execution locus of the thread. In the SPIN security implementation, we have associated the first of these, proxy-based models, with our support of user-level security requirements and the second, domain-based models, with our support of kernel-level security requirements. One area of investigation is the degree to which these models, when simultaneously active in a running system, potentially interfere with one another in the case of lattice-based policy models.

In this section, we provide an overview of specific security policy models for which we defined supporting policy language elements. These policy models include:

1. Multi-Level Security Model (MLS)
2. Role-Based Access Control Model (RBAC)
3. Chinese Wall Security Model
4. Access Control Lists (ACLs)
5. Domain and Type Enforcement (DTE).

We have chosen these security models based on their importance to the security community. We discuss the first four shortly, and distinguish these from the DTE policy model. While it has been characterized as user-level security support, we feel that the DTE policy model is more applicable to system security objectives such as confinement and least privilege. We will discuss the DTE policy model in the context of domain-based security.

3.3.1. Proxy-based policy models. Proxy-based security policies encompass those security policies associated with the security identity under which a user-associated proxy, i.e., a process or thread, executes. The security attributes of the proxy are relatively static in nature, having been acquired when the proxy was created. They are intended principally to control the user's access to security-sensitive containers of information. Proxy-based security is ultimately based on some aspect of user identity or user privilege delegated to create proxies that reflect the user's access rights in security mediation decisions. These might include the groups to which the user belongs, the clearance level at which he is operating, or the role that he has assumed.

A user's access privileges are often stated in terms of an allowed range of access, rather than a unique access level. Consequently, the delegation of access privilege to a user proxy, such as a thread may, for example, require fixing of the proxy's security label at one of several allowed labels contained within the full label range associated with the user's identity. For example, a subject created on behalf of a user will be only be given an MLS label allowed by his MLS maximum-read-level and minimum-write-level. Similarly, a subject created on behalf of a user will only be given a role that is permitted to the user. It is the specific label or roles attached to thread or process proxies rather than those potentially allowed by user identity, which we are dealing with in proxy-based policy support.

Multi-level security model. The multi-level security (MLS) policy model defines lattice-based information flow control based on the sensitivity of information. The goal of an MLS policy is to prevent information from higher sensitivity containers from flowing into less sensitive containers. MLS policy support requires the identification of security subjects, associated with users and their security clearance, and security objects, associated with containers of information. A (clearance, category-set) label is attached to each subject consistent with the read-max and write-min constraints of the initiating user and a (level, category-set) label to each object.

The classification levels establish a total ordering that represents the relative sensitivity of information identified with that level and the policy defines constraints on the flow of information between such levels.

The sets of categories that are assigned to objects are also used to order objects by their sensitivity. An object or data container labeled with a set of categories and a certain classification level is considered to be more sensitive than an object or container labeled with a *subset* of those categories and the same level (the former object's label is said to dominate the latter's). Therefore, the MLS policy model requires the assignment of sensitivity labels to subjects that represent the maximum sensitivity of information that they are allowed to read. The MLS policy states that a subject is only allowed to read information labeled with a sensitivity label that is dominated by the label of the subject (simple security property) and write information labeled with a sensitivity label that dominates the label of the subject (*-property).

Role-based access control model. RBAC establishes work roles in which the user may operate and controls access to information based upon those work roles. It is the role that the user has assumed for an interval of execution that determines what security objects he can access. Initially, RBAC policy models simply identified the collection of supported roles and the user's ability to assume a particular role or set of roles during a computational session and bound object access rights to these roles. More recent work, the RBAC96 model [15], goes beyond simple role identification in an attempt to support hierarchical role relationships as well as a family of RBAC models. Sandhu [16] identifies a three-tier architecture comprised of four separate but related models for specification and enforcement of role-based access control. $RBAC_0$ is comprised of users, roles, permissions, and sessions. $RBAC_1$ introduces a concept of role hierarchies to the $RBAC_0$ model analogous to object hierarchies in an object-oriented model. $RBAC_2$ is incomparable to $RBAC_1$ and adds a concept of constraints on roles to the $RBAC_0$ model. For example, constraints may indicate that the same user cannot belong to two distinct roles; i.e., they are mutually disjoint roles. $RBAC_3$ represents a consolidated model that combines $RBAC_1$ and $RBAC_2$ to provide both role hierarchies and constraints.

RBAC96 makes a distinction between users and sessions. Roles active in a session can be changed at the user's discretion. However, RBAC96 views the changing of roles as a security-sensitive act that constitutes a relatively heavyweight event that directly involves the user (although the model does acknowledge the concept of one session creating another session while not speaking further to that issue). We, too, view the transition between roles or activation of roles as a heavyweight event minimally associated with the creation of a new task, and not a control flow transition from one domain to another. In other words, we, at a minimum, see a new thread or process being created to represent the new role association, if not an even more substantial event such as a new logon action.

In the SPIN security architecture we restrict our policy language and security implementation to Sandhu's $RBAC_0$ model. Support for Sandhu's $RBAC_1$, $RBAC_2$, and $RBAC_3$ models is covered by role-specification and assignment mechanisms, of more of an administrative character, unrelated to the underlying mediation and enforcement mechanisms we are investigating for the SPIN system. They do not require any special run-time support beyond the role labeling established for $RBAC_0$ support. Consequently they are outside of the purview of this effort.

Chinese wall model. The Chinese Wall security model addresses the problem of upholding the confidentiality of information to which an analyst or consultant might be granted access in the course of working for his clients. It is characterized in the literature as the *code of practice* that must be followed by a consulting company. For example, a consulting company may have contracts with two companies that are in direct competition with one another. If no special policy is in place to handle insider knowledge, the two companies may both fear that the consulting company is a conduit of proprietary information.

The Chinese Wall model represents the application of this type of policy to computation. In this policy model, information is grouped in three different ways. At the lowest level, the individual containers of information are called objects. At the next level, objects are labeled with the company dataset to which they belong. These company datasets represent the information that is associated with a particular company. At the top level, company datasets are grouped together into conflict of interest classes (COIs). If two different company datasets are in the same conflict of interest class, then it is viewed as a conflict of interest for a user to access objects from both datasets. The statement of the Chinese Wall policy is that a user can obtain access to objects in only one company dataset for each conflict of interest class.

We have selected the model described in [14] as a basis for our policy language primitives for Chinese Wall. This model assigns a label to each user, his run-time-instantiated subjects, and all COI security objects in the system. The label is defined as a vector of identifiers associated with company datasets. Each position in the label vector corresponds to a particular conflict of interest class. In the case of the user label, a given position in the label vector contains the identifier for the company dataset accessed under that conflict of interest class, or NULL if no dataset has yet been accessed. The subject label indicates the access rights allowed during a given session, and is constrained by the access history recorded for the user label when the subject's label is created. The object label indicates the COI classes and associated datasets to which the object belongs (i.e., the object contains data from the indicated datasets associated with corresponding COI classes). Since the Chinese Wall policy model allows access to only one member of a COI class for a given user or his surrogates, the user label vector represents the history of user access to COI class datasets while the subject label constrains the run-time access rights to company datasets.

The intent is that a user be able to access the data associated with new companies within the constraints of other access controls provided that the user does not obtain access to two companies that have a conflict of interest. Thus a new user will be labeled as having never accessed any of the company datasets within the specified COI classes. A user's label floats upwards in the lattice with respect to dominance of subject/object labels as he is allowed access to datasets within COI classes.

A user interacts with the system by creating a subject to act on his behalf. When the user creates a subject, the user specifies a label for that subject that indicates what company datasets the subject is allowed to access, constrained by the user's access history and any discretionary security policy controls that apply to the user. As a subject attempts to access a dataset within a specific conflict of interest class, a check is made to see if and how a dominance relationship (defined on the subject's and object's label vectors) is satisfied with respect to the granting of specific combinations of read, write, and append access from the subject to the object.

ACL-based model. An ACL-based policy provides controls over how a subject can access a named object; control is at the discretion of users. An access control list is simply a list of principals that are authorized to have access to some object along with the allowed access rights. As stated in the Orange Book: *These access control lists shall be capable of including or excluding access to the granularity of a single user* [5]. Access control lists may specify *allowed* access, *denied* access, or a combination thereof. An implicit ordering exists as to the precedence of identity type entries. An implicit rights composition rule also exists as to how expressed *allowed* and *denied* access permissions are combined from multiple applicable ACL entries to determine the actual access permissions associated with a target object. Conflicts between allowed and denied access permissions require an ACL search rule or some other criteria that establishes how such conflicts are resolved. Actual interpretation of access rights depends on the kind of object with which the ACL is associated.

3.3.2. Domain-based policy models. Domain-based security policy support introduces an additional security constraint on the security subject beyond those we characterize as proxy-based constraints. Domain-based policies apply a dynamically variable security attribute in mediation decisions representative of the *domain of execution* in which a thread is executing. This domain-based security constraint is actually utilized in three different types of security decisions: the objects that a subject can access from a given domain (similar to traditional proxy-based security support except that it may involve access control to both system-level as well as user-level objects), the domains to which the subject can transition (i.e., other access rights it can acquire), and actions it can invoke against other domains.

Past as well as recent secure operating systems work [1,8,18,19] has utilized the domain concept as a mechanism for isolating system components from one another. We believe that the domain concept employed in older systems is especially suited to extensible operating system security to insure that misbehaving extensions cannot compromise other kernel or application functionality.

DTE policy model. The DTE policy model is based upon the Type Enforcement model first proposed by Boebert and Kain [3] for the Secure Ada Target (SAT) and subsequently implemented in the LOCK program [18]. Under Type Enforcement, each security object is assigned a *type*. Domains of privilege are established in which programs execute. Access to specified types is restricted to specified domains. DTE also supports the notion of control transfer between domains, defined at the object-module level, through specifically identified domain transition points associated with individual domains. These domain crossing points are identified statically in the DTEL language policy specification.

The TIS/NAI Labs DTE work [1] remedies a particular difficulty associated with Boebert and Kain Type Enforcement, while supporting the underlying policy model. The tables used under Type Enforcement to express domain-to-type and domain-to-domain access rights can grow quite large and cumbersome to represent and process on each access-mediation event. Recognizing that named, persistent objects, e.g., files and directories, make up the bulk of the security objects in traditional operating systems such as UNIX, DTE refines the techniques of Boebert and Kain by defining a sparse, shadow directory tree with parent-node-based default type inheritance. Explicit type inheritance can be used to override the default, reducing the storage and processing requirements associated with the technique. DTE also specifies a relatively simple language for specifying type-enforcement-based security policies.

Domain/type enforcement and SPIN. While DTE has been utilized to explicitly support proxy-based access control policies such as simple role-based access control, we find it most valuable for its ability to let us express and support confinement and least privilege requirements attached to extensions. Past experience with other secure operating systems work suggests that user extensions to the kernel should not all live and execute in the same protection domain even with software guarantees such as strong data typing. The domain concept is especially applicable to extensibility-oriented policy support and corresponds well to both the confinement of untrusted physical code segments, such as user-loaded extensions and the minimization of privilege required by specific functionality. If we associate kernel extensions (and possibly static kernel functionality) with particular domains utilizing type assignment to reflect domain-based access control requirements, DTE allows the imposition of a level of structuring on the core OS that isolates the effects of aberrant system behavior.

We implement security domains through the identification of gates that effect a transition from one security domain to another. However, we go further by associating a collection of *typed* kernel entities with security domains, specifically, software segments to which individual modules are assigned so as to establish desired system structuring and confinement properties. Software segments allow the compartmentalization of kernel code into separate sections with common functional and protection attributes. They allow the system to determine the domain in which a thread is executing before any security-relevant procedure call. They permit mediation and enforcement of access control decisions when a security-relevant procedure call is made.

4. Architectural enhancements

This section details the enhancements made to the original SPIN security architecture in order to support some specific user-based and system-based security policies. The discussion begins with a description of the original SPIN architecture, with an emphasis on the original security mechanisms, and continues with descriptions of the enhancements made to the SPIN policy manager to support the enforcement of specific user-based policies. The section concludes by detailing enhancements made to support the enforcement of system-based policies through software segment access control, which is also mediated by the policy manager.

4.1. Original SPIN architecture

The SPIN kernel is conceptually a single program written in the object-oriented programming language Modula-3, in addition to a stand-alone layer of code for low-level control of the target hardware. Higher-level kernel components include Modula-3 runtime support (strong type support, low-level linking, and garbage collection support) and several core services. The core services include memory management, an event dispatcher, and a strand scheduler, in addition to those that will be emphasized in this section--the dynamic linker and the enforcement and policy managers.

4.1.1. SPIN domains and dynamic linking. Because SPIN allows applications to reside in the same address space as the kernel, many mechanisms must be employed to protect kernel code and data from unwanted tampering and execution. At a low level, SPIN employs compiler restrictions on the extension code written in Modula-3 that preclude unsafe pointer operations such as pointer arithmetic and casting. This ensures that extension code cannot directly access data or code that has not been provided to the extension at link time.

The SPIN kernel also needs to control the access that extensions are given at link time. This is done through the concept of a SPIN domain [12]. The *SPIN domain* represents the unit of linkage in the SPIN system. It encapsulates a set of interfaces or a body of code that is to be linked against. The SPIN domain is represented as a Modula-3 object that cannot be forged by SPIN extensions because of the type-safety constraints. By treating a SPIN domain as a capability and controlling how SPIN extensions obtain access to SPIN domains, the SPIN kernel can restrict which extensions obtain access to privileged kernel calls.

It is important to observe that SPIN domains are different from security domains. SPIN domains are passed as capabilities. Once a SPIN domain is given to an extension the extension has full control over linking to the SPIN domain. Thus SPIN domains are not subject to any mandatory policy such as a domain/type policy.

4.1.2. Access control. SPIN's runtime access control mechanisms are the major responsibilities of two entities: the enforcement manager and the policy manager [9]. The role of the enforcement manager is to intercept every security-relevant access attempt and to request a mediation decision from the policy manager when necessary. The enforcement manager is policy neutral--it does not interpret subject or object identifiers or access modes. It simply provides access enforcement as requested by the policy manager and other policy-aware SPIN components.

When a SPIN domain requires access control on an exported procedure, it notifies the enforcement manager of this fact, along with the access modes that are required for procedure execution, access to the procedure arguments, and the result of the procedure. In addition, the SPIN domain may specify whether execution of the procedure signifies a SPIN domain transition for the subject, and, if so, the new SPIN domain.

When a procedure execution is attempted for the first time, the enforcement manager notifies the policy manager with the security identifiers (SIDs) of the subject (thread/domain pair) and the procedure to which access is to be mediated. The policy manager then returns the appropriate access modes for the subject/procedure pair to the enforcement manager. The subject/procedure call pair, along with the access modes, may be cached if requested by the policy manager. If the returned access modes include the mode required for executing the procedure, each of the procedure arguments is tested in a similar manner. If those tests pass, and a SPIN domain transition is specified for the procedure, the SPIN domain is changed and the procedure executes. On return from the procedure, the SPIN domain is possibly changed, then the return value tested.

4.2. Implementing a policy manager

The goal of the University of Washington SPIN security work was to produce a policy-neutral enforcement manager. Therefore, the policy manager provided by the University of Washington was intended for testing purposes. Part of the goal in our work was to build a prototype policy manager to mediate for several real-world policy models. Our policy manager utilizes a policy interpreter, which reads in the policy specification file and initializes the policy manager, and policy representations, which effect mediation of the policy. In addition to the description of the policy manager, the algorithm for mediating policy is described below.

4.2.1. Policy interpreter. When SPIN boots, the policy interpreter reads in the policy specification file. In the present implementation of SPIN, this file is obtained from the SPIN build machine. The policy specification file is a plain text file, and is transmitted to the SPIN crash machine via the trivial file transfer protocol (tftp). The policy interpreter first checks the syntax of the policy file. If an error is found in the syntax, the policy interpreter outputs a descriptive error message to the SPIN console and terminates. If the syntax is found to be correct, the policy interpreter populates the data elements of the policy manager that represent the security policy itself.

The policy interpreter is loaded at SPIN boot time as a trusted extension to the SPIN kernel. Parser and scanner functionality are written using the popular compiler/interpreter tools yacc and flex, and the actions that populate the policy manager data structures are written in C. There is a separate facility that allows checking the policy specification file for syntax errors before actually booting SPIN. The policy specification file is written in a language with elements that closely correspond to the descriptions of the individual policy representations described below.

4.2.2. Policy manager internals. Before discussing the implementation of the policy manager, it is instructive to describe the overall design of the elements interior to it, including the SID/attribute table, a mapping between the security identifiers, SIDs, of security-relevant entities, and the attributes associated with those entities. Also detailed in this section are the data structures that the policy manager uses to represent the policies to be mediated. In addition, we detail the attributes that are associated with objects' and subjects' SIDs in the SID/attribute table, and that are used in the policy representations to make policy decisions.

The design of the policy manager is shown in Figure 1. The data structures interior to the policy manager include a mapping between SIDs and attributes. This table is designed as a hash table with the SIDs as keys for quick lookups when the policy manager is queried with the SIDs of a subject and object.

In addition, the policy manager includes representations of five different policy models: MLS, Chinese Wall, RBAC, DTE, and ACL. These representations contain policy-specific information that is independent of the system security entities. The MLS representation contains a list corresponding to the security levels present in the system along with a list of categories to which datasets may belong.

The Chinese Wall representation contains the datasets belonging to the conflict of interest classes. In addition, both the MLS and Chinese Wall policy representations contain lists of access modes that correspond to read permission, write permission, and append permission. In

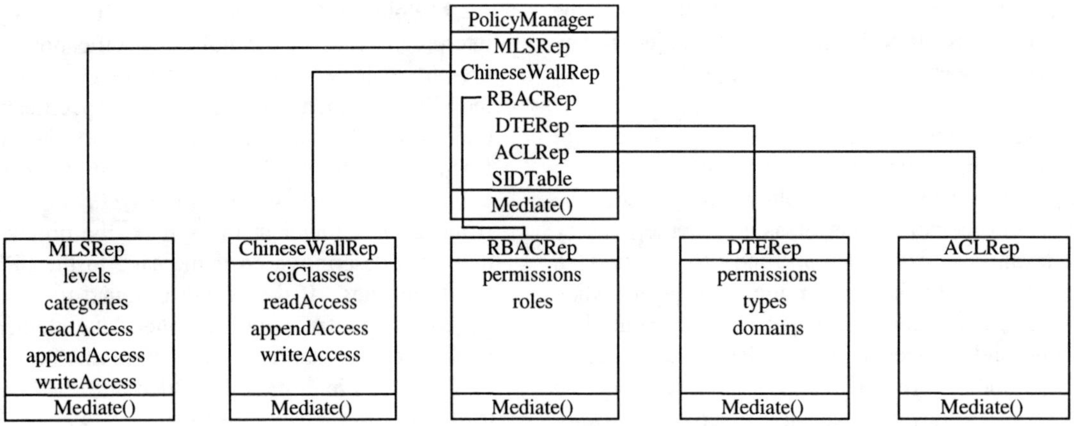

Figure 1. Design of the Policy Manager.

the MLS and Chinese Wall policies, these are the only access modes allowed.

The RBAC representation contains a table of roles. Each role associates the name of the role with a list of permissions, and each permission is a <object name, access mode list> pair. In addition, the RBAC representation keeps a list of permissions so that they may be accessed quickly.

The DTE representation also contains a list of permission objects for quick lookup, as well as a list of domains. Each domain consists of a name, a list of permissions corresponding to the name, a list of privileges, and a list of components. Each permission is a <type, access mode list> pair.

The ACL representation has no data structure associated with it because it needs none to represent its policy model. All policy-relevant information for the ACL model is contained in the ACL attributes.

Each of the policy representations makes mediation decisions based on the subject and object attributes that it is passed by the policy manager. The structures of these attributes depend on the policy representation in question, and on whether the attribute describes the subject or object involved in the mediation. Before describing mediation for each of the policy representations, then, it is necessary to describe in detail the structure of the attributes involved.

The MLS subject and object attributes keep the level and a list of categories to which the subject has access or to which the object belongs. The Chinese Wall subject and object attributes keep a vector of the datasets in the conflict of interest classes that the user has visited, or to which the object belongs. The first element of the vector corresponds to the first conflict of interest class, the second element to the second COI class, etc. An element of the subject attribute vector may contain a single dataset label if the user has visited the dataset, or it may contain a NULL if the user has not visited any dataset in the corresponding conflict of interest class. Similarly, an element of the object vector may contain either a single dataset or NULL.

The RBAC subject attribute contains the roles to which the subject belongs, and the RBAC object attribute contains a label corresponding to the name of the object (see the above discussion of the RBAC representation). The DTE subject attribute contains the domain in which the subject exists, and the DTE object attribute contains a label corresponding to the type of the object. Finally, the ACL subject attribute contains a label corresponding to the name of the subject, while the ACL object attribute contains the access-control list, a list of subject names and corresponding access modes.

4.2.3. Policy manager mediation. As detailed in the discussion of the policy/enforcement manager protocol, the main responsibility of the policy manager is to return a set of access modes based on the SIDs of the subject and object in question. In order accomplish this, the policy manager, when it receives a mediation request containing the subject and object SIDs, first looks up the attributes corresponding to the SIDs in an internal table. After obtaining the attributes, the policy manager then calls each active policy representation (one or more of the MLS, Chinese Wall, RBAC, DTE, and ACL representations) with the parts of the attributes relevant to them; i.e., it calls the MLS representation with the MLS subject and object attributes, the Chinese Wall representation with the Chinese Wall subject and object attributes, and so on. Each representation takes the subject and object attributes and returns a set of access modes reflecting the rights that the subject has to the object. The policy manager then collects the results and returns the intersection of the access modes. With this procedure in mind, then, it is instructive to describe the mediation of the individual policy representations.

The MLS and Chinese Wall policy representations have very similar implementations. They both utilize a dominance relation between subject and object attributes.

The details of these dominance relations are different for the two policies and are described below. However the manner in which the dominance relation is used for mediation is the same for the two policies. If the subject attribute dominates the object attribute, then the Read access mode is returned. If the object dominates the subject, the Append access mode is returned. If they are found to be of equal dominance, then the Read, Write, and Append access modes are returned. Note that these composite access modes are mapped to atomic access modes in the policy specification, so that a set of atomic access modes is returned in each case. It is possible that all of these tests fail. In that case, the attributes are incomparable, and an empty set of access modes is returned.

For the MLS policy representation, the subject dominates the object if the level of the subject is greater than or equal to the level of the object and the category set of the subject is an improper superset of that of the object. The object dominates the subject if, in the previous condition, the subject and object MLS attributes are reversed. If the subject and object have equal levels and categories then the subject and the object have equal dominance. If all three of these tests fail, the MLS attributes are incomparable.

For the Chinese Wall representation, the subject and object vector elements are tested one by one. The subject dominates the object if, for each vector element, the subject's element is equal to the object's element, or if the object's element is NULL. The object dominates the subject if, in the previous sentence, the roles of the object and subject are reversed. If all of the subject and object vector elements are equal, the subject and object have equal dominance. Again, if all of these tests fail, the attributes are incomparable.

The RBAC representation makes mediation decisions by taking the object's RBAC attribute, and searching through the subject's roles collecting the access modes corresponding to the object's label. After collecting the access modes, redundant modes are eliminated, and the resulting mode set is returned. The DTE representation works in similar fashion. Given the object's type label, a search through the subject's domain determines the access modes corresponding to the type. This set is returned.

The ACL representation takes the subject's attribute and searches for the label corresponding to the subject in the object's ACL. It returns the access modes corresponding to the subject's label.

4.3. Segmentation enhancements

As a complement to the policy manager enhancements described above, segments and associated access controls were added to SPIN in order to support least privilege and confinement. We view a segment as a SPIN security object, and a thread executing in a particular domain as a subject. By enforcing a security policy on the subject/object combination, segments can be protected against illegal or unwanted reads and writes. In addition, threads attempting to execute code in a segment can be prevented from doing so.

4.3.1. Segments. A *segment*, as we define the concept, is an association of related Modula-3 modules with common privilege and confinement mechanisms that are treated as a single entity for protection purposes. In implementing segments, we borrowed somewhat from the similar Multics [11] concept; however, we employ a segment as a logical subdivision of the kernel's address space based upon Modula-3 module lines. We do not require hardware support for segments, although we do exploit hardware features (described below) in order to facilitate the implementation of segments and associated access control, and in order to improve performance. With the proper hardware support (such as exists on the Intel x86 platform), segments could correspond directly to more traditional hardware segments. We chose to implement a more flexible solution while still retaining some performance advantages of using hardware segments.

4.3.2. Segment access caches. Segment access caches keep track of access decisions that have been made by the segment manager, described below. There is one segment access cache per subject in the system. At any given time, the current access cache is the segment access cache for the currently executing subject. The intention is that the segment access cache can be quickly accessed when an access operation is attempted. When an access operation is first attempted, the cache entry will be marked invalid and the operation will fault. As a result of the fault a mediation procedure is called. If the access is granted then the cache entry is filled in with the appropriate access modes and the operation is allowed to continue. If the access is denied then an access fault signal is sent to the calling thread. When a context switch occurs as a result of a gate crossing or if the scheduler runs a new thread, the current segment access cache is changed to be the segment access cache for the new subject. The use of the segment access cache is very similar to the use of the translation lookaside buffer in hardware. In fact, on certain hardware platforms it may be possible to implement the segment access cache as the hardware translation lookaside buffer.

In our extension of the SPIN system, each segment access cache is composed of an array of four-bit words. The elements of the array correspond to the segments existing in the system. For each segment, the bits in each array element are one or zero according to whether read, write, or execute access was previously granted to the subject/object pair. In addition, one bit in each element is zero or one according to whether the subject/object pair has been mediated. For fast convenient access, the segment access cache is held in a hardware segment.

4.3.3. Segment access intercept mechanism. The segment access intercept mechanism exists to ensure that a subject can read or write data, or execute code in a segment if it is permitted by policy. When a subject attempts to read or write data, or execute code, the segment access intercept mechanism extracts the appropriate segment access cache entry from the hardware segment register, and, if the valid bit is not set, a segment access fault occurs. If the entry is valid and the appropriate read, write, or execute permission is not given, the attempt fails and an exception is thrown. The segment access intercept code is inserted by the SPIN C/C++ compiler, which also serves as a back end for the SPIN Modula-3 compiler. Initially, when code is compiled, it is not known what segment the code is attempting to access. Only at link time will this be known; thus the linker patches the compiled code with the appropriate segment number, which is the offset into the hardware segment (the segment access cache entry).

4.3.4. Segment manager. The segment manager has two responsibilities: to hold a mapping from address ranges to segments, and to handle segment access faults. The linker, when it encounters code that is attempting an access, looks up the relevant segment, and patches the code with that segment. In order to look up the segment, the linker queries the segment manager with the address of the code or data in question, and the segment manager responds with the segment in which the code or data exists. The other responsibility of the segment manager is to mediate segment access when a segment access fault is generated. To do this, the segment manager queries the policy manager with the domain/segment pair and updates the bits in the segment access cache entry, along with the valid bit. The access is then retried.

5. Policy extension

In this section we explain the need for a mechanism that allows programs to extend the domain/type policy dynamically and describe how such a mechanism can be implemented. In addition, for illustration, we present two examples of the extension mechanism.

5.1. Approach

The domain/type policy has been proven to be very successful in confining a predefined set of programs to privileged but limited domains to enforce a least privilege requirement. For example, the DTE project [1] at TIS/NAI Labs has demonstrated how to confine programs that require root privilege so that they do not have all of the permissions associated with root privilege. For extensible operating systems, it is necessary to extend this capability to include policy constraints and protections for user programs whose needs were not foreseen at system build time. In order to satisfy the needs of such programs, we need a policy extension mechanism. In order for this extension mechanism to be useable by both untrusted applications and middleware, the extension mechanism must not allow the creation of new policies that are inconsistent with the intent of the old policy.

The major requirement of our approach is that the subject requesting the policy extension need not be trusted. Thus an extension is allowed only if the system can guarantee that the extension does not result in breaking guarantees that have been made to existing programs. For example, if a trusted program relies on the domain/type policy to prevent any other subjects from accessing a file, then this guarantee must still remain in place after any extension to the policy. We call such an extension *conservative*.

The technique that we use to meet this requirement is to require that both the original and the new policy constraints hold after a policy extension. When the policy is extended, every domain (type) in the new policy maps to a domain (resp., type) in the old policy. In addition, every event allowed by the new policy will map to some event that is allowed by the original policy.

5.2. Extending the domain/type policy

We now define how a domain/type policy can be extended in a conservative manner. We start with a notation for representing domains and types that leads naturally to a method of introducing new domains and types. We then describe the three types of extensions that can be made to a domain/type policy and define the mapping of the new policy to the old policy. Finally, we describe how this technology can be applied.

5.2.1. Representing domains and types. The domains and types that are defined in the policy language are all atomic domains and types. The *atomic domains and types* are exactly those domains and types that are available when the system is first started, before any policy extensions have taken place. *Composite domains and types* are those domains and types that are built out of existing domains and types when the system policy is extended. In order to define a notation for the composite domains and types, we assume that '/' is a character that does not occur in the representation of any atomic domain or type. A composite domain or type is represented through a path notation

$d_1/\ldots/d_n$

or

$t_1/\ldots/t_n$.

Here d_1 and t_1 are atomic domains and types respectively. For each $k>1$,

$d_1/\ldots/d_k$

represents a new domain that has been introduced through a policy extension. This new domain is based on the domain

$d_1/\ldots/d_{k-1}$

that existed before the extension. Similarly,

$t_1/\ldots/t_k$

represents a new type that has been introduced through a policy extension. This new type is based on the type

$t_1/\ldots/t_{k-1}$

that existed before the extension.

This notation has the property of maintaining a history of how new domains and types are introduced as the system policy is extended. At any given time, the domains and types that are available are those that either existed when the system was started or that were introduced by some policy extension. Initially, the only domains and types that are available are the atomic domains and types. Thus the domain, $d_1/\ldots/d_n$, is the result of n-1 extensions. In the original policy, the source of this domain was the atomic domain, d_1. The first extension adds a new domain of the form d_1/d_2. These extensions continue until the domain $d_1/\ldots/d_n$ is created.

Now we are ready to discuss the ways in which the system policy can be extended. There are three kinds of policy extensions that can occur. We can create new domains, new types, or new access modes.

5.2.2. Adding new domains. We now describe how a new domain is added to the policy. Such a policy extension is made by adding a new domain of the form

$d_1/\ldots/d_k$

to a system that already has a domain of the form

$d_1/\ldots/d_{k-1}$

We will also need a rule constraining how the access matrix will be modified. The rule that must hold is that the access modes that are allowed for the domain $d_1/\ldots/d_k$ must be a subset of the access modes that are allowed for the domain $d_1/\ldots/d_{k-1}$. Any change to the access matrix that respects this rule is allowed.

Another constraint on the extended policy involves the authorization checks that must occur when a subject is created in the new domain. These checks ensure that an untrusted program cannot simply load some code into a domain that confers privilege without some type of check. The rule that must be followed here is that the checks required for loading code into the domain $d_1/\ldots/d_k$ must be at least as stringent as the checks required for loading code into the domain $d_1/\ldots/d_{k-1}$.

To show that this policy extension is conservative, we must demonstrate a mapping of the new policy and all the allowed events into the existing policy and the events that were allowed there. The key point that makes the definition of this mapping possible is that the allowed access modes for the domain $d_1/\ldots/d_k$ are at least as restrictive as the allowed access modes for the domain $d_1/\ldots/d_{k-1}$. Thus the mapping that takes the domain $d_1/\ldots/d_k$ to the domain $d_1/\ldots/d_{k-1}$ will map the events in the extended policy to events in the existing policy.

New domains are added to the system in order to confine a program that would otherwise have more privilege than it needs or to allow a program to create a domain for itself that will allow it to avoid tampering by attackers.

5.2.3. Adding new types. New types are introduced in much the same way that new domains are introduced. Such a policy extension is made by adding a new type of the form

$t_1/\ldots/t_k$

to a system that already has a type of the form

$t_1/\ldots/t_{k-1}$

As with the extension that adds a new domain, we must have a constraint on the access matrix. The constraint is that the allowed access modes that domains have to the new type must be a subset of the allowed access modes that domains have to the type on which it is built.

New types are added to a system in order to provide a confined program with objects that it can access without damaging critical files or to give a program that is creating a protected subsystem objects that it can use that cannot be tampered with.

To show that this policy extension is conservative, we must demonstrate a mapping of the new policy and all the allowed events into the existing policy and the events that were allowed there. The mapping that we will use maps the type $t_1/\ldots/t_k$ to the type $t_1/\ldots/t_{k-1}$. The mapping is conservative because the access that subjects can have to a type $t_1/\ldots/t_k$ is a subset of the access that subjects can have to the type $t_1/\ldots/t_{k-1}$.

5.2.4. Adding new access modes. New access modes are introduced when a new kind of object is introduced. For example, suppose a new extension is introduced that implements a distributed file system. This extension introduces a new kind of object into the system. It may be that this extension introduces new access modes for the new kind of object representing different cache coherency policies for reading and writing. This new extension may have different versions of the read and write access modes representing the different cache coherency policies. The rule that this mapping has to follow is that the new domain/type access matrix must be the same as the old access matrix except that the new access modes may be added in arbitrarily throughout the matrix to regulate access to the new object kind.

To show that this policy extension is conservative, we must demonstrate a mapping of the new policy and all the allowed events into the existing policy and the events that were allowed there. For this policy extension, the new kind of object has no analogue in the existing policy. Access operations in the new policy merely map to inter-process communication in the existing policy. Therefore, accessing the new kind of object with the new access modes maps to a no-op in the existing policy. Thus the requirement that the access matrices differ only in that the new access modes are added to the access matrix is sufficient for the policy to be conservative.

5.3. Examples

We now give two examples of policy extensions. The first example involves the problem of confining a potentially hostile application to a domain where it cannot damage user files and applications. In order to keep the example simple, we start with a trivial domain/type policy that can be represented by the following access matrix:

	Initial Type
Initial Domain	rwx

This policy states that all subjects will live in a domain that has read, write, and execute access to any subject. We now have the problem that a new subject is going to be introduced that is potentially hostile. We do not want to put this new subject in the 'Initial Domain' because it would be able to corrupt arbitrary files in that domain. We therefore create a new domain and type.

We first create the new type 'Initial Type/Corruptible'.

	Initial Type	Initial Type/Corruptible
Initial Domain	rwx	rwx

All objects of the new type will be susceptible to attack from the hostile subject when it is created. Note that the each set of access modes to the new type for each domain is a subset of the set of access modes for each domain to an old type, in this case, the 'Initial Type'. Now we create the new domain in which the potentially hostile subject will live:

	Initial Type	Initial Type/Corruptible
Initial Domain	rwx	rwx
Initial Domain/Hostile	-	rwx

In this new policy, we have a domain 'Initial Domain/Hostile' where we can place the new subject. From this domain, the new subject will have no access to objects of type 'Initial Type'. It will in contrast have full access to objects of type 'Initial Type/Corruptible' which it can share with subjects in the domain 'Initial Domain'. The set of access modes for the new domain to each type is a subset of the access modes for an original domain, 'Initial Domain', to each type. Similar examples can be constructed to illustrate protection of middleware and least privilege.

As a slightly more complicated example, we consider a device driver that needs access to some, but not all, kernel resources. We start with a policy that models two levels of privilege:

	Protected	Unprotected
Trusted	rwx	rwx
Untrusted	-	rwx

We now want to introduce a device driver that needs some privilege but not all the privilege needed by trusted applications. After two extensions ('Protected' to 'Protected/Device Resource' and 'Trusted' to 'Trusted/Device Driver') we arrive at the following policy:

	Protected	Unprotected	Protected/ Device Resource
Trusted	rwx	rwx	rwx
Untrusted	-	rwx	-
Trusted/ Device Driver	-	rwx	rwx

Here the device driver will have access only to those protected objects that are identified as device driver resources.

6. Conclusions

The results of the EXOS project lead us to conclude that a significant degree of security functionality can be introduced into extensible operating systems that support the paradigm of user-oriented extensibility. This security functionality can be flexibly configured to the security needs associated with the dynamic introduction of code extensions to the kernel that may originate from untrustworthy user software. Just as these extensions may be introduced at the procedure call level, the granularity of security checking can be tailored to security objectives based upon interception and access mediation at that same level, including both procedure visitation as well as parameter validation.

We based our procedure visitation and parameter validation policy enforcement on the SPIN security work being done at the University of Washington. The University of Washington security work is based on a separation of the policy implementation into a policy-neutral Enforcement Manager and a replaceable Policy Manager that defines the policy for the system. The EXOS project validated this work by developing a very general policy manager that supports the MLS, RBAC, Chinese Wall, ACL, and DTE policy models. A conclusion of this phase of the project is that separation of policy mediation and enforcement in extensible systems is both important and practical, facilitating introduction of policies dynamically established over the lifetime of the system. Policy-specific mediation functionality can be separated from enforcement functionality via policy-neutral security protocols.

Procedure-call interposition can be augmented with confinement and least privilege controls based upon an implementation of security domains that further extends supported security functionality. The potential exposure of kernel-level system service functionality to misbehaving kernel extensions resulting from single address space execution environments such as SPIN suggests additional security support that regulates control transfer and data access across security domain boundaries. These confinement and least privilege controls define mandatory system-wide protection that complements the existing language-based protection mechanisms of the SPIN system.

To support confinement and least privilege, we designed and implemented a software segmentation support mechanism that we integrated into the SPIN system security architecture that could be used to provide a degree of kernel-component-to-kernel-component as well as system-to-extension functional isolation. We designed this mechanism to implement a form of software segments that would contain and confine language-distinguished programming elements (modules). This software segment mechanism was designed so as to be implementable with varying levels of dependence on hardware segmentation mechanisms for performance enhancement. In our implementation, we utilized a limited degree of hardware segmentation support for efficiency reasons to speed up the software segment lookup process, but could have extended our implementation to fully exploit hardware segmentation for kernel-level protection purposes had we so wished.

In our consideration of appropriate security policy support for extensible systems, we concluded that policy specification as well as implementation support for specific foundational security policy models is important to the expression, implementation, and eventual evaluation of security objectives and their realization in security software. In particular, the security policy should be specified by the security administrator in a format that is understandable with respect to the security objectives of a given policy. This is best accomplished if the policy can be stated in terms of the primitives of a specific policy model. Furthermore, it is advantageous to preserve the encompassing policy model context all the way through to the policy-dependent components of the security architecture, in our case, the Policy Manager. The advantages include the ease of implementing more complicated relationships between policy components such as the dominance relationships inherent in lattice-based polices such as MLS and the temporal relationships in policies such as Chinese Wall, as well as the facilitation of the assurance of the correctness of Policy Manager mediation algorithms. This distinguishes our approach from architectures where the policy is translated to some intermediate form in which the context of the original policy is lost. To this end, our policy language as well as our Policy Manager directly supports MLS, RBAC, Chinese Wall, ACL, and DTE policy models as opposed to an intermediate policy language such as Adage.

The increasing emphasis on extensible system software, of which the SPIN system is only one example, suggests a requirement for security policy extensibility beyond that which might be envisioned for a more traditional operating system. As systems become more dynamic in their composition and the functionality they export to user application software, the need for safe, dynamic policy extensibility becomes more evident. It is highly likely that with an extensible operating system such as SPIN as well as other extensible systems, the introduction of middleware or new object managers would be a frequent occurrence and that such software would require extension of existing policy to support the enhanced system functionality. Our work on the EXOS project demonstrates that a useful policy extensibility strategy can be devised that supports safe kernel security policy extensibility in support of untrusted software. We investigated the issue of safe policy extensibility and identified a conservative approach to adding new policy elements to existing policy for a domain-type policy model that we feel would be especially appropriate for extensible operating systems.

Acknowledgements

We would like to thank Keith Lee for his skill in implementing portions of the policy manager. This work was supported under DARPA contract #F30602-97-C-0258.

7. References

[1] L. Badger, D. Sterne, D. Sherman, K.Walker, and S. Haghihat, "Practical domain and type enforcement for UNIX", *Proc. of the 1995 IEEE Symposium on Security and Privacy*, Oakland, CA, May 1995, p. 66.

[2] Brian Bershad, Stefan Savage, Przemyslaw Pardyak, Emin Gun Sirer, David Becker, Marc Fiucyzynski, Craig Chambers, and Susan Eggers, "Extensibility, safety and performance in the SPIN operating system", *Proc. of the 15th ACM Symposium on Operating Systems Principles (SOSP '95)*, Copper Mountain Resort, CO, December 1995, pp. 267-284.

[3] W.E. Boebert and R.Y. Kain, "A practical alternative to hierarchical integrity policies", *Proc. of the 8th National Computer Security Conference*, 1985, pp. 18-27.

[4] A. Bricker, M. Gien, M. Guillemont, J. Lipkis, D. Orr, and M. Rozier, "A new look at micro-kernel-based UNIX operating systems: Lessons in performance and compatibility", *Proc. of the EurOpen Spring '91 Conference*, Tromsoe, Norway, May 1991.

[5] Department of Defense National Computer Security Center, *DoD Trusted Computer System Evaluation Criteria*, 1985 (Orange Book).

[6] Yasuhiro Endo, James Gwertzman, Margo Seltzer, Christopher Small, Keith A. Smith, and Diane Tang, "VINO: The 1994 Fall harvest", Harvard Computer Center for Research in Computing Technology Technical Report TR-34-94.

[7] Dawson R. Engler, M. Frans Kaashoek, and James O'Toole Jr., "Exokernel: An operating system architecture for application-level resource management", *Proc. of the 15th ACM Symposium on Operating Systems Principles (SOSP '95)*, Copper Mountain Resort, CO, December 1995, pp. 251-266.

[8] Paul Green, "Multics virtual memory - tutorial and reflections", Technical Report available on the Internet at: http://www.cs.berkeley.edu/~brewer/cs262/Green.ps.

[9] Robert Grimm and Brain N. Bershad, "Providing policy-neutral and transparent access control in extensible systems", University of Washington Department of Computer Science and Engineering Technical Report UT-CSE-98-02-02, February 1998.

[10] James G. Mitchell, Jonathan J. Gibbons, Graham Hamilton, Peter B. Kessler, Yousef A. Khalidi, Panos Kougiouris, Peter W. Madany, Michael N. Nelson, Michael L. Powell, and Sanjay R. Radia, "An overview of the Spring system", *Proc. of Compcon*, February 1994.

[11] Elliott I. Organick, *The Multics System: An Examination of Its Structure*, MIT Press, 1980.

[12] Przemyslaw Pardyak and Brian N. Bershad, "Dynamic binding for an extensible system", *Proc. of the Second USENIX Symposium on Operating Systems Design and Implementation (OSDI)*, Seattle, WA, October 1996, pp. 201-212.

[13] Richard Rashid, Robert Baron, Alessandro Forin, David Golub, Michael Jones, Daniel Julin, Douglas Orr, and Richard Sanzi, "Mach: A foundation for open systems", *Proc. of the Second Workshop on Workstation Operating Systems (WWOS2)*, September 1989.

[14] Ravi S. Sandhu, "A lattice interpretation of the Chinese Wall policy", *Proc. of the 15th NIST-NCSC National Computer Security Conference*, Baltimore, MD, October 1992, p. 221.

[15] Ravi S. Sandhu, Edward J. Coyne, Hal L. Feinstein, and Charles E. Youman, "Role-based access control: A multi-dimensional view", *Proc. of the Tenth Annual Computer Security Applications Conference*, Orlando, FL, December 1994, pp. 54-62.

[16] Ravi S. Sandhu, "Role-based access control", Laboratory for Information Security Technology Technical Report, September 1997.

[17] Stefan Savage and Brian N. Bershad, "Issues in the design of an extensible operating system", *Proc. of the First USENIX Symposium on Operating Systems Design and Implementation (OSDI)*, Seattle, WA, November 1994.

[18] O.S. Saydjari, J. Beckman, and J. Leaman, "LOCKing computers securely", *Proc. of the 10th DoD/NBS Computer Security Conference*, Gaithersburg, MD, September 1987, p. 129.

[19] K. Walker, D. Sterne, L. Badger, K.A. Oostendorp, M.J. Petkac, and D.L. Sherman, "Confining root programs with domain and type enforcement", *Proc. of the 1996 USENIX UNIX Security Symposium*, San Jose, CA, July 1996.

[20] Mary Ellen Zurko and Richard T. Simon, "User-centered security", *Proc. of New Security Paradigms Workshop*, September 1996.

Secure Virtual Enclaves: Supporting Coalition Use of Distributed Application Technologies

Deborah Shands, Richard Yee, Jay Jacobs
NAI Labs at Network Associates
{dshands, ryee, jjacobs}@nai.com

E. John Sebes
Kroll-O'Gara
ejs@securify.com

Abstract

The Secure Virtual Enclaves (SVE) collaboration infrastructure allows multiple organizations to share their distributed application objects, while respecting organizational autonomy over local resources. The infrastructure is transparent to applications, which may be accessed via a web server, or may be based on Java RMI, or Microsoft's DCOM. The SVE infrastructure is implemented in middleware, with no modifications to COTS operating systems or network protocols. The system enables dynamic updates to security policies to support changes in both coalition membership and participants' perception of risks. While the prototype demonstrates fine-grained access control for secure collaborative computing, we have identified significant issues that remain to be addressed, particularly in the area of policy development, before such collaboration will be convenient. The SVE infrastructure offers a platform and conceptual basis for further exploration of these issues and experimentation with new solutions.

1. Introduction

The need for mechanisms to allow organizations to collaborate securely is recognized in many environments. Military alliances and joint task forces are formed to accomplish a common goal and the participating organizations engage in some form of distributed collaborative planning. After a natural disaster, crisis management collaborations are formed from an often disjoint collection of disaster/incident response teams (e.g., medical personnel, local police, engineers). In a commercial environment, companies outsource some of their operations (e.g., payroll, data center operations), employ contractors to perform certain tasks, or offer some of their data to customers. They may also form consortiums to perform collaborative research, develop standards, or battle competitors. There are at least two common elements in any of the resulting scenarios: (1) the collaborating organizations have *limited trust* in one another, and (2) the coalitions are *dynamic*.

Because the organizations may have competitive or even adversarial relationships, they do not completely trust one another. They are, however, motivated by a common goal to share some of their resources. Their trust in one another and the limits of that trust are generally specified through some extra-technological means, such as contracts, treaties, or memoranda of agreement.

Coalitions are likely to be dynamic, in that organizations may join or leave over the lifetime of the collaboration. An organization's level of trust in its partners may also change with time, impacting the degree of resource sharing–local resources may be added or removed from the sharing arrangement. The mode of access to a particular resource may also change over time.

We believe that the degree of dynamism necessary to support coalition creation, evolution, and eventual dissolution preclude a hardware-intensive solution (e.g., setting up a new, joint network). Virtual Private Networks (VPNs) can authenticate individual users, but do not support access controls on fine-grained objects (e.g., a Java interface/method). VPN-based solutions are also relatively static, as adding new coalition members requires some manual reconfiguration. To support fine-grained access controls and dynamic changes to coalition membership, the SVE project focused on software solutions.

The goal of the Secure Virtual Enclaves (SVE) project was to develop software technology to enable multiple enclaves to engage in controlled collaborative computing using distributed applications, while retaining organizational autonomy over local resources. By *enclave*, we mean a collection of computers and networks managed by the same organization and subject to the same security policy. *Collaboration* occurs when principals in partner enclaves are permitted to access selected local resources. Security *controls* are necessary to ensure that collaborators get only the intended access to local resources. Local *autonomy* is exceedingly important, as an organization's willingness to share its resources with others is influenced by the degree of control it retains over those resources. An organization that knows it can

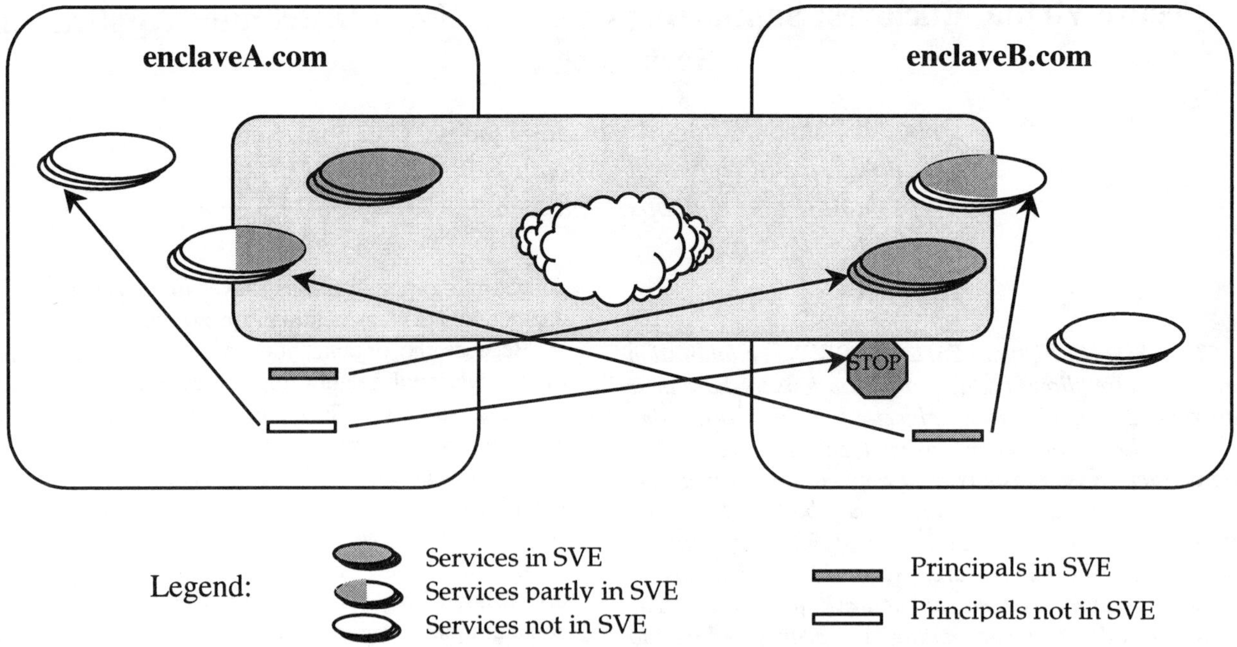

Figure 1: SVE concept of operation

unilaterally choose to withdraw its resources from a coalition at any time may be more willing to collaborate. Finally, by *distributed applications*, we mean applications built primarily on middleware infrastructures that support program and/or data object distribution. Examples include Java RMI, CORBA, Microsoft DCOM, and Enterprise JavaBeans. We may also include ordinary file system objects and resources accessible via a web server (e.g., HTML files).

Our approach was further bounded by the following constraints: the SVE infrastructure should be transparent to applications and based on commercially available operating systems and open networks. This forced us to work primarily in the realm of middleware, which, given our focus on distributed application technologies, was appropriate. We also emphasized the "Virtual" in "Secure Virtual Enclaves," deciding against solutions that replicate resources and synchronize multiple copies. Though these approaches can improve the fault tolerance of a system and availability of its resources, they introduce undesirable system complexity. SVE resources remain within, and under the protection of, their local enclaves, while mechanisms are introduced to control accesses to these resources by external subjects. There is one caveat to our assumption about application transparency: we must be able to authenticate the identity of a requesting principal, so we insist that application traffic use some authentication mechanism.

Based on our goals and constraints, we designed an SVE to work in the following way: two or more organizations decide, through extra-technological means, to collaborate by sharing some of their local resources. The administrator of one of those organizations begins the technical and administrative process of naming and creating an SVE, and noting which other enclaves are trusted to join the collaboration. The creating enclave becomes the sole member of the SVE. The administrator then creates a resource access policy to identify the local resources to be shared and the local principals that will be authorized to access SVE resources, both local and foreign. The administrators of the remaining enclaves follow an administrative process to request to join the established SVE. After a join request has been submitted to the local SVE system, the remainder of the process is automatic—SVE system components of one enclave communicate with SVE system components of the other enclaves to establish the desired coalition membership. Finally, clients (acting for authorized principals) may begin to access SVE resources residing in any of the member enclaves.

Figure 1 shows the general SVE concept of operations. Two enclaves are participating in an (already formed) SVE. The SVE, shown in the shaded area in the center of the figure, is a projection of access rights for principals belonging to the SVE. Thus, the SVE identifies a distributed collection of related resources, along with the principals that are authorized to access those resources.

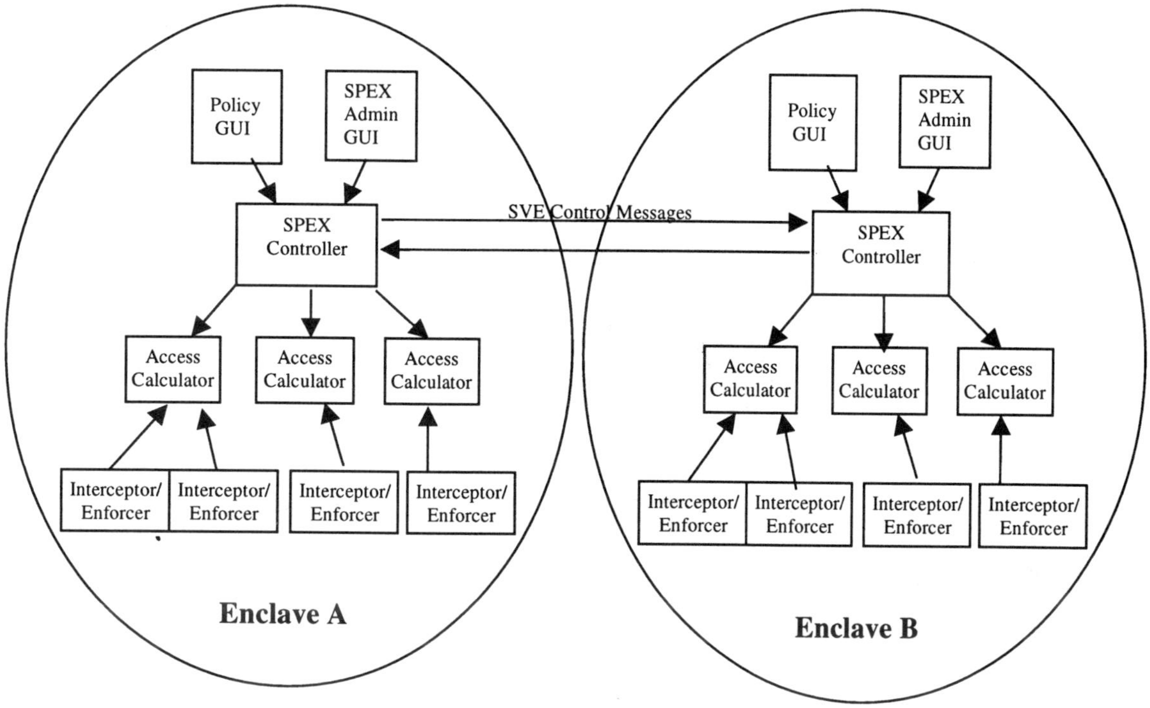

Figure 2: SVE component architecture

The resources and principals remain unchanged by the introduction of the SVE infrastructure.

2. Architecture

The SVE infrastructure consists of components that create, distribute, and enforce security policy, as shown in Figure 2. Each "egg" in the figure represents an enclave. The components in the upper halves of the eggs (policy GUI, administration GUI, and SPEX controller) are responsible for creating, maintaining, and distributing resource access policy, as well as administering SVE operations. The remaining components of the architecture interpret and enforce access policy.

The GUI components provide policy and configuration facilities to a local enclave administrator. The *Policy GUI* allows the administrator to develop and maintain access policy for local enclave resources. Through the policy tool, the administrator submits new policies or incremental updates. The *SVE Policy Exchange (SPEX) Controller* propagates policies within a local enclave to the SVE policy enforcement components, and to other SVE member enclaves. The SPEX controller also accepts SVE control commands from the *SPEX Administration GUI*, and participates in SVE control protocols (e.g., join, leave). The *Interceptor/Enforcers* capture client requests for server resources, query a local *Access Calculator* for an access decision, and enforce the decision by either allowing the request to proceed as usual, or dropping the request and returning an error message to the client. The access calculator encapsulates a local SVE resource access policy, and responds to access queries from local interceptor/enforcers. The SPEX controller provides asynchronous policy updates (either full or partial) to local access calculators.

Figure 3 shows a typical client-server application, with SVE interceptor/enforcers. The use of the SVE infrastructure is invisible to the application client, application server, and application developer. Interceptor/enforcers must, however, be installed between external clients and internal servers, via either gateways or server modifications. The remainder of the SVE infrastructure does not communicate with or affect the workings of the application.

2.1. Resource access policy

The administrator of an enclave creates and maintains the local resource access policy for the SVE by identifying the resources that may be shared and the principals that may participate. Since an enclave may choose to belong to multiple SVEs, there may be multiple local policies (one per SVE) in force at any given time. In fact, there may be several more policies lying dormant in

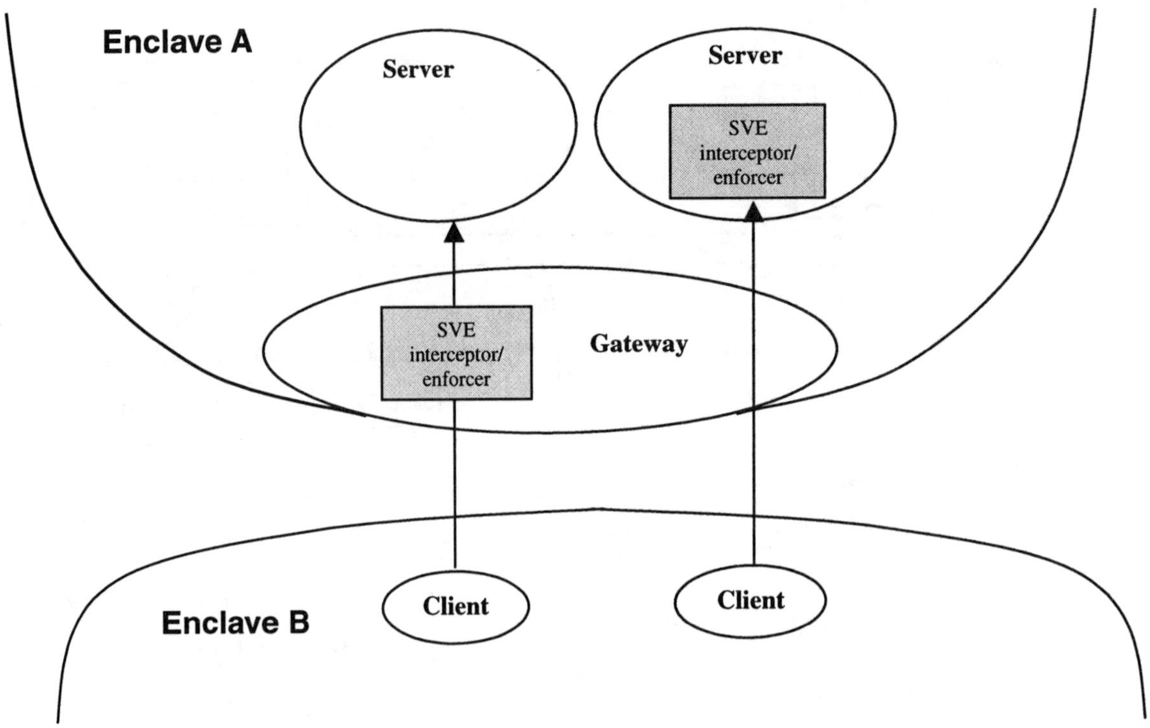

Figure 3: SVE client-server communication

case changing conditions (and levels of trust) compel the administrator to replace one of the active policies. In this section, we will describe a single SVE policy, local to the enclave in which it was created.

The SVE policy language uses concepts familiar from Domain and Type Enforcement (DTE) [5], which defines policy in terms of the access rights of equivalence classes of subjects to equivalence classes of objects. An object is a resource accessed by software. A subject is a software component that accesses resources on behalf of a principal. Principals are persons or persistent programs (such as servers). In DTE, objects are grouped into equivalence classes called *types*, and subjects are grouped into equivalence classes called *domains*.

SVE policies have four components: type definition rules, which map distributed objects into types; domain derivation rules, which map subjects into domains; a domain × type access matrix; and, potentially, a collection of additional access constraints. Figure 4 shows an example SVE resource access policy.

A type definition identifies a collection of resources that will be treated identically for access control purposes. For many object-oriented distributed systems, a resource is the pair (object interface, method). In systems that support distinguishable objects (e.g., Enterprise Java Beans), this could mean the more specific pair (named object, method). The example shows definitions for three types of resources: specification resources, source code resources, and financial resources. Thus, specification-related resources to be controlled include html files found in the `specs` directory at `eweb.com`, as well as the `view` method of the `SpecAdminI` interface. The "`_t`" appended to each type name is a mnemonic device, rather than a syntactic requirement.

Domain derivation rules uniquely identify principals on whose behalf a subject will be assigned to a particular domain. Domains are, essentially, role specifications. In the example, `frank` and `jane` (both of Acme, Inc.) are assigned to the engineering and accounting domains (roles) respectively. `sue` (of Toyco, Inc.) is assigned to the engineering domain. `frank` and `sue`, both engineers, will have identical access to resources as long as Acme and Toyco collaborate within an SVE. `jane` will also have access to SVE resources, but because she is an accountant, rather than an engineer, it is likely that her resource access permissions will be different from `frank`'s and `sue`'s. Thus, the semantics of "domain" within the context of Domain and Type Enforcement is very different from the "domain" addressed in network terminology. The example specifications show an email address (found within an X.509 certificate) prepended with a unique name for the user's home enclave (e.g., Acme). This name identifies the SPEX controller that issued the domain derivation rule. The "`_d`" appended to

Type Definitions

specifications_t = https://eweb.com/specs/*,
rmi://com.eweb.SpecAdminI/com.eweb.JavaSpec view(int)
source_code_t = rmi://com.eweb.sourceI/*
financials_t = dcom://123-456-789/321-654/1

Domain Derivation Rules (Principal Recognition Rules)

Acme!frank@acme.com = engineer_d
Acme!jane@acme.com = accountant_d
Toyco!sue@toyco.com = engineer_d

Access Matrix

engineer_d = specifications_t, source_code_t
accountant_d = financials_t+TimeInterval!900!1700!M!F

Figure 4: Example SVE resource access policy

each domain name is a mnemonic device, rather than a syntactic requirement.

The access matrix shows the types of resources to which subjects in a given domain are permitted access. In our example, engineers may access source code and specifications, while accountants may access financials.

Finally, an entry of the access matrix may be decorated with a *constraint*, identifying conditions that must be met before the subject (i.e., domain) can access the object (i.e., type). This allows us to extend our policies to address restrictions that may not be conveniently expressible in an ordinary access matrix. In the example, a time interval constraint is placed on accountants' access to financial data: subjects in the `accountant_d` domain may access `financials_t` type objects only during the hours between 9am and 5pm, Monday through Friday.

2.2. Policy distribution

A complete SVE policy, comprised of the four components described above, is necessary for an access control decision. An access calculator must, therefore, retain a complete policy and receive policy updates from its local SPEX controller. Some portions of the policy must also be shared among SVE member enclaves. In this section, we examine the need for inter-enclave policy distribution and discuss the role of the SVE Policy Exchange (SPEX) controller in the architecture.

An enclave defines its resource access policy, makes access control decisions, and enforces those decisions for each of its local resources, including resources it shares with other enclaves via an SVE. When a client from enclave **A** requests a resource on a server in enclave **B**, the principal responsible for the request must be identified by enclave **B** and the requestor mapped into an appropriate domain in order to determine whether the access should be permitted. **B** cannot recognize the principal behind the client without authentication and role information provided by **A**.

Information describing the requestor might be provided in a variety of ways. At one end of the spectrum, the client might present a credential at request time, containing the domain to which the requestor should be assigned. A certifying authority for **A** would have signed the credential. In this case, the client carries all of the data needed by **B**. At the other end of the spectrum, **A** might send all of its domain derivation rules to **B** in advance. When an **A** client makes a request, **B** would authenticate the identity of the responsible principal, then apply the appropriate rules to establish a domain for the requestor. In this case, the client carries almost none of the data needed by **B**. A range of hybrid solutions is possible, with some data delivered by the requesting client and some delivered in bulk, in advance.

There are advantages and disadvantages to any of these approaches. For the SVE architecture, we have chosen the approach in which an enclave sends domain derivation rules for its principals to other collaborating enclaves and only the identification and authentication data are carried by the client request. Because these rules allow enclaves to recognize "foreign" principals, we have renamed them *principal recognition rules*. Principal recognition rules are the only SVE policy component that must be exchanged among SVE member enclaves. All other policy components (resource-type mappings, access matrix, and constraints) remain within their local enclave. This provides the local administrator with the unilateral ability to control which of the local resources are shared – the administrator can change the resource-type mappings, the access matrix, or the constraints for the local resources without consulting other SVE member enclaves.

As a result, an enclave retains the right to authorize both local and foreign principals' access to local resources. In doing so, however, the enclave administrator does not authorize access directly to those principals. Rather, access is granted to a domain (role). Principal recognition rules (both local and foreign) provide the basis for determining membership in that domain. Thus, an enclave determines the roles that its principals assume within the coalition by providing appropriate principal recognition rules. This approach improves the scalability of the coalition-formation process, in that an organization need not have a priori knowledge of every foreign principal that might eventually participate. However, an enclave must trust its coalition partners to provide accurate and appropriate principal recognition rules. Note that the trust requirement for accurate domain placement does not diminish if clients carry domain designations or other forms of authorization in their credentials.

The SPEX component of the SVE architecture is responsible for distributing access policy to local access calculators, as well as communicating the local principal recognition rules to the SPEX controllers of other SVE members. Each enclave must have a SPEX controller. Since access calculators must be able to recognize principals from other enclaves, an aggregate of local and foreign principal recognition rules must be created by the local SPEX controller. The aggregate principal recognition rules, along with the local type definitions, access matrix, and constraints, are delivered to the access calculators by the SPEX controller. Local principal recognition rules are published to other SVE member enclaves via SPEX-to-SPEX communication when the enclave joins an SVE or when changes are made to the responsibilities of local personnel. When foreign enclaves update and publish their principal recognition rules, the SPEX controllers of SVE member enclaves deliver those updates to their local access calculators, without human administrative action.

2.3. Access calculation

The access calculators in the SVE architecture are responsible for deciding whether a given access is permissible. Each enclave contains one or more access calculators, though for performance reasons, we expect that an access calculator would be deployed on each host that supports SVE-sharable resources.

An access calculator presents two interfaces to components of the SVE infrastructure: an access decision interface to accept and respond to requests from interceptor/enforcers, and a policy update interface to the local SPEX controller. An interceptor/enforcer queries an access calculator for a policy decision, while the local SPEX controller pushes policy updates into the access calculator. The access policy is completely contained within the access calculator, so decision-making is accomplished strictly locally.

Access calculation is a four step process: domain derivation, type derivation, access matrix check, and constraint check. Domain derivation is accomplished using identity data, extracted from an authenticated credential, in combination with principal recognition rules, provided by the principal's "home" enclave. Type derivation is accomplished using resource request data, taken by the application interceptor, in combination with type derivation rules defined by the local SVE policy.

The premise underlying SVE access calculation is that each enclave cannot be expected to define access policy based on the individual identities of foreign principals. In order to grant access to foreign principals in a timely and scalable way, principals must be grouped into domains by their home enclaves. These groupings embody the roles of those individuals and the degree to which the individuals are trusted by their home enclaves.

In the SVE system, access authorization is granted equally to all principals (both local and foreign) represented by a domain, rather than to an individual principal. As in Figure 4, individuals acting in the `engineer_d` role will have access to the same SVE resources, regardless of their home enclaves or the location of the resources they access. In our prototype, these authorizations are represented by a domain-type access matrix. However, any policy representation that assigns access authorization to groups or roles could be used with the SVE system. If the access request is permissible according to the access matrix, then any constraints of the access are checked. The boolean result is returned to the interceptor/enforcer that initiated the query.

2.4. Request interception and policy enforcement

SVE interceptor/enforcers perform the tasks of capturing a request for a distributed system object, extracting data to identify both the target object and the requestor, forwarding this data to a local access calculator, and enforcing the access calculator's decision. One interceptor/enforcer may differ from another quite drastically, as their distributed application technologies can differ drastically from one another. We will discuss some of those implementation issues in Section 3, but will briefly identify two important classes of interceptor/enforcers, both of which are supported by the SVE architecture.

When server resources are to be protected, request interception and policy enforcement for distributed application technologies may be implemented via either a protocol gateway or a server-resident interceptor. Though the use of a gateway obviates the need for server modifications, information regarding a request is often incomplete "on the wire". In the case of distributed application technologies, this is often evident in that target identification (e.g., method call) is not resolved until the server receives the request. Gateways can sometimes be constructed to call out to a server for additional context data to circumvent this issue. When this is undesirable, when intra-enclave access control is necessary, or when client-server communication uses end-to-end encryption, server-resident policy enforcement may be the preferred approach. Note also that layered defenses can be built with combinations of gateway and server-resident interceptor/enforcers. The SVE infrastructure can support both types of interceptor/enforcer.

2.5. SVE administration

The administrator of an enclave is responsible not only for defining the enclave's resource access policies for each of the SVEs that the enclave joins, but also for performing local SVE administration tasks. In particular, the administrator must represent the enclave's trust relationships with foreign enclaves by identifying the list of enclaves with which it intends to collaborate in an SVE. This establishes an enclave-level trust policy that determines which foreign enclaves may have access to local resources. Both trust policy and resource access policy are managed by the enclave administrator via GUIs that communicate with the local SPEX controller. SVE administration commands for establishing trust policy, as well as creating, joining, and leaving an SVE, are initiated by an administrator via the GUI. Once initiated, these processes are carried out automatically by communicating SPEX controllers.

3. Implementation

The SVE infrastructure is primarily a Java-based architecture, tested on Sun Solaris, Windows NT 4.0, and Linux. Intra-enclave communication among SVE components is done via Java RMI. Inter-enclave communication is primarily accomplished using the group communication facilities provided by the Ensemble system [7], via the JavaGroups interface [3]. Except for the platform-specific binaries required by Ensemble, the SVE components are platform independent. Most of the components were engineered from scratch, using Java 1.1, with the considerable exception of the interceptor/enforcer code.

We chose to support distributed applications based on the following commonly used technologies: Java RMI, Microsoft's DCOM, and two web servers—Sun's Java Web Server, and Microsoft's IIS. In order to authenticate the identity of a requesting principal, application traffic must use an authentication mechanism. Since Java RMI and HTTP traffic can run over SSL, we use data from SSL-carried X.509 certificates to identify both web object requestors and Java RMI object requestors. The DCOM protocol does not currently run over SSL, but we can extract identity data from the Windows NT access token (created for the user at logon) which is carried with the DCOM request. If client identity cannot be established, (e.g., when a web or RMI application is not run over SSL, or a DCOM application allows an anonymous request) the SVE infrastructure will not permit access to resources it controls.

3.1. SVE Policy Exchange (SPEX) Controller

The SPEX controller is a multithreaded Java-based server that implements three interfaces for SVE administration and policy distribution: an administrative interface, which communicates with the administrative GUIs; an intra-enclave policy distribution interface; and an inter-enclave communication interface. The controller also implements repositories for resource access policies and SVE administrative data.

SVE administration (SVE policy updates and control requests) is handled through the SPEX controller's administrative interface. The administrative GUI communicates with the SPEX controller via Java RMI. At the administrator's request, the administrative GUI forwards SVE control requests (e.g., create a new SVE, join an existing SVE) to the SPEX controller. Changes in the enclave's state are pushed back to the administrative GUI for display to the administrator.

Intra-enclave policy distribution is handled through the policy service interface of the SPEX controller. Access calculators register via Java RMI as subscribers to the policy update mechanism. Policy changes originating at the policy GUI are accepted by the SPEX controller. Access calculators must enforce access policy for **all** of the SVEs to which the local enclave belongs. Thus, the SPEX controller must aggregate the local principal recognition rules, type mappings, access matrices, and constraints written for each SVE to which the local enclave belongs. Principal recognition rules received from foreign SVE member enclaves must also be aggregated with the local rules. This aggregate policy is propagated by either full or incremental updates to the subscribed calculators whenever the policy GUI gets an update from an administrator or when foreign members update their principal recognition rules.

Inter-enclave control messages and policy distribution are handled via the communication service interface of the SPEX controller. When an enclave attempts to join an existing SVE, its SPEX controller makes an RMI request to a liaison (a SPEX controller for an enclave already belonging to the SVE). The liaison launches a voting request by sending a message object through the JavaGroups interface to the Ensemble communication system. All of the current SVE members receive the voting request and consult their lists of trusted collaborators. The liaison tabulates the voting results – only a unanimous positive result will allow the prospective member to join. The result is returned to the prospective member as a response to the original RMI request. If accepted, the new member subsequently takes part in the group communication and submits its principal recognition rules for the SVE.

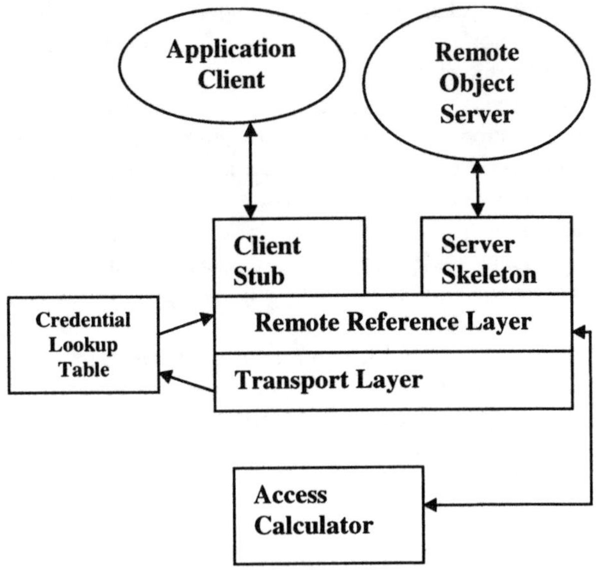

Figure 5: Java RMI interceptor

3.2. Interceptor/Enforcers

The SVE project implemented server-side (i.e., end-system-based, rather than gateway-based) interceptor/enforcers for Java RMI, Microsoft DCOM, and two web servers (Microsoft's IIS and Sun's Java Web Server). Both server-resident and gateway-based interception for CORBA requests were addressed by the Sigma project [13], a predecessor to the SVE project.

The implementation of interceptor/enforcers for a variety of distributed application technologies provided significant engineering challenges. Most of these technologies were not designed to allow for request filtering. Our interceptors are, therefore, highly implementation dependent, and vulnerable to version changes. Choosing to use gateway-based interception would have traded these problems for others. In particular, gateways must tolerate varying and evolving protocol implementations. For example, the developers of CORBA's GIOP/IIOP protocol did not take the need for boundary access control mechanisms into account when developing the original protocol specification. This decision created problems not only for firewall proxy development, but also for interoperability of different vendors' products when a gateway is involved. As mentioned in Section 2, identifying the requested resource before the request reaches the server is often exceedingly difficult as often only the server has sufficient contextual information to interpret the data carried by the protocol. We will now discuss some of the implementation challenges we faced in building server-resident interceptors for the various application technologies.

3.2.1. Java RMI. The first distributed application technology for which the SVE project implemented an interceptor/enforcer was Java RMI. Currently, Sun's RMI interface specification provides no defined application hook for intercepting client method invocation requests as they arrive at the RMI server. Consequently, providing a server-side interceptor for Java RMI required functional enhancements to Sun's RMI reference implementation.

The Sun specification describes the architecture of RMI in three layers. The topmost layer consists of the RMI stub and skeleton, which provide the client proxy and server dispatch functionality commonly found in distributed object models. The middle layer is designated as the remote reference layer and is responsible for providing specific remote invocation semantics, such as whether the remote server object will be a single object or part of a replicated object group. At the bottom is the transport layer, which is responsible for managing network connections and tracking remote server objects.

The RMI remote reference layer provides an ideal location for interceptor placement since it is considered part of the Java system API. In contrast, embedding an interceptor at the top layer would have required special skeletons to be generated for each application remote server implementation, while embedding an interceptor at the transport layer would not have provided adequate information about the invocation target. A consequence, however, of providing interception at the remote reference layer is that the interceptor is very specific to the RMI implementation. We chose to add interceptor capability specifically to Sun's RMI remote reference implementation for JDK 1.1. Figure 5 provides a high-level view of our design for RMI interception.

As shown in Figure 5, an additional complication to RMI interception was the inability to cleanly pass authentication credentials from the transport layer to the remote reference layer. The standard RMI interfaces in JDK 1.1 do not provide a means for passing credentials between layers. Thus, we were forced to implement a credential lookup table, which is shared by the transport and remote reference layers. We made an implementation-specific decision to use a thread identifier as the credential lookup key, since a single thread carries an invocation request through each layer in Sun's RMI implementation. Based on this design, the transport layer, upon receiving a client invocation request, will insert the associated credential into the lookup table using the current thread context identifier as the key. When the remote reference layer receives the same request, it will use its current thread context identifier as the lookup key for retrieving the credential associated with the request.

3.2.2. Microsoft DCOM. The second distributed application technology for which the SVE project implemented a interceptor/enforcer was Microsoft's DCOM. DCOM is the distributed specification for Microsoft's Component Object Model (COM) technology. A DCOM component can take any of three forms: a shared library (DLL), a binary executable (EXE), or a system service. A DCOM component is a collection of COM interfaces, each of which identifies methods, which are exported to applications. Using Windows NT 4.0, an access policy for DCOM components can be specified on a component-by-component basis. However, due to our goals of maximizing control and local autonomy, the SVE project required finer-grained constraints on interfaces and methods.

In order for a client to invoke a method on an interface offered by a remote DCOM component, the client must obtain an interface pointer to identify the requested object (i.e., instance of the component). Each interface of a DCOM component has a virtual table (v-table) data abstraction, which is a lookup table with pointers to method implementations. Pietrek, in [10], uses a custom DLL to modify DCOM v-tables and a kernel jump table to re-route DCOM method calls. The SVE system uses Pietrek's method of request interception. Other techniques have been developed which can be applied to the interception of DCOM requests. The method described by Balzer and Goldman in [4] replaces a portion of the application's assembly code, diverting the program to the interceptor. As we did not require the generality of the Balzer-Goldman approach, we chose to implement the simpler Pietrek method.

Once the interceptor has captured the DCOM request, it must request an access decision from an SVE access calculator. Assembly language code, such as that written for the kernel-based interceptor, cannot communicate directly with a Java-based server, such as an SVE access calculator. This problem forced us to build a bridge from the interceptor to the access calculator.

The bridge is based on COM. A COM client and a COM server are inserted in the communication path between the assembly code-based interceptor and the Java-based access calculator. The C++-based COM client is called by the interceptor code. The COM client calls the Java-based COM server. Since Microsoft's Java Virtual Machine (JVM) does not currently support RMI, we used a collection of RMI classes developed by IBM to patch the COM server. This server can then communicate with the SVE access calculator.

Our interceptor can currently handle DCOM applications in the binary executable (EXE) form. The interceptor may be extensible to shared library (DLL)-based and system service-based DCOM applications.

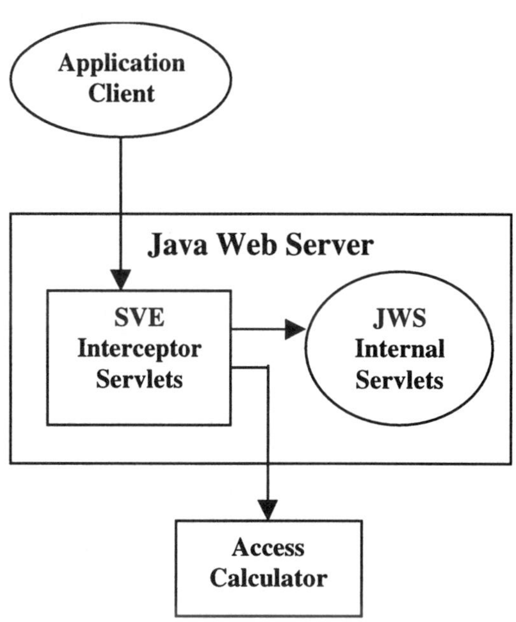

Figure 6: Java Web Server interceptor

3.2.3. Java Web Server. Web-based applications are another technology that SVE supports. We have developed two web server interceptor/enforcers, the first of which is Sun's Java Web Server. The Java Web Server (JWS) was designed with modular extensibility in mind and is built upon a server framework called the JavaServer Toolkit (JST). The JST allows developers to build network application services using the Java programming language. The types of services that can be built with the JST include established services such as HTTP and FTP, as well as application services that have yet to be created.

The JST supports the concept of a servlet, a service extension API that augments the capabilities of a particular service for customized application handling. Servlets run within a JST server as objects in support of a service and can be dynamically loaded on demand from any local or network source. The JWS uses servlets to manage all resources provided through its HTTP and HTTPS services. For example, the JWS uses a file servlet to provide its file-serving capabilities, a CGI servlet to execute any CGI-based scripts or programs, and an invoker servlet to execute custom application-specific servlets.

Providing an access control interceptor for the JWS was fairly straightforward. The JWS allows servlets to be chained together for the purpose of further augmenting service capabilities. Since the standard JWS utilizes internal servlets to manage all of its web resources, providing interception was merely a matter of inserting an interceptor servlet in front of each of the internal servlets. Thus, we were able to intercept all client requests for web resources managed by the JWS. Figure 6 provides a high-level view of our design for JWS interception.

3.2.4. Microsoft IIS Web Server. The final SVE interceptor/enforcer was developed for Microsoft's web server: the Internet Information Server (IIS). IIS uses Windows NT's Internet Server Application Programming Interface (ISAPI) as a customization interface. ISAPI is a server-side API with functionality similar to the Java servlet interface. An IIS *filter* conforms to the ISAPI, and is analogous to a Java servlet, used by the Java Web Server. Some filters are specially designated system filters, provided by Microsoft. Custom filters can be developed and added to the IIS server, as well. Both system and custom filters can be chained together to augment IIS services, with system filters being executed before any custom filters within the filter chain.

The SVE IIS interceptor is implemented within a custom filter, placed after the SSL system filter, but before other custom filters in the filter chain. The SVE interceptor makes use of an Active Server Pages (ASP) file, which should (according to the IIS documentation) allow the interceptor to forward the client request data to a local access calculator for an access decision.

Unfortunately, due to Microsoft's engineering problems with IIS 4, custom filters are unable to automatically execute ASP files before resource processing. Thus, the SVE filter is unable to automatically intercept resource requests. We worked around the problem by sacrificing transparency and manually modifying each of the IIS resources. During resource processing, the SVE filter first executes the SVE ASP file. Though this non-transparent approach would be unacceptable in an operational environment, the workaround was tolerable for experiments with SVE. We hope and expect that Microsoft will correct these problems with future releases of IIS.

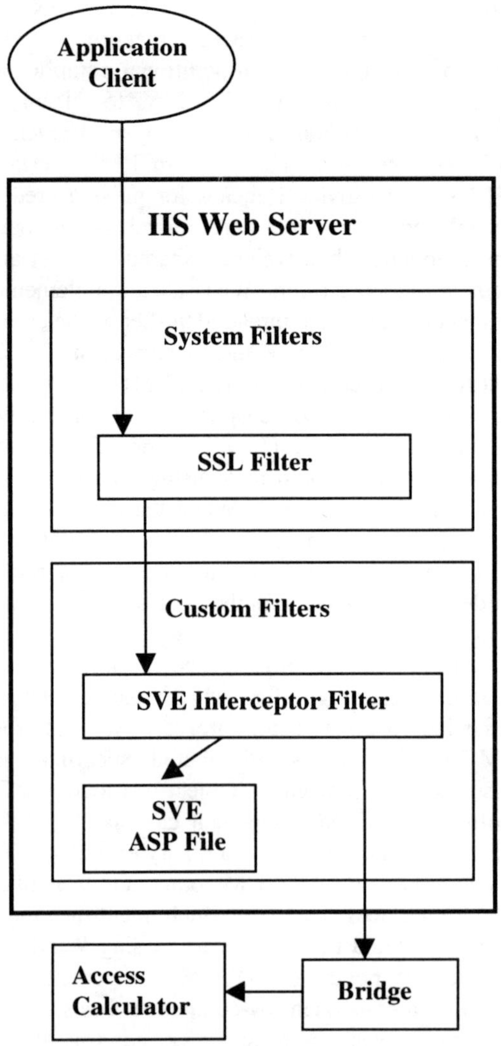

Figure 7: IIS Web Server interceptor

The lack of Microsoft JVM support for Java RMI once again forced us to develop a bridge between the ASP and the access calculator. We had hoped to use the same bridge that we built for the DCOM interceptor, but were unable to force the SVE interceptor filter to communicate directly with a COM client. We, therefore, prepended additional bridging components to enable the communication. Figure 7 provides a high-level view of IIS interception.

3.3. Communication security

To ensure that intra-enclave policy distribution takes place without danger of policy corruption or source spoofing, SVE Java RMI communication runs over SSL.

The SPEX controller authenticates the source of policy update commands and SVE administrative actions. Similarly, the access calculator ensures that the SPEX controller is, in fact, the source of a policy update. Communication between interceptor/enforcers and their access calculators has not yet been secured, due to project resource limits. To ensure that interceptor/enforcers cannot be fooled by imposters spoofing access decision results, these communications should be secured.

Inter-enclave communication has been secured to protect the integrity of policy data (in particular, principal recognition rules) and prevent source spoofing. The initial, RMI-based communication between a prospective enclave SVE member and its chosen liaison is secured using SSL. The remainder of inter-enclave SVE communication takes place using the group communication system. We had the option of using the communication security facilities provided by Ensemble (albeit not via JavaGroups); however, as we had earlier decided to insulate the SVE infrastructure from the choice of group communication facilities, we chose not to rely on Ensemble's security facilities. Instead, we chose to sign SVE message objects to provide source authentication and message data integrity.

To secure SVE communication, we used an implementation of the Java Cryptographic Extensions (JCE), produced by the Institute for Applied Information Processing and Communications (IAIK) from Graz, Austria. We used IAIK's Digital Signature Standard (DSS) implementation to ensure source authentication and data integrity for SVE messages carried over Ensemble. IAIK's iSaSiLk Java-based implementation of SSL was used to secure all of our RMI-based communication.

3.4. SVE policy

The implementation of SVE resource access policies facilitates both dynamism in policy updates and the use of a potentially broad range of policy models. An SVE policy is represented as a Java object that is created by an administrator, using the SVE policy GUI, stored and distributed via the SPEX controller, and, finally, lodged in an access calculator. The policy encapsulates both metadata (e.g., principal recognition rules, resource to type mappings, access matrix, and constraints) and rules (i.e., code) for interpreting that metadata. The policy object provides an update interface to allow the policy metadata to be changed, and an interface to respond to access queries from interceptor/enforcers. An active access policy resides in an access calculator. The calculator delegates any access decision requests to the policy object, which executes code to interpret its metadata and returns its decision to the calculator. The calculator responds to the interceptor/enforcer that initiated the query.

Because it is fully encapsulated, the policy can be passed within the SVE infrastructure and handled as an opaque object by most of the components. The access calculator provides an architectural placeholder in the SVE infrastructure to insulate interceptor/enforcers from details of policy-based decision making. The SPEX controller can push updated policy metadata into an active policy, or install a completely new policy in the access calculator without interrupting access decision requests from interceptor/enforcers. This approach ensures that the policy update process can be exceptionally dynamic.

The policy object's interface to an access calculator is quite simple: requestor and resource data are consumed and a boolean access decision is produced. Though the SVE policy language and GUIs reflect the OODTE influence, a policy **need not** be OODTE-based to be used within the SVE infrastructure. Any policy model implementation that respects the Java interface defined for the policy object could be substituted in the architecture. This offers us tremendous flexibility in selecting appropriate policy models. We hope to experiment with a variety of resource access policy models and use the SVE infrastructure to deliver and evaluate policies in working systems.

4. Discussion

The SVE project has developed a software infrastructure to enable collaborative distributed computing. During the analysis, design and implementation of this system, we identified several significant issues that impact the creation and use of secure virtual enclaves. Many of these issues (enclave autonomy, policy semantics, principal data representation and transmission, and trust policy) have both conceptual and implementation consequences. System design and implementation choices in each of these areas have significant conceptual repercussions on the relationships among coalition members, the protections offered by the system, and the complexity of establishing and maintaining a collaborative environment. Some of the issues, such as system implementation and performance and scalability, primarily impact the engineering of the SVE system and its fitness for use in particular collaborative environments. We will discuss each of these issues, identifying our results, limitations of the SVE approach, lessons learned, and some areas that deserve further attention.

4.1. Enclave autonomy

In this section, we describe how the SVE system enables enclave autonomy, and discuss security issues that arise from that autonomy. It is essential to facilitate as much organizational autonomy as possible because an infrastructure that required mutually suspicious organizations to cede significant control to potential competitors or adversaries would not encourage collaboration. The SVE infrastructure offers enclaves a great deal of autonomy in controlling access to their local resources.

The SVE infrastructure supports enclave autonomy within a coalition in two ways: (1) by ensuring that most resource access policy components are used only within the local enclave, and (2) by enabling an enclave to unilaterally withdraw from an SVE at any time.

An SVE member enclave retains full control over local resource access policy. Specifically, resource to type mappings, access matrices, and constraints are never propagated among enclaves. In the event that an enclave discovers a collaboration partner to be untrustworthy, the enclave may respond immediately by modifying any of these local policy components and updating its access calculators. This permits an enclave to unilaterally restrict access to its local resources by external entities. Alternatively, the enclave may withdraw from the SVE altogether, effectively removing all of the SVE members' principal recognition rules from the policy enforced by the local access calculator. This immediately prevents any principals in those foreign enclaves from accessing local resources. Leaving the SVE is a more drastic step as it denies local resource access to all of the SVE member enclaves.

As with any security mechanism, the protection offered by SVE policy enforcement cannot extend beyond the system itself. In particular, the design of the SVE infrastructure does not address the following two issues, both of which arise due to the autonomous operations of the collaborating enclaves: (1) despite an enclave's request to leave an SVE (or principal recognition rule update), its local principals may continue to access foreign SVE resources if other SVE members fail to update their access policies in a timely manner; and (2) a trusted SVE member enclave may inadvertently or intentionally share a copy of a resource with a non-SVE member.

In the first case, an enclave depends on other SVE members to correctly manage principal recognition rules. These rules are the only policy elements shared among SVE members, but they require careful handling by each of the SVE member enclaves to ensure that principals are granted appropriate access authorizations. For example, if principal "Alice" is transferred by her employer, enclave **A**, to a new position and no longer requires access to the "Alpha Project" SVE, enclave **A** depends on the other SVE members to remove Alice's principal recognition rules promptly. Alice may, otherwise, have continued access to Alpha Project resources held by foreign enclaves. We assume that **A** removed Alice's principal recognition rules promptly, so that she doesn't have access to the local Alpha Project resources. This problem also arises if enclave **A** leaves the Alpha Project SVE, due to new concerns about the trustworthiness of the other Alpha Project members. Alice may inadvertently continue to access Alpha Project resources held by foreign enclaves if **A**'s principal recognition rules are not purged from foreign Alpha Project members' active policies. This places Alice's, and, thus, **A**'s integrity at risk from compromised foreign resources.

In the second case, when a trusted collaboration partner shares resources (intentionally or unintentionally) with non-SVE members, we can't know which resources have been compromised, or how widely they may have been circulated. The compromise of an SVE member makes our local resources conveniently available to an attacker. We must, therefore, trust that our partners are not only honest and cautious, but also savvy about protecting their network infrastructures. These issues are not unique to the SVE system, but, rather, impact coalitions in general and are likely to be significant considerations in planning collaborative efforts.

4.2. Transmission of principal data

To enable principals to access resources in foreign enclaves, while satisfying the requirements for application-level transparency, requestor data must be transmitted to the resource owner's enclave. As discussed in Section 2, approaches to transmitting this data range from annotating principal certificates to supplying a list of all local principals' data. In this section, we discuss the advantages and disadvantages of the SVE approach.

The SVE approach to transmitting requestor's data to a resource owner's enclave is to propagate a collection of principal recognition rules in bulk before a request is made. This bulk propagation approach has the advantage that modifying a role specification (i.e., changing a principal recognition rule) is less expensive than reissuing a certificate. Certificates are usually issued off-line, so producing a new certificate can take a relatively long time. Furthermore, with each authorization change, the old certificate must be revoked and a certificate revocation list updated. While we can't avoid dealing with certificate revocation with respect to identity certificates, the use of authorization certificates might create even more significant certificate revocation issues because authorizations are likely to change more frequently than identities.

The use of bulk principal recognition rules has its disadvantages, however. When enclave **A** reorganizes its employees and changes their organizational roles, it must issue updates of its principal recognition rules to ensure that employees have access to the correct SVEs and that their subjects are mapped into the correct domains.

Whenever enclave **A** makes organizational changes, all other enclaves in the SVEs in which **A** is a member must update their systems to handle those changes. This misplaces the burden of accommodating local changes onto foreign entities.

It is possible that a hybrid approach to authorization transmission would be valuable: some role information is contained in an authorization certificate, a user manages multiple such certificates, and additional (more volatile) data is expressed in principal recognition rules.

4.3. Policy semantics

Policy semantics is a thorny issue when role/authorization data crosses enclave borders. In this section, we describe problems that may arise with regard to policy semantics, and their impact on the development of security policies for coalitions.

A major problem with policy semantics is the different interpretations of common entities across multiple enclaves. For example, though enclave **A** may designate a principal for the `manager` role, the semantics of `manager` are known only within **A**. `manager` might mean line management (e.g., matching staff to projects, reviewing salaries), or it might mean project management (e.g., project tasking, project budgets). In either case, the semantics of the `manager` designation are established within the **A** context.

Suppose, for discussion, that by `manager`, enclave **A** means project management. When the `manager` designation crosses into enclave **B**, either in an authorization certificate, or as a target domain in an SVE principal recognition rule, the critical context data is lost. At least three possible problems might arise: (1) If enclave **B** has no `manager` role, then it will not authorize resource access, as the principal designation is unknown in **B**. The foreign manager, therefore, will be denied access. (2) If enclave **B** locally defines a `manager` role with the semantics of line management, then the foreign manager may inadvertently gain access to some of **B**'s salary data. (3) If enclave **B** defines a project management role, but instead uses the spelling `project-manager`, then the local project managers may have access to different resources than the foreign manager, despite the intention to grant them identical resource access.

One might expect that this problem could be handled by treating this as a namespace issue, however, that approach is insufficient, as marking the `manager` role with an enclave qualification gives us something like "`enclaveA/manager`". This designation does not provide **B** with the information necessary to interpret the role in the **B** context. Nothing short of establishing the semantics of roles in advance of their use will completely solve this problem.

In the SVE project, we assumed that the domain names to be used in resource access policies were established by the SVE creator and agreed upon via extra-technological means by other SVE members in advance of any resource sharing. A system in which **A**'s and **B**'s roles could differ would require some translation mechanism, either directly between **A** and **B**, or into some agreed-upon intermediate specification. Translation mechanisms might be constructed by wrapping local role data with filters. Overall, the need to establish a common understanding of security policy semantics among collaborating organizations will prove crucial to the effective use of coalition enabling technologies.

4.4. Trust

Trust is the foundation of any collaboration. In this section, we consider three trust issues arising in an SVE context, and how the SVE infrastructure handles (or might be extended to handle) those issues.

The first issue pertains to the granularity of trust expressible within the system. In the SVE system, a principal's identification credentials are delivered to an interceptor/enforcer when an access request is made. The interceptor/enforcer must authenticate the credentials (e.g., check certificate signatures) and determine whether it trusts the signer before extracting the identification data and forwarding it to an access calculator. We either trust that the certificate accurately identifies the requestor or we don't. If we had used authorization certificates, we might have applied a trust designation to each authorization contained in the certificate. The certificate signer may or may not be trusted by the target enclave to certify each of the authorizations. A finer granularity of trust could be distinguished within an SVE policy, which could then influence access calculation.

The second issue pertains to the uniformity of trust relationships within an SVE. When joining an SVE, an administrator must specify to the local SPEX server the list of other enclaves that are trusted to join this SVE. These enclaves may share local SVE resources. The SVE system treats each of these member enclaves identically—each member is equally trusted. To create distinct trust relationships among enclaves, an administrator would form multiple SVEs with different resource access policies for each SVE. Extending the SVE trust model to better support asymmetric trust relationships could make the system applicable to a broader set of coalition environments.

The third issue pertains to the degree of symmetry required in SVE member relationships. SVE member enclaves are peers. For example, to admit a prospective member to the SVE, each current member must vote, and the votes must be unanimously positive to allow the prospective member to join. It would be worthwhile to consider non-peer relationships within the SVE context, to allow the concepts of "subordinate" or "subcontractor" to be represented. In an asymmetric relationship, some enclaves are primary SVE members, while, for example, other enclaves participate in the SVE because they manage resources at the direction of primary members. Secondary members that supply services to other enclaves may not require a full vote in this scheme. We have not determined how such asymmetries might be expressed within the SVE context.

4.5. Implementation

In this section, we briefly review our analysis of SVE implementation issues and identify alternative implementation options we might choose today, if the SVE project were just beginning. As mentioned in Section 3, most of the SVE infrastructure was implemented in Java 1.1, using RMI for intra-enclave communication and the Ensemble group communication system for inter-enclave communication. Though both Java and JavaGroups/Ensemble proved to be good choices for implementation, upcoming Java facilities and development tools would have eased our development task. For example, Sun Labs has recently developed a reference implementation of Java reliable multicasting (JRM) [12]. This might allow us to eliminate some of the platform-dependent multicasting facilities provided by Ensemble. Sun Labs plans to add the JRM API into future versions of the Java Development Kit (JDK).

By far, our most significant implementation problems arose during the development of the interceptor/enforcers. Neither Java RMI nor DCOM were built to allow the modular addition of security mechanisms. The problem was significantly compounded for DCOM due to its relative complexity and the lack of available internal specifications. Fortunately, the RMI interceptor was not quite as problematic, as RMI is simple and its specification is well-documented. However, both interceptors are implementation dependent and vulnerable to version changes.

The web servers were clearly designed to allow additional functions to be added. In the case of the Java Web Server, the servlet concept allowed us a simple means of encapsulating and installing SVE interception code. Microsoft's IIS web server offered a similar approach to extensibility via its filter concept. Unfortunately, several documented features of IIS did not work as specified. As a result, an administrator must manually prepare IIS resources before they can be controlled by the SVE system. We hope that Microsoft will correct the problems with IIS in a future release.

4.6. Performance and scalability

Project resources did not permit us to perform a quantitative performance evaluation of our system. We can, however, discuss our qualitative observations from the perspectives of an administrator and of a user client.

4.6.1. Performance. We developed the initial version of the system without any communication security for the SVE infrastructure. This initial version did require client authentication, however, to allow us to test our access decision-making and enforcement. We built a simple distributed application to test and demonstrate the SVE system, using Sun's JDK versions 1.1.6 and higher, with the Just in Time (JIT) compiler enabled. The application included two data repositories, one contained in a Microsoft access database and one contained in flat files. We built DCOM and RMI clients, and used web browsers to access the resources. We used a variety of machines ranging from a 166 MHz Pentium Pro with 32 MB RAM to a SPARC Ultra-1 with 128 MB RAM, up to a 450 MHz Pentium 2 with 128 MB RAM. All of these machines were connected via our local Ethernet.

When run with communication security turned on (SSL for RMI and the web browsers, and the Microsoft proprietary protocol for DCOM), users noticed a significant performance impact. When accessing Java-based resources via RMI over SSL, the delay between a request and the display of resource data was less than 2 or 3 seconds. The bridges we built to permit communication between the Microsoft interceptors and the access calculators degraded performance for these types of requests. In our system, access requests for Microsoft resources were significantly slower (perhaps on the order of 7 or 10 seconds) than requests for the Java-based resources. By inserting dummy access decisions into our enforcement code, we established that the bridges, rather than the interceptors, were at fault.

From an administrator's perspective, SVE control operations appeared to have reasonable performance (i.e., under 1 second for an SVE join operation), without communication security. Once we enabled message signing and verification, the performance degraded noticeably. An administrator might wait up to 5 seconds before receiving confirmation of a successful join operation. In summary, it appears that both intra-enclave and inter-enclave communication security introduce significant obstacles to attractive performance of the SVE system.

4.6.2. Scalability. Our system was tested with only a few SVEs (5 or fewer), and only a few member enclaves of each SVE (3 or fewer). The small size of the experiment was due primarily to the administrative overhead of developing security policies. Scalability was not, therefore, addressed by experiment. Roughly, the cost in administrative overhead and communication of joining and participating in two SVEs is twice the cost of joining only one. Thus, the scalability of the system largely hinges on how well we can support the growth of a single SVE.

SVEs grow as new member enclaves issue join requests, and are maintained as existing members issue policy update requests. The join operation has the highest messaging requirement of any of the SVE operations, as voting requests must be conveyed to each member, ballots collected from each member, and result notification sent to each member. All of these underlying operations are accomplished using Ensemble-based communication. Therefore, the scalability of inter-enclave SVE communication is heavily dependent on the scalability of the underlying Ensemble communication system, whose performance is addressed in [7].

5. Related work

Several research projects have addressed problems related to collaborative computing. Systems such as Ensemble [7], Rampart [11], Transis [1] and Enclaves [6] provide secure group communications, allowing collaborating entities to maintain privacy and message integrity in their communications. The SVE system, however, focuses on supporting policy-driven access controls for distributed object systems.

There is also research in the area of security policies for group communication. In [15], Srisuresh and Sanchez describe policy-based routing for IP security. Requirements are described for intermediate and end nodes to support security policies for packets crossing enclave boundaries. At this level, policies refer to packet forwarding rules or communication security mechanisms and their parameters. The Dynamic Cryptographic Context Management (DCCM) system [2] provides dynamically changeable mappings from low-level policy abstractions onto cryptographic mechanisms for secure group communications. DCCM offers a mechanism for inter-enclave policy negotiation. In this case, policies refer to cryptographic context specifications (e.g., ipsec 3des-cbc encryption, using sha-1 integrity verification).

Other work that has dealt with access controls for distributed objects include the Sigma project [13], which investigated the integration of security technologies into CORBA-based distributed computing environments. The Sigma project built prototypes of both gateway and server-resident (ORB plug-in) interceptor/enforcers for CORBA requests and developed the object-oriented version (OODTE) [14] of the Domain and Type Enforcement (DTE) [5] policy specification language. Sigma results were focused on policy definition (via OODTE and other policy specification languages) and request interception. The SVE project extended these results by introducing an infrastructure to support shared policy elements for collaborating organizations. The Multi-Protocol Gateway (MPOG) [9] extended the CORBA ORB gateway to enforce access policy for both CORBA and Java RM requests. We hope to extend the SVE system to provide dynamic policy updates for the MPOG to enable it to consistently enforce SVE resource access policies.

Each of the types of middleware resources (e.g., web resources, RMI resources, DCOM resources) for which the SVE infrastructure provides access controls has its own model of access policy specification and enforcement. For example, the default mechanism for specifying access policy for resources controlled by the Java Web Server is access control lists. Currently, there is no mechanism for controlling RMI access to distributed Java resources. In the future, however, the Java Authentication and Authorization Service (JAAS) [8] will be used to control this type of access. Access to DCOM applications can be controlled by Windows NT, but the finest granularity of control is at the per-application level (e.g., "launch," "configure"). Each of these middleware systems specializes in controlling access to its own distributed resources, rather than providing uniform access policy enforcement for multiple types of distributed object resources. None of these policy enforcement systems support notions of collaborative computing among multiple organizations. The SVE system focuses on supporting collaborative computing through uniform access controls on various distributed object resources.

The SVE system was engineered using many previously developed technologies. As mentioned in Section 3.2.2, the SVE approach to DCOM interception applied a method described by Pietrek in [10]. A more general approach to mediating calls into a Windows NT API is described by Balzer and Goldman in [4]. As we did not require the generality of Balzer/Goldman, we opted for the relative simplicity of the Pietrek method.

The SVE infrastructure made use of the Ensemble group communication system to provide reliable communication for SVE member enclaves. The JavaGroups toolkit [3], developed by Ban at Cornell, provided us with a Java-based interface to the Ensemble system. Protocols for SVE management make calls through JavaGroups to communicate over Ensemble.

6. Conclusions

Our primary goal was to study software mechanisms to support coalitions in collaborative computing efforts. Because coalition partners may have only limited trust in one another, a coalition support system must provide both the means for careful control of coalition partners' access to local resources and the ability to dynamically change those controls as trust relationships evolve. In addition, the need to share current application resources required that our system solutions be transparent to applications and based on commercially available operating systems and open networks.

To meet coalition requirements for a collaborative computing environment, we designed and implemented a prototype security infrastructure. The prototype addressed the problem of limited trust relationships by providing significant resource access policy definition and enforcement autonomy to individual member enclaves. To support the changing nature of a collaborative arrangement, we provided a dynamic policy update mechanism.

The SVE infrastructure allows multiple organizations to share their distributed application resources, while retaining organizational autonomy over local resources. While we demonstrated an approach to fine-grained access control for secure collaborative computing, we also identified significant problems that remain to be solved, particularly in the area of policy development, before such collaboration will be convenient. The SVE infrastructure offers a platform and conceptual basis for further exploration of these problems and experimentation with new solutions.

Acknowledgements

We would like to thank David Chieu for his patience and persistence in implementing the DCOM and Microsoft IIS interceptors. Yongjian Wang produced the first versions of our administrative GUIs, giving us a better understanding of SVE administrative issues. Debbie Sturdevant produced several of the diagrams for this paper. Bob Rice and Dennis Hollingworth reviewed drafts of the paper, and their suggestions have helped us improve the presentation. This work was supported under DARPA contract #F30602-97-C-0269.

References

[1] Y. Amir, D. Dolev, S. Kramer, and D. Malki. Transis: A communication sub-system for high availability. In *Proceedings of the 22nd International Symposium on Fault-Tolerant Computing*, pages 76-84, July 1992.

[2] D. M. Balenson, D. K. Branstad, P. Dinsmore, M. Heyman, and C. Scace. Dynamic cryptographic context management (DCCM) report 3: Cryptographic context negotiation protocol. Technical Report TISR #0757, TIS Labs at Network Associates, Inc., February 1999.

[3] B. Ban. Design and implementation of a reliable group communication toolkit for Java. http://www.cs.cornell.edu/home/bba/papers.html.

[4] R. Balzer and N. Goldman. Mediating connectors. In *Proceedings of the 19th IEEE International Conference on Distributed Computing Systems Workshop (ICDCS '99)*, pages 73-77, Austin, TX, May 1999.

[5] L. Badger, D. Sterne, D. Sherman, K. Walker, and S. Haghighat. Practical domain and type enforcement for UNIX. In *Proceedings of the 1995 IEEE Symposium on Security and Privacy*, pages 66-77, Oakland, CA, May 1995.

[6] L. Gong. Enclaves: Enabling secure collaboration over the Internet. In *Proceedings of the Sixth USENIX Unix and Network Security Symposium*, pages 149-159, San Jose, CA, July 1996.

[7] M. Hayden. The Ensemble System. Ph.D. dissertation. Cornell University, Ithica, New York. Available as technical report TR98-1662, 1998.

[8] C. Lai, L. Gong, L. Koved, A. Nadalin, and R. Schemers. User authentication and authorization in the Java platform. To appear in *Proceedings of the 15th Annual Computer Security Applications Conference*, Phoenix, AZ, December 1999.

[9] G. Lamperillo. Architecture and concepts of the MPOG. Technical Report NAI #0768, NAI Labs at Network Associates, Inc., June 1999.

[10] M. Pietrek. Learn system-level win 32 coding Techniques by writing an API spy program. In *Microsoft Systems Journal*, vol. 9, no. 12, pages 17-44, 1994.

[11] M. Reiter. Secure agreement protocols: Reliable and atomic group multicast in Rampart. *In Proceedings of the Second ACM Conference on Computer and Communication Security*, pages 68-80, Fairfax VA, November 1994.

[12] P. Rosenzweig, M. Kadansky, and S. Hanna. The Java reliable multicast service: a reliable multicast library. Technical Report #TR-98-68, Sun Microsystems, Inc., September 1998.

[13] E. J. Sebes and T. C. Vickers Benzel. Sigma: Security for distributed object interoperability between trusted and untrusted systems. In *Proceedings of the 12th Computer Security Applications Conference (ACSAC '96)*, pages 158-168, San Diego, CA, December 1996.

[14] D. Sterne, G. Tally, D. McDonnell, P. Pasturel, D. Sames, D. Sherman, and E.J. Sebes. Scalable access control for distributed object systems. In *Proceedings of the 8th USENIX Security Symposium*, August 1999.

[15] P. Srisuresh and L. A. Sanchez. Policy framework for IP security. *Internet Draft* draft-ietf-ipsec-policy-framework-00.txt, February 1999.

Domain and Type Enforcement Firewalls

Karen A. Oostendorp
Lee Badger
Christopher D. Vance
Wayne G. Morrison
Michael J. Petkac
David L. Sherman
Daniel F. Sterne
Trusted Information Systems, Inc.
3060 Washington Road Glenwood, Maryland 21738

Abstract

Internet-connected organizations often employ an Internet firewall to mitigate risks of system penetration, data theft, data destruction, and other security breaches. Conventional Internet firewalls, however, impose an overly simple inside-vs-outside model of security that is incompatible with many business practices that require extending limited trust *to external entities, for example, suppliers, bankers, accountants, advisors, consultants, partners, customers, and allies. Additionally, firewall security perimeters are somewhat weak: they provide no protection from inside attacks and do not protect sensitive data, which can be exported by tunneling through permitted protocols. As the Internet evolves towards applets, mobile agents, and object frameworks, these problems likely will worsen.*

This paper reports on our experience with an enhanced security firewall based on Domain and Type Enforcement (DTE), a strong but flexible form of access control. A DTE firewall provides several benefits. First, it runs application-level proxies in restrictive domains, thereby increasing security, and runs network services such as HTTP and FTP under DTE controls, thereby reducing risks that network-based attacks will compromise local resources. Second, a DTE firewall coordinates role-based security policies that span networks by passing DTE security attributes between DTE clients and servers. These attributes allow security policies at the endpoints to be coordinated; such coordination adds defense in depth *to the traditional firewall security perimeter: this permits safe exportation of normally risky services such as NFS. Finally, a DTE firewall interoperates with "non-DTE" systems and associates DTE security attributes with these systems so their interaction with DTE clients or servers can be controlled. We describe here the design of a prototype DTE firewall system and informally evaluate its security, compatibility, functionality, and performance.*[1]

1. Introduction

Internet-connected organizations often employ an Internet firewall to mitigate risks of system penetration, data theft, data destruction, and other security breaches. Conventional Internet firewalls, however, impose an overly simple inside-vs-outside model of security that is incompatible with many business practices that require extending *limited trust* to external entities, for example, suppliers, bankers, accountants, advisors, consultants, partners, customers, and allies. Additionally, firewall security perimeters are somewhat weak: they provide no protection from inside attacks and do not protect sensitive data, which can be exported by tunneling through permitted protocols. These problems likely will worsen in the next few years as applets [9], mobile agents [19], and object frameworks [12] provide opportunities for malicious programs to execute behind firewall security perimeters.

While firewall security perimeters are helpful, we believe that a fundamentally stronger mechanism will be required to protect "sacrificial lamb" network servers, to run network clients in environments that control damage inflicted by malicious servers (and applets),

[1]This research was supported by ARPA contract DABT63-95-C-0018.

and to provide a basis for controlled sharing of information between organizations. A number of computer operating system security controls [2, 3, 4, 7, 10] have been proposed, but most of these techniques either map poorly to commercial requirements or impose excessive administrative overheads. Domain and Type Enforcement (DTE) [1] is a relatively recent operating system access control mechanism that holds promise to provide needed security flexibility and strength while controlling administrative costs. We believe that a fundamental question for practical commercial security is whether strong-but-flexible access controls, such as DTE, can be combined with firewall defenses in a way that preserves the practicality of firewalls while substantially improving security.

This paper reports on our experience with a DTE-enhanced firewall prototype. Our preliminary results indicate that DTE access controls *can* be integrated with firewalls to cost-effectively control resources shared through firewalls and to protect enterprise resources from possibly malicious insider programs. After reviewing the primary concepts of DTE, this paper presents the design of a DTE firewall and the mechanisms it uses to control imported and exported services. This paper next presents our informal evaluation of DTE firewall security, functionality, compatibility, and performance characteristics. Finally, this paper reviews related work and future directions, and presents conclusions.

2. DTE review

DTE [1, 15, 18] is an enhanced form of type enforcement [4, 13, 14], which is a table-oriented mandatory access control mechanism. DTE has three main advantages over type enforcement. First, the security policy is specified in a high level language that reduces the burden of expressing, verifying, and maintaining security rules. Second, security attributes on objects are implicit, thereby allowing a file hierarchy to be typed concisely. Finally, DTE provides mechanisms for backward compatibility with existing software and with systems not running DTE.

As with a number of other access control mechanisms [2, 3, 4, 7, 11], DTE considers a system to be split logically into two categories: passive entities (e.g., files or network packets) and active entities (usually processes). A type is associated with each passive entity, or object; a domain and a DTE-protected user identifier (unchangeable even by the *root* user) is associated with each active entity, or subject. Using a language-based specification, DTE expresses allowed interactions between subjects and objects. Access control decisions are made by consulting a DTE database consisting of the "compiled" specification to determine if the subject's domain has the requested access (e.g., read or write) to the object's type. DTE also expresses allowed interactions between subjects. Access control logic consults the DTE database to determine if a subject A's domain has the requested access (e.g., execute or kill) to a subject B's domain.

To extend DTE protection across networks, DTE treats each network packet as an object with three associated attributes (carried in the IP option space of each datagram): the DTE type of the information, the domain (source domain) of the source process, and the DTE-protected UID of the source process. A process can send or receive a message object only if the process's domain has the appropriate access to the DTE type of the message. Communication with non-DTE systems also is mediated: when a message originates from a non-DTE system, the receiving DTE system assigns a type (and domain) to the message.[2] Similarly, a DTE system mediates a message before sending it to a non-DTE system to ensure that the domain associated with the non-DTE system can read the messages. If the associated domain cannot read the message's type, the message is not sent.

In our prototype, UNIX[3] kernel changes to enforce DTE mediation are localized to a relatively small subsystem; all system calls that represent process accesses to other processes or to objects are passed through this subsystem.[4] During the initialization phase of a DTE kernel (i.e., at boot-time), the DTE subsystem reads a security specification written in the DTE Language (DTEL) from the boot device, parses the specification into access control data structures, and then mediates the system calls of all processes (including the first system process) according to the specification, as described above.

3. DTE Firewall overview

A DTE firewall is an enhanced application gateway firewall. It runs application gateways in controlling

[2]A DTE policy currently associates attributes with packets received from non-DTE systems based on source IP address. While IP addresses can be spoofed in IPv4, the approach is adequate for prototyping. Stronger mechanisms could be provided using IP-layer cryptography, such as that proposed for IPv6.

[3]UNIX is a registered trademark in the United States and other countries, licensed exclusively through X/Open Company Ltd.

[4]Our prototype is based on BSD/OS 2.0 (and recently 2.1), a widely available PC UNIX. Excluding comments, DTE enhancements to the kernel represent approximately 14,000 lines of code, and DTE enhancements to the firewall proxies represent approximately 700 hundred lines of code.

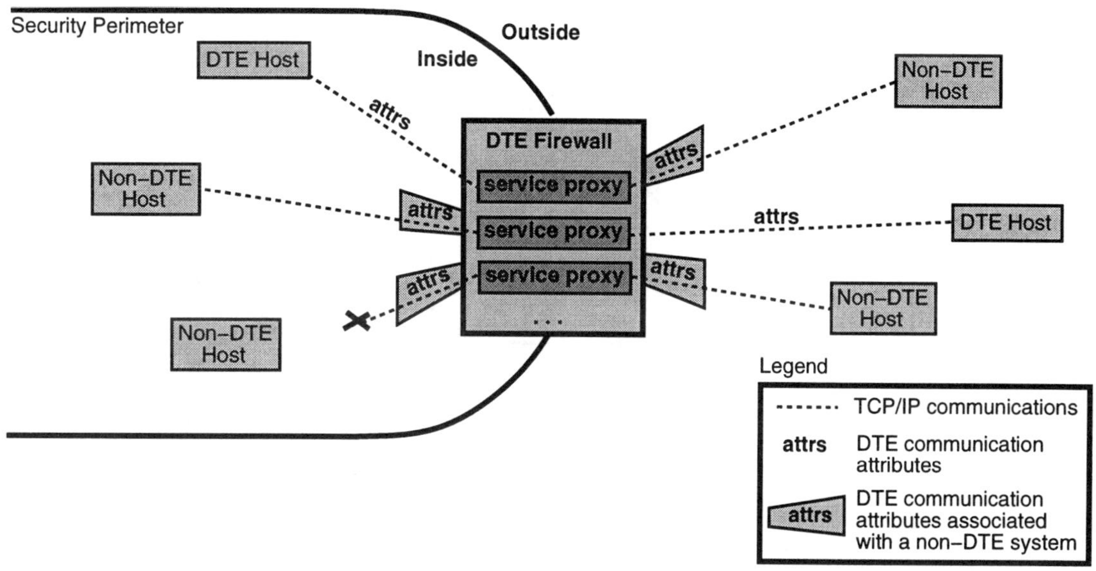

Figure 1. DTE Firewall Concept of Operations

DTE domains and mediates network communications based on DTE security attributes.

Figure 1 shows an overview of a network enclave protected by a DTE firewall. Some hosts behind the firewall are running DTE; there may or may not be DTE hosts outside the perimeter. As with "normal" firewalls, a DTE firewall intercepts and mediates all network traffic between internal and external hosts. Based on the DTE policy, the DTE firewall associates domains with non-DTE hosts and associates DTE communication attributes with data received from non-DTE hosts. As described in section 2, the DTE firewall and DTE hosts use the communication attributes to ensure that network messages are received only by domains that can appropriately control the type of data carried by the messages. If the end host is a DTE system, it carries the responsibility for confining network services: in this case, the DTE firewall's role is to coordinate the endpoints' security contexts by passing along the DTE communication attributes. In the case of non-DTE hosts, the DTE firewall performs access control on behalf of non-DTE hosts (interior or exterior) by mediating the network messages sent to each non-DTE host based on whether the host's associated domain grants access to the DTE attributes associated with the messages. By coordinating DTE policies between DTE endpoints (and on behalf of non-DTE endpoints), the DTE firewall is positioned both to protect and control exported services and to confine network clients that import services.

3.1. Controlling exported services

Figure 2 illustrates our general strategy for exporting services safely: run network server applications either on DTE firewalls or on DTE hosts behind the security perimeter and use DTE to control access to local resources. In figure 2, a DTE server system hosting a network server application (*service*) communicates through a DTE firewall to respond to service requests from a non-DTE external host (*Client System*). Since the client shown is not running DTE, the firewall associates type t and domain sd with messages received from it. The DTE firewall relays the DTE communication attributes to the server: these attributes establish DTE access controls consistent with the level of trust the firewall places in the client.

When a client attempts to initiate a connection with the server, the *inetd* daemon on the firewall runs the *netacl* program, which determines whether communication is allowed between the client and server hosts[5] and executes the *proxy* application for the specified protocol. The *netacl* program executes the *proxy* in the *proxy_d* domain, which is specific to the protocol (e.g., the HTTP proxy runs in the domain *fw_http_d*) and has DTE access permissions sufficient to pass DTE communication attributes between the client and the server.[6]

[5] This is standard application gateway firewall functionality; the check is based on IP addresses or host names.

[6] This paper assumes a homogeneous DTE policy (having the same domain and type definitions) for all hosts; future work will introduce interactions between hosts running different policies

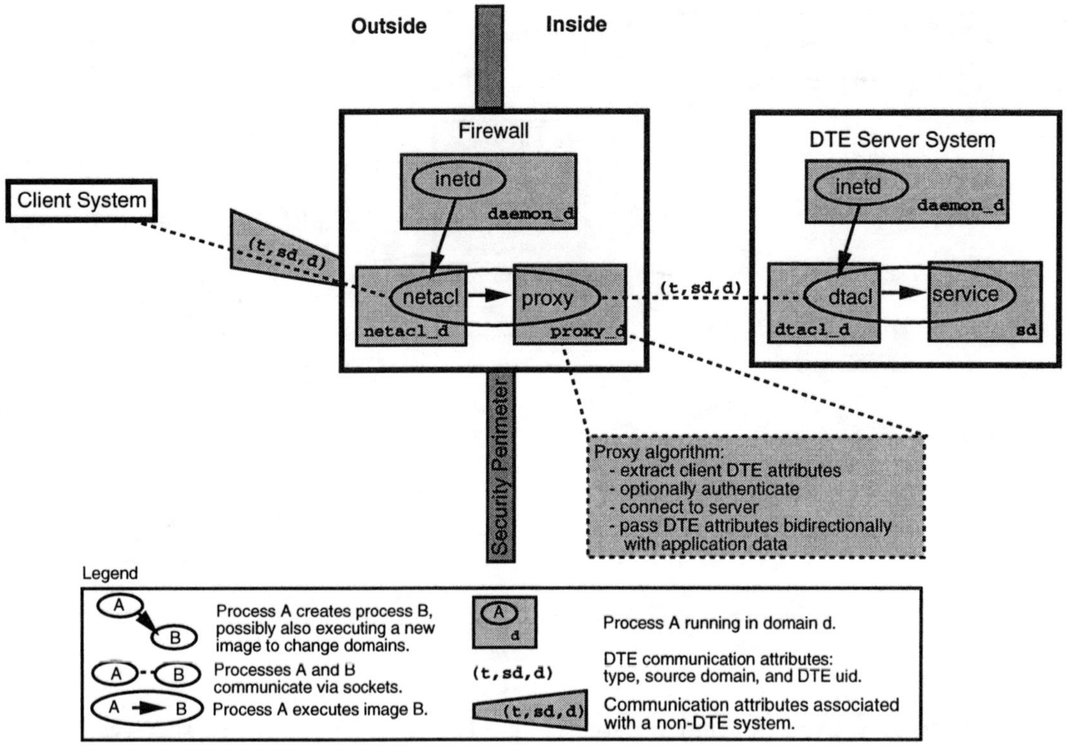

Figure 2. Exporting Services Safely

The *proxy_d* domain can be configured to control which clients may use the service.

At a high level, the *proxy*'s algorithm is simple:

1. **Extract client attributes** The DTE kernel's socket abstraction provides an interface to retrieve the client attributes *(t,sd,d)* each time data is read from the client socket. Because the attributes are carried in each IP message, they are available for both connection-oriented (e.g., TCP) and connectionless (e.g., UDP) protocols.

2. **Optionally authenticate** If the client is not a DTE system, the proxy authenticates the client using the method specified in the firewall's configuration. If the client is a DTE system, the proxy may choose to trust the DTE UID (*d*) in the communication attributes, which indicates the user's identity as authenticated by the client system. Using a DTE client's authentication attribute increases usability and performance by relieving the client of the need to authenticate for each service initiation (this has resulted in DTE firewall performance *increases* for some test cases).

and dynamic policy reconfiguration.

3. **Connect to server** The proxy connects to the DTE server system, passing the DTE attributes of the client.

4. **Pass data and DTE attributes bidirectionally** The DTE-enhanced proxy scans and copies data as in a conventional firewall except that it also passes the DTE communication attributes through to each endpoint. Also, for each message-receive and message-send operation, the proxy's domain is mediated (by the firewall's DTE kernel) with respect to the type of data being received or sent: the types of data allowed through the firewall therefore can be adjusted by adjusting the types of data that the proxies are allowed to read/write. Because DTE attributes are associated with remote non-DTE hosts by the firewall, the policy also can be adjusted so that some types of data are not sendable (or receivable) from specified hosts or networks. When the external host is running DTE, the control is more fine-grained and communications using specific types of data can be allowed or disallowed for individual domains (or processes) on the remote host.

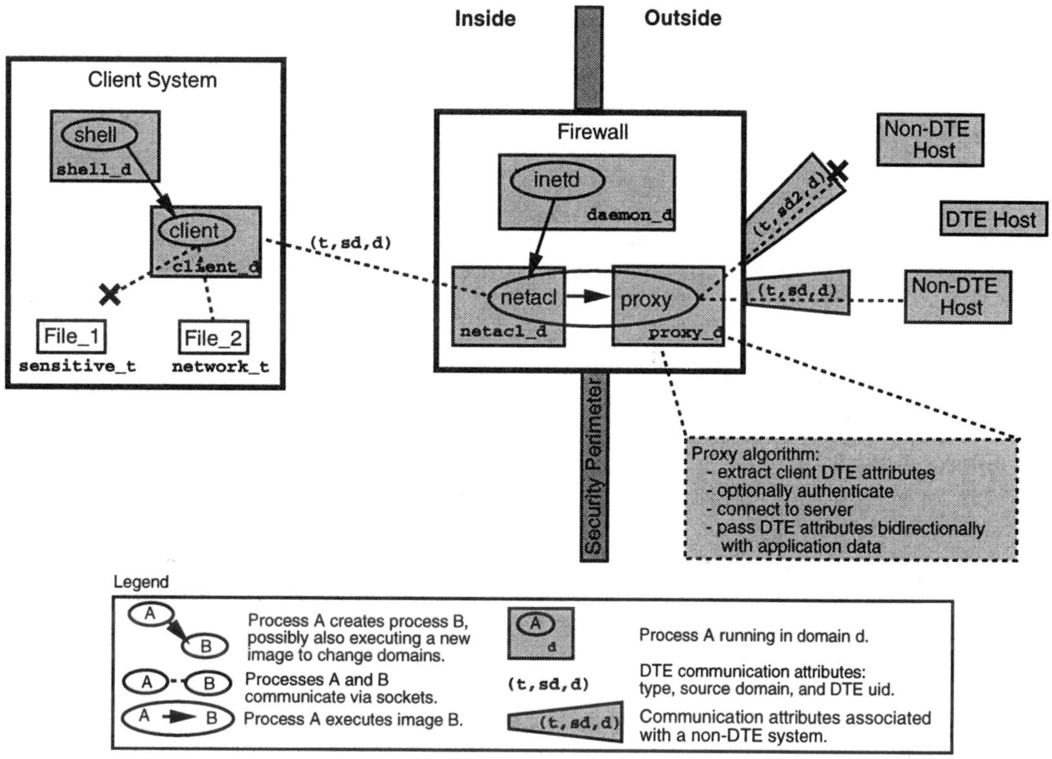

Figure 3. Importing Services Safely

As shown in figure 2, when the *proxy* application connects to the DTE server, the *inetd* program on the server runs the *dtacl* program. This program, like *netacl*, examines the DTE attributes sent in the connection request and then executes the server application *service* in domain *sd*, the domain of the client. By running the server application in the client's domain, the system ensures that the client and server run in a compatible security context: if the client is anonymous (e.g., anonymous FTP or HTTP), the server runs in a domain granting very limited access to data on the server system; if the client is a known quantity (perhaps administered by a business partner, or an authorized guest), the server runs in a domain granting access to more types of data.[7] However, for services that do not restart with each client request, such as NFS, a different strategy is needed. These services employ a kernel-supported "auxiliary" domain (i.e., the source domain of a received message) to further restrict the server when it is working on behalf of a particular client; DTE mediation then is performed using both domains. Because the service must ask the kernel to load and unload the auxiliary domain, this technique is available only to trusted servers.

[7] The access level granted to domains depends upon the DTE policy and is user-configurable.

The strategy of running the server in the client's domain does not necessarily result in least privilege, since the common domain may overstate access needs of the client or the server. However, the strategy is surprisingly simple and approximates least privilege (in our experience) closely enough to separate user roles and to protect the system. The DTE policy on the server host protects the host from the server application (using root confinement [18]) and also protects the server application from other programs that may run on the server host. Additionally, the types writable by the *sd* domain indelibly label data that originated from the exterior network. This labeling can alert users and programs that the data was received from the network and could be untrustworthy.

3.2. Controlling imported services

A DTE firewall controls imported services using the same mechanisms it uses to control exported services: the DTE firewall relays DTE communication attributes from an internal client system to potential server systems and performs mediation on behalf of non-DTE clients and servers. The overall effect of this mediation is to prevent a client from using a server unless either the server is running in a compatible domain or the server runs on a DTE host that is willing to start a copy of the server in the client's domain. As a consequence of these mechanisms, the same client operating in different domains may have differing levels of access to external services. Users on a DTE client, therefore, may choose restricted domains when accessing untrusted or unknown services (e.g., surfing the web), and choose more privileged domains when accessing important corporate data. Additionally, this constraint ensures, for example, that information cannot be accidentally sent to competitors (e.g., using mail) and that files cannot be accidentally imported from untrustworthy environments and executed.

In figure 3, a DTE client system runs a network program that communicates through the DTE firewall to access a service provided by a non-DTE external host. Figure 3 shows the user session running in domain *shell_d* and the client application running in domain *client_d*.[8] The domain *client_d* controls the client application in two important ways: 1) it prevents a successful attack on the client from damaging the client system, and 2) it labels data generated by the client application with a type that identifies the data's origin and indicates the amount of trust that can be placed in it. In addition, the DTE policy on the client system protects the client application from attacks by other running programs. This is particularly useful for preventing software of questionable quality (e.g., the most recent version of a free editor) from accessing the data streams of important client applications such as electronic commerce, banking, or Internet telephone programs.

As shown in figure 3, the client application communicates via sockets with the DTE firewall which then communicates with the external server. Since the serving host is not running DTE, the DTE firewall prevents communication with the serving host unless the DTE firewall associates a compatible domain with the host.

4. Network services evaluation

To assess the overall impact of DTE firewalls and selected hosts, we have evaluated informally the remote login (rlogin), TELNET, mail, FTP, NFS, and HTTP network services running through a DTE firewall. The evaluation considered the security, preservation of functionality, compatibility with non-DTE hosts, and performance of these services. For purposes of evaluation, we used the BSD/OS DTE prototype system and the TIS Firewall Toolkit (V1.3).

4.1. Security

Network attacks often exploit subtle weaknesses in programs that allow attackers to misuse program access rights, gain control over systems, steal or destroy data, and deny access to authorized users. The effectiveness of such attacks can be reduced if programs execute with the minimum access rights required to perform their functions. Our informal metric for evaluating the security of a DTE firewall, therefore, is the extent to which it restricts the access rights used by programs running on the firewall or on DTE systems that communicate with the firewall. We have identified three primary areas where program authorizations are reduced by DTE:

confined proxies A DTE firewall confines each network proxy in a separate domain. In general, the domains used to confine proxies prevent write access to system and administrative data and also prevent proxies from running other programs. This confinement protects the firewall from a possibly subverted proxy: for example, if the FTP proxy is compromised, it cannot alter system files.

Since proxies are trusted to propagate type labels of user data that passes through them, a faulty or subverted proxy may cause user data to be mislabeled, however this mislabeling can be limited: a proxy can attach a type label to an object only if the proxy's domain grants write access to the label. Consequently, the firewall's DTE policy can be configured to control the discretion afforded a proxy as well as to restrict which types of data are allowed to flow through different network service proxies.

protected servers on the firewall A DTE firewall can run network services in domains that protect their data and program files from network-based attack. By running services on the firewall (or, for load-sharing purposes, on companion DTE server hosts), the integrity of network services can be improved substantially even though the services are made available to the Internet. While running possibly vulnerable network services on a conventional firewall would pose an unacceptable security hazard, a DTE firewall can execute such services

[8] A DTE client system's DTE policy can be configured to give the user discretion in which domain the client application runs, to require that the client application run in the domain of the user's shell, or to specify automatically a domain in which the client application runs.

safely because their access rights are limited via DTE.

defense in depth As discussed in section 3.1, DTE firewalls coordinate security protections of DTE servers behind a security perimeter to restrict the access rights of programs that are run on behalf of external clients. When DTE servers are available, this strategy ensures that processing carried out on behalf of external clients is controlled in appropriate DTE domains that prevent clients from successfully tricking interior services into granting unauthorized access. This technique can be used to prevent unauthorized export of sensitive data through ubiquitous services, such as mail, and to export NFS hierarchies selectively to the Internet. Furthermore, this strategy typically does not rely on the correctness or security sophistication of services exported through the DTE firewall since the control is enforced by DTE on the serving hosts.[9]

The DTE firewall policy used to conduct our experiments consists of 105 lines of DTEL specification on the firewall and 122 lines of DTEL on supporting DTE hosts. The addition of a new network service or domain adds typically only 5 or 6 lines to these specifications. Additionally, small extensions to domains in the policy can be made that apply uniformly across the specification and expand or constrict access rights for all network services. As a result, DTE policy complexity can be controlled, resulting in enhanced assurance that protections are maintained as new network services are added. The combination of these techniques has significantly reduced the access rights used by software running on (or through) the firewall while allowing the software to function correctly. Based on our informal metric, we belive that these techniques have significantly increased network enclave security.

4.2. Functionality

For importing services, functionality is rarely affected. For services such as rlogin and TELNET, when the client is a DTE system, user authentication can be supplied automatically by the client DTE system and the proxy can accept and use this authentication instead of requiring additional authentication. Under some circumstances, this can increase usability; however, the DTE UIDs must be set up in the firewall configuration, adding a small amount of administrative overhead. Also, services such as HTTP and FTP can be made more widely available because of the reduced risk of malicious programs (e.g., applets).

Functionality for exported services has increased in several ways. With the additional security of running a server in a domain restricted according to trust level of the client, the server no longer needs to be located outside the firewall. Instead it can be on a system behind the firewall, or even on the firewall itself. Furthermore, with the server starting in the domain of the client, it is feasible to grant access to different classes of information based on that domain – something very useful for HTTP and FTP. In this way, for example, an anonymous FTP server can regulate access to files without resorting to multiple servers or hidden files. Also, a single HTTP server can provide sensitive data to users in trusted domains and general data to users in untrusted domains without fear of having the data compromised or altered. NFS requires slightly more administrative overhead, but becomes safely available through firewall security perimeters – something not possible before.

4.3. Compatibility

Overall, compatibility has been maintained: each service can interoperate either with DTE or non-DTE systems. Furthermore, the application-level proxies revert to standard Firewall Toolkit behavior when run on a non-DTE kernel. The use of IP options to carry DTE information removes the need for changing the protocols.

With the exception of the NFS server, which is kernel-resident in UNIX, few of the client or server applications have been changed to function with DTE firewalls. Mail final delivery agents were modified to be cognizant of the different types associated with users' mailboxes. Also, the rlogin server was modified to take advantage of DTE authentication mechanisms.

Some of the services running under DTE require changes in the administrative configuration. For example, external NFS clients must explicitly name the firewall host as the server whose file systems they wish to mount, since they cannot know the name of the server behind the firewall.

4.4. Performance

To evaluate the performance of DTE and DTE Firewalls, we have constructed a testbed consisting of three Pentium 166MHz machines on an isolated network running BSD/OS 2.0 and version "straw_19+" of the DTE prototype system.[10] We ran each test on a number of

[9] The exception is the NFS daemon, which for performance reasons is implemented as a trusted server.

[10] This is the 19th internal version of the BSD/OS-based DTE prototype with some performance enhancements incorporated;

Protocol		Data Transferred	Baseline (n,n,n)	Percentage Change (%) (n,y,n)	(n,y,y)	(y,y,n)	(y,y,y)
rlogin		0K	1.98	0	0	-15	-10
		200K	3.43	6	8	-16	-23
		500K	5.33	<1	<1	-14	-16
		5MB	32.21	<1	<1	-2	-2
TELNET		0K	6.79	<1	<1	-15	-12
		200K	7.79	<1	2	-10	-10
		500K	9.89	1	1	-9	-7
		5MB	41.36	<1	1	-2	-1
FTP		0K	1.98	9	13	6	11
		200K	3.98	4	6	4	5
		500K	4.59	3	4	3	4
		5MB	14.23	3	6	<1	3
		32MB	70.33	2	2	-1	<1
HTTP	Concurrency Level 4	1K	355.81	8	13	12	15
		50K	64.71	71	89	92	94
	Concurrency Level 8	1K	199.20	22	24	26	25
		50K	60.23	75	94	97	114

Figure 4. Raw Performance in Seconds and DTE Overheads

configurations where configuration is a triple (client, firewall, server) in which "y" indicates a system running DTE and "n" indicates a host not running DTE (so (n,y,n) is the configuration where only the firewall is running DTE). We did not measure the performance of mail since it is not interactive.

For rlogin, TELNET, and FTP, we used an Expect script to repeatedly authenticate and transfer various amounts of information through the service. Performance numbers were calculated by averaging results from 20 iterations of each test.

For HTTP, we used *ZeusBench*,[11] a standard benchmark that connects to the server via the HTTP proxy, retrieves a specified web page, and closes the connection. We varied the concurrency, the document length, and the number of requests (1000 requests for 1K documents, and, to save time, 32 requests for 50K documents).

As shown in figure 4, DTE overheads for rlogin, TELNET, and FTP are modest, with a maximal impact of 13% degradation in the worst case; with additional performance optimization, these probably could be reduced. As is shown in the table, performance actually increases (the negative numbers in the table) for rlogin and TELNET when the client is running DTE, because the DTE client passes a DTE UID which the firewall can accept instead of performing costly authentication. This performance increase does not manifest for FTP because the FTP daemon has its own, always invoked, built-in authentication which we did not disable.

Unlike rlogin, FTP, or TELNET, the HTTP service is approximately 50% slower in the worst case. After analysis, we believe this performance decline for HTTP is somewhat artificial, resulting from the low-performance implementation of the HTTP application gateway in the Firewall Toolkit which does a separate *read()* and *write()* system call for each byte transferred. This overstates DTE system call overheads because, in this test, the system spends most of its time dispatching system calls that each do very little work. More recent application gateways, such as Gauntlet's,[12] perform more efficient I/O; we expect that DTE performance for those gateways should approximate the DTE performance for rlogin, TELNET, and FTP.

For NFS, we used Iozone and NFSstones, two widely-used NFS benchmark packages. The Iozone package tests sequential file I/O by writing a 64 MB sequential file in 8 K chunks, then rewinds it, and reads it back (i.e., it measures the number of bytes per second that a system can read or write to a file). We

many more enhancements could be added in future versions.

[11] ZeusBench version 1.0 is copyright Zeus Technology Limited 1996.

[12] Gauntlet is a registered trademark of Trusted Information Systems, Inc.

		Baseline	Percentage Change (%)	
		(n,n,n)	(n,y,n)	(n,y,y)
Iozone	Bytes/Second Written	107,372.50	-4	-20
	Bytes/Second Read	409,430.50	-28	-38
NFSstones	NFSstones/Second	69.83	-18	-38

Figure 5. NFS Test Results (larger numbers indicate better performance)

chose a file size large enough to prevent the cache from dominating the results. NFSstones creates and deletes many directories, then does a variety of file accesses, including writes, sequential reads, and non-sequential reads. Using these results in a formula, it generates a single numeric indicator of relative NFS performance.

As shown by the results in figure 5, performance of writes under NFS is moderately affected by the addition of DTE to the firewall; adding DTE to the server produces a higher impact on performance, with a 20% performance hit. Reads under NFS, however, dominate NFS performance, with a slowdown of 38% when both the firewall and the server are DTE hosts. However, neither the NFS application-level gateway nor the DTE-enhanced NFS server is optimized. Two possible locations for performance degradation in the NFS server are the double mediation necessary because of the primary/auxiliary domain combination and the manipulation of additional file handles needed for DTE mediation in NFS-mounted files. (We believe this last issue dominated the results and can be largely ameliorated.)

5. Related work

DTE firewalls are related most closely to firewall techniques [5, 6], mandatory access controls [2, 3, 4, 7, 10], and type-enforcing systems [4, 13, 14, 16, 20].

There are three fundamental types of firewalls: packet-filtering, circuit gateway, and application gateway. Packet-filtering firewalls allow packets to pass through only if they satisfy a set of filtering rules based on packet direction, physical interface, and a variety of packet fields (e.g., source address, destination address, UDP/TCP port numbers, etc). Circuit gateway firewalls force all TCP connections to go through an intermediary process which performs initial access control and then copies (and perhaps audits) data streams. Application gateway firewalls force data streams penetrating a firewall to be processed on a per-protocol basis by a trustworthy application proxy program that audits protocol usage and possibly screens damaging data. DTE could be added to either packet-filters or circuit gateways; however, we have incorporated DTE into application gateway firewalls because they provide greater opportunities to explore interactions between DTE and individual protocols.

Mandatory access controls, sometimes called "rule-based" controls, are centrally configured by system managers and then enforced on ordinary users and their software. A variety of access control techniques [1, 2, 3, 4, 7, 10] provide centralized control over resources. In general, DTE policies are a proper superset of the DoD lattice model [2] and its integrity variation [3]: DTE can be configured to provide a lattice but also can enforce non-hierarchical security policies such as assured pipelines [4] that drive information through policy-specified pathways of arbitrary connectivity and complexity. DTE also can be configured to provide integrity categories as in [10] and to support the transformation procedures and constrained data items of the Clark/Wilson model [7].

A number of systems have implemented type enforcement, including the Secure Ada Target [4] (later renamed LOCK [14]), Trusted XENIX [16], and DTOS [8], which adds type enforcement to Mach port, task, and virtual memory abstractions. Type enforcement also has been integrated into at least one Internet firewall product, the SCC Sidewinder[13] system, [17] an embedded turnkey system employing a fixed, vendor-supplied access control configuration. DTE [1], an enhanced form of type enforcement, was first demonstrated on an OSF/1 UNIX prototype and later ported to BSD/OS.

6. Future directions

This paper discusses the first phase of our work: a manually-administered DTE firewall prototype. We plan two additional phases to increase the user-friendliness and flexibility of the prototype.

For the second phase, we have formulated simple DTEL modules for organizing policies into more main-

[13]Sidewinder is a trademark of Secure Computing Corporation, Inc.

tainable segments. We also have formulated enhancements that allow DTEL modules to express administrative agreements between network enclaves and to support interactions between compatible but not identical policies. Currently, we are experimenting with loosely-constrained DTEL modules that can be loaded and unloaded while the system runs, at the discretion of the administrator. Since the modules may affect running processes and existing DTE policy elements (in this phase), the administrator must exercise caution in their use. Additionally, we have integrated IP-layer cryptography into the DTE kernel in order to extend DTE protections across unprotected WANs. For this encryption, we chose IPSec in order to provide IPv6 protection mechanisms to IPv4 packets, and are basing our prototype on the Naval Research Laboratory's IPv6/IPSec Software Distribution. As IPv6 becomes more widely available, we will port the DTE network code to use its features.

In the third phase, we are designing facilities for central administration of security policies within enclaves, and for convenient configuration of policy elements that are shared between enclaves. In this phase, each DTE host will boot with a minimal security policy and then contact an enclave-specific Domain and Type Authority (DTA) server to obtain the remainder of its policy, which will be expressed as a set of dynamically loadable DTEL modules. The DTA for an enclave will be responsible for managing consistent and dynamic security policies for hosts within its enclave and also for configuring policies in the DTE firewall and specified hosts to support relationships with external entities.

7. Conclusions

As currently realized, firewall security perimeters are inexpensive but also inflexible and somewhat weak. A central question for practical commercial security is whether the advantages of firewalls can be preserved while adding additional security controls to protect against inside attacks, to protect sensitive data from export, and to fortify servers against corruption from the network. Adding DTE to firewalls appears to address many security concerns because DTE supports role-based policies that relate resource access to individual responsibilities within the enterprise; and because roles provide an intuitive framework for expressing controlled sharing of resources between organizations. We believe the primary question for enhanced-security firewalls is whether useful security features can be added while preserving the functionality of existing applications and protocols, interoperating with non-enhanced systems and programs, imposing minimal performance overheads, and imposing acceptable administrative costs.

This paper reports our results with a prototype DTE firewall system that runs application gateways in DTE domains, exports network services that run in DTE-protected environments either on the firewall itself or on protected servers, and controls interactions with non-DTE systems. In general, our prototype DTE firewall coordinates DTE security protections of communicating endpoints and ensures that clients and corresponding servers execute in the same DTE domains. This relatively simple strategy has allowed us to demonstrate controlled sharing through the firewall, protection of sensitive information, confinement of client applications, and increased protection of network servers from client-based attacks. We have found that security can be increased significantly if DTE runs on the firewall and selected server systems. In addition, running DTE on client systems enables support for user roles and confinement of client applications. For rlogin, TELNET, FTP, HTTP, and mail, we have found no significant decrease in functionality or compatibility. The functionality of the NFS protocol has increased since it previously could not be exported safely. For many protocols, performance is not significantly affected by DTE; we believe that significant performance impacts for NFS and HTTP can be eliminated through optimization techniques. Finally, administrative costs, while still an open issue, appear to be manageable since useful DTE policies can be expressed concisely.

References

[1] L. Badger, D. F. Sterne, D. L. Sherman, and K. M. Walker. A Domain and Type Enforcement UNIX Prototype. In *Usenix Computing Systems Volume 9*, Cambridge, Massachusetts, 1996.

[2] D. E. Bell and L. LaPadula. Secure Computer System: Unified Exposition and Multics Interpretation. Technical Report ESD-TR-75-306, Electronics Systems Division, AFSC, Hanscom AF Base, Bedford, Massachusetts, 1976.

[3] K. J. Biba. Integrity Considerations for Secure Computer Systems. Technical Report ESD-TR-76-372, USAF Electronic Systems Division, Bedford, Massachusetts, 1977.

[4] W. E. Boebert and R. Y. Kain. A Further Note on the Confinement Problem. In *IEEE 1996 Carnahan Conference on Security Technology*, pages 198–203, 1996.

[5] D. B. Chapman and E. D. Zwicky. *Building Internet Firewalls.* O'Reilly and Associates, Inc., 1995.

[6] W. R. Cheswick and S. M. Bellovin. *Firewalls and Internet Security: Repelling the Wily Hacker.* Addison-Wesley, 1994.

[7] D. D. Clark and D. R. Wilson. A Comparison of Commercial and Military Computer Security Policies. In *Proceedings of the 1987 IEEE Symposium on Security and Privacy*, Oakland, California, 1987.

[8] T. Fine and S. E. Minear. Assuring Distributed Trusted Mach. In *Proceedings of the 1993 IEEE Computer Society Symposium on Research in Security and Privacy*, Oakland, California, 1993.

[9] J. Gosling and H. McGilton. The Java Language Environment: A White Paper. Technical report, JavaSoft, Mountain View, California, May 1996.

[10] S. B. Lipner. Non-Discretionary Controls for Commercial Applications. In *Proceedings of the 1982 IEEE Symposium on Security and Privacy*, Oakland, California, 1982.

[11] National Computer Security Center. *A Guide to Understanding Discretionary Access Control in Trusted Systems*, September 1987.

[12] The Object Management Group, Inc., Framingham, Massachusetts. *The Common Object Request Broker Architecture and Specification*, revision 2.0 edition, July 1995.

[13] R. O'Brien and C. Rogers. Developing Applications on LOCK. In *Proceedings of the 14th National Computer Security Conference*, pages 147–156, Washington D.C., October 1991.

[14] O. S. Saydjari, J. M. Beckman, and J. R. Leaman. LOCK Trek: Navigating Uncharted Space. In *Proceedings of the 1989 IEEE Symposium on Security and Privacy*, Oakland, California, 1989.

[15] D. L. Sherman, D. F. Sterne, L. Badger, S. L. Murphy, K. M. Walker, and S. A. Haghighat. Controlling Network Communication with Domain and Type Enforcement. In *Proceedings of the 1995 National Information Systems Security Conference*, Baltimore, Maryland, 1995.

[16] D. Sterne. A TCB Subset for Integrity and Role-Based Access Control. In *Proceedings of the 15th National Computer Security Conference*, pages 680–696, Baltimore, Maryland, 1992.

[17] D. J. Thomsen. Sidewinder: Combining Type Enforcement and UNIX. In *Proceedings of the 11th Computer Security Applications Conference*, Orlando, Florida, December 1995.

[18] K. W. Walker, D. F. Sterne, M. L. Badger, M. J. Petkac, D. L. Sherman, and K. A. Oostendorp. Confining Root Programs with Domain and Type Enforcement. In *Proceedings of the 6th Usenix Security Symposium*, San Jose, California, 1996.

[19] J. Williams. *Bots and Other Internet Beasties.* Sams.net Publishing, 1996.

[20] S. Wiseman. A Secure Capability Computer System. In *Proceedings of the 1986 IEEE Symposium on Security and Privacy*, Oakland, California, 1986.

Secure Execution of Mobile Programs *†

Raju Pandey Brant Hashii Manoj Lal
Parallel and Distributed Computing Laboratory
Computer Science Department
University of California, Davis, CA 95616
{pandey, hashii, lal}@cs.ucdavis.edu

Abstract

There is increasing interest in computing models that support extensibility of systems through code migration. Although appealing both from the system design and extensibility points of view, extensible systems are vulnerable to an external program's aberrant execution behaviors. In this paper, we examine the problems of resource access control and resource consumption. We propose solutions for these problems and analyze their effectiveness.

1 Introduction

On today's Internet, users routinely and often unknowingly download and run programs, such as Java applets. Some web servers permit users to upload foreign programs and execute them. These practices already have the sanction of widespread use, but their security implications have not yet been systematically addressed. In the brief, dynamic history of the Internet, this situation is not unusual. The implementation of new communication mechanisms and computing paradigms has often preceded a rigorous analysis of the security issues they engender.

*This work is supported by the Defense Advanced Research Project Agency (DARPA) and Rome Laboratory, Air Force Materiel Command, USAF, under agreement number F30602-97-1-0221. The U.S. Government is authorized to reproduce and distribute reprints for Governmental purposes notwithstanding any copyright annotation thereon. The views and conclusions contained herein are those of the authors and should not be interpreted as necessarily representing the official policies or endorsements, either expressed or implied, of the Defense Advanced Research Project Agency (DARPA), Rome Laboratory, or the U.S. Government.

†Portions reprinted, with permission, from (IEEE Internet Computing; volume 2, number 6, pages 35-45; Nov/Dec 1998. ©1998 IEEE), from (13th Conference on Object-Oriented Programming. ECOOP'99; June 1999; published in Lecture Notes in Computer Science number 1628, pages 449-473 by Springer-Verlag) and from (Proceedings of the First International Symposium on Agent Systems and Applications; ©Oct. 1999 IEEE)

In computing models that support program migration, a host provides a set of services to an *external program* by loading it and executing it within a local execution environment such as an operating system or a run time system. We call such computing models *extensible computing models*. Two examples are extensible operating systems and mobile programming systems.

Extensible operating systems allow the collocation of external programs by loading them directly into the kernel's address space. Thus, the external program shares its address space with the kernel and all other loaded external programs. SPIN [2] is an example of an extensible operating system that lets users download user-level extensions into the kernel.

Mobile programming systems also support migration by letting users upload programs to a remote host. Further, external programs can stop in mid-execution and then migrate to another host while retaining their state and data.

In addition, Web browsers such as Netscape support extensibility by letting applets be downloaded and executed within the browser. All of these systems provide an execution environment that loads externally defined user programs and executes them within its local name space. We refer to these execution environments as runtime systems.

The extensible computing model is appealing, first, because it lets operating system kernels implement only basic, core functionality, which can then be extended through external programs. This facilitates customization and efficient implementation of specific services and policies, such as application specific memory management or caching policies.

Second, external programs are sometimes far more efficient at utilizing network bandwidth than traditional programming paradigms, such as remote procedure call (RPC). For example, if the external program encodes an application that must filter huge amounts of data, it can migrate to the host with the data, execute there, and then return with its results to the originating host. In this case, the network load incurred by migrating the external program is insignificant

0-7695-0490-6/99 $10.00 © 1999 IEEE

compared with the cost of migrating data and processing it locally.

Although appealing both from system design and extensibility points of view, runtime systems are extremely vulnerable to misbehaving external programs. Since external programs run within the same name space as the runtime system, many of the traditional protection mechanisms, such as address space containment, no longer apply. As a result, an external program can maliciously disrupt the execution behavior of extensible systems by interfering with the runtime system's execution or with the execution of other programs within the name space, by using unauthorized resources, by over-using resources. It might also access unauthorized resources or use more than its fair share of resources, denying them to other programs.

In this paper, we focus on two problems: resource access control and resource consumption control. Note that our focus is protecting a runtime system against external programs. We do not address the problem of protecting the communication medium [14] or protecting an external program from runtime systems. Furthermore, we do not address the problem of correctly identifying the source of a mobile program (authentication). The rest of this paper is organized as follows: In Section 2, we describe the two security problems in detail. In Section 3, we present our solution for resource access control. We then describe our approach to resource consumption control in Section 4. We provide a brief summary in Section 5.

2 Security Problems

To help classify security issues, we distinguish two types of resources that a runtime system provides to external programs:

- System resources are those implicitly allocated to external programs, such as memory and CPU.
- Conceptual resources are those explicitly defined and managed by a host. They have well-defined interfaces that external programs use to access resources and request services. For example, a host might provide an interface to a database repository.

This distinction highlights fundamental differences in the mechanisms used to control resource access:

- Because system resources are implicitly allocated and managed by runtime systems, the mechanisms for access control are usually implemented within the runtime system. They also depend on the resource model that runtime systems create. For example, CPU resource access control is traditionally implemented in runtime systems through CPU scheduling algorithms.
- Because conceptual resources are accessed by explicit calls from external programs, access control is generally based on trapping these calls in software. For example, in the Java runtime system (JRTS) [5] calls to protected resources are trapped when these resources call a reference monitor, which the host uses to track and control accesses.

2.1 Resource-centric Security Problems

Security problems result out of conflicts between how a host wants external programs to access its resources and how the external programs access them. There are two kinds of resource access constraints:

- *Access control* refers to restricting a program's access to resources. We focus on limiting access to conceptual resources with explicit interfaces.
- *Consumption control* refers to restricting how much of a given resource a program can consume.

Our primary focus here is on those aspects of resource access and consumption control that are specific to the extensible computing model.

2.2 Resource Access Control

Access control prevents an external program from using unauthorized resources. Because external programs execute within the runtime system's name space, they can directly access any resource by naming it. Naming involves getting a resource handle and using it to invoke operations on the resource. An external program can use the handle to read sensitive files and send the information to remote hosts by accessing network resources. It can also disrupt the operations of a computer system by accessing local resources in an unintended manner. For example, the "Ghost of Zealand" Java applet misuses the ability to write to the screen: It turns areas of the desktop white, making the machine practically useless until it is rebooted. [1]

Once downloaded on a machine, external programs can read private files from local disks or copy proprietary information by accessing databases. Indeed, Hamburg's Chaos Computer Club demonstrated on German television how to use ActiveX, Microsoft's external programming system, to steal funds. In this exploit, the victim uses Internet Explorer to visit a Web page that downloads an ActiveX control. The control checks to see if Quicken, a financial management software, is installed. If it is, the control adds a monetary transfer order to Quicken's batch of transfer orders. When

[1] For full details, see http://www.finjan.com/applet_alert.cfm or http://www.internetworld.com/print/1998/05/11/webdev/19980511-hostile.html

the victim next pays the bills, the additional transfer order is performed. All of this goes unnoticed by the victim.[2]

Although the resource access control problem has been studied within the context of traditional systems, the problems are different for extensible systems [6] in several ways.

First, in extensible systems, authorization is more complex. In traditional operating systems, programs run on behalf of principals who are given certain rights. Once a program attains these rights, they usually remain valid during the program's execution. In extensible systems, however, an external program contains individual components that might have different rights and permissions. Hence, the level of granularity at which access rights must be checked and enforced is much finer – sometimes at the level of individual objects and functions.

Second, security mechanisms should be independent of the site's resources. Most traditional operating systems manage a fixed set of resources such as memory, CPU, files, and the network. Extensible systems, however, must manage resources that can vary from site to site.

Finally, most traditional operating systems implement an access control model that either allows or denies access. Extensible systems can allow conditional resource access based on runtime or program state [10]. For example, a database vendor can specify that if there are more than 20 external programs in the system, each external program can only access its database up to 10 times.

2.3 Resource Allocation Problem

When external programs use more than their share of resources, they leave the system vulnerable to attack. For example, external programs can stage denial-of-service attacks by intentionally over-using CPU, thereby denying CPU to other programs and the runtime system. Consumption control is required not only for system resources, but also for conceptual resources. For example, by opening multiple socket connections and flooding the network with data, an external program can deny network resources to other programs.

The resource-consumption problem is similar to the CPU-scheduling problem in that both require the runtime system to control the resources allocated to requesting programs. The two problems differ in the type of control that runtime systems need to exercise. The primary goal in CPU scheduling algorithms is to allocate CPU so it provides some quality of service (QoS) to executing programs. The QoS requirements can be specified as constraints, such as optimal resource utilization, response time, lower bounds on resource allocation, and deadline. In mobile systems, we can classify resource usage constraints both according to how mobile programs want to consume resources, and how a host wants to facilitate as well as protect usage of these resources. This results in two different views of resource allocation – client (mobile program) view and server (host system) view.

Notation: Let the term $\sigma_t(P)$ denote the amount of CPU allocated to a program P in a *schedulable unit*[3] t.

2.3.1 Client View

From a client program's perspective, requests for resources are driven by how the program demands and uses resources. Programs want to use as much CPU as possible so that they can perform their job quickly. They specify such requirements in terms of several constraints such as the following:

- **Lower bound**: A lower bound constraint, l, associated with a program P specifies that in case of contention, P be allocated at least $l\%$ of CPU in each schedulable unit.

$$\langle \forall\, t : \sigma_t(P) \geq (\frac{l}{100}) \times t \rangle$$

- **Weight**: A weight constraint, w, associated with a program P specifies that P be allocated $w\%$ of CPU in each schedulable unit.

$$\langle \forall\, t : \sigma_t(P) = (\frac{w}{100}) \times t \rangle$$

- **Share**: Shares are closely related to weights in that relative amounts of shares owned by a program define the program's weight constraint. Thus, if program P has s shares and the total number of shares in the system is S, then

$$\langle \forall\, t : \sigma_t(P) = (\frac{(\frac{s}{S})}{100}) \times t \rangle$$

- **Deadline**: Real-time constraints in the form of $<S,E,T>$ for a program P specify that between times S and E, P must get T amount of CPU.

$$\langle \sum_{t \geq S}^{t \leq E} \sigma_t(P) = T \rangle$$

2.3.2 Server View

From the server's perspective, two concerns govern the allocation of resources – meeting client's needs and tightly controlling allocation of resources. Two factors unique to mobile environments accentuate the latter concern. First, a level of distrust exists between a host

[2]For more on this, see http://www.iks-jena.de/mitarb/lutz/security/activex.hip97.html or http://www.iksjena.de/mitarb/lutz/security/activex.en.html

[3]We divide the time line into schedulable units and specify resource usage constraints in terms of these schedulable units.

and mobile programs since mobile programs and hosts typically belong to different administrative domains. Mobile programs can maliciously disrupt a host by using unauthorized resources, by over-using resources, and by denying resources to other programs. Second, in distributed systems such as the web [1], hosts may provide varying degree of services to clients. A host may differentiate requests from different clients and allocate resources to these requests in accordance with the kinds of services the host wants to provide. For instance, a host may reserve 85% of its resources to mobile programs that originate from the paying customer sites and allocate the rest for other programs.

In mobile programming models [3, 13], the resource allocation problem has an additional dimension: A mobile program can circumvent resource control by migrating to another host and then returning to its previous host for more resources. Resource usage constraints, thus, apply not only to specific executions but to all executions of a program. We refer to such constraints as *lifetime constraints*.

Thus, a host must differentiate and categorize mobile programs if it intends to impose resource constraints on them. In addition to the lower bound, shares and weight constraints, a host may also specify the following constraints:

- **Priority**: A priority constraint specifies that, given a set of programs to schedule, a host selects the program with the highest priority.

- **Upper bound**: An upper bound constraint, u, associated with a program P specifies that, within each schedulable unit, P be allocated at most $u\%$ of CPU.

$$\langle \forall\, t : \sigma_t(P) \leq (u/100) \times t \rangle$$

  A host can also specify upper bounds in absolute form.

- **Life-time constraints**: A lifetime constraint, l associated with a program P specifies that P be allocated at most l units of CPU time over all executions of P.

2.3.3 The Scheduling Problem The CPU resource control problem is, therefore, one of developing a scheduling scheme that, given client and server resource usage constraints, constructs a schedule that enforces the constraints.

3 Access Control Model

Out access control model contains two parts: a resource model for representing resources and an access constraint specification language. We describe the two in detail below.

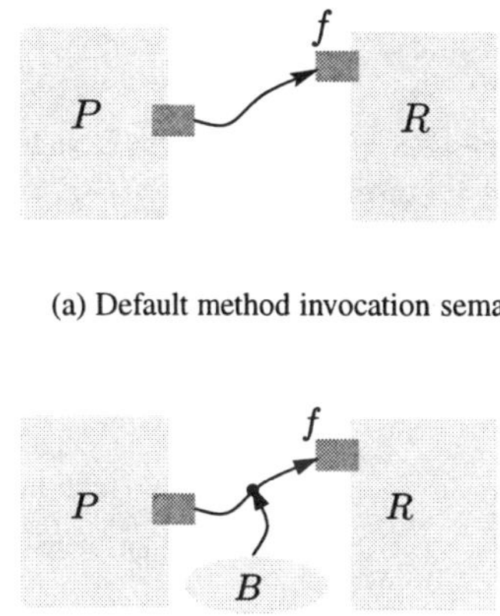

(a) Default method invocation semantic

(b) Security constraints on method invocations

Figure 1. Method invocation semantics

3.1 Resource Model

A site provides many resources to a mobile program. These resources include classes for utility libraries, accessing files, networks, and interfaces to other resources such as a proprietary database. For instance, a site providing access to a weather database exports a set of interfaces that specify how the database can be accessed. In our security model, each Java class or method represents a resource and, thus, is a unit of protection. Our access control mechanism does not differentiate between system classes and user-defined classes, or between locally defined classes and classes downloaded from remote hosts. The model also allows the definition of class-subclass relationships among resources using the Java's inheritance model.

3.2 Access Constraint Specification Language

The access constraint specification language contains two parts: a notation for constraining accesses to resources and its inheritance model.

3.2.1 Access Constraints We first describe the motivation behind our access control language. A Java program uses a resource by invoking its methods. In Fig. 1(a), we show that program P invokes a method f to access resource R. During an execution of P, the control jumps to f, executes f, and returns back to P upon termination. The Java compiler implements a simple access semantics in which there are no constraints on accesses to R through f.

Our approach is to allow a host to make the access relationship between P and R *conditional* by adding a constraint, B (see Fig. 1(b)). The access constraint is specified *separately* from both P and R and has the effect of imposing the constraint that P can invoke f on R only if condition B is true. A site, thus, restricts accesses to specific resources by enumerating a set of access constraints, which forms a site's access control policy.

Below, we present the core aspects of the language. The following EBNF shows how a site can specify access constraints:

```
Constraints       ::= { AccessConstraint | EnableStatement }
AccessConstraint::= deny '('[Entity]Relationship Entity')'
                        [when Condition]
EnableStatement ::= enable '(' Entity
                        [Relationship Entity] ')'
Relationship      ::= ↦ | ⊣
Entity            ::= ClassIdentifier | MethodIdentifier
                        | GroupName
Condition         ::= BooleanExpression
GroupStatement    ::= define group GroupName
                        '{' Entity ';' { Entity } '}'
```

A site controls accesses to different resources (Java objects) by defining a set of AccessConstraints. We describe the various terms in the grammar informally below:

- Entity: An entity denotes objects and method invocations of Java programs. A ClassIdentifier, thus, identifies the set of objects to which a given access relationship applies. Similarly, a MethodIdentifier denotes a set of invocations of a method. In addition, an entity can be a group of entities.
- Relationship: The composition mechanisms of a programming language allow one to define various relationships (data composition through aggregation and inheritance, and program composition through method invocations) among the entities of a program. We are primarily interested in the following two access relationships here:
    1. Instantiate ($\dashv$): A relation $E \dashv R$ exists if an entity E creates an instance of class R.
    2. Invoke ($\mapsto$): A relation $E \mapsto R$ exists if an entity E invokes an entity R.
- Condition: The term Condition denotes a boolean expression that can be defined in terms of object states, program state (global state), runtime system state, security state, and parameters of methods.
- Enable: In addition to access constraints, we support an enable statement that allows a host to certain accesses, thereby overriding these constraints. This is needed in cases when a host wants to override the default principles of least privileges. For example, assume that a security policy specifies that an applet cannot access the file system. The security infrastructure implements a default policy of least privilege, which ensures that the applet cannot access the file system directly or indirectly by calling other methods that access the file system. However, in many cases, this may not be desirable [15]. For instance, suppose the applet can write to the screen using the font files stored on the disk. In such cases, we want to enable the display manager to be able to access these files, regardless of the calling program. The `enable` statement allows one to override the default policy. This is similar to the `enablePrivileged` command of the JDK1.2 security model [5].
- Group: The current implementation defines a group based on its name. However, this can be extended to define an entity on the basis of its source, signature, or behavior pattern.

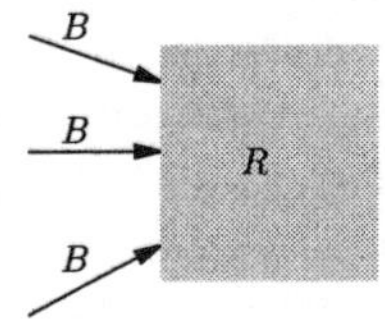

(a) Global constraints

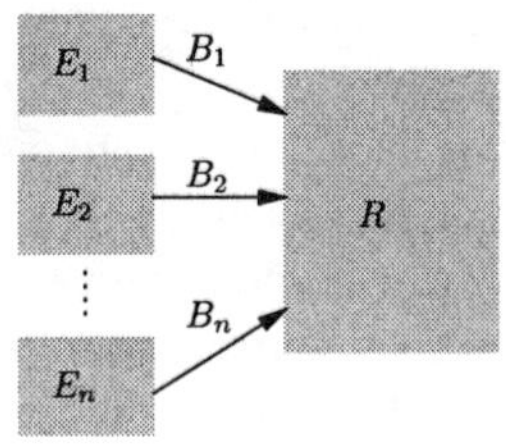

(b) Selective access constraints

Figure 2. Category of access constraints

3.2.2 Semantics

An access constraint of the form

$$\texttt{deny } (E\ \xi\ R) \texttt{ when Condition}$$

specifies that entity E cannot access R through relationship ξ if Condition is true. Since E is optional, there are two kinds of access constraints: *global constraints* and *selective access constraints*. Global constraints denote those constraints that do not depend on the initiator of the access relationship. As shown in Fig. 2(a), no program can access R when B is true. For example, a host may specify the constraint that no Java code can access a set of proprietary files.

Selective access constraints denote those constraints that depend on the initiator of the access relationship. For instance, as shown in Fig. 2(b), each entity E_i's access to R is constrained by a separate and possibly different B_i. A site can use selective access constraints to associate different security policies with different Java programs that come from different sites.

3.2.3 Inheritance of Access Constraints We now present an inheritance model that describes what the denial of a resource means to its subclasses.

Assume that a site defines two resources, R_c and R_s:

```
class R_c {
    public void f();
    public void g();
    public void h();
}

class R_s extends R_c {
    ⋮
}
```

R_s is a subclass of R_c. R_s inherits methods f, g, and h from R_c. Assume that the site defines the following constraints on the resources:

$$\begin{array}{l} \texttt{deny } (E \not\mapsto R_c.f) \texttt{ when } B_{cf} \\ \texttt{deny } (E \not\mapsto R_c.g) \texttt{ when } B_{cg} \\ \texttt{deny } (E \not\mapsto R_s.f) \texttt{ when } B_{sf} \\ \texttt{deny } (E \not\mapsto R_s.h) \texttt{ when } B_{sh} \end{array}$$

There are two components to the inheritance model:

- Inheritance of access constraints: A subclass inherits all access constraints from its superclasses. Hence, the resulting access constraint on invocations of g on an instance of R_s is defined by the following expression:

  $$\texttt{deny } (E \not\mapsto R_s.g) \texttt{ when } B_{cg}$$

  Access constraints are not inherited from subclasses to superclasses. Hence, although the access constraint on h in R_s is B_{sh}, there are no access constraints on h in R_c.

- Strengthening of access constraints: A subclass cannot override its inherited constraints. Specification of additional constraints in the subclass only strengthen the constraints defined in its superclasses. Hence, the resulting access constraint on invocations of f on an instance of R_s is:

  $$\texttt{deny } (E \not\mapsto R_s.f) \texttt{ when } B_{cf} \vee B_{sf}$$

  In other words, method $R_s.f$ cannot be invoked from E if either B_{cf} or B_{sf} is true.

This model of inheritance ensures that a mobile program cannot override access constraints on methods by defining a subclass and by weakening the access constraints. Also, the above inheritance model applies for access constraints on $\dashv$ as well. That is, if a class R_c cannot be instantiated, none of its subclasses can be instantiated.

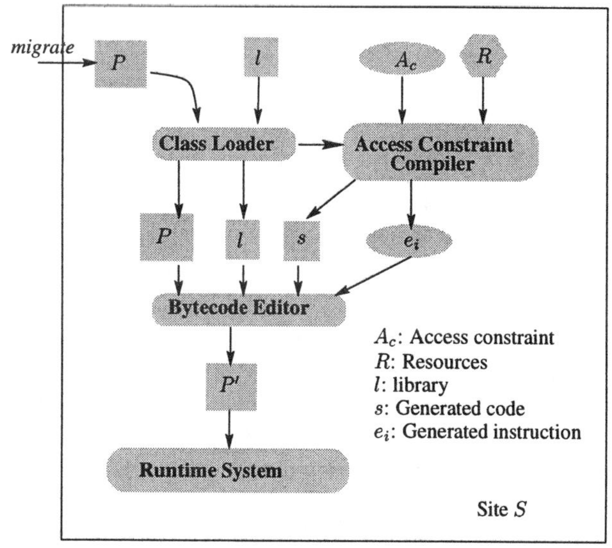

Figure 3. Security policy enforcement of mobile programs

3.3 Access Constraint Enforcement

An enforcement of access constraints on a resource involves placing interposition code between the resource access code and resource definition code. The interposition code checks if a specific resource access is allowed. It can be inserted *manually* by site managers, generated by the compiler, or defined by the runtime systems or operating systems through special system calls. For instance, in the Java runtime system [4, 5], resource developers manually insert calls to a reference monitor in the resources they want to protect. The reference monitor consults access control policies to check if a specific resource access is allowed.

We use an alternate approach for generating interposition code. In this approach, a set of tools generates the interposition code and integrates them within mobile programs and resources before they are loaded in the JVM. In this approach, there are no reference monitors. In essence, the approach generates reference monitors on the fly and integrates them within the relevant Java programs and resources. The approach, thus, eliminates the need to manually include calls to reference monitors in resource definitions.

In Fig. 3, we describe our implementation for enforcing access control policies on Java programs. We show a Java program P that migrates to a site S. R denotes resources that the site makes available to mobile programs; and l denotes local libraries linked into P.

During class name resolution and dynamic linking, the Java class loader [8] retrieves R and l and passes them to a tool, called the *access constraint compiler*. The access constraint compiler examines P, R, and l to determine the resource access relationships that must be constrained in order to implement the access constraint A_c. It then generates interposition code s that implement the specific access constraints. It also generates a set of editing instructions e_i for the bytecode editor. The bytecode editor uses e_i to integrate s within P, R and l. The transformed programs and resources are then loaded into the JVM and executed.

3.4 Code Generation and Binary Editing

We now describe the nature of the code that is generated and its integration within mobile programs. Our code generation and editing involves modifying class definitions in order to add runtime state to classes and to insert runtime checks into methods.

An access constraint of the form

deny $(E \ \xi \ R)$ when B

is implemented by generating the following code:

```
if (B)
    then error(); // raise exception
else
    access R
```

and patching it into classes and methods. The nature of the editing depends on the nature of the access constraints. Global constraints of the form

deny $(\xi \ R)$ when B

specify constraints on accesses to R without any regard to objects or methods that may access R. The generated code is, thus, integrated into the methods of R. On the other hand, selective access constraints of the form

deny $(E \ \xi \ R)$ when B

imposes conditions on accesses to R from E. In this case, R must determine if it had been called by E. We implement this using stack inspection [15, 16], which we will discuss in the following sections.

We also support addition of security states to specific Java classes in order to monitor site-specific behavior. This mechanism allows a site to customize its security policies, especially if the policies cannot be represented directly by the policy language. Security state objects are added to a class definition by using the statement:

```
add SecurityStateType SecurityStateObject
    to R
```

The constraint compiler generates code for initializing these objects.

3.5 Discussion

In this section, we analyze the proposed technique for its suitability as an access constraint enforcement mechanism and discuss its performance behavior.

3.5.1 Characteristics of the Approach In our approach, a site specifies access constraints separately from mobile programs, resources, and other class definitions. Further, the access constraint enforcement mechanism is not part of either the Java runtime system or the compiler. This impacts how access control code is managed and enforced at a site:

- Both access constraints and resource definitions can be modified independently. This makes it easy for a site to specify different access constraints for different mobile programs for the same resource. For instance, a site may specify that mobile program P can access R under condition B_p whereas mobile program Q can access R under condition B_q.

- The same set of access constraints can be applied to different resources without requiring one to copy it from one resource to another. For example, if a single access constraint B applies to multiple resources, it can be defined once and used for all resources.

- An important advantage of the separation is that our approach can be used for enforcing security on resources that were not designed with security in the first place. In other words, the security component can be added to a resource after it has been designed and implemented. Thus, it frees a library or resource designer from worrying about security concerns when designing and implementing the library.

Despite these advantages, there are several limitations of our approach.

3.5.2 Performance Analysis In this section, we describe the performance behavior of the access constraint enforcement mechanism. Specifically, we analyze the following:

- What are the time and space overheads associated with our approach?
- How does our approach perform with respect to the Java runtime system's approach for enforcing access control?

We performed our experiments on a 266 MHz Pentium II running Red Hat Linux 5.0. The results show that both the time and space overheads of the approach are moderate. Further, the approach performs better than the Java runtime system in certain cases.

Overhead Measurements: We measured both the time and space costs of modifying resources.

There are four factors that affect the execution time associated with access constraint check code generation and editing:

- the cost associated with reading a method
- the number of access constraints
- the types of constraints
- the number of occurrences of restricted methods in a program

We do not consider the cost of reading class files in our measurements since the run-time system must perform this operation anyway.

In the first experiment, we looked at how the size of the method being modified affects the cost of editing. In this experiment, only a single method invocation must be wrapped. The cost of editing here is minimally affected by the size of the method. The cost varied between 0.08 and 0.16 seconds for methods ranging from 0 to 3200 instructions. In the second experiment, we looked at how the cost of editing changes when the number of method calls that needs to be wrapped changes. We found the cost to be proportional to number of methods that are wrapped.

We have also calculated the increase in size caused by adding code to class definitions. While the amount of code that is added to a class is independent of the size of the class, it depends on the number of method invocations that need to be wrapped and the complexity of the boolean portion of the constraint. For one wrapper, the minimum addition size (for a true boolean constraint), is 56 bytes. For two simple boolean expressions, it is about 206 bytes.

Performance Comparison: We now compare the performance behavior of our approach with the runtime system approach, as implemented in the JDK 1.1.3.

For this experiment we created a small program to test the performance of implementing security checks around one method invocation. Since the actual amount of work a particular site must perform depends on both the complexity of the access control policy and the number of restricted method invocations in a program, implementing a single policy statement once forms a good basis for comparison. We based our comparisons on an access control policy limiting the number of times a particular resource can be accessed. The complete code for our approach is shown in Fig. 4. We implemented the same policy using Java's security manager as shown in Fig. 5. The test program calls the constrained method variable number of times. The access policy is that the method cannot be called more than 1000000 times.

```
class SecState {
  public SecState() {count = 0;}
  public int check()
    { count++; return count; }
  private int count;
}
```

(a) Security object

```
add SecState SecurityState to R
deny ↦ R.f()V when
  #1.SecurityState.check() > 1000000
```

(b) Control access constraints

Figure 4. The binary editing approach

Figure 6 shows the execution times of our approach and the Java's runtime system approach. In our approach, there is an initial overhead of about 0.08 seconds for code editing, which does not occur in the Java runtime system. However, after about 100000 method calls, our approach performs better than the Java runtime system. This is because our approach inlines the access control check code, whereas in case of the Java runtime system approach, each access constraint check involves making two method calls: one to the system, to get the security manager, and another to the security manager itself. We can reduce our cost even further by pre-editing the methods if we know that only a single access constraint will be applied to the method, as is the case in the Java runtime system approach. Our approach, in this case, will then always outperform the Java runtime system approach.

In the second experiment, we ran the same program with no policy implemented. As shown in Fig. 7, the Java runtime system is always less efficient that our approach. This is because in the Java runtime system approach, a method must always call the runtime system to check if there is a se-

```
class newSecMan
  extends SecurityManager {
  public newSecMan() {count = 0;}
  public void checkf()
      throws SecurityException {
    count++;
    if (count > 1000000)
      throw new SecurityException();
  }
  int count;
}
```

(a) Security Manager

```
class R {
  public void f() {
    newSecMan security;
    security =
      System.getSecurityManager();
    if (security != null)
      security.checkf();
  }
}
```

(b) Resource definition

Figure 5. The Java Runtime System-based approach

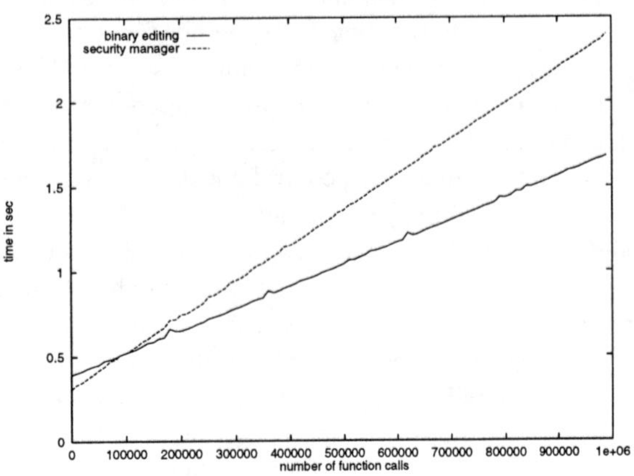

Figure 6. Comparison of execution times with a policy

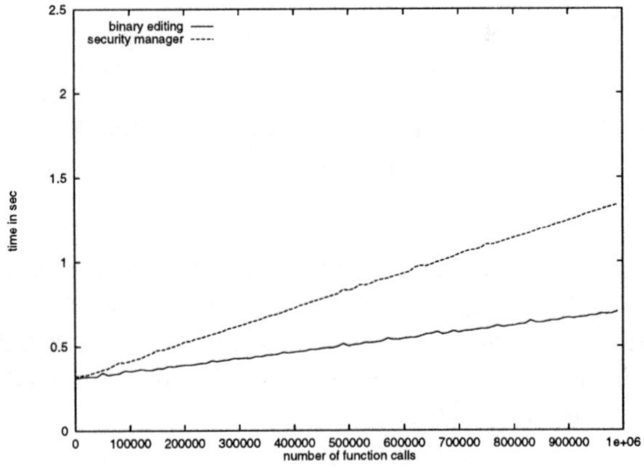

Figure 7. Comparison of execution times without a policy

curity manager installed, incurring the overhead of this call. Our approach does not incur any overhead since it does not add any code to methods that do not need to be constrained.

4 Resource Consumption Control

In this section, we describe our approach to resource consumption control. We have developed a runtime interface and a specification language that clients and servers use to specify resource usage constraints. We briefly describe them below.

We organize mobile programs into groups and subgroups. For instance, a group `ucdavis.edu` denotes all mobile programs that originate from this domain. This group may contain subdomains such as `cs.ucdavis.edu` and `ece.ucdavis.edu`. The notion of groups and subgroups results in a hierarchical partitioning of mobile programs. Clients can define their own groups and determine constraints for individual jobs within the group. Hosts can specify resource constraints on groups, subgroups, or individual mobile programs.

Mobile programs and hosts can specify the following sets of constraints:

- **Client constraints:** Real-time constraints and shares.
- **Host constraints:** Shares, priority, absolute upper bound, and lifetime constraints.
- **Dynamic constraints:** Both mobile programs and hosts can change their constraints dynamically. The constraints can be specified as functions of the state of the system. Currently, we consider constraints that can vary on only two state variables: the number of mobile programs belonging to a group and the time of day.

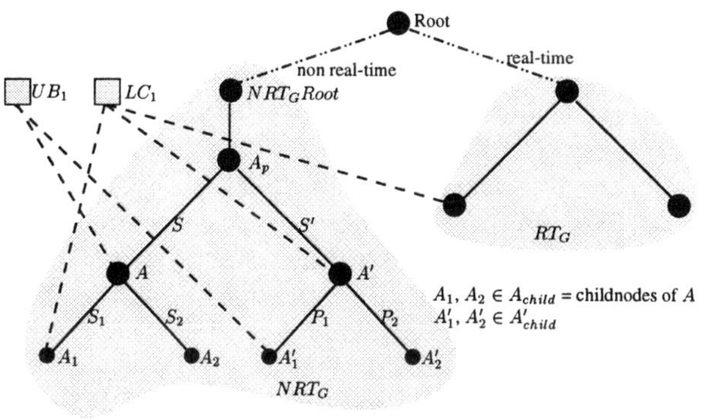

Figure 8. The scheduling graph

We omit the description of the constraint specification language here due to lack of space.

We now describe the overall approach for scheduling resources to mobile programs.

- Construction of scheduling graph: The scheme partitions mobile programs into real-time and non real-time programs. It captures the group-subgroup relationships along with the various constraints to construct a scheduling graph.

- Application of algorithms: The scheme applies three algorithms to the scheduling graph: (i) an algorithm to enforce upper bound and lifetime constraints; (ii) an algorithm to enforce share and priority constraints; and (iii) an algorithm to enforce real-time deadline based constraints.

- Monitoring of system state: Since the host can specify constraints as a function of state variables, the scheduling scheme monitors the state of the system and adapts to the changes in the resource constraints by modifying the scheduling graph.

In the following sections, we describe the individual algorithms and how they are composed to build the scheduling scheme.

4.1 Construction of Scheduling Graph

The scheduling scheme first builds a *scheduling graph* (Fig. 8) from resource usage constraints. The scheduling graph is made of three subgraphs: (i) Real-time, (ii) Non real-time, and (iii) upper-bound. The real-time subgraph is a single node (a real-time guarantee group, RT_G) containing all programs that specify deadline based constraints.

The non real-time subgraph, NRT_G in Fig. 8, is a hierarchical graph, where each node denotes a group and each edge the group-subgroup relationship. Mobile programs are at the leaves of NRT_G. The edges of NRT_G are annotated with share or priority constraints. For instance, in Fig. 8, the label on edge (A_p, A) specifies that group A has S shares. Similarly, the label on (A', A'_1) specifies that mobile program A'_1 has priority P_1. The amount of shares or priorities allocated to a group node is relative to its parent group. For instance, group A_1's share S_1 of CPU resources are with respect to the CPU resources allocated to its parent group, A. This results in a modular allocation of CPU: Any changes in the share allocations to A_1 or A_2 do not affect the CPU allocations to programs that belong to different groups, for instance A'.

The upper bound subgraph represents the security constraints. Nodes in this subgraph denote specific upper bound and lifetime constraints. Edges link these constraints to the relevant groups and mobile programs. Upper bound and lifetime constraints are general in that they can encapsulate more than one node in the scheduling graph. Moreover, the nodes encapsulated by a particular constraint need not be at the same level. For instance, as shown in Fig. 8 the upper bound constraint UB_1 applies to group A and mobile program A'_1. There is a need for such constraints so that the host can control mobile programs belonging to different levels in the hierarchy. For example, suppose a site wants to impose an upper bound constraint that mobile programs accessing a particular database be allocated at most 10% of CPU. Such mobile programs may exist in different groups and may span multiple subtree domains.

Hosts define the backbone of the scheduling graph – a graph consisting of empty RT_G and possibly non-empty NRT_G. When a new client program arrives, the host can create a new group node (A_p) for the client, specify constraints for A_p, and add A_p at an appropriate place in the scheduling graph. The client program can create subgroups under A_p and define resource usage constraints for the subgraph under A_p. Note that the scheduling graph is dynamic; it changes whenever mobile programs arrive and when client-specific and host-specific resource usage constraints change.

4.2 Application of Algorithms

The scheduling scheme operates on the scheduling graph to allocate CPU to mobile programs. An important aspect of a scheduling scheme is how conflicts between the mobile program and host specific resource usage constraints are resolved. A scheduling scheme must include a set of policies, called *algorithm composition policies*, that resolves any conflicts. Our scheduling scheme implements a policy that always resolves conflicts in favor of host constraints. The policy is summarized as follows:

- The scheme first ensures that constraints related with the security aspects of the host are satisfied. Thus, it

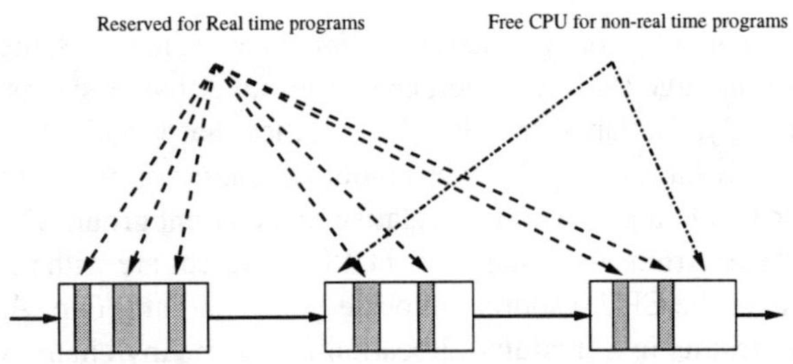

Figure 9. List of quantum time chunks with reservations for real-time programs

always applies the upper bound algorithm first to enforce the upper bound and lifetime constraints even if it means that other programs do not get their requested CPU allocation or that some deadlines are missed.

- Next, the scheme enforces host-specified priority and share constraints in order to implement host's preferences.
- Finally, the scheme schedules non real-time jobs according to their relative shares, whereas it schedules real-time jobs so that their deadlines constraints are met.

The scheme partitions the continuous time line into small *quantum time chunks* (Fig. 9). Within each quantum time chunk, it schedules mobile programs from RT_G according to their reservations. The reservations fix the times when the scheme allocates CPU to real-time programs. This is shown as shaded parts in a single quantum time chunk. The scheme then allocates the remaining time to non real-time programs. The scheduling of non real-time mobile programs starts from the root node of NRT_G graph ($NRT_G Root$). The scheme traverses from $NRT_G Root$ to one of the leaves of the graph.

In the next sections we describe the individual algorithms for non real-time and real time programs.

4.2.1 Scheduling of Non Real-time Programs The crux of the algorithm for non real-time programs is the decision associated with the children nodes of a node. If the children nodes have priority based constraints, the algorithm selects the child node with the highest priority. If the children nodes have share based constraints, the algorithm selects a child node on the basis of the share allocations of the children nodes.

The algorithm to allocate CPU on the basis of share based constraints extends the ideas in the SMART scheduling algorithm [11] to a hierarchy. We define three quantities: upper virtual time (UVT), virtual finish time (VFT) and lower virtual time (LVT) for each node in the hierarchy. The reason we require UVT and LVT is that in NRT_G, each internal node is both a child node and a parent node. UVT of the internal node is compared with the LVT of the parent node to select the child node that should be scheduled.

Assume that the algorithm has reached a particular internal node A and the children nodes of A have share based constraints associated with them (Fig. 8). Let the parent of A be A_p, and let A own S shares under A_p. Let A_{child} be the set of children nodes of A. Also, let each A_i in the set A_{child} own shares S_i.

- UVT: When A joins the hierarchy for the first time at time t:

$$UVT_A(t) = LVT_{A_p}(t) \tag{1}$$

Later, if a mobile program from the subtree within A was initiated for execution at time τ and is currently (t) executing:

$$UVT_A(t) = UVT_A(\tau) + \frac{t - \tau}{S} \tag{2}$$

- LVT: The lower virtual time at A selects one of A's children. Initially, when A joins the hierarchy,

$$LVT_A(t) = 0 \tag{3}$$

Later, if a mobile program from the subtree within A was initiated for execution at time τ and is currently (t) executing,

$$LVT_A(t) = LVT_A(\tau) + \frac{t - \tau}{\sum_{a \in A_{child}} S_a} \tag{4}$$

The UVT of a node measures the degree to which the node has received its proportional share of CPU from the parent node. The difference between UVT of a node and LVT of the parent node gives a measure of whether the node has received its share-based allocation. If the node's UVT is less then the parent node's LVT, the node has received less than its share and vice-versa. UVT advances at a rate inversely proportional to the number of shares the node holds. If a node has a large number of shares, its UVT will increase at a smaller rate, and therefore it will be selected more often to make its UVT same as LVT of the parent node.

The virtual finish time of a node refers to its UVT, had the node been selected for the current quantum for execution.

- VFT: The VFT of a node A is its UVT had A been selected for the current quantum. When A joins the hierarchy for the first time at time t:

$$VFT_A(t) = UVT_A(t) + \frac{Q}{S} \tag{5}$$

where Q is the quantum size. Later, when a mobile program from within A was initiated for execution at time τ and now (t) some other program is going to be scheduled:

$$VFT_A(t) = VFT_A(\tau) + \frac{Q}{S} \tag{6}$$

A property of the virtual finish time is that it does not change while the application is executing. It changes only when the task is rescheduled. The algorithm selects the child node with the earliest virtual finish time (VFT). To summarize: The scheduling of non real-time programs starts at the root of the non real-time programs (Fig. 8). The algorithm starts at the root and traverses the tree till it reaches a leaf, which represents a mobile program. At an internal node A, the algorithm examines the constraints associated with the children nodes of A. If the children nodes have priority based constraints associated with them, the algorithm selects the child node with the highest priority. If children nodes have share based constraints, the algorithm selects the child node with the earliest VFT. If the node selected is a mobile program, it is scheduled for execution, otherwise the process is repeated.

4.2.2 Scheduling of Real-time Programs Real-time mobile programs are members of RT_G. The scheduling of mobile programs in RT_G is based on the scheduling algorithm in Rialto [7]. Rialto uses a *precomputed scheduling graph* to implement continuously guaranteed CPU reservations with application defined periods, and to guarantee time constraints. Applications make *CPU reservations* in the form of "reserve X units of time out of every Y units". Real-time applications request CPU resources by specifying *time constraints* of the form <S,E,T>. On the basis of the CPU reservations, Rialto constructs a `Rialto scheduling graph`. The nodes in the Rialto scheduling graph indicate either reserved time periods for applications or free time not reserved for any application. The time constraints for threads are then satisfied from the reserved time periods and from any free time that might be available.

Our real-time scheduling problem differs from the problem solved in Rialto in the following ways: First, we use simpler real-time constraints. We don't consider continuous *CPU reservations* of the form "reserve X units ....". Instead, we define CPU reservations over discrete base periods, i.e., quantum time chunks. With this simplification, there is no need to compute the *Rialto scheduling graph*. However, the RT_G algorithm is now more general, as CPU reservations can be carried out from any place in the base period rather than from some fixed locations in the Rialto scheduling graph. Further, our real-time scheduling algorithm must satisfy additional constraints in the form of upper bounds.

Resource allocation for real-time programs is done on the basis of a set of constraints of the form:

- RT_G`.upperbound = val1`: An upper bound on the time reserved for RT_G within each quantum time chunk. This prevents starvation of non real-time programs.
- `group.`RT_G`-bandwidth = val2`: Groups can reserve bandwidth within RT_G so that deadline based constraints for member mobile programs can be satisfied from the reserved bandwidth.
- `mobileprogram.deadline = `$<S,E,T>$: A mobile program within a group can request that its time constraints be satisfied by utilizing the bandwidth reserved for its parent group. If there is no bandwidth reserved for the parent group, the program will get only unreserved RT_G bandwidth to satisfy its constraints.

The scheduling algorithm allocates time within the quantum time chunks to satisfy reservation requests. The use of quantum time chunks is similar to the notion of *slot lists*[12]. While the slot list method considers only real-time applications, in our case the amount of CPU time available in each quantum time chunk is constrained by the upper bound on RT_G.

The real-time algorithm first reserves the bandwidth for each group in each quantum time chunk. For each <S,E,T> constraint, the scheduling algorithm makes reservations in the quantum time chunks (Fig. 9) that fall within times S and E. The algorithm reserves the computation time T from within the parent group's reserved bandwidth, if any, and any free unreserved RT_G bandwidth that might be available within the quantum chunk. It does so by creating reservation nodes in each quantum time chunk. The reservation nodes specify the start time, the time reserved, and the mobile program for which the time has been reserved.

When a new real-time program arrives, the algorithm performs a feasibility check to determine if the deadline request can be met. The algorithm goes through the list of quantum time chunks, reserving any available RT_G time for the request. If the program's deadline cannot be met, any reservation made for the program is freed. In the process of carrying out the feasibility checks, the algorithm performs a rearrangement of any reservations already made for earlier programs so that the deadline based constraints are added in the Earliest Deadline First (EDF) [9] order. Using EDF for adding new reservations increases the number of reservations satisfied.

In Fig. 10, we show how the algorithm makes reservations for two requests: reservation A (<150,280,40>) and reservation B (<150,170,5>) in that order. We assume that the size of a quantum time chunk is $40ms$ and the host specifies an upper bound of 40% ($16ms$) for RT_G. Also, assume

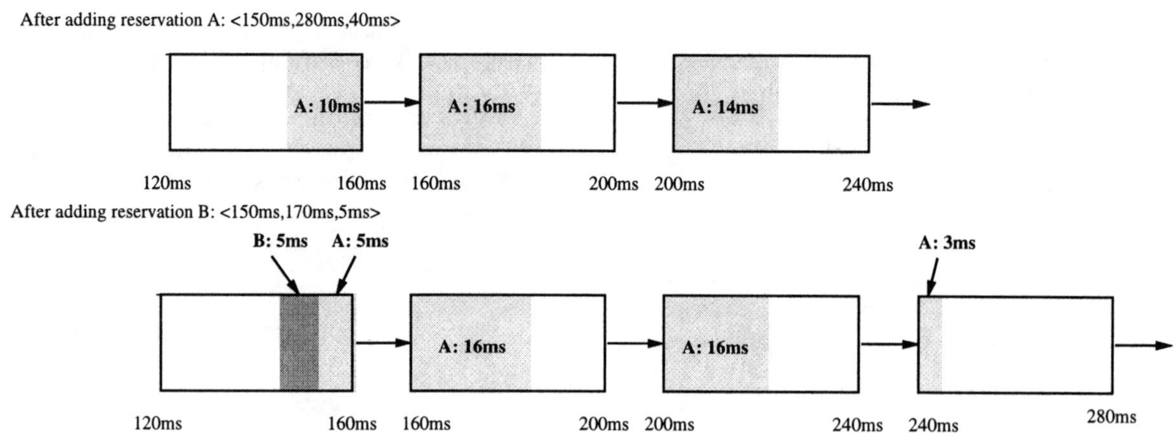

Figure 10. List of quantum time chunks for two reservations

that the entire CPU time ($16ms$) for RT_G is available to the mobile programs. When request A is made, the algorithm greedily reserves any RT_G time available to A. When request B arrives, the algorithm rearranges the reservation for A so that the constraints of B can also be satisfied. This is done because B has an earlier deadline. If there were no rearrangement, B cannot be guaranteed since all RT_G time will already be used by A.

4.2.3 Resource Usage Control The upper bound subgraph captures the upper bound and lifetime constraints on groups and mobile programs. Each security node in the graph maintains the usage information for the groups and programs that the node monitors. As the scheduling scheme traverses the scheduling graph, it checks the security node associated with a node before it applies any scheduling algorithm to the node. If selecting a program from within that node will cause an upper bound or a lifetime constraint to be violated, the particular internal node is not selected. For example, assume that the scheme decides to schedule a program in the subtree under A_p in Fig. 8. Before it decides between nodes A and A', the scheme checks with the security nodes that control A and A' (UB_1 for A and LC_1 for A') to ensure that the two nodes do not violate any constraints. The scheme then employs the selection algorithm as described earlier to select one of the two.

4.3 Implementation and Performance Analysis

To assess the behavior of the scheduling scheme, we first implemented the scheme as part of a simulation engine and conducted several experiments using the simulation engine to analyze the performance behavior of the scheme. Once we were satisfied with the scheme, we then implemented the scheduling scheme within the Java virtual machine (JVM).

We conducted several experiments on the simulation engine and the JVM. The goals of these experiments were to examine the effectiveness of the scheme in (i) satisfying both real-time and non real-time constraints; (ii) enforcing upper bound and lifetime constraints; and (iii) satisfying constraints that change dynamically. Due to lack of space, we only show two results, one from the JVM and another from the simulation environment. In all the experiments, the time quantum for a program is $5ms$.

4.3.1 General scheduling behavior The first experiment demonstrates how the scheme schedules groups of mobile programs that are constrained by shares, priorities and upper bounds. Further, it shows how the upper bound constraints interact with shares and priority constraints. We performed this experiment on the modified JVM. In Fig. 11(a), we show the hierarchy constructed from the client and host resource usage constraints. In Fig. 11(b), we show the relative CPU allocations of groups G_0, G_1 and G_2. We also show the relative CPU allocations of mobile programs MP_9, MP_{10}, and MP_{11}.

Between times $(0, 35)$, G_0, G_1 and G_2 get 30%, 60% and 10% of the CPU respectively which matches their share allocations. At time $35s$, G_0 reaches its upper bound. This results in relative allocation for G_1 and G_2 to increase to 86% and 14% respectively that corresponds to the share ratio of $60 : 10$. When the upper bound of G_2 is reached, G_1 is the only group and it gets all the CPU resources till its upper bound is achieved as well.

Within G_0, the relative allocations of mobile programs MP_9 and MP_{10} are 20% and 80% respectively, accordingly to their share allocations. MP_{11} is not scheduled in the beginning because it belongs to a lower priority group. At time $25s$, the upper bound for PG_3 is achieved and then mobile programs from PG_4 are scheduled till the upper

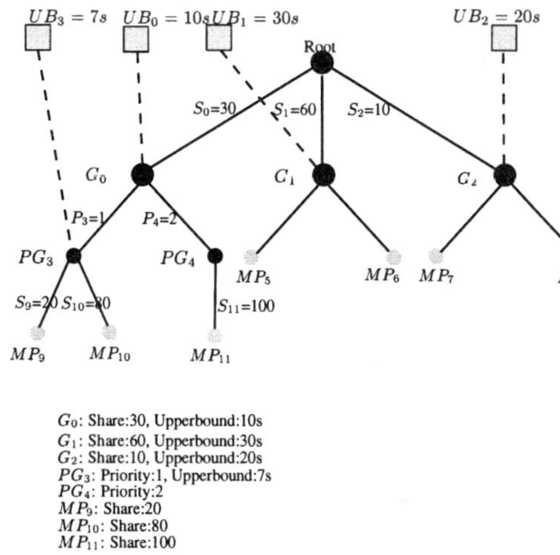

(a) The scheduling graph

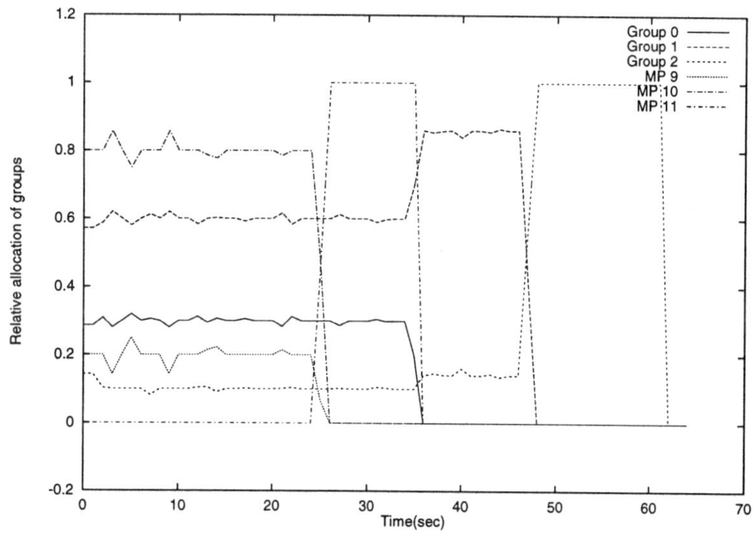

(b) Relative allocation of CPU for groups and mobile programs

Figure 11. General scheduling behavior of the scheme

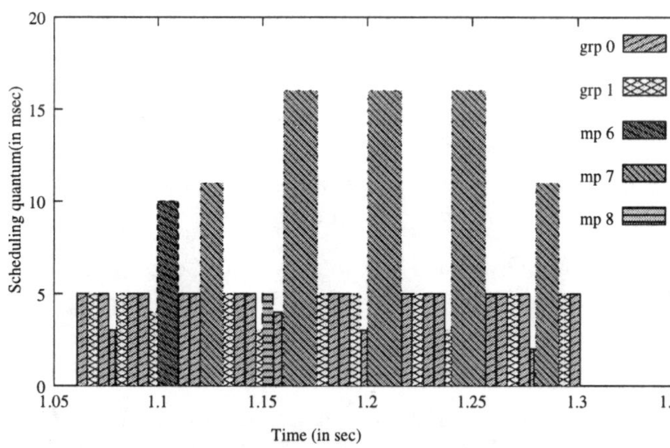

Figure 12. Scheduling for a combination of real-time and non real-time programs

bound for G_0 is reached.

The scheme, thus, effectively implements relative allocations of resources within hierarchies of groups. Further, it enforces upper bounds constraints as well. Note that changes in CPU allocation to MP_9, MP_{10} and MP_{11} (programs in G_0) does not affect the allocation to G_1 or G_2. This highlights the modularity of the scheme.

4.3.2 Real-time Programs In this experiment (Fig. 12), we demonstrate how the scheduling of real-time programs takes place in the presence of non-real time programs. We conducted the experiments within the simulation engine.

There are two non-real time groups: $Group0$ and $Group1$ have shares 40 and 20, respectively. There are three programs with real-time reservations:

MP6: <1.1s,1.15s,10ms>
MP7: <1.12s,1.5s,70ms>
MP8: <1.15s,1.18s,5ms>

The host specifies an upper bound of 40% ($16ms$) on the RT_G group for each quantum time chunk of $40ms$. The plot shows that the real-time programs are allocated according to their reservations. At the same time non-real time programs are allocated according to their shares. Also, since there is an upper bound on RT_G group, real-time programs cannot starve the non-real time programs, even though real time programs are scheduled for more than $5ms$ (the default time quantum) at a given time.

5 Summary

In this paper, we have presented schemes for enforcing both resource access and resource usage control.

We have described a mechanism for implementing general access control policies on mobile programs. There are two components of our approach. The first is a simple declarative access constraint language that allows a site to restrict accesses to the objects and methods of the system. The declarative nature of the language makes it easy to specify policies while still allowing a hook to express procedural policies if necessary. The second is a set of tools that enforce the specified constraints by editing mobile programs and resources. Our approach's appeal is that a site can specify access constraints separately from both mobile program definitions and resource definitions. This separation of concerns has a number of benefits. Both access constraints and resource definitions can be modified independently. Sites can easily specify different access constraints for different mobile programs for the same resource. Finally, our approach can enforce security on systems that were not originally designed with security in mind.

We control usage of CPU by developing a CPU scheduling scheme that addresses the security and quality of service requirements of a host. The scheme presents an environment for specifying resource usage constraints. Mobile programs specify shares, priority and deadline constraints. Hosts specify shares, priority, upper bound and lifetime constraints. The scheme constructs a scheduling hierarchy to apply a set of algorithms that enforce the various constraints. The non-real time algorithm enforces share and priority based constraints. The real time algorithm enforces deadline constraints. The upper bounds algorithm enforces security constraints specified by the host. Any conflicts between the client and server constraints are resolved by our algorithm composition policy that always favors the server constraints.

References

[1] T. Berners-Lee, R. Cailliau, A. Luotonen, H. Frystyk Nielsen, and A. Secret. The World-Wide Web. *CACM*, 37(8):76–82, August 1994.

[2] B. N. Bershad, S. Savage, P. Pardyak, E. G. Sirer, M. Fiuczynski, D. Becker, S. Eggers, and C. Chambers. Extensibility, safety and performance in the SPIN operating system. In *15th ACM Symposium on Operating Systems Principles*, pages 267–284, Copper Mountain Resort, CO., December 1995.

[3] D. Chess, C. Harrison, and A. Kreshenbaum. Mobile Agents: Are They a Good Idea? In *Second International Workshop on Mobile Object Systems: Towards the Programmable Internet.*, pages 25–47. Springer-Verlag, 1996.

[4] L. Gong. Java security: Present and near furture. *IEEE Micro*, 17(3):14–19, May-June 1997.

[5] L. Gong, M. Mueller, H. Prafullchandra, and R. Schemers. Going beyond the sandbox: An overview of the new security architecture in the Java Development Kit 1.2. In *Proceedings of the USENIX Symposium on Internet Technologies and Systems*, Monterey, California, December 1997. Available to Usenix members at http://www.usenix.org/publications/library/proceedings/usits97/gong.html.

[6] T. Jaeger, J. Liedtke, and N. Islam. Operating system protection for fine-grained programs. In *Proceedings of the 7th USENIX Security Symposium*, San Antonio, Texas, Jan. 1998.

[7] M. B. Jones, D. Rosu, and M-C. Rosu. CPU Reservations and Time Constraints: Efficient, Predictable Scheduling of Independent Activities. *16th ACM Symposium on Operating Systems Principles*, October 1997. St. Malo, France.

[8] S. Liang and G. Bracha. Dynamic class loading in the Java Virtual Machine. In C. Chambers, editor, *Object-Oriented Programming Systems, Languages and Applications Conference,* in *Special Issue of SIGPLAN Notices*, number 10, Vancouver, October 1998. ACM.

[9] C. L. Liu and J. W. Layland. Scheduling Algorithms for Multiprogramming in a Hard Real-Time Environment. *Journal of the ACM*, 20(1), January 1973.

[10] D. V. Miller and R. W. Baldwin. Access control by boolean expression evaluation. In *Fifth Annual Computer Security Applications Conference*, pages 131–139, Tucson, AZ, 1990. IEEE, IEEE Comput. Soc. Press.

[11] J. Nieh and M. Lam. The Design, Implementation and Evaluation of SMART: A Scheduler for Multimedia Applications. *16th ACM Symposium on Operating Systems Principles*, October 1997.

[12] K. Schwan and H. Zhou. Dynamic Scheduling of Hard Real-Time Tasks and Real-Time threads. *IEEE Transactions on Software Engineering*, 18(8):736–748, August 1992.

[13] T. Thorn. Programming Languages for Mobile Code. *ACM Computing Surveys*, 29(3), September 1997.

[14] J. Vitek, M. Serrano, and D. Thanos. Security and communication in mobile object systems. In *Mobile Object Systems. Towards the Programmable Internet. Second International Workshop, MOS '96*, number 1222 in Lecture Notes in Computer Science, pages 177–199, Linz, Austria, July 1997.

[15] D. S. Wallach, D. Balfanz, D. Dean, and E. W. Felten. Extensible security architecture for Java. In *16th ACM Symposium on Operating Systems Priciples*, volume 31(5) of *Operating System Review*, pages 116–128, Saint Malo, France, Oct. 1997.

[16] D. S. Wallach and E. W. Felton. Understanding Java stack inspection. In *1998 IEEE Symposium on Security and Privacy*, pages 52–63, Oakland, CA, USA, May 1998. IEEE, IEEE Comput. Soc.

Security Agility for Dynamic Execution Environments

Mike Petkac
NAI Labs
mpetkac@nai.com

Lee Badger
NAI Labs
lbadger@nai.com

Wayne Morrison
NAI Labs
wmorrison@nai.com

Abstract

The evolution of dynamic execution environments increasingly requires security policies that are also dynamic in nature to address such events as process migration, changes in personnel, shifts in alliances, and detected intrusion that cannot be well anticipated or addressed by static policies. However, traditional software components designed without (dynamic) security in mind may fail in various degrees when a policy change takes place, including termination of critical processes. This paper reports on techniques developed by the Security Agility for Dynamic Execution Environments project[1] that are encapsulated in a prototype toolkit for integration with software components so they may function effectively in a dynamic security environment. These techniques include methods for making software components aware of their dynamic security policy environment, helping components adapt to policy changes, and adding security policy enforcement mechanisms to software components to mediate client actions when the underlying system software cannot.

1. Introduction

The evolution of dynamic execution environments requires software that adapts to changing security policies during runtime with minimal loss of functionality and with little or no manual assistance. Quorum, a Defense Advanced Research Project Agency (DARPA) Information Technology Office (ITO) sponsored program, exemplifies evolving dynamic execution environments. Quorum describes a mobile distributed automated information processing system of heterogeneous nodes (hardware platforms and software components) which provide a Quality of Service (QoS) architecture. Such systems have two dynamic aspects that make it difficult for traditional software components that are either unaware of their security environment in general or its dynamic nature to support the system mission. The first aspect is the ability to reconfigure the global security policy (by reconfiguring the security policy on one or more hardware platforms) at any time to address events such as shifts in alliances, changes in personnel, and crisis situations. The second aspect is component migration - the ability to move components from one node to another at runtime to meet QoS guarantees. These dynamic aspects of evolving systems bring two general shortcomings of traditional software components to light: the inability of software components to withstand resource revocation and reinstatement and inconsistent security mechanisms across non-homogeneous sets of nodes.

Implementation of dynamic security policies in support of our prior research [1] has shown that the changes to resource availability caused by security policy updates can wreak havoc among running components. When processes find themselves unable to access resources they expect will always be available, they may hang, crash, abandon further access attempts, or exhibit undefined behaviors. Additionally, since most components have little or no knowledge of security, they typically do not share in security responsibilities or have knowledge of the underlying policy governing their operation. This lack of knowledge contributes to occurrences that can also negatively impact dynamic system operations, including continuation of processes invalidated by a new security policy and loss of intended access mediation when a component transitions to another node whose security policy requirements differ from its previous host.

This paper reports on techniques developed by the Security Agility for Dynamic Execution Environments project ([2], [3]) that are encapsulated in a prototype toolkit to create *agile* software components that function effectively in a dynamic security environment. These techniques include methods for making software components aware of the dynamic nature of their security policy environment, techniques for helping them adapt to changes, and methods for adding security policy

[1] This research is supported by DARPA contract F30602-97-C-0225.

0-7695-0490-6/99 $10.00 © 1999 IEEE

enforcement mechanisms to software components to mediate the actions of their clients when the underlying system software components cannot. The toolkit is designed to minimize the cost and development burden associated with creating agile components. We first describe security agility concepts and strategies. We then present the design, implementation, and application of the Security Agility Toolkit. Finally, we discuss future directions and present our conclusions.

2. Security agility

Security agility is a software flexibility technique to address security properties and their dynamic evolution. An agile software component is cognizant of the security environment in which it executes, is aware of its responsibilities for enforcing "its part" of a more global policy, and contains internal mechanisms that adapt its functionality in coordination with authorized external policy changes.

The heterogeneous nature of a dynamic execution environment presents some significant obstacles to developing general security adaptiveness. The first such impediment is the wide range of possible security semantics. A variety of access control policies might be employed, for example, including information disclosure policies ([4], [11], [12]), integrity policies ([5], [6], [7], [8]), firewall security perimeters ([13], [14]), and role-based policies [15]. Likewise, many cryptographic algorithms (e.g., DES, RSA, Skipjack) and certificate management strategies (e.g., Microsoft CryptoAPI, SunCA) may be implemented. The variety of software components employed in a heterogeneous environment, including operating systems, system software, and mission-specific software requires the development of flexible techniques that are not bound to a single environment. Finally, the availability of operating system and component source code (e.g., freely available, unavailable, or license restricted) must also be considered when developing adaptive security techniques.

To help overcome the heterogeneity obstacles, security agility employs two primary strategies to address security properties and their dynamic evolution. First, it embeds software components with preformulated security policy models and mechanisms to provide policy awareness in support of security reconfiguration. Second, it provides a flexible software component architecture that allows dynamic code extensions for adding new security semantics or changing security-relevant behavior to maintain compatibility with new security rules. Using these strategies to constrain the problem space and utilizing methods common to many programming environments, this project has packaged security agility techniques in a prototype toolkit that can be customized to the security framework and execution environment in which it is deployed.

2.1. Embedded models and mechanisms

Although the literature documents a large number of security policies and models, many important models, including the Bell and LaPadula confidentiality model [4], the Biba integrity model [5], the Clark/Wilson model [6], and Domain and Type Enforcement [7] (an elaboration of the type enforcement model [8]), share a common underlying organization. That organization consists of:

a subject/object framework that views a system as a collection of active subjects (e.g., processes, tasks) and a collection of passive data containers (e.g., files, messages),

security attributes that form equivalence classes of subjects and objects, for example, grouping subjects according to their classification (e.g., Secret, Top Secret), or grouping their files according to their type (e.g., system file, Project X), and

transition rules that define allowed operations on subjects and objects based on their associated attributes (e.g., the "*" property for nondisclosure policies).

This common organization of key security models was a prime factor in the decision to constrain the problem space by leveraging our Domain and Type Enforcement (DTE) technology as the policy model of this research. The DTE prototype system is a highly flexible and robust UNIX[2] kernel-based access control mechanism that provides tools for dynamically changing a system's security policy. The policy changes immediately effect all running programs, providing an ideal testbed to rapidly prototype and validate security agility techniques.

DTE also provides an implicit attribute management framework that allows security and reliability attributes to be associated easily with hierarchically organized objects (e.g., files) and active entities (e.g., processes) that can greatly assist in security agility implementation. By extracting this attribute framework and combining it with

[2] UNIX is a registered trademark of The Open Group.

access decision functionality, an application-level policy specification and enforcement model was constructed that can be embedded in components in an unobtrusive manner. The application-level policy model, referred to as the Access Decision Function (ADF), adheres to many aspects of the General Framework for Access Control (GFAC) [10]. An ADF policy can be implemented in conjunction with or independent of additional policy enforcement engines, such as DTE, providing a mechanism for software to mediate the actions of their clients when the underlying system software components cannot.

2.2. Policy-aware components

Embedded security policy models allow a component to be aware of the security policy governing their operation, including security attributes associated with other subjects and objects, security rules, and algorithms. This awareness can be used to help components survive in dynamic security environments. For example, in a DTE environment, which uses a security attribute known as a *type* to label every interprocess message, a client component must communicate with a server component using a type the server is permitted to read. If the security policy were static, it could be configured so the type associated with the policy-unaware client was readable by the server. However, in a dynamic security environment, access permissions granted to the server process may change such that it can no longer read the client's default type, resulting in the non-aware client losing the ability to communicate with the server. On the other hand, if the client were policy aware, the client could select one of the other types it can write after determining that the new server permissions allow the server to read the chosen type, assuming such a type exists. In this way, the client and server may be able to continue communications under a new security policy.

One of the most important considerations of policy awareness in dynamic execution environments is the likely fluctuation in resource availability resulting from security policy reconfiguration. When processes find themselves unable to access resources they expect will always be available, undesirable events may occur, including termination of critical processes. Files and directories are typically thought of first when resources are mentioned. However, resources might also include other elements, such as memory segments or interprocess communications, as described in the previous paragraph. Process dependencies, such as client/server relationships, are often subtle, or even unknown, and may be overlooked when dynamic policy changes are implemented, particularly in time-critical situations such as in response to an intrusion detection event.

Another important aspect of policy awareness is the ability to revisit security decisions that are typically evaluated once, such as those enforced to validate access to a connection-oriented service. For instance, TCP wrappers [16], commonly employed on UNIX hosts, consult an access control policy to validate remote service requests prior to connection establishment and DTE adds additional access mediation to initialization routines of the services themselves. Though a change in the TCP wrapper policy or the DTE policy may prohibit further connections, the grandfathered decisions of existing connections allow them to continue in violation of the new security policy. This could result, for instance, in an intruder continuing an attack on the system until the offending processes are manually terminated.

Once components have been made aware of the security policy changes that could affect their execution, they can be extended with adaptive functionality that reacts positively to these changes, rather than failing in some manner. Positive responses might include terminating connections invalidated by policy changes, temporarily suspending or reducing normal operations until lost resources become available once again, reacquiring lost resources, or switching to alternate algorithms to produce equivalent results. The security agility toolkit architecture facilitates the coordination of policy awareness and adaptive behavior functionality.

2.3. Security agility architecture

Although the UNIX system-based DTE technology was chosen as the security model basis for this research, a primary goal in developing security agility was to structure the toolkit to maximize portability to other security policy representations and execution contexts. These goals heavily influenced the design of the architecture and its resulting implementation.

A number of computer security models and policies (e.g., confidentiality and integrity policies) view security as a safety property because security rules constrain the system states. Some security policies (e.g., key recovery and auditing), however, are viewed more accurately as liveness properties because the policies add security requirements to carry out additional actions, such as writing audit records. To support the full spectrum of security semantics, the security agility architecture

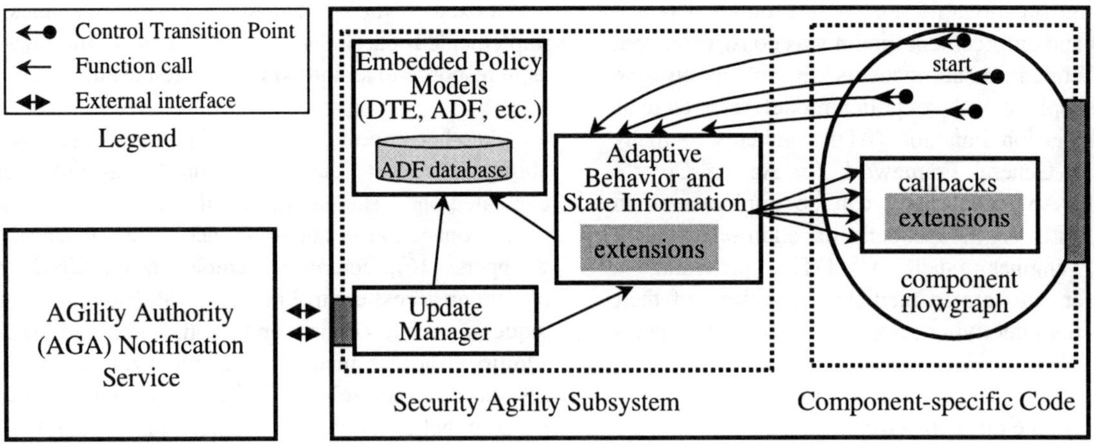

Figure 1: Security agile component architecture

includes facilities to not only constrain behavior according to security rules but also for dynamically adding behavior.

Figure 1 shows the general architecture of a security *agile* system component. The component-specific code represented by the oval in Figure 1 implements the component's non-security responsibilities. The security-specific functionality is carried out by the agility subsystem, a collection of extensible cooperative libraries. In this architecture, the component-specific code also includes a number of Control Transition Points (CTPs), which are interfaces that conditionally transfer control to the security agility subsystem. In general, CTPs bracket Application Program Interfaces (APIs) to libraries, including system call entry points to the UNIX kernel, to intercept external resource access. CTPs are used to maintain up-to-date component security state information and/or augment processing based on embedded policy decisions attained by consulting the underlying policy enforcement implementation.

An agile system component is able to receive notification of dynamic changes to the system security policy and directives to modify the application-level security policy. In addition to updating its internal security policy database in response to such a request, the security agility subsystem provides the mechanisms for updating component state data to bring a component into conformance with policy changes. Although the security agility subsystem is able to accomplish some reconfiguration internally, (e.g., closing files, opening resources previously denied) it may rely on help from customized component-specific code (callbacks in Figure 1). Callback functions can be used, for example, to switch to an alternate algorithm, re-evaluate connection criteria, or write component-specific file formats.

Dynamic code extension techniques present significant opportunities for adding flexibility beyond that afforded by embedded models and mechanisms and the limited adaptiveness of static control transition points and security-specific callback functions. Paralleling the need to dynamically change the security policy of a system, it may be desirable or necessary to change the *agile behavior* carried out by a component's agility subsystem. Agile behaviors define the adaptive functionality an agile component exhibits to adhere to policy changes. An agile system component able to accept code extensions could also add entire security feature suites, such as auditing or cryptographic services, through modified or additional control transition points and callback functions.

Finally, the security agility architecture shown in Figure 1 shows the AGility Authority (AGA) notification service that interfaces with an agile component. The AGA informs agile components of security policy changes and dynamic code extensions.

3. The security agility toolkit

The prototype Security Agility Toolkit is the first step in implementing cost-effective agility. It focuses on implementation of security agility on a single host, and is the basis for development of a Distributed Security Agility Toolkit that will manage coordinated security agility in distributed systems.

The toolkit was initially implemented on the BSD/OS[3] UNIX system-based operating system using the DTE and ADF embedded policy models. However, our development goals consisted not only of minimizing the cost and development burden associated with creating agile software components, but also of providing extensibility to other security semantics, including additional policy models and operating systems. The toolkit's implementation of the security agility architecture accomplishes these goals using techniques that are generally available in all programming environments. The agility subsystem encapsulates embedded policy models and adaptive functionality in extensible libraries that utilize an object-oriented class framework for customization with other security semantics. Coupling this framework with standard C++ language and compiler techniques and dynamic linking, the toolkit provides a transparent runtime implementation of the underlying security agility subsystem functionality to support agile components.

The standard techniques employed by the toolkit facilitated its port to the FreeBSD[4] (UNIX system-based) and the Windows NT[5] environments. DTE development concluded on the now outdated BSD/OS 2.1 operating system, whose license restriction prohibited free dissemination of source code. However, NAI Labs is currently porting the DTE technology to the up-to-date freely available FreeBSD 3.x operating system that permits public (re)distribution and will be the host of future UNIX system-based security agility development[6]. Additionally, the ADF application-level policy model was ported and demonstrated in the Windows NT environment, which also hosts the initial distributed security agility development effort through a fully ported toolkit that relies on the ADF embedded policy model. Thus, the toolkit is a development tool that can be customized for the security policy environment in which it is deployed, though our presentation of the stand-alone toolkit focuses on the (BSD/OS) UNIX environment.

The Security Agility Toolkit consists of three main elements: 1) the agile policy, 2) the AGility Authority (AGA), and 3) the agility subsystem. The security environment and agile behavior, or adaptive functionality, of a component are described by its agile policy. The agility authority is a multi-instantiated server that provides the toolkit's agile and security policy notification services. The agility subsystem libraries, linked and compiled with component-specific source code, cooperate with the AGA to implement the agile behavior of the component specified by its agile policy.

3.1. Agile policy

A component's agile policy defines its security environment and the behavior the component will implement to adapt to environment changes that affect its operation. Currently, the overall agile policy of a system consists of a collection of individual agile policies defined for each agile component of the system. The agility authority maintains the initial agile policy for each agile component in a configuration file and conveys them to the components' agility subsystem at process initialization. A component's agile policy information includes the security environment governing the process (e.g., DTE and/or ADF), the control transition points specifying the agile behavior it will exhibit, and specific files the component is dependent on, such as configuration files, that the AGA monitors for changes on behalf of the component. The AGA-monitored files can be viewed as part of the overall security policy of the system.

Like the initial security policy of a system, the initial agile policy of a component may be insufficient to characterize its agile qualities during the runtime lifecycle of the component. For example, a shift in the overall security policy strategy due to a detected intrusion may require tightening or extending a component's agile behavior or modifying ADF policy rules. Therefore, the agile policy is dynamic in the same sense that a system's security policy may be dynamic. The ADF policy rules and the agile behavior of a component can be dynamically modified during runtime. The agile behavior is modified by changing the CTPs (which may additionally introduce new callback functions) implemented by a component.

3.2. AGility Authority (AGA)

The agility authority is the notification service of the Security Agility Toolkit and is the central point of security-specific control for agile processes. The AGA is implemented in user space to support portability to other platforms, customization for varying security policy enforcement techniques, and coordination for distributed security agility. It is based on a model consisting of a

[3] BSD/OS is a trademark of Berkley Software Design, Inc.

[4] FreeBSD is a registered trademark of FreeBSD, Inc. and Walnut Creek CDROM.

[5] Windows NT is a registered trademark of Microsoft Corporation.

[6] The initial FreeBSD Security Agility Toolkit (without DTE) is available at ftp://ftp.tislabs.com/pub/agility.

network of cooperative agility authorities, with each node hosting a user-space agility authority process, which will be employed in the Distributed Security Agility Toolkit. A configuration file specifies all the initial agile policy information necessary for the AGA to manage agile processes in the stand-alone toolkit.

During process initialization, an agile process registers with its AGA to receive its agile policy. Each agile process has a transparent interface, provided by the agility subsystem, to its own dedicated child AGA that is spawned by the master (parent) AGA after verifying the requesting process is a known agile component. All security and agility policy changes implemented on an agile system are coordinated with the AGAs, then passed to the agile processes using asynchronous signal notification. Security policy changes are also funneled through the AGA, enabling it to efficiently notify agile processes of updates as they occur. This frees components from having to conduct periodic queries of the policy enforcement engine(s). In the distributed toolkit, policy changes will propagate through an AGility Authority Manager (AGAM), which will define the community of nodes affected by a policy change, and provide the concurrency control needed to ensure that distributed policy changes are safe.

3.3. Agility subsystem

The security agility subsystem provides the primary implementation of the security agility architecture presented in Figure 1. It is a collection of libraries that incorporates embedded policy models and implements a component's adaptation to security policy changes. The agility subsystem libraries are linked with a component's source code and any other libraries the component depends on at compile time. The UNIX security agility subsystem contains libraries that provide the application-level policy specification and enforcement capability, a modified version of the standard C library for control transition point implementations, and a central library, referred to as the agility library, that provides the primary agile functionality and event coordination.

The agility library offers an extensible object oriented class framework to represent resource and security policy abstractions. The framework coordinates the interaction between security policies and resource accesses as well as the dynamic code implementation library and communicates with external control components such as the toolkit's policy notification service, the agility authority. The agility library is the key component of the subsystem that reduces the development burden associated with creating agile software by providing transparent implementation of the underlying functionality that can be integrated with little or no modification of a component's source code. In particular, the agility library's framework supports:

resource and embedded policy model abstractions: Resources are designed as abstract data types whose concrete implementations can represent any resource used by an agile component that are subject to security policy restrictions. Abstract resources can be derived to represent files, network connections, internal data structures, etc. Security policies are also represented as abstractions intended to support the derivation of specific policies. Examples of concrete policies are DTE and the ADF application-level embedded policy specification and enforcement model. Additional concrete policies can be derived and incorporated, including the UNIX and Windows NT forms of discretionary access control.

dynamic resource state and callback functions: Resource state information is retained in a dynamic repository of abstract instances of resource usage from creation to access to destruction. An asynchronous policy change notification is essentially a message sent to resources in the repository, which consult the embedded policy model(s) to update their current state under the new policy. Each concrete resource instance is responsible for processing state changes through an associated callback function. The agility library API provides routines for resource registration and callback specification.

control transition points: Control Transition Points (CTPs) are the set of transfer points bracketing library Application Program Interfaces (APIs) that conditionally transition control to dynamically extended code segments. Library APIs normally provide the means for components to access resources. For example, on UNIX systems, all system calls and Domain Name Service resolver routines are accessed through the standard C library while Remote Procedure Call (RPC) clients and servers access RPC services through the RPC library. CTPs are generally used to intercept these external resource accesses for resource state registration and/or augmentation of processing based on security decisions. However, CTPs need not be limited to this purpose. The agility subsystem provides a basic set of CTPs associated with many of the standard C library resource access routines and

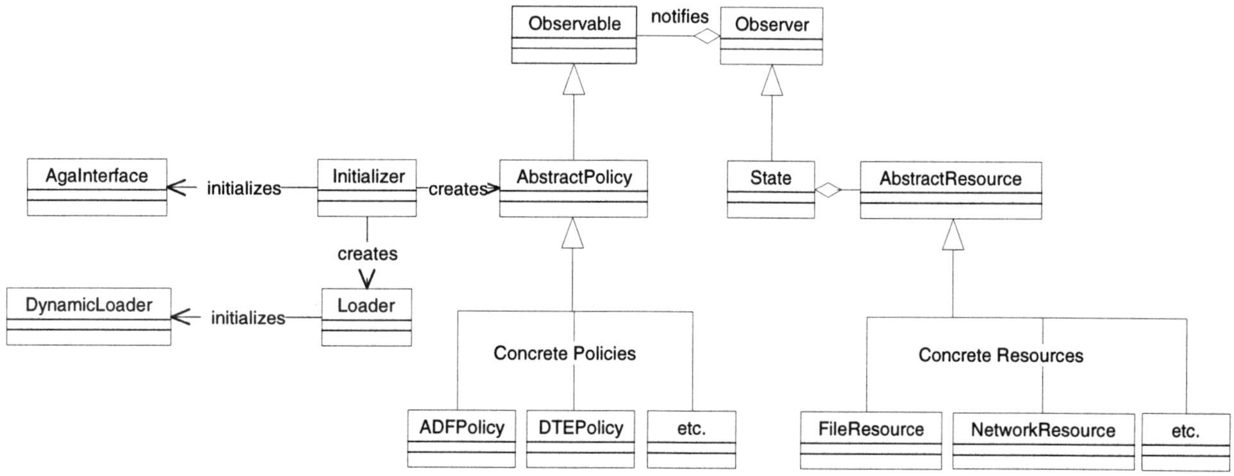

Figure 2: Agile library class diagram

build-time utilities to develop CTPs for additional library routines or component-specific functions to support the needs of the execution environment. Existing CTPs can also be eliminated to reduce the overhead (one additional function call and boolean evaluation) associated with unnecessary interception.

security policy and agile behavior change notification: The agility library implements an asynchronous mechanism for notifying resources of changes in the external security policy. In addition, mechanisms for updating CTPs and code extensions have been implemented using simple concurrency techniques.

Ideally, security agility would provide pre-packaged adaptive behavior for all components. However, though some adaptive functionality can be identified and applied, some components must be able to specify finer-grained adaptive behavior to fulfill their missions. Therefore, the toolkit does furnish a limited set of predefined callback functions and CTPs to implement *default agile behaviors* that include waiting until all resources are available, reconfiguring for a change in the domain name server(s), and re-establishing DTE IPC attributes. However, it also supplies the interface to register component-specific callback functions and the templates to develop CTPs that provide *custom agile behaviors.*

Figure 2 shows a class diagram that illustrates the relationships between policy and resource abstractions and their concrete implementations. Using standard C++ language and compiler techniques, process control for an agile component begins in the agility library initialization routine before executing any of the original component code. The initialization routine creates the interface to the AGA then requests and receives the agile policy for the component via the AGA interface. The agility subsystem uses the agile policy to configure the component by creating the local infrastructure for control transition points and dynamic code extensions and creating an application level policy database when an ADF policy is specified. A single concrete in-process policy object is created from the agile policy information for the governing policy enforcement engine(s). The abstract part of the policy object is an "observable" object that maintains a list of "observers" that register interest in any changes to the policy. The list of observers is the resource state repository for the framework. Observers are derived as state objects held by the abstract resource. Instantiating a concrete resource through the agility library API registers the resource for interest in changes to the policy object. Releasing a concrete resource de-registers the resource through the observer. The AGA sends security policy change event notification and data to the policy object. These notifications pass through the abstract policy to each of the registered state objects via their callback functions parameterized with a reference to the changed policy. A policy class-scope callback is triggered to handle any global responses to policy changes that are independent of specific resources (e.g., resources that a component independently maintains). While some agile behaviors can be accomplished transparently through CTPs without resource registration, the list of observers and their associated callbacks help support customization of agile behaviors to address component-specific needs.

The agility library implements control transition points in conjunction with slightly modified system libraries and/or component-specific code. To implement specific instances of CTPs, the agility library uses a form of dynamic dispatch which replaces a direct call to an API with a function pointer dereference. With this technique (and some concurrency control), the library can easily change to a different implementation at runtime, making the agile policy and the behavior it specifies truly dynamic. All original calls are saved as function pointers and the new calls overlay those pointers. At any time the intercepted call can be restored back to the original call or modified with another implementation.

Security policy change and dynamic code extension events, including CTPs and callback functions, may update component-internal data structures as well as the security agility subsystem: in these cases, updates must be synchronized with component-specific processing. While we have thus far only implemented elementary concurrency control techniques, providing more robust synchronization is a primary focus of the distributed toolkit development.

The agility subsystem allows an agile component to respond to changes in resource availability that were not anticipated when the component was originally coded. There are some assumptions placed on the actual usage of these techniques. Most importantly, control transition points and callback functions must be carefully constructed and validated. No additional techniques, such as fault tolerance, are implemented in the toolkit to guard against code that may be corrupted.

3.4. Toolkit application

A variety of strategies are available for adding agility to software components. Figure 3 displays a continuum of application strategies offered by the toolkit's agility subsystem for integration with components. At one end of the spectrum, the agility subsystem can simply be compiled into an application to make use of default behaviors provided by the subsystem. At the other end of the spectrum, a component can be re-engineered to make use of the class structure of the agility subsystem libraries. As you move down the continuum, the cost of toolkit integration increases.

Our experiments integrating the DTE toolkit with existing components have concentrated on the lower cost strategies of linking the agility subsystem libraries and using default or custom agile behaviors. However, we anticipate that using the toolkit to build new components rather than re-engineering existing components would substantially reduce the cost of fully integrating the object framework of the agility subsystem and may be beneficial for agile behavior specification and runtime modification.

In general, default agile behaviors apply to many processes while custom agile behaviors apply to a single process or a small set of processes. While there are many system calls and library routines offered on a UNIX system, a more limited set are commonly used by processes to perform fundamental tasks such as file manipulation or Interprocess Communication (IPC). Default agile behaviors use CTPs to intercept the common system calls and library routines and augment their processing. Custom agile behaviors, the middle ground in Figure 3, address the specific needs of a process and can give functionality above what some default agile behaviors may provide. Since knowledge of the component source code is necessary to develop the CTPs to implement custom agile behaviors, they add implementation costs if the developer is unfamiliar with the process.

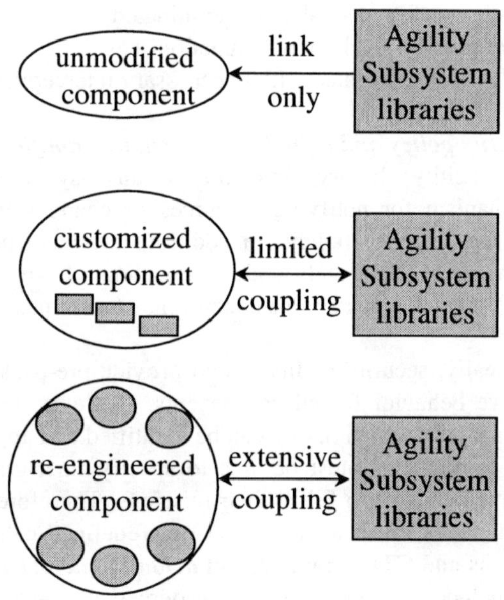

Figure 3: Toolkit Application Strategies

We have integrated the toolkit with several UNIX system processes. Those processes include the system logging daemon, *syslogd*, the remote shell execution daemon, *rshd*, the Internet name query service, *nslookup*, the Apache HTTP server, *httpd*, and the daemon to execute scheduled commands, *cron*. While integrating the toolkit with these components and experimenting with others, it became evident that the time to integrate the toolkit with existing components, though not excessive, lessened with subsequent integrations due to familiarization with the toolkit application techniques. Even CTPs that provide custom agile behaviors for a specific process can be quickly modified to address the similar needs of another process.

We will present a detailed look at toolkit application to the *syslogd*, *rshd*, *nslookup*, and *httpd* processes to illustrate how the toolkit can be applied to implement security agility techniques. *Syslogd* and *nslookup* demonstrate the security policy awareness and adaptive techniques enabled by the toolkit. *Rshd* and *httpd* additionally demonstrate policy enforcement. The security policy which governs DTE components is the one enforced by the DTE BSD/OS operating system's two access control mechanisms: POSIX-style discretionary access control and DTE non-discretionary access control. Since the *syslogd* and *rshd* server components run under the special POSIX *root* user identity, they are exempt from all POSIX-style access control mediation. We also assume the *nslookup* client, which runs with the logged-in user's identity, has the POSIX-style access required for proper implementation, which simply requires read access to a single configuration file available to all processes in a properly configured UNIX platform. This leaves only the DTE security policy to determine access control for these three components. Additionally, the *httpd* web server discussion concentrates on the ADF embedded policy model, which was derived from DTE. A quick presentation of DTE is therefore helpful to set the stage for discussion of the agile processes.

As with a number of other policy models, DTE considers a system to be logically split into two categories: active entities (usually processes) and passive entities (e.g. files or network packets). A security attribute known as a *type* is associated with each passive entity, or object; a *domain* is associated with each active entity, or subject. In addition to a type label, DTE network packets carry the domain of the source process in the IP option space of each message. Using a language-based specification (policy), DTE expresses allowed interactions between subjects and objects. Access control decisions are made by consulting a DTE database consisting of the "compiled" specification to determine if the domain has the requested access (e.g., read or write) to the object. DTE also expresses the allowed interactions between different subjects to determine if subject A has the requested access (e.g., execute or kill) to subject B.

3.4.1. System Logging Component. The UNIX system logging daemon, *syslogd*, is the primary mechanism for recording important system events, such as failed login attempts. System and user processes convey messages to *syslogd* via a UNIX socket created at *syslogd* initialization. *Syslogd* processing is implemented transparently with no indication of failure should a policy change or misconfigured policy cause access failures that result in lost log files or process termination. Without modifying the original source code, the agile *syslogd* component uses default and custom CTPs combined with resource state registration and callbacks to implement agile behaviors.

Toolkit integration with *syslogd* exemplifies construction of an agile component: identify the component's primary points of failure and introduce functionality to reliably adapt mission processing should a failure occur. *Syslogd* is typical of most non-trivial processes since it depends on underlying library APIs to support access to critical resources. In particular, the UNIX logging service depends on the C library interfaces to the *bind()*[7], *open()* and *write()* system calls to carry out its mission. These three system calls are the potential failure points we considered in the *syslogd* service caused by a misconfigured DTE policy or a dynamic policy change. A failure in the *bind()* system call results in process termination, while failure in the *open()* and/or *write()* system calls causes further logging attempts to the denied files to be abandoned, even if a subsequent policy change reinstates access to the file(s). Until the system administrator notices that *syslogd* is not running or is not fully functional and restarts the process, log messages, which may be critical in nature, will be lost.

Linking in the agility subsystem and using the 'wait until all resources are available' default CTPs for the *bind()*, *open()*, and *write()* system calls, results in an agile *syslogd* process that will sleep until the resource becomes available rather than exiting or ignoring inaccessible log files. While the 'wait until all resources are available' default agile behavior applied to *syslogd* is an advance over the original *syslogd* performance, it does not attain the level of robustness achieved by introducing simple custom CTPs. For example, while *syslogd* waits to regain access to its log files, incoming messages will accumulate in its socket buffer. When the capacity of the socket buffer is surpassed, any additional messages intended for logging will be discarded.

[7] For UNIX sockets, the *bind()* system call attempts to create a new socket entry in the file system.

We constructed component-specific CTPs to implement two sets of additional, more robust, agile behaviors for *syslogd.*[8] Both additional sets of agile behaviors are expressed through an agile policy that retains the 'wait' default behavior for the *bind()* system call, but augments the *open()* and *write()* system calls with custom behaviors. Rather than returning failures when trying to open or write to log files *syslogd* cannot access, the custom CTPs return success to avoid *syslogd* abandonment of future logging attempts. Both sets of custom behaviors implement an *open()* system call that registers a file resource for each unavailable log file with a callback function that will open the file if a DTE policy change allows future access. The *write()* CTP of the first set was not modified beyond returning success to *syslogd* regardless of the outcome of the write attempt. However, the first custom set will still result in a loss of messages, since errors are simply ignored. The second set of custom CTPs adds additional customization to the *write()* CTP to internally buffer messages during loss of log file access. If a subsequent policy change then restores log file access, the buffered data is written to the log file so data is not lost.

There are many other agile behaviors that could be implemented by *syslogd*. For example, *syslogd*'s agile policy could specify that alternate log files be used during periods of lost resource access so data is always available in a timely manner. On access restoration, the files could then be merged to restore seemingly uninterrupted *syslogd* operation. Since the agile policy implemented is dynamic, agile behaviors selected at process startup can be modified during runtime without restarting the process, a property demonstrated in the *syslogd* component.

Access to concrete resources, such as *syslogd*'s file system entries, are not the only points of failure that an agile component can address. The next agile component we review, *rshd*, demonstrates security agility with respect to another resource affected by dynamic security policy changes: connection criteria.

3.4.2. Remote execution server component. When a security policy changes, existing services must honor the new security policy, revoking client access to resources if necessary. However, this revocation is often absent, particularly in connection-oriented components that provide remote access to a system. The UNIX remote execution server, *rshd,* which allows a client to remotely execute a command, provides a clear example of a grandfathered security decision that may allow connections to continue in violation of a new security policy.

[8] Note that though we classify these CTPs as custom or component-specific, the strategies they implement could certainly benefit other components.

The DTE system implements a DTE-aware *rshd* that supplements normal UNIX *rshd* access checks with DTE requirements. A DTE-aware *rshd* reads the security attributes (the DTE domain and type) from the client's request and attempts to execute the command in the client's domain.[9] If the client domain does not have sufficient permissions to execute the command or the user issuing the request does not have access to the requested domain on the local DTE system, the command will fail. Also, if *rshd* does not have permission to transition into the client's domain so it can run the command on the client's behalf, the remote execution of the command will fail.

DTE-awareness, however, is not sufficient to make remote execution services agile. If a change in the security policy results in the client no longer having access to remote execution services, any commands that are already executing will continue to run. This grandfathering prevents the DTE-aware *rshd* from honoring the new security policy. To address this shortcoming, 'suspend' and 'terminate' global callback CTPs, referred to as policy class scope callbacks in section 3.3, were constructed that re-examine DTE access criteria for all open agile *rshd* connections when notified of a change in the security policy. If the agile policy indicates the connection is to be suspended or terminated, the CTP specified by the agile policy will suspend or terminate the connection and all child processes[10] spawned by the connection. If a subsequent security policy once again validates the connection while it is suspended, the suspend CTP will reinstate the connection and its spawned processes. If, however, an agile policy change occurs specifying termination of invalid connections while the connection and child processes are suspended, the new CTP will terminate the invalid connection and child processes. In this way, *rshd* ensures that it and its clients honor the new security policy as specified by the agile policy in place.

[9] DTE employs the principle of least-privilege. *Rshd* typically runs a system domain that has more access than granted to user domains, so the DTE-aware rshd process transitions the command to run in the weaker user domain.

[10] The suspend and terminate actions are currently implemented using the *killpg()* system call. In the future, we intend to implement more robust process tracking functionality.

3.4.3. **Internet name query client component.** The *nslookup* component is a Domain Name System (DNS) query client used to obtain domain and host information. *Nslookup* contains functionality common in client/server relationships that cause failure in both DTE and non-DTE environments. The agility policy implemented for the *nslookup* client shows how three additional issues can be addressed via the toolkit. First, how changes to the DTE policy that would otherwise cause client/server communication to be interrupted or lost completely can be overcome. Second, how CTPs can also be used to dynamically extend or modify component specific code. Finally, how the toolkit can adjust to changes in the DNS server specified by the /etc/resolv.conf configuration file.

In our experiences developing non-trivial DTE policies, we have found it difficult to configure the policy correctly to allow all intended client/server message exchanges to occur, as alluded to in section 2.2. If either client or server cannot read the DTE type labeling the message from its peer, a failure will occur in attempts to read data. For the *nslookup* client, this will simply cause a failure of the lookup. However, in many client/server applications, such an error would cause unwanted termination, perhaps of a critical process. There are many client/server interactions on a system that can't be anticipated without a comprehensive knowledge of each system process. These interactions can make security policy changes difficult to properly implement.

However, by implementing CTPs for the *connect()* and *send()* socket-related system calls, communication failures caused by DTE policy configurations and dynamic changes can be overcome in many cases. If a failure to communicate with the default output type of the domain is detected, the corresponding CTP will try to communicate with other types the domain can write until success is attained or the possibilities are exhausted. The type of the first successful communication is then retained to label future messages. It will only be recalculated if a policy change indicates a change in the access rights of the client or server domain.

A component-specific CTP was added to the original *nslookup* source code to demonstrate that CTPs can also extend or reconfigure component functions in addition to system calls and library routines. This allows component-specific algorithms to be modified during runtime, for example, to change the cryptographic algorithm used to encrypt data files. To implement the component-specific CTP, the only required change to the original source was to rename the function to be intercepted (providing the indirection necessary to implement the toolkit's CTPs). The toolkit includes automated methods to construct the intermediate code that supports dynamic replacement of the original routine, requiring the developer to simply construct the augmented or replaced function. Our CTP simply replaced the primary client request routine to retain a connection to the name server for subsequent queries rather than the default functionality to initiate a new connection with each request.

The agile *nslookup* process also demonstrates the ability of the AGA to monitor files changes on behalf of agile processes, in this case the /etc/resolv.conf file. When the periodic AGA monitoring routine notices a change to the modification time of the file, it notifies *nslookup*, which then invokes a global callback CTP to transition to the updated domain nameserver(s) during the life of the process, an action that is not performed by typical UNIX components. This static initialization of the DNS servers can be looked upon as another example of a grandfathered decision that could potentially have catastrophic effects in the event the primary name server(s) were to be disabled for any number of reasons, from machine failure to hostile intrusion.

3.4.4. Policy-enforcing HTTP server component. In a dynamic execution environment, a server component *C* could migrate to a new node whose lower-level server components (such as the operating system) have insufficient policy enforcement mechanisms compared to *C*'s original node. However, with embedded policy enforcement mechanisms, *C* can begin enforcing policy itself, effectively providing its clients with the same security policy environment they had before.

The *httpd* daemon, which integrates the ADF policy model for access control to HTTP resources, has been demonstrated on UNIX and Windows NT platforms. Like DTE, the ADF policy specification associates a domain with the IP address of the incoming request and checks the domain's access rights to the types associated with the files being served. Using the UNIX toolkit, only seven lines of code modification to the original source code were required to implement ADF *httpd* access control along with one CTP, *fork*(), to associate an AGA with each *httpd* server. The ADF policy of a running component is dynamic and can be modified during component runtime.

4. Future directions

This paper primarily focuses on the results attained through the second of three phases of our security agility research. The first phase investigated UNIX system components to identify potential shortcomings of operating in a dynamic policy environment and experimented with techniques to address those shortcomings. The second phase of the project packaged these techniques in the Security Agility Toolkit. The third and final phase will extend security agility to a distributed environment focusing on two primary activities: developing the mechanisms to support a Distributed Security Agility Toolkit, and demonstrating the potential of the distributed toolkit in conjunction with Intrusion Detection Systems (IDSs).

Distributed toolkit development is currently underway with initial efforts focusing on the Windows NT environment to expand the heterogeneous environment supported by security agility. The FreeBSD operating system will host the phase three UNIX development that will include DTE and non-DTE versions. The mechanisms that must be developed to support distributed security agility include synchronization, communication, and concurrency control techniques. An AGility Authority Manager (AGAM) component that is being constructed to coordinate security policy and agile policy updates with each node's AGA will provide much of these capabilities. Additionally, the concept of an agility policy will be extended and redefined to increase its scalability, manageability, and generalized applicability for the distributed heterogeneous environment. Finally, we will strive to optionally implement some agile behaviors more transparently by eliminating the need for source code recompilation extending agile behaviors to many components. This has already been successfully demonstrated on a small scale in the Windows NT environment, where we have customized Dynamic Link Library (DLL) interface interception techniques to specify control transition points and combined them with our first attempt at the new agile policy specification. We intend to develop parallel mechanisms for the Executable and Linking Format (ELF) binaries employed in the FreeBSD environment by replacing standard shared libraries with agile library versions.

We also intend to explore integration of the toolkit with an intrusion detection system to use agility techniques to respond to intrusions. Our current intent with respect to intrusion detection driven agility is to conduct initial validation testing by integrating the distributed toolkit with intrusion detection wrappers produced by our Generic Software Wrappers sister project. The Common Intrusion Detection Framework (CIDF) [9], initiated by DARPA ITO, will be used ultimately provide the security agility interface to intrusion detection systems.

5. Conclusions

Distributed execution environments are becoming increasingly dynamic in nature. Software components that execute in a dynamic environment must be able to respond flexibly and quickly to changes in resources and security policies to accommodate such occurrences as shifting coalitions, changing states of readiness, and administrative security directives. Conventional software, however, is extremely security rigid: dynamic policy changes may lead to incidences such as abortion of critical missions, violation of new security rules, or destruction of communication channels.

This paper reports on security agility concepts and implementation. Security agility is a software flexibility technique to address security properties and their dynamic evolution by making software components aware of their security environment and allowing them to adapt to its changing nature during runtime. Security agility techniques are packaged in a prototype toolkit that consists of reusable code that can be customized for integration into software components in the environment they are deployed. The toolkit offers two primary mechanisms for adaptive behavior: embedded security models and mechanisms and dynamic code extensions to modify security related processing. We have experimentally integrated the toolkit with a variety of system components in a DTE-BSD/OS (UNIX) environment to show how dynamic policy changes that would normally negatively impact running components can be positively addressed by agile versions of the components. The behaviors exhibited by an agile process can also be updated during the life cycle of the running process to add increasing flexibility.

Though we have not constructed formal benchmarks to measure the toolkit's ease of use, we found integrating the toolkit with selected DTE-BSD/OS components to be relatively unobtrusive. The toolkit's object oriented framework provides a 'plug-in' capability for other policy engines and its use of standard techniques offer portability to other operating systems. Through toolkit ports to the Windows NT and FreeBSD environments, we discovered that though there were some minor changes required for compilation, the primary porting effort was in customizing interception techniques to match the new environments. However, the interception techniques used in the three toolkit environments should map well into other architectures, such as LINUX or Windows 98.

Finally, the toolkit's integrated software-based approach is just one of the possible solutions to dynamic security adaptability. Other approaches, such as software wrapper techniques, are certainly viable alternatives. However, we believe that the toolkit approach provides several advantages over other adaptive methods, particularly when coupled with existing policy models and enforcement engines, such as DTE.

The ability to formulate control transition points for all library and component level routines provides additional flexibility to isolate code segments for security relevant augmentation or even algorithm changes. For example, several security policy models (e.g., Bell-La Padula's "trusted subjects", Clark-Wilson's "transformation procedures") rely on the use of special components which are guaranteed by their design and implementation never to perform actions disallowed by the (static) security policy. Therefore, they are allowed to operate without the usual oversight or mediation. Unfortunately, in the dynamic policy environment, these rules may change, making these components useless, even dangerous, without reconfiguration or reimplementation. However, the code that provides a trusted component with its essential security properties may be (re)structured into a single subsystem within the component that acts as a governor over the rest of the component's operation to ensure that it avoids security violations. Since the toolkit's CTP/dynamic code extension techniques allow the subsystem to be dynamically reconfigured, it can force the trusted component to change its security-relevant behavior in response to security policy updates.

The toolkit's software-based approach also allows those most familiar with the processes' functionality, such as a developer or a system analyzer, to address problems that might be introduced by dynamic security policy changes in a familiar source code environment. This eliminates the need to learn additional specification semantics that might be required by other alternatives.

References

[1] Karen A. Oostendorp, Lee Badger, Christopher D. Vance, Wayne G. Morrison, Michael J. Petkac, David L. Sherman, and Daniel F. Sterne. "Domain and Type Enforcement Firewalls," In *Proceedings of the 13th Computer Security Applications Conference*, San Diego, California, December 1997.

[2] Karen A. Oostendorp, Timothy J. Fraser, Michael J. Petkac, John Grillo, Benjamin L. Uecker, and Lee Badger. *Security Agility for Dynamic Execution Environments Initial Prototype Evaluation Report*, Technical Report 0741, TIS Labs, June 1998.

[3] Mike Petkac, Eric Cantori, Wayne Morrison, and Lee Badger. *Security Agility for Dynamic Execution Environments Security Agility Toolkit Evaluation Report*, Technical Report 0765, NAI Labs, June 1999.

[4] D. E. Bell and L. LaPadula. *Secure Computing System: Unified Exposition and Multics Interpretation,* Technical Report EDS-TR-75-306, Electronics Systems Division, AFSC, Hanscom AF Base, Bedford, Massachusetts, 1976.

[5] K. J. Biba. *Integrity Considerations for Secure Computing Systems*, Technical Report ESD-TR-76-372, AFSC Electronics Systems Division, Bedford, Massachusetts, 1977.

[6] D. D. Clark and D. R. Wilson. "A Comparison of Commercial and Military Computer Security Policies," In *Proceedings of the 1987 IEEE Symposium on Security and Privacy*, Oakland, California, 1987.

[7] Lee Badger, Daniel F. Sterne, David L. Sherman, and Kenneth M. Walker. "A Domain and Type Enforcement UNIX Prototype," USENIX Computing Systems Vol. 9, No. 1, Winter 1996, p. 47.

[8] W. E. Boebert and R. Y. Kain. "A Practical Alternative to Hierarchical Integrity Policies," In *Proceedings of the 8th National Computer Security Conference*, Gaithersburg, MD, p. 18, 1985.

[9] Stuart Staniford-Chen, Brian Tung, and Dan. Schnackenburg. "The Common Intrusion Detection Framework," Position paper accepted to the Information Survivability Workshop, Orlando, Florida, 1998.

[10] M. D. Abhams, K. W. Eggers, L. J. LaPadula, I. M. Olson, "A Generalized Framework for Access Control: an Informal Description," In *Proceedings of the 13th National Computer Security Conference*, October 1990.

[11] D. McCullough, "Specifications for Multi-Level Security and a Hook-Up Property," Proceedings of the 1987 IEEE Symposium on Security and Privacy, 1987.

[12] D. Sutherland, "A Model of Information". In *Proceedings of the 9th National Computer Security Conference*, September, 1986, p. 175.

[13] D. B. Chapman and E. D. Zwicky, *Building Internet Firewalls*, O'Reilly and Associates, Inc., 1995.

[14] W.R. Cheswick and S. M. Bellovin, *Firewalls and Internet Security: Repelling the Wily Hacker*, Addison-Wesley, 1994.

[15] D. Sterne, "A TCB Subset for Integrity and Role-Based Access Control," In *Proceedings 15th National Computer Security Conference*, pp. 680-696, Baltimore, MD, 1992.

[16] W. Venema, "TCP Wrapper: Network Monitoring, Access Control, and Booby Traps," in *Proceedings of the 3rd Unix Security Symposium*, Baltimore, MD, September 1992.

Software Architecture and Logic for Secure Applications

Myong H. Kang and Judith N. Froscher

Information Technology Division
Naval Research Laboratory
{mkang, froscher}@itd.nrl.navy.mil

Abstract

Today's military supports many different missions as well as the traditional role of national defense. This diversity means that these systems must be easily configurable to accommodate integration with coalition partners all over the world and with local law enforcement and emergency response organizations, and secure and survivable to ensure the success of the military mission. Commercial workflow and enterprise application integration products provide solutions to similar problems in the commercial world. However these products cannot satisfy the military because they lack security and survivability features.

DoD needs multilevel secure (MLS) and survivable system and tools that allow users to program multilevel mission logic, securely coordinate globally distributed users and existing applications, and monitor the progress of the distributed program across classification domains. This paper summarizes an effort to provide a methodology and tools to support mission critical operations.

1. Introduction

Streamlining of current business and information processes has brought about significant increases in productivity and the ability for US companies to compete in the global market place. Software developers have been challenged to streamline the software development process and to produce software that can be reused, that separates concerns, that supports autonomy and heterogeneity in a distributed computing environment, that allows extensibility, etc. The information technology (IT) community has developed distributed object computing standards, like CORBA and DCOM, that provide a basic level of interoperability among distributed applications.

The next step is to build application specific software architectures that encode business logic for coordinating widely distributed manual and automated tasks in achieving enterprise level objectives. Commercial enterprise application integration (EAI) tools and workflow management system (WFMS) products address these issues. The potential benefits of these technologies are enormous because of their broad reach in managing business process productivity.

However, commercial products do not support distributed mission critical applications. Because they neither provide recovery mechanisms from failure nor do they enforce access control policies (certainly not multilevel secure (MLS) access control policies). Even though there is a great need for this technology in DoD, DoD cannot rely on commercial products to protect national security information and perform mission critical business processes.

Today's military has to respond to new threats in completely different environments. For example, today's military supports disaster relief, drug interdiction, peace-keeping missions in worldwide regional skirmishes, treaty enforcement, as well as the traditional role of national defense against aggression and weapons of mass destruction. At no other time in the nation's history has the military relied so heavily on IT systems for all of its operations, including command and control, logistics, surveillance and reconnaissance, personnel management, finances, etc. Challenges to DoD's information systems include,

- The organizations that participate in a dynamic coalition may be located in different classification domains.
- The guidelines for sharing and exchanging information among organizations in different classification domains are stricter than those for sharing information in the same classification domain.
- DoD information systems are increasingly exposed to great risks (i.e., cyber and physical attacks).

To address those problems, the Naval Research Laboratory (NRL) has embarked upon a research project to investigate software architecture and convenient way to build logic for secure applications. Two main technical objectives that need to be addressed are *multilevel security* and *survivability*. In other words, we need

- secure access to the right information at the right place at the right time,
- reduced development cycle for distributed applications and maximum use of commercial software/hardware,
- high confidence in the application behavior,
- survivability from anomalies as well as information warfare attacks, and
- enterprise-level information management and monitoring.

Even though there may be many ways to achieve these goals, we have started with the WFMS paradigm, to which we can add new capabilities such as multilevel security, distributed scheduling, recovery, etc. We have chosen the workflow paradigm for several reasons. First, because it provides a design environment, it allows us to address distributed security engineering issues during the design phase. In other words, it allows a designer to consider the security issues in conjunction with functional issues. Second, because it provides a separate runtime environment, it allows us to distribute some functionality, such as scheduling and monitoring, for survivability. Third, this separation of design and runtime environments allows us to add other validation functionality such as structural validation for correctness of mission logic and security validation (see section 5) without affecting design and runtime environments.

The direction of the project is to make use of COTS components as much as possible. When we cannot find the suitable components due to lack of industry interests (e.g., MLS) or progress (e.g., interoperability), we then develop tools and security critical components that allow DoD to harvest emerging commercial off-the-shelf (COTS) technology and still rely on legacy resources with reduced risk.

In this paper, we summarize the first phase (i.e., 18 months) effort to investigate and prototype MLS WFMS, and lessons learned. In this first phase, we focus on the aspect of multilevel security. However, all decisions are made with survivability in mind. We document the reasons behind each decision.

2. A Strategy for MLS Workflow

The MLS WFMS that we are building will provide equivalent functionality to a single-level WFMS and hooks into an MLS infrastructure for enforcing the MLS security policy. Tasks that may be single-level individually, but located in different classification domains, have to cooperate to achieve a higher-level MLS mission objective. Therefore, we need to provide an MLS distributed programming (design) tool that allows programmers to specify their distributed program logic (sometimes called *mission logic* in this paper). This design tool allows MLS workflow designers to

- divide a design area into multiple domains,
- specify dominance relationships among domains (e.g., Top Secret > Secret > Unclassified),
- describe the tasks in each domain,
- specify information flow, dependency, and the condition of the dependency among tasks that are in the same or different domains, and
- specify exception conditions and recovery strategies for different exceptions.

On the other hand, the runtime engine needs to provide MLS services in a distributed and heterogeneous computing environment. The MLS runtime system must enforce the following information flow requirements:

- High users may have access to low data and low resources,
- High processes may have access to low data, and
- High data must not leak to low systems or users.

An MLS WFMS should obey this MLS policy. Atluri *et. al.* have investigated MLS workflow in general [1]. The development of high-assurance software, necessary to provide separation between unclassified and TS/SCI information, such as MLS workflow systems, has proven to be both technically challenging and expensive. Today's fast paced advances in technology and the need to use COTS products make the traditional MLS approach untenable. Therefore, the approach we have chosen for building MLS workflow is to integrate multiple single-level workflows with an MLS distributed architecture [3]. This is in line with modern distributed computing paradigms that support autonomy and heterogeneity.

To implement an MLS WFMS using the architectural method, the following technical approach has been established:

- Implement the necessary design tool for supporting MLS workflow because there is no COTS MLS workflow design tool available. Even though this tool allows workflow designers to specify information and control flow among tasks in different domains, the operational environment of the tool will be system-high (i.e., the workflow design tool neither accesses sensitive data in multiple domains nor passes it around). Hence, we can implement this tool without too much concern for multilevel security issues (e.g., information leakage across classification domain boundary). This tool will be run on a single-level system.
- Choose an MLS distributed architecture where an MLS workflow can be executed. We chose the DARPA IA architecture [9], which is sometimes called the multiple security levels (MSL) (see section 6.1), as our underlying MLS infrastructure.
- Choose a strategy for dividing an MLS workflow that has been specified into multiple single-level workflows (see section 5.2).
- Choose a single-level WFMS to execute a single-level workflow in each classification domain.
- Extend the workflow interoperability model to accommodate cooperation among workflows at different classification domains. This extension was necessary because commercial workflow products support very limited interoperability model (see section 6.2).
- Extend the single-level workflow enactment service (i.e., runtime engine) to accommodate communication among tasks in different classification domains. This extension of the runtime engine is necessary to support extended workflow interoperability model. The communication among workflows at different classification domains is performed through the extended runtime workflow interoperability mechanism.

In the following section, we describe the overall structure of the system.

3. System Overview

We believe that an MLS WFMS is a good starting point to meet the requirements that were described in section 1. In general, a WFMS consists of two main components: a design tool and runtime tools. Our requirements contain an unusual requirement, which is MLS. However, all other requirements are generic enough for use by corporate environments. We believe the MLS requirement and the way we solve this MLS problem actually helps us to look into workflow interoperability from a fresh perspective [4].

We have developed generic platform independent distributed programming tools based on an object-oriented paradigm; hence, Java was chosen as our development language. We also want our design tool to be not only independent of the runtime engine, but also independent of the underling MLS infrastructure. Based on these requirements, we have developed the system organization, shown in figure 1.

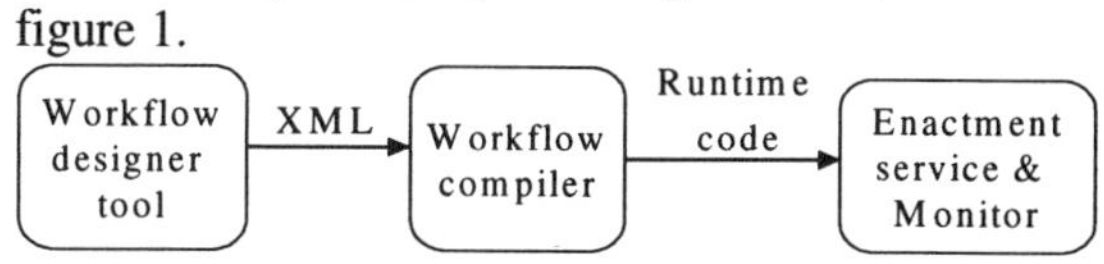

Figure 1: Internal structure of MLS METEOR

A workflow designer specifies mission logic through the workflow design tool (i.e., various workflow editors). The workflow design tool saves the design in XML (eXtensible Markup Language) [7]. When a workflow design is completed, the workflow compiler reads the XML representation of the design and performs the necessary design analysis and validation. Finally it generates runtime code for enactment services. Even though there are many workflow runtime engines, there are very few distributed runtime engines that are necessary for survivability. We believe that OrbWork, a distributed runtime workflow engine from the University of Georgia, is a good starting point. This distributed scheduling capability is vital for the survivability. To satisfy MLS and other requirements, OrbWork has been extended. In the following sections, we will explain each component in detail.

4. Workflow Design Tool

The workflow design tool is a generic distributed programming tool that can express programming logic through GUI-based editors. We have developed the workflow design tool based on the MLS distributed computing model.

4.1. MLS workflow model

MLS distributed programming adds another dimension of complexity to single-level distributed programming, which itself is not a trivial task. Therefore, we need a new programming model for MLS distributed computing that

- eases the burden of MLS distributed programming, especially in the context of large system integration,
- promotes the re-use of existing components,
- facilitates the specification of security requirements (e.g., roles),
- enables secure cooperation among autonomous systems at different classification levels, and
- provides a global picture of the whole mission and a proper view of the mission to users at different levels of abstraction,

In the MLS workflow model, a *task* represents an abstraction of an activity. A task can be regarded as a unit of work that is performed by a variety of processing entities, depending on the nature of the task. A task can be performed by (*realized by*) a human, a computerized activity that executes a computer program, a database transaction, or possibly a network of interconnected tasks. Hence, a task provides one level of abstraction (view) and its realization provides a lower level of abstraction (view). Since the realization of a task may contain many tasks at different levels of abstraction, a task is a recursive reference in our workflow model. In other words, one task from a particular user's point of view may be a network of many tasks from another user's perspective.

There are two types of tasks in the model:

- *Foreign task:* A task whose realization (i.e., strategy for implementation) is unknown to the workflow designer. It represents a task that is a part of cooperating independent system. It is required for a designer to declare a foreign task explicitly to provide a hint to the runtime code generator. A foreign task should have a minimal information set (e.g., where to send the request, how to receive output).
- *Native task:* A task for which the realization is known or the realization will be provided before runtime code generation (i.e., all other tasks except the foreign tasks).

For example, foreign tasks can be used to define communication and synchronization with a task in other classification domains. If an MLS workflow is created at the highest classification domain, then the complete MLS workflow with realizations of all its tasks can be specified. However, if the workflow designer creates an MLS workflow that requires input from (or output to) higher classification domains, then he may know only the interfaces to the tasks at the higher levels but not the detailed workflow process at higher levels.

A native task can be either a simple task or a network task. A simple task is a task that cannot be broken down further from a workflow designer's point of view. A *network task* represents the core of the workflow activity specification. Since a network task is one of the realizations of a task, it is always associated with a task called its *parent task.* A single network of tasks defines a relationship among workflow tasks, transferred data, exception handling, and other relevant information. It is a collection of either foreign or native tasks and transitions from one task to another. Figure 2 shows a simplified version of two levels of abstractions (views) where Task2 is the parent task of the projected workflow W_l, which contains tasks 4, 5, 6, and 7, and transition t_j represents a transition from Task1 to Task2. In Figure 2, Task1, Task2, and Task3 may belong to different classification domains. Hence, the MLS workflow model can be thought of as follows: along the *xy*-surface, there are tasks in different domains and along the *z*-axis, there are different levels of abstraction.

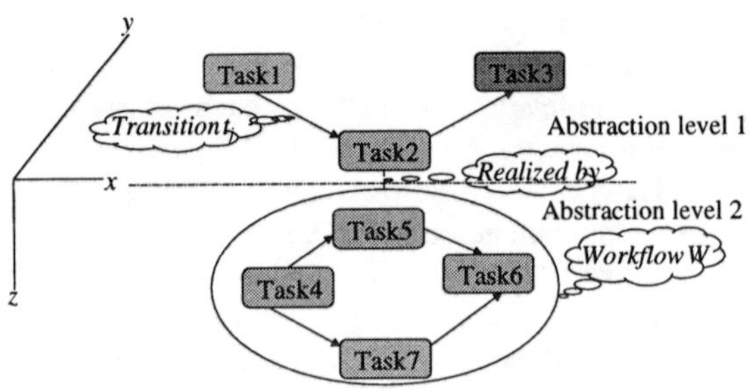

Figure 2: MLS workflowModel

A task may play the role of a source task or a destination task (e.g., Task1 is the source task and Task2 is the destination task of the transition t_j in Figure 2) for a number of transitions. All of the transitions for which a task is the destination task are called the *input transitions* for that task (e.g., transition t_j is an input transition for Task2). Likewise, all the transitions for which a task is the source task are called its *output transitions* (e.g., transition t_j is an output transition of Task1). A transition may have an associated Boolean condition called its *guard.* A transition may be activated only if its guard is true. When there is a transition from task T_i to task T_j where T_i and T_j are in different classification domains, we call this an MLS *transition* from T_i to T_j.

An *external transition* is a special type of a transition in which the two participating tasks (source and destination) are not in the same workflow (i.e., transition to and from a foreign task). An external transition may lead to a start task of another workflow. Similarly, an implied transition leads from the final task and is used to notify the external entity that the network has terminated. Note that an MLS transition is turned into an external transition when an MLS design is divided into multiple single-level workflows for runtime.

External transitions are also used to specify synchronization points with some external events. Typically, external transitions may be used to specify communication and synchronization between two independent workflows. Here, an external transition leading into a task in the workflow is assumed to have an implied source task (outside the workflow). Similarly, an external transition leading out of a task in the workflow is considered to have an implied destination task (outside the workflow). External transition is a cornerstone of our strategy to support MLS workflow through workflow interoperability.

The classes (i.e., types of objects) that are associated with an input transition to a task are called the task's *input classes*, and those appearing on an output transition are called *output classes* of that task. If an output class is also not an input class then the class is *created* by the task. Specifically, an object instance of the specified class is created by the workflow runtime. An input class that is not an output class is *dropped (consumed)*. When input classes are unused by the task, they are transferred to the task's successor(s).

A group of input transitions is called an *AND-join* if all of the participating transitions must be activated for the task to be *enabled* for execution. An AND-join is called *enabled* if all of its transitions have been activated. All the input transitions of a task may be partitioned into a number of AND-joins. A group of input transitions is called an *OR-join* if the activation of one of the participating transitions enables the task.

A group of transitions is said to have a *common source* if they have the same source task and all lead either from:

- Its success state or
- Its fail state.

A group of common source transitions may form one of the following:

- *AND-split*: Each of the transitions in the group has the condition set to `true`. This means that all of the transitions in the group are activated once the task is completed.
- *OR-split* (selection): An ordered list of transitions where all but the last transition may have arbitrary conditions (i.e., the last transition on the list has the condition set to `true`). The first transition whose condition is satisfied will be activated.
- *Loop*: A special case of an OR-split, where the list is composed of exactly two transitions: continue and break. Continue implies branch taken and break implies branch not taken (i.e., fall through).

All tasks that we define in this paper are single-level tasks. What we mean by single-level is that the task receives input from one classification domain and produces output at the same classification domain. There are four special tasks: *begin, success, failure,* and *synchronization*. The synchronization tasks represent external transitions to and from other workflows. In general, workflow designers do not manipulate synchronization nodes directly. They are automatically generated by the system based on the specification of foreign tasks and input and output transitions to and from the foreign tasks.

An MLS *workflow* is a network of interconnected single-level (foreign or native) tasks from more than one classification domain. Note that we call a task single-level from one particular level of abstraction (view). Since a single-level task may be realized by an MLS workflow at a lower level of abstraction, it may have side-effects on different classification domains at lower abstraction levels. Hence, our distinction between single-level and multilevel is purely from the perspective of a specific abstraction level.

Let $CL(T_i)$ represent the classification domain of task T_i. The relationships between the classification domains form a lattice. An MLS workflow that is the realization of task T_i where $CL(T_i) = S_a$ must obey the following constraints:

- The *begin, success*, and *fail* nodes of the MLS workflow must be CL(*begin*) = CL(*success*) = CL(*failure*) = S_a.
- It may have tasks in other classification domains; however, if the $CL(T_j) = S_b$ where S_a does not dominate S_b, then T_j must be a foreign task. In other words, only tasks in S_c where $S_a \geq S_c$ may have realizations.

In the following section, we present workflow design editors though which users can design a specific workflow. These design editors are our reference implementation of the MLS model that we presented in this section.

4.2. Design Editors

We provide various platform independent GUI-based editors to support the MLS distributed computing model. There are two starting points into a specific design process (see figure 3). They are the task and network editors, both of which will create an initial top-level component (task). The arc and operator editors are mainly used in conjunction with the network and task editors to specify data and control flow. There are three additional editors that aid MLS workflow design: data, domain, and role editors. These three editors can be used independent of the specific workflow. For example, domain structure, especially in MLS, may be predefined based on physical separation. Also role hierarchy [8] may be predefined by organizations. This semi-independence enables a workflow designer either

- to use predefined data, and domain and role structures from a previous design or
- to define necessary data, domain and role structures during a workflow design.

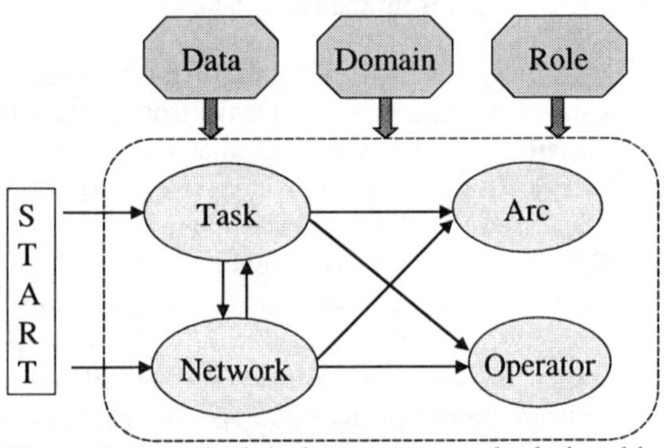

Figure 3: A typical design process and relationship among editors

The description of each editor follows.

Network Editor

A network editor, alternatively called a map editor, is a graphical programming tool that allows users to lay out the control flow of the intended mission logic. Hence, at the highest abstraction level, it provides a global picture of a mission. In this editor, the designer can

- divide the drawing area into many domains,
- drop tasks in the different domains, and
- draw arcs that represent control flow between tasks.

A designer can traverse the different abstraction levels to observe or specify different workflow logic with this editor. It also provides links to all other editors to refine a design. For example, if a designer wishes to specify operators (e.g., AND-split, loop) for a specific task, then he can do so by accessing the operator editor.

An ability to prescribe recovery routes and alternative tasks in case of failure is an important feature for an MLS WFMS. Our designer provides this capability through various editors. First, the network editor supports two types of arcs that represent transitions: one is success arc and the other is fail arc. Second, the MLS workflow also supports system and user-defined exceptions that can be specified through the task editor. Using exceptions and fail arc, a workflow designer can specify a recovery strategy for predictable failures. Figure 4 shows a snapshot of the network editor.

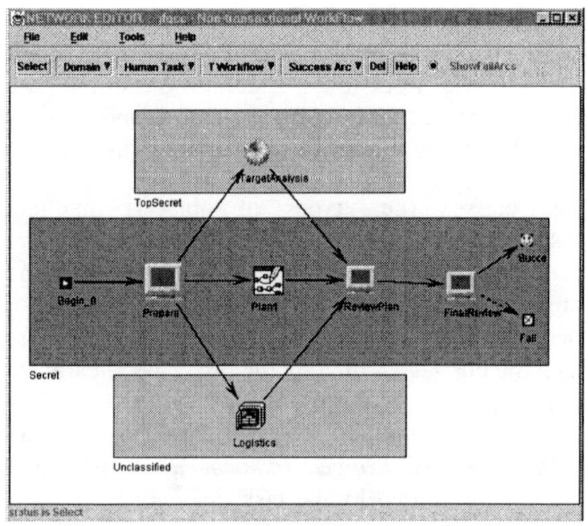

Figure 4: A snapshot of the network editor

Task Editor

The task editor provides the designer with a tool for describing the interface, operating environment, and implementation of the task. Since tasks can be connected together in the implementation of another task (i.e., network task), the task editor also provides information about connections to other tasks and their corresponding editors. In other words, the task editor provides a task-centric view of the workflow.

To describe the task interface, the designer provides a unique name for identifying the task, along with a textual description. The task's type specifies how the task is implemented. The two primary task types are network and simple, where simple tasks include: human, transactional and non-transactional. To complete the interface to a task the designer must specify the input data objects necessary to invoke the task along with the task's output data objects. The designer can specify multiple invocations, but during runtime all the data objects for one of the invocations must be available for a task to start. If the designer wishes to specify failure states for a task then special data objects, called exceptions, are used.

The designer must also describe the environment within which the task has been designed to operate. Of critical importance to an MLS design is what classification domain the task will be operating in. The designer can use compartments (e.g. data restrictions) to further restrict access to the task and its data. In addition the designer can specify what organization owns the task and what roles (which will map to a list of users at runtime) are allowed to perform the task. The designer can also specify the host where a task should be located and specify any system and operational constraints (e.g., allocated time to completion) for the task. During workflow design it is important that the designer considers constraints that have been specified in parent tasks.

The designer must also specify the task implementation (realization). The task's realization is highly dependant upon the type of the task. For a network task, the designer can use the network editor to describe the underlying workflow. For a simple human task, the workflow design tools will generate a generic HTML page based upon the inputs and outputs of the task. The designer can then specify an html editor and viewer that he can use to customize the HTML page for the desired result. When implementing a simple transactional task, the designer will be able to enter the database query commands that are necessary to carry out the commands. For a simple non-transactional task, the designer can enter the code necessary for invoking the task (e.g., executable code, CORBA invocation) or can enter a description of what needs to be done, and the runtime designer can actually implement the functionality. Both transactional and non-transactional tasks can be connected to existing legacy applications.

There are two other special tasks, which the task editor can edit. They are the abstract and the foreign tasks. A designer uses an abstract task to describe the interface and security of a task that some other designer will complete before runtime code generation. But a foreign task is used to describe the interface to a task that will be implemented by another designer and will be available only at runtime.

To achieve a task-centric view of the workflow, the task editor provides the necessary connectivity to look at the entire design. If the task is used within a workflow, the task editor provides the designer with a view of all the task's connections (arcs) whether input, output, or failure arcs. For each connection, the editor provides the designer with quick access to the associated arc, operator, and task editors. And the task editor provides the ability to view down to the implementation details.

Arc Editor

An arc in the network editor represents a transition from a source task to a destination task. In our implementation, arcs specify the data transferring from the source task to the destination task (i.e., input and output classes). The arc editor provides an easy way to map outputs from one task to inputs of another task in a given workflow. For example, one task has three outputs, type1, type1, and type2. Another task has three inputs type1, type1, and type2. Since there is an ambiguity in matching two type1 outputs to two type1 inputs, an arc editor provides a handy way for the designer to specify which output of a task corresponds to the input of another task.

Operator Editor

The new model uses operators to specify the input and output transitions for a task. Hence, a designer needs a capability to edit the structure of these operators. Due to the complexity of workflow design for most applications, it does not seem practical to attach complex operator structures to each task (i.e., three operator structures per task; input, success output, and failure output) in the network editor. So we provide a separate editor to organize the input transition operator and two output transition operators. The input operators are organized using a structure of AND-joins and OR-joins to combine transitions from other tasks. The two output transitions (one for success and one for fail) are organized using a structure of AND-splits, OR-splits, and a LOOP to distribute transitions to other tasks.

Domain Editor

The domain editor allows a designer to specify attributes of each domain (e.g., name, description). As mentioned earlier, the dominance relationship among classification domains form a lattice. The domain editor allows the designer to specify the dominance relationship among classification domains. It also lets users change the GUI properties of classification domains (e.g., color). This editor provides a convenient place to specify receive and release policies between pairs of domains. This policy information can be used as a view into a complete list of policies that are described in a more comprehensive policy definition and enforcement tool.

Data Editor

Data for the workflow design tool is specified as an object interface. The data editor provides a graphical interface for a designer to specify new data and access already defined data. All data objects must extend an existing workflow data object, since the root workflow data object implements functionality required by the runtime of data object management. This is similar to the Java concept where the "Object" class is the root of the entire Java class hierarchy. The data editor allows the designer to specify the package, class name, what class the current class extends (single inheritance only), fields, and methods.

Workflow data is used for task invocations, outputs, and exceptions. In the case of exceptions, the data must extend an existing user exception or the root "UserException". Data is also used for the guards (conditional statements in operators). In a conditional statement, the designer will have access to all the fields and methods defined in the interface of the data object. The relationship among workflow data is shown in figure 5.

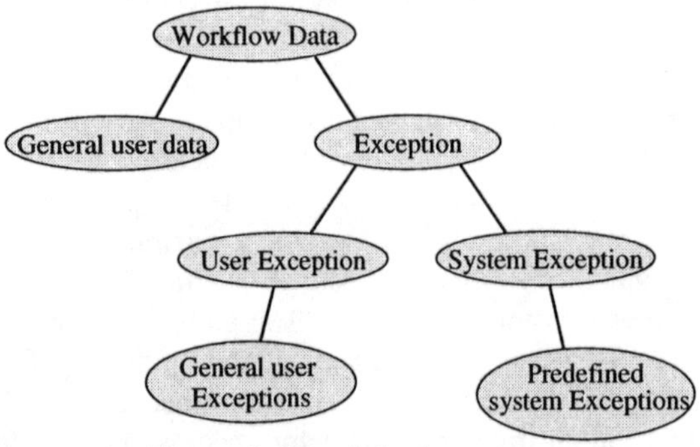

Figure 5: Data inheritance tree

Role editor

The role editor is similar to the classification domain editor in the sense that it allows a designer to define roles and the relationships among them (i.e., role hierarchies). In defining a role the designer can specify the name of the role, its description, and the privileges associated with the role. In general, role hierarchies reflect an organization's line of authority and responsibility. For example, if role A is higher than role B, then role A may have all the permissions that role B has and more. A designer can specify which role is more privileged than another in a given organization through this editor. The role editor will generate XML files on a per organization basis. These XML files can be used by an external application to assign users to roles and enforce permissions in the runtime system.

Coordination among Editors

Editors share common workflow related information and several of them might be displayed at the same time. If an object is modified after an editor displays information, then the editor needs to know about the changes so that it can refresh its display. We use a very simple scheme to ensure consistent display. There is an editor registry that maintains a list of active editors. When a user opens an editor, it registers itself to the registry. When a user closes an editor, it drops itself from the registry. When an editor modifies an internal workflow object, it notifies other active editors. It is each editor's responsibility either to update display if the editor uses the modified object or ignores the notification if the editor does not use the object.

In the next section we present the modules that bridge the gap between the workflow design tool and enactment services.

5. Workflow Compiler

There are three main reasons that we decided to separate the workflow compiler from the workflow design tool. First, the design tool becomes a fairly complex system because it has to handle various inputs and outputs (e.g., mouse movement, context sensitive menu display). Since we do not want to make the design tool any more complex than is necessary, it is logical to create another module, the workflow compiler, to handle some functions that do not require much user interaction. Second, even though the workflow design tool performs limited local design validation (e.g., task name conflict), it is logical to move global design analysis and code generation out of the workflow design tool for maintainability and extendibility reasons. Third, the workflow design tool is a generic distributed programming and documentation tool that can be used for many different purposes. Therefore, a different user community may have different analysis and validation requirements (e.g., vulnerability analysis). Also a different user community may prefer different enactment services. Hence we need to create a simplified (i.e., no GUI components) data structure that can be used by other people to write a new analyzer or runtime code generator. In our case, the extended version of OrbWork [6] is used as our target runtime engine. However, the structure of the workflow compiler allows other runtime engines to be easily incorporated.

5.1. Structure of MLS workflow compiler

Based on the above requirements, we structured the workflow compiler to perform three important tasks. They are:

- Analysis and validation of the design,
- Splitting an MLS workflow into multiple single-level workflows, and
- Generation of runtime code for enactment services.

The workflow compiler is organized in such a way that new analyzers or runtime code generators can be easily integrated. The internal structure of the workflow compiler is as shown in figure 6.

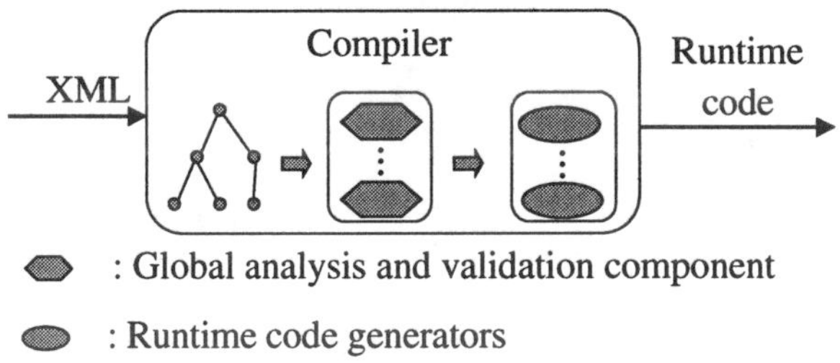

Figure 6: the structure of the MLS workflow compiler

The workflow compiler reads an XML representation of a design and converts it to an internal tree data structure. The analyzer and runtime code generator receive the internal data structure and perform their tasks. Runtime code generation largely depends on

- the specific runtime engine that will be used and
- the MLS infrastructure (see section 6.1) where the runtime engine is executed.

The process of splitting an MLS workflow design into multiple single-level workflows that will be described in the next subsection precedes runtime code generation.

5.2. MLS Dependence and MLS Workflow Decomposition

One of the essential functions of the MLS workflow compiler is to divide an MLS workflow into multiple single-level workflows. These multiple single-level workflows will be executed on the underlying MLS distributed architecture that will be described in section 6. Multiple workflows work together through the extended workflow interoperability model that will be described in section 6.2.

5.2.1. MLS Dependence

An MLS workflow design tool should allow the same kind of intertask dependence as a single-level workflow design tool (e.g., guards, input and output classes). However, some dependence in an MLS workflow may be specified across classification boundaries (i.e., MLS dependence). In other words, workflow state information and some values may have to move across classification boundaries during workflow execution. Hence, it is important to understand what the vulnerability is, whether it is easily exploitable and how to reduce it.

In our MLS WFMS, all information that has to move across classification domains must go though information release and receive policy servers (see section ***) and a high-assurance flow controller (e.g., Pump [2] or downgrader). We divide a transition between two tasks in different classification domains into a series of transitions. For example, if there is a transition from a task in domain 1 (T_{D1}) to a task in domain 2 (T_{D2}) as in Figure 7, then

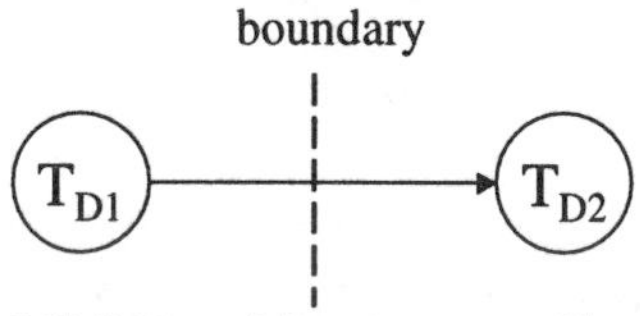

Figure 7: MLS Transition Across a Classification Boundary

this transition will be divided into transitions from T_{D1} to P_{D1}, P_{D1} to P_{D2}, and P_{D2} to T_{D2} as shown in Figure 8. Note that there is the flow controller between P_{D1} and P_{D2} where P_{D1} and P_{D2} are proxies that combine flow controller wrappers and policy servers (i.e., P_{D1} combines a flow controller wrapper and information release policy server, and P_{D2} combines a flow controller wrapper and information receive policy server).

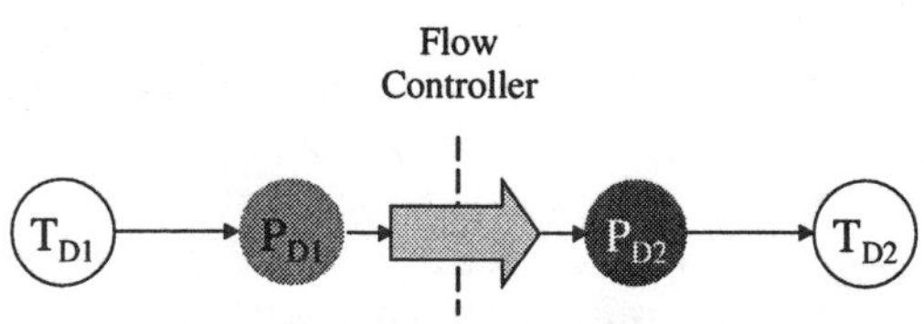

Figure 8: Indirect Transition Through a Flow Controller and Policy Servers

Note that domain policies may not allow combining flow controller wrappers and information policy servers. In that case, we have to break down transitions further. Also note that since T_{D1} and T_{D2} belongs to two different workflows, the P_{Di} serve as synchronization points that we discusses in section 4.1.

5.2.2. Decomposition of an MLS Workflow

Using the method that we described above, an MLS workflow will be separated into multiple single-level workflows. The single-level workflows neither communicate directly nor recognize single-level workflows from other classification domains. The number of single-level workflows that will be generated is equal to the number of distict domains in the MLS workflow (except the domains that contain only foreign tasks). When the breakdown is performed, direct transition (see Figure 7) becomes indirect transition as in Figure 8, and security depends on the underlying MLS architecture. Before we show an example of the division from an MLS workflow to multiple single-level workflows, we will formalize the MLS workflow model and the decomposition of an MLS workflow.

Formalism

An abstraction represents how we view a workflow. At the highest level of abstraction a workflow should be just one network task. However, as we traverse down the hierarchy of abstractions, workflows become more and more concrete with both network and simple tasks making up connected components via the flows. At the bottom of this hierarchy of abstractions, there should be only simple tasks.

There exists a partially ordered set of classification domains $\{D_i, \leq\}$ and a function CL:$\{Tasks\} \cup \{W\} \rightarrow \{D_i\}$.

There exists a set *Tasks.* (To distinguish the elements from the set itself, we capitalize the set name.) The set *Tasks* can be distinguished into the disjoint union of *Network Tasks* and *Simple Tasks.* The set of *Simple Tasks* can be further decomposed into TT ∐ NT ∐ HT. A simple task can be thought of as the smallest unit of work from the workflow's point of view because it is only associated with only one level of abstraction. TT represents transactional tasks, NT are non-transactional tasks, and HT are human tasks. One can think of a network task as a convenient way to group tasks that are associated with multiple levels of abstractions.

There also exists a set *Flows.* A flow can be thought of as the order of program execution (control and data flows). We are interested in tasks and flows that can be formed into directed graphs with tasks being the vertices and flows being the edges.

A **workflow** W is a directed graph made up of tasks and flows. {W} is the set of workflows. The goal of this section is to formalize the ideas about what workflows represent in the rest of this paper, thus ensuring rigor. In this section the concept of "MLS" is implicit in a workflow or tasks. We express this by the following definition.

There exists a set of *Abstractions* $\{z_1, z_2, \ldots, z_n\}$, the abstraction levels, with an ordering $\leq$, such that $z_1 < z_2 < \cdots < z_n$.

A workflow is always associated with an abstraction level z_j. The notation W^j signifies this association. The tasks of W^j are $\{N^j_1, N^j_2, \ldots, N^j_{\alpha 1}, S^j_1, S^j_2, \ldots, S^j_{\beta 1}\}$ where N^j_i is a network task of W^j and S^j_k is a simple task of W^j. For j = n, W^n consists of a single network task N^n_1 and For j = 1, W^1 consists only of simple tasks.

For each abstraction z_j, $j > 1$ there is a projection function π_{j-1} which takes an N^j_i of some W^j and gives a workflow W^{j-1} (associated with z_{j-1}). Furthermore we have the following:

Non-Increasing Security Projection Condition: $CL(\pi_{j-1}(N^j_i)) \leq CL(N^j_i)$, and this comparison is well-defined.

In other words the projection of a network task of a workflow will never give a workflow of a higher classification domain. Note that there is no concept of projection of a simple task because there is no lower level of abstraction for the simple task.

Starting with a workflow W^j (associated with z_j) we may form the *family of* W^j, $\mathcal{F}(W^j)$. The family of W^j gives a view of the workflow at its present, and all lower, levels of abstraction. Before making this definition precise we first discuss some other concepts.

Given W^j we define $P_{j-1}(W^j)$ as
$P_{j-1}(W^j) = \{\pi_{j-1}(N^j_i), \forall N^j_i \in W^j\}$, we may then form $P_{j-2}(W^j)$ as

$P_{j-2}(W^j) = \{\pi_{j-2}(N^{j-1}{}_i),\ \forall\ N^{j-1}{}_i \in P_{j-1}(W^j)\}$; thus we may recursively define
$P_{j-3}(W^j),\ldots, P_1(W^j)$. Now we are ready to define the family of W^j.

$$\mathcal{F}(W^j) = \{W^j, P_{j-1}(W^j),\ldots, P_1(W^j)\}$$

Given any classification domain D_k we may form the filter function F_{D_k} as:
$F_{D_k}(\mathcal{F}(W^j)) = \{\forall\ N^j{}_i \in W^j \ni: CL\ (N^j{}_i) = D_k\} \cup \{\ \forall\ N^{j-1}{}_i \in P_{j-1}(W^j) \ni: CL\ (N^{j-1}{}_i) = D_k\} \cup \cdots$
$\cup \{\forall\ N^1{}_i \in P_1(W^j) \ni: CL\ (N^1{}_i) = D_k\}$

We may also view the filter as a vector $\mathbf{F}_{D_k}(\mathcal{F}(W^j))$ where the (j-h)-component is
$\{\forall\ N^{j-h}{}_i \in P_{j-h}(W^j) \ni: CL\ (N^{j-h}{}_i) = D_k\}$.

We have the following obvious result: $\cup_k F_{D_k}(\mathcal{F}(W^j)) = \mathcal{F}(W^j)$.

Similarly, we may also form F_{bD_k} and $\mathbf{F}_{bD_k}$ by using network tasks whose classification is equal to or below D_k, instead of simply equal to D_k. Note that it is possible that some components are empty. The filter function provides a way to decompose multilevel workflow into single-level workflows, or a specified classification domain and below. Hence, the filter function provides a different view of multilevel workflows for users at different classification domains. For example, the view of a multilevel workflow at the unclassified level and the view of the same multilevel workflow at the secret level are different.

An Example Decomposition

For example, a workflow designer should be able to specify an MLS workflow as in Figure 9 where ovals represent tasks and arrows represent control and data flow. In the figure, B (begin), S (success), and F (failure) represent special tasks.

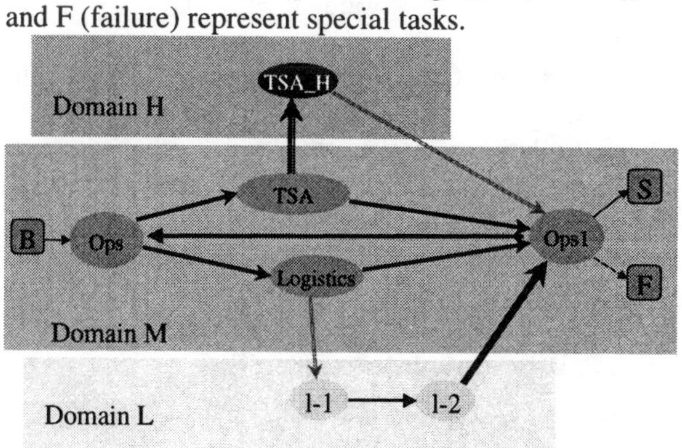

Figure 9: An Example of an MLS Workflow Design

Let us consider the simple MLS workflow design shown in Figure 9 with classification domains L < M < H. Since this particular workflow is designed at domain M (special nodes, B, S, and F signify where the particular workflow was designed), all tasks in domain M and L, which is dominated by domain M, may be native tasks. Since the designer of the workflow in domain M may not know the detailed structure of the workflow in domain H, which dominates M, he can declare the TSA_H a foreign task. The specification of foreign task expresses interfaces (i.e., invocation methods and outputs) and where to send requests. The transitions, TSA to TSA_H, TSA_H to Ops1, Logistics to l-1, and l-2 to Ops1, and input and output classes that are associated with each transition define when and what kind of data will be passed to other workflows at different classification domains.

To execute this MLS workflow on MSL architecture, we have to decompose the MLS workflow design into multiple single-level workflows. This decomposition can be achieved by applying filter functions to the MLS workflow design. After applying filter functions F_M and F_L, the runtime code generator generates the two workflows as shown in Figures 10-a and 10-b. The shaded proxies in Figure 10 represent the combination of policy server, flow controller wrapper and synchronization nodes that represent external transitions.

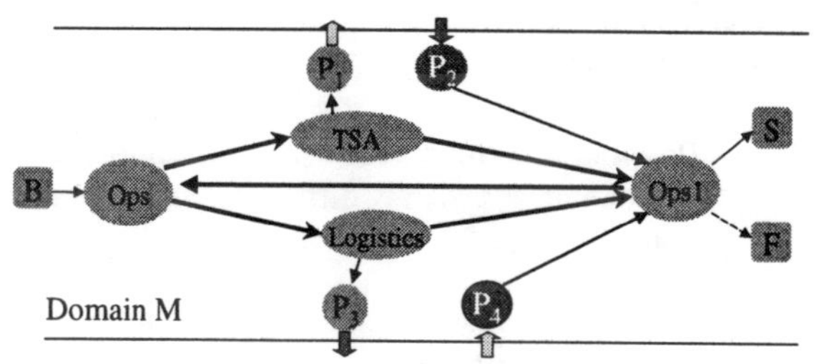

Figure 10-a: An Outcome of the M workflow after applying filter functions F_M

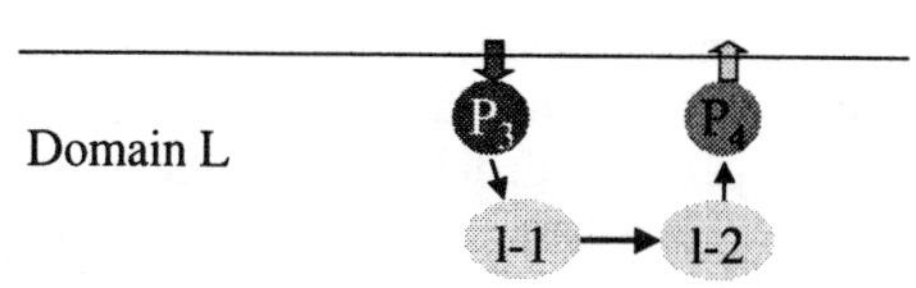

Figure 10-b: An Outcome of the L workflow after applying filter functions F_L

Note that workflows in domains M and L are two independent workflows after the filter function has been applied. The concept of cooperative processes (see section 6.2) and the synchronization node that we defined earlier are useful for interoperability between these two independent workflows. Also note that applying the filter function to the original MLS workflow does not generate the workflow in domain H because there is only a foreign task in domain H.

6. Runtime-system

An MLS workflow runtime management system accesses information in many classification domains. Therefore, the MLS requirement needs to be addressed by the runtime system. Our strategy for providing an MLS workflow management capability that can reduce MLS trust requirements is through composing multiple single-level WFMSs on a particular MLS infrastructure. We first describe the MLS infrastructure that will host single-level WFMSs. Then we describe a way to achieve the MLS workflow management capability from multiple single-level WFMSs.

6.1. MLS Infrastructure

Composing an MLS workflow from multiple single-level workflows is the only practical way to construct a high-assurance MLS WFMS today. In this approach, the multilevel security of our MLS workflow does not depend on a single-level WFMS, but rather on the underlying MLS distributed architecture. The MLS distributed architecture will:

- Host multiple single-level workflows to be executed and
- Provide conduits for passing information among tasks in different classification domains.

Our MLS distributed architecture is based on a security engineering philosophy: a few trusted devices in conjunction with information release and receive policy servers enforce the information flow policy of the classification domains, and single-level systems and single-level engineering solutions provide other functionality. This is in line with DARPA's information assurance architecture [9] and some people call this architecture as multiple single-level (MSL) architecture. A generic MLS distributed architecture is shown in Figure 7.

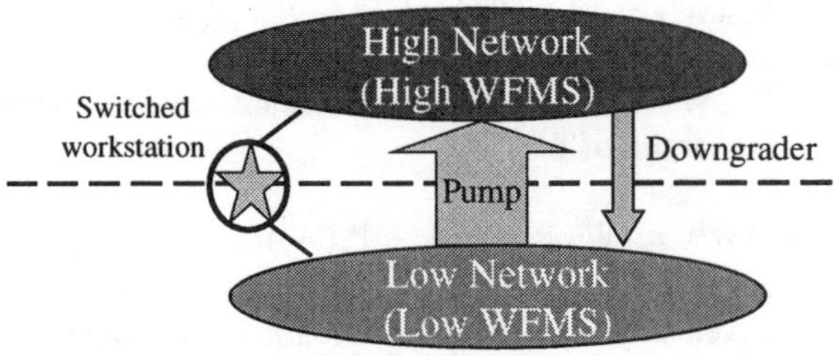

Figure 11: An MLS Distributed Architecture

In this architecture, switched workstations (e.g., "Starlight") enable a user to access resources in multiple classification domains and create information in domains that the user is authorized to access. One-way devices (e.g., a flow controller such as "A Network Pump") together with information release and receive policy servers provide a secure way to pass information from one classification domain to another. An information release policy server resides in a classification domain where the information is released, and an information receive policy server resides in a classification domain where the information is received as shown in Figure 8.

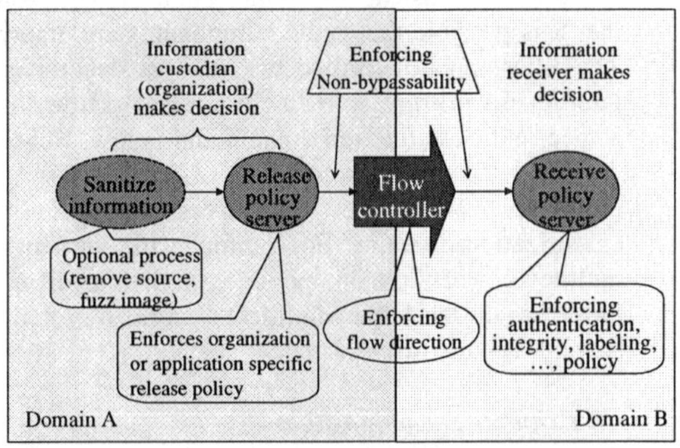

Figure 12: Information Release and Receive Policies in Conjunction with a Flow Controller

When COTS software passes data to a flow controller, in general, a *wrapper* translates the protocol of COTS software to that of the flow controller because COTS software and flow controllers communicate with other software through their own protocols. Hence, a wrapper can be considered a protocol translator. The detailed description of the architecture and the MLS services are in [3].

6.2. Workflow Interoperability

As we described earlier, an MLS workflow becomes multiple single-level workflows that will be run on the MSL architecture. These independent workflows should work together to achieve the intended mission. Therefore, workflow interoperability is an important issue.

There are two aspects of workflow interoperability:

1. The interoperability protocol between independent WMFSs.
2. The ability to model the interoperability in a workflow process definition tool (i.e., workflow builder).

The first aspect can be handled by a standards body such as OMG (e.g., jFlow). However, the second aspect should be handled by each WFMS.

OMG's jFlow introduces two models of interoperability. They are nested sub-process and chained processes as shown in Figure 13.

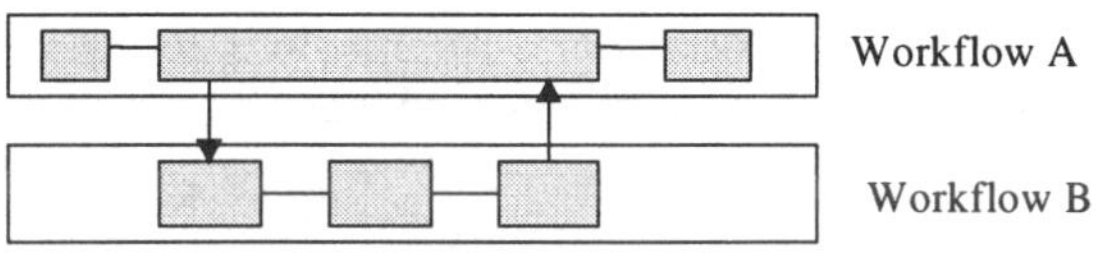

Figure 13-a: Nested sub-process

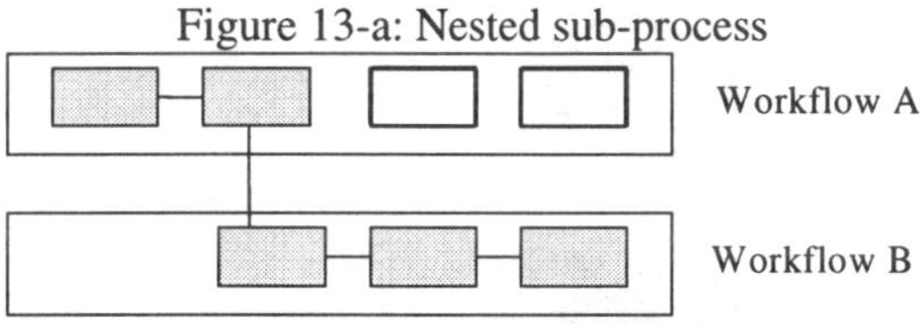

Figure 13-b: chained processes

In nested sub-process workflow structures, a task in workflow A may invoke workflow B as the performer of a task and then wait for it to complete. Hence, the task in workflow A is a requester, and the task that is realized by the sub-processes can serve as the synchronization point for interaction of the two workflows.

In chained workflow structures, one task may invoke another, then carry on with its own business logic. The workflows terminate independently of each other; in this case, the task registered with the sub-process would be another entity that is interested in the results of the sub-process.

The above two models provide powerful mechanisms for interoperability. However, these two models are not sufficient for supporting our purposes (i.e., decomposed workflows to work together). Therefore, we would like to extend the above two models to support a richer interoperability model: *cooperative processes.* Consider two independent autonomous workflows that need to cooperate. Let's assume that there are agreements among organizations that participate in the cooperation. Organization A is in charge of workflow A and Organization B is in charge of workflow B. Tasks in workflow A and workflow B can communicate and synchronize with each other as shown in Figure 14. In this example, two workflows may have independent starting and ending points.

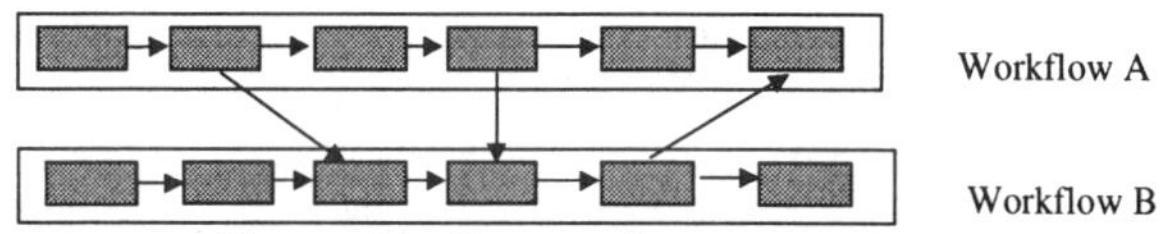

Figure 14: Cooperative processes

There is another situation that we want to support in the context of cooperative processes. In general, the designer of workflow A does not need to know the structure of workflow B and vice versa. This may be because organization A does not want organization B to know the structure of its workflow process and vice versa. In conjunction with the MLS principles, the designer of a workflow may not be allowed to know the detailed workflow structure of a higher-level workflow. For example, in Figure 9, the designer of the workflow, whose classification level is M, may not be allowed to know the workflow structure that contains TSA_H task.

However, there is a minimal set of information that is required for communication and synchronization among tasks in cooperating autonomous workflows. These include:

1. where and how to send/receive requests (i.e., the location and invocation method of tasks) and
2. how and where to receive replies (i.e., expected outputs and the location to return them).

Therefore, the above specification has to appear in the workflow design so that proper runtime code can be generated. Hence, we need a primitive that represents this situation in the design tool. The foreign task, introduced in section 4.1, can accommodate this need.

6.3. Single-level Enactment Services and Monitor

There can be many enactment services for a given workflow design. Currently we are using modified OrbWork [6] for our runtime engine. OrbWork is a single-level distributed workflow engine implemented in Java. It does not have a central scheduler for the whole workflow, rather there is a distributed scheduler per task that the workflow designer defines in the network editor. Each scheduler only knows its immediate predecessors and successors. Most of the EAI and WFMS have centralized engine structure. We chose OrbWork as our baseline implementation because of its distributed scheduling capability. This distributed capability is a useful characteristic for survivability that we continue to investigate.

Briefly, OrbWork works as follows. OrbWork schedulers are CORBA servers. Hence, they communicate with each other through CORBA's IIOP. OrbWork schedulers are also HTTP servers. When a human operator has to interact with a scheduler (e.g., human task), he can do so through the HTTP protocol. Also when a human workflow manager needs to intervene for some reasons, he can do so through the HTTP protocol.

OrbWork has two other CORBA servers: data servers and workflow monitor servers. Data servers act as a repository for data that needs to be passed among schedulers. The workflow monitor server receives progress reports from schedulers. Simplified communication paths among different components in OrbWork are shown in figure 15.

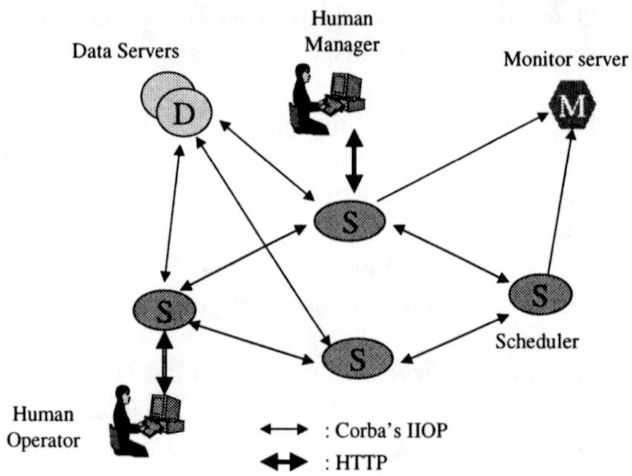

Figure 15: Communication among OrbWork components

6.4. MLS Enactment Services and MLS Monitor

As was presented in section 2, our strategy for achieving MLS workflow is through the interoperability of single-level workflows that were generated from an MLS workflow spefication. For interoperability, the workflow runtime engine should be able to pass and receive the necessary information across domain boundaries. We extend OrbWork with the *synchronization node*. We can categorize synchronization nodes into release and receive synchronization nodes following our MLS infrastructure shown in figure 8. The responsibilities of synchronization nodes are as follows:

- To act as proxies for a task in another domain,
- To serve as exit and entry points to pass necessary information from one domain to another domain, and
- To ensure only proper information is passed to another domain (i.e., make sure release and receive policies are enforced).

For example, if there is an MLS transition from task A to task B as in figure 16, the release synchronization node acts as a proxy for task B and the receive synchronization node acts as a proxy for task A. Also release synchronization nodes serve as an exit point of the release domain and the receive synchronization node serves as an entry point for the receive domain. Therefore, release and receive policies can be enforced there.

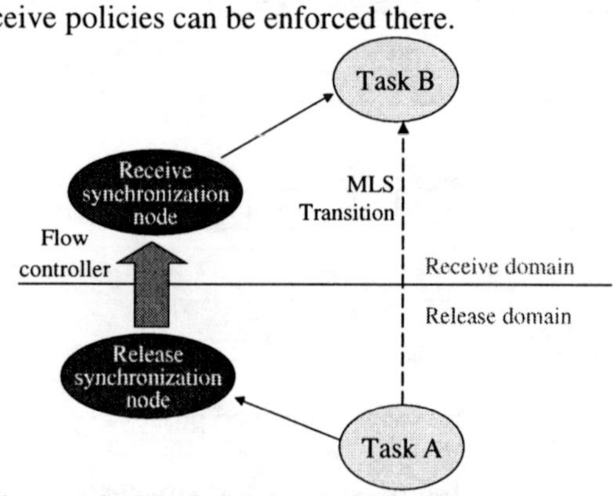

Figure 16: Break down of an MLS transition into multiple transitions

In our MLS workflow system, information exists as objects. Hence, there is the possiblility that objects that contain variables and methods have to be passed across domain boundaries. Passing an object across classification domains can cause integrity and security violations. To mitigate risks associated with passing entire objects (i.e., state and definition of objects) across domain boundaries, currently, we only pass the necessary state information so that the objects can be reconstructed in the receiving domains. When a release synchronization node receives an object, it extracts the necessary information from the object and sends it through a flow controller. When a receive synchronization node receives, it instantiates the object in the receiving domain.

An MLS WFMS requires an MLS monitor. Workflow managers in a classification domain may need to know the progress of work in their classification domain and other domains that they are authorized to access. In other words, users of

MLS workflow in different classification domains may have different views of the workflow they are running. Hence, an MLS WFMS should provide the capability to monitor activities in all domains the workflow manager is authorized to access. We are providing MLS monitoring capabilities that use similar techniques to those for data transfer from one domain to another. In other words, we place a monitoring proxy that receives monitoring information in a lower domain. This monitoring proxy corresponds to a release synchronization node and transfers the monitored information to a higher domain. In the higher domain, there is another monitoring proxy which corresponds to a receive synchronization node. This monitoring proxy relays the lower-level monitoring message to a higher-level monitor server. As in the case of data transfer, information must satisfy release and receive policy and must go though a boundary controller.

7. Summary

In this paper, we presented the results of 18 months effort to investigate and prototype MLS WFMS. To provide security and survivability, a system must be designed from ground up. We presented the strategy for implementing multilevel secure WFMS, system organization and the rationale behind the organization. MLS workflow is an example of an MLS application that can run on the MLS infrastructure. We believe this is one of the first MLS applications that run on the MSL architecture [3, 9]. In this approach, multilevel security enforcement does not depend on the single-level WFMS but rather on the underlying MLS infrastructure and a few security critical components (i.e., synchronization nodes that enforce release and receive policies, and information boundary controllers). Therefore, MLS workflow allows a workflow designer to concentrate on the functionality of the system he is building.

Our effort, thus far, was focused on the foundation of MLS workflow. The remaining work needs to be done to make the MLS WFMS more useful:

- Support richer data type and method evaluation capability at the workflow level,
- Support workflow repositories so that pre-designed tasks can be reused,
- Support more powerful validation for workflow design,
- Support adaptive workflow so that workflow can be altered at runtime, and
- Support survivability.

References

1. V. Atluri, W-K. Huang and E. Bertino, "A Semantic Based Execution Model for Multilevel Secure Workflows," Journal of Computer Security, To appear.

2. M. Kang , I. Moskowitz, and D. Lee, "A Network Pump," *IEEE Transactions on Software Engineering,* Vol. 22, No. 5, pp. 329 - 338, 1996.

3. M. H. Kang, J. Froscher, and B. Eppinger, "Toward an Infrastructure for MLS Distributed Computing," 14th Annual Computer Security Applications Conference, Scottsdale, AZ, 1998.

4. M. H. Kang, J. N. Froscher, B. J. Eppinger, and I. S. Moskowitz, "A Strategy for an MLS Workflow Management System" 13th IFIP Conference on Database Security, Seattle, WA, 1999.

5. M. H. Kang, B. Eppinger, and J. Froscher, "Tools to Support Secure Enterprise Computing," 15th Annual Computer Security Applications Conference, Scottsdale, AZ, 1999.

6. K. Kochut, A. Sheth, and J. Miller, "ORBWork: A CORBA-Based Fully Distributed, Scalable and Dynamic Workflow Enactment Service for METEOR," UGA-CS-TR-98-006, Technical Report, Department of Computer Science, University of Georgia, 1998.

7. Extensible Markup Language (XML) 1.0," World-wide-Web Consortium, http://www.w3.org/TR/1998/REC-xml-19980210.html

8. R. Sandhu, E. Coyne, H. Feinstein and C. Youman, "Role-Based Access Control Models," IEEE Computer, Vol. 29, No. 2, 1996.

9. Security Architecture for the AITS Reference Architecture, http://web-ext2.darpa.mil/iso/ia/Arch/SecArch1/SecArc1.htm

Automated Techniques for Provably Safe Mobile Code*

Karl Crary, Robert Harper, Peter Lee, and Frank Pfenning
Department of Computer Science
Carnegie Mellon University
Pittsburgh, PA 15213-3891

Abstract

We present a general framework for provably safe mobile code. It relies on a formal definition of a safety policy and explicit evidence for compliance with this policy which is attached to a binary. Concrete realizations of this framework are proof-carrying code (PCC), where the evidence for safety is a formal proof generated by a certifying compiler, and typed assembly language (TAL), where the evidence for safety is given via type annotations propagated throughout the compilation process in typed intermediate languages. Validity of the evidence is established via a small trusted type checker, either directly on the binary or indirectly on proof representations in a logical framework (LF).

1. Introduction

Integrating software components to form a reliable system is a long-standing fundamental problem in computer science. The problem manifests itself in numerous guises:

1. How can we dynamically add services to an operating system without compromising its integrity?

2. How can we exploit existing software components when building a new application?

3. How can we support the safe exchange of programs in an untrusted environment?

4. How can we replace components in a running system without disrupting its operation?

*This research was sponsored by the Advanced Research Projects Agency CSTO under the title "The Fox Project: Advanced Languages for Systems Software", ARPA Order No. C533, issued by ESC/ENS under Contract No. F19628-95-C-0050. The views and conclusions contained in this document are those of the authors and should not be interpreted as representing official policies, either expressed or implied, of the Defense Advanced Research Projects Agency or the U.S. Government.

These are all questions of *modularity*. We wish to treat software components as "black boxes" that can be safely integrated into a larger system without fear that their use will compromise, maliciously or otherwise, the integrity of the composite system.

Three main techniques have been proposed to solve the problem of safe component integration:

1. **Run-time checking**. Untrusted components are monitored at execution time to ensure that their interactions with other components are strictly limited. Typical techniques include isolation in separate hardware address spaces and software fault isolation [25]. These methods impose serious performance penalties in the interest of safety. Moreover, there is often a large semantic gap between the low-level properties that are guaranteed by checking (*e.g.*, address space isolation) and the high-level properties that are required (*e.g.*, black box abstraction).

2. **Source-language enforcement**. All components are required to be written in a designated language that is known, or assumed, to ensure "black box" abstraction. These techniques suffer from the requirement that all components be written in a designated, safe language, a restriction that is the more onerous for lack of widely-used safe languages. Moreover, one must assume not only that the language is properly-defined, but also that its implementation is correct, which, in practice, is never the case.

3. **Personal authority**. No attempt is made to enforce safety, rather the component is underwritten by a person or company willing to underwrite its safety. Digital signature schemes may be used to authenticate the underwritten code. In practice few, if any, entities are willing to make assurances for the correctness of their code.

What has been missing until now is a careful analysis of *what* is meant by safe code exchange, rather than yet another proposal for *how* one might achieve a vaguely-defined

0-7695-0490-6/99 $10.00 © 1999 IEEE

notion of safe integration. Our contention is that safe component integration is fundamentally a matter of *proof.* To integrate a component into a larger system, the code recipient wishes to know that the component is suitably well-behaved — that is, compliant with a specified *safety policy.* That is, it must be apparent that the component satisfies a *safety specification* that governs its run-time behavior. Checking compliance with such a safety specification is a form of *program verification* in which we seek to prove that the program complies with the given safety policy.

When viewed as a matter of verification, the question arises as to who (the code producer or the code recipient) should be responsible for checking compliance with the safety policy. *The problem with familiar methods is that they impose the burden on the recipient.* The code producer insists that the recipient employ run-time checks, or comply with the producer's linguistic restrictions, or simply trust the producer to do the right thing. But, we argue, this is exactly the wrong way around. To maximize flexibility we wish to exploit components from many different sources; it is unreasonable to expect that a code recipient be willing to comply with the strictures of each of many disparate methods. Rather, we argue, *it is the responsibility of the code producer to demonstrate safety*. It is (presumably) in the producer's interest for the recipient to use its code. Moreover, it is the producer's responsibility (current practices notwithstanding) to underwrite the safety of its product. In our framework we shift the burden of proof from the recipient to the producer.

Having imposed the burden of proof on the producer, how is the consumer to know that the required obligations have been fulfilled? One method is to rely on trust — the producer signs the binary, affirming the safety of the component. This suffers from the obvious weakness that the recipient must trust not only the producer's integrity, but also must trust the tools that the producer used to verify the safety of the component. Even with the best intentions, it is unlikely that the methods are foolproof. Consequently, few producers are likely to make such a warrant, and few consumers are likely to rely on the code they receive. A much better method is that one that we propose here: require the producer to provide a formal representation of the *proof* that the code is compliant with the safety policy. After all, if the producer did carry out such a proof, it can easily supply the proof to the consumer. Moreover, the recipient can use its own tools to check the validity of the proof to ensure that it really is a genuine proof that the given code complies with the safety specification. Importantly, *it is much easier to check a proof than it is to find a proof in the first place.* Therefore the code recipient need only trust its *own* proof checker, which is, if the method is to be effective, much simpler than the tools required to find the proof in the first place.

The message of this paper is that this approach can, in fact, be made to work in practice. We are exploring two related techniques for implementing our approach to safe component exchange: *proof-carrying code* and *typed object code.* In both cases mobile code is annotated with a formal warrant of its safety, which can be easily checked by the code recipient. To produce such a warrant, we are exploring the construction of *certifying compilers* that produce suitably-annotated object code. Such a compiler could be used by a code producer to generate certified object code. Two points should be kept in mind when reading this paper:

1. The tools and techniques of logic, type theory, and semantics are indispensable.
2. These methods have been implemented and are available *now.*

2. Safety Infrastructures

The first component in a system for safe mobile code is the *safety infrastructure.* The safety infrastructure is the piece of the system that actually ensures the safety of mobile code before execution. It forms the *trusted computing base* of the system, meaning that all consumers of mobile code install it and depend on it, and therefore it must work properly. Any defect in the trusted computing base opens a possible security hole in the system.

A fundamental concern in the design of the trusted computing base is that it be small and simple. Large and/or complicated code bases are very likely to contain bugs, and those bugs are likely to result in exploitable security holes. For us to have confidence in our safety infrastructure, its trusted components must be small and simple enough that they are likely to be correct.

The design of the safety infrastructure consists of three parts. First, one must define a safety policy. Second, one specifies what will be acceptable as evidence of compliance with the safety policy. Suppliers of mobile code will then be required also to supply evidence of compliance in an acceptable form. Third, one must build software that is capable of automatically checking whether purported evidence of safety is actually valid.

2.1. Safety Policies

The first task in the design of the safety infrastructure is to decide what properties mobile code must satisfy to be considered safe. In this paper we will consider a relatively simple safety policy, consisting of *memory safety, control-flow safety,* and *type safety.*

1. Memory safety is the property that a program never dereferences an invalid pointer, never performs an unaligned memory access, and never reads or writes any

memory locations to which it has not been granted access. This property ensures the integrity and privacy of all data not available to the program, and also ensures that the program does not crash due to memory accesses.

2. Control-flow safety is the property that a program never jumps to an address not containing valid code, and never jumps to any code to which it has not been granted access. This property ensures that the program does not jump to any code to which it is not allowed (*e.g.,* low-level system calls), and also ensures that the program does not crash due to jumps.

3. Type safety is the property that every operation the program performs is performed on values of the appropriate type. Strictly speaking, this property subsumes memory and control-flow safety (since memory accesses and jumps are program operations), but it also makes additional guarantees: For example, it ensures that all (allowable) system calls are made using appropriate values, thereby ruling out attacks such as buffer overruns on other code in the system. The additional guarantees provided by type safety are often very expensive to obtain using dynamic means, but the static means we discuss in this paper can provide them at no additional cost.

Stronger safety policies are also possible, including guarantees of the integrity of data stored on the stack [9], limits on resource consumption [14, 4], and policies specified by allowable traces of program operations [26]. However, for policies such as these, the evidence of compliance (which we discuss in the next section) can be more complicated, thereby requiring greater expense both to produce and to verify that evidence, and possibly reducing confidence in the system's correctness. Thus the choice of safety policy in a practical system involves important trade-offs.

It is also worth observing that stronger policies are not always better if they rule out too many programs. For example, a policy that rejects *all* programs provides great safety (and is cheap to implement), but is entirely useless for a safety infrastructure. Therefore, it is important to design safety policies to allow as many programs as possible, while still providing sufficient safety.

2.2. Evidence of Compliance

The safety policy establishes what properties mobile programs must satisfy in order to be permitted to execute on a host. However, it is impossible in general for a code consumer to determine whether an arbitrary program complies with that policy. Therefore, we require that suppliers of mobile code assist the consumer by providing evidence that their code complies with the safety policy. This evidence, which we may think of as a *certificate of safety,* is packaged together with the mobile program and the two together are referred to as *certified code.* Upon obtaining certified code, the code consumer (automatically) verifies the validity of the evidence before executing the program code.

The second task in the design of the safety infrastructure is to decide what form the evidence of compliance must take. This decision is made in the light of several considerations:

1. Since evidence of safety must be transferred over the network along with the program code they certify, we wish the evidence to be as small as possible, in order to minimize communication overhead.

2. Since evidence must be checked before running any program code, we desire verification of evidence to be as fast as possible. Clearly, smaller evidence can lead to faster checking, but we can also speed evidence verification by careful design of the form of evidence.

3. As discussed above, the evidence verifier is an essential part of the trusted computing base; it must work properly or there will be a potential security hole in the system. For us to have confidence that the verifier works properly, it must be simple, which means that the structure of the evidence it checks must also be simple. Thus, not only is simplicity desirable from an aesthetic point of few, but it is also essential for the system to work.

4. Finally, to have complete confidence that our system provides the desired safety, we must prove with mathematical rigor that programs carrying acceptable evidence of safety really do comply with the safety policy. This proof is at the heart of the safety guarantees that the system provides. For such proofs to be feasible, the structure of the evidence must be built on mathematical foundations.

In light of these considerations, we now discuss two different forms that evidence of compliance may take: explicit proofs, which are employed in the Proof-Carrying Code infrastructure [13], and type annotations, which are employed in the Typed Assembly Language infrastructure [11, 9].

Explicit Proofs

The most direct way to provide evidence of safety is to provide an explicit formal proof that the program in question complies with the safety policy. This is the strategy employed by Proof-Carrying Code (PCC). It requires a formal language in which safety proofs can be expressed. Any such language should be designed according to the following criteria.

Effective Decidability: It should be efficiently decidable if a given object represents a valid safety proof.

Compactness: Proofs should have small encodings.

Generality: The representation language should permit proofs of different safety properties. Ideally, it should be open-ended so that new safety policies can be developed without a change in the trusted computing base.

Simplicity: The proof representation language should be as simple as possible, since we must trust its mathematical properties and the implementation of the proof checker.

Our approach has been to use the LF logical framework [5] to satisfy these requirements. A *logical framework* is a general meta-language for the representation of logical inferences rules and deductions. Various logics or theories can be specified in LF at a very high-level of abstraction, simply by stating valid axioms and rules of inference. This provides *generality*, since we can separate the theories required for reasoning about safety properties such as arithmetic modulo 2^{32} or memory update and access from the underlying mechanism of checking proofs. It is also *simple*, since it is based on a pure, dependently typed λ-calculus whose properties have been deeply investigated [6].

Proofs in a logic designed for reasoning about safety properties are represented as terms in LF. Checking that a proof is valid is reduced to checking that its representation in the logical framework is well-typed. This can be carried out effectively even for very large proof objects. Experiments in certifying compilation [16] and decision procedures [23] yield proofs whose representation is more than 1 MB, yet can still be checked. On the other hand, proofs in LF are not compact without additional techniques for redundancy elimination. Following some general techniques [7], Necula [17] has developed optimized representations for a fragment of LF called LF_i which is sufficient for its use in PCC applications. The experimental results obtained so far have validated the practicality of this proof compression technique [16] for the safety policies discussed here. Current research [20] is aimed at extending and improving these methods to obtain further compression without compromising the simplicity of the trusted computing base.

Type Annotations

A second way to provide evidence of safety is using type annotations. In this approach a typing discipline is imposed on mobile programs, and the architects of the system prove a theorem stating that any program satisfying that type discipline will necessarily satisfy the safety policy as well [11]. However, determining whether a program satisfies a type discipline involves finding a consistent type scheme for the values in the program, and such a type scheme cannot be inferred in general. Therefore, in this approach programs are required to include enough type annotations for the type checker to reconstruct a consistent type scheme. Such type annotations constitute the evidence of safety, provided they are taken in conjunction with a theorem the well-typed programs comply with the safety policy (as above).

A principal advantage of the type annotation approach over the explicit proof approach is that the size of the evidence can be much smaller. This is because many facts about program that must be established individually (and included) in explicit proofs can, in a type annotation approach, be established once and for all as invariants of well-typed programs, and therefore those invariants can be omitted from the evidence.[1] The drawback of type annotations is related to this, any program that violates a type system's invariants will not be typeable under that type system, and therefore cannot be accepted by the safety infrastructure, even if it is actually safe; in contrast, with explicit proofs such invariants are not built in, so it is possible to work around cases in which they do not hold.

The idea of using types to guarantee safety is by no means new. Many modern high-level languages (*e.g.,* ML, Modula-3, Java) rely on a type system to ensure that all legal programs are safe. Such languages have even been used for safety infrastructures; for example, the SPIN operating system [3] required that operating system extensions be written in Modula-3, thereby ensuring their safety. The drawback to using a high-level language to ensure safety is that programs are checked for safety before compilation, rather than after, thereby requiring that the entire compiler be included in the trusted computing base. As discussed above, the confidence that one can have in the safety architecture is inversely related to the size of its trusted computing base.

The Typed Assembly Language (TAL) infrastructure resolves this problem by employing a type safe *low-level* language. In TAL, a type discipline is imposed on executable code, and therefore the program code being checked for safety is the exact code that will be executed. There is no need to trust a compiler, because if the compiler is faulty and generates unsafe executables, those executables will be rejected by the type checker.

The principal exercise in developing a type system for executable code is to isolate low-level abstractions satisfying two conditions:

- The abstractions should be independently type checkable; that is, to whatever extent type checking of the abstractions depends on surrounding code and data, it should only depend on the *types* of that code and data.

[1] In current implementations, the TAL architecture has an average 45% evidence overhead, whereas the PCC architecture has a 100% overhead or greater.

- The atomic operations on the abstractions should be single machine instructions.

For example, consider function calls. High-level languages usually provide a built-in notion of functions. Functions can certainly be type checked independently, but they are not dealt with by a single machine instruction. Rather, function calls are processed using separate call and return instructions and the intervening code is by no means atomic: the return address is stored in accessible storage and can be modified or even disregarded. To satisfy the second condition, TAL's corresponding abstraction is the *code block,* and code blocks are invoked using a simple jump instruction. Functions are then composed from code blocks by writing code blocks with an explicit extra input containing the return address. (This decomposition corresponds to the well-known practice in high-level language of programming in continuation-passing style [22].) The first condition is satisfied by requiring code blocks to specify the types of their inputs, just as functions in high-level languages specify the types of their arguments and results. Without such specifications, it would be impossible to check the safety of a jump without inspecting the body of the jump's target.

As an alternative, one can also strike a compromise between high- and low-level languages by exploiting *typed intermediate languages* for safety [10]. Using typed intermediate languages enlarges the trusted computing base, since some part of the compiler must be trusted, but it loosens the second condition on type systems for executable code. This provides a spectrum of possible designs, the closer an intermediate language is to satisfying the second condition, the lesser the amount of the compiler that needs to be trusted. Moreover, as we discuss in Section 3, typed intermediate languages are valuable for automated certification, even if the end result is a typed executable.

2.3. Automated Verification

Since it plays such a central role for provably safe mobile code, we now elaborate on the mechanisms for verifying safety certificates.

Explicit Proofs

As discussed above, the Proof-Carrying Code infrastructure employs the LF logical framework. In the terminology of logical frameworks, a *judgment* is an object of knowledge which may be evident by virtue of a *proof.* Typical safety properties require only a few judgments, such as the truth of a proposition in predicate logic, or the equality of two integers.

In LF, a *judgment* of an object logic is represented by a *type* in the logical framework, and a *proof* by a *term.* If we have a proof P for a judgment J, then the representation $\overline{P}$ has type $\overline{J}$, where we write $\overline{()}$ for the representation function. The adequacy theorem for a representation function guarantees this property and its inverse: Whenever we have a term M of type $\overline{J}$ then there is a proof P for J. Both directions are critical, because together they mean that we can reduce the problem of checking the validity of a proof P to verifying that its representation $\overline{P}$ is well-typed.

So in PCC, checking compliance with a safety policy can be reduced to type checking the representation of a safety proof in the logical framework.

But how does this technique allow for different safety policies? Since proofs are represented as terms in LF, an *inference rule* is represented as a function from the proofs of its premises to the proof of its conclusion. To represent a complete logical system we only need to introduce one type constant for each basic judgment and one term constant for each inference rule. The collection of these constant declarations is called a *signature.*[2] So a particular safety policy consists of a verification condition generator, which extracts a proof obligation from a binary, and signature in LF, which expresses the valid proof principles for the verification condition. This means that different policies can be expressed by different signatures, and that the basic engine that verifies evidence (the LF type checker) does not change for different policies. However, we do have to trust the correctness of the LF signature representing a policy—an inconsistent signature, for example, would allow arbitrary code to pass the safety check.

Type checking in LF is syntax-directed and therefore in practice quite efficient [7], especially if we avoid checking some information which can be statically shown to be redundant [17]. Currently, the Touchstone compiler for PCC discussed in Section 3.1 uses a small, efficient type checker for LF terms written in C. Related projects on proof-carrying code at Princeton [1, 2] and certifying decision procedures at Stanford [23] use the Twelf implementation [21, 19, 18] developed over many years by researchers in logical frameworks at Carnegie Mellon University.

Type Annotations

In case the safety policy is expressed in the form of typing rules, checking compliance immediately reduces to type checking. In this case we have to carefully design the language of annotations so that type checking is practical. Generally, the more complicated the safety property the more annotations are required. Once the safety property is fixed, there is a trade-off between space and time: the more type annotations we have, the easier the type checking problem. One extreme consists of no type annotations at all, which means that type checking is undecidable. The

[2]This should not be confused with a digital signature used to certify authenticity.

other extreme is a full typing derivation (represented, for example, in the logical framework) which is quite similar to a proof in the PCC approach. For the safety policies we have considered so far, it has not been difficult to find appropriate compromises between these extremes that are both compact and permit fast type checking [8].

It is worth noting that in both cases (explicit proofs and type annotations) the verification method is type checking. For proof-carrying code, this is always type checking in the LF logical framework with some optimizations to eliminate redundant work. For typed assembly language the algorithm for type checking varies with the safety property that is enforced, although the basic nature of syntax-directed code traversal remains the same.

3. Automated Certification

How are certificates of safety to be obtained? In principle we may use any means at our disposal, without restriction or limitation. This freedom is assured by the checkability of safety certificates — it is always possible to determine mechanically whether or not a given certificate underwrites the safety of a given program. Since the code recipient can always check the validity of a safety certificate, there is no need to rely on the means by which the certificate was produced.

Two factors determine how hard it is to construct a safety certificate for a program:

1. The strength of the assurances we wish to make about a program. The stronger the assurances, the harder it is to obtain a certificate.

2. The complexity of the programming language itself. The more low-level the language, the harder it is to certify the safety of programs.

As a practical matter, the utility of code certification for safe mobile code varies directly with the ease with which we may construct safety certificates.

The main technique we have considered for building safety certificates is to build a *certifying compiler* for a safe, high-level language such as ML or Java or "safe C". A certifying compiler generates object code that is comparable (and often superior) in quality to that of an ordinary compiler. A certifying compiler goes beyond conventional compilation methods by augmenting the object code with a checkable safety certificate warranting the compliance of the object code with the safety properties of the source language. In this way we are able to exploit the safety properties of semantically well-defined high-level languages without having to trust the compiler itself or having to ensure the integrity of code in transit from producer to consumer.

The key to building a certifying compiler is to propagate *safety invariants* from the source language through the intermediate stages of compilation to the final object code. This means that each compilation phase is responsible for the preservation of these invariants from its input to its output. Moreover, to ensure checkability of these invariants, each phase must annotate with enough information for a code recipient to reconstruct the proof of these invariants. In this way the code recipient can check the safety of the code, without having to rely on the correctness of the compiler. In the (common) case that the compiler contains errors, the purported safety certificate may or may not be valid, but the recipient can detect the mistake. Since each compilation phase can be construed as a recipient of the code produced by the preceding stage, the compiler can check its own integrity by verifying the claimed invariants after each stage. This has proved to be an invaluable aid to the compiler writer [16, 12].

3.1. Constructing Evidence of Safety

We have explored two main methods for propagating safety invariants during compilation:

1. **Type-directed translation between typed intermediate languages** [24]. Safety invariants are captured by a type system for the intermediate languages of the compiler. The type system is designed to ensure that well-typed expressions are safe, and enough type information attached to intermediate forms to ensure that we may mechanically check type correctness. The typed intermediate forms are "self-certifying" in the sense that the attached type information serves as a checkable certificate of safety.

2. **Compilation to proof-carrying code** [16]. Safety invariants are directly expressed as logical assertions about the execution behavior of conventional intermediate code. The soundness of the logic ensures that these assertions correctly express the required safety properties of the code. The safety of the object code is checked by a combination of verification condition generation and automatic theorem proving. By equipping the theorem prover with the means to generate a formal representation of a proof, we may generate checkable safety certificates for the object code.

These two methods are not mutually exclusive. We are currently exploring their integration using *dependent types* which allow assertions to be blended with types in a single type-theoretic formalism. This technique is robust and can applied to high-level languages [29, 30] as well as low-level languages [28, 27], thereby providing an ideal basis for their use in certifying compilers.

3.2. Typed Intermediate Languages

To give a sense of how type information might be attached to intermediate code, we give an example derived from the representation of lists. At the level of the source language, there are two methods for creating lists:

1. `nil`, which stands for the empty list;
2. `cons` (h, t), which constructs the non-empty list with head h and tail t.

These values are assigned types according to the following rules:[3]

1. `nil` has type `list`;
2. If h has type `int` and t has type `list`, then `cons` (h,t) has type `list`.

There are a variety of operations for manipulating lists, including the `car` and `cdr` operations, which have the following types:

1. If l has type `list`, then `car` (l) has type `int`;
2. If l has type `list`, then `cdr` (l) has type `list`.

The behavior of these operations is governed by the following transitions in an operational semantics for the language:

1. `car(cons(`h,t`))` reduces to h;
2. `cdr(cons(`h,t`))` reduces to t.

One task of the compiler is to decide on a representation of lists in memory, and to generate code for `car` and `cdr` consistently with this representation. A typical (if somewhat simple-minded) approach is to represent a list by

1. A *pointer* to ...
2. ...a *tagged* region of memory containing ...
3. ...a *pair* consisting of the head and tail of the list.

The tag field distinguishes empty from non-empty lists, and the pointer identifies the address of the node in the heap. This representation can be depicted as the following compound term:

`ptr(tag[cons](pair(`h`, `t`)))`

What is interesting is that each individual construct in this expression may be thought of as a primitive of a typed intermediate language. Specifically,

1. `ptr(`v`)` has type `list` if v has type `[nil:void,cons:int*list]`. The bracketed expression defines the tags (`nil` and `cons`), and the type of their associated data values (none, in the case of `nil`, a pair in the case of a `cons`).
2. `tag[t](`v`)` has type `[t:`τ`,...]` if v has type τ. In particular, `tag[cons](pair(`h,t`))` has type `[nil:void,cons:int*list]` if h has type `int` and t has type `list`.
3. `pair(`l,r`)` has type τ_l`*`τ_r if l has type τ_l and r has type τ_r. In particular, `pair(`h,t`)` has type `int*list` if h has type `int` and t has type `list`.

Corresponding to this representation we may generate code for, say, `car` (l) that behaves as follows:

1. Dereference the pointer l. The value l must be a pointer because its type is `list`.
2. Check the tag of the object in the heap to ensure that it is `cons`. It must be either `cons` or `nil` because the type of the dereferenced pointer is `[nil:void,cons:int*list]`.
3. Extract the underlying pair and project out its first component. It must have two components because the type of the tagged value is `int*list`.

When expressed formally in a typed intermediate language, the generated code for the `car` operation is defined in terms of primitive operations for performing these three steps. The safety of this code is ensured by the typing rules associated with these operations — a type correct program cannot misinterpret data by, for example, attempting to retrieve the head of an empty list.

A type-directed compiler [24] is one that performs transformations on typed intermediate languages, making use of type information to guide the translation, and ensuring that typing is preserved by each transformation stage. In a type-directed compiler each compilation phase translates not only the program code, but also its type, in such a way that the translated program has the translated type. How far this can be pushed is the subject of ongoing research. In the TILT compiler we are able to propagate type information down to the RTL (register transfer language level), at which point type propagation is abandoned. The recent development of Typed Assembly Language (TAL) [11, 9] demonstrates the feasibility of propagating type information down to x86-like assembly code. The integration of TILT and TAL is the subject of ongoing research.

What does the propagation of type information have to do with safety? A well-behaved type system is one for which we can prove a *soundness theorem* relating the execution behavior of a program to its type. One consequence

[3]For simplicity we consider only lists of integers.

of the soundness theorem for the type system is that it is impossible for well-typed programs to incur type errors, memory errors, or control errors. That is, well-typed programs are safe. Of course not every safe program is well-typed — typing is a sufficient condition for safety, but not a necessary one. However, we may readily check type correctness of a program using lightweight and well-understood methods. The technique of type-directed compilation demonstrates that a rich variety of programs can be certified using typed intermediate languages. Whether there are demands that cannot be met using this method remains to be seen.

3.3. Logical Assertions and Explicit Proofs

Another approach to code certification that we are exploring [15, 16] is the use a combination of logical assertions and explicit proofs. A certifying compiler such as Touchstone works by salting intermediate code with logical assertions tracking the types and ranges of values. Checking the validity of these assertions is a two-step process:

1. Verification condition generation (vcgen). The program is "symbolically evaluated" to propagate the implications of the logical assertions through each of the instructions in the program. This results in a set of logical implications that must hold for the program to be considered properly annotated.

2. Theorem proving. Each of the implications generated during vcgen are verified using a combination of automatic theorem proving techniques, including constraint satisfaction procedures (such as simplex) and proof search techniques for first-order logic.

In this form the trusted computing base must include both the vcgen procedure and the theorem prover(s) used to check the verification conditions.

To reduce the size of the trusted computing base, we may regard the combination of vcgen and theorem proving as a kind of "post-processing" phase in which the validity of the annotated program is not only checked, but a formal representation of the proof of the validity of the verification conditions is attached to the code. This is achieved by using *certifying theorem provers* [16, 23] that not only seek to prove theorems, but also provide an explicit representation of the proof whenever one is found. Once the proofs have been obtained, it is much simpler to check them than it is to find them. Indeed, only the proof checker need be integrated into the trusted computing base; the theorem provers need not be trusted nor be protected from tampering.

To gain an understanding of what is involved here, consider the array subscript operation in a safe language. Given an array A of length n and an integer i, the operation sub(A, i) checks whether or not $0 \leq i < n$ and, if so, retrieves the ith element of A. At a high level this is an atomic operation, but when compiled into intermediate code it is defined in terms of more primitive operations along the following lines:

```
if (0 <= i && i <= *A)
   return A[i+1] /* unsafe access */
 else
   ... signal an error ...
```

Here we assume that an integer array is represented by a pointer to a sequence of words, the first of which contains the array's length, and the rest of which are its contents.

Annotating this code with logical assertions, we obtain the following:

```
/* int i, array A */
if (0 <= i && i <= *A)
   /* 0 <= i < length(A) */
   return A[i+1]
 else
   ... signal an error ...
```

The assertion that A is an array corresponds to the invariants mentioned above; in practice, a much lower-level type system is employed [16]. It is a simple matter to check that the given assertions are correct in this case.

Observe that the role of the conditional test is the enable the theorem prover to prove that the index operation `A[i+1]` is memory-safe — it does not stray beyond the bounds of the array. In many cases the run-time test is redundant because the compiler is able to *prove* that the run-time test must come out true, and therefore can be eliminated. For example, if the high-level code were a simple loop such as the following, we can expect the individual bounds checks to be elided:

```
int sum = 0;
for (i=0; i<length(A); i++)
  /* 0<=i<length(A) */
  sum += sub(A,i);
```

At the call site for sub the compiler is able to prove that $0 \leq i < n$, where n is the length of A. Propagating this through the code for sub, we find that the conditional test can be eliminated because the compiler can prove that the test must always be true. This leads to the following code (after further simplification):

```
int sum = 0;
for (i=0; i<*A; i++)
  /* 0<=i<length(A) */
  sum += A[i+1];
```

Given this annotation, we can now perform verification condition generation and theorem proving to check that the required precondition on the unsafe array subscript operation is indeed true, which ensures that the program is safe to execute. However, rather than place this additional burden on the programmer, we can instead attach a formal representation π of the proof of this fact to the assertions:

```
int sum = 0;
for (i=0; i<*A; i++)
  /* π : 0 <= i < length(A) */
  sum += A[i+1]
```

The proof term π is a checkable witness to the validity of the given assertions that can be checked by the code recipient. In practice this witness is a term of the LF λ-calculus for which proof checking is simply another form of type checking (see Section 2.3).

4. Experimental Results

As mentioned earlier, we have implemented several systems to test and demonstrate the ideas of certified code, typed intermediate languages, certifying compilers, and certifying theorem provers. The results of our experiments with these systems confirm several important claims about the general framework for safety certification of code that we have presented in this paper.

1. Approaches to certified code such as PCC and TAL allow highly optimized code to be verified for safety. This means that few if any compromises need to be made between high performance and safety.

2. The various approaches to certifying compilers that we have explored, such as typed intermediate languages and logical assertions, can be scaled up to languages of realistic scale and complexity. Furthermore, they provide an automatic means of obtaining code that is certified to hold standard safety properties such as type safety, memory safety, and control safety.

3. The need to include annotations and/or proofs with the code is not undue burden. Furthermore, checking these certificates can be performed quickly and reliably.

In order to support these claims and give a better feel for the practical details in our systems, we now present some results of our experiments.

4.1 The Touchstone Certifying Compiler

Touchstone is a certifying compiler for a safe language based on the C programming language. Given a source program, it generates a highly optimized native code binary for the DEC Alpha architecture, with an attached proof of its type, memory, and control safety.

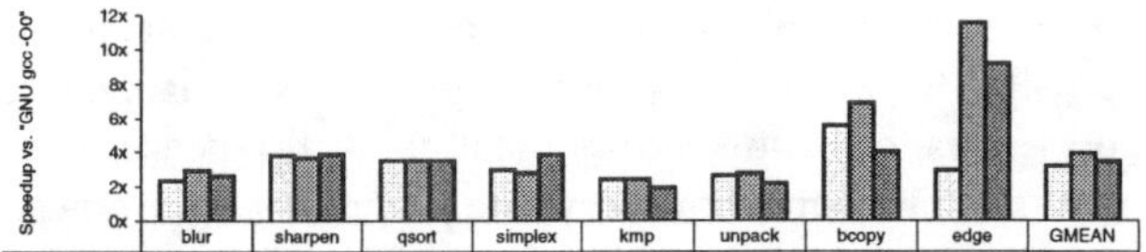

	blur	sharpen	qsort	simplex	kmp	unpack	bcopy	edge	GMEAN
GNU gcc -O4	2.33	3.82	3.51	2.97	2.44	2.62	5.50	2.92	3.17
DEC cc -O4	2.92	3.68	3.52	2.79	2.44	2.76	6.88	11.52	3.92
Cert Comp	2.64	3.89	3.52	3.86	1.93	2.20	4.00	9.16	3.48

Figure 1. Comparison of generated object-code performance between the Touchstone, GCC, and DEC CC optimizing compilers. The height of the bars shows the speedup of the object code relative to unoptimized code as produced by gcc.

Figure 1 shows the results of a collection of benchmark programs when compiled with Touchstone, the Gnu gcc C compiler, and the DEC cc C compiler. The benchmark programs were obtained from standard Unix utility applications (such as the xv and gzip programs) and then edited to replace uses of pointer arithmetic with array-indexing syntax. (The Safe-C language compiled by Touchstone does not support pointer arithmetic.) The bars in the figure were generated by first compiling each program with the Gnu gcc compiler with all optimizations turned off. Then, Touchstone, Gnu gcc, and DEC cc were used to compile the programs with all optimizations turned on. The bars in the figure show the relative speed improvements produced by each optimizing compiler relative to the unoptimized code.

The figure shows that the Touchstone compiler generates object code which is comparable in speed to that produced by the gcc and cc compilers, and in fact is superior to gcc overall. This result is particularly surprising when one considers that Touchstone is obligated to guarantee that all array accesses and pointer dereferences are safe (that is, Touchstone must sometimes perform array-bounds and null-pointer checks), whereas the gcc and cc compilers do not do this. In fact, Touchstone is able to optimize away almost all array-bounds and null-pointer checks, and generates proofs that convince any code recipient that all array and pointer accesses are still safe.

In Figure 2 we provide a breakdown of the time required to compile each benchmark program into a PCC binary. Each bar in the figure is divided into four parts. The bottommost part shows the "conventional" compile time. This is the time required to generate the DEC Alpha assembly code plus invariant annotations required by the underlying PCC system. Because Touchstone is a highly aggressive optimizing compiler, it is a bit slower than typical compilers. However, on average it is comparable in compiling times to the DEC cc compiler with all optimizations enabled. The second part shows the time required to generate the verification conditions. Finally, the third and fourth parts show

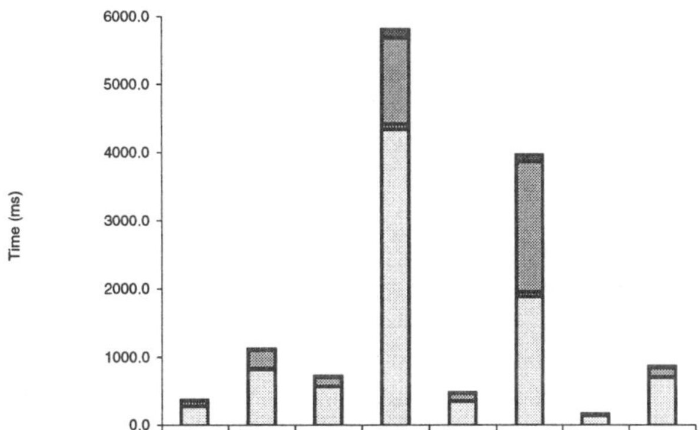

	blur	sharpen	qsort	simplex	kmp	unpack	bcopy	edge
Proof Checking (ms)	7.7	23.5	16.3	120.3	9.6	92.5	3.6	14.7
Proving (ms)	81.0	257.0	127.0	1272.0	108.0	1912.0	25.0	143.0
VC Generation (ms)	6.3	20.9	11.5	73.9	8.4	70.6	3.5	9.7
Code Generation (ms)	271.0	818.0	560.0	4340.0	348.0	1885.0	136.0	697.0

Figure 2. Breakdown of time required to generated the proof-carrying code binaries.

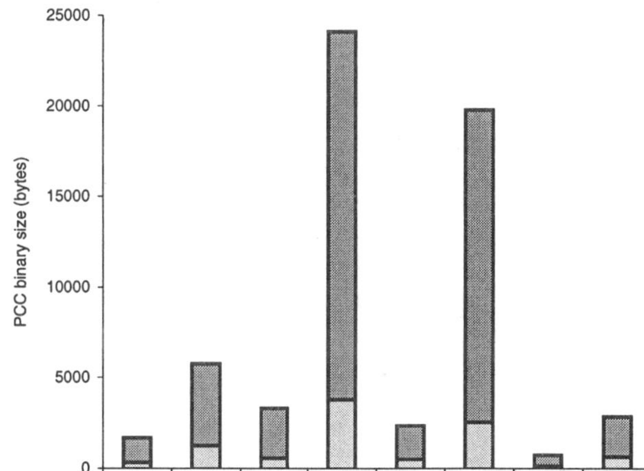

	blur	sharpen	qsort	simplex	kmp	unpack	bcopy	edge
Proof	1372	4532	2748	20308	1874	17260	620	2218
Code	320	1248	576	3792	496	2528	128	640

Figure 3. Early measurement of proof sizes, relative to code sizes.

the times required for proof generation and proof checking, respectively.

One can see that very little time is required for the verification-condition generation and proof checking. This is important because it is these two steps that must also be performed by any recipient of the generated code. The fact that these two parts are so small is an indication that the code recipient in fact has very little work to do.

Figure 3 shows the size of the PCC binaries, divided into the code and proof. The figure shows that the proofs are about 3 to 4 times larger than the code. We note that since the time that these experimental results were obtained, we have made considerable progress on reducing the size of the proofs, without increasing the time or effort required to check them. Currently, we observe proof sizes on the order of 10% to 40% of the size of the code. We are now devising new experiments in order to confirm these observations. We hope to be able to describe our techniques for reducing proof size and show their effects in a future report.

4.2 The Cedilla Systems SpecialJ Compiler

Recently, we have "spun off" a commercial enterprise to build an industrial-strength implementation of a proof-carrying code system. This enterprise, called Cedilla Systems Incorporated, is essentially an experiment in technology transfer, in the sense that it is attempting to take ideas and results directly out of the laboratory and into commercial practice. While it is too early to determine the success of this enterprise, Cedilla Systems has already shown that the ideas presented in this paper can be scaled up to full-scale languages. This is shown most clearly in an optimizing native-code compiler for the Java programming language, called SpecialJ.

The operation of SpecialJ is similar to Touchstone, in that SpecialJ produces optimized target code annotated with invariants that make it possible to construct a proof of safety. A verification-condition generator is then used to extract a verification condition, and a certifying theorem prover generates the proof which is attached to the target code.

To see a simple example of this process, consider the following Java program:

```
public class Bcopy1 {
  public static void bcopy(int[] src,
                           int[] dst)
  {
    int l = src.length;
    int i = 0;
    for(i=0; i<l; i++) {
      dst[i] = src[i];
    }
  }
}
```

This source program is compiled by SpecialJ into a target program for the Intel x86 architecture. Included in this target program are numerous data structures to support Java's object model and run-time system. The core of this output, however, is the native code for the bcopy method shown above.

```
ANN_LOCALS(_bcopy__6arrays6Bcopy1AIAI, 3)
.text
.align 4
.globl _bcopy__6arrays6Bcopy1AIAI
_bcopy__6arrays6Bcopy1AIAI:
        cmpl  $0, 4(%esp)
        je    L6
        movl  4(%esp), %ebx
        movl  4(%ebx), %ecx
        testl %ecx, %ecx
        jg    L22
        ret
L22:
        xorl  %edx, %edx
        cmpl  $0, 8(%esp)
        je    L6
        movl  8(%esp), %eax
        movl  4(%eax), %esi
L7:
ANN_INV(ANN_DOM_LOOP,
        LF_(/ (csubneq ebx 0)
           (/ (csubneq eax 0)
           (/ (csubb edx ecx)
               (of rm mem))))_LF,
        RB(EDI,EDX,EFLAGS,FFLAGS,RM))
        cmpl  %esi, %edx
        jae   L13
        movl  8(%ebx, %edx, 4), %edi
        movl  %edi, 8(%eax, %edx, 4)
        incl  %edx
        cmpl  %ecx, %edx
        jl    L7
        ret
ANN_INV(ANN_DOM_LOOP,
        LF_true_LF,
        RB(FFLAGS))
        ret
L13:
        call  __Jv_ThrowBadArrayIndex
ANN_UNREACHABLE
        nop
L6:
        call  __Jv_ThrowNullPointer
ANN_UNREACHABLE
        nop
```

This code is largely conventional except for the insertion of several invariants, each of which is marked with a special "ANN_" macro. These annotations are "hints" from the compiler that help the automatic proof generator do its job. They do not generate code, and they do not constrain the object code in any way. However, they serve an important engineering purpose, as we will now describe.

The `ANN_LOCALS` annotation simply says that the compiled method uses three locals. In this case, the register allocator did not need any spill space on the stack, so the only locals are the two formal parameters and the return address. This hint is useful for proving memory safety. The prover could, in principle, analyze the code itself to reverse-engineer this information; but it is much easier for the compiler to communicate what it already knows. Since one of our engineering goals is to simplify as much as possible the size of the trusted computing base, it is better to have the compiler generate this information, leaving only the checking problem to the PCC infrastructure.

The `ANN_UNREACHABLE` annotations come from the fact that the safety policy specifies that array accesses must always be in bounds and null pointers must never be dereferenced. In Java, such failures result in run-time exceptions, but the safety policy in our example requires a proof that these exceptions will never be thrown. Therefore, the compiler points out places that must never be reached during execution so that the proof generator does not need to reverse-engineer where the source-code array accesses and pointer dereferences ended up in the binary.

The first `ANN_INV` annotation is by far the most interesting of all the annotations. Note that the SpecialJ compiler has optimized the tight loop:

- Both required null checks are hoisted. (Note that the null check on dst cannot be hoisted before the loop entry because the loop may never be entered at all; but it can be hoisted to the first iteration.)

- The bounds check on src is hoisted. (Note that hoisting the bounds check on dst would be a more exotic optimization, because in the case that dst is not long enough, the loop must copy as far as it can and then throw an exception.)

The proof generator must still prove memory safety, so it must prove that inside the loop there are no null-pointer dereferences or out-of-bounds memory accesses. Essentially, the proof generator needs to go through the same reasoning that the compiler went through when it hoisted those checks outside the loop. Therefore, to help the proof generator, the compiler outputs the relevant loop invariants that it discovered while performing the code-hoisting optimizations. In this case, it discovered that:

- `src` (in register `ebx`) is not null: `(csubneq ebx 0)`.

- `dst` (in register `eax`) is not null: `(csubneq eax 0)`.

- `i` (`edx`) is unsigned-below `src.length` (`ecx`): `(csubb edx ecx)`.

The "csub" prefix denotes the result of a Pentium comparison. Other things in the loop invariant specify:

- which registers are modified in the loop: `RB(...)`,

IEEE Computer Society Publications

The world-renowned IEEE Computer Society publishes, promotes, and distributes a wide variety of authoritative computer science and engineering texts. These books are available from most retail outlets. Visit the Online Catalog, *http://computer.org*, for a list of products.

IEEE Computer Society Proceedings

The IEEE Computer Society also produces and actively promotes the proceedings of more than 141 acclaimed international conferences each year in multimedia formats that include hard and softcover books, CD-ROMs, videos, and on-line publications.

For information on the IEEE Computer Society proceedings, send e-mail to *cs.books@computer.org* or write to Proceedings, IEEE Computer Society, P.O. Box 3014, 10662 Los Vaqueros Circle, Los Alamitos, CA 90720-1314. Telephone +1 714-821-8380. FAX +1 714-761-1784.

Additional information regarding the Computer Society, conferences and proceedings, CD-ROMs, videos, and books can also be accessed from our web site at *http://computer.org/cspress*

Revised 9 November 1999

- that memory safety is a loop invariant: `(of rm mem)`

Pseudo-register `rm` denotes the computer's memory, and `(of rm mem)` means that no unsafe operations have been performed on the memory.

After this target code is generated by SpecialJ, the Cedilla Systems proof generator reads it and outputs a proof that the code satisfies the safety policy. The first step to doing this is to generate a logical predicate, called a *verification condition* (or simply VC), whose logical validity implies the safety of the code. It is important that the same VC be used by both the producer and the recipient of the code, so that the recipient can guarantee that the "right" safety proof is provided, as opposed to a proof of some unrelated or irrelevant property.

As we explained earlier, both the proofs and the verification conditions are expressed in a language called the Logical Framework (LF). Space prevents us from including the entire VC for our bcopy example. However, the following excerpt illustrates the main points:

```
(=> (csubb X3 (sel4 X2 (add X1 4)))
 (=> (csubneq X0 0)
  (=> (csubneq X1 0)
   (=> (csubb X3 (sel4 X2 (add X0 4)))
    (/\ (saferd4
          (add X1 (add (imul X3 4) 8)))
     (/\ (safewr4
          (add X0 (add (imul X3 4) 8))
          (sel4 X4
           (add X1 (add (imul X3 4) 8))))
      (/\
       (=> (csublt (add X3 1)
                   (sel4 X2 (add X1 4)))
        (/\ (csubneq X1 0)
         (/\ (csubneq X0 0)
          (/\ (csubb (add X3 1)
                     (sel4 X2 (add X1 4)))
```

This excerpt of the VC says that, given the loop-invariant assumptions

- `(csubb X3 (sel4 X2 (add X1 4)))` (i.e., `src` is in bounds),
- `(csubneq X0 0)` (i.e., `dst` is non-null), and
- `(csubneq X1 0)` (i.e., `src` is non-null),

and given the bounds check that was emitted for `dst`:

- `(csubb X3 (sel4 X2 (add X0 4)))`

as well as some additional assumptions outside the loop (not shown in this snippet), proofs are required to establish the safety of the read of the `src` array and the write to the `dst` array. Furthermore, given the additional loop-entry condition

- `(csublt (add X3 1) (sel4 X2 (add X1 4)))`

proofs are required to reestablish the loop invariants.

Here, `X0` corresponds to `eax` (`dst` in the source), `X1` to `ebx` (`src` in the source), `X2` to `rm` (the memory pseudo-register), and `X3` to `edx` (`i` in the source). Note that `src.length` is `(sel4 X2 (add X1 4))`, because the length is stored at byte-offset 4 in an array object. The safety policy, and hence the VC, specifies and enforces these requirements on data-structure layout.

The proof generator reads the VC and outputs a proof of it. A tiny excerpt of this proof is shown below:

```
(impi
 ([ASS10: pf (csubb X3
               (sel4 X2 (add X1 4)))]
  (impi
   ([ASS11: pf (csubneq X0 0)]
    (impi
     ([ASS12: pf (csubneq X1 0)]
      (impi
       ([ASS13: pf (csubb X3
                    (sel4 X2
                     (add X0 4)))]
        (andi
         (rdArray4 ASS4 ASS3
                   (sub0chk ASS12)
                   szint
                   (aidxi 4
                    (belowl ASS10)))
   ...
```

The proofs are shown here in a concrete syntax for LF developed by Pfenning in the Elf system. In this very small snippet of the proof, one can see that assumptions (marked with the "`ASS...`" identifiers) are labeled and then used in the body of the proof. Logical inference rules such as "`impi`", which in this case stands for the "implication-introduction" rule, are specified in declaratively in the LF language, and included with the PCC system as part of the definition of the safety policy.

Finally, a binary encoding of the proof is made and attached it to the target code. The proof is included in the data segment of a standard binary in the COFF format. In this case, the proof takes up 7.1% of the total object file. We note that we currently use an unoptimized binary encoding of the proof in which all proof tokens are 16 bits long. Huffman encoding produces an average token size of 3.5 bits, and so a Huffman-encoded binary should be expected to be about 22% of the size of a non-Huffman-encoded binary. In this case, that would make the size of the proof approximately 45 bytes, or less than 2% of the object file. In addition to Huffman-encoding, several other optimizations are currently under development.

5. Conclusion and Future Work

We have presented a general framework for the safety certification of code. It relies on the formal definition of a safety policy and explicit evidence for compliance attached to mobile code. This evidence may take the form of formal safety proofs (in *proof-carrying code*) or type annotations (in *typed assembly language*). In both cases one can establish with mathematical rigor that certified code is tamper-proof and can be executed safely without additional run-time checks or operating system protection boundaries. Experience with the approaches has shown the overhead to be acceptable in practice, both in the time to validate the certificate and the space to represent it, using advanced techniques from logical frameworks and type theory.

We also sketched how certificates can be obtained automatically through the use of certifying compilers and theorem provers. The approach of *typed intermediate languages* propagates safety properties which are guaranteed for the high-level source language throughout the compilation process down to low-level code. Safety remains verifiable at each layer through type-checking. A *certifying compiler* such as Touchstone uses logical assertions throughout compilation in a similar manner, except that the validity of the logical assertions must be assured by theorem proving. This is practical for the class of safety policies considered here, since the compiler can provide the information necessary to guarantee that a proof can always be found. Finally, a *certifying theorem prover* does not need to be part of the trusted computing base since it produces explicit proof terms which can be checked independently by an implementation of a logical framework.

The key technology underlying our approaches to safety is type theory as used in modern programming language design and implementation. The idea that type systems guarantee program safety and modularity for high-level languages is an old one. We see our main contribution in demonstrating in practical, working systems such as Touchstone, the TILT compiler, and the Twelf logical framework, that techniques from type theory can equally be applied to intermediate and low-level languages down to machine code in order to support provably safe mobile code for which certificates can be generated automatically.

References

[1] A. W. Appel and E. W. Felten. Proof-carrying authentication. In G. Tsudik, editor, *Proceedings of the 6th Conference on Computer and Communications Security*, Singapore, Nov. 1999. ACM Press. To appear.

[2] A. W. Appel and A. P. Felty. A semantic model of types and machine instructions for proof-carrying code. Submitted, July 1999.

[3] B. Bershad, S. Savage, P. Pardyak, E. Sirer, M. Fiuczynski, D. Becker, C. Chambers, and S. Eggers. Extensibility, safety and performance in the SPIN operating system. In *Fifteenth ACM Symposium on Operating Systems Principles*, pages 267–284, Copper Mountain, Dec. 1995.

[4] K. Crary and S. Weirich. Resource bound certification. In *Twenty-Seventh ACM Symposium on Principles of Programming Languages*, Boston, Jan. 2000. To appear.

[5] R. Harper, F. Honsell, and G. Plotkin. A framework for defining logics. *Journal of the Association for Computing Machinery*, 40(1):143–184, Jan. 1993.

[6] R. Harper and F. Pfenning. On equivalence and canonical forms in the LF type theory. In A. Felty, editor, *Proceedings of the Workshop on Logical Frameworks and Meta-Languages (LFM'99)*, Paris, France, Sept. 1999.

[7] S. Michaylov and F. Pfenning. An empirical study of the runtime behavior of higher-order logic programs. In D. Miller, editor, *Proceedings of the Workshop on the λProlog Programming Language*, pages 257–271, Philadelphia, Pennsylvania, July 1992. University of Pennsylvania. Available as Technical Report MS-CIS-92-86.

[8] G. Morrisett, K. Crary, N. Glew, D. Grossman, R. Samuels, F. Smith, D. Walker, S. Weirich, and S. Zdancewic. TALx86: A realistic typed assembly language. In *Second Workshop on Compiler Support for System Software*, Atlanta, May 1999.

[9] G. Morrisett, K. Crary, N. Glew, and D. Walker. Stack-based typed assembly language. In *Second Workshop on Types in Compilation*, volume 1473 of *Lecture Notes in Computer Science*. Springer-Verlag, Mar. 1998. Extended version published as CMU technical report CMU-CS-98-178.

[10] G. Morrisett, D. Tarditi, P. Cheng, C. Stone, R. Harper, and P. Lee. The TIL/ML compiler: Performance and safety through types. In *Workshop on Compiler Support for Systems Software*, Tucson, Feb. 1996.

[11] G. Morrisett, D. Walker, K. Crary, and N. Glew. From System F to typed assembly language. *ACM Trans. Prog. Lang. Syst.*, 1999. To appear. An earlier version appeared in the 1998 Symposium on Principles of Programming Languages.

[12] J. G. Morrisett. *Compiling with Types*. PhD thesis, School of Computer Science, Carnegie Mellon University, Pittsburgh, PA, December 1995. (Available as Carnegie Mellon University School of Computer Science technical report CMU–CS–95–226.).

[13] G. Necula and P. Lee. Safe kernel extensions without run-time checking. In *Second Symposium on Operating Systems Design and Implementation*, pages 229–243, Seattle, Oct. 1996.

[14] G. Necula and P. Lee. Safe, untrusted agents using proof-carrying code. In *Special Issue on Mobile Agent Security*, volume 1419 of *Lecture Notes in Computer Science*. Springer-Verlag, Oct. 1997.

[15] G. C. Necula. Proof-carrying code. In N. D. Jones, editor, *Conference Record of the 24th Symposium on Principles of Programming Languages (POPL'97)*, pages 106–119, Paris, France, Jan. 1997. ACM Press.

[16] G. C. Necula. *Compiling with Proofs*. PhD thesis, Carnegie Mellon University, Oct. 1998. Available as Technical Report CMU-CS-98-154.

[17] G. C. Necula and P. Lee. Efficient representation and validation of logical proofs. In *Proceedings of the 13th Annual Symposium on Logic in Computer Science (LICS'98)*, pages 93–104, Indianapolis, Indiana, June 1998. IEEE Computer Society Press.

[18] F. Pfenning. Logic programming in the LF logical framework. In G. Huet and G. Plotkin, editors, *Logical Frameworks*, pages 149–181. Cambridge University Press, 1991.

[19] F. Pfenning. Elf: A meta-language for deductive systems. In A. Bundy, editor, *Proceedings of the 12th International Conference on Automated Deduction*, pages 811–815, Nancy, France, June 1994. Springer-Verlag LNAI 814. System abstract.

[20] F. Pfenning and C. Schürmann. Algorithms for equality and unification in the presence of notational definitions. In T. Altenkirch, W. Naraschewski, and B. Reus, editors, *Types for Proofs and Programs*. Springer-Verlag LNCS 1657, 1998. To appear.

[21] F. Pfenning and C. Schürmann. System description: Twelf — a meta-logical framework for deductive systems. In H. Ganzinger, editor, *Proceedings of the 16th International Conference on Automated Deduction (CADE-16)*, pages 202–206, Trento, Italy, July 1999. Springer-Verlag LNAI 1632.

[22] J. C. Reynolds. Definitional interpreters for higher-order programming languages. In *Conference Record of the 25th National ACM Conference*, pages 717–740, Boston, Aug. 1972.

[23] A. Stump and D. L. Dill. Generating proofs from a decision procedure. In A. Pnueli and P. Traverso, editors, *Proceedings of the FLoC Workshop on Run-Time Result Verification*, Trento, Italy, July 1999.

[24] D. Tarditi, G. Morrisett, P. Cheng, C. Stone, R. Harper, and P. Lee. TIL: A type-directed optimizing compiler for ML. In *ACM SIGPLAN Conference on Programming Language Design and Implementation*, pages 181–192, Philadelphia, PA, May 1996.

[25] R. Wahbe, S. Lucco, T. E. Anderson, and S. L. Graham. Efficient software-based fault isolation. In *14th ACM Symposium on Operating System Principles*, pages 203–216, December 1993.

[26] D. Walker. A type system for expressive security policies. In *Twenty-Seventh ACM Symposium on Principles of Programming Languages*, Boston, Jan. 2000. To appear.

[27] H. Xi and R. Harper. A dependently typed assembly language. Technical Report OGI-CSE-99-008, Computer Science Department, Oregon Graduate Institute, July 1999.

[28] H. Xi and R. Harper. A dependently typed assembly language (extended abstract). (To appear), July 1999.

[29] H. Xi and F. Pfenning. Eliminating array bound checking through dependent types. In K. D. Cooper, editor, *Proceedings of the Conference on Programming Language Design and Implementation (PLDI'98)*, pages 249–257, Montreal, Canada, June 1998. ACM Press.

[30] H. Xi and F. Pfenning. Dependent types in practical programming. In A. Aiken, editor, *Conference Record of the 26th Symposium on Principles of Programming Languages (POPL'99)*, pages 214–227. ACM Press, Jan. 1999.

Author Index